Contemporary Enzyme Kinetics and Mechanism

THIRD EDITION

Reliable Lab Solutions

Contemporary Enzyme Kinetics and Mechanism

THIRD EDITION

Reliable Lab Solutions

Edited by

Daniel L. Purich

Professor of Biochemistry and Molecular Biology
University of Florida, College of Medicine
Gainesville, Florida, USA

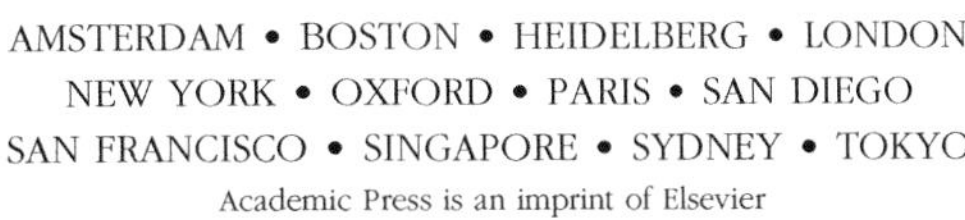

Academic Press is an imprint of Elsevier
Linacre House, Jordan Hill, Oxford OX2 8DP, UK
30 Corporate Drive, Suite 400, Burlington, MA 01803, USA
525 B Street, Suite 1900, San Diego, CA 92101-4495, USA
32 Jamestown Road, London NW1 7BY, UK

Third edition 2009

Portions of this work have previously appeared in Methods in Enzymology (Academic Press., 1979, 1995, and 1999 and Elsevier Science (USA)., 2002)

ISBN: 978-0-12-378608-1

For information on all Academic Press publications
visit our website at www.elsevierdirect.com

Printed and bound by CPI Group (UK) Ltd, Croydon, CR0 4YY

Transferred to Digital Print 2012

CONTENTS

PART II Inhibitors as Probes of Enzyme Catalysis

PART III Detection of Enzyme Reaction Intermediates

PART IV Isotopic Probes of Enzyme Processes

CONTRIBUTORS

Numbers in parentheses indicate the pages on which the authors' contributions begin.

R. Donald Allison (35, 451), Department of Biochemistry & Molecular Biology, University of Florida College of Medicine, Gainesville, Florida 32610-0245, USA

Bennett W. Baugher (55), Department of Biochemistry Rice University, Houston Texas 77001

Robert S. Beissner (55), Department of Rice University, Houston Texas 77001

Paul J. Berti (571), Department of Chemistry, Department of Biochemistry & Biomedical Science McMaster University, Hamilton, Ontario, Canada

Benjamin B. Braunheim (609), Department of Physiology and Biophysics, Albert Einstein College of Medicine, Bronx, New York 10461

Henry B. F. Dixon✠ (123), Department of Biochemistry, Trinity College, Dublin 2, Ireland

Donald J. Douglas (433), Department of Chemistry, University of British Columbia, Vancouver, British Columbia, Canada V6T 1Z1

Ronald G. Duggleby (95), Department of Biochemistry, Centre for Structure, Function, and Engineering, University of Queensland, Brisbane, Queensland 4072, Australia

Carol A. Fierke (373), Department of Biochemistry, College of Medicine, Duke University Medical Center, Durham, North Carolina 27710

Herbert J. Fromm (279), Department of Biochemistry and Biophysics, Iowa State University, Ames, Iowa 50011

Gordon G. Hammes (373), Department of Biochemistry, Duke University Medical Center, Durham, North Carolina 27710

Charles Y. Huang (1), Laboratory of Biochemistry, National Heart, Lung, and Blood Institute, National Institutes of Health, Bethesda, Maryland 20892, USA

Kenneth A. Johnson (407), Department of Chemistry and Biochemistry, Institute for Cell and Molecular Biology, University of Texas, Austin, TX 78735, USA

Lars Konermann (433), Department of Chemistry, University of Western Ontario, London, Ontario, Canada N6A 5B7

Keith J. Laidler✠ (177), Department of Chemistry, University of Ottawa, Ottawa, Ontario KIN 9B4, Canada

Bengt Mannervik (75), Department of Biochemistry and Organic Chemistry, Uppsala University, Biomedical Center, Box576, SE-75123 Uppsala, Sweden

✠Deceased

Andrew G. McDonald (123), Department of Biochemistry, Trinity College, Dublin 2, Ireland

Leisha S. Mullins (493), Department of Chemistry, Texas A&M University, College Station, Texas 77843

Kenneth E. Neet (227), Department of Biochemistry and Molecular Biology, Chicago Medical School, Rosalind Franklin University of Medicine and Science

Branko F. Peterman (177), Department of Chemistry, University of Ottawa, Ottawa, Ontario KIN 9B4, Canada

Bryce V. Plapp (199, 301), Department of Biochemistry, Carver College of Medicine, The University of Iowa, Iowa City, IA 52242-1109, USA

Daniel L. Purich (35, 451), Department of Biochemistry & Molecular Biology, University of Florida College of Medicine, Gainesville, Florida 32610-0245, USA

Frank M. Raushel (493), Department of Chemistry, Texas A&M University, College Station, Texas 77843

Frederick B. Rudolph✠ (55), Department of Biochemistry Rice University, Houston Texas 77001

Vern L. Schramm (519), Department of Biochemistry, Albert Einstein College of Medicine, Bronx, New York 10461

Steven D. Schwartz (609), Departments of Physiology and Biophysics, and Biochemistry, Albert Einstein College of Medicine, Bronx, New York 10461

Richard B. Silverman (331), Department of Chemistry, Northwestern University, Evanston, Illinois 60208

Keith F. Tipton (123), Department of Biochemistry, Trinity College, Dublin 2, Ireland

✠Deceased

PREFACE

The elucidation of kinetic models and chemical mechanisms is deeply rooted in a keen interest in the design and execution of appropriate rate experiments. Those familiar with enzymology immediately appreciate both the challenge and satisfaction of designing enzyme experiments. Experienced enzymologists invariably convey an excitement and fascination for the details of a well-conceived experiment as well as the task of learning about a particular enzyme's "chemical personality." And in enzyme kinetics, nothing is quite as helpful as a useful kinetic theory and reliable kinetic techniques with which to test theoretical predictions. The chapters presented in this monograph were originally published in the *Enzyme Kinetics and Mechanism* subseries, comprising volumes 63, 64, 87, 249, 308, and 354 of *Methods in Enzymology*. These chapters represent some of the most useful and enduring sources for theory and best-practice advice for the systematic kinetic examination of enzyme catalysis and control. In fact, many of the authors actually played pivotal roles in originating the kinetic theories and/or in establishing their far-reaching utility. By offering these chapters in this newly organized compendium, I am confident that students of enzyme kinetics as well as those interested in broader aspects of the molecular life sciences will benefit from the wisdom and experience embodied in each and every chapter.

Daniel Purich

PART I

Initial Rate Theory and Methods

CHAPTER 1

Derivation of Initial-Velocity and Isotope-Exchange Rate Equations

Charles Y. Huang
Laboratory of Biochemistry
National Heart, Lung, and Blood Institute
National Institutes of Health
Bethesda, Maryland 20892, USA

A rate equation for an enzymic reaction is a mathematical expression that depicts the process in terms of rate constants and reactant concentrations. It serves as a link between the experimentally observed kinetic behavior and a plausible model or mechanism. The characteristics of the rate equation permit tests to be designed to verify the mechanism. Conversely, the experimental observations provide clues to what the mechanism may be, hence, what form the rate expression shall take.

Derivation of rate equations is an integral part of the effective usage of kinetics as a tool. Novel mechanisms must be described by new equations, and familiar ones often need to be modified to account for minor deviations from the expected pattern. The mathematical manipulations involved in deriving initial-velocity or isotope-exchange rate laws are in general quite straightforward, but can be tedious. It is the purpose of this chapter, therefore, to present the currently available methods with emphasis on the more convenient ones.

 DOI: 10.1016/B978-0-12-378608-1.00001-3

I. Derivation of Initial-Velocity Equations

The derivation of initial-velocity equations invariably entails certain assumptions. In fact, these assumptions are often conditions that must be fulfilled for the equations to be valid. Initial velocity is defined as the reaction rate at the early phase of enzymic catalysis during which the formation of product is linear with respect to time. This linear phase is achieved when the enzyme and substrate intermediates reach a steady state or quasi-equilibrium. Other assumptions basic to the derivation of initial rate equations are as follows:

1. The enzyme and the substrate form a complex.
2. The substrate concentration is much greater than the enzyme concentration, so that the free substrate concentration is equivalent to the total concentration. This condition further requires that the amount of product formed is small, such that the reverse reaction or product inhibition is negligible.
3. During the reaction, constant pH, temperature, and ionic strength are maintained.

In the past decades, it was not uncommon to hear some presumably knowledgeable persons question the value of steady-state kinetics on the ground that the conditions stated in point 2 above often do not exist *in vivo*. To clarify this recurring doubt, one should look at these conditions as requirements for initial rate experiments in order to obtain the desired kinetic parameters. One can ask how closely the isolated enzyme retains its *in vivo* catalytic characteristics. However, whether steady state is ever reached in a certain metabolic pathway is a separate issue. In fact, kinetic measurements remain the best way to study partially purified enzymes and their effectors.

A. Steady-State Treatment

During the steady state, the concentrations of various enzyme intermediates are essentially unchanged; that is, the rate of formation of a given intermediate is equal to its rate of disappearance. This assumption was first introduced to the derivation of enzyme kinetic equation by Briggs and Haldane (1925).

To derive a rate equation, the first step is to write a reaction mechanism. The nomenclature used by Fromm in volume [63] of *Methods in Enzymology* will be adopted here with the exception that rate constants in the forward and reverse directions will be denoted by positive and negative subscripts. For example, the simplest one substrate-one product reaction can be written as:

$$\mathrm{E} + \mathrm{A} \underset{k_{-1}}{\overset{k_1}{\rightleftharpoons}} \mathrm{EA} \xrightarrow{k_2} \mathrm{E} + \mathrm{P} \tag{1}$$

or

$$\mathrm{E} \underset{k_{-1}}{\overset{k_1\mathrm{A}}{\rightleftharpoons}} \mathrm{EA} \xrightarrow[\quad]{k_2 \nearrow \mathrm{P}} \mathrm{E}$$

Since both the k_{-1} and k_2 steps (or branches) lead from EA to E, the two branches, as has been shown by Volkenstein and Goldstein (1966), can be combined into a single branch. This simplification procedure will be used whenever feasible.

$$\mathrm{E} \underset{k_{-1}+k_2}{\overset{k_1\mathrm{A}}{\rightleftharpoons}} \mathrm{EA}$$

The initial rate is given by

$$v = d\mathrm{P}/dt = k_2(\mathrm{EA})$$

Applying the steady-state assumption, we have

$$d(\mathrm{EA})/dt = k_1\mathrm{A}(\mathrm{E}) - (k_{-1} + k_2)(\mathrm{EA}) = 0 \tag{2}$$

To obtain an expression for (EA), the enzyme conservation equation

$$\text{Total enzyme} = \mathrm{E}_0 = \mathrm{E} + \mathrm{EA} \tag{3}$$

is required. Substitution of (E) = (E_0 − EA) into Eq. (2) yields

$$(\mathrm{EA}) = \frac{\mathrm{E}_0\mathrm{A}}{[(k_{-1} + k_2)/k_1] + \mathrm{A}}$$

and

$$v = k_2(\mathrm{EA}) = \frac{k_2\mathrm{E}_0\mathrm{A}}{[(k_{-1} + k_2)/k_1] + \mathrm{A}} = \frac{V_1\mathrm{A}}{K_\mathrm{m} + \mathrm{A}} \tag{4}$$

where V_1 is the maximum velocity in the forward direction (caution: in most cases, the notation V_m is recommended because the capitalized V is often misprinted as a small v, causing tremendous confusion in complex situations for unsuspecting readers) and K_m is the Michaelis constant.

It should be noted that the validity of the steady-state method does not depend on the assumption $d(\mathrm{EA})/dt = 0$. Without setting Eq. (2) equal to zero, one can obtain the following expression from Eqs. (2) and (3):

$$(\mathrm{EA}) = \frac{k_1\mathrm{A}\mathrm{E}_0 - d(\mathrm{EA})/dt}{k_1\mathrm{A} + k_{-1} + k_2}$$

Wong (1975) has pointed out that the steady-state approximation only requires that $d(\mathrm{EA})/dt$ be small compared with $k_1\mathrm{A}\mathrm{E}_0$. In the early phase of the reaction, if $\mathrm{A} \gg \mathrm{E}_0$, the rate of change of EA due to diminishing A will be relatively slow. It is clear that the validity of steady state is intimately tied to the condition of high substrate to enzyme ratio.

1. The Determinant Method

For a mechanism involving several enzyme-containing species, derivation of the rate equation can be done by solving the simultaneous algebraic equations by the determinant method. Consider the mechanism described by Eq. (1) with the addition of an EP intermediate.

$$\mathrm{E} \underset{k_{-1}}{\overset{k_1\mathrm{A}}{\rightleftharpoons}} \mathrm{EA} \underset{k_{-2}}{\overset{k_2}{\rightleftharpoons}} \mathrm{EP} \xrightarrow{k_3} \mathrm{E} + \mathrm{P} \tag{5}$$

The three simultaneous equations are given in the following form:

$$\begin{array}{r|ccc|l} & \mathrm{E} & \mathrm{EA} & \mathrm{EP} & \\ d\mathrm{E}/dt = & -k_1\mathrm{A} & k_{-1} & k_3 & = 0 \\ d\mathrm{EA}/dt = & k_1\mathrm{A} & -(k_{-1}+k_2) & k_{-2} & = 0 \\ d\mathrm{EP}/dt = & 0 & k_2 & -(k_{-2}+k_3) & = 0 \end{array}$$

The determinant, or distribution term, for E, for example, can be calculated from the coefficients listed above, after deleting the E column. For a mechanism of n intermediates, only $n-1$ equations are needed. Thus, by leaving out the $d\mathrm{EP}/dt$ row, we can write

$$(\mathrm{E}) = \begin{vmatrix} k_{-1} & k_3 \\ -(k_{-1}+k_2) & k_{-2} \end{vmatrix} = k_{-1}k_{-2} + k_3(k_{-1}+k_2)$$

If the $d\mathrm{E}/dt$ row is omitted instead, we have

$$\begin{aligned}(\mathrm{E}) = \begin{vmatrix} -(k_{-1}+k_2) & k_{-2} \\ k_2 & -(k_{-2}+k_3) \end{vmatrix} &= k_{-1}(k_{-2}+k_3) + k_2(k_{-2}+k_3) - k_2k_{-2} \\ &= k_{-1}k_{-2} + k_3(k_{-1}+k_2)\end{aligned}$$

Note that deletion of different equations often leads to different amounts of algebraic manipulations. Application of the same operations to EA and EP yields

$$(\mathrm{EA}) = k_1(k_{-2}+k_3)\mathrm{A}$$
$$(\mathrm{EP}) = k_1k_2\mathrm{A}$$

The rate equation is readily obtained as

$$\begin{aligned}\frac{v}{\mathrm{E}_0} &= \frac{k_3(\mathrm{EP})}{(\mathrm{E})+(\mathrm{EA})+(\mathrm{EP})} \\ &= \frac{k_1k_2k_3\mathrm{A}}{k_{-1}k_{-2}+k_3(k_{-1}+k_2)+k_1(k_{-2}+k_3)\mathrm{A}+k_1k_2\mathrm{A}}\end{aligned}$$

or

$$v = \frac{k_2k_3\mathrm{E}_0\mathrm{A}/(\mathrm{k}_2+k_{-2}+k_3)}{\{[k_{-1}k_{-2}+k_3(k_{-1}+k_2)]/[k_1(k_2+k_{-2}+k_3)]\}+\mathrm{A}} = \frac{V_1\mathrm{A}}{K_\mathrm{m}+\mathrm{A}}. \tag{6}$$

Equation (6) is identical in form with Eq. (4). In fact, if $k_3 \gg k_2, k_{-2}$, Eq. (6) reduces to Eq. (4). Although Eq. (5) is a more realistic mechanism compared with Eq. (1), especially when the rapid-equilibrium treatment is applied to the reversible reaction, the information obtainable from initial-rate studies of such unireactant system remains nevertheless the same: V_1 and K_m. This serves to justify the simplification used by the kineticist; that is, the elimination of certain intermediates to maintain brevity of the rate equation (provided the mathematical form is unaltered). Thus, the *forward* reaction of an ordered Bi Bi mechanism is generally written as diagrammed below.

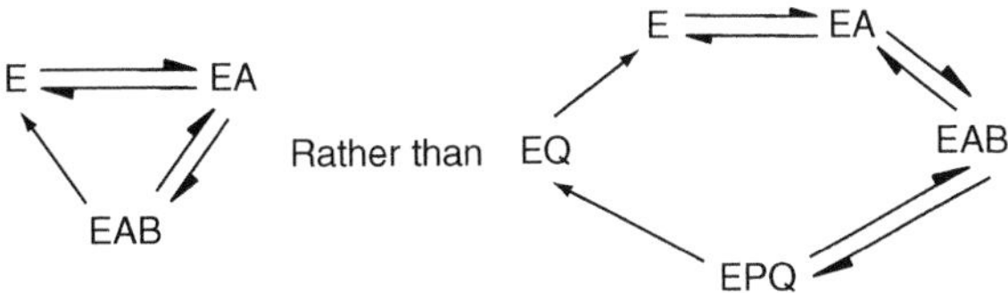

The use of the determinant method for complex enzyme mechanisms is time-consuming because of the stepwise expansion and the large number of positive and negative terms that must be canceled. It is quite useful, however, in computer-assisted derivation of rate equations. For example, Fromm and Fromm (1999) have developed a two-step computer-assisted procedure utilizing the readily accessible *Mathematica* software.

2. The King and Altman Method

King and Altman (1956) developed a schematic approach for deriving steady-state rate equations, which has contributed to the advance of enzyme kinetics. The first step of this method is to draw an *enclosed* geometric figure with each enzyme form as one of the corners. Equation (5), for instance, can be rewritten as:

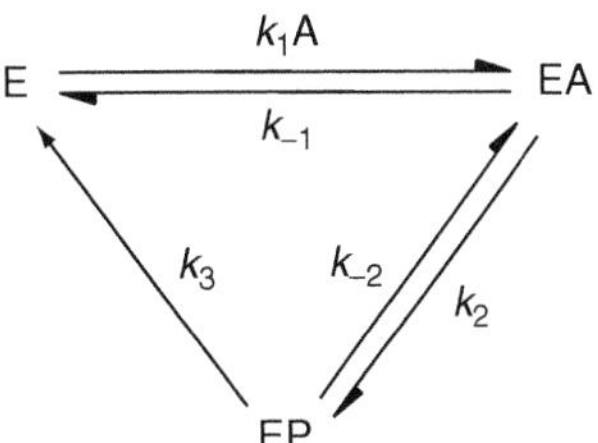

The second step is to draw all the possible patterns that connect all the enzyme species without forming a loop. For a mechanism with n enzyme species, or a figure with n corners, each pattern should contain $n - 1$ lines. The number of valid patterns for any single-loop mechanism is equal to the number of enzyme forms. Thus, there are three patterns for the triangle shown above:

The determinant for a given enzyme species is obtained as the summation of the product of the rate constants and concentration factors associated with all the branches in the patterns *leading toward* this particular enzyme species. The same patterns are used for each species, albeit the direction in which they are read will vary. However, when an irreversible step is present, e.g., the EP → E step, some patterns become invalid for certain enzyme forms.

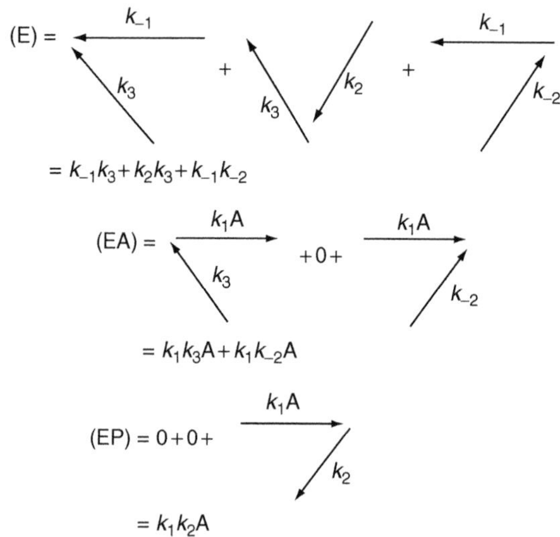

The rate equation is obtained as

$$\frac{v}{\text{E}_0} = \frac{k_3(\text{EP})}{(\text{E}) + (\text{EA}) + (\text{EP})},$$

where (E), (EA), and (EP) are the determinants for E, EA, and EP, respectively.

The presence of an enzyme intermediate(s) that is not part of a loop will not affect the number of King-Altman patterns. For instance, the addition of a competitive inhibitor, I, to the above system will result in the same number of patterns.

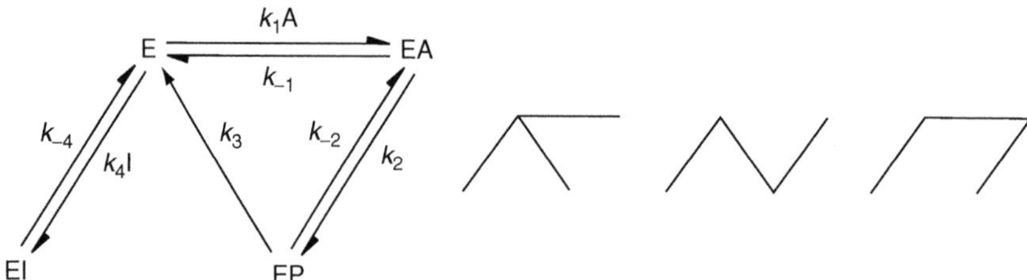

The additional E $\leftrightarrow$ EI branch is present in *all* the diagrams. Thus, in calculating the number of valid King-Altman patterns, only the *closed loops* need be considered. The determinants of E, EA, EAB, and EI can be obtained by the method just described:

$$\begin{aligned}
(\mathrm{E}) &= k_{-4}(k_{-1}k_3 + k_2k_3 + k_{-1}k_{-2}),\\
(\mathrm{EA}) &= k_{-4}(k_1k_3\mathrm{A} + k_1k_{-2}\mathrm{A}),\\
(\mathrm{EP}) &= k_{-4}(k_1k_2\mathrm{A}),\\
(\mathrm{EP}) &= k_4\mathrm{I}(k_{-1}k_3 + k_2k_3 + k_{-1}k_{-2}).
\end{aligned}$$

It is more convenient, however, to treat this case by first considering only the loop portion (ignoring the additional E $\leftarrow$ EI step for the time being).

$$\begin{aligned}
(\mathrm{E}) &= k_{-1}k_3 + k_2k_3 + k_{-1}k_{-2},\\
(\mathrm{EA}) &= k_1(k_{-2} + k_3)\mathrm{A},\\
(\mathrm{EP}) &= k_1k_2\mathrm{A}.
\end{aligned}$$

The determinant for EI is then obtained as

$$\begin{aligned}
(\mathrm{EI}) = (\mathrm{E})k_4\mathrm{I}/k_{-4} &= (k_{-1}k_3 + k_2k_3 + k_{-1}k_{-2})k_4\mathrm{I}/k_{-4}\\
&= (k_{-1}k_3 + k_2k_3 + k_{-1}k_{-2})\mathrm{I}/K_i,
\end{aligned}$$

where $K_i = k_{-4}/k_4$.

The King-Altman method is most convenient for single-loop mechanisms. In practice, there is no need to write down the patterns. One can use an object, say, a paper clip, to block one branch of the loop, write down the appropriate term for each enzyme species, then repeat the process until every branch in the loop has been blocked once.

For more complex mechanisms having alternative pathways that form several closed loops, the precise number of valid King-Altman patterns must be calculated to avoid omission of terms. To illustrate the various situations that may occur in such calculation, let us consider Scheme 1.

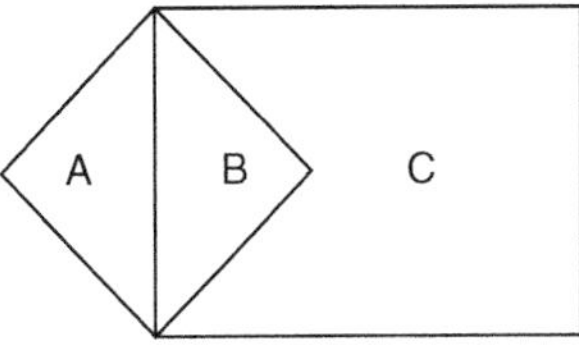

Scheme 1

The total number of patterns with $n-1$ lines is given by the equation

$$\frac{m!}{(n-1)!(m-n+1)!},$$

where m = the number of lines in the complete geometric figure. In the above scheme, $m = 8$ and $n = 6$, and the total number of patterns with 5 lines is

$$\frac{8!}{5!3!} = \frac{(8 \times 7 \times 6 \times 5 \times 4 \times 3 \times 2)}{(5 \times 4 \times 3 \times 2)(3 \times 2)} = 56.$$

This number, however, includes patterns that contain the following loops, which must be subtracted from the total:

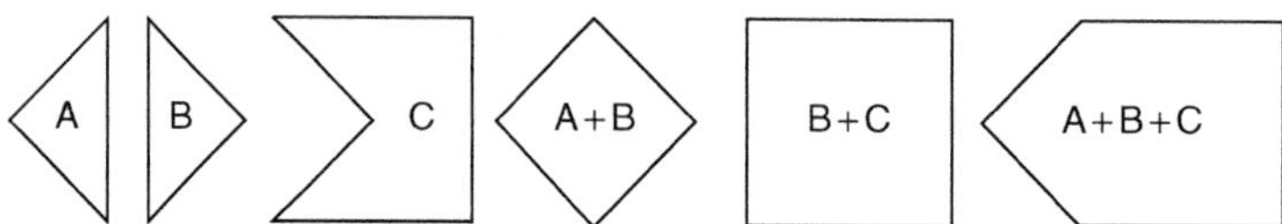

The number of patterns for a loop with r lines is given by

$$\frac{(m-r)!}{(n-1-r)!(m-n+1)!}.$$

According to this equation, for loops A and B, $r = 3$, we have 10 patterns each; for loops A + B and B + C, $r = 4$, we have 4 patterns each; and for loops C and A + B + C, $r = 5$ (note that $0! = 1$), we have 1 pattern each. The total number of loop-containing patterns to be subtracted is 30. One of the patterns, however, occurred three times in the above calculations, but should be discarded only once. This pattern involves both loop A and loop B (solid lines indicate the loop that gives rise to this pattern).

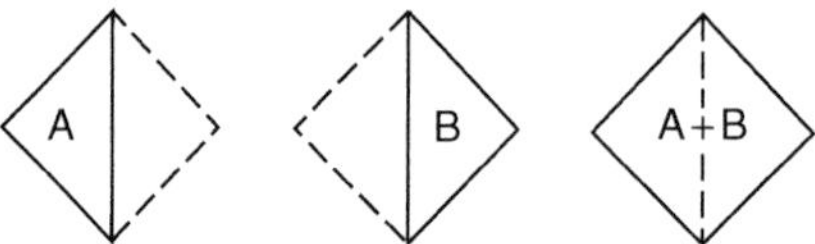

Thus, the total of loop-containing patterns is 28, and the total number of valid patterns is $56 - 28 = 28$.

The 28 5-lined patterns are shown below.

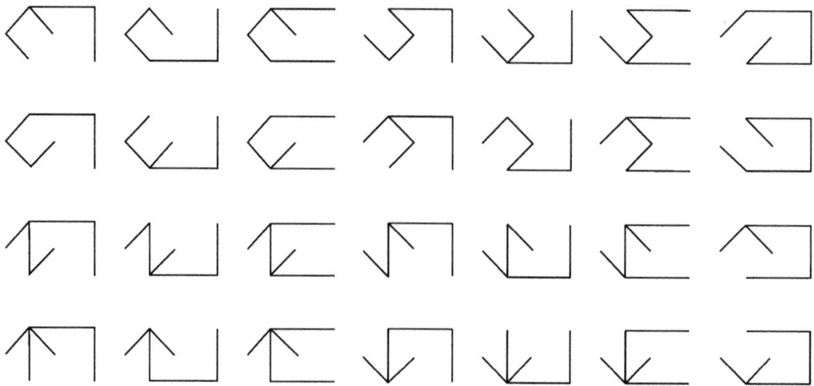

The conventional way of computing the valid King-Altman patterns is rather tedious. A set of formulas developed by the author allows the calculation of the desired number in a very short time. Each of these formulas is applicable to a particular geometric arrangement. For any figure consisting of three subfigures arrayed in sequence like the one shown in Scheme 1, the general formula for calculating the exact number of the valid King-Altman pattern, π, is

$$\pi = a \cdot b \cdot c - (l_{AB}^2 \cdot c + l_{BC}^2 \cdot a),$$

where a, b, and c = the number of lines in subfigures A, B, and C; l_{AB} and l_{BC} = the number of lines in the common boundaries between A and B, and B and C, respectively. For $a = 3, b = 3, c = 5, l_{AB} = 1$, and $l_{BC} = 2$ (Scheme 1), we have

$$\pi = 3 \times 3 \times 5 - (1^2 \times 5 + 2^2 \times 3) = 45 - (5 + 12) = 28.$$

The general equation for a scheme containing three subfigures is

$$\pi = a \cdot b \cdot c - ({l^2}_{AB} \cdot c + {l^2}_{AC} \cdot b + {l^2}_{BC} \cdot a + 2 l_{AB} \cdot l_{AC} \cdot l_{BC})$$

In the case of two subfigures A and B sharing a common boundary as shown in Scheme 2,

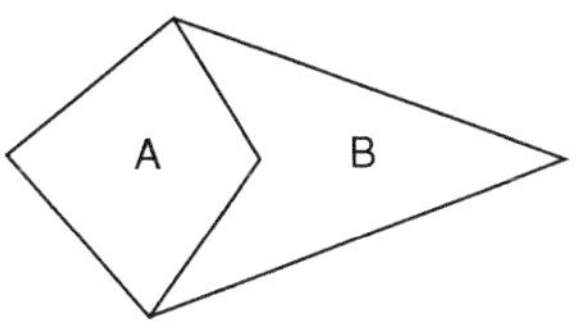

Scheme 2

the formula is given by

$$\pi = a \cdot b - l_{AB}^2 = 4 \times 4 - 2^2 = 12.$$

General formulas for calculating 4- and 5-subfigure schemes have been established.

3. The Method of Volkenstein and Goldstein

Volkenstein and Goldstein 2 have applied the theory of graphs to the derivation of rate equations. Their approach has three main features: the use of an auxiliary "node," the "compression" of a path into a point, and the addition of parallel branches. These can be best explained by an example (Scheme 3).

Each enzyme-containing species is assigned a number and referred to as a node. Suppose we want to calculate the determinant for EA (node 2). First, we choose another node, say node 3, as the auxiliary node (a reference starting point). The choice of the auxiliary node is arbitrary; it will not affect the outcome of the derivation, but may affect the amount of work involved. All the possible

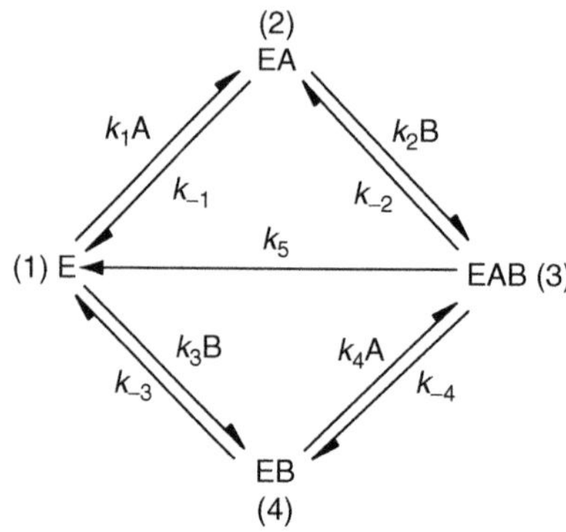

Scheme 3

pathways (flow patterns) leading from (3) to (2) are then written (marked by solid branches).

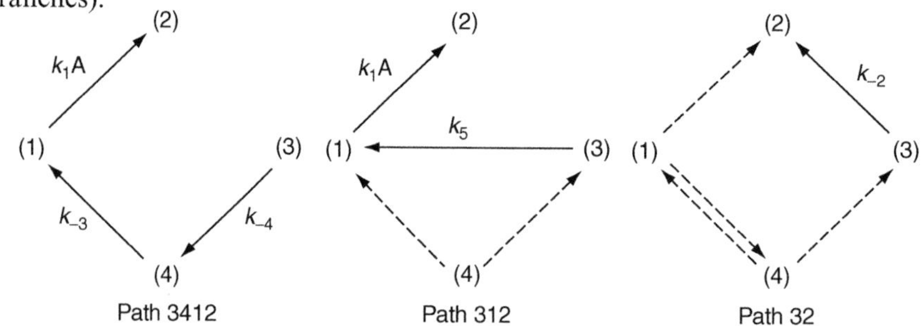

The nodes not included in the pathways retain the branches leading *away* from them (dashed branches). Since path 3412 flows through all the nodes, it is one of the terms of the determinant with a path value of $k_1k_{-3}k_{-4}A$. Path 312 ($= k_1k_5A$) and path 32 ($= k_{-2}$) are now compressed into points.

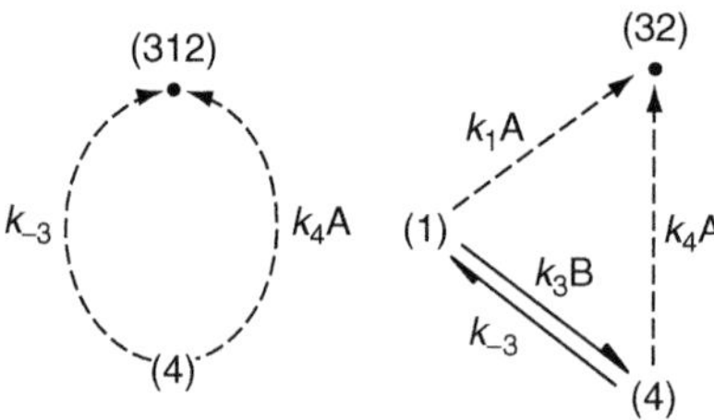

The two parallel branches leading from (4) to the compressed point (312) can be added together to yield

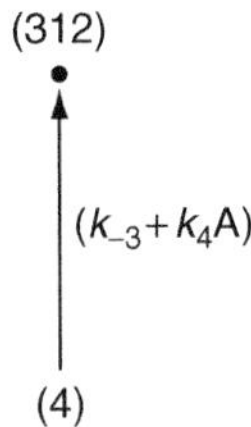

and the expression for this part is $(\mathrm{P312})(k_{-3} + k_4 A) = k_1 k_5 A(k_{-3} + k_4 A)$. The part containing point (32) can be treated by selecting a secondary auxiliary node, say node (4), and repeating the procedure described at the onset.

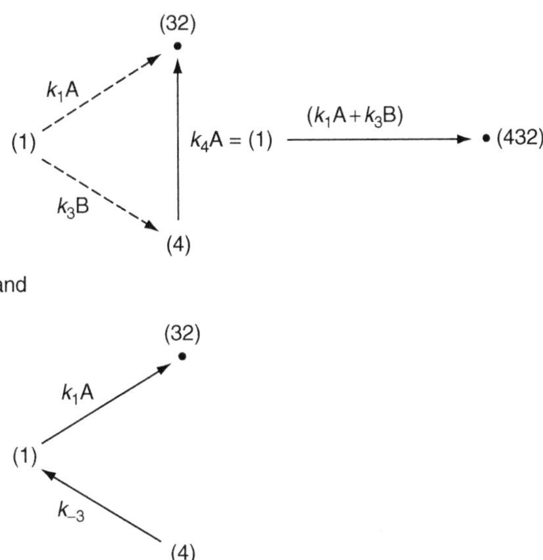

The contribution from this part is $(\mathrm{P432})(k_1\mathrm{A} + k_3\mathrm{B}) = k_{-2}k_4 A(k_1 A + k_3\mathrm{B})$ and $(\mathrm{P32})(k_1 k_{-3}\mathrm{A}) = k_1 k_{-2} k_{-3} A$. Adding the terms together, we obtain the determinant for EA

$$\begin{aligned}(\mathrm{EA}) &= k_1 k_{-3} k_{-4}\mathrm{A} + k_1 k_5 \mathrm{A}(k_{-3} + k_4\mathrm{A}) + k_{-2}k_4\mathrm{A}(k_1\mathrm{A} + k_3\mathrm{B}) + k_1 k_{-2} k_{-3}\mathrm{A} \\ &= k_1 k_{-3}(k_{-2} + k_{-4} + k_5)\mathrm{A} + k_{-2}k_3k_4\mathrm{AB} + k_1 k_4(k_{-2} + k_5)\mathrm{A}^2.\end{aligned}$$

The determinants for E, EB, and EAB can be obtained in a similar fashion. The complete rate equation is given by

$$\frac{v}{\mathrm{E}_0} = \frac{k_5(\mathrm{EAB})}{(\mathrm{E}) + (\mathrm{EA}) + (\mathrm{EB}) + (\mathrm{EAB})}.$$

Rate equations for more complex mechanisms can be derived by repeating the procedure described above as many times as necessary. The choice of the auxiliary point becomes important for reaction schemes containing several loops. The process is analogous to deciding which row (equation) should be omitted from the matrix in the determinant method. In general, one should choose, by inspection of the geometric structure of the mechanism, a node such that, if one removes from the figure the auxiliary node and the node whose determinant is desired, the remaining nodes do not form a closed loop. In addition, one should select a node situated in a symmetrical position with respect to the desired node. For instance, node (4) is a better choice as an auxiliary node for the calculation of the determinant for node (2). Node (3) was chosen for the sole purpose of illustrating the use of secondary auxiliary nodes.

4. The Systematic Approach

The systematic approach for deriving rate equations was first devised by Fromm (1970) based on certain concepts advanced by Volkenstein and Goldstein (1966). The procedure to be described here (Huang, 1978) is a modified method that includes the contributions from the aforementioned workers and from Wong and Hanes (1962).

Let us use as an example the ordered Bi Bi mechanism, in which an alternative substrate, A$'$, for the first substrate, A, is present (Scheme 4).

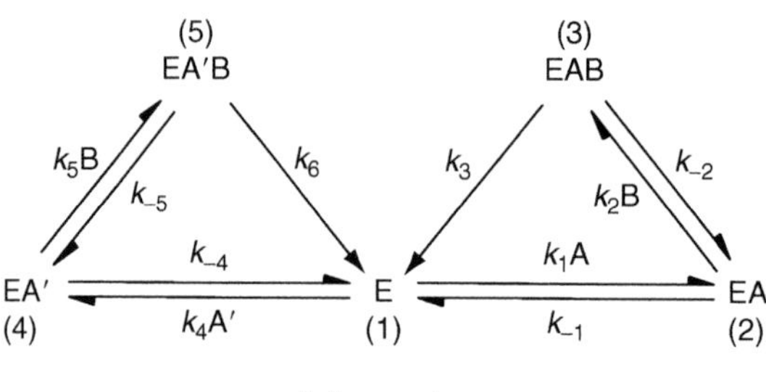

Scheme 4

Each enzyme-containing species is assigned a node number as previously described in the Volkenstein and Goldstein method. For each node, a node value is written, which is simply the summation of all branch values (rate constant and concentration factor) leading *away* from the node (Fromm, 1970):

$$
\begin{aligned}
(1) &= k_1\mathrm{A} + k_4\mathrm{A}', \\
(2) &= k_{-1} + k_2\mathrm{B}, \\
(3) &= k_{-2} + k_3, \\
(4) &= k_{-4} + k_5\mathrm{B}, \\
(5) &= k_{-5} + k_6.
\end{aligned}
$$

The determinant of a given enzyme species is equal to the noncyclic terms generated by multiplying together all the node values, excluding its own (Wong and Hanes, 1962). For example,

$$\begin{aligned}(\mathrm{E}) &= (2)(3)(4)(5)\\ &= (k_{-1} + \underline{k_2\mathrm{B}})(\underline{k_{-2}} + k_3)(k_{-4} + \underline{k_5\mathrm{B}})(\underline{k_{-5}} + k_6)\end{aligned}$$

and

$$\begin{aligned}(\mathrm{EAB}) &= (1)(2)(4)(5)\\ &= (\underline{k_1\mathrm{A}} + \underline{\underline{k_4\mathrm{A}'}})(\underline{k_{-1}} + k_2\mathrm{B})(\underline{k_{-4}} + \underline{\underline{k_5\mathrm{B}}})(\underline{k_{-5}} + \underline{k_6})\cdot\end{aligned}$$

The cyclic terms—terms that contain the products of reversible steps (underlined: $k_1\mathrm{A} \cdot k_{-1}$, $k_2\mathrm{B} \cdot k_{-2}$, $k_4A' \cdot k_{-4}$, and $k_5\mathrm{B} \cdot k_{-5}$; readily identified by their subscripts differing only in positive and negative signs)—or the product of a closed loop [underscored by dashed lines; e.g., $k_4A' \cdot k_5\mathrm{B} \cdot k_6$, constituting the (1) (4) (5) loop] are to be deleted to obtain the correct expression. The "reversible-step" terms can be eliminated during expansion:

$$\begin{aligned}(\mathrm{E}) &= (k_{-1} + \underline{k_2\mathrm{B}})(\underline{k_{-2}} + k_3)(k_{-4} + \underline{k_5\mathrm{B}})(\underline{k_{-5}} + k_6)\\ &= \left[\left(k_{-1} + \frac{k_2\mathrm{B}}{\mathrm{X}}\right)\underline{k_{-2}} + (k_{-1} + k_2\mathrm{B})k_3\right]\left[\left(k_{-4} + \frac{k_5\mathrm{B}}{\mathrm{X}}\right)\underline{k_{-5}} + (k_{-4} + k_5\mathrm{B})k_6\right]\\ &= k_{-1}k_{-4}(k_{-2} + k_3)(k_{-5} + k_6) + [k_2k_3k_{-4}(k_{-5} + k_6) + k_{-1}k_5k_6(k_{-2} + k_3)]\\ &\quad \mathrm{B} + k_2k_3k_5k_6\mathrm{B}^2.\end{aligned}$$

Note that the canceled terms are marked by an X (in practice, these terms are simply crossed out). Also note that $k_2\mathrm{B}$ and $k_5\mathrm{B}$, instead of k_{-2} and k_{-5}, are crossed out because the products $k_{-1}k_{-2}$ and $k_{-4}k_{-5}$ are not "cyclic."

The "loop terms" can be eliminated prior to expansion by a "branching" technique. From the expression

$$(\mathrm{EAB}) = (1)(2)(4)(5)$$

a quick glance at Scheme 4 will reveal that (1) (4) (5) form a closed loop. Thus, the "one-branch" approach of Fromm (1970) is first applied. With this approach, the determinant of an enzyme species, e.g., EAB, is obtained as the summation of the products of the nearest branch values leading to it (for EAB, there is only one nearest branch, EA $\rightarrow$ EAB or 23) and the remaining node values

$$(\mathrm{EAB}) = (23)(1)(4)(5).$$

Similarly

$$(\mathrm{EA}) = (12)(3)(4)(5) + (32)(1)(4)(5).$$

Note that the (1) (4) (5) loop is not eliminated by the one-branch approach. We now apply the "consecutive-branch" technique (Huang, 1978)

$$(\text{EAB}) = (23)(12)(4)(5),$$
$$(\text{EA}) = (12)(3)(4)(5) + (32)\text{X}.$$

The procedure for using this technique is as follows: (a) Only the *loop-containing* terms require further branching; e.g., since (3) (4) (5) do not form a loop, the term (12) (3) (4) (5) remains unchanged. *Unnecessary consecutive-branching may result in omission of terms.* (b) For the loop-containing terms, the first branch(es) is followed by its nearest branches and the remaining nodes not involved in these branches; e.g., the 23 branch is followed by the 12 branch. This is done by inspection of the reaction scheme. (c) In the case of (32) (1) (4) (5), since there is no branch leading from other nodes to (3), the whole term is deleted (marked by X).

When all the loops have been removed, the resultant expression is expanded to obtain the desired determinant. Thus,

$$(\text{EAB}) = k_2\text{B}\cdot k_1\text{A}(k_{-4} + \underline{k_5\text{B}})(\underline{k_{-5}} + k_6) = k_1k_2k_{-4}(k_{-5} + k_6)\text{AB} + k_1k_2k_5k_6\text{AB}^2,$$
$$(\text{EA}) = k_1\text{A}(k_{-2} + k_3)(k_{-4} + \underline{k_5\text{B}})(\underline{k_{-5}} + k_6) = k_1k_{-4}(k_{-2} + k_3)(k_{-5} + k_6)\text{A} + k_1k_5k_6(k_{-2} + k_3)\text{AB}.$$

Note that the product of reversible steps, $k_5\text{B} \cdot k_{-5}$, is canceled in the expanding process, as has been previously described.

The determinants of EA′ and EA′B can be obtained by the same approach, and the complete rate equation is expressed as v/E_0. For the example given here, if the common product P is measured, we have

$$\frac{v}{\text{E}_0} = \frac{k_3(\text{EAB}) + k_6(\text{EA}'\text{B})}{(\text{E}) + (\text{EA}) + (\text{EAB}) + (\text{EA}') + (\text{EA}'\text{B})}.$$

The first rule for using the systematic approach, broadly stated, is as follows:

Rule 1:. The determinant of a given enzyme-containing species is equal to the product of the node values of the other enzyme species, minus the reversible-step terms. When the nodes form one or more closed loops, apply the one-branch approach. Apply the consecutive-branch approach to any *remaining* loop-containing terms until all loops are eliminated.

It should be noted that unnecessary application of the one-branch approach may lead to needless algebraic manipulations. Furthermore, the branching technique often results in "redundant terms" that must be searched out and deleted. As an example, let us write the determinant for E in Scheme 4 by the one-branch method:

$$(\text{E}) = (21)(3)(4)(5) + (31)(2)(4)(5) + (41)(2)(3)(5) + (51)(2)(3)(4)$$
$$= k_{-1}(k_{-2} + k_3)(k_{-4} + k_5\text{B})(k_{-5} + k_6) + k_3(k_{-1} + k_2\text{B})(k_{-4} + k_5\text{B})(k_{-5} + k_6) + k_{-4}(k_{-1} + k_2\text{B})(k_{-2} + k_3)(k_{-5} + k_6) + k_6(k_{-1} + k_2\text{B})(k_{-2} + k_3)(k_{-4} + k_5\text{B}).$$

Expansion of the above equation will generate many redundant terms; e.g., $k_{-1}k_3k_{-4}k_6$ can be found in all four terms above, but only one is needed. These redundant terms can be eliminated by using Rule 2.

Rule 2:. The nearest branch values cannot appear in *subsequent* node values. When the consecutive-branch approach is used, the *product* of the consecutive branches cannot appear in *subsequent* terms.

Consequently, the 21 term, k_{-1}, is crossed out from all subsequent terms; the 31 term, k_3, from all subsequent node (3) terms, but not from the node (3) term preceding it; and the 41 term, k_{-4}, from the subsequent node (4) term:

$$\begin{aligned}(\mathrm{E}) = {} & k_{-1}(k_{-2}+k_3)(k_{-4}+k_5\mathrm{B})(k_{-5}+k_6) \\ & + k_3\left(\underset{\mathrm{X}}{k_{-1}} + k_2\mathrm{B}\right)(k_{-4}+k_5\mathrm{B})(k_{-5}+k_6) \\ & \left.\begin{aligned} & + k_{-4}\left(\underset{\mathrm{X}}{k_{-1}} + \underline{k_2\mathrm{B}}\right)\left(\underline{k_{-2}} + \underset{\mathrm{X}}{k_3}\right)(k_{-5}+k_6) \\ & + k_6\left(\underset{\mathrm{X}}{k_{-1}} + \underline{k_2\mathrm{B}}\right)\left(\underline{k_{-2}} + \underset{\mathrm{X}}{k_3}\right)\left(\underset{\mathrm{X}}{k_{-4}} + k_5\mathrm{B}\right) \end{aligned}\right\} \text{Canceled.}\end{aligned}$$

After elimination of the redundant terms, the last two terms are canceled because they all contain the $k_2\mathrm{B} \cdot k_{-2}$ reversible-step product. The remaining terms are identical with the expression obtained from (E) = (2) (3) (4) (5), demonstrating the fact that unnecessary branching may lead to wasteful algebraic exercise.

There are situations where certain redundant terms are difficult to eliminate. The following example serves to illustrate a procedure useful for the complicated cases. Consider the hypothetical mechanism shown in Scheme 5.

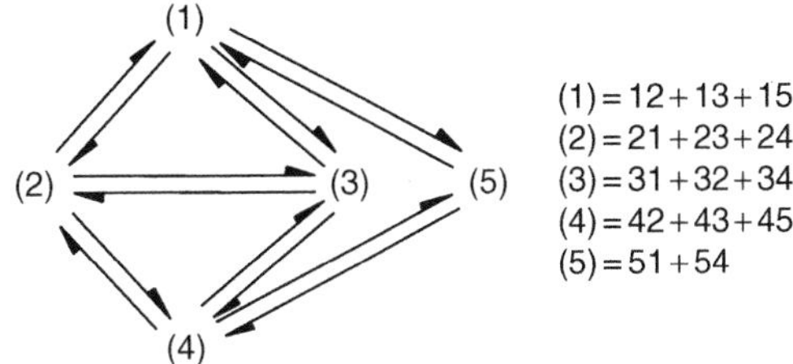

Scheme 5

The numerical notation used for Scheme 5 has several advantages. (a) The tedium of writing down all the rate constants and concentration factors is avoided. The constants can be substituted into the final expression (after canceling all the loop terms, redundant terms, and reversible-step terms) to obtain the desired determinant. (b) When two or more steps are assigned the same rate constant, e.g., k_1 and k_{-1}, this method prevents the cancellation of $k_1 \cdot k_{-1}$ product formed from different steps. (c) The reversible-step terms can still be easily identified and eliminated by their numbers, e.g., 12 · 21, 35 · 53, and etc.

Suppose we want to calculate the determinant for (5), we can write $D_5 = (1)\,(2)\,(3)\,(4)$. Applying the consecutive-branch technique, we have

$$\begin{aligned} D_5 &= (15)(2)(3)(4) + (45)(1)(2)(3) \\ &= 15[21(3)(4) + 31(2)(4)] + 45[24(1)(3) + 34(1)(2)] \\ &= 15[21(31 + 32 + 34)(42 + 43 + 45) + 31(\underline{21} + 23 + 24)(42 + 43 + 45)] \\ &\quad + 45[24(12 + 13 + \underline{15})(\underline{31} + 32 + 34) + 34(12 + 13 + \underline{15})(\underline{21} + 23 + \underline{24})]. \end{aligned}$$

Note that within the same brackets, the redundant terms (underlined by solid lines) are readily removed, but the products of consecutive branch terms, $15 \cdot 21$ and $15 \cdot 31$ (underlined by dashed lines) within the second set of brackets cannot be canceled without further manipulation. One can either eliminate them during further expansion of the equation or repeat the consecutive-branch approach until *all* the consecutive branches end with a branch leading away from a *common* node. This "common-node consecutive branching" approach is always valid—even when all the loop-containing terms have been eliminated. Using this approach, we can write

$$\begin{aligned} D_5 &= 15[21(3)(4) + 31(2)(4)] + 45[24(1)(3) + 34(1)(2)] \\ &= 15\{21(3)(4) + 31[23(4) + 43 \cdot 24]\} + 45\{24(1)(3) + 34[23(1) + 13 \cdot 21]\}. \end{aligned}$$

In the above operation, we chose node (2) as the common node; we could have chosen node (3) and obtained the same determinant. The principle involved here is analogous to the "auxiliary node" used by Volkenstein and Goldstein. It is not routinely used because the extra operations are not needed under most circumstances.

5. Comparison of Different Steady-State Methods

For relatively simple mechanisms, all the diagrammatic and systematic procedures illustrated in the foregoing sections are quite convenient. The King-Altman method is best suited for single-loop mechanisms, but becomes laborious for more complex cases with five or more enzyme forms because of the work involved in the calculation and drawing of valid patterns. With multiloop reaction schemes involving four to five enzyme species, the systematic approach requires the least effort, especially when irreversible steps are present, since it does away with pattern drawing. When the number of enzyme forms reaches six or more in a mechanism with several alternate pathways, all the manual methods become tedious owing to the sheer number of terms involved.

B. The Rapid-Equilibrium Treatment

The first rate equation for an enzyme-catalyzed reaction was derived by Henri and by Michaelis and Menten, based on the rapid-equilibrium concept. With this treatment it is assumed that there is a slow catalytic conversion step and the combination and dissociation of enzyme and substrate are relatively fast, such that they reach a state of quasi-equilibrium or rapid equilibrium.

The derivation of initial-velocity equations for any rapid-equilibrium system is quite simple. When the equilibrium relationship among various enzyme-substrate complexes are defined, the rate equation can be written simply by inspection. Consider the one-substrate system

$$\mathrm{E} + \mathrm{A} \xrightleftharpoons{K_\mathrm{a}} \mathrm{EA} \xrightarrow{k_2} (\mathrm{EA}) = [(\mathrm{E})(\mathrm{A})/K_\mathrm{a}],\ v = k_2(\mathrm{EA}).$$

We can write

$$\frac{v}{\mathrm{E}_0} = \frac{k_2(\mathrm{EA})}{(\mathrm{E}) + (\mathrm{EA})} = \frac{k_2(\mathrm{E})\mathrm{A}/K_\mathrm{a}}{(\mathrm{E}) + (\mathrm{E})\mathrm{A}/K_\mathrm{a}} = \frac{k_2(\mathrm{A}/K_\mathrm{a})}{1 + \mathrm{A}/K_\mathrm{a}} = \frac{k_2\mathrm{A}}{(K_\mathrm{a} + \mathrm{A})}$$

It is clear that there is no need to write down (E), and one can obtain the rate expression by replacing (E) with 1. Thus, for the Random Bi Bi mechanism shown below

$$\mathrm{E} + \mathrm{A} \xrightleftharpoons{K_\mathrm{ia}} \mathrm{EA},\ \mathrm{EA} + \mathrm{B} \xrightleftharpoons{K_\mathrm{b}} \mathrm{EAB} \xrightarrow{k_5} \mathrm{E} + \mathrm{P},$$

$$\mathrm{E} + \mathrm{B} \xrightleftharpoons{K_\mathrm{ia}} \mathrm{EB},\ \mathrm{EB} + \mathrm{A} \xrightleftharpoons{K_\mathrm{a}} \mathrm{EAB} \xrightarrow{k_5} \mathrm{E} + \mathrm{P},$$

we can quickly write

$$\frac{v}{\mathrm{E}_0} = \frac{k_5\mathrm{AB}/(K_{ia}K_\mathrm{b})}{1 + \mathrm{A}/K_{ia} + \mathrm{B}/K_{ib} + \mathrm{AB}/K_{ia}K_\mathrm{b}}.$$

Note that although two equations describe the formation of EAB, only one of them is used (because $K_{ia}K_\mathrm{b} = K_{ib}K_\mathrm{a}$).

In using the equilibrium treatment, one should bear in mind that the rate laws so obtained are generally different in form from those derived by the steady-state assumption for the same mechanism.

C. The Combined Equilibrium and Steady-State Treatment

There are a number of reasons why a rate equation should be derived by the combined equilibrium and steady-state approach. First, the experimentally observed kinetic patterns necessitate such a treatment. For example, several enzymic reactions have been proposed to proceed by the rapid-equilibrium random mechanism in one direction, but by the ordered pathway in the other. Second, steady-state treatment of complex mechanisms often results in equations that contain many higher-order terms. It is at time necessary to simplify the equation to bring it down to a manageable size and to reveal the basic kinetic properties of the mechanism.

The procedure to be described here was originally developed by Cha (1968). Northrop has successfully applied the method to treat the two-site ping pong mechanism of transcarboxylase (Northrop (1969)). The basic principle of cha's approach is to treat the rapid-equilibrium segment as though it were a single enzyme species at steady state with the other species. Let us consider the hybrid Rapid-equilibrium Random-ordered Bi Bi system:

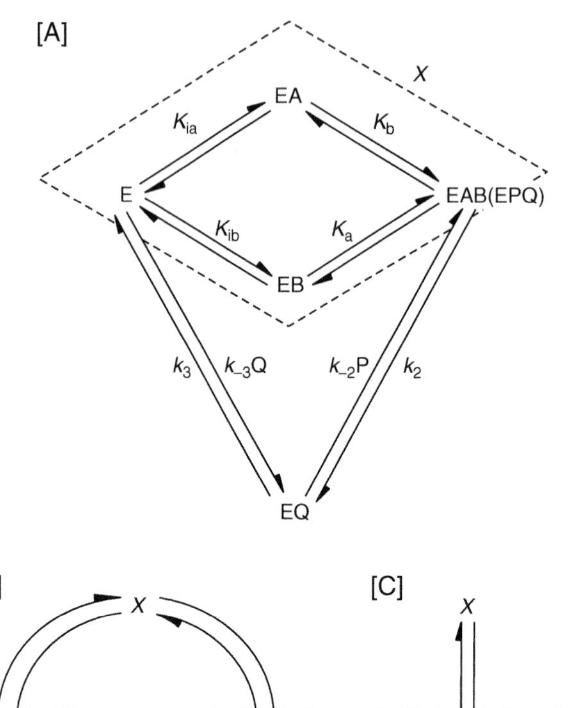

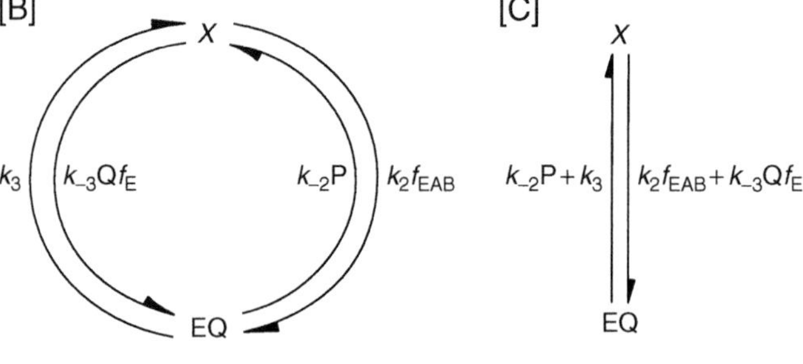

Scheme 6

The area enclosed by dashed lines in Scheme 6A is the rapid-equilibrium random segment. The segment is called *X*, and Scheme 6 is reduced to a basic figure consisting of *X* and EQ connected by two pathways (Scheme 6b):

Scheme 6b can be further condensed into one line (Scheme 6c) according to the principle of "addition of parallel branches." The rate constants k_2 and k_{-3}Q are multiplied by the fractional concentration of the enzyme species within the equilibrium segment that participates in that pathway, f_{EAB} and f_E.

$$f_{\mathrm{EAB}} = \frac{\mathrm{EAB}}{\mathrm{E + EA + EB + EAB}} = \frac{\mathrm{AB}/K_{ia}K_{\mathrm{b}}}{1 + \mathrm{A}/K_{ia} + \mathrm{B}/K_{ib} + \mathrm{AB}/K_{ia}K_{\mathrm{b}}},$$

$$f_{\mathrm{E}} = \frac{1}{1 + \mathrm{A}/K_{ia} + \mathrm{B}/K_{ib} + \mathrm{AB}/K_{ia}K_{\mathrm{b}}}.$$

From Scheme 6C, we have

$$(EQ) = k_2 f_{EAB} + k_{-3} Q f_E,$$
$$(X) = k_{-2} P + k_3.$$

Thus, we obtain the rate equation

$$\frac{v}{E_0} = \frac{k_2 f_{EAB}(X) - k_{-2} P(EQ)}{(X) + (EQ)} = \frac{k_2 k_3 f_{EAB} - k_{-2} k_{-3} PQ f_E}{k_{-2} P + k_3 + k_2 f_{EAB} + k_{-3} Q f_E}$$
$$= \frac{k_2 k_3 (AB/K_{ia} K_b) - k_{-2} k_{-3} PQ}{(k_{-2} P + k_3)(1 + A/K_{ia} + B/K_{ib} + AB/K_{ia} K_b) + k_2 (AB/K_{ia} K_b) + k_{-3} Q}.$$

This equation reveals atypical product inhibition patterns for a random mechanism: P is noncompetitive with both A and B; Q is competitive with both A and B. Whenever abnormal product inhibition patterns are observed, therefore, partial equilibrium treatment of the mechanism may be considered.

In the case of a rate-limiting step *within* a rapid-equilibrium segment, it is necessary to include such a rate-limiting step in the velocity equation. Scheme 7a serves as an example for this type of mechanism:

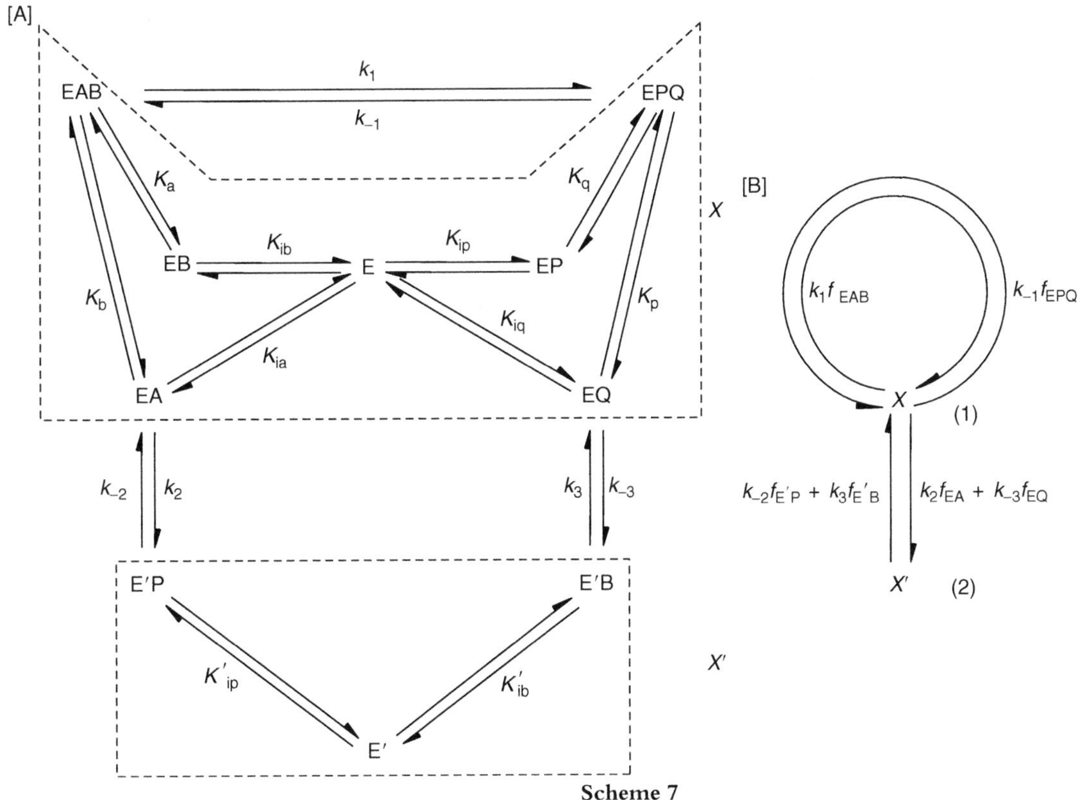

Scheme 7

The two rapid-equilibrium segments X and X' are again indicated by areas enclosed by dashed lines. Within X, there is a rate-limiting step involving the interconversion of EAB and EPQ. By treating X and X' as though they were two enzyme species, adding parallel branches together, and multiplying the rate constants with appropriate fractional enzyme concentrations, we arrive at Scheme 7b.

To obtain the determinants of X and X', the internal rate-limiting step in X is not included since it leads both away from and toward X. This can be readily shown by applying the systematic method described in the derivation of steady-state equations. Recall that the node values for X and X' are the summation of branch values leading *away* from them, as follows:

$$(1) = k_2 f_{\mathrm{EA}} + k_{-3} f_{\mathrm{EQ}},$$
$$(2) = k_{-2} f_{\mathrm{E'P}} + k_3 f_{\mathrm{E'B}}.$$

The determinants for X and X' are defined as the product of other node values excluding their own; that is, $X = (2) = k_{-2} f_{\mathrm{E'P}} + k_3 f_{\mathrm{E'B}}$ and $X' = (1) = k_2 f_{\mathrm{EA}} + k_{-3} f_{\mathrm{EA}}$.

The fractional enzyme concentrations are obtained as their fractions in the appropriate equilibrium segments:

$$f_{\mathrm{EA}} = \frac{A/K_{ia}}{1 + \mathrm{A}/K_{ia} + \mathrm{B}/K_{ib} + \mathrm{AB}/K_{ia}K_{\mathrm{b}} + \mathrm{P}/K_{ip} + \mathrm{Q}/K_{iq} + \mathrm{PQ}/K_{ip}K_{\mathrm{q}}},$$

$$f_{\mathrm{EQ}} = \frac{\mathrm{Q}/K_{iq}}{1 + \mathrm{A}/K_{ia} + \mathrm{B}/K_{ib} + \mathrm{AB}/K_{ia}K_{\mathrm{b}} + \mathrm{P}/K_{ip} + \mathrm{Q}/K_{iq} + \mathrm{PQ}/K_{ip}K_{\mathrm{q}}},$$

$$f_{\mathrm{EAB}} = \frac{\mathrm{AB}/K_{\mathrm{ia}}K_{\mathrm{b}}}{1 + \mathrm{A}/K_{ia} + \mathrm{B}/K_{ib} + \mathrm{AB}/K_{ia}K_{\mathrm{b}} + \mathrm{P}/K_{ip} + \mathrm{Q}/K_{iq} + \mathrm{PQ}/K_{iq}K_{\mathrm{q}}},$$

$$f_{\mathrm{EPQ}} = \frac{\mathrm{PQ}/K_{\mathrm{ip}}K_{\mathrm{q}}}{1 + \mathrm{A}/K_{ia} + \mathrm{B}/K_{ib} + \mathrm{AB}/K_{ia}K_{\mathrm{b}} + \mathrm{P}/K_{ip} + \mathrm{Q}/K_{iq} + \mathrm{PQ}/K_{ip}K_{\mathrm{q}}},$$

$$f_{\mathrm{E'B}} = \frac{\mathrm{B}/K'_{ib}}{1 + \mathrm{B}/K'_{ib} + \mathrm{P}/K'_{ip}},$$

$$f_{\mathrm{E'P}} = \frac{\mathrm{P}/K'_{ip}}{1 + \mathrm{B}/K'_{ib} + \mathrm{P}/K'_{ip}}.$$

In writing an expression for the initial velocity for Scheme 7b, however, one must include the internal rate-limiting step. Thus, we have

$$v = (k_1 f_{\mathrm{EAB}} - k_{-1} f_{\mathrm{EPQ}})X + k_2 f_{\mathrm{EA}} X - k_{-2} f_{\mathrm{E'P}} X'.$$

Note that in steady-state treatment, any of the pathways can be used to write the velocity expression. Either one of the two pathways linking X and X' (see Scheme 7a) will yield the same expression:

$$\begin{aligned} k_2 f_{\mathrm{EA}} X - k_{-2} f_{\mathrm{EP}} X' &= k_2 f_{\mathrm{EA}}(k_{-2} f_{\mathrm{E'P}} + k_3 f_{\mathrm{E'B}}) - k_{-2} f'_{\mathrm{EP}}(k_2 f_{\mathrm{EA}} + k_{-3} f_{\mathrm{EQ}}) \\ &= k_2 f_{\mathrm{EA}} k_3 f_{\mathrm{E'B}} - k_{-2} f_{\mathrm{E'P}} k_{-3} f_{\mathrm{EQ}}, \end{aligned}$$

$$\begin{aligned} k_3 f_{\mathrm{E'B}} X' - k_{-3} f_{\mathrm{EQ}} X &= k_3 f_{\mathrm{E'B}}(k_2 f_{\mathrm{EA}} + k_{-3} f_{\mathrm{EQ}}) - k_{-3} f_{\mathrm{EQ}}(k_{-2} f_{\mathrm{E'P}} + k_3 f_{\mathrm{E'B}}) \\ &= k_2 f_{\mathrm{EA}} k_3 f_{\mathrm{E'B}} - k_{-2} f_{\mathrm{E'P}} k_{-3} f_{\mathrm{EQ}}. \end{aligned}$$

The complete rate equation is given by

$$\begin{aligned} \frac{v}{\mathrm{E}_0} &= \frac{(k_1 f_{\mathrm{EAB}} - k_{-1} f_{\mathrm{EPQ}})X + k_2 f_{\mathrm{EA}} X - k_{-2} f_{\mathrm{E'P}} X'}{X + X'} \\ &= \frac{(k_1 f_{\mathrm{EAB}} - k_{-1} f_{\mathrm{EPQ}})(k_{-2} f_{\mathrm{E'P}} + k_3 f_{\mathrm{E'B}}) + k_2 f_{\mathrm{EA}} k_3 f_{\mathrm{E'B}} - k_{-2} f_{\mathrm{E'P}} k_{-3} f_{\mathrm{EQ}}}{k_{-2} f_{\mathrm{E'P}} + k_3 f_{\mathrm{E'B}} + k_2 f_{\mathrm{EA}} + k_{-3} f_{\mathrm{EQ}}}. \end{aligned}$$

For mechanisms involving three or more rapid-equilibrium segments, once the segments are properly represented as "nodes" in a scheme, the rate equation can be obtained by the usual "systematic approach." For example, consider the case of one substrate-one product reaction in which a modifier M is in rapid equilibration with E, EA, and EP.

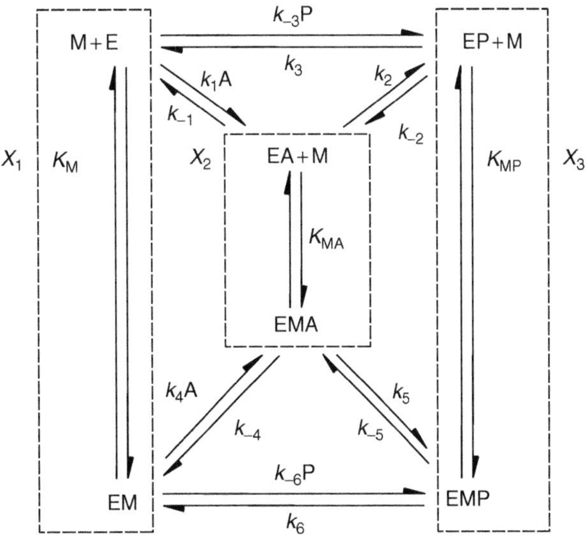

Treating X_1, X_2, and X_3 as nodes and adding parallel branches together, we obtain

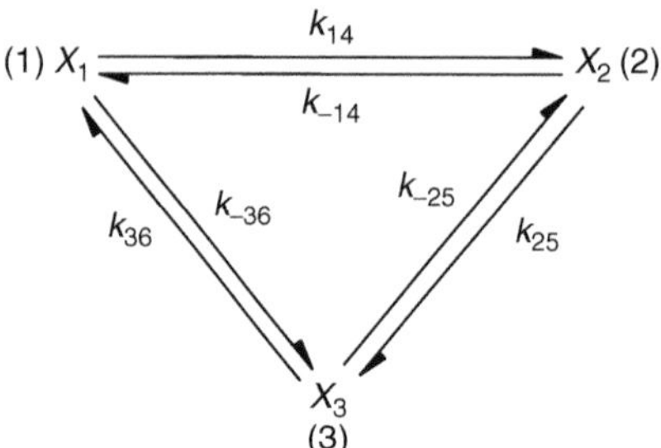

where $k_{14} = k_1\mathrm{A}f_{\mathrm{E}} + k_4\mathrm{A}f_{\mathrm{EM}}$, $k_{-14} = k_{-1}f_{\mathrm{EA}} + k_{-4}f_{\mathrm{EMA}}$, $k_{25} = k_2f_{\mathrm{EA}} + k_5f_{\mathrm{EMA}}$, $k_{-25} = k_{-2}f_{\mathrm{EP}} + k_{-5}f_{\mathrm{EMP}}$, $k_{36} = k_3f_{\mathrm{EP}} + k_6f_{\mathrm{EMP}}$, and $k_{-36} = k_{-3}\mathrm{P}f_{\mathrm{E}} + k_{-6}\mathrm{P}f_{\mathrm{EM}}$. Note that $v = k_{36}X_3 - k_{-36}X_1$.

II. Derivation of Isotope-Exchange Rate Equations

Isotope-exchange rate equations can be classified into two types: exchange at equilibrium and at steady state. The theory and technique of equilibrium exchange was pioneered by Boyer (1959). Most applications of isotope-exchange methods to enzyme systems have been of this type. Under equilibrium conditions there are no net chemical changes; whereas under steady-state conditions there is a net catalysis. Although the derivation of these two types of rate equations differs in the assumptions involved, an equation derived by the steady-state approach can be readily converted into one for equilibrium exchange. Therefore, procedures intended for steady-state exchange are equally applicable to the derivation of equilibrium-exchange rate equations.

A. Equations for Exchanges at Equilibrium

The derivation of equilibrium isotope-exchange equations for enzymic reactions was first formulated by Boyer (1959). Others have subsequently contributed to its development. Yagil and Hoberman (1969) and Flossdorf and Kula (1972) have devised generalized approaches that treat the flux of a label in a chemical reaction in a way analogous to the flow of charge in an electrical circuit. In this approach, (a) the equilibrium velocity of a reaction proceeding through n parallel steps is equal to the sum of the n individual rates; (b) when proceeding through n consecutive steps, the reciprocal of equilibrium velocity is equal to the sum of the n reciprocals. To demonstrate the use of this method, let us consider the B $\rightarrow$ P exchange in an ordered Bi Bi mechanism (Scheme 8).

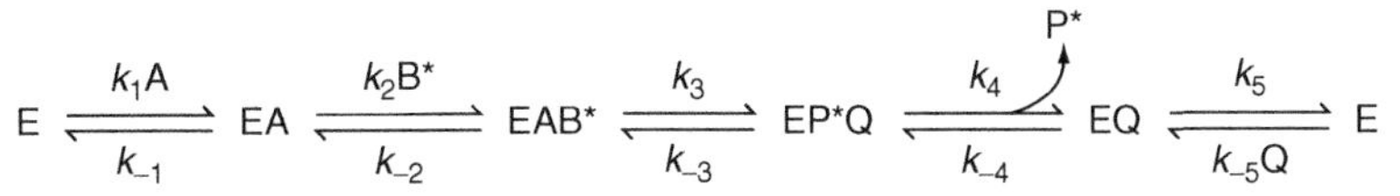

$$\mathrm{E} \underset{k_{-1}}{\overset{k_1\mathrm{A}}{\rightleftharpoons}} \mathrm{EA} \underset{k_{-2}}{\overset{k_2\mathrm{B}^*}{\rightleftharpoons}} \mathrm{EAB}^* \underset{k_{-3}}{\overset{k_3}{\rightleftharpoons}} \mathrm{EP}^*\mathrm{Q} \underset{k_{-4}}{\overset{k_4}{\rightleftharpoons}} (\mathrm{P}^*) \; \mathrm{EQ} \underset{k_{-5}\mathrm{Q}}{\overset{k_5}{\rightleftharpoons}} \mathrm{E}$$

Scheme 8

The asterisks mark the labeled species. Since the system is at equilibrium and the exchange involves only three steps, we can write a new scheme in the direction of isotopic flux:

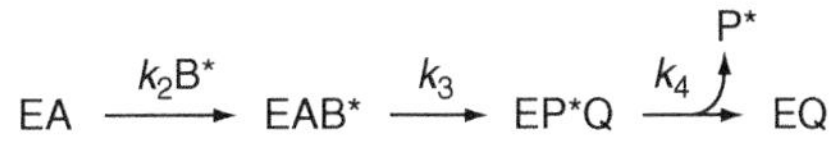

$$\mathrm{EA} \xrightarrow{k_2\mathrm{B}^*} \mathrm{EAB}^* \xrightarrow{k_3} \mathrm{EP}^*\mathrm{Q} \xrightarrow{k_4} (\mathrm{P}^*) \; \mathrm{EQ}$$

The reverse steps are not shown because they are included in the equilibrium relationships to be substituted into the equation later.

Using the rule governing consecutive steps, we have

$$\frac{1}{v^*_{\mathrm{B}\to\mathrm{P}}} = \frac{1}{k_2\mathrm{B}^*(\mathrm{EA})} + \frac{1}{k_3(\mathrm{EAB}^*)} + \frac{1}{k_4(\mathrm{EP}^*\mathrm{Q})}. \tag{7}$$

From the equilibrium relationships

$$(\mathrm{EAB}^*) = \frac{\mathrm{k}_2\mathrm{B}^*(\mathrm{EA})}{k_{-2}},$$

$$(\mathrm{EP}^*\mathrm{Q}) = \frac{\mathrm{k}_3(\mathrm{EAB}^*)}{k_{-3}} = \frac{k_2k_3\mathrm{B}^*(\mathrm{EA})}{k_{-2}k_{-3}},$$

we can substitute the expressions of (EAB*) and (EP*Q) into Eq. (7) to obtain

$$\frac{1}{v^*_{\mathrm{B}\to\mathrm{P}}} = \frac{k_3k_4 + k_{-2}(k_{-3} + k_4)}{k_2k_3k_4\mathrm{B}^*(\mathrm{EA})} \tag{8}$$

or

$$\frac{v^*_{\mathrm{B}\to\mathrm{P}}}{\mathrm{E}_0} = \frac{k_2k_3k_4\mathrm{B}^*(\mathrm{EA})/\mathrm{E}_0}{k_3k_4 + k_{-2}(k_{-3} + k_4)}, \tag{9}$$

where

$$\frac{(\mathrm{EA})}{\mathrm{E}_0} = \frac{(\mathrm{EA})}{(\mathrm{E}) + (\mathrm{EA}) + (\mathrm{EAB}) + (\mathrm{EPQ}) + (\mathrm{EQ})}.$$

Hence,

$$\frac{v^*_{\mathrm{B}\to\mathrm{P}}}{\mathrm{E}_0} = \frac{k_1k_2k_3k_4\mathrm{AB}^*}{k_{-1}[k_3k_4 + k_{-2}(k_{-3} + k_4)][1 + k_1\mathrm{A}/k_{-1} + k_1k_2\mathrm{AB}/k_{-1}k_{-2} + k_{-5}\mathrm{Q}/k_5 + k_{-4}k_{-5}\mathrm{PQ}/k_4k_5]}.$$

The derivation of exchange-rate equations for mechanisms with branched pathways requires the use of rules governing consecutive and parallel steps. Consider as an example the A → P exchange in the Random Bi Bi mechanism (Scheme 9):

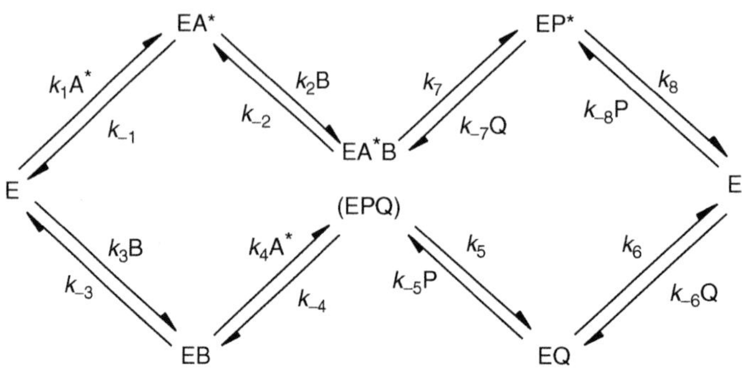

Scheme 9

Again, only the flux in one direction need be considered, and the steps not involved in the flux are ignored (Scheme 10).

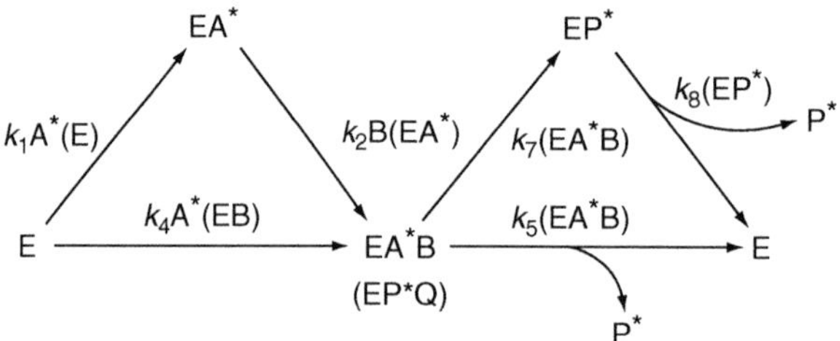

Scheme 10

The first step is to obtain the expressions for the E → EA* → EA*B and EA*B → EP* → E pathways. Let us call them k_{12} and k_{78}:

$$\frac{1}{k_{12}} = \frac{1}{k_1\mathrm{A}^*(\mathrm{E})} + \frac{1}{k_2\mathrm{B}(\mathrm{EA})}; \quad k_{12} = \frac{k_1\mathrm{A}^*(\mathrm{E}) \cdot k_2\mathrm{B}(\mathrm{EA})}{k_1\mathrm{A}^*(\mathrm{E}) + k_2\mathrm{B}(\mathrm{EA})},$$

$$\frac{1}{k_{78}} = \frac{1}{k_7(\mathrm{EA}^*\mathrm{B})} + \frac{1}{k_8(\mathrm{EP}^*)}; \quad k_{78} = \frac{k_7(\mathrm{EA}^*\mathrm{B}) \cdot k_8(\mathrm{EP}^*)}{k_7(\mathrm{EA}^*\mathrm{B}) + k_8(\mathrm{EP}^*)}.$$

Scheme 10 can now be represented by Scheme 11.

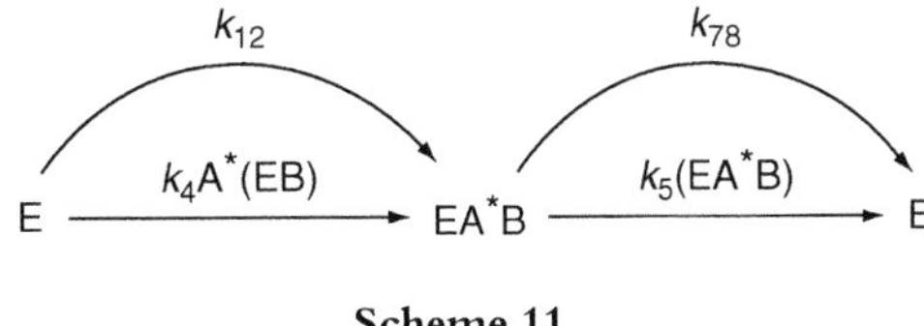

Scheme 11

Since parallel steps are additive, we have

$$\mathrm{E} \xrightarrow{k_4\mathrm{A}^*(\mathrm{EB})+k_{12}} \mathrm{EA^*B} \xrightarrow{k_5(\mathrm{EA^*B})+k_{78}} \mathrm{E}$$

and finally

$$\frac{1}{v^*_{\mathrm{A}\rightarrow\mathrm{P}}} = \frac{1}{k_4\mathrm{A}^*(\mathrm{EB}) + k_{12}} + \frac{1}{k_5(\mathrm{EA^*B}) + k_{78}}. \tag{10a}$$

The exchange rate equation is obtained by substituting the expressions for k_{12}, k_{78}, and the equilibrium expressions for the enzyme intermediates into Eq. (10a).

From the following equilibrium relationships

$$(\mathrm{EA}^*) = k_1\mathrm{A}^*(\mathrm{E})/k_{-1},$$
$$(\mathrm{EB}) = k_3\mathrm{B}(\mathrm{E})/k_{-3},$$
$$(\mathrm{EA^*B}) = k_4\mathrm{A}^*k_3\mathrm{B}(\mathrm{E})/k_{-4}k_{-3},$$
$$(\mathrm{EP}^*) = k_7(\mathrm{EA^*B})/k_{-7}\mathrm{Q},$$

and

$$(k_1\mathrm{A}k_2\mathrm{B}/k_3\mathrm{A}k_4\mathrm{B}) = (k_{-1}k_{-2}/k_{-3}k_{-4})$$

it can be shown that

$$\begin{aligned}\frac{1}{v^*_{\mathrm{A}\rightarrow\mathrm{P}}} &= \frac{1}{k_4\mathrm{A}^*(\mathrm{EB}) + [k_1\mathrm{A}^*(\mathrm{E}) \cdot k_2\mathrm{B}(\mathrm{EA}^*)]/[k_1\mathrm{A}^*(\mathrm{E}) + k_2\mathrm{B}(\mathrm{EA}^*)]} \\ &+ \frac{1}{k_5(\mathrm{EA^*B}) + [k_7(\mathrm{EA^*B}) \cdot k_8(\mathrm{EP}^*)]/[k_7(\mathrm{EA^*B}) + k_8(\mathrm{EP}^*)]} \\ &= \frac{1}{(\mathrm{E})}\frac{[(k_{-1} + k_2\mathrm{B})(k_{-7}\mathrm{Q} + k_8)(k_{-4} + k_5) + k_7k_8(k_{-1} + k_2\mathrm{B}) + k_{-1}k_{-2}(k_{-7}\mathrm{Q} + k_{-8})]}{[k_1\mathrm{A}k_2\mathrm{B} + k_4\mathrm{A}k_3\mathrm{B}(k_{-1} + k_2\mathrm{B})/k_{-3}][k_5(k_{-7}\mathrm{Q} + k_8) + k_8k_7]}.\end{aligned} \tag{10b}$$

B. Derivation by the Steady-State Method

Britton (1964) first derived isotope flux equations under steady-state rather than equilibrium conditions. To illustrate his procedure, we shall again use Scheme 8, the B $\rightarrow$ P exchange in ordered Bi Bi mechanism, as an example, so that the results can be compared (Scheme 8A).

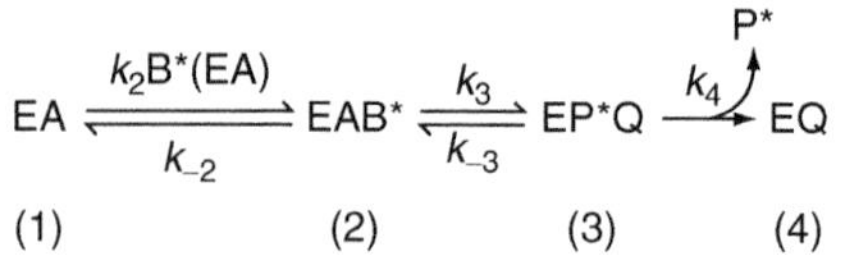

$$\mathrm{EA} \underset{k_{-2}}{\overset{k_2\mathrm{B}^*(\mathrm{EA})}{\rightleftharpoons}} \mathrm{EAB}^* \underset{k_{-3}}{\overset{k_3}{\rightleftharpoons}} \mathrm{EP}^*\mathrm{Q} \xrightarrow{k_4} \mathrm{EQ} + \mathrm{P}^*$$

(1) (2) (3) (4)

Scheme 8 (A)

Note that (a) the reverse steps are needed for steady-state treatment; (b) the k_{-4} step is not shown because the initial rate of exchange is being measured; and (c) only the unlabeled enzyme concentration is included in the derivation because the concentration factors of labeled enzyme forms will cancel out during the derivation. Also, each enzyme form is assigned a number for reference purposes.

The procedure is to calculate the B $\rightarrow$ P or 1 $\rightarrow$ 4 isotope transfer in a stepwise manner using partition theory:

$$\text{Flux } 1\rightarrow 2 = k_2\mathrm{B}^*(\mathrm{EA}),$$

$$\text{Flux } 1\rightarrow 3 = (1\rightarrow 2)\cdot\frac{(2\rightarrow 3)}{(2\rightarrow 3)+(2\rightarrow 1)} = \frac{k_2\mathrm{B}^*(\mathrm{EA})k_3}{k_3+k_{-2}},$$

$$\begin{aligned}\text{Flux } 1\rightarrow 4 &= (1\rightarrow 3)\cdot\frac{(3\rightarrow 4)}{(3\rightarrow 4)+(3\rightarrow 1)}\\ &= \frac{k_2\mathrm{B}^*(\mathrm{EA})k_3}{k_3+k_{-2}}\cdot\frac{k_4}{k_4+[k_{-2}k_{-3}/(k_3+k_{-2})]}\\ &= \frac{k_2k_3k_4\mathrm{B}^*(\mathrm{EA})}{k_4(k_3+k_{-2})+k_{-2}k_{-3}}\\ &= v^*_{\mathrm{B}\rightarrow\mathrm{P}}.\end{aligned} \tag{11}$$

Note that Eq. (11) is identical with Eq. (8). More recently, Cleland (1975) has developed an approach that is more convenient. His procedure starts with the release of labeled product and works backward as follows:

$$\text{Flux } 3\rightarrow 4 = k_4,$$

$$\text{Flux } 2\rightarrow 4 = (2\rightarrow 3)\cdot\frac{(3\rightarrow 4)}{(3\rightarrow 4)+(3\rightarrow 2)} = \frac{k_3k_4}{k_4+k_{-3}},$$

$$\begin{aligned}\text{Flux } 1\rightarrow 4 &= (1\rightarrow 2)\cdot\frac{(2\rightarrow 4)}{(2\rightarrow 4)+(2\rightarrow 1)}\\ &= \frac{k_2\mathrm{B}^*(\mathrm{EA})k_3k_4/(k_4+k_{-3})}{[k_3k_4/(k_4+k_{-3})]+k_{-2}}\\ &= \frac{k_2k_3k_4\mathrm{B}^*(\mathrm{EA})}{k_3k_4+k_{-2}(k_4+k_{-3})}.\end{aligned}$$

These procedures, however, are more suitable for deriving exchange-rate equations for mechanisms without branched pathways. For more complex mechanisms, the schematic method of Cleland (1967) based on the approach of King and Altman is quite convenient. This method can be adapted to the systematic approach and combined with the deletion of steps linking unlabeled species previously described to further reduce the amount of work involved (Huang, 1978). Consider the A → P exchange in the Random Bi Bi mechanism (cf. Scheme 9; the figure has been redrawn in a folded geometric form) shown in Scheme 12.

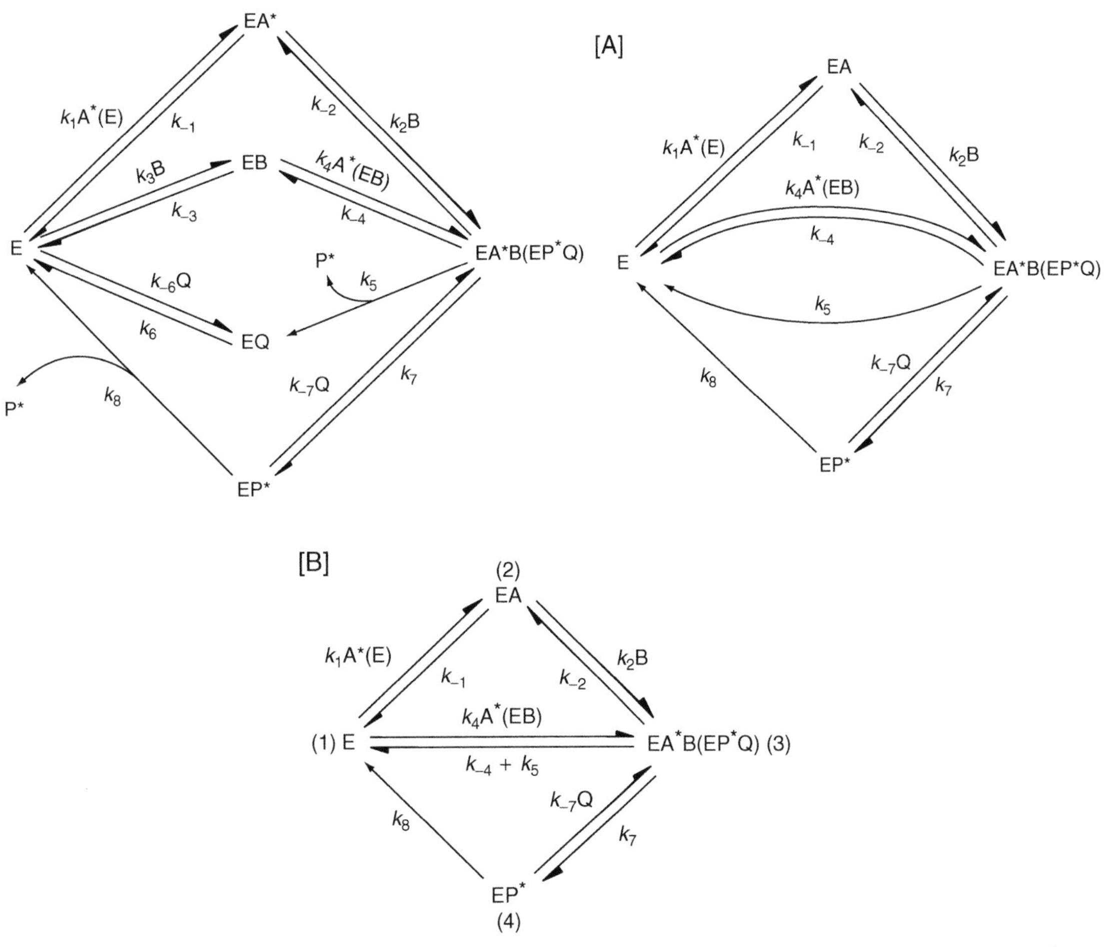

Scheme 12

The pathways connecting the unlabeled enzymes species, E $\leftrightarrow$ EB and E $\leftrightarrow$ EQ, as has been shown in Scheme 10, can be eliminated; and the EB $\leftrightarrow$ EAB and EAB $\rightarrow$ EQ steps can be directly linked to E (Scheme 12A).

By combining the k_{-4} and k_5 branches, Scheme 12A can be further reduced to Scheme 12B.

Now we apply the systematic method to Scheme 12B by assigning node numbers to the enzyme species and write down the node values:

$$\begin{aligned}
(1) &= k_1\mathrm{A}^*(\mathrm{E}) + k_4\mathrm{A}^*(\mathrm{EB}),\\
(2) &= k_{-1} + k_2\mathrm{B},\\
(3) &= k_{-2} + k_{-4} + k_5 + k_7,\\
(4) &= k_{-7}\mathrm{Q} + k_8.
\end{aligned}$$

We then use Cleland's procedure to write the expression for the initial rate of formation of labeled P from A.

$$v^* = d\mathrm{P}^*/dt = k_5(\mathrm{EA^*B}) + k_8(\mathrm{EP}^*) = \frac{k_5 N_{\mathrm{EA*B}} + k_8 N_{\mathrm{EP*}}}{D},$$

where $N_{\mathrm{EA*B}}$, $N_{\mathrm{EP*}}$, and D are the "determinants" of EA*B, EP*, and E in Scheme 12B. The N terms vary with the mechanism, depending on the enzyme species from which the labeled product is released, but the D term is always obtained as the "determinant" of E.

Thus, we can write

$$\begin{aligned}
N_{\mathrm{EA*B}} &= (1)(2)(4) = [k_1\mathrm{A}^*(\mathrm{E}) + k_4\mathrm{A}^*(\mathrm{EB})](k_{-1} + k_2\mathrm{B})(k_{-7}\mathrm{Q} + k_8)\\
&= k_1\mathrm{A}^*(\mathrm{E})(k_2\mathrm{B})(k_{-7}\mathrm{Q} + k_8) + k_4\mathrm{A}^*(\mathrm{EB})(k_{-1} + k_2\mathrm{B})(k_{-7}\mathrm{Q} + k_8),\\
N_{\mathrm{EP*}} &= (34)(1)(2) = k_7[k_1\mathrm{A}^*(\mathrm{E}) + k_4\mathrm{A}^*(\mathrm{EB})](k_{-1} + k_2\mathrm{B})\\
&= k_7 \cdot k_1\mathrm{A}^*(\mathrm{E}) \cdot k_2\mathrm{B} + k_7k_4\mathrm{A}^*(\mathrm{EB})(k_{-1} + k_2\mathrm{B}),\\
D &= (2)(3)(4) = (k_{-1} + k_2\mathrm{B})(k_{-2} + k_{-4} + k_5 + k_7)(k_{-7}\mathrm{Q} + k_8)\\
&= (k_{-4} + k_5)(k_{-1} + k_2\mathrm{B})(k_{-7}\mathrm{Q} + k_8) + k_{-1}k_{-2}(k_{-7}\mathrm{Q} + k_8)\\
&\quad + k_7k_8(k_{-1} + k_2\mathrm{B}).
\end{aligned}$$

The complete exchange equation is obtained as

$$\frac{v^*}{\mathrm{E}_0} = \frac{(k_5 N_{\mathrm{EA*B}} + k_8 N_{\mathrm{EP*}})/D}{(\mathrm{E}) + (\mathrm{EA}) + (\mathrm{EB}) + (\mathrm{EAB}) + (\mathrm{EP}) + (\mathrm{EQ})}.$$

It should be noted that (E), (EA), (EB), (EAB), (EP), and (EQ) are now the *normal* determinants of E, EA, EB, EAB, EP, and EQ obtained from the steady-state treatment of the intact reaction scheme. If the equilibrium-exchange rate equation is desired, (E), (EA), (EB), (EAB), (EP), and (EQ) should be obtained from the equilibrium relationships.

The following derivations demonstrate that an isotope exchange rate equation obtained by the steady-state treatment can be converted to one of exchange at equilibrium:

$$v^* = \frac{k_5[k_1A^*k_2B(k_{-7}Q + k_8)(E) + k_4A^*(k_{-1} + k_2B)(k_{-7}Q + k_8(EB)] + k_8[k_7k_1A^*k_2B(E) + k_7k_4A^*(k_{-1} + k_2B)(EB)]}{(k_{-4} + k_5)(k_{-1} + k_2B)(k_{-7}Q + k_8) + k_{-1}k_{-2}(k_{-7}Q + k_8) + k_7k_8(k_{-1} + k_2B)}.$$

Substitution of $(EB) = k_3\,B(E)/k_{-3}$ into the above expression yields

$$v^* = \frac{[k_5(k_{-7}Q + k_8) + k_8k_7][k_1A^*k_2B + k_4A^*k_3B(k_{-1} + k_2B)/k_{-3}](E)}{(k_{-4} + k_5)(k_{-1} + k_2B)(k_{-7}Q + k_8) + k_{-1}k_{-2}(k_{-7}Q + k_8) + k_7k_8(k_{-1} + k_2B)},$$

which is identical with Eq. 10b derived by the equilibrium treatment.

The systematic method is equally convenient for the derivation of rate equations for simple mechanisms. Scheme 8, for example, can be redrawn as an enclosed figure after deleting the pathways between unlabeled enzyme forms.

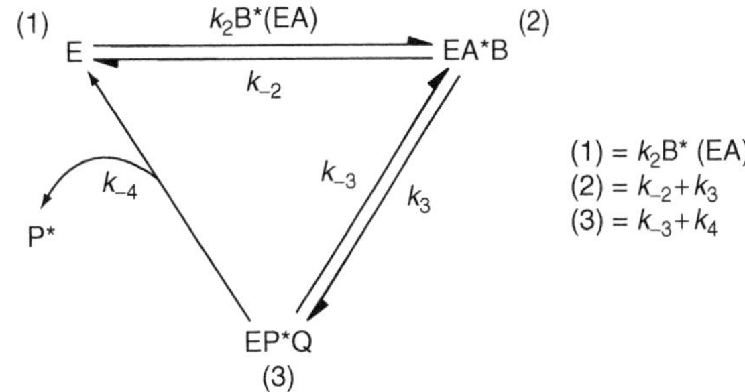

$$v^*_{B\to P} = \frac{k_4 N_{EP*Q}}{D}$$

$$N_{EP*Q} = (1)(2) = k_2B^*(EA)(k_{-2} + k_3) = k_2k_3B^*(EA)$$

$$D = (2)(3) = (k_{-2} + k_3)(k_{-3} + k_4) = k_{-2}(k_{-3} + k_4) + k_3k_4$$

Thus

$$v^* = \frac{k_2k_3k_4B^*(EA)}{k_{-2}(k_{-3} + k_4) + k_3k_4}.$$

III. Concluding Remarks

In this chapter, the derivation of initial-velocity equations under steady-state, rapid-equilibrium, and the hybrid rapid-equilibrium and steady-state conditions has been covered. The derivation of isotope exchange rate laws at chemical equilibrium and steady state has also been dealt with. Derivation of initial velocity equation for the quasi-equilibrium case is quite straightforward once the equilibrium relationships among various enzyme-containing species are defined. The combined rapid equilibrium and steady-state treatment can be reduced to the steady-state method by treating the equilibrium segments as though they were enzyme intermediates. Rate equations for isotope exchange at equilibrium can be obtained by substitution of equilibrium expressions for unlabeled enzyme forms into equations derived by the steady-state procedure. Clearly, different types of rate equations are within reach when one has mastered the steady-state approach.

Albeit several methods for deriving steady-state rate equations have been described, one has to be proficient at only one of them since, for a given reaction scheme, they all lead to the same equation. The method to be recommended here is the *systematic* approach. It requires the least amount of time and work, especially when irreversible steps are present in the reaction scheme. Furthermore, no pattern drawing is needed in this approach. The initial setup is done by systematic inspection of the reaction diagram, and the risk of omission of terms (due to omission of certain patterns) is minimized.

It should be emphasized that the validity of a rate equation depends also on whether the reaction diagram is properly constructed; that is, whether the number and types of enzyme species involved, the pathways linking these species, the sequence of reaction steps, etc., are carefully considered and appropriately arranged. While the procedures presented in this chapter allow one to obtain correct algebraic expressions, the rate equations so obtained are only as valid as the assumptions made in constructing the reaction scheme.

Addendum

Thirty years have passed since this chapter was first published in 1979. No major improvement in deriving rate equations has emerged, most likely due to (1) that the existing methods are already well developed and (2) that research in this area has not been active. Thanks to the advance in technology, however, computer-assisted procedures are available to ease the tedium of obtaining complex equations. In the meantime, it has become clear to this author that people must be reminded to observe several basic physicochemical principles. For instance, in a given reaction cycle, when unconsumed ligands are involved or equilibrium is attained, the product of all the rate constants in one direction must equal the product of those in the other direction. Thus, in Scheme 9, $k_1k_2k_{-4}k_{-3}$ must equal $k_{-1}k_{-2}k_4k_3$ and $k_1k_2k_5k_6$AB must equal $k_{-1}k_{-2}k_{-5}k_{-6}$PQ during equilibrium exchange, and so on. These relationships dictate that certain terms in an equation should be eliminated or cancelled to reach a correct "final" equation. Failure to do so often lead to wrong conclusions based on improper derivations. Furthermore, in simulation analyses, the size of each constant must also obey this restriction for each cycle. In addition, for a given condition, the chemical equilibrium constant cannot change. The Haldane relationship requires that one cannot change one constant, e.g., a larger k_{cat} as a result of activation, without compensating for it elsewhere. One should also keep in mind that the equations are applicable only to data collected by initial rate measurements. Non-initial rate assays and ligands present at levels comparable to that of the enzyme must be treated by taking these factors into account.

In this updated chapter, a few references and equations have been given and a few misprints have been corrected. As a tip, it is worthwhile to re-emphasize two points to those who derive rate equations by the conventional methods. (With simple mechanisms, there is no need to employ the computer-assisted method since the time needed to set up the differential equations and input the constants is sufficient to allow one to do it by "long hand.") First, for any mechanism involving only a single "loop," the King-Altman method is the best. Second, for mechanisms containing more than one loop, use the "systematic" method by using numerical notations to express the node values at the beginning and replacing them with rate constants and concentration factors to obtain the final equation.

References

Boyer, P. D. (1959). *Arch. Biochem. Biophys.* **82,** 387.
Briggs, G. E., and Haldane, J. B. S. (1925). *Biochem. J.* **19,** 338.
Britton, H. G. (1964). *J. Physiol. (Lond.)* **170,** 1.
Cha, S. (1968). *J. Biol. Chem.* **243,** 820.
Cleland, W. W. (1967). *Annu. Rev. Biochem.* **36,** 77.
Cleland, W. W. (1975). *Biochemistry* **14,** 3220.
Flossdorf, J., and Kula, M. (1972). *Eur. J. Biochem.* **30,** 325.
Fromm, H. J. (1970). *Biochem. Biophys. Res. Commun.* **40,** 692.
Fromm, S. J., and Fromm, H. J. (1999). *Biochem. Biophys. Res. Commun.* **265,** 448.
Huang, C. Y. (1978). *Fed. Proc., Fed. Am. Soc. Exp. Biol.* **37,** 1423.
King, E. L., and Altman, C. (1956). *J. Phys. Chem.* **60,** 1375.
Northrop, D. B. (1969). *J. Biol. Chem.* **244,** 5808.
Volkenstein, M. V., and Goldstein, B. N. (1966). *Biochim. Biophys. Acta* **115,** 471.
Wong, J. T. (1975). "Kinetics of Enzyme Mechanisms." Academic Press, New York.
Wong, J. T., and Hanes, C. S. (1962). *Can. J. Biochem. Physiol.* **40,** 763.
Yagil, G., and Hoberman, H. D. (1969). *Biochemistry* **8,** 352.

CHAPTER 2

Practical Considerations in the Design of Initial Velocity Enzyme Rate Assays

R. Donald Allison and Daniel L. Purich

Department of Biochemistry & Molecular Biology
University of Florida College of Medicine
Gainesville, Florida 32610-0245, USA

I. Introduction & Update

Developing a reliable initial velocity enzyme assay procedure is of prime importance for achieving a detailed and faithful analysis of any enzyme. This objective is quite different from the use of enzyme assays in enzyme purification or clinical chemistry, where the focus is on estimates of the enzyme content of various samples. In this case, one is particularly concerned with optimizing assay conditions by including substrates, cofactors, and activators at optimal (often saturating) levels and with minimizing interfering agents. Thus, the emphasis is on determining enzyme concentration in a routine, easy, and reproducible fashion. On the other hand, the enzyme kineticist must often work at subsaturating

CONTEMPORARY ENZYME KINETICS AND MECHANISM

DOI: 10.1016/B978-0-12-378608-1.00002-5

substrate and effector levels to evaluate the rate-saturation behavior. When two or more substrates are involved, the problem of obtaining initial velocity data becomes more considerable. This chapter treats of the practical aspects of initial rate enzyme assay.

Although written nearly three decades ago, this chapter continues to define the underlying logic and best-practice approaches for designing and implementing reliable initial-rate enzyme assays. The following comments are offered with the goal of providing additional guidance to the interested reader.

As noted in the chapter, the defining feature of all initial-rate assays is that the observed reaction rate v must exhibit a linear dependence on the total enzyme concentration $[E_{Total}]$, irrespective of: (a) the substrate concentration, (b) the type of assay (e.g., stopped-time or continuous), or (c) the method of detection (e.g., spectral, radiometric, etc.). This linear relationship is an implicit property of the Michaelis-Menten ($v = V_m/\{1 + K_m/[S]\} = k_{cat}[E_{Total}]/\{1 + K_m/[S]\}$) as well as any similar rapid-equilibrium and steady-state initial rate equation for reactions requiring one or more substrates. The only exception to this statement is that self-associating enzymes may fail to yield a linear dependence, if the enzyme's activity depends on its state of oligomerization.

With the burgeoning use of mutant enzymes produced by site-directed mutagenesis, there is always a need for the experimenter to confirm that the enzyme rate assay still yields a linear dependence of v on $[E_{Total}]$. This is especially true for continuous assays employing one or more auxiliary enzymes, because the lag-phase in coupled enzyme assays depends on the kinetic parameters of wild-type and mutant enzyme forms.

Finally, those interested in applying initial-rate measurements in enzyme kinetic investigations are strongly encouraged to master the methods for deriving rate equations, as described in Chapter 1.

II. General Experimental Design

The initial rate phase of an enzymic reaction typically persists for 10 s to several hundred seconds. Thus, various methods including spectrophotometry, radioactive assay, and pH-stat procedures may be used along with manual mixing and manipulation of samples. Prior to addition of the enzyme (or one of the substrates) to initiate the reaction, the assay sample (usually in 0.05–3.0 ml volumes) is preincubated at the reaction temperature for several minutes to achieve thermal equilibration, and a small aliquot of enzyme is added to initiate the reaction. The increase in the product concentration or the drop in substrate concentration may then be measured. The basic goals are to initiate the reaction in a manner that leads to immediate attainment of the initial velocity phase and to obtain an accurate record of the reaction progress.

For most enzyme rate equations to apply to real systems, one must be certain that the conditions placed upon the mathematical derivation are satisfied in the experiment. Since the rate equations become quite complex as product accrual becomes significant, the initial rate assumption is frequently taken to linearize the equations. Experimentally, one draws the tangent to the reaction progress curve as shown in Fig. 1. The best estimates of the slope of this line will be obtained from the most complete record of the initial rate phase, and continuous assays are thus preferable to single-point assays. The duration of the linear initial rate phase depends upon many factors, including the equilibrium constant, the fractional saturation of the enzyme with substrate(s) and product(s), the buffering capacity of the medium, and the concentration ratio of the least abundant substrate relative to the enzyme. Below a $[S]_{total}/[E]_{total}$ value of 100, the steady state may not persist for long, and nonlinear initial rates will frequently be observed. In some cases, the rate may appear linear, but virtual linearity should not be the only criterion used in establishing reaction conditions. With conditions such as a very favorable equilibrium constant, no product inhibition, and a high $[S]_{total}/[E]_{total}$ value, the initial reaction velocities can be maintained for a considerable period of time.

III. The Initial Rate Condition

As a general guideline, one assumes that the initial rate persists for a period of time during which the substrate(s) concentration is within 10% of the initial value. This is probably true only for reactions that are thermodynamically quite favorable, and even so it is best to choose an assay method that is safely within this range. Nonetheless, there is no *a priori* guarantee that product inhibition will not account for a significant error in the estimation of initial rates, and tests should be made for even the most favorable reactions. Since the equilibrium constant for the particular reaction will presumably be known, one may estimate the extent of the reaction by the following simple expression:

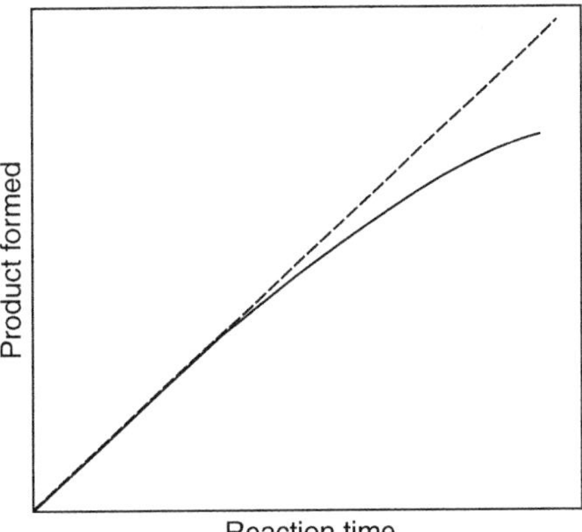

Fig. 1 Plot of product formation versus reaction time for an enzyme-catalyzed reaction. The solid line represents the reaction progress, and the dashed line is the tangent to the curve at low product formation. This tangent is the initial velocity, and it is expressed in units of molarity per minute.

$$K'_t = \frac{(P_0 + x)(Q_0 + x)(R_0 + x)\ldots}{(A_0 - x)(B_0 - x)(C_0 - x)\ldots} \quad (1)$$

where K'_t is the apparent product: substrate ratio (or "mass-action ratio") at time t, x is the concentration change measured at the midpoint of the experimental assay (i.e., time t), and A_0, B_0, C_0, P_0, Q_0, and R_0 are the initial substrate and product concentrations. If it is found that K'_t is not much different from the apparent equilibrium constant (K') for the reaction, then one must reduce x by use of a more sensitive assay. Let us consider the case of yeast hexokinase, where the apparent equilibrium constant ($K' = 4.9 \times 10^3$ at pH 7.5) is quite favorable. Assuming we have both the glucose and ATP concentrations at 0.1 mM (i.e., near their Michaelis constants), then a 5% conversion would yield a K'_t of 2.8×10^{-3}, suggesting that the system is quite far from equilibrium. On the other hand, we may consider the acetate kinase reaction (written in the direction of acetyl phosphate formation where $K' = 3.3 \times 10^{-4}$ at pH 7.4). At an acetate concentration of 10 mM and an ATP level of 1 mM, a 5% conversion of substrate to products would yield a K'_t of 2.7×10^{-4}, not far away from the equilibrium value. In this case, it would be advisable to reduce the percentage of substrate conversion in the rate assay. An obvious extension of these comments is that the deviation from initial rates may be a greater problem in some product inhibition studies.

It is not certain that product accumulation during the "initial rate" will lead to insignificant error even when substrate conversion is quite low. Indeed, for some systems the inhibition constants for a particular product may be quite low. Consider the brain hexokinase reaction where $K_{\text{glucose-6-P}}$ is approximately 10^{-5} M (Ning *et al.*, 1969) but the K_{glucose} is about fivefold higher. Another case in point is the PRPP:ATP phosphoribosyltransferase from *Salmonella typhimurium*; here N^1-phosphoribosyl-ATP has a dissociation constant of 3.8×10^{-6} M, but the affinity for either substrate is considerably lower (Kleeman and Parsons, 1976). The reduced coenzyme in NAD^+-dependent dehydrogenases is frequently a potent inhibitor as well. One strategy around the problem of product accumulation is to remove the product by use of an auxiliary enzyme system (see Chapter 3, this volume). This can be especially useful when the auxiliary system also serves to regenerate one of the substrates. For example, the pyruvate kinase/lactate dehydrogenase-coupled assay for kinases maintains the initial ATP concentration, and it also provides for a convenient method of assay. Another example is provided by the use of adenylate kinase (AK) and AMP deaminase (AD) in the assay for nucleoside diphosphate kinase (NDPK):

$$\begin{aligned} \text{ATP} + \text{GDP} &\xrightleftharpoons{\text{NDPK}} \text{ADP} + \text{GDP} \\ 2\ \text{ADP} &\xrightleftharpoons{\text{AK}} \text{ATP} + \text{AMP} \\ \text{AMP} &\xrightleftharpoons{\text{AD}} \text{IMP} + \text{NH}_3 \end{aligned} \quad (2)$$

Here, ATP is partially regenerated by the auxiliary enzyme system.

Finally, one may be forced to restrict the assay period to prevent the occurrence of competitive reactions. This is particularly true with many transglycosylases, where a number of competing processes may proceed in parallel. In other cases, there are kinases that catalyze slow hydrolysis of ATP in the absence of a phosphoryl acceptor substrate. For this reason, it is advisable to rigorously establish the reaction stoichiometry by characterization and quantitation of product formation.

When the initial rate phase is particularly short, the observed velocity will fall off quickly. This can lead to a situation that yields false cooperativity in plots of velocity versus substrate concentration. The reason for this is that the duration of the initial rate period is short at low substrate concentration and longer at high levels of substrate. Thus, the velocities will be depressed at low concentrations and a sigmoidal curve will be observed. This problem becomes most serious with single-point determinations of velocity where a continuous record of the reaction progress is not available. For example, glutamine synthetase may be assayed by the γ-glutamyl transfer reaction.

$$\text{Glutamine} + NH_2OH \underset{Me^{2+}}{\overset{ADP,As_I}{\rightleftharpoons}} \gamma\text{-glutamyl hydroxamate} + NH_3 \tag{3}$$

The hydroxamate concentration may be estimated by a colorimetric assay with ferric chloride in strongly acidic solutions. The assay is fast and convenient, but, as shown in Fig. 2, care must be exercised to validate the linear formation of hydroxamate with time. In this particular case, the reaction was linear for a short time (10 min), and the plots are based upon values plotted at 10- and 20-min

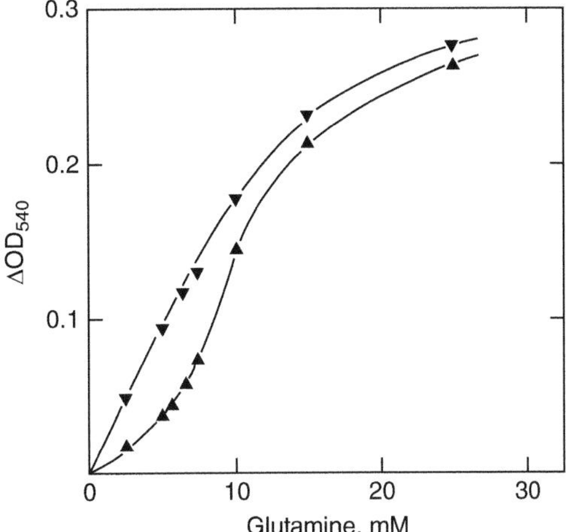

Fig. 2 Plot of the formation of γ-glutamyl hydroxamate (measured at 540 nm) versus the concentration of glutamine. ADP, hydroxylamine, arsenate, magnesium chloride, HEPES (pH 7.8), and KCl were present at 1.5 mM, 100 mM, 30 mM, 30 mM, 40 mM, and 100 mM, respectively (temperature 37 °C). Glutamine was varied from 2.5 to 25 mM. The final concentration of *Escherichia coli* glutamine synthetase (state of adenylylation, 1.5) was 4 μg/ml, and the velocity was measured by single-point assays at 10 (▼) and 20 min (▲), respectively (D. L. Purich, unpublished findings).

incubations. A false cooperativity with respect to L-glutamine becomes pronounced and evident because the initial rate phase has disappeared.

IV. Enzyme Purity and Stability

Ideally, one wants a pure, stable enzyme preparation. Although enzyme purity permits one to evaluate numerically various kinetic parameters that depend on the knowledge of the enzyme concentration, a more important aspect of purity deals with elimination of contaminating activities. It is imperative to demonstrate that such minor activities do not unfavorably interfere with the particular assay under consideration. For multisubstrate reactions, the stability of each substrate, product, and effector may be examined separately in the absence and in the presence of the enzyme preparation. The enzyme's presence should not cause any change in the substrate, product, or effector concentrations in the absence of the complete reaction mixture. For example, one might consider the liver pyruvate kinase reaction, which is activated by fructose 1,6-bisphosphate. ADP should be unaffected by the enzyme in the absence of phosphoenol pyruvate (PEP), and this may be confirmed by chromatographic analysis. Likewise, the PEP should not be affected by the enzyme in the absence of ADP, and the fructose 1,6-bisphosphate should be stable with the enzyme in the absence or presence of ADP and PEP. In this case, the presence of contaminating AK, enolase, fructose-1,6-bisphosphatase, or phosphofructokinase activity can easily be determined. AK is a frequent contaminant in nucleotide-dependent phosphotransferases; fortunately, in this case, one may purchase specific and potent inhibitors including P^1,P^4-di(adenosine-5′)-tetraphosphate and P^1,P^5-di(adenosine-5′)-pentaphosphate to block this activity (Lienhard and Secemski, 1973; Purich and Fromm, 1972).

Sometimes, it is impossible to eliminate a side reaction that arises from the intrinsic catalytic properties of a particular enzyme. For example, there are several transglycosylation reactions that are catalyzed by sucrose phosphorylase, and it was such evidence that provided insight into the possible role of a glucosyl-enzyme intermediate. Likewise, *Escherichia coli* acetate kinase also catalyzes a purine nucleoside-5′-diphosphate kinase reaction in the absence of acetyl phosphate. In this case, it was possible to use inactivation studies to demonstrate parallel loss of both activities when the enzyme was first treated with acetyl phosphate and subsequently submitted to hydroxaminolysis (Webb *et al.*, 1976). Unfortunately, the loss of some enzymic activities by such treatment does not prove that a common active site was responsible for both activities.

Another check of enzyme purity may be the use of coupled dye assays with polyacrylamide gels. The presence of isozymes may be frequently uncovered by such procedures. This is a powerful probe that is recommended wherever practicable. One problem with using coupled assays to generate NADH or NADPH is that the auxiliary enzymes must diffuse into the gel. Newer methods, especially flat-bed isoelectric focusing, permit the use of starch, agarose, and very fine dextrans. Such support materials allow uniform entry of auxiliary enzymes into the matrix of the support.

Enzyme instability is also a problem often encountered in enzyme assays. One must search out conditions that minimize or eliminate time-dependent changes in the enzyme's catalytic power. Often the presence of stabilizing agents such as glycerol, certain salts, one substrate, or an effector may render the enzyme more stable. Indeed, instability is common with very pure enzyme preparations where adsorption to glass walls, the absence of protective effects by other proteins, or high dilution into the assay mixture may be encountered. With regulatory enzymes this may be a particular problem since protein conformational mobility may lead to conformational states that are kinetically controlled. Simple heat treatments during enzyme purification or cold exposure during storage may lock the enzyme into an inactive (or less active) conformation.

If inactivation of enzyme during the course of a series of assays is simply due to loss of activity, one may correct for this by establishing an activity-decay curve. Here, a standard reference assay is intermittently used to measure the variation in enzyme activity during the experiment. But, one must be certain that the inactivation is just the result of a decrease in the fraction of total active enzyme. If there is a change to a less-active state, either the Michaelis constant or the maximum velocity will change and might jeopardize the validity of the kinetic study. (This possibility is frequently ignored in many experiments.) A change in Michaelis constant with time can be detected by using two reference assays: one at subsaturating substrate levels and another at saturating substrate levels.

Another useful approach is to examine the effect of preincubation of the enzyme with one of the substrates in a multisubstrate enzyme-catalyzed reaction. Substrate binding frequently acts to stabilize an enzyme to thermal denaturation and proteolysis. Likewise, some enzymes have remarkable conformational mobility, and reversibly inactive forms may occur in the absence of a stabilizing effect from a substrate. Incidentally, it might be added that there are situations where an inhibitor might actually "snap" an inactive enzyme into its active conformation. The *Salmonella typhimurium* PRPP:ATP phosphoribosyltransferase (Bell *et al.*, 1974) provides one such an example. Histidine is a potent allosteric feedback inhibitor of this enzyme, and it was observed that prior incubation with the inhibitor can eliminate lags in the assay. In this case, the investigators made good advantage of the cooperativity of histidine binding by exposing the concentrated enzyme to 0.4 mM histidine and then diluting the enzyme by a factor of 315 to a final histidine level where its inhibitory effect on the initial rate was negligible. Under these conditions, the active enzyme form could be studied without a lag occurrence. Likewise, the *E. coli* acetate kinase enzyme undergoes reversible cold denaturation, and brief incubation with ATP, a product, can restore the activity (Webb *et al.*, 1976).

Some enzymes require reduction of critical thiol groups for activity, and the inclusion of 2-mercaptoethanol, dithiothreitol (DTT), or dithioerythritol (DTE) is useful to restore activity. With *E. coli* coenzyme A-linked aldehyde dehydrogenase, it was observed that omission of 2-mercaptoethanol resulted in a lag in the activity in both directions. The lag was eliminated by prior incubation for 15 min with enzyme, 2-mercaptoethanol, NAD^+, and either CoA or acetaldehyde. In this case, the reaction was then initiated by addition of the substrate omitted in the preincubation. It was possible to demonstrate a requirement for the thiol in both directions of the reaction, and this eliminates a trivial explanation of the thiol effect in terms of reducing coenzyme A. Other cases of lags are described in Chapter 8 of vol. 64 in *Methods in Enzymology* series. Generally, DTT and DTE are preferred as reducing agents (Cleland, 1964). The intramolecular displacement of reduced enzyme is more facile than intermolecular reduction of mixed disulfides, and DTT and DTE may be used at lower concentrations.

Finally, the enzyme dilution should be minimized. The presence of 1–2 mg/ml of serum albumin or another protein may frequently afford the enzyme greater stability, and this approach has been gainfully exploited in many cases. It is advisable to test several different proteins if this method is used, especially in light of the tendency of serum albumin to bind fatty acids and other metabolites.

V. Substrate Purity

Many biochemical substances are fairly unstable, and impurities in each substrate must be considered as a possible source of experimental error. Although chromatographic and spectral analyses are among the best tools for establishing substrate purity, enzymic analysis is probably one of the most powerful tests of purity. Here, advantage is made of the stereochemical specificity of certain enzymes, but contaminating alternative substrates might give misleading results.

Since substrates and competitive inhibitors are generally sufficiently similar to bind to the same active site, it is not surprising that the similar physical properties of some substrates and inhibitors prevent facile and complete purification. With a competitive inhibitor present in a constant ratio to substrate, the observed kinetic parameters may be affected. This can be shown by rearranging the competitive inhibition expression [Eq. (4)] to account for this contamination.

$$\frac{1}{v} = \frac{1}{V_m} + \frac{K_m}{V_m}\frac{1}{[S]}\left(1 + \frac{[I]}{K_i}\right) \tag{4}$$

when an inhibitor is present in the substrate (i.e., $[I] = \alpha[S]$), we get

$$\frac{1}{v} = \frac{1}{V_m} + \frac{K_m \alpha}{V_m K_i} + \frac{K_m}{V_m}\frac{1}{[S]} \tag{5}$$

The form of this equation is indistinguishable from the Michaelis-Menten equation, and the double-reciprocal plots will yield the wrong estimates of the kinetic parameters. The occurrence of vanadate ions in some commercial yeast

ATP preparations is a notable example of this situation. Another similar situation may occur when substrates are contaminated by alternative substrates. Depending on whether the alternative substrate has a different V_{max} or K_m, or both, a variety of nonlinear reactions may be observed. A corollary situation also occurs with inhibitors or effectors containing substrates, and incomplete inhibition may be observed in such cases. For example, the validity of product inhibition studies of NDPK may be compromised if the nucleoside triphosphate contains significant levels of the corresponding nucleoside diphosphate. In this respect, the stability of substrates and effectors can be extremely important.

Sometimes, substrate and inhibitor instability is so serious that detectable decay may occur during the initial rate assay. This may be especially true for impure enzyme preparations containing contaminating enzyme activities. For multisubstrate cases, it is fairly easy to examine the stability of each substrate separately to check for this possibility. By using identical concentrations of two substrates containing different isotopic labels, one may verify the stoichiometry of bisubstrate reactions by the maintenance of the ratio of radioactivity. For example, [^{3}H] glutamate and [^{14}C]ATP may be used in the glutamine synthetase reaction to examine the presence of contaminating enzymes acting on either substrate. If the identical concentrations of ATP and glutamate are used, then the [^{3}H]:[^{14}C] ratio of the products should be identical to the same ratio of the substrates. While such tests appear a bit tedious, there is much merit in preventing the accumulation of false data. When effectors are added to an enzyme system, it may also prove to be advantageous to demonstrate their stability by reisolation or direct assay.

VI. Range of Substrate Levels

For many one-substrate systems, choice of the substrate concentration range is not particularly a problem. In preliminary trials, one merely chooses the widest range about the Michaelis constant with due care to avoid substrate inhibition. A rough value of K_m may then be estimated, and the range can be refined. Since the greatest velocity change occurs in the region of the K_m, it is frequently satisfactory to vary the concentration from about 0.2 to 5.0 times this constant. This changes the fractional attainment of maximal velocity from 0.14 to 0.83, and reasonable estimates of K_m can be made. (See Chapter 6 vol. 64 in *Methods in Enzymology* series for statistical treatment of rate data.) Since reciprocal plots are commonly used to analyze the rate data, it is best to choose substrate concentrations that yield an equal spacing across the reciprocal plot. By initially making up the most concentrated sample, one may dilute with buffer to get dilutions of 1/1, 1/3, 1/5, 1/7, 1/9, 1/11, etc., and these will give values of 1, 3, 5, 7, 9, 11, etc. on the abscissa of a Lineweaver-Burk plot.

For one-substrate systems requiring a nonconsumed cofactor (e.g., monovalent or divalent metal ions, an essential activator) and for all multisubstrate cases, the choice of a suitable reactant concentration may be more tedious. Let us consider the case of a bisubstrate enzyme to illustrate the problem. Now, the velocity is a function of two components, and the relative contribution of each to velocity is determined by the value of the rate constants in the initial rate expression. If both substrates are below one-third to one-half of the corresponding substrate dissociation constants, the fractional attainment of V_m will be small, and velocity measurements may have considerable error. Above concentrations of substrates corresponding to five times their respective dissociation constants, the change in velocity will be relatively small. Some investigators work in a very narrow range, but then the chance for obscuring slope changes can be sizable. Fromm (1975) outlined a useful method for obtaining rate data using a five-by-five matrix of substrate concentrations. Three solutions (A, containing one substrate at the highest level to be employed; B, containing the second substrate also at the highest level; and C, containing buffer, nonvaried cofactors, and auxiliary enzymes) are prepared. Then, five dilutions of solution A (A/1, A/3, A/5, A/7, and A/9) and five dilutions of solution B (as solution A was prepared) are combined with solution C added to each. Thus, 25 velocity measurements are made in a single experimental trial, and one may make plots of v^{-1} versus $[A]^{-1}$ at five constant [B] levels and likewise, v^{-1} versus $[B]^{-1}$ at five constant [A] levels.

For three-substrate enzymic systems, the problems of data gathering become more cumbersome. To use the approach outlined above, 125 velocity measurements for each experiment would be required if A, B, and C were each varied at five concentrations. Even with a continuous-assay protocol, the amount of time needed to carry out a single experiment would be considerable. Thus, two basic procedures have been used in such investigations. In the Frieden method (Frieden, 1959), one substrate is constant while the other two are treated like solutions A and B described in the previous paragraph. In this way, the three-substrate system reduces to a pseudo-two-substrate mechanism, but care must be taken to keep the nonvaried substrate at a level above its respective Michaelis constant but nonsaturating. As noted by Fromm (1975), there is always the possibility that a high concentration of the nonvaried substrate can lead to apparent parallel-line data in double-reciprocal plots. The careful investigator will repeat experiments that yield parallel-line plots, but at a lower level of the nonvaried substrate. An example of the Frieden protocol is presented in Fig. 3 for the sheep brain glutamine synthetase system (Allison *et al.*, 1977). A second method was proposed and implemented by Fromm and coworkers (Fromm, 1967, 1975; Rudolph and Fromm, 1969; Rudolph *et al.*, 1968). Here, one substrate is varied while the other two substrates are maintained constant in the general concentration range of their Michaelis constants. The experiment is then repeated; however, a different concentration of fixed substrates is chosen, care being exercised to maintain the ratio of fixed substrates constant in both experiments. This procedure is then repeated until all substrates are varied. It is noteworthy that all mechanisms involving quaternary complexes (i.e., EABC complexes) will give Lineweaver-Burk plots that intersect to the left of the v^{-1} axis. On the other hand, a Ping-Pong mechanism will yield one or more sets of data that result in parallel-line plots.

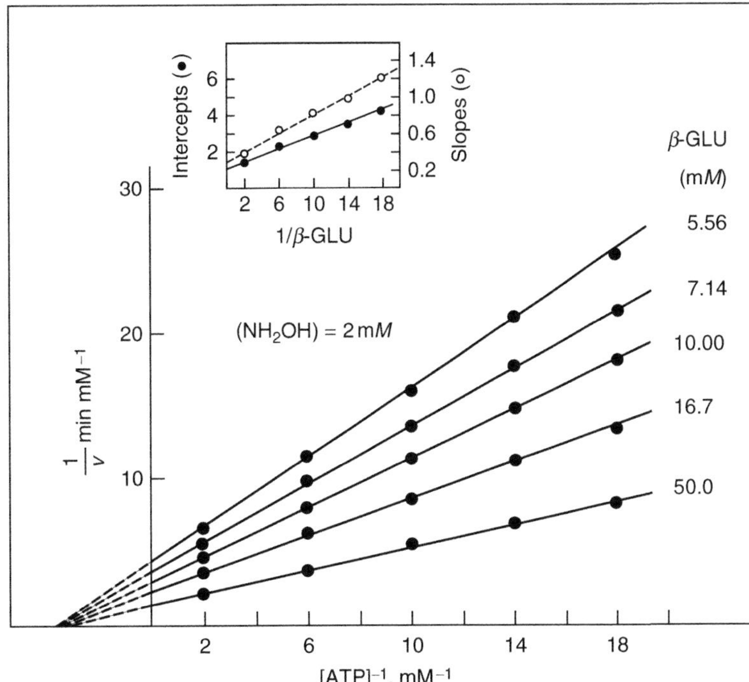

Fig. 3 Plot of the reciprocal of the initial reaction velocity (in units of min/mM) versus the reciprocal of ATP concentration at varying levels of β-glutamate (β-GLU) and a constant level of NH_2OH (2 mM). The amount of ovine brain glutamine synthetase used was 0.2 μg. Assays were performed at 37 °C in 50 mM HEPES (pH 7.2), 100 mM KCl, with free, uncomplexed Mg^{2+} held at 1 mM and a final reaction volume of 0.135 ml. *Inset:* A replot of the slopes and intercepts versus the reciprocal of β-GLU concentration. From Allison *et al.* (1977).

This procedure is illustrated in Fig. 4 by experiments of Rudolph *et al.* (1968) on the *E. coli* CoA-linked aldehyde dehydrogenase reaction. The major limitation of the latter approach is that values for kinetic parameters are difficult to obtain from such plots.

Finally, a common assumption in many experiments is that the concentration of enzyme-bound substrate is negligible relative to the total substrate concentration, and this eliminates various quadratic terms from rate expressions. In most cases, this assumption will be valid, but it is advisable to determine that the total true substrate concentration exceeds the active enzyme concentration by 50- to 100-fold. The emphasis on a true "substrate" refers to the fact that many enzymes act on only one conformer of the substrate, which if in low concentration may violate the above assumption. For example, this may occur in cases where the open chain of a monosaccharide is the true substrate. With very low K_m substrates, some problems may also be encountered. Many enzymes acting on macromolecules (e.g., protein and polynucleotide kinases, various polymerases and depolymerases)

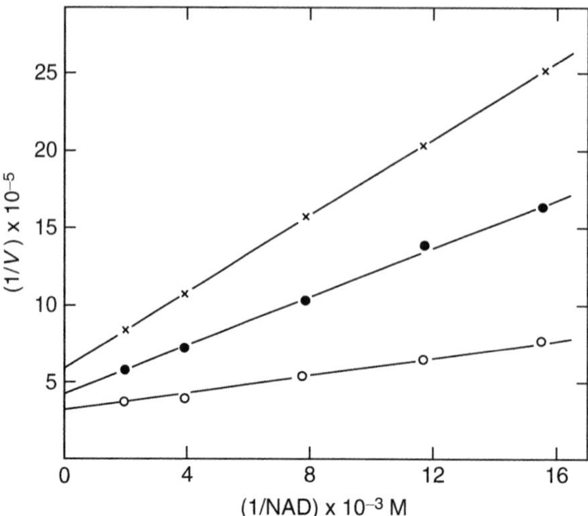

Fig. 4 Plot of the reciprocal of the initial reaction velocity (in units of min/mM) versus the reciprocal of NAD^+ concentration (in units of M^{-1}). The respective concentrations of CoA and acetaldehyde were: ○, 9.75 μM and 5.50 μM; •, 4.87 μM and 2.76 mM; ×, 2.44 μM and 1.38 mM. The NAD^+ concentration was varied from 5.2×10^{-4} M to 6.5×10^{-5} M. For other experimental conditions for the assay of the *Escherichia coli* coenzyme A-linked aldehyde dehydrogenase, see Rudolph *et al.* (1968).

may have exceptionally low Michaelis constants for the macromolecules, and their turnover numbers may not be large. In such cases, the substrate and enzyme concentrations required to yield an adequate rate determination may be in the same neighborhood, and the velocity dependence will be complex. Here, the initial rate assumption becomes impractical, and computer fitting may be required.

VII. Analytical Methods

Every activity assay relies upon an analytical method to properly represent the progress of reaction. The possibility of side reactions during the analysis must be adequately controlled to maintain experimental validity. With enzymic systems, it is convenient to distinguish two different analytical methods: continuous and periodic sampling. The choice depends upon the availability of some reliable method to continuously follow the extent of reaction, and such methods generally rely upon some spectral change attending the reaction. Spectrophotometry, radioactivity methods, polarimetry, fluorimetry, polarography, and pH measurements represent common methods for analytically determining concentration changes. Only the first is discussed here since they are probably the most frequently applied methods. Periodic sampling is frequently utilized when the number of assays is quite large or when an adequate continuous assay is unavailable. Basically, one estimates product formation over a fixed period of reaction.

The spectrophotometric method thrives as a result of the wide variety of naturally occurring chromophores (including nucleotides, coenzymes, thiol esters, and hematoporphyrins) and many synthetic alternative substrates containing any of the variety of chromophoric groups. The most common assays involve NAD^+-linked dehydrogenases by taking advantage of the large molar extinction of NADH at 340 nm ($\varepsilon_M = 6.22 \times 10^3\ M^{-1}\ cm^{-1}$). Their use as auxiliary enzyme assay systems is described in Chapter 3 of this volume.

By far the best spectrophotometers use a double monochromator system for "purifying" the light. This is especially important at high absorbances, where deviation from the Beer-Lambert relation may occur. It cannot be overemphasized that one must work under conditions where the absorbance change is a *linear* measure of the concentration change of the absorbing reaction component. In the presence of a significant stray light component leaking through the monochromator:

$$A = \log\left(\frac{I_0 + I_s}{I + I_s}\right) \neq \varepsilon cl \tag{6}$$

where A, I_0, I, I_s, ε, c, and l are absorbance, incident light intensity, transmitted light intensity, stray light intensity, the molar absorptivity ($M^{-1}\ cm^{-1}$), molar concentration, and light path length (cm), respectively. At high sample absorbance, the value of I may approach I_s and the error will be considerable. Other sources of nonlinearity may result from micellar formation, stacking of aromatic chromophores, and sample turbidity. The first two may be relieved by addition of certain nonionic detergents, provided no unfavorable effects on the enzyme occur. The latter may be prevented by centrifugation or filtration. For highest sensitivity of measurement, the wavelength should be adjusted to yield the greatest difference between substrate and product absorbances. This may not necessarily be the so-called λ_{max} for the product, since the spectra of substrate and product may overlap. If the λ_{max} can be conveniently used, it is probably the region that best obeys Beer-Lambert's law. Of course, a record of the entire spectrum would be desirable in some special cases, but single-wavelength monitoring is probably the most common method. In some cases, changes at several wavelengths may be followed, and a simultaneous equation treatment of the rate data can be applied.

The most sensitive direct assay of product accumulation can be made with radionuclides. Assays can be made continuously or by periodic sampling, and the latter is most frequently employed. Product formation, rather than substrate depletion, provides the most sensitivity and accuracy. In periodic sampling protocols, it is absolutely necessary to verify that the reaction is thoroughly quenched at the time of sampling. Likewise, the stability of labeled substrate and product during the quenching of the reaction's progress must be demonstrated. These two problems are frequently the source of published errors in enzyme rate data. Periodic sampling requires one to stop reactions completely at desired times. Many agents (including ethanol, other organic solvents, mercurials, phenol, acids, bases,

EDTA, and EGTA) may be employed to quench the progress of enzyme-catalyzed reactions. In addition, heating can often affect complete denaturation of the enzyme. The problem that is frequently encountered is the failure to rapidly achieve complete quenching. This is especially true for low-molecular-weight enzymes which have very stable protein structures (e.g., ribonuclease or AK). Nonetheless, one may generally rely on pH changes to stop many reactions, provided the conditions are suited to substrate and product stability. The chief advantage of acid or base additions is the rapidity of the inactivation, and storage on ice can further diminish enzymic activity and ensure product stability. For each new experimental condition, it is useful to demonstrate the efficiency of the stopping procedure, and one convenient method involves the use of radioisotopes. Here, the enzyme sample with substrates, cofactors, and buffer, as used in the assay, is subjected to the stopping protocol. Isotopically labeled substrate is then added and any conversion to product resulting from incomplete quenching may be measured. It is important to mimic the reaction conditions since substrates, cofactors, and buffers are known to affect the susceptibility of many enzymes to inactivation.

Some techniques for substrate and product separation include ion exchange, chromatography, simple extraction or adsorption, electrophoresis, derivatization, precipitation, isotope dilution, and the deliberate labilization of tritium or carbon as tritiated water or labeled CO_2. Undoubtedly, medium- and high-pressure liquid chromatography will be increasingly exploited as a high-resolution method. Indeed, poor resolution is the single most important source of high blanks in such assays, and the high sensitivity of the assay is dependent upon the minimization of the radioactivity of blank samples. The occurrence of labeled impurities in the substrate resulting from nonenzymic processes can also plague sensitive measurements. Nonetheless, the case of substrate and product resolution has been remarkably increased through the availability of DEAE-, ECTEOLA-, PEI-, and CM-cellulose papers and thin-layer plates. In addition, several types of paper-loaded polystyrene ion exchangers have been advantageously employed. The sample may be added quickly without regard to the size of the dampened spot because the ions will adsorb near the point of addition. Then, by addition of absolute ethanol or another volatile solvent, one can affect complete arrest of the reaction. With ion-exchange papers, the capacity is generally 1–10 μeq/cm^2; thus, it is best to consider the amount of total ions added to each spot. For complete nucleotide adsorption, the magnesium ion level should be below 1 mM. For this reason, EDTA may be required in the buffer used to develop the chromatogram (Morrison, 1968; Roberts and Tovey, 1970).

Sample counting can also be the source of problems, especially if the radioactive samples are counted directly on the paper support. With ^{3}H and ^{14}C the amount of quenching of the radioactivity can be sizable, and it is preferable to elute ^{3}H samples and to count them as liquid samples. In any case, the properties of the scintillation counting fluid should be considered to prevent slow release of counts from the paper support into the scintillant. It is advisable to recount one sample for a number of hours to ascertain that this is not a source of possible error. Even with dissolved compounds such as ATP, the low solubility in the scintillation fluor may require use of thixatropic agents for gel suspensions.

The radioactive substrate should be scrupulously examined for impurities that result from radiolysis. This is particularly important with highly labeled materials when ethanol or some other stabilizing agent cannot be present in the sample. The position of labeling should also be such that no primary or secondary kinetic isotope effects can lead to apparently reduced activity. Thus, isotopic substitution should be preferably away from the atoms undergoing reaction and in a position that is chemically inert. For certain cases where the position of the isotopic substitution is somewhat labile, it may be best to distill away the solvent from a small aliquot of stored radionuclide and count the distillate. In all cases, it is advisable to learn about special handling problems intrinsic to particular substances. For example, even in 50% ethanol, the stability of ^{32}P-labeled metabolites is not very great, and repurification just prior to use is often required. The availability of HPLC-purified compounds from commercial vendors of radioactive materials has increased the reliability of most preparations. Even so, the investigator should verify this.

VIII. Determination of Reaction Equilibrium Constants

Equilibrium constants for biochemical reactions are frequently more complex than one might suppose at first sight. Substrates and products are almost always ionic, and hydrogen ions and metal ions can substantially displace the so-called mass-action ratio. For example, consider the AK reaction

$$\mathrm{ADP^{3-} + MgADP^{-} \rightleftharpoons AMP^{2-} + MgATP^{2-}} \tag{7}$$

The mass-action ratio is written as

$$K = \frac{[\mathrm{AMP}]_0[\mathrm{ATP}]_0}{[\mathrm{ADP}]_0^2} \tag{8}$$

where the subscripts indicate total substrate and product concentrations. Yet, even Eq. (8) is an inadequate representation of the reaction in that ATP^{4-} and $HATP^{3-}$ differ in affinity for metal ions, and the value of K will show both pH and metal ion concentration dependences. Even the buffer can affect the equilibrium constant if it interacts with metal ions, as is the case with Tris and imidazole buffers. The important point is that one must know the value of the equilibrium constant under the reaction conditions to be used in experiments, and many literature values may only serve as a rough guide. Thus, it is best to take the time and effort to obtain an accurate estimate for the equilibrium constant. This is especially important if one is interested in using Haldane relationships.

Basically, one should determine the equilibrium constant by a number of methods, and each should give a self-consistent value. Starting with substrates only (or products only) one may add a sufficient amount of enzyme to equilibrate the reaction. One may periodically sample the reaction mixture for substrate(s) and product(s) concentrations, and the system should reach a time-independent state. (The equilibrium value should also be independent of the amount of enzyme added if no competing side reactions occur.) One problem with this method is that the catalytic efficiency of the enzyme may be decreased by product accumulation, and the approach to equilibrium may be sluggish. If the enzyme is irreversibly inactivated by denaturation over a long incubation period, a false equilibrium value may be obtained. Likewise, the pH may change if protons are taken up or released as the reaction progresses, and the buffering capacity of the system must be sufficient to accommodate this change. For this reason, the pH should be carefully monitored or one may use a pH-stat to maintain the pH. It is also advisable to demonstrate that the same equilibrium constant is obtained by starting from either side of the reaction.

Another approach, which obviates some of these difficulties, involves the preparation of a group of reaction samples with different substrate(s):product(s) ratios roughly in the range of the estimated equilibrium constant. Provided one has a sensitive measure of changes in product or substrate levels, one may follow each reaction and plot the deviation in concentration attained after equilibration in the presence of the enzyme. The sample giving zero deviation must correspond to the equilibrium ratio of substrates and products. If there are several substrates and several products, [A] and [P] may be set at several constant values, and the [B]:[Q] ratio may be changed (or vice versa). The zero deviation position on the graph should be always at the same point. If it is not, careful scrutiny of the free metal ion concentration, the pH, and the enzyme's activity is indicated. This method is further illustrated in Chapter 16 of this volume.

Probably the best analytical measure of substrate and product levels can be achieved by use of radioisotopically labeled substrates. At isotopic equilibrium, the isotope will be distributed in the substrate or product pool in strict accord with the analytical concentrations of the substrate and product. Thus, one may rigorously demonstrate the occurrence of equilibrium by double-label methods. If ^{14}C-labeled substrate and ^{3}H-labeled product were incubated with enzyme, the equilibrium condition requires that the ^{3}H:^{14}C ratio in substrate and product be identical.

IX. Choice of Buffer Agents

With few exceptions, studies of enzyme-catalyzed systems require that a buffer agent be present in the reaction solutions. Unfortunately, the choice of the buffer substance cannot be made strictly by matching closely the pK_a value and the desired pH of the reaction medium. Too many buffers are inappropriate for enzyme studies as a result of undesired interactions with the enzyme or some reaction component. For example, few investigators recognize the oxidizing potential of arsenate and cacodylate (dimethylarsinic acid) under acidic conditions. Likewise, many phosphate esters, phosphate itself, and a variety of carboxylic acids are also natural metabolites, and they may bind to special enzyme sites that affect the catalytic activity. Borate buffers are also often unacceptable as a result of complexation with many polyols, ribonucleotides, and carbohydrates. Thus, care must be exercised in choosing from the many available buffering agents.

One solution to the difficult problem of selecting a buffer is to examine as many buffers as possible. Choose several that appear to yield the greatest activity. To determine whether the buffer interacts with the enzyme, buffer dilutions at constant ionic strength can be made (and solutions readjusted to the desired pH). Buffer dilution should not affect the activity of the enzyme provided pH and ionic strength are maintained. The use of buffers that cause activity changes should be questioned. It might also be noted that several counterions should be used to discover the best counterion. (Additional information on pH effects can be obtained in Chapter 6 of this volume.)

It cannot be overemphasized that control of divalent metal ion concentrations requires the correct choice of buffers, especially in nucleotide-dependent reactions (Perrin and Dempsey, 1974). The reader should see Chapters 11 and 12 of vol. 63 in the *Methods in Enzymology* series.

Finally, the equilibrium position of many reactions is pH dependent, and it is wise to fully consider the scope of the planned experiments. If the kinetics of a reaction are to be studied in both the forward and reverse directions, one is well advised to select a pH that permits this to be readily accomplished. For example, the mass-action ratio for the hexokinase reaction, [glucose-6-P][ADP]:[ATP][glucose], is 490 at pH 6.5 but around 4900 at pH 7.5. Measuring the reverse reaction rate at the higher pH may be quite difficult. The same is true for many dehydrogenase reactions, and a little prior consideration may eliminate considerable work.

X. Temperature Control

Valid enzyme assays will be obtained only when the temperature is carefully maintained. Usually, this requires a good temperature-regulated bath and circulator with a variability of less than 0.1 °C. Samples should be sufficiently immersed to allow full thermal equilibration, which may require several minutes with glass and plastic vessels. With spectrophotometers, the entire sample compartment is commonly thermally isolated by water circulation through the hollow walls, but the equilibration of the cuvette is subject to a rather inefficient convective heat transfer by air from the warmed or chilled walls. Thus, one may have considerably less temperature control than indicated by the temperature regulator. A more efficient system uses a brass block fabricated to mount directly around the cuvettes. This block should be designed to permit sufficient circulation of fluid from the regulated water bath.

The problem of adequate temperature control is most serious when reactions are monitored at more than 10 °C above or below ambient temperature. In such cases, sample removal from the bath for manual mixing alone may be sufficient to disturb the temperature. The sample tube and cuvette should be thermally preequilibrated, and a minimum of sample handling is desirable. It may be desirable to prewarm the enzyme solution prior to initiating the reaction. Even a 0.1 ml aliquot of ice-cold enzyme solution can perturb a 3.0 ml assay by 1° or 2°, and the reaction velocity may be altered by 10% or 15%. One approach to ensure rigorous temperature control is to use a spring-loaded plunger mounted on the sample compartment lid. The end of the plunger may contain a Teflon "spoon" holding up to 0.1 ml of solution, and mixing is readily achieved by rapidly depressing the plunger several times. The "spoon" can be fashioned to have small jets that permit complete mixing in a few seconds. This method has the added advantage of permitting mixing while the photomultiplier tube is already operating. With the Cary 118C, for example, opening and reclosing of the sampling compartment leads to a 4-s delay before the opaque safety shutter to the photomultiplier tube is reopened. With the above apparatus, mixing can be complete in 3 s, and the spectrophotometer can give a record of the early progress of the reaction.

XI. Reporting Initial-Rate Data

As a recommended course of action in publishing rate data, the following statements are offered. Velocity should always be expressed in terms of molarity changes per unit time. Other terms, which are proportional to molarity, are frequently reported, but the use of an intensive variable such as molarity is preferable. The conditions of the assay should be fully described, and any special treatments required for linearity should be thoroughly detailed. The specific activity and amount of enzyme used in each experiment should be stated in each figure legend, especially when these undergo change during the course of the kinetic study. Reviewers are especially grateful if the investigator provides estimates of the maximum percentage of substrate depletion during the course of rate assays. Statements regarding the number of replicate samples that were assayed and the variation in observed constants are also helpful. Unfortunately, some investigators fail to report such statistical data, and this leads to confusion for those interested in repeating the work.

XII. Concluding Remarks

The development of a reliable initial-rate assay may appear to be an insurmountable task involving the interplay of many variables. Yet, for those interested in enzymology, this activity frequently presents the opportunity to uncover new aspects of the behavior of a particular enzyme. Many fascinating details of enzyme mechanism and metabolic regulation have evolved from such exercises, and a valid assay represents a powerful tool to probe further.

References

Allison, R. D., Todhunter, J. A., and Purich, D. L. (1977). *J. Biol. Chem.* **252,** 6046.

Bell, R. M., Parsons, S. M., Dubravac, S. A., Redfield, A. G., and Koshland, D. E., Jr. (1974). *J. Biol. Chem.* **249,** 4110.

Cleland, W. W. (1964). *Biochemistry* **3,** 480.

Frieden, C. (1959). *J. Biol. Chem.* **234,** 2891.

Fromm, H. J. (1967). *Biochim. Biophys. Acta* **139,** 221.

Fromm, H. J. (1975). "Initial Rate Enzyme Kinetics." Springer-Verlag, Berlin and New York.

Kleeman, J. E., and Parsons, S. M. (1976). *Arch. Biochem. Biophys.* **175,** 687.

Lienhard, G. E., and Secemski, I. I. (1973). *J. Biol. Chem.* **248,** 1121.

Morrison, J. F. (1968). *Anal. Biochem.* **24,** 106.

Ning, J., Purich, D. L., and Fromm, H. J. (1969). *J. Biol. Chem.* **244,** 3840.

Perrin, D. D., and Dempsey, B. (1974). Buffers for pH and Metal Ion Control, pp. 176, Chapman and Hall, London.

Purich, D. L., and Fromm, H. J. (1972). *Biochim. Biophys. Acta* **276,** 563.

Roberts, R. M., and Tovey, K. C. (1970). *Anal. Biochem.* **34,** 582.

Rudolph, F. B., and Fromm, H. J. (1969). *J. Biol. Chem.* **244,** 3832.

Rudolph, F. B., Purich, D. L., and Fromm, H. J. (1968). *J. Biol. Chem.* **243,** 5539.

Webb, B. C., Todhunter, J. A., and Purich, D. L. (1976). *Arch. Biochem. Biophys.* **173,** 282.

CHAPTER 3

Techniques in Coupled Enzyme Assays

Frederick B. Rudolph,[*,✠] **Bennett W. Baugher,**[*] **and Robert S. Beissner**[†]

[*]Department of Biochemistry Rice University
Houston Texas 77001

[†]Department of Rice University
Houston Texas 77001

The major problem in initial-rate kinetic studies is often the method of assay. If possible, an assay should be accurate, sensitive, and convenient. In addition, the ability to continuously monitor a reaction process is of great value. Unfortunately, many reactions do not produce changes in the spectral or other properties of the reactants and cannot be directly measured. To allow continuous assay of such reactions, the formation of a product can be measured by addition of an auxiliary enzyme that produces a measurable change. Such methods are sensitive and convenient, but have certain disadvantages, the most important being a lag period before the steady state is reached, which will be detailed in this chapter. Various theoretical analyses of the use of consecutive enzyme reactions for assay systems have been made and will be considered here along with practical aspects of their use, precautions to be observed, and examples of such assays.

✠ Deceased

CONTEMPORARY ENZYME KINETICS AND MECHANISM
Reprinted from *Methods in Enzymology*, Volume 63 (Academic Press, 1979).

DOI: 10.1016/B978-0-12-378608-1.00003-7

I. Theory

A. Models and Analysis of Coupled Systems

A number of approaches describing ways to ensure valid coupled assays have appeared in recent years (Barwell and Hess, 1970; Bergmeyer, 1953; Easterby, 1973; Gutfreund, 1965; Hess and Wurster, 1970; Kuchel *et al.*, 1974; McClure, 1969; Storer and Cornish-Bowden, 1974). The basic problem is to determine that auxiliary enzymes will react at a rate that allows monitoring only of the steady-state concentration of the product(s) (P) of the reaction being studied, not of the rates of the auxiliary enzymes. The systems have an inherent lag time prior to the steady state that must be analyzed and minimized. The simplest example of such a system is

$$\mathrm{A} \underset{\mathrm{E_1}}{\rightarrow} \mathrm{P} \underset{\mathrm{E_2}}{\rightarrow} \mathrm{Q} \tag{1}$$

where E_1 is the enzyme whose activity is being measured (primary enzyme), E_2 is the auxiliary enzyme, and k_1 and k_2 are the rate constants for the respective enzymes. The general approach to such assays has been to use a large excess of the auxiliary enzymes to assure steady-state conditions. The behavior of such a system was first treated quantitatively by McClure (1969) using the following assumptions. (1) k_1 is the rate constant for E_1, which is assumed to be an irreversible zero-order step. To meet this criterion, all substrates for E_1 must be saturating or only a small fraction of the substrates can react. Irreversibility is assumed since P is continuously removed by E_2 during the assay. (2) The second reaction is irreversible and first order with respect to P (rate constant k_2). This necessitates that $P \ll K_p$ (the Michaelis constant for P for E_2) and that the other substrates for E_2 be nearly saturating. If the equilibrium for the reaction catalyzed by E_2 lies to the right or only a small amount of reaction occurs so that the other substrates for E_2 are not depleted, irreversibility can be assumed. Proper choice of the auxiliary enzyme will satisfy this condition. With these assumptions, it is possible to calculate the amount of E_2 required for a theoretically correct assay. For the sequence in Eq. (1)

$$\frac{d\mathrm{P}}{dt} = k_1 - k_2\mathrm{P} \tag{2}$$

which integrates, using the limits $t = 0 \rightarrow t$ and $\mathrm{P} = 0 \rightarrow \mathrm{P}$, to

$$\mathrm{P} = \left(\frac{k_1}{k_2}\right)(1 - e^{-k_2 t}) \tag{3}$$

as $t \rightarrow \infty$, P approaches its steady-state concentration ($\mathrm{P_{ss}}$) or

$$\mathrm{P_{ss}} = \frac{k_1}{k_2} \tag{4}$$

Equation (3) can be rearranged to

$$\ln\left(1 - \frac{k_2\mathrm{P}}{k_1}\right) = -k_2 t \tag{5}$$

These equations allow calculation of the amount of E_2 required in an assay to reach the steady state for P (P_{ss}) in a given period of time. Combining Eqs. (4) and (5) results in

$$\ln\left(1 - \frac{P}{P_{ss}}\right) = -k_2 t \tag{6}$$

If $K_p \gg P_{ss}$ as assumed initially, then from the Michaelis-Menten equation for E_2,

$$v_2 = \frac{V_2 P_{ss}}{K_p + P_{ss}} \simeq \frac{V_2 P_{ss}}{K_p} = k_2 P_{ss} \tag{7}$$

This expression can be substituted into Eq. (6) to give an expression for the amount of E_2 (or V_2 which is the maximal velocity for E_2) required to attain a given fraction of the steady-state phase of the coupled reaction at any given time t.

$$V_2 = \frac{-\ln(1 - P/P_{ss})K_p}{t} \tag{8}$$

If F_p is the fraction of P_{ss} desired at time t, the equation can be expressed as

$$V_2 = \frac{-2.303 \log(1 - F_p)K_p}{t} \tag{9}$$

The time required to each F_p (t^*) can be calculated from

$$t^* = \frac{-\ln(1 - F_p)}{k_2} \tag{10}$$

To use expression (9), one decides on the F_p desired with a given lag period and calculates V_2. Only K_p has to be known to make the calculation.

Certain features of the Eqs. (7)–(9) are apparent (McClure, 1969): (1) k_2 is influenced by the ratio V_2/K_p and is not just dependent on a large excess of E_2; this point is illustrated in Fig. 1. Both the enzyme concentration and K_p will influence the lag time. Either an increase in K_p or decrease in E_2 will cause an increase in the lag time. These factors should be considered if isozymes or similar enzymes are available for the same assay. (2) The time required for establishment of the steady state (P_{ss}) is independent of k_1. The system is therefore suitable for assaying any activity of the primary enzyme as long as $P_{ss} \ll K_p$. This is an important factor and suggests that a lower limit for K_p exists and should be considered when a choice of auxiliary enzymes is made. (3) P_{ss} is a function of both k_1 and k_2.

The attainment of the steady-state concentration of P is shown in Fig. 2. In this hypothetical example sufficient E_2 is present to give the steady state within 5 s, after

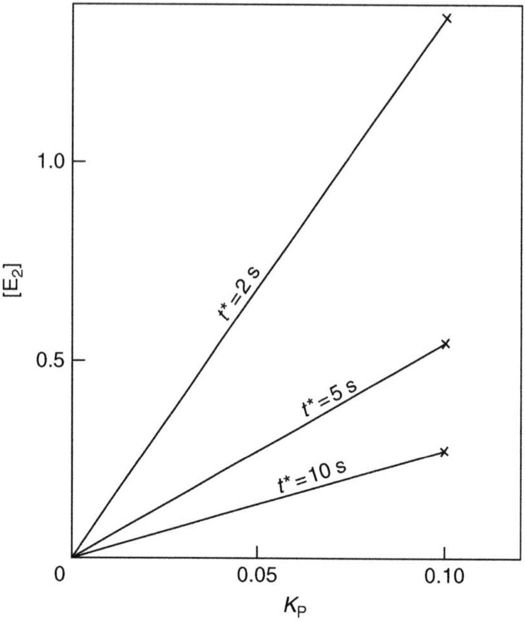

Fig. 1 A plot of amount of auxiliary enzyme, E_2, versus the K_p for various indicated lag times (t^*).

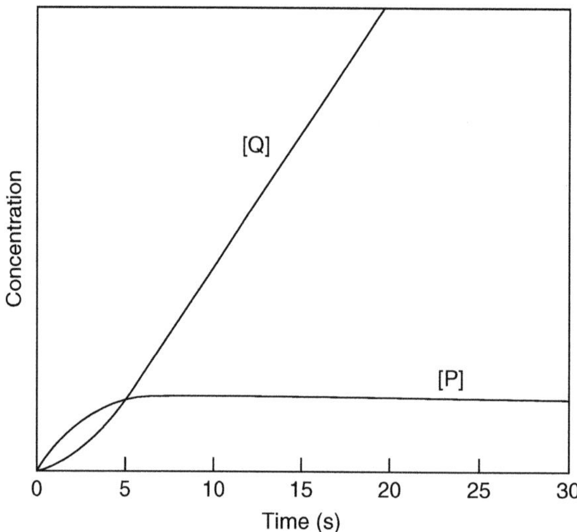

Fig. 2 A plot of the concentration of the intermediate P, product, and the final product, Q, as a function of time.

which the formation of Q is constant (dQ/dt = C and −dP/dt = 0) and initial velocity conditions are achieved.

Assay systems utilizing more than one auxiliary enzyme are more difficult to analyze. For the system

$$\mathrm{A} \underset{\mathrm{E_1}}{\rightarrow} \mathrm{P} \underset{\mathrm{E_2}}{\rightarrow} \mathrm{Q} \underset{\mathrm{E_3}}{\rightarrow} \mathrm{R} \tag{11}$$

where E_1 is the primary enzyme and E_2 and E_3 are the auxiliary enzymes, McClure has applied an analysis similar to that described above for Eq. (1). An expression for [Q] analogous to Eq. (3) is

$$\mathrm{Q} = \frac{k_1}{k_3} - \frac{k_1}{k_3 - k_2}\left[\mathrm{e}^{-k_2 t} - \left(\frac{k_2}{k_3}\right)\mathrm{e}^{-k_3 t}\right] \tag{12}$$

This equation contains the following features: (1) The first term is equal to the steady-state concentration of Q (Q_{ss}) and the negative term is the time dependence of steady-state attainment. (2) The rate of steady-state attainment is symmetrical with respect to k_2 and k_3, allowing values of k_2 and k_3 to be interchanged without affecting the lag time. (3) The time required to achieve a given fraction of Q_{ss} is not a function of k_1.

Equation (12) cannot be solved as easily for t or t^* as the two-enzyme case, but some conclusions were drawn by McClure (1969). If either k_2 or k_3 is large compared to the other rate, then the lag time is primarily dependent on the smaller value. The equation reduces to

$$t^* = \frac{-\ln(1 - F_\mathrm{Q})}{k_2}, \text{ if } k_3 \gg k_2 \tag{13}$$

or

$$t^* = \frac{-\ln(1 - F_\mathrm{Q})}{k_3}, \text{ if } k_2 \gg k_3 \tag{14}$$

In fact, the difference between the two constants need to be only four- to fivefold for satisfactory prediction of t^*. When neither auxiliary enzyme is in excess, there is no single solution for t^*. McClure (1969) solved Eq. (12) numerically, and the results are presented in Fig. 3 as nomograms showing the time required to reach 99% of Q_{ss} for different values of k_2 and k_3.

A similar approach to dealing with the lag time of a coupled assay has been presented by Easterby (1973), and his approach allows calculation of some useful values. A resultant equation from his treatment of Eq. (1) is

$$[\mathrm{P}] = v_1(t + \tau \mathrm{e}^{-t/\tau} - \tau) \tag{15}$$

where τ, the transient (lag) time = K_p/V_2 and $P_{ss} = v_1\tau$. For this treatment the plot of Q formation versus t is shown in Fig. 4. The P_{ss} concentration is found from the y intercept, and the transient time (τ) from the x intercept. This information allows

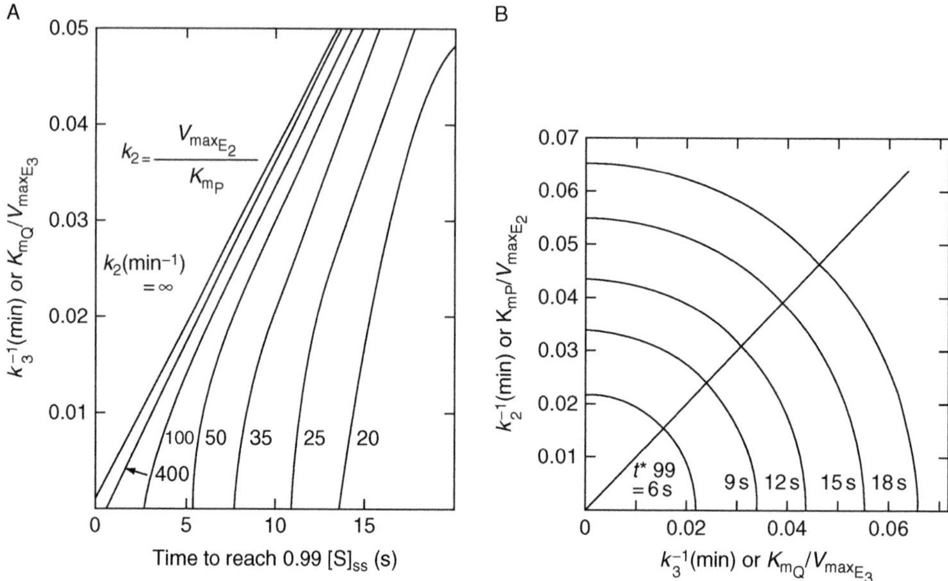

Fig. 3 Nomograms of McClure showing the time required for 99% attainment of the steady state (t^*) in a coupled assay using the auxiliary enzymes. R_2 and R_3 represent the first-order rate constants (V_{max}/K_m) for the two coupling enzymes. Adapted with permission of McClure (1969) and the American Chemical Society (copyright holder).

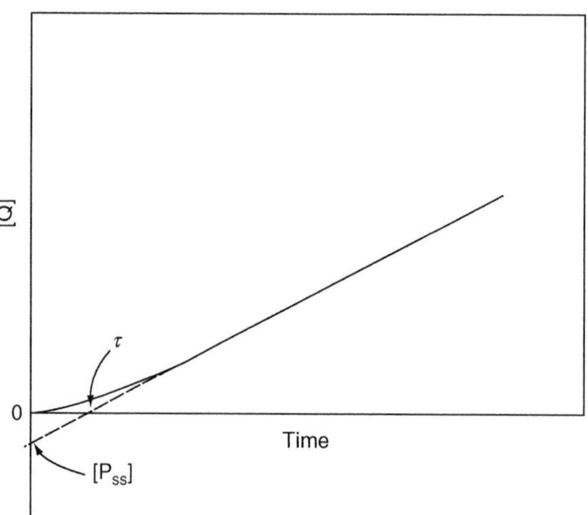

Fig. 4 A plot of the formation of the final product Q of E_2 as a function of time. The x intercept represents the transient time (τ) as discussed in the text, and the y intercept represents the steady-state concentration of intermediate P (P_{ss}).

calculation of v_1, the velocity of the reaction being studied. Determination of v_1 is done at a series of A concentrations, and the velocity dependence on substrate concentration is plotted. The transient time approximates K_p/V_2 only if $V_2/K_p \gg V_1/K_A$. The ability to determine P_{ss} allows a check on the assumptions used in McClure's derivations. If $P_{ss} \ll K_p$, then the methods of calculation of the amount of coupling enzymes required are more valid. It also allows monitoring if product inhibition is a potential problem in the assay.

Easterby (1973) also treated assays using two or more auxiliary enzymes. He found the transient time for reaching the steady state for the assay to be the sum of the transients for the individual coupling enzymes. Thus, the lag time increases as the number of auxiliary enzymes increases. Of importance is that this treatment also indicates that the transient time is independent of the initial rate-limiting enzyme. Changing the activity of E_1 does not change the lag. Easterby (1973) also showed that the minimum time for accurate measurement of initial velocity (steady-state intermediate) has to be at least five times the transient time.

Storer and Cornish-Bowden (1974) considered coupled assays without assuming that the second and subsequent reactions followed first-order kinetics, as was done by McClure and Easterby. This more general treatment for the reaction sequence

$$A \xrightarrow{v_1} P \xrightarrow{v_2} Q \quad (16)$$

where v_1 and v_2 are the experimental velocities, gives an equation for the time (transient time) required for v_2 to reach any given value as

$$t^* = \frac{\phi K_p}{v} \quad (17)$$

where ϕ is defined as

$$\phi = \frac{V_2 v_1}{(V_2 - v_1)^2} \ln\left[\frac{v_1(V_2 - v_2)}{V_2(v_1 - v_2)}\right] - \frac{v_1 v_2}{(V_2 - v_2)(V_2 - v_1)} \quad (18)$$

ϕ is a dimensionless number and a function of the ratios v_2/v_1 and v_1/V_2 only. V_2 is the maximal velocity of the auxiliary enzyme. Values of ϕ can be calculated at various values of v_1/V_2 and v_2/v_1 and are given in Table I.

The value of v_2/v_1 depends on the accuracy desired. In a precise assay it is desirable that it be at least 0.99. The amount of E_2 required for a desired accuracy can be estimated for a given v_1 and the lag time calculated from Eq. (17).

The analysis of Storer and Cornish-Bowden (1974) also considered multiple coupled systems. They found that, by treating each step individually, the lag time was bounded by $t_1, \ldots, t_n < t_{lag} < \sum_{i=1}^{n} t_1$. That is, it was longer than any individual time but less than the sum of all. Assuming that the lag time is the sum gives a useful upper limit similar to the method of Easterby (1973). The experimental study by Storer and Cornish-Bowden (1974) illustrated that linearity of an assay cannot be assumed unless the apparently linear portion is 10 times as long as the obvious lag period.

Table I
Analysis of Lag Times for Coupled Assays[a,b]

	ϕ	
V_1/V_2	$v_1/v_2 = 0.95$	0.99
0	0	0
0.05	0.16	0.25
0.10	0.35	0.54
0.15	0.56	0.89
0.20	0.81	1.31
0.25	1.11	1.81
0.30	1.46	2.42
0.35	1.88	3.18
0.40	2.39	4.12
0.45	3.02	5.32
0.50	3.80	6.86
0.55	4.79	8.91
0.60	6.08	11.7
0.65	7.80	15.6
0.70	10.2	21.4
0.75	13.6	30.3
0.80	18.7	45.5
0.85	27.2	74.3
0.90	42.8	141
0.95	77.9	377

[a]Adapted from Storer and Cornish-Bowden (1974).

[b]The time required for v_2 (for the auxiliary enzyme) to reach the specified percentage of the steady-state velocity (v_1) of the primary enzyme (E_1) is given by $t = \phi K_p/v_1$.

The techniques discussed so far have assumed that the steady-state reaction of the primary enzyme can be measured with accuracy. The lag period has been described only as to its duration and how to shorten it. Kuchel *et al.* (1974) solved the basic equations describing Eq. (1) by formulation as a set of Maclaurin polynomials. This treatment leads to the following equation:

$$[\mathrm{P}] = \frac{V_1 V_2 \mathrm{A}_0 t^2}{2K_\mathrm{p}(K_\mathrm{A} + \mathrm{A}_0)} \tag{19}$$

where A_0 is the initial A concentration.

A plot of P versus t^2 gives an initially linear plot with a slope $V_2/2K_\mathrm{p} \cdot V_1\mathrm{A}_0/(K_\mathrm{A} + \mathrm{A}_0)$, which describes a rectangular hyperbola. The slopes of a number of A_0 concentrations are plotted in double reciprocal form to give an x intercept of $-1/K_\mathrm{A}$ and a y intercept of $2K_\mathrm{p}/V_1V_2$. Prior analysis of the V_2 and K_p allows calculation of both V_1 and K_A. This analysis can be extended to any number of auxiliary enzymes by similar plots (Kuchel *et al.*, 1974). This treatment will apply best when the activities of the primary and auxiliary enzymes are similar. A long lag time is needed to obtain the initial linear slope of the P versus t^2 plot. If the lag is short, the data are readily analyzed by the techniques described by McClure (1969), Easterby (1973), and Storer and Cornish-Bowden (1974). This technique is of use particularly if the coupling enzymes are not available in sufficient quantities to saturate the system and does avoid some difficulties inherent in coupled systems. Additional information concerning individual rate constants can be obtained from presteady-state studies as described by Roberts (1977). These are beyond the scope and intent of this report but are available for reference.

The major conclusions of all the theoretical treatments on coupled enzyme assays are that they can be designed so that lag times are known and minimal and the measured velocity accurately represents the steady-state velocity of the primary enzyme. The calculations will usually provide the correct answers, but certain precautions and limitations must be considered and will be discussed in the following sections.

II. Practical Aspects

The basic question in designing a coupled enzyme assay is how much auxiliary enzyme(s) to add to ensure a short lag time and accurate measurement of the enzyme being studied. It would be convenient if one could simply say to use a 10-fold excess of the first coupling enzyme and 100-fold excess of the second enzyme if it is necessary, which has been a convention among some enzymologists. However, the systems may not always work well even with these excesses of coupling enzymes. The expense of the enzymes, contamination with other proteins, and other factors necessitate a more defined approach to the problem. The theories discussed above have the limitations that they are based on models requiring certain assumptions but they generally are experimentally applicable.

The choice of auxiliary enzymes is obviously dictated by the products formed in the reaction. A number of examples of coupled assays are listed in Tables II and III. This listing is not an exhaustive compilation of such assays but generally represents recently developed methods. Included in Table III are a number of stopped-time assays, in which auxiliary enzymes are present during the reaction to allow formation of an identifiable product. The reactions are stopped at various time intervals, and the product is measured. The use of the coupled system requires that the assay times be done after the steady-state condition is achieved (McClure, 1969). The excellent series entitled "Methods of Enzymatic Analysis" edited by Bergmeyer (1974) should be the primary source for finding suitable systems for coupled enzyme assays. The series details assays for both enzymes and metabolites by using enzymic analysis. Such methods can be generally adapted to kinetic assays. Choices between systems should be made by comparing the K_m's for the measured substrates and activity of the auxiliary enzymes. Different isozymes can be more useful in coupled systems as illustrated by the fact that the H_4 isozyme of

Table II
Examples of Coupled Enzyme Assays

Product measured	Species monitored	Coupling enzymes	Examples
Assays for nucleotides			
ATP	Disappearance of NADH	Hexokinase, glucose-6-phosphate dehydrogenase	Adenylate kinase, creatine kinase (Oliver, 1955)
ADP	Disappearance of NADH	Pyruvate kinase (PK), lactate dehydrogenase (LDH)	Fructokinase (Sabater *et al.*, 1972), γ-glutamyl-cysteine synthetase (Rathbun and Gilbert, 1973), mannokinase (Sabater *et al.*, 1972), ATPase (Horgan *et al.*, 1972; Monk and Kellerman, 1976)
ADP	Disappearance of NADH	PK, LDH, 3′-nucleotidase	Adenosine-5′-phosphosulfate kinase (Burnell and Whatley, 1975)
AMP	Disappearance of NADH	Adenylate kinase	Adenosine phosphotransferase (Park *et al.*, 1973), GMP synthetase (Spector *et al.*, 1974)
GDP	Disappearance of NADH	PK, LDH	Adenylosuccinate synthetase (Spector and Miller, 1976)
GMP	Disappearance of NADH	GMP kinase, PK, LDH	Hypoxanthine-guanine phosphoribosyltransferase (Giacomello and Salerno, 1977)
Inosine	Formation of uric acid	Nucleoside diphosphorylase, xanthine oxidase	Adenosine deaminase (Yasmineh *et al.*, 1977)
UDP	Disappearance of NADH	PK, LDH	Glycogen synthetase (Passonneau and Rottenburg, 1973)
dUMP	Formation of dihydrofolate	Thymidylate synthetase	Adenosine phosphotransferase (Spector *et al.*, 1974)
S-Adenosyl homocysteine	Disappearance of *S*-adenosyl methionine	Adenosine deaminase	Catechol-*O*-methyltransferase (Coward and Wu, 1973)
Assays for carbohydrates and glycolytic intermediates			
Cellobiose	O_2 consumption	β-1,4-Glucosidase, mutarotase, glucose oxidase, catalase	Cellulase (Green *et al.*, 1977)

Fructose 6-phosphate	Formation of NADH	Glucosephosphate isomerase, glucose-6-phosphate dehydrogenase	Fructokinase (Sabater *et al.*, 1972)
Galactose	Formation of NADH	Galactose dehydrogenase	Lactase (Asp and Dahlqvist, 1972)
β-d-Glucose	O_2 consumption	β-Fructofuranosidase, glucose oxidase, catalase	Aldose-1-epimerase (Weibel, 1976)
β-d-Glucose	O_2 consumption	Glucose oxidase	Mutarotase (Miwa, 1972)
Glucose 1-phosphate	Formation of NADH	Phosphoglucomutase, glucose-6-phosphate dehydrogenase, 6-phosphogluconate dehydrogenase	Galactose-1-phosphate uridylyltransferase (Pesce *et al.*, 1977)
Maltose	Formation of NADH	α-Glucosidase, hexokinase, glucose-6-phosphate dehydrogenase	α-Amylase (Guilbault and Rietz, 1976; Wilson and Barret, 1975)
Mannose 6-phosphate	Formation of NADH	Mannose-6-phosphate isomerase, glucose-6-phosphate dehydrogenase	Mannokinase (Sabater *et al.*, 1972)
3-Phosphoglycerate	Disappearance of NADH	Phosphoglyceromutase, enolase, pyruvate kinase, LDH	Phosphoglycerate kinase (Ali and Brownstone, 1976)
Pyruvate	Disappearance of NADH	LDH	Neuraminidase (Ziegler and Hutchinson, 1972)
Assays for TCA intermediates			
Acetyl-CoA	Disappearance of *p*-nitroaniline	Arylamine acetyltransferase	Acetyl-CoA synthetase, ATP-citrate lyase (Hoffman *et al.*, 1978)
Fumarate	Formation of NADH fluorimetrically	Fumarase, malate dehydrogenase	Argininosuccinate lyase (Ali and Brownstone, 1976)
Oxaloacetate	Disappearance of NADH	Malate dehydrogenase	Phosphoenolpyruvate carboxykinase (Hansen *et al.*, 1976)
Assays for amino acids and related compounds			
Aspartate	Disappearance of NADH	Aspartate aminotransferase, malate dehydrogenase	Asparaginase (Jayaram *et al.*, 1974)
Elastin fragments	Solubilization of Congo red-elastin	Trypsin	Elastase (Gnosspelius, 1977)

(continues)

Table II (*continued*)

Product measured	Species monitored	Coupling enzymes	Examples
Histidinol	Formation of NADH	Histidinol dehydrogenase	Histidinolphosphate phosphatase (Brady and Houston, 1972)
Assays for various other organic compounds			
Cystamine aminoaldehyde	Disappearance of NADH	Alcohol dehydrogenase	Diamine oxidase (Dupré *et al.*, 1977)
Phenol	O_2 consumption	Tryosinase	Alkaline phosphatase (Kumar and Christian, 1976)
R-CHO	O_2 consumption	Alcohol dehydrogenase	Glycerol dehydratase (Yakusheva *et al.*, 1974)
Assays for inorganic compounds			
H_2O_2	Formation of 2′,7′-dichlorofluorescein	Peroxidase	Monoamine oxidase (Köchli and von Wartburg, 1978)
H_2O_2	Oxidation of *O*-dianisidine	Peroxidase	Peroxisomal oxidases (Duley and Holmes, 1975)
NH_3	Disappearance of NADH	Glutamate dehydrogenase	D- and L-Amino acid oxidases (Holme and Goldberg, 1975), asparaginase (Ferguson, 1974)
PP_i	Disappearance of NADH	Pyrophosphate: fructose-6-phosphotransferase, aldolase, triosephosphate isomerase, glycerol-3-phosphate dehydrogenase	Argininosuccinate synthetase (O'Brien, 1976)

Table III
Miscellaneous Coupled Assays

Product measured	Species monitored	Coupling enzymes	Examples
Stop-time coupled enzyme assays			
ATP	Formation of [^{14}C]glucose 6-phosphate	Hexokinase	Pyruvate kinase (Preller and Ureta, 1976)
NAD-^{3}H	Formation of [^{3}H]lactate	LDH	9-Hydroxyprostaglandin dehydrogenase (Tai and Yuan, 1977)
[^{14}C]Biliverdin	Formation of [^{14}C]bilirubin	NADPH-Dependant biliverdin reductase	Heme oxygenase (Tenhunen, 1972)
Octopamine	Formation of [^{14}C]synephrine	Phenylethanolamine *N*-methyltransferase, adenosylho-cysteinase, adenosine deaminase	Dopamine-β-hydroxylase (Karahasonoglu *et al.*, 1975)
P_i groups from end RNA fragments	Formation of phosphomolybdate	Alkaline phosphatase	Ribonuclease (Stein and Wilczek, 1973)
Other assays			
FMN	Disappearance of riboflavin fluorescence; binding of FMN to apoflavodoxin quenches its fluorescence	Apoflavodoxin	Flavokinase (Mayhew and Wassnik, 1977)
[^{3}H]*S*-Adenosylmethione	Production of [^{3}H]melatonin, which can be separated from the other radioactive compounds in the assay by extraction with $CHCl_3$	Hydroxyindole-*O*-methyltransferase	Methionine adenosyltransferase (Mathysse *et al.*, 1972)
Phosphoenol pyruvate (PEP)	Disappearance of oxaloacetate; PK converts PEP to pyruvate, which does not absorb at 280 nm, where OAA is monitored	PK	PEP carboxykinase (Hatch, 1973)

lactate dehydrogenase is much better than the M_4 isozyme on coupling systems (Chang and Chung, 1975).

An interesting suggestion has been presented by Marco and Marco (1974), who proposed using an alternative substrate, 3-acetyl-NAD, for dehydrogenase coupling enzymes. This analog has a more favorable equilibrium for 3-acetyl-NADH formation, allowing assays monitoring the production of 3-acetyl-NADH to be done in the pH range of 7–8. Normal dehydrogenase assays are not as effective in this range. Also, the extinction change is higher than for NADH so the assay is even more sensitive. In general, alternative substrates can offer ways to develop better assays for a given reaction. The number of coupling enzymes should be minimized, but if the enzymes are readily available the sensitivity of the given assay should be the most important factor. Spectrophotometric assays are normally sufficiently sensitive for use if an adequate change in extinction coefficient is involved. A change of at least 0.05–0.1 OD unit is necessary for good analytical measurement. For NADH, a change in concentration of 8 μM will generate an OD change of 0.05 in a 1-cm cuvette. If the coupled assay contains a species that has a high absorbance, care must be taken to avoid stray light errors (Cavalieri and Sable, 1974). A single-beam spectrophotometer is not accurate at an absorbance above 2 owing to stray light effects. Beer's law should be verified for the spectrophotometer being used to ensure accurate measurements. Changes in substrate concentration or addition of inhibitors that absorb at the assay wavelength can cause serious errors and lead to incorrect interpretations. Fluorescence assays are 2–3 orders of magnitude more sensitive than spectrophotometric assays if a fluorescing species is involved. Self-quenching can be a problem when disappearance of fluorescence is being measured (Dalziel, 1961). Reactions can be coupled to enzymes that release or take up protons and assayed by use of a pH stat. This method is probably not as sensitive as spectral assays but would be useful in some cases. The use of isotopes in coupled assays as illustrated in Table III can be very sensitive, limited only be separation methods and the specific radioactivity of the substrate.

Once the choice of coupling system and method of assay is made, the parameters of the auxiliary enzymes must be evaluated. The V_{max} and K_m for each auxiliary enzyme should be determined with identical buffer, ionic strength, temperature, and other conditions as will be used in the actual measurements. It may be necessary to consider the inhibitory effects of substrates or inhibitors for each enzyme on the others. For example, glucose-6-phosphate dehydrogenase is significantly inhibited by free ATP, producing a long lag time (Avigad, 1966) and limiting its use for measurement of glucokinase and hexokinase activity in the presence of high free ATP. The measured kinetic parameters are likely to be different from manufacturer or literature values but are the proper values for the system being studied. In addition, the solution in which the auxiliary enzyme is kept must not cause inhibition of the primary enzyme. Ammonium sulfate is a particular problem requiring dialysis of the auxiliary enzymes prior to use. Even albumin, which is often added to stabilize proteins, can bind many substrates and cofactors and cause inhibition in that manner. If possible, the coupling enzymes should be in the same buffer system as the assay. Use of small molecular sieve columns or rapid dialyzers will help to avoid stability problems encountered during normal dialysis.

The amount of auxiliary enzymes needed can then be calculated as described either by McClure (1969) or Storer and Cornish-Bowden (1974). The parameters to be specified are the ratio of v_2/v_1 or the fractional attainment (F_p) of the steady-state rate and the lag time before F_p is reached. To ensure accurate assays,

F_p should usually be specified as 0.99 so that the measured rate represents 99% of the steady-state rate of the primary enzyme. Lag time will depend on the system being studied. A lag of 30 s, or longer, may not be a problem in some assays if a long linear region is then observed. If the primary enzyme is suspected of being hysteretic (Frieden, 1970), the lag time should be minimal to ensure actual measurement of the primary enzyme's behavior.

The two methods can be compared in the following examples. Storer and Cornish-Bowden (1974) used the assay of glucokinase with glucose-6-phosphate dehydrogenase for illustration of their method. The K_m for glucose-6-phosphate was found to be 0.11 mM under assay conditions. They specified that the lag time prior to $v_2 = 0.99\ v_1$ would be 1 min and v_1 would not exceed 0.04 mM/mi. Substitution of these values into Eq. (17), $t^* = \phi K_2/v_1$, gives $\phi = 0.36$. Using Table I and interpolating for this value of ϕ shows $v_1/V_2 = 0.08$ or $V_2 = 0.5$ mM/min or 0.5 IU/ml. Using McClure's analysis with Eq. (9) gives

$$\begin{aligned} V_2 &= \frac{-2.303 \log(1 - F_p)K_p}{t} \\ &= -2.303 \log(1 - 0.99)(0.11) \\ &= 0.51\text{mM}/\min \text{ or } 0.51\text{IU}/\text{ml} \end{aligned}$$

So, for a simple system, the two treatments will give similar answers. As v_1 decreases, the amount of E_2 added in the Storer and Cornish-Bowden (1974) treatment will be somewhat less, but not proportionally since both ϕ and V_2 are related to v_1.

An example of a two-auxiliary enzyme couple was presented by McClure (1969) for a generalized kinase assay using pyruvate kinase and lactate dehydrogenase. The values used were $k_1 = 0.05$ mM/min, $K_{ADP} = 0.21$ mM, and $K_{pyruvate} = 0.14$ mM. The transient time (t^*) was chosen to be 12 s. From the nomogram in Fig. 3(b), if t^* is 12 s then k_2 and k_3 must be larger than 1/0.043 or 23 min^{-1}. Assuming that k_2 and k_3 are approximately equal and that sufficient values are $k_1 = k_3 = 33$ min^{-1}, then $k_2 = V_2/K_{ADP}$ and $k_3 = V_3/K_{pyruvate}$ and $k_2 = 6.9$ IU/ml of pyruvate kinase and $k_3 = 4.6$ IU/ml of lactate dehydrogenase. Using equal amounts of k_2 and k_3 allows the minimum total units of enzyme to be used (McClure, 1969).

Using the Storer and Cornish-Bowden (1974) analysis, necessitates separation of the transient time for each couple and calculation of the ϕ for each reaction as was done in the one auxiliary enzyme case. If the t_1 and t_2 are assumed to each equal to 6 s, then the ϕ for pyruvate kinase is equal to

$$\frac{(t^*)(v_1)}{K_p} = \frac{(0.1)(0.05)}{0.21} = 0.024$$

The ratio of v_1/V_2 from Table I is 0.005, so $V_2 = 10$ IU/ml. For lactate dehydrogenase v_3 is assumed $\simeq v_2$, so the calculation can be made. $\phi = (0.1)$

$(0.05)/(0.14) = 0.036$, and from Table I $v_2/V_3 = 0.007$. Thus, the lactate dehydrogenase concentration should be 7.1 IU/ml. The two methods once again give similar results although McClure's (1969) method is probably easier to handle for a two-auxiliary enzyme system. The treatment of Storer and Cornish-Bowden (1974) gives high values for V_2 and V_3 because of the assumption that the lag times are independent. The actual observed lag time will be shorter than assumed in the calculations. The treatment of Storer and Cornish-Bowden (1974) has the advantage that it can be adapted to three or more coupled enzymes.

Calculations of transient time for a given assay using defined levels of coupling enzymes can be done using the methods of McClure (1969), Easterby (1973), and Storer and Cornish-Bowden (1974). The analysis of Easterby (1973) illustrated in Fig. 4 allows easy determination of the lag time and the steady-state concentration of the intermediate product. Knowledge of that concentration allows confirmation of the validity of the assumptions in the models presented in Section I and analysis of product inhibition effects by the intermediate product.

Making the calculation as to the amount of enzyme to be added does not free the investigator from making sure that the coupled assay really measures the velocity of the primary enzyme. Test assays should be run with levels of the coupling enzymes 2- to 10-fold higher than calculated, and the rates should all be the same. If an increase in rate is observed, further checking of the assays is required.

The calculation of required enzyme levels allows economical use of the auxiliary enzyme and will avoid many problems. McClure (1969) listed the information that should be evaluated for a coupled assay and presented as part of the experimental methods. This includes (1) the units of auxiliary enzyme added per milliliter of assay as described under the experimental conditions; (2) the apparent Michaelis constants for the measured products under the experimental conditions; (3) t^* and the F_p chosen on the basis of the calculations; and (4) the effect of the additional auxiliary substrates on the activity of the primary enzyme. Most assays currently used meet these criteria, having been determined by trial and error. The treatment described here allows a more rational approach to future design of coupled assays.

If a coupling enzyme is not available in sufficient quantity or some circumstance limits its use, the treatment of Kuchel *et al.* (1974) will allow analysis of a system where the activities of the primary and auxiliary enzymes are similar. This technique would obviously not work with hysteretic enzymes but can be used with allosteric proteins as the primary enzyme (Kuchel *et al.*, 1974).

III. Precautions

Once one is sure that sufficient auxiliary enzyme(s) is present to allow accurate assays, there are a few other problems to be dealt with. The reaction mixture for the primary enzyme should be preincubated with the assay enzymes to determine whether any change occurs. Often the substrates will be contaminated with a small amount of product, such as ADP in ATP, which can react with the coupling system. The endogenous product should be exhausted prior to actual assay. Often this results in actually better kinetic experiments, since no product will be present initially in the reaction mixture. Also the system should be checked to see whether there is a reaction observed in the absence of the assay enzymes. The presence of other enzymes, particularly during purification, can give rise to a blank rate that has to be considered and accounted for.

Another problem is contamination of the auxiliary enzymes by the primary enzyme. Often commercial enzymes will have a low contamination of the activity being studied, but even a 0.1% contamination is often a serious problem. This is more critical if the studies are being done with auxiliary enzyme based on the calculations presented above. A related problem is that the auxiliary enzymes will sometimes react with one of the primary substrates. Glucose-6-phosphate dehydrogenase has a small but detectable reaction with glucose (Storer and Cornish-Bowden, 1974), which limits the amount of enzyme that can be used to couple enzymes such as hexokinase.

The final problem deals with interpretation of inhibition experiments on the primary enzyme. If the inhibitor does not inhibit the auxiliary enzymes, no problems occur, since the activity of the primary enzyme will be accurately measured. Often, however, an inhibitor that is a structural analog of a substrate will be an inhibitor of the auxiliary enzymes and will cause an observed increase in the lag time. Generally, the advice has been simply not to use coupled enzyme assays with inhibitors. The model systems presented in Section I do make useful predictions regarding such effects. The inhibitor should be tested with the auxiliary enzymes and the K_i and type of inhibition be determined. If the inhibitor is a competitive inhibitor of the auxiliary enzyme with respect to the measured product, both McClure (1969) and Storer and Cornish-Bowden (1974) suggested that simply increasing the coupling enzyme will reduce the lag. In some cases, the lag may be too long to be able to reduce to a reasonable value, but the addition of more auxiliary enzyme will usually allow the measurements to be made. In general, similar considerations can be made for a noncompetitive inhibitor. McClure's treatment suggests that an uncompetitive inhibitor will not alter the lag time since K_p and V_2 change in a constant ratio, but Storer and Cornish-Bowden (1974) showed that this is not always true. It will be approximately true only if V_2 is at least 10 times v_1.

The effects of inhibitors on the auxiliary enzymes can be readily handled using the techniques described here. The inhibitor can be determined and compensated for by adding more auxiliary enzyme, and the lag time can be analyzed on the actual assay so that the validity of the assay is assured.

IV. Summary

The amount of auxiliary enzyme to be added to an assay system utilizing a single-coupled reaction can be calculated from

$$V_2 = \frac{-2.303 \log(1 - F_p)K_p}{t}$$

based on McClure's (1969) analysis where V_2 is the number of units of E_2, F_p is the desired fraction of the steady-state reaction of the primary enzyme to be measured, K_p is the Michaelis constant for P for E_2, and t is the desired lag time. Alternatively, the method of Storer and Cornish-Bowden (1974) used Eq. (17)

$$t^* = \frac{\phi K_p}{v_1}$$

where t^* is the transient time, K_p is as above, and v_1 is the highest velocity of the primary enzyme to be measured. A value for ϕ is calculated at a specified time, and from Table I the ratio v_1/V_2 is determined for a specified v_2/v_1 ratio.

A plot of product (Q) appearance versus time allows evaluation of the steady-state intermediate product level (P_{ss}) and the lag time as a check on the assumptions for the above equations. A check should always be done to assure that the assays are linear with a reasonable lag time.

Storer and Cornish-Bowden's (1974) treatment can be applied to systems with several coupling enzymes, and the nomogram of McClure (Fig. 3) is useful for a two-auxiliary enzyme system.

A check should always be done by adding a small amount of the primary enzyme product and determining that it reacts very rapidly with the assay enzymes, ensuring that the assay system is actually working.

V. Concluding Remarks

The use of coupled enzyme assays affords a convenient, reliable method of measuring the steady-state activity of an enzyme. Certain precautions must be taken to assure accuracy, and the system used should be well documented as described above. Under proper conditions even inhibition studies can be done with assurance of accurate measurements. New assays can be designed with confidence avoiding trial-and-error determinations.

Acknowledgments

This work was supported by Grants CA14030 awarded by the National Cancer Institute and C-582 from the Robert A. Welch Foundation. B. W. B. and R. S. B. are Robert A. Welch Foundation Predoctoral Fellows.

References

Ali, M., and Brownstone, Y. S. (1976). *Biochim. Biophys. Acta* **445,** 74.
Asp, N., and Dahlqvist, A. (1972). *Anal. Biochem.* **47,** 527.
Avigad, G. (1966). *Proc. Natl. Acad. Sci. USA* **56,** 1543.
Barwell, C. J., and Hess, B. (1970). *Hoppe-Seyler's Z. Physiol. Chem.* **351,** 1531.

Bergmeyer, H. U. (1953). *Biochem. Z.* **324,** 408.
Bergmeyer, H. U. (ed.) (1974). <?show “?>Methods of Enzymatic Analysis,<?show ”?> 2nd edn, vols. 1–4. Academic Press, New York.
Brady, D. R., and Houston, L. L. (1972). *Anal. Biochem.* **48,** 480.
Burnell, J. N., and Whatley, F. R. (1975). *Anal. Biochem.* **68,** 281.
Cavalieri, R. L., and Sable, H. Z. (1974). *Anal. Biochem.* **59,** 122.
Chang, M., and Chung, T. (1975). *Clin. Chem.* **21.**
Coward, J. K., and Wu, F. Y. (1973). *Anal. Biochem.* **55,** 406.
Dalziel, K. (1961). *Biochem. J.* **80,** 440.
Duley, J., and Holmes, R. S. (1975). *Anal. Biochem.* **69,** 164.
Dupré, S., Solinas, S. P., Guerrieri, P., Federici, G., and Cavallini, D. (1977). *Anal. Biochem.* **77,** 68.
Easterby, J. (1973). *Biochim. Biophys. Acta* **293,** 552.
Ferguson, D. A. (1974). *Anal. Biochem.* **62,** 81.
Frieden, C. (1970). *J. Biol. Chem.* **245,** 5788.
Giacomello, A., and Salerno, C. (1977). *Anal. Biochem.* **79,** 263.
Gnosspelius, G. (1977). *Anal. Biochem.* **81,** 315.
Green, T. R., Han, Y. W., and Anderson, A. W. (1977). *Anal. Biochem.* **82,** 404.
Guilbault, G. G., and Rietz, E. B. (1976). *Clin. Chem.* **22,** 1702.
Gutfreund, H. (1965). “An Introduction to the Study of Enzymes,” pp. 302–306. Blackwell, Oxford.
Hansen, R. J., Hinz, H., and Holzer, H. (1976). *Anal. Biochem.* **74,** 576.
Hatch, M. D. (1973). *Anal. Biochem.* **52,** 280.
Hess, B., and Wurster, B. (1970). *FEBS Lett.* **9,** 73.
Hoffman, G., Weiss, L., and Weiland, O. H. (1978). *Anal. Biochem.* **84,** 441.
Holme, D., and Goldberg, D. M. (1975). *Biochim. Biophys. Acta* **377,** 61.
Horgan, D. J., Tume, R. K., and Newbold, R. P. (1972). *Anal. Biochem.* **48,** 147.
Jayaram, H. N., Cooney, P. A., Jayaram, S., and Rosenblum, L. (1974). *Anal. Biochem.* **59,** 327.
Karahasonoglu, A. N., Ozand, P. T., Diggs, D., and Tildon, J. T. (1975). *Anal. Biochem.* **66,** 523.
Köchli, H., and von Wartburg, J. P. (1978). *Anal. Biochem.* **84,** 127.
Kuchel, P. W., Nichol, L. W., and Jeffery, P. D. (1974). *J. Theor. Biol.* **48,** 39.
Kumar, A., and Christian, G. D. (1976). *Anal. Biochem.* **48,** 1283.
Marco, E. J., and Marco, R. (1974). *Anal. Biochem.* **62,** 472.
Mathysse, S., Baldessarini, R. J., and Vogt, M. (1972). *Anal. Biochem.* **48,** 410.
Mayhew, S. G., and Wassnik, J. H. (1977). *Biochim. Biophys. Acta* **482,** 341.
McClure, W. R. (1969). *Biochemistry* **8,** 2782.
Miwa, I. (1972). *Anal. Biochem.* **45,** 441.
Monk, B. C., and Kellerman, G. M. (1976). *Anal. Biochem.* **73,** 187.
O'Brien, W. (1976). *Anal. Biochem.* **76,** 423.
Oliver, L. T. (1955). *Biochem. J.* **61,** 116.
Park, W. D., Tischler, M. E., Dunlop, R. B., and Fisher, R. R. (1973). *Anal. Biochem.* **54,** 495.
Passonneau, J. S., and Rottenburg, D. A. (1973). *Anal. Biochem.* **51,** 528.
Pesce, M. A., Bodian, S. H., Harris, R. C., and Nicholson, J. F. (1977). *Clin. Chem.* **23,** 1711.
Preller, A., and Ureta, T. (1976). *Anal. Biochem.* **76,** 416.
Rathbun, W. B., and Gilbert, H. D. (1973). *Anal. Biochem.* **54,** 153.
Roberts, D. V. (1977). “Enzyme Kinetics.” Cambridge University Press, London and New York.
Sabater, B., Sebastián, J., and Asensio, C. (1972). *Biochim. Biophys. Acta* **284,** 406–414.
Sherwin, J. F., and Natelson, S. (1975). *Clin. Chem.* **21,** 230.
Spector, T., and Miller, R. L. (1976). *Biochim. Biophys. Acta* **445,** 509.
Spector, T., Miller, R. L., Fyfe, J. A., and Krenitsky, T. A. (1974). *Biochim. Biophys. Acta* **370,** 585.
Stein, R., and Wilczek, J. (1973). *Anal. Biochem.* **54,** 419.
Storer, A., and Cornish-Bowden, A. (1974). *Biochem. J.* **141,** 205.
Tai, H., and Yuan, B. (1977). *Anal. Biochem.* **78,** 410.
Tenhunen, R. (1972). *Anal. Biochem.* **45,** 600.

Weibel, M. K. (1976). *Anal. Biochem.* **70,** 489.

Wilson, C. S., and Barret, M. J. (1975). *Clin. Chem.* **21,** 947.

Yakusheva, M. I., Malakhov, A. A., Poznanskaya, A. A., and Yakovlev, V. A. (1974). *Anal. Biochem.* **60,** 293.

Yasmineh, W. G., Byrnes, K., Lum, C. T., and Abbasnezhad, M. (1977). *Clin. Chem.* **23,** 2024.

Ziegler, D. N., and Hutchinson, H. D. (1972). *Appl. Microbiol.* **23,** 1060.

CHAPTER 4

Regression Analysis, Experimental Error, and Statistical Criteria in the Design and Analysis of Experiments for Discrimination Between Rival Kinetic Models

Bengt Mannervik

Department of Biochemistry and Organic Chemistry
Uppsala University
Biomedical Center
Box576, SE-75123 Uppsala, Sweden

I. Basic Concepts

The fitting of rate equations to kinetic data in enzymology is an application of the treatment of experimental data in general and the use of mathematical models for quantitative description. By using statistical methods, a certain degree of objectivity is ascertained insofar as all investigators should get the same analytical results once they have agreed on the techniques to use. However, it should be borne in mind that the choice of a statistical method (such as the least-squares algorithm) is not necessarily unbiased and will often affect the results. Certain statistical fitting procedures also provide quantitative measures of goodness of fit and of the reliability of the kinetic constants estimated, facilitating evaluation of the results

DOI: 10.1016/B978-0-12-378608-1.00004-9

and testing of hypotheses. This is the case for nonlinear regression analysis based on the principle of least squares, which was originally used in enzymology in 1961 to fit the Michaelis-Menten equation to experimental data (Johansen and Lumry, 1961; Wilkinson, 1961). Cleland (1963, 2009) has made a major contribution by encouraging and facilitating the use of regression analysis of enzyme kinetic data and has reviewed some of its principles. Even if many of the fundamental contributions to enzyme kinetics have been based adequately on graphical analysis, the current interest in more complicated rate laws often requires curve-fitting by computer for quantitative evaluation. Some of the early applications of computer analysis to enzyme kinetics have been reviewed (Garfinkel *et al.*, 1977). Various aspects of both the theoretical background and the applications of statistical methods are covered in a monograph (Endrenyi, 1981a).

In the following treatment it will be assumed that a mathematical model (rate equation) should be fitted by nonlinear regression analysis to a set of experimental data. Most procedures use the principle of least squares, according to which the regression function Y is minimized:

$$Y = \sum_{i=1}^{n} w_i(v_i - \hat{v}_i)^2 \tag{1}$$

where v_i and $\hat{v}_i$ are the observed and calculated velocity values and w_i is a weighting factor in the ith experimental point ($i = 1, 2, \ldots, n$). The difference $v_i - \hat{v}_i$ is called the *residual* in the ith point. The calculated (predicted) value is based on the model (e.g., the Michaelis-Menten equation), the setting of the independent variable(s) (e.g., substrate concentration) in the ith measurement, and the current estimates of the parameter values of the model (e.g., K_m and V). The computer program minimizes Y by searching for the optimal combination of parameter values. The weighting factors are used to compensate for unequal experimental variance in the different experimental points. After convergence, the value of Y is referred to as the *residual sum of squares*, SS, and is a measure of the goodness of fit of the mathematical model to the data set. A model with p parameters has $(n - p)$ degrees of freedom and a *mean sum of squares*, Q^2, for n data

$$Q^2 = \frac{\text{SS}}{n - p}$$

which is expected to equal the experimental variance σ^2. In addition to the residual sum of squares, it is desirable that the output from the computer includes *parameter values* and their *standard deviations*, *predicted velocities*, and *residuals* for each experimental point, and the variance-covariance matrix. Most modern programs (Jennrich and Ralston, 1979) achieve this and also offer graphics that facilitate evaluation of the results.

II. Discrimination Between Rival Mathematical Models (Rate Equations)

It is assumed that two rival mathematical models (here rate equations) have been fitted by the method of least squares to a kinetic data set. Common pairs of rival models are the steady-state rate laws for linear competitive and noncompetitive inhibition,

$$v = \frac{V[\mathrm{A}]}{K_{\mathrm{m}}(1 + [\mathrm{I}]/K_{\mathrm{is}}) + [\mathrm{A}]} \tag{2}$$

$$v = \frac{V[\mathrm{A}]}{K_{\mathrm{m}}(1 + [\mathrm{I}]/K_{\mathrm{is}}) + [\mathrm{A}](1 + [\mathrm{I}]/K_{\mathrm{ii}})} \tag{3}$$

and the rate laws for simple Ping-Pong and sequential two-substrate mechanisms:

$$v = \frac{V[\mathrm{A}][\mathrm{B}]}{K_{\mathrm{m}}^{\mathrm{B}}[\mathrm{A}] + K_{\mathrm{m}}^{\mathrm{A}}[\mathrm{B}] + [\mathrm{A}][\mathrm{B}]} \tag{4}$$

$$v = \frac{V[\mathrm{A}][\mathrm{B}]}{K_{\mathrm{m}}^{\mathrm{B}}[\mathrm{A}] + K_{\mathrm{m}}^{\mathrm{A}}[\mathrm{B}] + [\mathrm{A}][\mathrm{B}] + K_{\mathrm{s}}^{\mathrm{A}}K_{\mathrm{m}}^{\mathrm{B}}} \tag{5}$$

The questions that arise are

1. Do the models adequately describe the data?
2. Is one model better than the other?

The first question may be answered by evaluating the results of the regression by the criteria for goodness of fit discussed below. If both models are adequate and fit the data equally well, the simplest model is chosen. However, independent information, obtained by additional kinetic studies or by completely different experimental methods, should, when available, be included in the discrimination procedure. By use of such supplementary input, the more complex model may be the most realistic alternative even if the simpler model would have been adequate according to regression analysis. However, in the following discussion such extraneous input, albeit relevant and obligatory when available, will not be considered. Instead, focus will be put on a procedure for discrimination between rival models, based on the results of the regression analysis (Bartfai and Mannervik, 1972a,b; Bartfai *et al.*, 1973; Mannervik and Bartfai, 1973).

The following *criteria* have been formulated and been found valuable for evaluating *goodness of fit* as well as for *discrimination*:

1. A good model is expected to give convergence in the regression analysis.
2. A good model should give meaningful parameter values with low standard deviations.

3. A good model should give residuals showing random distribution about the zero level and lacking correlation with any of the dependent or independent variables.
4. A good model should give a low residual sum of squares that is compatible with the experimental variance.

If two models fulfill all the above criteria and a difference between their mean sum of squares exists, the model giving the lowest value is chosen if the difference can be considered meaningful in comparison with the experimental variance. When no significant difference between the two models exists, additional information must be collected.

In the following, let us consider in some detail the four discrimination criteria (1)–(4) (Mannervik, 1981):

1. *Convergence* is an indicator of the success of the fitting procedure. Occasionally, convergence is not obtained, because of unnoticed relationships between variables that result in a singular design matrix (Bartfai *et al.*, 1973; Mannervik, 1981). More commonly, numerical problems cause the lack of convergence. Several sets of initial values for the parameters to be estimated should be tested. A new set of data may also be tried. In some cases, a more powerful optimization algorithm may help. Often reparametrization of the model or tranformation of the variables (e.g., log v fitted instead of v) is worthwhile. If one model has poorer convergence properties than the rival model, the explanation may be that the poor model is intrinsically inferior. However, discrimination should not be based on this criterion alone. In most cases, convergence of the fitting procedure is obtained for both of the rival models and the results of the regression should be examined from all possible points of view. The predicted (calculated) and the experimentally determined velocity values should be plotted versus each of the independent variables (substrate and inhibitor concentrations, etc.). The overall impression of such plots should be that predicted values closely follow the experimentally determined values (cf. Fig. 1A). Plots of the dependent variable (v) versus the different independent variables are available in the output of many computer program packages for regression analysis. Such diagrams can also be constructed manually. Lack of proper convergence as well as inadequate models are often clearly detected by examination of the plotted data. In particular, it should be noted that simple mathematical models can only give rise to certain "canonical" curve shapes: a rational 1:1 function, 1 shape; a 2:2 function, 4 shapes; a 3:3 function, 27 shapes (cf. Solano-Muñoz *et al.*, 1981). If the fit to the experimental data is poor, further analysis of the results of the regression may give clues to the formulation of a better model, but the estimates of the parameter values have no physical meaning.

2. *Parameter values* should be meaningful in relation to the model. An inhibition constant (K_i), for example, should have a finite, real, positive value. If unacceptable parameter values are obtained (such as $K_i < 0$), constraints on some or all of the parameters may be introduced in the regression program.

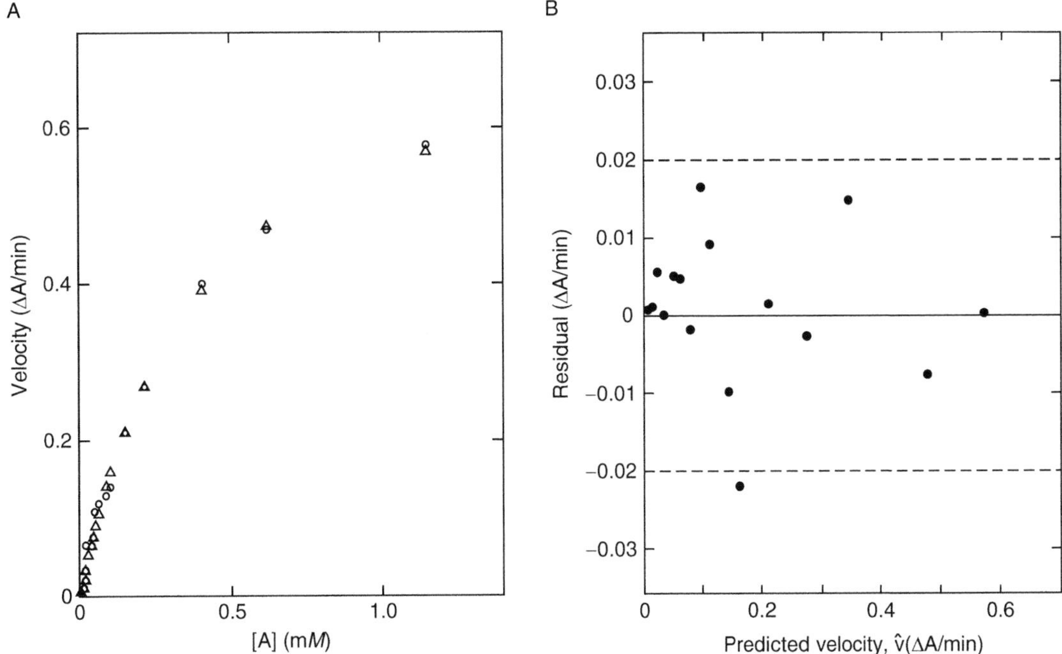

Fig. 1 Successful fit of the Michaelis-Menten equation to a set of experimental data by use of nonlinear regression. (A) Convergence to the proper minimum of the regression function is indicated by the close correspondence between observed (○) and calculated (Δ) velocity values. (B) Residuals of the fit plotted versus predicted velocity appear to have a random distribution. The dashed lines mark the 95% confidence levels ($0 \pm t_{0.05}\sigma$) calculated from the t-statistic and an estimate of the experimental error (σ). (From Mannervik (1981) with permission of the publishers.)

Physically meaningless parameter values may occur owing to numerical problems, use of an erroneous model, or improper design of the experiment being analyzed.

The regression analysis allows calculation of standard deviations of the parameters. In a good model they should be low in comparison with their corresponding parameter values. A standard deviation of less than 50% of the parameter value can usually be considered satisfactory, but for models containing many parameters it may sometimes be necessary to accept larger standard deviations.

In classical model building, the parameters should be independent. If they are, the covariance between any two of the parameters should be nil. This prerequisite cannot be fulfilled in the analysis of equations that contain intrinsically correlated constants. For example, according to Briggs and Haldane (1925), V and K_m are equal to $k_3[E]_0$ and $(k_2 + k_3)/k_1$, respectively, and contain k_3 as a common element. Thus, the estimates of V and K_m may show covariance even if the Michaelis-Menten (Briggs-Haldane) equation is the "true" model. In fact, the variance-covariance matrix or the corresponding normalized correlation matrix of the parameters of a good model is expected to reflect the occurrence of common rate constants in the coefficients of a steady-state rate equation (Mannervik, 1981; Mannervik and Bartfai, 1973). Intrinsic correlation between the parameters can be avoided if the steady-state model is expressed in terms of independent elementary rate constants, but in most cases such models contain more constants than can be fitted.

3. *Residuals* are the differences between the observed values (v_i) and the corresponding calculated values ($\hat{v}_i$) obtained by regression analysis:

$$q_i = v_i - \hat{v}_i$$

In a good fit of a model, nonvanishing residuals represent experimental error only. In general, the residuals are expected to be distributed about the zero level with approximately equal numbers of positive and negative values and without significant correlation with the dependent (v) or independent variables (substrate and inhibitor concentrations, etc.). If the experimental error of the experimental data has a Gaussian (normal) distribution, the residuals of the regression are expected to have a Gaussian distribution and a mean of zero. When (independent) information about the experimental variance (σ^2) is available, appropriate confidence levels can be defined for the residuals. Figure 1B shows a representative plot of residuals versus predicted velocity for a good fit; all but one of the residuals fall within the 95% confidence limits. The experimental variance can be estimated by replicate measurements at different substrate concentrations. (In the case shown in Fig. 1B, the estimated variance was found to be constant under the experimental conditions used.) The confidence limits are defined as $0 \pm t\sigma$, where t is the t-statistic (here $t_{0.05,\, r-1}$ for r measurements to determine σ^2) and

$$\sigma = \left(\sum_{i=1}^{r} \frac{(v_i - \bar{v})^2}{r - 1} \right)^{1/2} \tag{6}$$

$\bar{v}$ is the mean of v_i.

Systematic deviations from a random distribution of the residuals are often revealed in plots of residuals versus the different variables. Figure 2B shows a case in which residuals are clearly correlated with the concentration of a chemical component (G) of the reaction system. The nonrandom distribution of signs in a map of residuals in a two-dimensional plane defined by the (total) concentrations of two components (M and G) in the system also distinctly demonstrates the inadequacy of the fit (Fig. 2A).

A useful and simple *test for randomness* of the residuals can be performed when they are ordered as a function of one of the variables. In any array of this kind, the signs of the residuals are expected to be positive and negative at random. For example, in Fig. 1B, the residuals have the following grouping of signs as a function of predicted velocity:

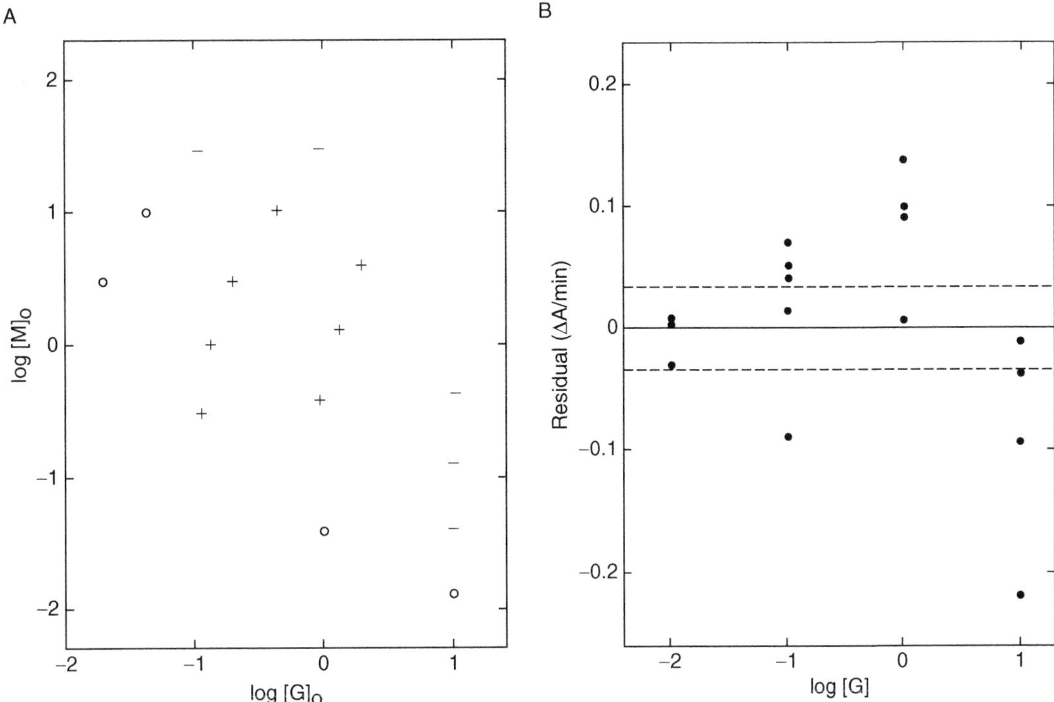

Fig. 2 Results of fitting the Michaelis-Menten equation to data that require a more complex rate law for adequate description. Lack of fit is shown by plots of residuals. (A) Map of residuals in the plane of the independent variables (total concentrations of G and M in mM): ○, residuals smaller than the experimental error (σ); ±, residuals $>\sigma$. −, residuals $<-\sigma$. (B) Residuals plotted versus [G] (mM); the dashed lines mark the 95% confidence levels. Note the trend in the points, demonstrating that a significant influence of [G] on velocity has not been accounted for in the model fitted. (From Mannervik (1981) with permission of the publishers.)

$$(+++)(-)(++)(-)(++)(--)(+)(-)(+)(-)(+)$$

The smallest number of signs of one kind (m) is 6, and the largest (n) is 10. The probability of 11 "runs" in any random sequence of groups with equal signs with $m = 6$ and $n = 10$ can be calculated as 0.958 (Swed and Eisenhart, 1943). Thus, there is no reason to believe that the residuals in Fig. 1B have a nonrandom distribution as plotted versus velocity. In Fig. 2B, on the other hand, the grouping of residuals is

$$(-)(++)(-)(+++++++)(----)$$

with 5 runs and $m = 6$ and $n = 10$. The probability for this arrangement is 0.047—that is, the null hypothesis that the arrangement is random can be excluded at the 95% confidence level. This indicates that the fit of the model to the data is inadequate. Table I gives numbers for testing randomness of groupings of any two kinds of different objects at the 95% confidence level.

Table I
Numbers for Testing Randomness of Groupings of Two Kinds of Objects at the 95% Confidence Level[a,b]

	m																		
n	2	3	4	5	6	7	8	9	10	11	12	13	14	15	16	17	18	19	20
2																			
3																			
4			2																
5		2	2	3															
6		2	3	3	3														
7		2	3	3	4	4													
8	2	2	3	3	4	4	5												
9	2	2	3	4	4	5	5	6											
10	2	3	3	4	5	5	6	6	6										
11	2	3	3	4	5	5	6	6	7	7									
12	2	3	4	4	5	6	6	7	7	8	8								
13	2	3	4	4	5	6	6	7	8	8	9	9							
14	2	3	4	5	5	6	7	7	8	8	9	9	10						
15	2	3	4	5	6	6	7	8	8	9	9	10	10	11					
16	2	3	4	5	6	6	7	8	8	9	10	10	11	11	11				
17	2	3	4	5	6	7	7	8	9	9	10	10	11	11	12	12			
18	2	3	4	5	6	7	8	8	9	10	10	11	11	12	12	13	13		
19	2	3	4	5	6	7	8	8	9	10	10	11	12	12	13	13	14	14	
20	2	3	4	5	6	7	8	9	9	10	11	11	12	12	13	13	14	14	15

[a]The table contains the largest integral numbers for groupings ("runs") of m and n ($m \leq n$) objects of two kinds, for which the probability $P \leq 0.05$ for a random array. For example, the entry at $m = 6$, $n = 10$ is 5, showing that any sequence of 6 objects of one kind and 10 objects of another kind containing $\leq$5 groups ("runs") can be regarded as nonrandom at the 95% confidence level.

[b]Calculated probabilities as well as numbers for additional confidence levels can be found in Swed and Eisenhart (1943), from which the numbers of this table are compiled.

The test for randomness has valuable use not only in the examination of the goodness of fit of a model, but also in the search for systemic errors in the experimental data. If residuals are ordered in the sequence in which the experimental points were recorded, time-dependent changes in the experimental conditions could be detected. Such changes are not infrequent and may be caused by declining enzyme activity, altered reactant concentrations, perturbations of physical parameters (temperature, pressure, pH, etc.), and drifting instrument responses.

4. The *residual sum of squares* of an adequate fit of a model to experimental data should reflect the experimental variance only, the expected value of which is σ^2. The mean sum of squares, Q^2, (defined under Section I) should be compared with an independent estimate, s^2, of σ^2. The quotient Q^2/s^2 can be compared with the appropriate F-statistic to evaluate the possibility that any difference between Q^2 and s^2 is significant. A model giving a Q^2 value significantly higher than the experimental error does not provide a satisfactory fit. However, before the model

is rejected, it must be considered that only the estimate s^2, and not the true value of σ^2, is available, and that only for models which are linear in the parameters (unlike most rate equations) are the significance levels calculated by use of the F-statistic fully reliable. The uncertainty about significance levels for models that are nonlinear in parameters applies to Student's t-test as well (Draper and Smith, 1966; Mannervik, 1981), but the rigor of the use of these statistics increases if some measure demonstrates that the nonlinearity in the appropriate region in the parameter space is small or negligible (Mannervik and Bartfai, 1973).

When two models j and k with p_j and p_k parameters are fitted (separately) to the same data set, the model giving the lowest Q^2 value should normally be regarded as giving the "best" fit. Often two similar models give residual sums of squares, SS_j and SS_k, that differ only by a small amount. Thus, the question arises of whether the difference is large enough for discrimination between the alternative models. Usually one model (k) is an extension of a simpler model (j) obtained by addition of terms to the latter equation. The significance of the improvement obtained by addition of the $(p_k - p_j)$ new parameters can be tested by comparing the quotient

$$\frac{(SS_j - SS_k)(n - p_k)}{(p_k - p_j)SS_k}$$

with the F-statistic $F(p_k - p_j, n - p_k)$ at the desired level of probability. If the quotient is larger than the F-value, the choice of the more complex model is warranted (keeping in mind the limitations of the F-test for nonlinear models). If no significant difference exists, the simpler model should normally be preferred. In some cases, the discrimination involves completely different models, which are not interconvertible by addition or elimination of terms. An example is the comparison of a one-substrate model with a two-substrate model for the kinetics of glyoxalase I (Bartfai and Mannervik, 1972b; Bartfai *et al.*, 1973; Mannervik, 1981; Mannervik and Bartfai, 1973). In these cases, the test described earlier is not applicable, and quantitative statistical measures based on the residual sum of squares are lacking. However, experience in our laboratory suggests that, if the number of data is not too small ($n > 15$), a quotient of the mean sums of squares for the rival models exceeding 1.5, that is,

$$\frac{Q_j^2}{Q_k^2} > 1.5$$

can be used as a reliable criterion for a better fit of model k.

III. Design of Experiments

The design of an experiment determines and strictly limits the information contained in a data set. An enzyme kinetic experiment is often simply planned to find out, for example, whether a two-substrate reaction follows a Ping-Pong or a sequential mechanism and what the values of the kinetic constants (K_m, V, etc.) are.

The fact that such an investigation involves two separate problems: (1) model discrimination and (2) parameter estimation, each requiring different designs of the experiment for optimal results, is usually not recognized.

When the model is known, it has been shown that maximal precision in the determination of the p parameter values is obtained if measurements are made only in p experimental points (Duggleby, 1981; Endrenyi, 1981b). The measurements should be replicated at least five times for each point. Thus, if the Michaelis-Menten equation is known to be the "true" model and the relative error is constant, the optimal design for estimation of K_m and V is an equal number (≥ 5) of measurements at the highest and lowest substrate concentrations feasible (Endrenyi, 1981b). Measurements at more than two experimental points will not improve the precision of the estimates of K_m and V ($p = 2$). The location of the p experimental points that are optimal for parameter estimation is determined by the model and the nature of the experimental error (Duggleby, 1981; Endrenyi, 1981b). Proper designs for some of the common steady-state rate equations in enzymology have been derived for systems with constant absolute error (Duggleby, 1979).

However, usually the model is not known with certainty and, in fact, the finding of the "best" model may be the primary goal of the investigation. Therefore, it is advisable to make measurements that cover all independent variables (substrate and inhibitor concentrations, etc.) at low, intermediate, and high values. If wide ranges of the variables are experimentally accessible, it is usually recommendable that the points be spaced geometrically (i.e., constant ratio of successive concentrations): 1:10:100:1000: . . . . Evidently, the number of experimental points must at least equal the number of parameters of the most complex model considered, but it is normally necessary to map the entire space of the independent variables by adequately distributed measurements in order to fully explore the response (v). An example showing how inadequate spacings of experimental points conceal the intricacies of the rate behavior of an enzyme is found in Mannervik (1981).

When an experimental data set is analyzed, it is often the case that two models pass the goodness-of-fit criteria (1)–(4) and that the difference in their residual sums of squares is too small for discrimination between the rival models. In this case, supplementary information must be acquired to make possible the choice of the "best" model. If new experiments are feasible, additional experimental data may be combined with the original data set, provided that the basic conditions for the measurements can be accurately reproduced. However, enzyme and substrate solutions are often unstable, and it is often worthwhile to perform a complete experiment with a new design to facilitate discrimination.

The reason that the residual sum of squares for a model j is larger than that of the "best" model k is that model j has a certain amount of lack of fit (bias error) in addition to the experimental variance of the data (pure error). If experimental conditions are selected that stress the lack of fit of model j, an increased mean sum of squares (Q_j^2) and a nonrandom distribution of the residuals are expected. The "true" model should be well behaved under all experimental conditions.

Since the "true" model is unknown, the current "best" model (k) has to be used as its substitute. A *discrimination function*, g, has been introduced that helps find the optimal experimental conditions for discrimination between two rival models (Bartfai and Mannervik, 1972a,b; Bartfai *et al.*, 1973; Mannervik, 1981; Mannervik and Bartfai, 1973).

The discrimination function can be formulated as the absolute difference between the residuals or between the predicted velocities for the two rival models, j and k:

$$g_i = |q_{i\mathrm{j}} - q_{i\mathrm{k}}| = |\hat{v}_{i\mathrm{j}} - \hat{v}_{i\mathrm{k}}|$$

where the index i denotes the ith experimental point. Since the residuals of both models contain the same contribution of experimental error (pure error), g expresses only the bias error (lack of fit) of the inferior model, provided that one of the alternatives is the "true" model and the errors are additive. The square of the value of the discrimination function for a point is proportional to the information content for discrimination. Accordingly, the optimal experimental conditions are those that maximize g. By differentiation of g with respect to the independent variables (reactant concentrations, etc.), the maximum can be located in the space on which the dependent variable (v) is defined (Bartfai and Mannervik, 1972a,b; Mannervik and Bartfai, 1973). Alternatively, g can be mapped by plotting the difference between the predicted values of the two rival models in each experimental point (Mannervik, 1975; Mannervik *et al.*, 1973). When new data can be combined with the original data set, these should be obtained from additional measurements in the region which maximizes g. Otherwise, a new complete experiment should be made that includes experimental points that define both the general outline of the response function (v) and the optimal conditions for discrimination. For optimal overall discrimination, the sum of the values of g in all experimental points should be maximized:

$$\sum_{i=1}^{m} g_i = \max$$

where m is the number of experimental points [$m \geq \max(p_\mathrm{j}, p_\mathrm{k})$], and p_j and p_k are the number of parameters in models j and k, respectively. The minimum number of experimental settings (design points) is $\max(p_\mathrm{j}, p_\mathrm{k})$, and if this design is chosen, replicates (≥ 5) should be made in each point. A combination of the experimental settings that optimize parameter estimation for each of the rival models, as well as those giving maximal discrimination, may be the best design. Evaluation of the discrimination function requires prior estimates of the parameter values of the two rival models. When the "best" model has been selected, its parameter values should be refined by use of experimental points that afford maximal precision.

As a simple illustration of the discrimination function, the alternative models for the data in Fig. 3 are presented. Here, the Michaelis-Menten equation and its extension with an $[\mathrm{A}]^2$ term for substrate inhibition are considered. The discrimination function is

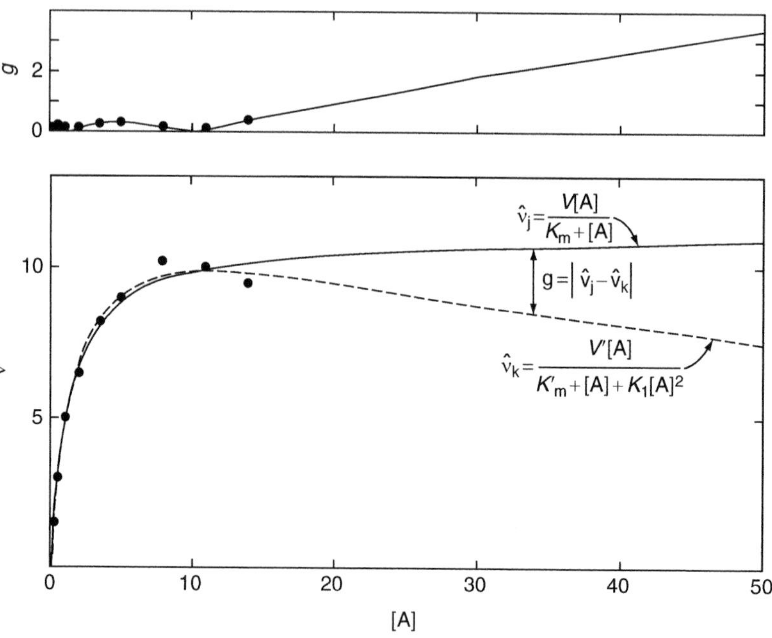

Fig. 3 The discrimination function g for the Michaelis-Menten equation and the corresponding rate law describing inhibition by high substrate concentration, [A]. The two rival rate equations were fitted (separately) by nonlinear regression to the data points (lower panel). The discrimination function (upper panel) is the absolute value of the difference between the respective calculated v values of the rival models [cf. Eq. (7)]. Note that both models afford reasonable fits to the data points and that only for high [A], where g is large, could clear discrimination between the models be expected. (From Mannervik (1981) with permission of the publishers.)

$$g = \left| \frac{V[A]}{K_m + [A]} - \frac{V'[A]}{K'_m + [A] + K_I[A]^2} \right| \tag{7}$$

The global maximum of g is V, which is approached when the substrate concentration [A] is extrapolated to infinity. Thus, the optimal design point for discrimination is at the highest substrate concentration attainable. Evidently, physical and chemical parameters—such as limitations of solubility, changes of ionic strength, and dielectric constant of the system, etc.—constrain the ranges of reactant concentrations that can be explored. The discrimination function, nevertheless, demonstrates in which region additional data should be collected.

Discrimination between the rate laws for a Ping-Pong and a sequential two-substrate mechanism [Eqs. (4) and (5)] presents an example where the optimal conditions for discrimination are less obvious than in the previous example (Bartfai and Mannervik, 1972b; Mannervik, 1975; Mannervik and Bartfai, 1973). By analytical differentiation of the discrimination function, it can be shown that the partial derivatives with respect to the substrate concentrations ([A], [B]) vanish under the following conditions:

$$\frac{\partial g}{\partial[\mathrm{A}]} = 0 \quad \text{for}[\mathrm{A}] = \frac{\sqrt{K_{\mathrm{m}}^{\mathrm{A}}[\mathrm{B}](K_{\mathrm{m}}^{\mathrm{A}}[\mathrm{B}] + K_{\mathrm{m}}^{\mathrm{B}}K_{\mathrm{s}}^{\mathrm{A}})}}{K_{\mathrm{m}}^{\mathrm{B}} + [\mathrm{B}]}$$

and

$$\frac{\partial g}{\partial[\mathrm{B}]} = 0 \quad \text{for}[\mathrm{B}] = \frac{\sqrt{K_{\mathrm{m}}^{\mathrm{B}}[\mathrm{A}](K_{\mathrm{m}}^{\mathrm{B}}[\mathrm{A}] + K_{\mathrm{m}}^{\mathrm{B}}K_{\mathrm{s}}^{\mathrm{A}})}}{K_{\mathrm{m}}^{\mathrm{A}} + [\mathrm{A}]}$$

When the substrate concentrations approach zero or infinity, the discrimination function approaches zero and no discrimination is possible. Optimal conditions for discrimination are found at the point where both partial derivatives equal zero for finite positive values of [A] and [B]. This point, at which g has a maximum, can be found numerically or graphically by plotting the curves corresponding to $\partial g/\partial[\mathrm{A}] = 0$ and $\partial g/\partial[\mathrm{B}] = 0$ (Bartfai and Mannervik, 1972a,b; Mannervik and Bartfai, 1973).

The optimal conditions for discrimination between the rate laws for linear competitive and linear noncompetitive inhibition [Eqs. (2) and (3)] can be found in a similar manner by analysis of the corresponding discrimination function. The partial derivative with respect to inhibitor concentration ([I]) vanishes for

$$[\mathrm{I}] = \frac{K_{\mathrm{m}} + [\mathrm{A}]}{\sqrt{(K_{\mathrm{m}}[\mathrm{A}]/K_{\mathrm{is}}K_{\mathrm{ii}}) + (K_{\mathrm{m}}/K_{\mathrm{is}})^2}} \tag{8}$$

Further, it can be shown that [A] should be as high as possible for optimal discrimination, and when $[\mathrm{A}] \gg K_{\mathrm{m}}K_{\mathrm{ii}}/K_{\mathrm{is}}(K_{\mathrm{ii}} > K_{\mathrm{is}}$ when the problem of discrimination arises), it follows that the optimal setting of [I] (for $\partial g/\partial[\mathrm{I}] = 0$) is simplified to

$$[\mathrm{I}] = \sqrt{\frac{K_{\mathrm{is}}K_{\mathrm{ii}}[\mathrm{A}]}{K_{\mathrm{m}}}} \tag{9}$$

As a final example, the optimal conditions to demonstrate the putative effect of a compound believed to act as a linear competitive inhibitor on an enzyme following Michaelis-Menten kinetics are given (Mannervik, 1981; Mannervik and Bartfai, 1973). The inhibitor should be used at the highest possible concentration, $[\mathrm{I}]_{\mathrm{max}}$, and the substrate concentration should be

$$[\mathrm{A}] = K_{\mathrm{m}}\sqrt{\frac{1 + [\mathrm{I}]\,\mathrm{max}}{K_{\mathrm{i}}}} \tag{10}$$

In all of the above examples, the analytical expressions for optimal conditions are based on the assumption that the expected parameter values for corresponding terms are equal in the rival models (e.g., $V = V'$ and $K_m = K'm$ in Fig. 3). When experimental data are fitted by computer, the corresponding estimates of the parameter values may differ to various extents. The parameter values obtained are determined by the fits for the respective models and the numerical properties of the alternative rate equations. Differences in pairs of corresponding estimated parameter values of two good alternative models are usually not too large to prevent the use of the above analytical expressions, and, even if they are, do not limit the use of the discrimination function. However, it should be noted that meaningful use of the discrimination function is based on the prerequisite that at least one of the rival models gives a reasonable fit to the experimental data. The difference between the fits of two inadequate models cannot be expected to display information leading to the identification of the "best" model.

It may be added that for discrimination between complicated models involving many terms, it may be useful to confine the analysis to a restricted domain of the independent variables, highlighting the difference between the rival models. This approach was used successfully to distinguish a random (nonequilibrium) sequential mechanism from a hybrid Ping-Pong/sequential mechanism (Mannervik, 1981; Mannervik and Askelöf, 1975).

IV. Experimental Error

Optimal design of experiments as well as proper analysis of experimental data is critically dependent on knowledge of the experimental variance. Design points for which the experimental error is large compared with the measured value of the dependent variable (v) contain less reliable information than those points in which the error is small. As a general rule, the dependent variable should be studied in a more restricted range when the absolute value of the experimental error tends to be constant than when it increases with the measured values. As a preliminary study, it may be useful to make replicate (≥ 5) measurements at the lowest and at the highest values considered for the dependent variable in the investigation. From these measurements, the variance, σ^2, can be estimated [cf. Eq. (6)] in the two domains and the values compared by use of the F-statistic (Mannervik, 1975). If the variance is constant, the relative error increases with decreasing values of the dependent variable. The upper limit of the relative error that can be accepted may be considered to define the natural lower constraint on future measurements. Such information about the experimental error should be combined with considerations about optimal conditions for parameter estimation and model discrimination in the design of experiments.

When experimental data are analyzed by the method of least squares, it is assumed that the independent variables have negligible error in comparison with the dependent variable. This condition is probably often fulfilled and has been put to experimental test (Askelöf *et al.*, 1976). Furthermore, the data fitted should have equal variance (constant absolute error), which is probably seldom the case when wide ranges of the dependent variable are investigated. As a remedy for unequal variances, each term in the regression equation [Eq. (1)] can be multiplied with a weighting factor w_i, inversely proportional to the experimental variance of the dependent variable (v_i):

$$w_i \propto \frac{1}{\mathrm{Var}(v_i)} \tag{11}$$

The problem is to find the values of the weighting factors. The weights can be based on estimates of the variance, which are calculated from replicates in each experimental point (Mannervik, 1975; Ottaway, 1973). However, a better approach taken by several investigators (see Mannervik, 1981) is to express the experimental variance as a function of the dependent variable. In a detailed analysis of enzyme kinetic data, it was found that the error distribution, as determined by replicates ($r = 10$), was not significantly different from a Gaussian distribution and that the variance could be expressed as a constant term plus one or several terms which increased with velocity (Askelöf *et al.*, 1976). However, only two parameters could be estimated with significant values, and it was therefore found useful to approximate the variance by a power function

$$\mathrm{Var}(v) = K_1 v^{\alpha} \tag{12}$$

where K_1 and α are empirical constants. A similar error function was independently proposed (Siano *et al.*, 1975). The constants can be estimated by replicate measurements of v in the range under investigation. Constant absolute error corresponds to $\alpha = 0$ and constant relative error to $\alpha = 2$. Thus, the weighting factors can be derived from a weighting function:

$$w_i \propto v_i^{-\alpha} \tag{13}$$

In principle, the velocity values in Eq. (13) used in weighting should be the predicted values ($\hat{v}_i$), but before an adequate model has been selected, the experimental values may be better; usually, any of the choices will be adequate in the application to regression analysis.

An alternative to the use of replicate measurements for defining the weighting function is based on the principle that the residuals of a good fit should reflect the experimental error only (Mannervik *et al.*, 1979). Figure 4 shows an example of a residual plot of such a good fit. It is evident that the experimental error (residuals) increases with velocity. If the neighboring residuals are ordered in groups of five to six residuals, the local variance can be estimated as the mean of the squared residuals,

$$\frac{1}{m}\sum_{i=1}^{m} q_i^2$$

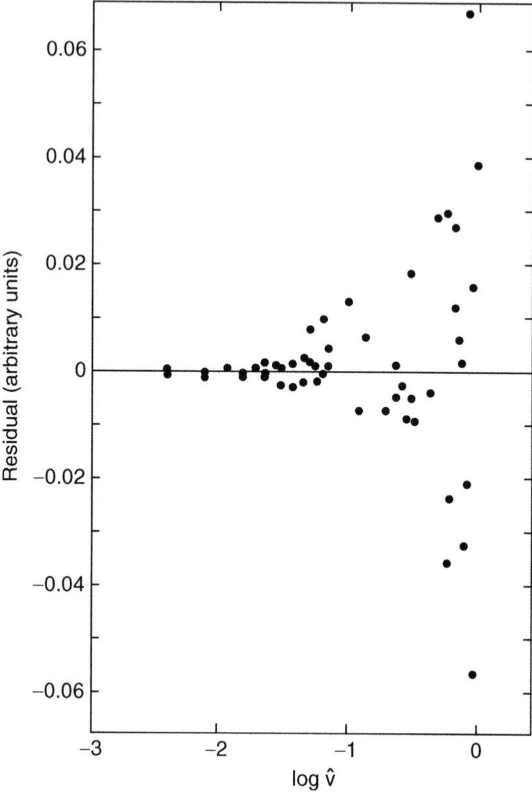

Fig. 4 Dependence of experimental error on velocity. The error is expressed as unweighted residuals of a good fit of a rational function of 3:3 degree to a set of experimental data. Groups of residuals along the v-axis could be used to estimate the local variance as a function of velocity (Mannervik *et al.*, 1979). (Further details are given in Mannervik (1981), from which the figure has been taken with permission of the publishers.)

where m is the number of residuals in the group. By treating the different estimates of the local variance as a function of velocity (using the mean or median to represent velocity in each group, the choice depending on the spacing of the experimental data), the constants of Eq. (12) can be estimated by weighted regression analysis (weighting factors inversely proportional to the local variances). Alternatively, the constants can be estimated from a logarithmic plot. For further details and considerations of this method (see Mannervik, 1981; Mannervik *et al.*, 1979).

The derivation of weighting factors from residuals has been found to be a reliable method, eliminating the need for the numerous replicates required in earlier procedures. The method can be applied to every data set ($n \geq 10$) and consequently can account for differences in error structure from design to design, day to day, or experimenter to experimenter. Furthermore, it appears likely, in general, that the variance of replicates may be smaller than the local population variance of a kinetic data set in the same experimental point. The analysis of the residuals is believed to give the latter variance, which is the one wanted for the regression analysis.

Although Eq. (13) has been found useful in defining weighting factors as a function of velocity, it should be remembered that the velocity is a dependent, stochastic variable. The true independent variables are substrate and inhibitor concentrations, etc. In several experimental systems investigated in our laboratory (Mannervik *et al.*, 1982), it has been found that the experimental variance is more correctly and accurately expressed in terms of the independent variable(s) than in velocity (cf. Mannervik, 1981). Therefore, the nature of the experimental error should not be taken for granted, but should be subject to investigation. Examination of residuals as a function of each of the variables is useful for this purpose.

In some data sets it may be found that one or a few isolated points deviate considerably more from the predicted value(s) than expected for a "normal" experimental error. Such points, referred to as *outliers*, have large residuals and consequently have a very important influence on the residual sum of squares in regression analysis. It is not safe to simply eliminate such data, because they may be explained and fitted by an alternative model not yet considered and, furthermore, may reflect the true error behavior of the data set. On the other hand, the least-squares fitting of the data set may not be unbiased if outliers are included in the analysis. The best way to handle the outliers is to perform the analysis both with and without the aberrant values, and to make sure that their elimination does not influence the model discrimination. In cases where omission of an outlier appears desirable, the criterion for elimination could be that its residual exceeds 2σ, where σ is estimated as $\sqrt{Q^2}$. It must be emphasized that only one point at a time and only very few data, if any, should be eliminated from a data set. (For a Gaussian distribution, the probability is 0.05 that 1 residual of 20 will exceed 2σ.)

V. Guidelines for Design and Analysis of Kinetic Experiments

Proper design of an experiment optimizes the information content of a data set. Proper analysis extracts a maximum of the information that is implicit in a data set. A strategy that highlights the sequence of events, including design and analysis, in the mathematical modeling of enzyme kinetic data is shown in Fig. 5. (For further details, see Mannervik, 1981.)

1. The *problem* should be *identified* and accurately *defined*—for example, the dependence of velocity on reactant concentrations.

2. The problem should be formulated by use of one or some alternative *mathematical models*—for example, rate equations.

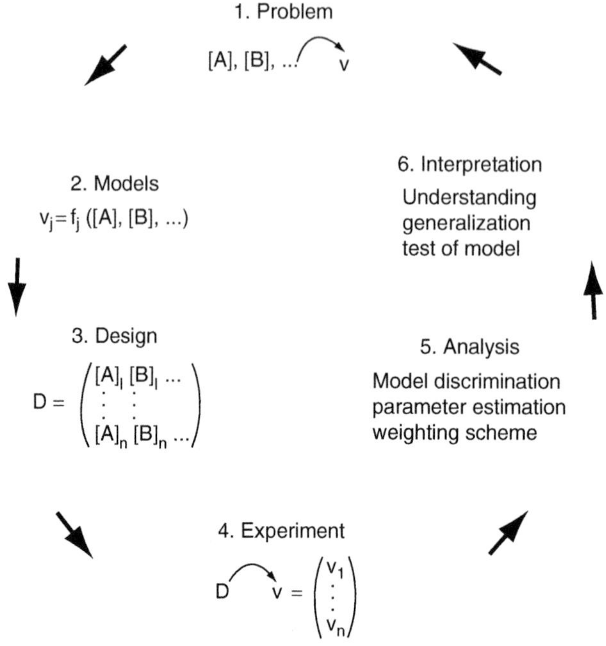

Fig. 5 Sequential strategy of design and analysis in mathematical modeling of experimental data. (From Mannervik (1981) with permission of the publishers.)

3. A suitable *design of experiments*, optimal for the immediate goal of the study, should be made. Goals may be parameter estimation or model discrimination. Optimal design requires prior knowledge about the error structure of the experimental data.

4. *Experimental data* should be collected as outlined in the design. The record of the experiment should show the order in which the data were acquired, possible differences in stock solutions used for subsets of data, and any other information that may reveal systematic errors.

5. The *analysis of data* should consider goodness of fit, discrimination between rival models, parameter estimation, and the influence of experimental error.

6. Finally, an *interpretation of the results* of the investigation should be attempted. An understanding of the system investigated is desirable, and possible generalizations of the conclusions should be considered. An important test of the model is the prediction of the outcome of new experiments.

Completion of the cycle in Fig. 5 usually does not constitute the end of the investigation. If the primary problem is the discrimination between rival models, a new cycle of steps emphasizing parameter estimation may be required. In fact, the proper execution of the steps in Fig. 5 could not be performed without some prior information about possible mathematical models, parameter values, and the experimental error. Preliminary information on all these issues could be obtained if replicate measurements were made in $(p + 1)$ experimental points. For example, if the Michaelis-Menten equation were considered $(p = 2)$, replicates (≥ 5) could be obtained at low, intermediate, and high substrate concentrations. If measurements at the three levels appear to belong to a rectangular hyperbola, the model seems appropriate; otherwise, an alternative model must be considered. The data also permit calculations of preliminary values of the parameters, and from the replicate measurements the local variance of the data can be estimated in the points of experimentation. On the basis of these variance estimates, the design of further experiments can be made and a preliminary value of α (cf. Mannervik *et al.*, 1979) be calculated and used in the weighting function [Eq. (13)]. Consequently, a typical investigation is characterized by a *sequential design* in which the cycle of steps in Fig. 5 may be repeated with successive shifts of emphasis from model to model, and from model identification to parameter estimation. In this manner, the analysis proceeds by consecutive steps of refinements until a satisfactory result has been obtained.

Acknowledgments

The work from the author's laboratory described in this chapter was supported by the Swedish Natural Science Research Council and the Swedish Cancer Society.

References

Askelöf, P., Korsfeldt, M., and Mannervik, B. (1976). *Eur. J. Biochem.* **69,** 61.

Bartfai, T., and Mannervik, B. (1972a). *FEBS Lett.* **26,** 252.

Bartfai, T., and Mannervik, B. (1972b). *In* "Analysis and Simulation of Biochemical Systems" (H. C. Hemker and B. Hess, eds.), p. 197. North-Holland/Elsevier, Amsterdam.

Bartfai, T., Ekwall, K., and Mannervik, B. (1973). *Biochemistry* **12,** 387.

Briggs, G. E., and Haldane, J. B. S. (1925). A note on the kinetics of enzyme action. *Biochem. J.* **19,** 338–339.

Cleland, W. W. (1963). *Nature (London)* **198,** 463.

Cleland, W. W. (2009). Statistical analysis of enzyme kinetic data. *Methods Enzymol.*, **63,** Article [6].

Draper, N. R., and Smith, H. (1966). "Applied Regression Analysis," p. 263. Wiley, New York.

Duggleby, R. G. (1979). *J. Theor. Biol.* **81,** 671.

Duggleby, R. G. (1981). *In* "Kinetic Data Analysis. Design and Analysis of Enzyme and Pharmacokinetic Experiments." (L. Endrenyi, ed.), p. 169. Plenum, New York.

Endrenyi, L. (ed.) (1981). "Kinetic Data Analysis. Design and Analysis of Enzyme and Pharmacokinetic Experiments". Plenum, New York.

Endrenyi, L. (1981b). *In* "Kinetic Data Analysis. Design and Analysis of Enzyme and Pharmacokinetic Experiments" (L. Endrenyi, ed.), p. 137. Plenum, New York.

Garfinkel, L., Kohn, M. C., and Garfinkel, D. (1977). *CRC Crit. Rev. Bioeng.* **2,** 329.

Jennrich, R. I., and Ralston, M. L. (1979). *Annu. Rev. Biophys. Bioeng.* **8,** 195.

Johansen, G., and Lumry, R. (1961). *C. R. Trav. Lab. Carlsberg.* **32,** 185.

Mannervik, B. (1975). *BioSystems* **7,** 101.

Mannervik, B. (1981). *In* "Kinetic Data Analysis. Design and Analysis of Enzyme and Pharmacokinetic Experiments" (L. Endrenyi, ed.), p. **235**. Plenum, New York.
Mannervik, B., and Askelöf, P. (1975). *FEBS Lett.* **56,** 218.
Mannervik, B., and Bartfai, T. (1973). *Acta Biol. Med. Ger.* **31,** 203.
Mannervik, B., Górna-Hall, B., and Bartfai, T. (1973). *Eur. J. Biochem.* **37,** 270.
Mannervik, B., Jakobson, I., and Warholm, M. (1979). *Biochim. Biophys. Acta* **567,** 43.
Mannervik, B., Jakobson, I., and Warholm, M. (1982). *Fed. Proc. Fed. Am. Soc. Exp. Biol.* **41,** 524 (Abstr. 1483).
Ottaway, J. H. (1973). Normalization in the fitting of data by iterative methods. Application to tracer kinetics and enzyme kinetics. *Biochem. J.* **134,** 729–736.
Siano, D. B., Zyskind, J. W., and Fromm, H. J. (1975). *Arch. Biochem. Biophys.* **170,** 587.
Solano-Muñoz, F., McGinlay, P. B., Woolfson, R., and Bardsley, W. G. (1981). Deviations from Michaelis-Menten kinetics. Computation of the probabilities of obtaining complex curves from simple kinetic schemes. *Biochem. J.* **193,** 339–352.
Swed, F. S., and Eisenhart, C. (1943). *Ann. Math. Stat.* **14,** 66.
Wilkinson, G. N. (1961). Statistical estimations in enzyme kinetics. *Biochem. J.* **80,** 324–332.

CHAPTER 5

Analysis of Enzyme Progress Curves by Nonlinear Regression

Ronald G. Duggleby

Department of Biochemistry
Centre for Structure, Function, and Engineering
University of Queensland
Brisbane, Queensland 4072, Australia

CONTEMPORARY ENZYME KINETICS AND MECHANISM
Reprinted from *Methods in Enzymology*, Volume 249 (Academic Press, 1979).

DOI: 10.1016/B978-0-12-378608-1.00005-0

I. Introduction

The usual way in which an enzyme-catalyzed reaction is monitored is by measuring the *amount* of reactant remaining or product formed at one or more times. By contrast, most kinetic models are formulated in terms of *rates* of reaction. There is, therefore, a fundamental incompatibility between the data and the underlying kinetic model. This incompatibility can be resolved in two ways. Either the model can be integrated to give a description of the time course of the reaction or the data can be differentiated to determine rates.

Traditionally, enzyme kinetic studies have employed the second approach and have focused on determining rates, especially initial rates, by measuring tangents to the reaction progress curves. There are several reasons for this preference for initial rates. (1) The substrate concentration is exactly equal to that which is added. (2) There is a vanishingly small concentration of potentially inhibitory reaction products present, unless these are added deliberately. (3) There is no opportunity for the enzyme to have undergone partial inactivation.

There are, however, several advantages in studying progress curves, and, ironically, the first two of these are restatements of what were mentioned above as reasons for preferring initial rates. (1) The substrate concentration is automatically varied as the reaction proceeds, providing information about the dependence on substrate concentration of the enzyme. (2) Products of the correct stereochemistry and of complete purity are formed automatically as the reaction proceeds, providing information about the dependence on product concentration of the enzyme. (3) As a consequence of the above, considerably more information is obtained from each assay so a complete description of the kinetic properties of the enzyme can be obtained from fewer experiments.

Given these advantages, one may inquire why all enzyme kinetic studies do not exploit progress curves. Again there are several reasons. (1) Frequently, we may be interested only in a limited subset of the kinetic properties, the Michaelis constant for a particular substrate, for example. Progress curves contain information about the complete rate equation, and there is often no simple way to dissect out a subset of this information unless the enzyme has fortuitous properties, such as a lack of significant inhibition by the products. (2) The data analysis is considerably more complex than that for rate measurements, and it is only over the last few years that reliable methods have been developed that cover most of the common types of reactions. (3) The shape of the progress curve may not depend solely on the variations in reactant concentrations that are due to the catalyzed reaction. The enzyme may be subject to progressive inactivation, and substrates and products may undergo uncatalyzed side reactions. There is a reasonable prospect that it will be possible to handle such complications as methods of data analysis are developed further.

This chapter focuses mainly on the second point immediately above, the methods of data analysis. Because these are still being developed, the presentation is largely chronological and concentrates on the methods and experimental systems that have been developed and used in our laboratory. This personal perspective is in no way intended to diminish the contributions that others have made to the area of progress curve analysis. Later I shall indicate the type of developments that can deal with the complications mentioned in the third point. For a discussion of much of the earlier literature, see a previous chapter in this series by Orsi and Tipton (1979).

It is worth emphasizing at this point that analyses based on progress curves are not preferable to those based on rates in all circumstances. Indeed, there are many instances where rate measurements are the method of choice. Most commonly this will be where one is interested solely in characterizing the kinetics toward the substrate(s) for an enzyme that either (a) has cosubstrates whose kinetics are not of immediate interest, (b) catalyzes a readily reversible reaction, or (c) shows significant inhibition by its product(s). Applying progress curve analysis leads to unnecessary complications as an accurate description of the data requires that the integrated form of the full rate equation is used. For example, a simple ordered reversible reaction with two substrates and products has an integrated rate equation involving eight independent kinetic constants (maximum velocities for the forward and reverse reactions, Michaelis constants for all four reactants, and inhibition constants for the first substrate bound in each direction). The experiments needed to characterize the system completely will be excessive if all one is interested in is the Michaelis constant for one of the substrates.

II. Simple Michaelian Enzyme

A. Fitting Equations

In any situation where an equation is to be fitted to some experimental data there must be some sense in which the data lie close to the line or surface predicted by the equation. Fitting the equation will involve manipulating certain constants whose values are not fixed by the nature of the experiment; these constants are referred to as the parameters of the equation. If the data are perfect and the equation represents a theoretically correct description of the experimental system, it should be possible to find values of the parameters for which the line or surface agrees exactly with the data. In practice this does not occur because the data are inexact.

It is common practice to regard one of the coordinates of each datum as a measured or dependent variably (y) and the others as controlling or independent variables (x_1, x_2, x_3, etc.). Under this convention, each of the independent variables is regarded as being known exactly, while any experimental uncertainties are associated with the dependent variable. Thus, to fit an equation to the data, the degree of closeness to the line or surface is assessed from the distances along the y coordinate only.

Unless one is simply fitting by eye, there has to be some objective measure of closeness, and this is usually taken as the sum of the squares of these distances. The justification for this is that if the experimental errors follow a Gaussian distribution, minimizing the sum of squares yields the most likely values of the fitted parameters. In practice, however, the distribution of errors is rarely determined, and the independent variables may also contain some uncertainties. The widespread use of the least squares criterion is more a reflection of its mathematical convenience than a firm belief in its assumptions.

B. Integrated Michaelis-Menten Equation

The basic principles, and the difficulties, of progress curve analysis can be illustrated by considering the simplest situation, namely, that of an enzyme catalyzing an irreversible reaction in which one substrate is converted to a noninhibitory product.

The rate equation for the reaction is given as Eq. (1), where V_{m} is the maximum velocity, K_{a} is the Michaelis constant, and $[\mathrm{A}]_t$ is the substrate concentration at any time:

$$\frac{\mathrm{d}[\mathrm{A}]_t}{\mathrm{d}t} = \frac{-V_{\mathrm{m}}[\mathrm{A}]_t}{K_{\mathrm{a}} + [\mathrm{A}]_t} \tag{1}$$

It is useful to rewrite Eq. (1) in terms of the initial substrate concentration ($[\mathrm{A}]_0$) and a variable y, the amount of product formed by reaction:

$$[\mathrm{A}]_0 = [\mathrm{A}]_t + y \tag{2}$$

Combining Eqs. (1) and (2) yields Eq. (3), which then may be integrated to give Eq. (4):

$$\frac{\mathrm{d}y}{\mathrm{d}t} = \frac{V_{\mathrm{m}}([\mathrm{A}]_0 - y)}{K_{\mathrm{a}} + [\mathrm{A}]_0 - y} \tag{3}$$

$$V_{\mathrm{m}}t = y - K_{\mathrm{a}} \ln\left(1 - \frac{y}{[\mathrm{A}]_0}\right) \tag{4}$$

In Eq. (4), the quantity that is normally measured experimentally is y itself, or some linear function of y. However, Eq. (4) is an implicit function of y, meaning that it cannot be rearranged to express y as an explicit function of the other variables. As a result, direct calculation of y is impossible, and fitting Eq. (4) to obtain values for V_{m} and K_{a} from an enzyme progress curve is not straightforward. There are, of course, simple rearrangements of Eq. (4) that put it into a form which V_{m} and K_{a} can be obtained, such as Eq. (5):

$$\frac{y}{t} = V_{\mathrm{m}} + K_{\mathrm{a}}\left(\frac{\ln(1 - y/[\mathrm{A}]_0)}{t}\right) \tag{5}$$

A plot of y/t versus $\ln(1 - y/[\mathrm{A}]_0)/t$ should be a straight line from which V_{m} and K_{a} are easily obtained by linear regression. There are two objections to this approach. The first is a statistical one; y is the measured variable that is considered to contain any experimental error. However, y appears on both sides of Eq. (5), and estimates of V_{m} and K_{a} obtained by least squares fitting will be statistically biased. The second objection is more fundamental, although not evident for this simple single-substrate model. As we move to more complex enzymatic reactions that may involve multiple substrates or inhibitory products, or where the back reaction must be taken into account, no linear transformation may be possible.

To overcome these difficulties it is necessary to fit Eq. (4) to an enzyme progress curve by nonlinear regression. This will involve minimizing the sum of the squares of the difference between the measured value of y and that expected from Eq. (4); thus, it is necessary to solve Eq. (4) for y, given particular values for V_m, K_a, $[A]_0$, and t.

C. Solving Integrated Rate Equations

A practical method for solving integrated rate equations was first mentioned by Nimmo and Atkins (1974) and described more fully by Duggleby and Morrison (1977). This method, the Newton-Raphson procedure, is a standard mathematical algorithm for solving implicit equations and is described briefly.

Equation (4) may be rearranged to give Eq. (6), in which the problem now becomes one of finding a value of y for which $F(y) = 0$

$$F(y) = y - K_a \ln\left(1 - \frac{y}{[A]_0}\right) - V_m t \tag{6}$$

Equation (6) describes a curve with the general shape shown in Fig. 1A. That the curve will have this shape is readily seen when it is understood that Eq. (6) describes a progress curve that has the axes interchanged, followed by a downward displacement of the curve by an amount $V_m t$.

The solution (i.e., the point where the curve crosses the abscissa) may be found using the differential of $F(y)$ with respect to y (Eq. (7)) and an initial estimate of

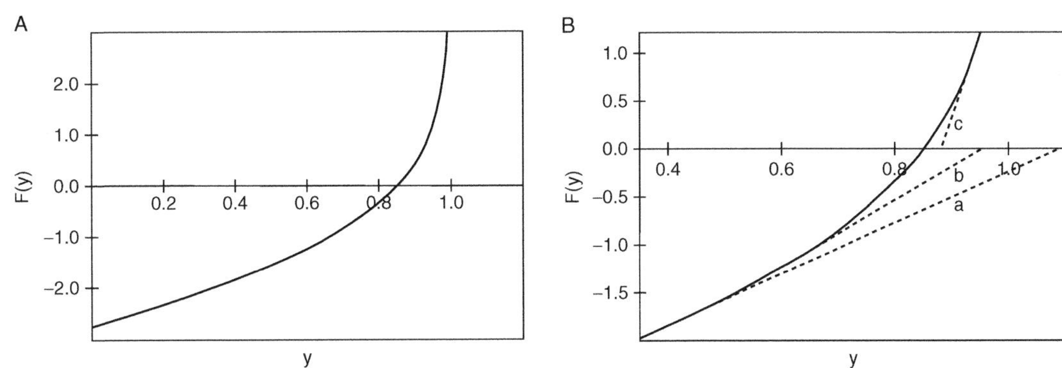

Fig. 1 Solving the integrated Michaelis-Menten equation by the Newton-Raphson method. Given a value of $K_a = [A]_0$, the time to reach 85% of $[A]_0$, was calculated from Eq. (4). Using this value ($t = 2.7471/V_m$), $F(y)$ was calculated using Eq. (6) for a range of values of y. (A) Graph over the entire range of y and (B) enlargement of the central region. Also shown in (B) are the results of the first cycle in the Newton-Raphson solution for initial estimates of 0.4 (line a), 0.6 (line b), and 0.94 (line c).

$y(y_i)$, as illustrated in Fig. 1B. The dashed lines are tangents to the curve obtained with different initial estimates. Each allows a "refined" estimate (y_{i+1}) to be calculated from the point where the line has $F(y) = 0$. This point may be obtained using Eq. (8):

$$F'(y) = \frac{\mathrm{d}F(y)}{\mathrm{d}t} = \frac{1}{1 + K_\mathrm{a}/([\mathrm{A}]_0 - y)} \tag{7}$$

$$y_{i+1} = y_i - \frac{F(y^i)}{F'(y_i)} \tag{8}$$

Several repetitions of this procedure using the successive refined estimates in place of the initial estimate will, in most cases where a reasonable starting value has been chosen, yield the required solution to any desired degree of numerical accuracy.

The problem with this procedure is that there is no guarantee that the refined estimate is closer to the solution than initial estimate, nor any automatic constraint to prevent it from exceeding $[\mathrm{A}]_0$ (Fig. 1B, curve a). This is a computational disaster as the logarithmic term in Eq. (6) becomes undefined. Although this can never occur if the initial estimate is greater than the solution (Duggleby, 1986a) (Fig. 1B, curve c), there is no simple method of choosing an initial estimate in the appropriate range if the solution is unknown.

Duggleby (1986a) proposed an alternative definition of $F(y)$ that contains an exponential term rather than algorithmic term (Eq. (9)) and a corresponding definition (Eq. (10)) of $F'(y)$:

$$F(y) = \frac{y}{[\mathrm{A}]_0} - 1 + \exp\left[\frac{y - V_\mathrm{m}t}{K_\mathrm{a}}\right] \tag{9}$$

$$F'(y) = \frac{1}{[\mathrm{A}]_0} + \exp\left[\frac{(y - V_\mathrm{m}t)/K_\mathrm{a}}{K_\mathrm{a}}\right] \tag{10}$$

As shown in Table I, this variation provides an almost completely fool-proof method for solving Eq. (4), and it is usually faster than the method using Eqs. (6)–(8). Moreover, it can be extended in a simple manner to more complex integrated rate equations. However, it is not completely general and cannot be applied to solving the integrated rate equations that describe many reactions with two substrates.

A method that is virtually immune to failure was suggested by Boeker (1987) and generalized further to more complex reactions by Duggleby and Wood (1989). The value of y that solves any integrated rate equation clearly lies between certain limits; it cannot be less than 0 nor greater than the equilibrium value (y_∞), which is easily calculated. If $F(y)$ is calculated at a value of y midway between these limits,

Table I
Newton-Raphson Solution of Integrated Michaelis-Menten Equation[a]

	Eqs. (6) –(8)			Eqs. (8) –(10)		
y_i	y_{i+1}	$F(y)$	Cycles	y_{i+1}	$F(y)$	Cycles
0.00	1.3736	−2.7471	Fail	0.8795	−0.9359	3
0.05	1.3390	−2.6458	Fail	0.8769	−0.8826	3
0.10	1.3040	−2.5418	Fail	0.8743	−0.8291	3
0.15	1.2686	−2.4346	Fail	0.8718	−0.7755	3
0.20	1.2329	−2.3240	Fail	0.8693	−0.7217	3
0.25	1.1970	−2.2094	Fail	0.8669	−0.6677	3
0.30	1.1608	−2.0904	Fail	0.8646	−0.6135	3
0.35	1.1246	−1.9663	Fail	0.8624	−0.5590	3
0.40	1.0886	−1.8363	Fail	0.8603	−0.5044	3
0.45	1.0530	−1.6993	Fail	0.8584	−0.4495	3
0.50	1.0180	−1.5540	Fail	0.8566	−0.3943	3
0.55	0.9841	−1.3986	6	0.8550	−0.3389	3
0.60	0.9517	−1.2308	6	0.8536	−0.2832	3
0.65	0.9215	−1.0473	5	0.8523	−0.2272	3
0.70	0.8946	−0.8431	5	0.8514	−0.1709	3
0.75	0.8722	−0.6108	5	0.8506	−0.1143	2
0.80	0.8563	−0.3377	5	0.8502	−0.0573	2
0.85	0.8500	0.0000	1	0.8500	0.0000	1
0.90	0.8586	0.4555	4	0.8502	0.0577	2
0.95	0.8929	1.1986	5	0.8507	0.1158	2
0.99	0.9618	2.8480	6	0.8514	0.1625	3
0.999	0.9938	5.1596	7	0.8516	0.1731	3
0.9999	0.9992	7.4630	8	0.8516	0.1742	3

Given a value of $K_a = [A]_0$, the time to reach 85% of $[A]_0$ was calculated from Eq. (4) (the integrated Michaelis-Menten equation). Using this value ($t = 2.7471/V_m$) and initial estimates (y) shown in the first column, values of y_{i+1} and $F(y)$ were calculated using Eqs. (6)–(8). This process was repeated until y_{i+1} was equal to the correct solution (i.e., 85% of $[A]_0$), and the number of iterations required was counted. A similar procedure was applied using Eqs. (8)–(10).

the sign will indicate whether the estimate is above or below the solution. If it is below ($F(y)$ is negative), this value of y is taken as a new lower limit and the process repeated until $F(y)$ is positive. The value is then refined using the standard Newton-Raphson procedure. The computer code necessary to implement this method in BASIC is shown in Fig. 2 for case $y_\infty = [A]_0$.

D. Nonlinear Regression

Fitting an integrated rate equation to a progress curve involves manipulating the parameters so as to minimize the sum of the squares of the difference between the measured value of y and that expected from the equation. In Eq. (4), the

```
5100 Y = 0: GOSUB 5180: IF F = 0 THEN GOTO 5190
5110 Y = (Y + A0) / 2: IF Y > .99999 * A0 THEN GOTO 5190
5120 GOSUB 5180: IF F = 0 THEN GOTO 5190
5130 IF F < 0 THEN GOTO 5110
5140 FI = 1 + KA / (A0 - Y)
5150 DY = F / F1: IF DY / A0 < .00001 THEN GOTO 5190
5160 Y = Y - DY: GOSUB 5180: IF F = 0 THEN GOTO 5190
5170 GOTO 5140
5180 F = Y - KA * LOG(1 - Y / A0) - VM * T: RETURN
5190 'end
```

Fig. 2 BASIC code to solve the integrated Michaelis-Menten equation by the modified Newton-Raphson method. It is assumed that on entry to this routine, values of V_m, K_a, $[A]_0$, and time have been defined. First (line 5100) the solution $y = 0$ is tested by calculating $F(y)$ in line 5180. If this is not a solution then successive values of $[A]_0/2$, $3[A]_0/4$, $7[A]_0/8$, etc., are tried (lines 5110-5130) until the solution is found or $F(y)$ is positive. In the latter case, the standard Newton-Raphson method is then used to refine the solution; $F'(y)$ is calculated (line 5140) then the correction $F(y)/F'(y)$ is compared to $[A]_0$ to ensure that the correction is not insignificant. The correction is made (line 5160) and the new value used in the next cycle unless it is already an exact solution.

parameters that may be varied are V_m and K_a, although $[A]_0$ may be included as well if there is some uncertainty in its true value. Often it may be judicious to apply weighting factors to the true value. Often it may be judicious to apply weighting factors to the individual squared differences if there is reason to believe that all measurements are not equally accurate.

Nonlinear regression methods fall into two classes: search methods fall into two classes: search and gradient methods (Swann, 1969). In search methods, the parameters are varied systematically and at each set of values the sum of squares is calculated. This process continues until the minimum (or at least a local minimum) in the sum of squares surface is located. Gradient methods use information about the slope of the sum of squares surface at a particular point to predict the approximate location of the minimum. The gradient is then recalculated at this new location and used to improve the prediction of the location of the minimum.

Generally, gradient methods are faster than search methods. Moreover, they include information that will yield estimates of the standard errors of the fitted parameters. The principal disadvantage is that the gradients represent additional information that must be supplied. For most enzyme kinetic studies, including progress curve analysis, gradient methods have been used almost exclusively.

Perhaps the most well-known gradient procedure is the Gauss-Newton method that was first introduced to enzyme kineticists by Wilkinson (1961) and the reader wishing to understand nonlinear regression will find that chapter well worthwhile studying. A brief outline of this method is presented in Appendix A at the end of this chapter. The technique was applied to progress curve analysis by Nimmo and Atkins (1974), and much of this chapter is simply an extension of their work.

The gradient information is in the form of the differential of the fitted variable with respect to each of the fitted parameters. In the case of the integrated

Michaelis-Menten equation, these partial differentials with respect to V_m and K_a are given by Eqs. (11) and (12), respectively, where $F'(y)$ is defined by Eq. (7):

$$\frac{\delta y}{\delta V_m} = \frac{t}{F'(y)} \tag{11}$$

$$\frac{\delta y}{\delta K_a} = \frac{\ln(1 - y/[A]_0)}{F'(y)} \tag{12}$$

Sometimes one may wish to treat $[A]_0$ as a third parameter to be estimated from the data, and to do so requires the partial differential with respect to $[A]_0$, which is given as Eq. (13):

$$\frac{\delta y}{\delta [A]_0} = \frac{K_a y}{([A]_0 - y)F'(y)} \tag{13}$$

Although these partial differential equations are not hard to obtain for the integrated Michaelis-Menten equation, with more complex mechanisms their derivation becomes quite cumbersome. However, they can be approximated by numerical differentiation with little loss of accuracy in the final results. This process of numerical differentiation involves calculating y, changing the value of a parameter (say, K_a) by a small fraction of its value, then recalculating y. The differential $\delta y/\delta K_a$ is then approximated by $\Delta y/\Delta k_a$. It is more accurate if the parameter is perturbed both up and down by equal amounts and the partial derivative calculated from the total changes in y and K_a.

The effectiveness of numerical differentiation is illustrated by an analysis of data on the hydrolysis of phenyl phosphate by human prostate acid phosphatase (taken from Schønheyder, 1952). Fitting the data using the algebraic expressions for the partial derivatives gave $V_m = 0.4562 \pm 0.0139$ mM/min and $K_a = 6.146 \pm 0.235$ mM. Using numerical differentiation, identical values and very similar standard errors were obtained: $V_m = 0.4562 \pm 0.0140$ mM/min and $K_a = 6.146 \pm 0.236$ mM. The fitted line gives an excellent description of the data (Fig. 3).

The results just described were obtained using DNRP53, a general nonlinear regression computer program written in our laboratory (Duggleby, 1984). Most of the results to be described here are based on analyses with this program, copies of which may be obtained by writing to the author.

As noted earlier, it is sometimes desirable to treat $[A]_0$ as a another fitted parameter to be estimated from the data rather than as a constant. The reason for this is that the fitted values of V_m and K_a are sensitive to quite small inaccuracies in the value of $[A]_0$, particularly when $[A]_0/K_a$ is not high (Newman *et al.*, 1974). This point is illustrated in Table II in which the data of Schønheyder (1952) are further analyzed. Even small changes of $[A]_0$ lead to very large errors in the values of V_m and K_a. These uncertainties can be eliminated by including $[A]_0$ as a third fitted parameter; although the results obtained are not necessarily exactly correct, they will not be biased by an incorrect value of $[A]_0$.

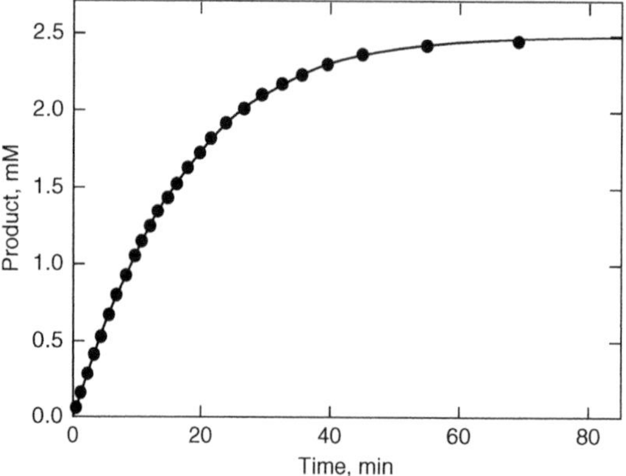

Fig. 3 Fit of the integrated Michaelis-Menten equation to a progress curve. The data represent the hydrolysis of 2.494 mM phenyl phosphate catalyzed by human prostate acid phosphatase, and were taken from Schønheyder (1952). The line is the fit by Eq. (4), with $V_m = 0.4562$ mM/min and $K_a = 6.146$ mM.

Table II
Effect of Variations in $[A]_0$ on the Fitted Values of V_m and K_a

Variation (%)	Concentration (mM)	V_m (mM/min)	K_a (mM)
−2.0	2.4441	0.3410 ± 0.0172	4.024 ± 0.278
−1.0	2.4691	0.3894 ± 0.0132	4.912 ± 0.218
−0.5	2.4815	0.4199 ± 0.0121	5.475 ± 0.202
Fitted	2.4891 ± 0.0047	0.4412 ± 0.0191	5.868 ± 0.338
Nominal	2.4940	0.4562 ± 0.0139	6.146 ± 0.235
+0.5	2.5065	0.5001 ± 0.0207	6.958 ± 0.352
+1.0%	2.5189	0.5541 ± 0.0332	7.960 ± 0.571
+2.0%	2.5439	0.7102 ± 0.0851	10.864 ± 1.493

The expected value of $[A]_0$, taken from Schønheyder (1952) was 2.4940 mM. The data were fitted using Eq. (4) (the integrated Michaelis-Menten equation) using this nominal concentration and values that differed by the amounts shown in the first column. The data were also fitted using Eq. (4) treating $[A]_0$ as a third fitted parameter.

III. Effects of Inhibitors

A. Dead-End Inhibition

A compound that is neither a substrate nor a product may combine with either the free enzyme or the enzyme-substrate complex, diverting a fraction of the enzyme into an inactive form. The effect of such dead-end inhibitors will be reduce V_m/K_a or V_m by a constant factor of $1 + [I]/K_i$, leading to competitive and uncompetitive inhibition, respectively. These effects can be analyzed using progress curve analysis to determine V_m and K_a in the presence of one or more concentrations of the inhibitor. However, it is the view of this author that these analyses do not properly fall within the ambit of progress curve analysis insofar as the inhibitor is not a compound whose concentration varies as the reaction proceeds. In much the same way, interactions of the enzyme with solvent ions and molecules (including H^+) may well affect the kinetics but will not be considered further in this chapter. The reader is referred to the review of Orsi and Tipton (1979) for a discussion of the effects of dead-end inhibitors on progress curves.

B. Product Inhibition

So far it has been assumed that the accumulating product is noninhibitory, which in many cases is an unsafe assumption. The most common situation is where the product is a competitive inhibitor with respect to the single substrate. In this case, the integrated rate equation is identical to Eq. (4), except that V_m and K_a are now apparent values (V'_m and K'_a) that depend on a product inhibition constant (K_{ip}), $[A]_0$, and the concentration ($[P]_0$) of any product that is initially present:

$$V'_m = \frac{V_m K_{ip}}{K_{ip} - K_a} \tag{14}$$

$$K'_a = \frac{K_a(K_{ip} + [A]_0 + [P]_0)}{K_{ip} - K_a} \tag{15}$$

Equations (14) and (15) lead to a number of interesting points. First, because any given progress curve obtained with a particular combination of initial substrate and product concentrations can be defined by two parameters (V'_m and K'_a), it is clearly impossible to estimate all three of V_m, K_a, and K_{ip} from such a curve. Second, V'_m and K'_a must have the same sign, but both can be negative if the product is a strong inhibitor ($K_{ip} < K_a$). Third, K'_a is a linear function of the sum $[A]_0$ and $[P]_0$, whereas V'_m is independent of these concentrations. This then leads to the fourth point; it should be possible to estimate V_m, K_a, and K_{ip} using several progress curves, provided that they are obtained over a range of $[A]_0$ (or, less usually, of $[P]_0$).

The last point was illustrated by Duggleby and Morrison (1977) in an analysis of five progress curves for the reaction catalyzed by prephenate dehydratase from which were obtained values for V_m (18.0 ± 0.8 U/mg), K_a (0.472 ± 0.008 mM), and K_{ip} (4.59 ± 0.62 mM). This reaction has a single product (apart from water), but reactions with two or more products can be analyzed in a similar manner when only one is inhibitory. Such is the case for the hydrolysis of *p*-nitrophenyl phosphate by potato acid phosphatase (Duggleby and Morrison, 1977) which is not inhibited by *p*-nitrophenol (or by the *p*-nitrophenolate ion) but is strongly affected by phosphate ion. From eight progress curves, values for V_m (117 ± 8 U/mg), K_a (2.12 ± 0.20 mM), and K_{ip} (0.262 ± 0.01 mM) were obtained.

Note that because K_{ip} (0.262 mM) is less than K_a (2.12 mM), it would be predicted that both V'_m and K'_a from the analysis of a single curve should be negative. This prediction is confirmed by a reanalysis of the data obtained at

0.988 mM substrate, where the best fit progress curve (Fig. 4A) has values of $V'_m = -15.58 \pm 0.31$ U/mg and $K'_a = -1.443 \pm 0.014$ mM. The linear dependence of K'_a on ($[A]_0 + [P]_0$) that is predicted from Eq. (15) is illustrated in Fig. 4B.

In the case of potato acid phosphatase, the inhibitory product is competitive with the substrate. However, if the first product to leave was the stronger inhibitor, or if the enzyme has an iso-mechanism (Rebholz and Northrop, this volume), more complex inhibition is observed. For these cases, the integrated rate equation is a function of two inhibition constants (K_{is} and K_{ii}). In addition to the linear and logarithmic terms in y seen in Eq. (4), there is now a term in y^2 as well (Eq. (16)):

$$\rho t = y + \frac{\gamma y^2}{2} - \delta \ln\left(1 - \frac{y}{[A]_0}\right) \tag{16}$$

$$\rho = \frac{V_m K_{is} K_{ii}}{K_{is} K_{ii} - K_a K_{ii} + [P]_0 K_{is}} \tag{17}$$

$$\gamma = \frac{K_{is}}{K_{is} K_{ii} - K_a K_{ii} + [P]_0 K_{is}} \tag{18}$$

$$\delta = \frac{K_a K_{ii}(K_{is} + [A]_0 + [P]_0)}{K_{is} K_{ii} - K_a K_{ii} + [P]_0 K_{is}} \tag{19}$$

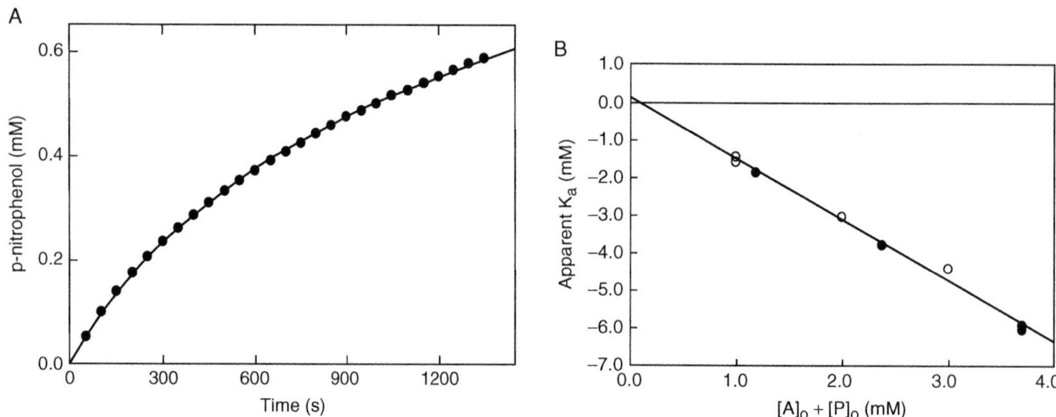

Fig. 4 Progress curve analysis of the reaction catalyzed by potato acid phosphatase. (A) Hydrolysis of 0.988 mM *p*-nitrophenyl phosphate and the fit of Eq. (4) to the data with the apparent values $V'_m = -15.58$U/mg and $K'_a = -1.443$mM. (B) Linear dependence of the apparent values on the total initial concentration of substrate plus product (phosphate), where the filled symbols represent reactions that contained added phosphate.

C. Reactions with Two Inhibitory Products

The next level of complexity beyond those exemplified by prephenate dehydratase and acid phosphatase is where there is one substrate but two inhibitory products. Despite this added complexity, the integrated rate equation has the same form as Eq. (16) although the definitions of ρ, γ, and δ are altered (Duggleby and Morrison, 1979).

Cox and Boeker (1987) studied the enzyme-catalyzed decarboxylation of arginine to agmatine as an example of such a system. Their initial analysis was based on Eq. (20) which is identical in form to Eq. (16):

$$[\mathrm{E}]_0 t = C_1 y + \frac{C_2 y^2}{2} - C_\mathrm{f} \ln\left(1 - \frac{y}{[\mathrm{A}]_0}\right) \qquad (20)$$

By separately fitting each progress curve to Eq. (20), they concluded that the C_2 term did not contribute significantly to the shape of the progress curve. By analyzing the relationship between C_1, C_f, and the initial concentrations of arginine and agmatine, they were able to determine several of the kinetic constants. Although the approach seems to have been successful, there is some doubt as to whether such a procedure would be useful in general. This is because fits to individual curves can give parameter values that vary wildly as a result of random fluctuations in the data.

A simple example illustrates this point. An exact progress curve was simulated for a single-substrate, single-product, irreversible reaction showing uncompetitive production inhibition, using $V_\mathrm{m} = 1$, $K_\mathrm{a} = 1$, $K_\mathrm{ii} = 0.3$, and $[\mathrm{A}]_0 = 3$. Nineteen points, equally spaced on the concentration axis, were selected from the curve, and these were perturbed by a random error with a standard deviation of 1% $[\mathrm{A}]_0$. The appropriate progress curve equation was then fitted to the simulated curve; the entire process was repeated 100 times. Despite the fact that the simulated noise was quite small, 15 of the curves could not be fitted at all. For the remaining 85, V_m ranged from 0.68 to 42, K_a from −0.16 to 206.2, and K_ii from 0.12 to 0.42. In only 19 cases were all three parameters within 50% of the correct values, and in only four cases were they within 10%.

Thus, individual fits are not expected to give very reliable estimates, and we would recommend combining the entire set of progress curves and analyzing them using the complete integrated rate equation for the system. An example of a reaction that behaves kinetically as a one-substrate and two-product reaction and that was analyzed by overall fitting of the complete set of progress curves is described in the next section.

IV. Complex Reactions

A. Reactions with Two Substrates

The oxidation of NADH catalyzed by lactate dehydrogenase has been studied (Duggleby and Morrison, 1977) as an example of a reaction with two substrates. To simplify the analysis, the enzyme can be made to behave as a single-substrate system by raising the concentration of pyruvate to a constant value that is well in excess of that of NADH. The integrated rate equation is identical to Eq. (16) but with new definitions of ρ, γ, and δ (Eqs. (21)–(25)) where A, B, P, and Q refer to NADH, pyruvate, lactate, and NAD^+, respectively:

$$\rho = \frac{V_m}{\Omega} \tag{21}$$

$$\gamma = \frac{1/K_{ip} - K_a/K_p^* K_{iq}}{\Omega} \tag{22}$$

$$\delta = \frac{K_a\left\{1 + ([A]_0 + [Q]_0)/K_{iq} + ([A]_0 + [P]_0)([A]_0 + [Q]_0)/K_p^* K_{iq}\right\}}{\Omega} \tag{23}$$

$$K_p^* = \frac{K_a K_{ib} K_p}{K_{ia} K_b} \tag{24}$$

$$\Omega = 1 + \frac{[P]_0}{K_{ip}} - \left(\frac{K_a}{K_{iq}}\right)\frac{1 + ([A]_0 + [P]_0 + [Q]_0)}{K_p^* K_{iq}} \tag{25}$$

As it later turned out, the kinetic constants for lactate (K_p^* and K_{ip}) are quite high relative to those for NAD^+ and NADH, which are in the micromolar range. Consequently, progress curves obtained in the absence of lactate will not accumulate enough of this product for it to have a significant influence. In effect, the reaction behaves exactly like acid phosphatase with one inhibitory product. If some reactions are conducted in the presence of added lactate, however, then the inhibition will be expressed; using this approach, all five of the kinetic constants for this reaction could be estimated.

For two-substrate reactions, using one substrate at a much higher concentration than the other provides a general means of simplifying the analysis, by eliminating some terms from the integrated rate equation. However, the choice of the substrate that is used in excess is constrained by the kinetic properties of the enzyme and the assay system. In the case of lactate dehydrogenase, pyruvate was used at a constant concentration of 4 mM while NADH never exceeded 0.15 mM. However, if one wanted to use NADH as the excess substrate some difficulties would arise. Because the Michaelis constant for pyruvate is approximately 0.2 mM under the reaction conditions employed, it would be necessary to run at least some of the progress curves with an initial pyruvate concentration well above this value, perhaps as high as 1 mM. Consequently, the initial NADH concentration, if it was to be considered constant as the reaction proceeded, would need to be 10 mM or higher. Although the resulting high absorbencies could be overcome by measurements away from the λ_{max}, one would still have to measure small absorbance changes against a large background, which would certainly introduce some noise into the data. Moreover, these high NADH concentrations might well have adverse effects on the enzyme.

Another two-substrate reaction that has been manipulated experimentally to simplify the progress curve analysis is the reaction catalyzed by aspartate aminotransferase (Duggleby and Morrison, 1978). The reaction is readily reversible, and the complete integrated rate equation contains kinetic constants for all four reactants. The system was simplified by using glutamate and oxaloacetate as substrates and by coupling the reaction to the oxidation of NADH by the addition of glutamate dehydrogenase and NH_4Cl. This coupling reaction has three functions: (a) it provides a means of monitoring the reaction; (b) it recycles all formed 2-ketoglutarate back to glutamate, causing the concentration of this substrate to remain constant at its initial value; and (c) it prevents the accumulation of 2-ketoglutarate, making the reaction irreversible and eliminating any kinetic constants associated with this reactant and with the back reaction from the integrated rate equation. The third point has one consequence that may not always be desirable; there is no way that the kinetic properties with respect to 2-ketoglutarate or the back reaction can be determined from the data.

The integrated rate equation for this reaction is identical to Eq. (16) where ρ is V_m/Ω [Eq. (21)], γ, δ, and Ω are defined by Eqs. (26)–(28), and A, B, and P refer to oxaloacetate, glutamate, and aspartate, respectively:

$$\gamma = \frac{K_b}{[B]_0 K_{ip} \Omega} \tag{26}$$

$$\delta = \frac{K_a + K_{ia} K_b([A]_0 + [P]_0)/[B]_0 K_{ip}}{\Omega} \tag{27}$$

$$\Omega = 1 + K_b \frac{1 + ([P]_0 - K_{ia})/K_{ip}}{[B]_0} \tag{28}$$

All five kinetic parameters were estimated from the data obtained from 20 progress curves containing varying amounts of oxaloacetate, glutamate, and aspartate; two further repetitions of the entire experiment showed that each parameter was estimated reliably.

This recycling assay offers a general means to simplify progress curve analysis of any aminotransferase that uses glutamate as an amino donor, which includes most of the enzymes of this type. Similar types of recycling assays could be devised for kinases to study them in the direction of ATP utilization (using pyruvate kinase) or ATP synthesis (using hexokinase).

Even when recycling is not feasible, it still may be possible to more products using a coupling reaction. The potential of these methods for product removal do not appear to have been exploited in practice.

B. Reversible Reactions

Very few reversible reactions have been studied by progress curve analysis; the principal exception is that catalyzed by fumarase (fumarate hydratase), which has been investigated by several groups (Darvey *et al.*, 1975; Duggleby and Nash, 1989; Taraszka and Alberty, 1964). The integrated rate equation for this one-substrate and one-product reaction contains four kinetic parameters: a maximum velocity for each direction (V_f and V_r) and a Michaelis constant for each reactant (K_f and K_r). The equation is identical in form to Eq. (4) except that V_m and K_a are replaced with apparent values (V'_m and K'_a) while the logarithmic term is $\ln(1 - y/y_\infty)$; that is, instead of the reaction proceeding until y equals $[A]_0$, it stops when y reaches the equilibrium value. The definitions of V'_m and K'_a are given as Eqs. (29) and (30):

$$V'_m = \frac{V_f K_r + V_r K_f}{K_r - K_f} \tag{29}$$

$$K'_a = K_f \left\{ K_r + ([A]_0 + [P]_0)\left(1 + \frac{V_r}{V'_m}\right)\right\} (K_r - K_f) \tag{30}$$

The value of y_∞ depends on the equilibrium constant which, in turn, can be expressed in terms of the four kinetic parameters through the Haldane relationship. One such expression for y_∞ is given as Eq. (31):

$$y_\infty = \frac{[A]_0 V_f K_r - [P]_0 V_r K_f}{V_f K_r + V_r K_f} \tag{31}$$

Using 10 progress curves containing a range of fumarate concentrations but no added malate, it was shown that all four kinetic parameters could be estimated, although those associated with the substrate (fumarate) were determined with more accuracy than those associated with the product (malate). A similar result was obtained using progress curves measured with range of malate concentrations but no added fumarate; the kinetic parameters associated with the substrate had smaller standard errors than those associated with the product. Overall, somewhat better results were obtained by using fumarate as the starting reactant, and this is probably because the equilibrium constant is approximately 4.0 in this direction. The reaction proceeds further so more information is obtained from each progress curve. Similar conclusions have been drawn from a study of the reaction catalyzed by enolase.

V. General Approach to Deriving Integrated Rate Equations

The complexity of integrated rate equations escalates rapidly as one move from one to two substrates, from one to two products, and from irreversible reactions. Sometimes, owing to a fortuitous combination of kinetic parameters, the reaction can be treated as a simpler system as for the acid phosphatase example mentioned earlier. In other case, the experimental conditions can be manipulated to achieve a similar result by maintaining one substrate at a constant concentration, as for lactate dehydrogenase (Duggleby and Morrison, 1977) and asparatate aminotransferase (Duggleby and Morrison, 1978).

In general one would like to be able to deal with any reaction, and Boeker (1984a,b, 1985) provided a suitable theoretical framework for reactions with up to two substrates and products. This framework was further developed by Duggleby and Wood (1989) as follows.

Unbranched kinetic mechanisms with two or fewer substrates and products have a kinetic mechanism that is no more complex than Eq. (32):

$$\frac{dy}{dt} = \frac{V_f J_{ab}([A]_t[B]_t - [P]_t[Q]_t/K_{eq})}{\begin{aligned} &J_0 + J_a[A]_t + J_b[B]_t + J_p[P]_t + J_q[Q]_t + J_{ab}[A]_t[B]_t + J_{ap}[A]_t[P]_t \\ &+J_{aq}[A]_t[Q]_t + J_{bp}[B]_t[P]_t + J_{bq}[B]_t[Q]_t + J_{pq}[P]_t[Q]_t \\ &+J_{abp}[A]_t[B]_t[P]_t + J_{abq}[A]_t[B]_t[Q]_t + J_{apq}[A]_t[P]_t[Q]_t \\ &+J_{bpq}[B]_t[P]_t[Q]_t + J_{abpq}[A]_t[B]_t[P]_t[Q]_t \end{aligned}} \tag{32}$$

In Eq. (32), the various J terms represent combinations of the usual kinetic constants (Michaelies and inhibition constants), and for most mechanisms several of these terms will be absent from the rate equation. Except for the unusual case where $[A]_0$ and $[B]_0$ are exactly equal, the integrated rate equation is given by Eq. (33):

$$Ct = C_1Cy + \frac{C_2Cy^2}{2} + \frac{C_3Cy^3}{3} - C_fC\ln\left(1 - \frac{y}{y_\infty}\right) + C_sC\ln\left[1 - \frac{y}{y_\infty + D}\right] \tag{33}$$

Boeker (1984a,b, 1985) has given expressions that define all the variables of Eq. (33) in terms of combinations of initial concentrations and the various J constants (Eq. (32)), although it should be noted that there are some typographical errors that have been corrected (Duggleby and Wood, 1989). There is a common factor (c) that could be canceled throughout, but this term is included because it simplifies the definitions (e.g., C_1C is a less complicated expression than C_1).

Thus, to obtain the integrated rate equation for a specific kinetic mechanism, the J terms and the Haldane relationship are defined, then substituted into the appropriate expressions for C, C_1, C_2, C_3, C_f, y_∞, and D. This is a purely mechanical process, and manipulation of the symbols is readily automated. Duggleby and Wood (1989) have written a computer program (AGIRE) to achieve this automation. Given definitions of each of the J constants, the program generates computer code that defines both $F(y)$ and $F'(y)$. This code is directly compatible with the DNRP53 (Duggleby, 1984) nonlinear regression program. Thus, fitting of progress curves requires only a knowledge of the appropriate differential rate equations, and these have been catalogued extensively (Segal, 1975).

As noted earlier, Eqs. (32) and (33) may not be applicable to random mechanisms as the branches can introduce squared terms in substrate and product. This will make the general rate equation somewhat more complex, and the definitions of Boeker (1984a,b, 1985) may not apply. The principles remain the same, however, and it should not be overwhelmingly difficult to extend this method to random mechanisms.

The use of the AGIRE program is illustrated with lactate dehydrogenase reaction considered earlier. Under conditions where the pyruvate concentration is high, the rate equation simplifies to the particular from (Eq. (34), where A, P, and Q refer to NADH, lactate, and NAD^+, respectively) of the general equation for an irreversible, one-substrate, two-product reaction that is given as Eq. (35):

$$\upsilon = \frac{V_m[A]_t}{K_a + [A]_t + [Q]_t K_a/K_{iq} + [A]_t[P]_t/[K]_{ip} + [P]_t[Q]_t K_a/K_p^* K_{iq}} \tag{34}$$

$$\frac{dy}{dt} = \frac{V_f V_a[A]_t}{J_0 + J_a[A]_t + J_p[P]_t + J_q[Q]_t + J_{ap}[A]_t[P]_t + J_{aq}[A]_t[Q]_t + J_{pq}[P]_t[Q]_t + J_{apq}[A]_t[P]_t[Q]_t} \tag{35}$$

This leads to the following definition: $J_0 = K_a$; $J_a = 1$; $J_p = 0$; $J_q = K_a/K_{iq}$; $J_{ap} = 1/K_{ip}$; $J_{aq} = 0$; $J_{pq} = K_a/K_p^* K_{iq}$; and $J_{apq} = 0$. As illustrated in Fig. 5, it is now a simple matter to create a nonlinear regression program to fit the integrated from of Eq. (34) to the progress curve data. All that needs to be done to create the fitting program is to merge the code generated by AGIRE (which is stored in a file called 121.MDL in this example) with the DNRP53 program. AGIRE is available from the author.

VI. Practical Considerations

A. Selwyn's Test

As noted earlier, progress curve analysis may be applied only when the shape of the progress curve depends solely on the variations in reactant concentrations that are due to the catalyzed reaction. It is important to show that the enzyme does not undergo a progressive inactivation in the assay and that the reactants do not participate in any uncatalyzed side reactions. There is a simple method to show whether the system is suitable for progress curve analysis. This method, known as Selwyn's test (Selwyn, 1965) relies on the fact that integrated rate equations have the general form given by Eq. (36):

$$[E]_0 t = f(y, k, [X]_0) \tag{36}$$

where k represents the various rate constants, $[X]_0$ is the initial reactant concentrations, and $[E]_0$ is the total enzyme concentration. Plots of y versus $[E]_0 t$ obtained

```
                 P R O G R A M   A G I R E

Automatic Generation of Integrated Rate Equations

Please specify the type of reaction catalysed.

Enter the number of substrates (1 or 2): 1                 ←
Enter the number of products   (1 or 2): 2                 ←
Reversible/Irreversible reaction (R/I): I                  ←

The reaction has 1 substrate, 2 products and is irreversible

Is this all correct (Y/N): Y                               ←

The differential form of the rate equation is taken to be:

                         VfJaA
V = ________________________________________________________
     Jo + JaA + JpP + JqQ + JapAP + JaqAQ + JpqPQ + JapqAPQ

You must supply the combinations of kinetic constants which
go to make up these J terms and the Haldane relationship.

  Term        Combination
 ______      ____________

( 1) Jo:     Ka                                            ←
( 2) Ja:     1                                             ←
( 3) Jp:                                                   ←
( 4) Jq:     Ka/Kiq                                        ←
( 5) Jap:    1/Kip                                         ←
( 6) Jaq:                                                  ←
( 7) Jpq:    Ka/KpKiq                                      ←
( 8) Japq:                                                 ←

Terms for correction (0 if all OK): 0                      ←

The following 5 kinetic constants have been used
Vf Ka Kiq Kip Kp
Are you happy with the analysis so far (Y/N): Y            ←

In the integrated rate equation the various kinetic
constants will be referred to as follows.

B( 1) represents Vf
B( 2) represents Ka
B( 3) represents Kiq
B( 4) represents Kip
B( 5) represents Kp

BASIC code generation in progress
Completed and saved in 12I.MDL
```

Fig. 5 Use of the AGIRE computer program. The lactate dehydrogenase reaction behaves as a one-substrate, two-product, irreversible system when pyruvate is added at a high concentration. Comparison of the general rate equation [Eq. (35)] with the specific rate equation [Eq. (34)] allows the J terms to

at several enzyme concentrations will be superimposable unless there is enzyme, substrate, or product instability. The results of Selwyn's test on the reaction catalyzed by fumarase (Duggleby and Nash, 1989) are illustrated in Fig. 6.

B. Preliminary Fitting

Integrated rate equations are formulated in terms of product formed or substrate utilized. However, what is measured experimentally is usually a linearly related quantity such as an absorbance, and conversion to concentration requires the measurements to be multiplied by a constant factor, frequently after subtraction of a zero-time value. The zero-time values have to extremely reliable a small inaccuracy can introduce a systematic error that can have quite substantial effects on subsequent analysis (Newman *et al.*, 1974).

One approach to this problem is to fit a generalized progress curve to the individual curves, and a suitable from might be a modified from of Eq. (33) in which y and y_∞ include an offset term (y_0) that is estimated from the data. In practice, Eq. (33) contains far too many parameters to be estimated from a single progress curve, and we have found empirically that Eq. (37) describes almost any progress curve very well, despite the fact that it lacks terms in y^2 and y^3 and has only a single-logarithmic term:

$$V_m' t = \{y - y_0\} - K_a' \ln\left(1 - \frac{y - y_0}{y_\infty - y_0}\right) \tag{37}$$

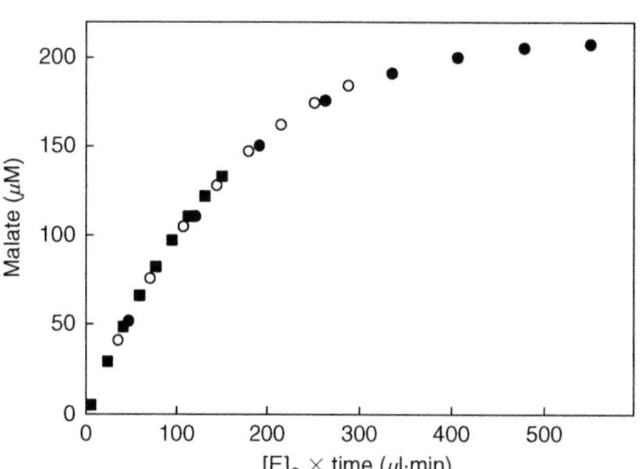

Fig. 6 Selwyn's test on the reaction catalyzed by fumarase. Assays contained 253 μM fumarate and 6 (■), 12 (○), or 24 μl (●) of fumarase, where 1 μl corresponds to a final assay concentration of 18 pM.

---

be defined. Using this information, the AGIRE program generates appropriate computer code for solving and fitting the integrated rate equation. User responses follow the colon on the lines indicated arrows; where there is nothing following the colon (e.g., for the J_p term), the "enter" key was pressed without any other response.

If we are to use Eq. (37) to estimate y_0 we must fit this to the data, which implies that we will also be estimating V'_m and K'_a, and perhaps y_∞ as well, from the same progress curve. The difficulty here is that nonlinear regression requires initial estimates of the fitted parameters and, although it should not be difficult to make reasonable guesses about the values of y_0 and y_∞, initial values for V'_m and K'_a may not be so easy to obtain. For example, it was shown earlier that the progress curve illustrated in Fig. 4A has values of V'_m and K'_a of -15.58 U/mg and -1.443 mM, respectively, but it would be almost impossible to guess at these values by visual inspection on the curve.

Duggleby (1985) reparameterized Eq. (37) in terms of quantities that have a more straightforward relationship to the shape of a progress curve. These are the initial velocity (υ_0) and the time for the reaction to reach half-completion ($t_{1/2}$). The resulting expression is shown as Eq. (38):

$$t = \frac{y - y_0}{\upsilon_0} + \frac{2\upsilon_0 t_{1/2} - (y_\infty - y_0)}{\upsilon_0(1 - \ln 4)}\left[\frac{y - y_0}{y_\infty - y_0} + \ln\left(\frac{y_\infty - y}{y_\infty - y_0}\right)\right] \tag{38}$$

The value of υ_0 so calculated can be used in conventional initial velocity analysis, and it was for this purpose that Eq. (38) was first developed and shown to perform well even when the integrated rate equation is not strictly valid owing to enzyme inactivation. It was later used to estimate y_∞ in an analysis of progress curves of an unstable enzyme (Pike and Duggleby, 1987). If required, V'_m and K'_a can be calculated using Eqs. (39) and (40):

$$V'_m = \frac{\upsilon_0(1 - \ln 4)}{(2\upsilon_0 t_{1/2}/y_\infty) - \ln 4} \tag{39}$$

$$K'_a = \frac{y_\infty(1 - 2\upsilon_0 t_{1/2}/y_\infty)}{(2\upsilon_0 t_{1/2}/y_\infty) - \ln 4} \tag{40}$$

The reason for performing a fit to Eq. (38) is to estimate, and correct for, y_0; however, the value of y_∞ is often useful as well. It has been mentioned previously that systematic errors in the substrate concentration can cause substantial errors in the subsequent analysis; estimating y_∞ from the data offers a solution to this problem.

C. Experimental Design

It is relatively easy to collect and analyze individual progress curves. If the exercise is to be at all useful, however, then the initial reactant concentrations for each progress curve must be considered carefully. Except in the simplest of cases, this question of experimental design has relied more on intuition than on detailed analysis.

When we are dealing with a system that obeys Eq. (4) then there is just one reactant, and the question to be addressed is what initial substrate concentration should be used. Defining $R = y/[\mathrm{A}]_0$ (i.e., the fractional reaction, which will range from zero at $t = 0$ to unity as $t \rightarrow \infty$), Eq. (4) is rewritten as Eq. (41):

$$t = \frac{R[\mathrm{A}]_0}{V_\mathrm{m}} - \left(\frac{K_\mathrm{a}}{V_\mathrm{m}}\right)\ln(1 - R) \tag{41}$$

If $[\mathrm{A}]_0$ is high, the first term in Eq. (41) will dominate, and the curve will contain information about V_m only. For example, if $[\mathrm{A}]_0$ is five times K_a then the first term will exceed the second for more than 99% of the progress curve. In contrast, if $[\mathrm{A}]_0$ is low, the second term will dominate and the curve will contain information about $K_\mathrm{a}/V_\mathrm{m}$ only. Clearly, $[\mathrm{A}]_0$ must take an intermediate value, and R must subtend a sufficient range that both terms have a sizable influence. Suppose that at the end of the reaction we want the first and second terms to be equal; it is easy to show that Eq. (42) is true:

$$\frac{[\mathrm{A}]_0}{K_\mathrm{a}} = -\left(\frac{1}{R}\right)\ln(1 - R) \tag{42}$$

Values calculated from this relationship are shown in Table III, from which it is seen that if the reaction is stopped when it is 80% complete, the starting substrate concentration should be twice K_a. If the reaction is allowed to proceed to 95% completion, $[\mathrm{A}]_0$ should be three times K_a. Very similar results were obtained by Duggleby and Clarke (1991) (Table III), based on calculations of the expected standard error of K_a. Moreover, the correctness of this design was verified

Table III
Effect of Extent of Reaction on Optimal Initial Substrate Concentration

	$[\mathrm{A}]_0/K_\mathrm{a}$	
R	Eq. (42)	Optimal
0.50	1.39	1.38
0.55	1.45	1.44
0.60	1.53	1.52
0.65	1.62	1.61
0.70	1.72	1.70
0.75	1.85	1.82
0.80	2.01	1.96
0.85	2.23	2.14
0.90	2.56	2.38
0.95	3.15	2.71

The extent of reaction, R, represents the total fraction of substrate utilized at the end of the progress curve. Using the values of R shown in the first column, $[\mathrm{A}]_0/K_\mathrm{a}$ was obtained using Eq. (42), or the optimal value was calculated as described by (Duggleby and Clarke (1991).

experimentally using pyruvate kinase under conditions where it behaves as a single-substrate (phosphoenolpyruvate) reaction that is not inhibited by products.

A quantitative analysis of the experimental designs appropriate for more complex reactions has not been published. However, some general remarks can be made from an examination of the relevant integrated rate equations. For an irreversible single-substrate reaction showing competitive product inhibition, it has already been pointed out that it is necessary to measure the progress curves obtained over a range of concentrations of $[A]_0 + [P]_0$. This is evident from Eqs. (14) to (15); the simultaneous estimation of V_m, K_a, K_{ip} requires that K_a' is not a constant, and this will occur only when $[A] + [P]_0$ is varied. It will also be evident that these concentrations must be varied over a range that encompasses the value of K_{ip}. Normally, this would be done by varying $[A]_0$ only, but if K_{ip} is much greater than K_a, addition of product may be advisable. For a reversible reaction involving one substrate and one product (Eqs. (29) and (30)), a similar design is applicable although the range of $[A]_0 + [P]_0$ is required is such that $([A]_0 + [P]_0)(1 + V_r/V_m')$ spans the value of K_r. Clearly, more work is needed to quantify the experimental designs for progress curve analysis.

VII. Unstable Enzymes

When an enzyme fails Selwyn's test because of instability, the advantages of progress curve analysis are lost. This is unfortunate, and we have tried previously to overcome this limitation (Duggleby, 1986b; Pike and Duggleby, 1987). It has been shown that it is possible, from the residual substrate concentration after prolonged incubation, to determine the inactivation rate constants of various complexes along the catalytic pathway. However, the normal kinetic constants (maximum velocity, Michaelis constants, and inhibition constants) are not obtained readily from the data. Using an analysis that includes provision for enzyme inactivation, it has now been shown (Duggleby, 1994) that progress curves can be used for kinetic characterization of an unstable enzyme.

Consider an enzyme that obeys Eq. (1) but where the free enzyme and the enzyme-substrate complex are each unstable, being converted to inactive forms with rate constants of j_1 and j_2, respectively. The kinetics of this system are described by two differential equations defining the rate of loss of active enzyme $[E']_t$ [Eq. (43)] and the rate of utilization of substrate $[A]_t$ (Eq. (44), where $k_3\,[E']_t$ equals V_m at $t = 0$):

$$\frac{d[E']_t}{dt} = \frac{-[E']_t(j_1 + j_2[A]_t/K_a)}{1 + [A]_t/K_a} \tag{43}$$

$$\frac{d[A]_t}{dt} = \frac{-k_3[E']_t[A]_t/K_a}{1 + [A]_t/K_a} \tag{44}$$

Applying progress curve analysis to this system requires calculation of $[A]_t$ at any time, but these differential equations are coupled; that is, to obtain an expression for $[E']_t$ from Eq. (43) requires a knowledge of $[A]_t$ while to obtain an expression for $[A]_t$ from Eq. (44) requires a knowledge of $[E']_t$. Except in some special cases (e.g., when j_1 is equal to zero or to j_2), Eqs. (43) and (44) cannot be solved algebraically. They can, however, be solved numerically; given particular values of the parameters, as well as defined starting conditions ($[E']_0$ and $[A]_0$), Eqs. (43) and (44) can be evaluated. These rates are applied over a small time interval to calculate $[E']_t$ and $[A]_t$ after this time. The process is then repeated as often as required to obtain the entire progress curve for substrate utilization.

This is a relatively crude method, and more sophisticated algorithms have been described (Stoer and Bulirsch, 1980). The third/fourth-order Runge-Kutta-Fehlberg numerical integration algorithm with automatic step size control has been incorporated into the DNRP53 computer program to analyze reaction systems described by coupled differential equations. The experimental system that has been used to evaluate the method is bovine intestinal mucosal alkaline phosphatase. This is a slightly more complex system than that described by Eqs. (43) and (44) because the enzyme is inhibited by one of its products, phosphate ion. Although the enzyme is normally quite stable, the essential zinc ion can be removed by EGTA, leading to inactivation. This instability is illustrated by applying Selwyn's test (Fig. 7A) where the curves are seen clearly not to be superimposable. It was shown previously (Pike and Duggleby, 1987) that only the free enzyme is susceptible to inactivation; in the enzyme-substrate complex the zinc is locked in, as it is in the enzyme-phosphate complex. The system is described by Eqs. (45) and (46):

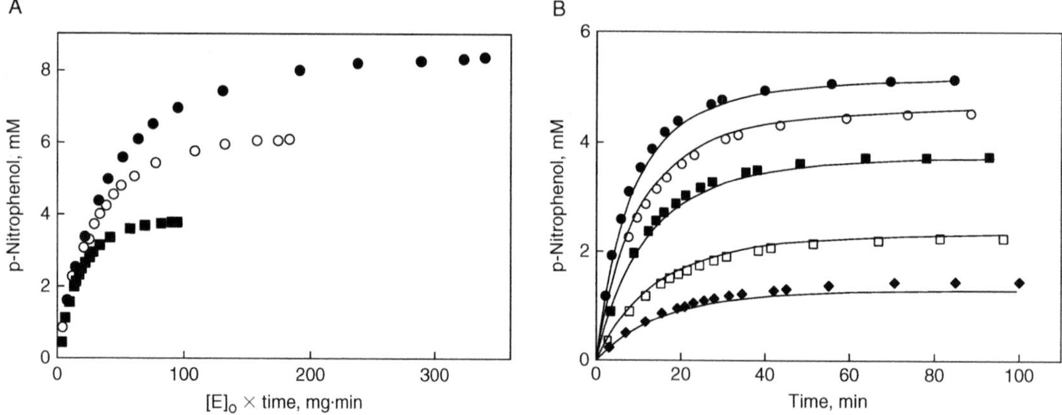

Fig. 7 Progress curve analysis of an unstable enzyme, alkaline phosphatase. (A) Results of Selwyn's test. Each reaction contained 10 mM *p*-nitrophenyl phosphate and 0.846 (■), 1.692 (○), or 3.384 mg (●) of enzyme. (B) Data obtained with 6 mM substrate and 0.423 (◆), 0.846 (□), 1.692 (■), 2.538 (○), or 3.384 mg (●) of enzyme. The curves are the results of fitting Eqs. (45) and (46) to the data shown plus similar data obtained at 10, 8, 4, and 2 mM substrate.

$$\frac{d[E']_t}{dt} = \frac{-j_1[E']_t}{1 + [A]_t/K_a + [P]_t/K_p} \tag{45}$$

$$\frac{d[P]_t}{dt} = \frac{k_3[E']_t[A]_t/K_a}{1 + [A]_t/K_a + [P]_t/K_p} \tag{46}$$

Some representative progress curves for the hydrolysis of *p*-nitrophenyl phosphate are shown in Fig. 7B. Fitting Eqs. (45) and (46) to the data gave kinetic parameters that are in excellent agreement with those obtained previously, and, as shown in the curves in Fig. 7B, describe the data quite well.

More complex schemes than that described by Eqs. (45) and (46) can be analyzed in much the same way. For example, the added possibility of instability of both the enzyme-substrate and enzyme-product complexes could be included. Further, there are a number of other possible situations where progress curve analysis should be useful but has not be applied since the underlying model leads to coupled differential equations. For example, it is known that a number of compounds act as slow-binding enzyme inhibitors (Szedlacsek and Duggleby, this volume). The slow onset of inhibition is best studied by following progress curves, but the analyses currently available require that this is be done under conditions where substrate depletion is insignificant. This may necessitate high concentrations of substrate that tend to obscure the very inhibition one wishes to observe. The present approach offers a facile solution to this problem that will allow the progress curves for slow-binding enzyme inhibitors to be analyzed even when there is considerable substrate depletion.

VIII. Conclusions

The progress curves of enzyme-catalyzed reactions are a rich source of kinetic information. Provided suitable methods are used to analyze the data, relatively few such curves can allow a full quantitative description of the dependence of activity on the concentrations of substrates and products. Although difficulties in data analysis have impeded the development of this technique, there is now little reason why it could not be applied to a wide variety of enzyme kinetic studies. Problems of enzyme instability that distort progress curves form the shape dictated purely form the effects of substrate depletion and inhibitory product accumulation can now be overcome by including enzyme decay in the kinetic scheme.

Straightforward methods are available for analyzing all unbranched reaction mechanisms with two or fewer substrates and products, and there is no reason to suppose that more complex situations, such as random mechanisms, will not become amenable to progress curve analysis in the near future.

Appendix A: Regression Analysis

The fundamentals of regression analysis are not difficult to understand and are best considered in relation to linear regression. The basic problem is to fit an equation to experimental data by minimizing the sum of the squares (SSQ) of the differences between a series of experimental values (y_i) and those predicted (μ_i) by the fitted line:

$$\mathrm{SSQ} = \sum(y_i - \mu_i)^2 \tag{A1}$$

Suppose that the equation for a fitted line is a function of a single parameter (β) and some combination (X) of independent variables:

$$\mu_i = f(\beta, X) \tag{A2}$$

Equation (A2) is said to be *linear* if β appears as a constant multiplier; if this is so, we may partition Eq. (A2) as shown by Eq. (A.3):

$$\mu_i = \beta f_1(X_1) + f_2(X_2) \tag{A3}$$

Substituting Eq. (A3) into Eq. (A1), introducing $y_i' = y_i - f_2(X_2)$, and expanding leads to Eq. (A.4):

$$\mathrm{SSQ} = \sum(y_i')^2 - 2\beta\sum[y_i' f_1(X_1)] + \beta^2\sum[f_1(X_1)]^2 \tag{A4}$$

From simple calculus we can find the value of β that yields the minimum SSQ by differentiating with respect to β and setting the resulting expression equal to zero:

$$0 = -2\sum[y_i' f_1(X_1)] + 2\beta\sum[f_1(X_1)]^2 \tag{A5}$$

$$\beta = \frac{\sum[y_i' f_1(X_1)]}{\sum[f_1(X_1)]^2} \tag{A6}$$

This procedure is easily generalized if Eq. (A2) has several linear parameters to be estimated; instead of a single linear equation (Eq. (A5)) we obtain a set of linear simultaneous equations that may be solved readily.

In the Gauss-Newton method of nonlinear regression we start with an expression of the same form as Eq. (A2), but, because it is nonlinear, it cannot be rewritten in the same form as Eq. (A3). However, suppose we have an approximate value (B) of the nonlinear parameter β that differs from the best fit value by an unknown amount (b). We may rewrite Eq. (A2) as Eq. (A.7):

$$\mu_i = f(B+b, X) \tag{A7}$$

From Taylor's theorem, the right-hand side of Eq. (A7) may be expanded as an infinite series (Eq. (A8)) involving successive derivatives (f', f'' f''', etc.) of the function evaluated at the current value B:

$$\mu_i = f(B, X) + bf'(B, X) + \frac{b^2 f''(B, X)}{2!} + \frac{b^3 f''(B, X)}{3!} + \tag{A8}$$

Provided b is small (i.e., B is a reasonably good estimate of β), the terms in b^2 and higher powers can be ignored, giving Eq. (A9):

$$\mu_i \approx f(B, X) + bf'(B, X) \tag{A9}$$

Subtracting the experimental values (y_i) from both sides and defining the residual r_i as the different between y_i and the value calculated from the equation at the current value of the parameter [$f(B, X)$], we get Eq. (A10):

$$y_i - \mu_i = r_i - bf'(B, X) \tag{A10}$$

An expression for SSQ can now be written as before, then differentiated and solved for b (Eqs. (A11)–(A13)):

$$\mathrm{SSQ} = \sum (r_i)^2 - 2b\sum [r_i f'(B, X)] + b^2 \sum [f'(B, X)]^2 \tag{A11}$$

$$0 = -2\sum [r_i f'(B, X)] + 2b\sum [f'(B, X)]^2 \tag{A12}$$

$$b = \frac{\sum [r_i f'(B, X)]}{\sum [f'(B, X)]^2} \tag{A13}$$

From the relationship $\beta = B + b$, we would be able to use Eq. (A13) to calculate the best fit value, except for the fact that an approximation was made in going from Eqs. (A8) to (A9). Provided the approximation is good, $B + b$ will be a better estimate of β than was B, and we can repeat the whole process using the refined estimate. The process of refinement is continued until no further improvement is obtained.

The procedure can be generalized to several nonlinear parameters by observing that Eq. (A13) has the same form as Eq. (A6) with *b substituted for* β, r_i replacing y_i', and the derivative $f'(B, X)$ used instead of $f_1(X_1)$. Thus, if the original function has several parameters we will obtain a system of linear simultaneous equations instead of the single equation (Eq. (A12)). These may be solved to yield correction for each of the parameters, and, as before, repeated application of the process will yield the best fit values.

Acknowledgments

This work supported by the Australian Research Council. The author is indebted to Prof. Dexter Northrop for useful comments on this chapter.

References

Boeker, E. A. (1984a). *Biochem. J.* **223,** 15.
Boeker, E. A. (1984b). *Experientia* **40,** 453.
Boeker, E. A. (1985). *Biochem. J.* **226,** 29.
Boeker, E. A. (1987). *Biochem. J.* **245,** 67.
Cox, T. T., and Boeker, E. A. (1987). *Biochem. J.* **245,** 59.
Darvey, I. G., Shrager, R., and Kohn, L. D. (1975). *J. Biol. Chem.* **250,** 4696.
Duggleby, R. G. (1984). *Comput. Biol. Med.* **14,** 447.
Duggleby, R. G. (1985). *Biochem. J.* **228,** 55.
Duggleby, R. G. (1986a). *Biochem. J.* **235,** 613.
Duggleby, R. G. (1986b). *J. Theor. Biol.* **123,** 67.
Duggleby, R. G. (1994). *Biochim. Biophys. Acta* **1205,** 268.
Duggleby, R. G., and Clarke, R. B. (1991). *Biochim. Biophys. Acta* **1080,** 231.
Duggleby, R. G., and Morrison, J. F. (1977). *Biochim. Biophys. Acta* **481,** 297.
Duggleby, R. G., and Morrison, J. F. (1978). *Biochim. Biophys. Acta* **526,** 398.
Duggleby, R. G., and Morrison, J. F. (1979). *Biochim. Biophys. Acta.* **568,** 357.
Duggleby, R. G., and Nash, J. C. (1989). *Biochem. J.* **257,** 57.
Duggleby, R. G., and Wood, C. (1989). *Biochem. J.* **258,** 397correction. *Biochem. J.* **270,** 843.
Newman, P. F. J., Atkins, G. L., and Nimmo, I. A. (1974). *Biochem. J.* **143,** 779.
Nimmo, I. A., and Atkins, G. L. (1974). *Biochem. J.* **141,** 913.
Orsi, B. A., and Tipton, K. F. (1979). *Methods Enzymol.* **63,** 159–183.
Pike, S. J., and Duggleby, R. G. (1987). *Biochem. J.* **224,** 781.
Rebholz, K. L., and Northrop, D. B. (1995). *Methods Enzymol.* **249,** Chapter 9.
Schønheyder, F. (1952). *Biochem. J.* **50,** 378.
Segal, I. H. (1975). "Enzyme Kinetics." Springer-Verlag, Berlin.
Selwyn, M. J. (1965). *Biochim. Biophys. Acta* **105,** 103.
Stoer, J., and Bulirsch, R. (1980). "Introduction to Numerical Analysis." Springer-Verlag, New York.
Swann, W. H. (1969). *FEBS Lett.* **2**(Suppl.), S39.
Szedlacsek, S. E., and Duggleby, R. G. (1995). *Methods Enzymol.* **249,** Chapter 6.
Taraszka, M., and Alberty, R. A. (1964). *J. Phys. Chem.* **68,** 3368.
Wilkinson, G. N. (1961). *Biochem. J.* **80,** 324.

CHAPTER 6

Effects of pH on Enzymes*

Keith F. Tipton, Andrew G. McDonald, and Henry B. F. Dixon✠

Department of Biochemistry
Trinity College, Dublin 2, Ireland

The effects of variations of the hydrogen ion concentration on the activity of enzymes have close similarities to the effects of activators and inhibitors, and the same kinetic methods and theory can be applied to both types of system. These

* Dedicated to the memory of the late Dr. Henry B.F. Dixon (Hal) who contributed so much.
✠ Deceased

DOI: 10.1016/B978-0-12-378608-1.00006-2

close similarities are often obscured, however, by the fact that a logarithmic scale (pH) is usually used for hydrogen ion concentration. Treatment of the effects of hydrogen ion concentration in the same way as other effectors can yield valuable information on the nature of the kinetic mechanism obeyed by the enzyme; in addition, the characteristic ionization constants of amino acid side-chain groups has led to the use of such studies in attempts to identify specific groups as playing a role in the reaction. This latter approach has frequently been regarded as being the most important function of these studies. These two aspects of the subject, however, are complementary, and attempts to identify groups from pH studies without considering the kinetic aspects not only will miss a great deal of valuable information, but also can frequently lead to erroneous conclusions. In this chapter, we will consider the effects of hydrogen ion concentration on the activity of enzymes both in terms of kinetic analysis with the hydrogen ion concentration as the variable and in terms of the use of the effects of pH to identify specific ionizing groups.

I. Theory

Although most enzymes contain many ionizing groups, the variation of initial velocity with pH often gives a "bell-shaped" curve like that shown in Fig. 1. Michaelis and Davidsohn (1911) first interpreted this in terms of a model in which the enzyme contained only two ionizable groups that were essential for activity. At first sight, this interpretation may appear oversimplistic, but it merely indicates that, of the many ionizable groups that may be needed in a particular ionic form for activity, it is only the first to protonate or deprotonate as the pH is moved up or down from the optimum pH, whose ionization can be detected by measurement of enzyme activity. The ionizations of other groups at pH values farther from the pH optimum will be undetectable because they will affect only the equilibrium between inactive forms of the enzyme. If the simple model is correct,

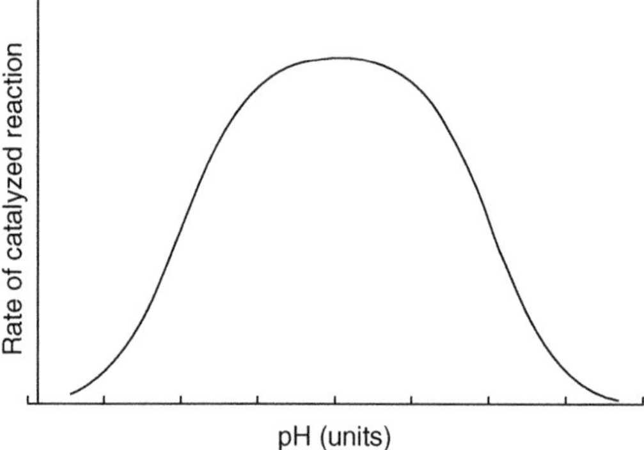

Fig. 1 Typical dependence on pH of the rate of an enzyme-catalyzed reaction. Here and in Figs. 2, 3, 8–10 the marks on the horizontal scale indicate pH units.

the enzyme can be considered as being a dibasic acid, and we therefore start by giving the equations that describe the ionization of such compounds.

A. The Ionization of Dibasic Acids (Adams, 1916)

For the scheme (Eq. (1)) where the concentrations of the different species are represented by the symbols w, x, y, and z, these can be related in terms of the dissociation constants shown in Eq. (1) by the equations

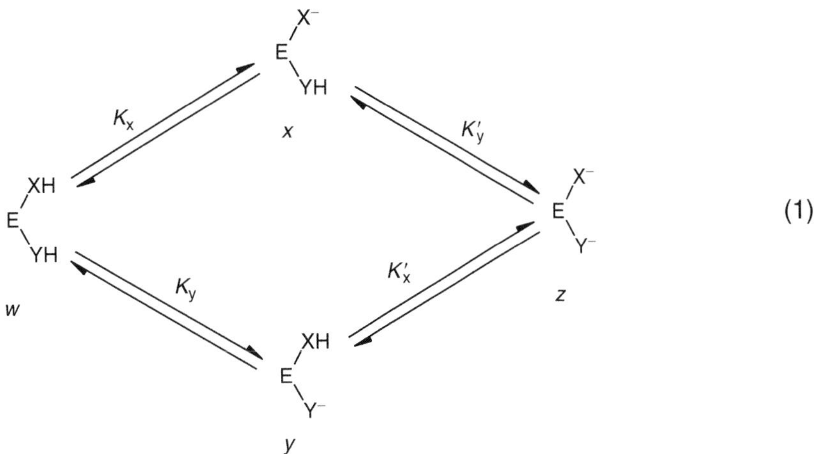

(1)

$$x = wK_x/[H^+] \quad (2)$$

$$y = wK_y/[H^+] \quad (3)$$

$$z = xK'_y/[H^+] \quad (4)$$

$$z = yK'_x/[H^+] \quad (5)$$

where $[H^+]$ represents the hydrogen ion concentration. Only three of the dissociation constants of Eq. (1) are required to define the concentrations, because Eqs. (2) and (4) give

$$z = wK_xK'_y/[H^+]^2 \quad (6)$$

and Eqs. (3) and (5) give

$$z = wK'_xK_y/[\mathrm{H}^+]^2 \tag{7}$$

so that from Eqs. (6) and (7)

$$K_xK'_y = K'_xK_y \tag{8}$$

and thus fixing any three of these constants defines the fourth.

Although K_x and K'_x both refer to dissociation of a proton from the same group XH, they are not in general identical. Usually K_x will be larger than K'_x because a negative charge on the group Y^- can help to hold the proton on group XH. Nevertheless K'_x can occasionally exceed K_x: loss of a proton from YH may facilitate loss from XH by, for example, a consequent conformational change or by permitting binding of a multivalent metal ion whose higher positive charge may repel the proton on XH. Similar arguments apply to K_y and K'_y, since from Eq. (8) $K_x/K'_x = K_y/K'_y$.

The total concentration of the dibasic acid, e, will be given by

$$e = w + x + y + z \tag{9}$$

and the concentrations of each ionic form can be written as

$$w = \frac{e}{1 + [(K_x + K_y)/[\mathrm{H}^+]] + K_xK'_y/[\mathrm{H}^+]^2} \tag{10}$$

$$x = \frac{eK_x/[\mathrm{H}^+]}{1 + [(K_x + K_y)/[\mathrm{H}^+]] + K_xK'_y/[\mathrm{H}^+]^2} \tag{11}$$

$$y = \frac{eK_y/[\mathrm{H}^+]}{1 + [(K_x + K_y)/[\mathrm{H}^+]] + K_xK'_y/[\mathrm{H}^+]^2} \tag{12}$$

$$z = \frac{eK_xK'_y/[\mathrm{H}^+]^2}{1 + [(K_x + K_y)/[\mathrm{H}^+]] + K_xK'_y/[\mathrm{H}^+]^2} \tag{13}$$

It can be seen from Eqs. (11) and (12) that the ratio of the concentrations of the two singly protonated species is given by

$$x/y = K_x/K_y \tag{14}$$

and that this ratio is independent of $[\mathrm{H}^+]$; thus any change in the concentration of one of these species with pH will be accompanied by a proportional change in the concentration of the other. Because of this there is no way of telling how much of a given effect is due to one of these two species in isolation. It is therefore more convenient to treat them as a single species whose concentration is given by

$$x + y = \frac{e}{1 + [\mathrm{H}^+]/K_\mathrm{A} + K_\mathrm{B}/[\mathrm{H}^+]} \tag{15}$$

where

$$K_A = K_x + K_y = [H^+](x + y)/w \tag{16}$$

and

$$K_B = K_x K_y'/(K_x + K_y) = [H^+]z/(x + y) \tag{17}$$

(thus from Eqs. (8), (16), and (17) $K_A K_B = K_x K_y' = K_x' K_y$).

The constants K_A and K_B are termed *molecular* dissociation constants to distinguish them from the *group* dissociation constants shown in Eq. (1). Only the molecular constants can be measured experimentally, although evidence can be obtained by argument from analogy on the magnitude of group constants (Dixon, 1976). Any effect that is due to the singly protonated species, to whatever degree each of them contributes to it, increases as the sum of the two and obeys equations whose only parameters are the molecular constants. The constant K_A is the dissociation constant for the first proton to dissociate, regardless of the fractions of it that are derived from groups XH and YH, and K_B is that for the second proton. Because it is only molecular constants that can be determined experimentally, it is useful to express the concentrations of the other two species shown in Eq. (1) in terms of them (Eqs. (18) and (19)).

$$w = \frac{e}{1 + K_A/[H^+] + K_A K_B/[H^+]^2} \tag{18}$$

$$z = \frac{e}{1 + [H^+]/K_B + [H^+]^2/K_A K_B} \tag{19}$$

A theoretical curve showing the variations in the concentrations of each of the species is given in Fig. 2. If the dibasic acid under consideration is an enzyme that is active in one or more ionic forms and is converted into inactive forms by protonation and deprotonation, then it follows from Eq. (15) that the initial rate of the catalyzed reaction, v, varies with $[H^+]$ according to the equation

$$v = \frac{ke}{1 + [H^+]/K_A + K_B/[H^+]} \tag{20}$$

When K_A greatly exceeds K_B, so that the terms $[H^+]/K_A$ and $K_B/[H^+]$ cannot both be appreciable at any value of $[H^+]$, then K_A and K_B represent the values of $[H^+]$ that give half the maximum velocity. Since rates are often plotted against pH rather than against $[H^+]$, it is convenient to use pK values ($-\log K$) rather than K_A and K_B themselves. Thus if p$K_B \gg$ pK_A, pK_A and pK_B are the pH values at which the activity is half maximal. Figure 3 shows curves of activity (i.e., of the concentration of the mono-protonated species of a dibasic acid) against pH. It can be seen that, as pK_B approaches and then falls below pK_A, the pH values at which half-maximal velocity is reached no longer approximate to the true pK values. The maximum of the curve is always at a pH equal to (pK_A + pK_B)/2, and this value will also be given by the mean of the two pH values that give half the maximal rate.

In analyzing the effects of pH on enzyme activity it is important to remember the two points developed above: (a) the pH values that give half-maximal activity

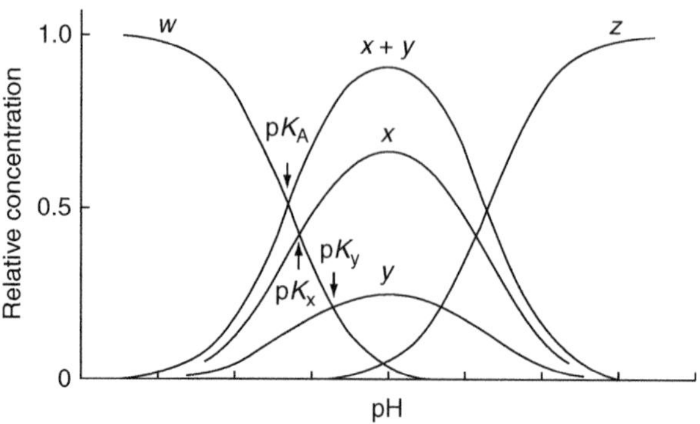

Fig. 2 pH Dependence of the concentrations of the forms of a dibasic acid. The species are labeled in Eq. (1). Note that $w = x$ when pH = pK_x and that $w = x + y$ when pH = pK_A.

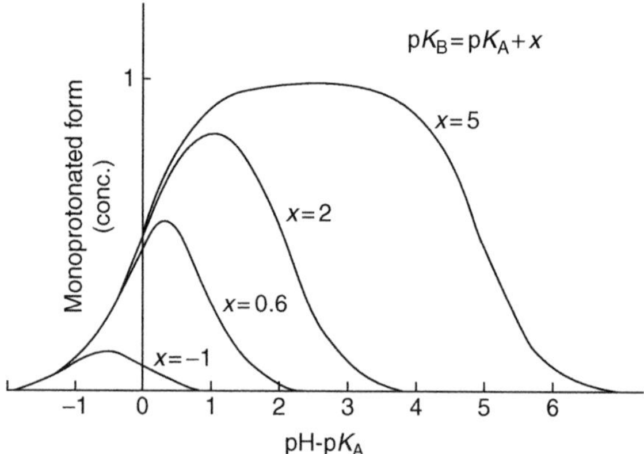

Fig. 3 pH Dependence of the concentration of the monoprotonated form of a dibasic acid. Curves are shown for various values of pK_B − pK_A. The curve for pK_B = pK_A + 0.6 is the narrowest that can be achieved without positively cooperative proton binding. When pK_B − pK_A is less than 0.6, there is very little of the monoprotonated form, because binding of one proton facilitates binding of the second.

correspond to pK values only if they are far apart (over about 3.5 units); (b) the pK values obtained will be the molecular constants derived from Eqs. (16) and (17) rather than group constants.

A simple method of estimating whether the two pH values that give half-maximal activity are far enough apart to give pK values is to measure the difference between them and to derive the pK difference from this by the relationship shown in Fig. 4.

Although it is molecular constants that are obtained, often they can be interpreted as group constants. Groups so closely linked (by proximity or indirect interactions) that both affect activity are unlikely not to affect each other's ionization; hence we cannot assume that $K_x = K'_y$. Nevertheless K_x and K_y may often differ appreciably, say that $K_x \gg K'_x$. Then $x \gg y$ and K_A will approximate to K_x and K_B to K'_y. A typical α-amino acid exemplifies this (Eq. (21)).

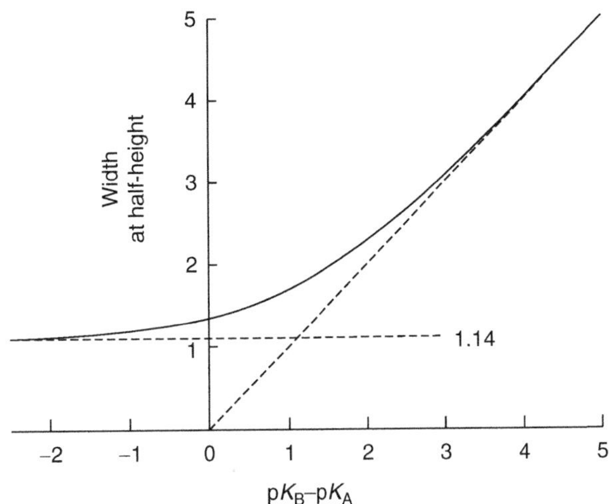

Fig. 4 Dependence of the width of a bell-shaped curve at half its height on the pK difference that determines it. If the pH values at which the concentration of the monoprotonated form of a dibasic acid is half its maximal value are called p$K^* \pm \log q$, then the molecular pK values pK_B and pK_A are given by p$K^* \pm \log (q - 4 + 1/q)$.

Much evidence shows that the zwitterion greatly predominates over the uncharged form. Hence the observed pK_A of 2 for the first proton to dissociate

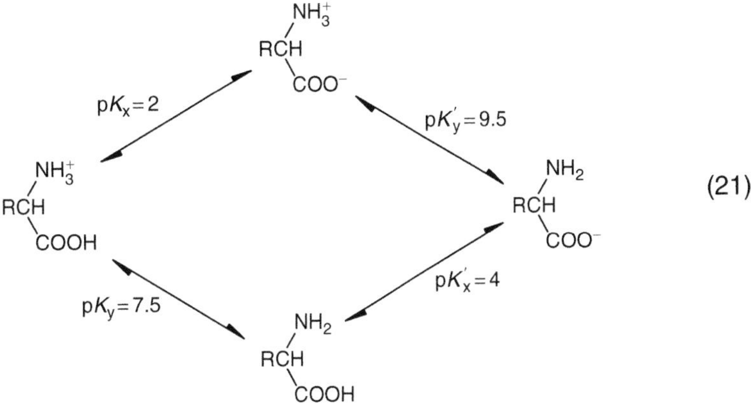

(21)

can be identified with pK_x for the dissociation of the carboxyl group in the form with a protonated amino group. Likewise the pK_B of 9.5 for the second proton can be identified with pK'_y for the amino group in the form with a dissociated carboxyl group. By analogy with other acids, we can assign to pK'_x a value of about 4. This fixes pK_y at about 7.5, which is reasonable for an amino group in view of the electron attraction of a neighboring undissociated carboxyl group.

Two points should be noted. One is that the possibility of identifying the molecular constants with group constants depends on the difference between pK_x and pK_y; it does not depend on whether or not pK_A and pK_B differ. If pK_x and pK_y were similar, then both groups would contribute to pK_A, and it would be close to both pK_x and pK_y. The second point is that pK values of groups in minor forms (here pK'_x and pK'_y) are unimportant, because even if an effect such as enzyme activity depends on a minor form, its concentration depends on pH as governed by the parameters pK_A and pK_B (i.e., largely pK_x and pK'_y), not pK_y and pK'_x, as derived above.

Determination of the pK values that govern the variation of the initial rate of an enzyme-catalyzed reaction does little in itself to show the roles of specific groupings in the function of the enzyme. Even in the simplest analysis, changes in the pH may affect either the Michaelis constant (K_m) or the maximum velocity (V) of the reaction. Where both of these are affected by pH, the apparent pK values seen at an arbitrary substrate concentration are unlikely to correspond to any individual ionization. Thus, to obtain useful information, it is necessary to investigate the effects of pH on the kinetic parameters of the reaction catalyzed.

B. Simplified System

The effects of pH on enzyme kinetic parameters are often interpreted in terms of a highly simplified kinetic mechanism. The theory will be extended later to cover more complicated, but perhaps more realistic, systems, but even in such extensions a number of assumptions that may not be easy to justify will remain, so that uncertainties are introduced into the interpretations obtained. We will first consider a simple system to illustrate the general approach that is usually applied.

Consider the system shown in Eq. (22) where K_A^E and K_B^E represent the molecular dissociation constants for the free enzyme, and K_A^{ES} and K_A^{ES} represent those for the enzyme-substrate complex. The constants K_A^E and K_A^{ES} may differ either because substrate binding changes the dissociation constant of the relevant group or

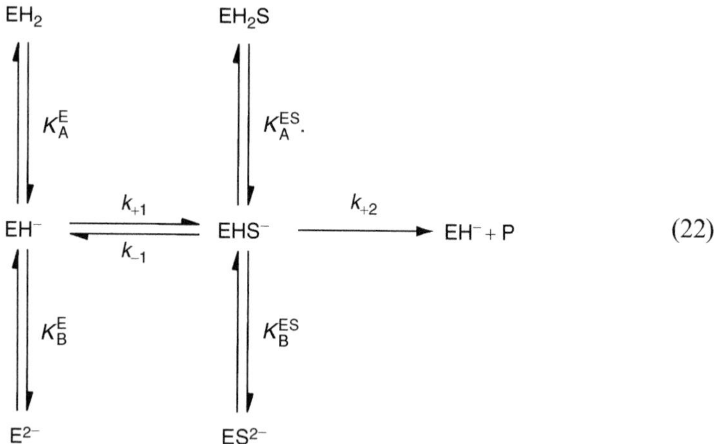

(22)

because the constants refer to different ionizations, since it needs not to be the protonation of the same group that inhibits substrate binding by the free enzyme and reaction of the enzyme-substrate complex once formed. Steady-state treatment of this mechanism yields the initial velocity equation (Alberty and Massey, 1955; Waley, 1953).

$$v = \frac{\tilde{V}}{(1 + [\mathrm{H}^+]/K_{\mathrm{A}}^{\mathrm{ES}} + K_{\mathrm{B}}^{\mathrm{ES}}/[\mathrm{H}^+]) + \tilde{K}_m/s(1 + [\mathrm{H}^+]/K_{\mathrm{A}}^{\mathrm{E}} + K_{\mathrm{B}}^{\mathrm{E}}/[\mathrm{H}^+])} = \frac{\tilde{V}/(1 + [\mathrm{H}^+]/K_{\mathrm{A}}^{\mathrm{ES}} + K_{\mathrm{B}}^{\mathrm{ES}}/[\mathrm{H}^+])}{1 + (\tilde{K}_m/s)[(1 + [\mathrm{H}^+]/K_{\mathrm{A}}^{\mathrm{E}} + K_{\mathrm{B}}^{\mathrm{E}}/[\mathrm{H}^+])/(1 + [\mathrm{H}^+]/K_{\mathrm{A}}^{\mathrm{ES}} + K_{\mathrm{B}}^{\mathrm{ES}}/[\mathrm{H}^+])]} \quad (23)$$

where $\tilde{V} = k_{+2}[\mathrm{E}]_t$, $\tilde{K}_{\mathrm{m}} = (k_{-1} + k_{+2})/k_{+1}$, $[\mathrm{E}]_t$ represents the total enzyme concentration, and s represents the concentration of unbound substrate, which can usually be taken to be the total substrate concentration. Thus $\tilde{V}$ and $\tilde{K}_{\mathrm{m}}$ are pH-corrected parameters: they are the limits to which V and K_{m} tend at pH values between the relevant pK_{A} and pK_{B} values if these pK values are sufficiently far apart.

Comparison of Eq. (23) with the simple Michaelis equation (Eq. (24))

$$v = \frac{V_s}{s + K_{\mathrm{m}}} = \frac{V}{1 + K_{\mathrm{m}}/S} \quad (24)$$

shows that the hydrogen ion concentration will affect both the apparent K_{m} and V values; thus hydrogen ions can be regarded as mixed (or noncompetitive) effectors of the enzyme. The apparent maximum velocity will be given by

$$V = \frac{V}{1 + [\mathrm{H}^+]/K_{\mathrm{A}}^{\mathrm{ES}} + K_{\mathrm{B}}^{\mathrm{ES}}[\mathrm{H}^+]} \quad (25)$$

It will therefore depend only on ionizations of the enzyme-substrate complex (Eq. (22)). The apparent K_{m} value will be given by

$$K_{\mathrm{m}} = \tilde{K}_{\mathrm{m}} \frac{1 + [\mathrm{H}^+]/K_{\mathrm{A}}^{\mathrm{E}} + K_{\mathrm{B}}^{\mathrm{E}}[\mathrm{H}^+]}{1 + [\mathrm{H}^+]/K_{\mathrm{A}}^{\mathrm{ES}} + K_{\mathrm{B}}^{\mathrm{ES}}[\mathrm{H}^+]} \quad (26)$$

and will therefore be affected by ionizations of both the free enzyme and the enzyme-substrate complex. Combination of Eqs. (25) and (26) gives

$$\frac{V}{K_{\mathrm{m}}} = \frac{\tilde{V}/\tilde{K}_{\mathrm{m}}}{1 + [\mathrm{H}^+]/K_{\mathrm{A}}^{\mathrm{E}} + K_{\mathrm{B}}^{\mathrm{E}}[\mathrm{H}^+]} \quad (27)$$

Equation (24) shows that V/K_{m} is the limit approached by v/s at low substrate concentrations when $s \ll K_{\mathrm{m}}$. Hence v varies with pH under these conditions in a way that depends only on ionizations of the free enzyme. Methods of finding the various ionization constants will now be given.

C. Methods of Obtaining Ionization Constants

1. Double-Reciprocal Plots, H^+ as Effector

The following discussion is presented in terms of the double-reciprocal plot. This is known to be a most inaccurate procedure for estimating kinetic parameters. Direct fitting of the initial-rate data to a rectangular hyperbola is much more accurate and is now the method of choice, since there are many readily available computer programs that will fit the data, with error estimates obtained from replicate values, to such curves to determine the kinetic parameters and their associated errors. Although the double-reciprocal plot might be expected to be of mere historical interest, such plots provide a clear illustration of inhibition type are frequently used to illustrate enzyme kinetic behavior. They are used only for that purpose here.

If K_A and K_B for both E and ES differ enough, there will be pH values where terms containing one of them can be neglected and terms containing the other are appreciable. Thus at low concentrations of H^+ ($[H^+] \ll K_A^E$ and $[H^+] \ll K_A^{ES}$), it will act as a kinetically mixed essential activator of the enzyme according to Eq. (28) (derived from Eq. (23)).

$$
\begin{aligned}
v &= \frac{\tilde{V}}{1 + K_B^{ES}/[H^+] + (\tilde{K}_m/s)(1 + K_B^E/[H^+])} \\
&= \frac{V}{1 + \tilde{K}_m/s + K_B^{ES}/[H^+] + K_B^E\tilde{K}_m/[H^+]s}
\end{aligned} \tag{28}
$$

Figure 5 shows the type of double-reciprocal plots to be expected for such a system. At higher concentrations of hydrogen ions the inhibition terms ($[H^+]/K_A^E$ and $[H^+]/K_A^{ES}$) will no longer be negligible, and this results in the inhibition at high $[H^+]$ shown in Fig. 5A and in the deviation of the lines from the simple intersecting pattern in Fig. 5B.

The situation at high concentrations of hydrogen ions where $K_B^E/[H^+]$ and $K_B^{ES}/[H^+]$ are negligible is described by Eq. (29).

$$
v = \frac{\tilde{V}/(1 + [H^+]/K_A^{ES})}{1 + (\tilde{K}_m/s)[(1 + [H^+]/K_A^E)/(1 + [H^+]/K_A^{ES})]} \tag{29}
$$

which shows that the hydrogen ion will function as a mixed inhibitor of the enzyme, the situation becoming more complicated at lower concentrations, where the terms $K_B^E/[H^+]$ and $K_B^{ES}/[H^+]$ are no longer negligible, as shown in Fig. 6.

Just as with other activators and inhibitors (Segel, 1993; Cook and Cleland, 2007), the values of the individual ionization constants can be obtained from replots of the slopes and intercepts of plots of $1/V$, against $1/s$, that is, of K_m/V and of $1/V$, against 1/[activator] or against [inhibitor]. Thus the slopes and

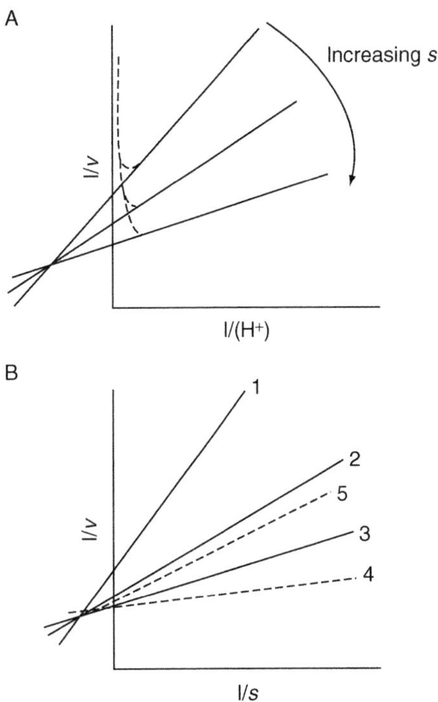

Fig. 5 Reciprocal plots of the effects of hydrogen ions on enzyme activity for a mechanism obeying Eq. (23). The solid lines correspond to the situation where $[H^+] \ll K_A^E$ and $[H^+] \ll K_A^{ES}$, and thus Eq. (28) is obeyed. The dashed lines describe the situation where these inequalities no longer hold. The solid lines intersect at a point to the left of the $1/v$ axis, which will be above the horizontal axis (as shown here) if $K_B^E > K_B^{ES}$, on that axis if $K_B^E = K_B^{ES}$, and below it if $K_B^E < K_B^{ES}$. In (B) the hydrogen ion concentration increases in the order 1–5.

intercepts of Fig. 5B are plotted against $1/[H^+]$ in Fig. 7A and B, and those of Fig. 6B $[H^+]$ in Fig. 7C and D. The values of the intercepts on the horizontal axes of these replots allow the values of the ionization constants to be read off as shown.

Inspection of Eq. (23) and the mechanism of Eq. (22) shows that if hydrogen ions bind only to the free enzyme ($K_A^{ES} \to \infty, K_B^{ES} \to 0$) the kinetic patterns will be competitive (the slope K_m/V is affected, but not V). Conversely, if only the enzyme-substrate complex can ionize ($K_A^E \to \infty, K_B^E \to 0$) the effects will appear uncompetitive with V and K_m changed in the same proportion and the slopes of plots of $1/v$ against $1/s$ unchanged. If substrate binding does not affect the ionization constants, so that $K_A^{ES} = K_A^E$ and $K_B^{ES} = K_B^E$, the effect will be truly noncompetitive, with V and the slope changed but K_m unaffected, so that either kind of replot can be used to derive the ionization constants.

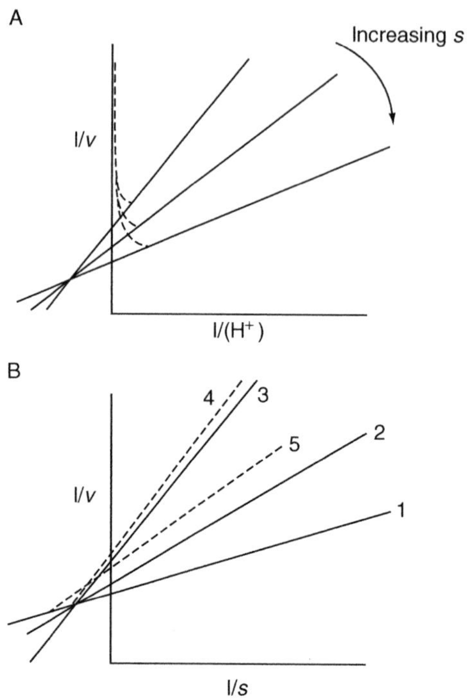

Fig. 6 Reciprocal plots of the effects of hydrogen ions on enzyme activity for a mechanism obeying Eq. (23) when $[H^+] \gg K_B^E$ and $[H^+] \gg K_B^{ES}$ (solid lines) and when this inequality breaks down (dashed lines). The solid lines intersect at a point to the left of the $1/v$ axis, which will be above the horizontal axis if $K_A^E < K_A^{ES}$, on that axis if $K_A^E = K_A^{ES}$, and below it if $K_A^E > K_A^{ES}$. In Fig. 5B the hydrogen ion concentration increases in the order 1–5.

2. Plots of Kinetic Constants Against pH

An alternative method of finding pK values is to use direct curve fitting of the appropriate parameter against pH, V to give the pK values of the enzyme-substrate complex and V/K_m to give those of the free enzyme. Most curve fitting programs will calculate the pK values automatically. However, accurate values will only be obtained if pK_B is far above pK_A, when the bell-shaped curve will have a flat top, and the pK values correspond to the pH values at which half the maximal velocity is reached (Fig. 3). If the pK values are close together or reversed, then the relationship given in Fig. 4 can be used to obtain the pK values. This is derived from the treatment of Alberty and Massey (1955), who showed that

$$K_A = [H^+]_A + [H^+]_B - 4[H^+]_{opt} \qquad (30)$$

where $[H^+]_A$ and $[H^+]_B$ are the hydrogen ion concentrations that give half the maximal velocity, and $[H^+]_{opt}$ is the concentration that gives the maximal velocity. From this it follows that if pH_B and pH_A are called $pH_{opt} \pm \log q$ (note that pH_{opt} is

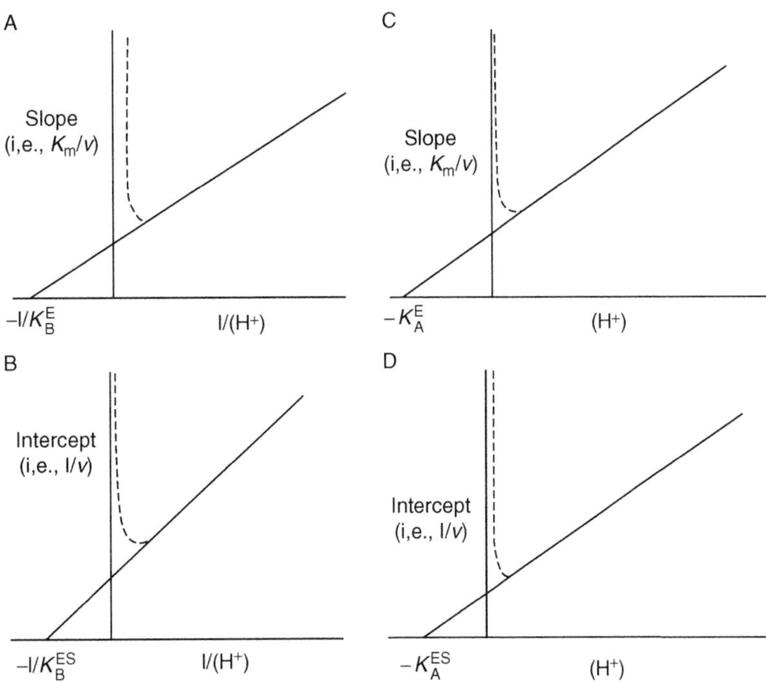

Fig. 7 Determination of ionization constants from secondary plots of the data shown in Figs. 5 and 6. Replots of the slope and vertical axis intercepts of the lines in Fig. 5B are shown in (A) and (B), respectively; those for Fig. 6B are shown in (C) and (D). The dashed lines correspond to deviations due to the breakdown of the simplifying inequalities described in the legends to Figs. 5 and 6.

the average of pH_A and pH_B, and also of pK_A and pK_B), then pK_A and pK_B are given by $pH_{opt} \pm \log (q - 4 + 1/q)$ (Dixon, 1979). The values thus found for pK_A and pK_B can be checked from the fit of the results over the whole of the bell-shaped curve by using its simplest expression $(\alpha + 1)/(\alpha + \cosh x)$ where $x = 2.303\ (pH - pH_{opt})$ and $\alpha = (K_A/4K_B)^{1/2}$ (so that $pK_B - pK_A = 2 \log 2\alpha$) (Dixon, 1974).

The accuracy of determination of pK_A and pK_B becomes very low when pK_B falls below pK_A (i.e., when there is strongly cooperative proton binding). This is because the width of the curve at half its maximal height becomes insensitive to changes in $pK_B - pK_A$ when the latter is small or negative (Fig. 4). Thus this width changes only from 1.53 pH units when $pK_B - pK_A = 0.6$ (the lowest it can be without positively cooperative proton binding) to 1.14 pH units as $pK_B - pK_A$ approaches minus infinity. In terms of Eq. (30), K_A becomes a small difference between large numbers as $[H^+]_A$ and $[H^+]_B$ approach $(2 \pm \sqrt{3})$ $[H^+]_{opt}$.

At the time Eq. (3) was derived it was not generally realized that it applies only to molecular ionization constants. If a group of high pK needs to be deprotonated, this does *not* cause pK_B to fall below pK_A, since these are the values for the second and first protons to dissociate, from whatever groups these come. It merely represents the situation of Eq. (1) in which $x \gg y$ and activity depends on y. Kinetics can tell us that the monoprotonated form is needed, but cannot tell us where the proton must be. A fall of pK_B below pK_A represents positively cooperative proton binding, so that the second proton to dissociate does so more readily than the first. Positive cooperativity at the group level is required, that is, K'_x of Eq. (1) must exceed K_x, if K_B is to exceed K_A, since $K_A/K_B = (2 + K_x/K_y + K_y/K_x)$ K_x/K'_x (Dixon and Tipton, 1973).

Although K_m varies with pH in a manner determined by the K values of both free enzyme and enzyme-substrate complex (Eq. (26)), a plot of K_m against pH gives only the pK values of the enzyme-substrate complex as the pH values of the midpoints of sigmoid waves in the curve. The pK values from the numerator of Eq. (26) determine the heights of these waves (as a consequence of determining where they start) but not the positions of their centers (Brocklehurst and Dixon, 1976; Fig. 3). Hence the pK values of the enzyme-substrate complex can be determined from this plot if they are well separated. If $1/K_m$ is plotted against pH, the pK values represented by the midpoints of the sigmoid waves will be those of the free enzyme (Fersht, 1977), since these terms will be in the denominator of the reciprocal form of Eq. (26), whereas the pK values of the enzyme-substrate complex will determine the heights of the waves.

3. Plots of Logarithms of Kinetic Constants Against pH

A convenient method of treating the effects of pH on enzyme activity is that of Dixon (1953), which involves plotting the logarithm of the kinetic constant against the pH (Fig. 8). It is treated in detail here because it has been widely used, although many would now prefer to fit the data directly to the titration curves or the bell-shaped curve represented by Eq. (20). For the maximum velocity (Fig. 8A), taking the logarithm of Eq. (25) gives

$$\log V = \log \tilde{V} - \log(1 + [\mathrm{H}^+]/K_A^{ES} + K_B^{ES}/[\mathrm{H}^+]) \qquad (31)$$

Since $\log \tilde{V}$ is independent of pH, the dependence will be determined by the second term. At low pH values where $[\mathrm{H}^+] \gg K_A^{ES}$ and $[\mathrm{H}^+] \gg K_B^{ES}$, the equation will simplify to

$$\log V = \log \tilde{V} - \log \frac{[\mathrm{H}^+]}{K_A^{ES}} = \log \tilde{V} + (\mathrm{pH} - \mathrm{p}K_A^{ES}) \qquad (32)$$

and thus the slope of the graph will be +1. As the pH is raised there will be a range over which two terms contribute to the slope, and provided that $K_A^{ES} \gg K_B^{ES}$ there will then be a range where $K_A^{ES} \gg [\mathrm{H}^+] \gg K_B^{ES}$. Over this range Eq. (31) simplifies to

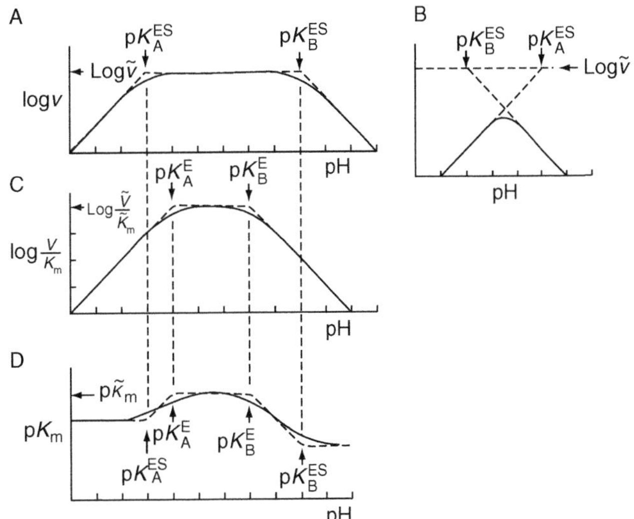

Fig. 8 Plots of logarithms of kinetic constants against pH for a system obeying Eq. (22). (A) Plot of $\log V$ (when $pK_B^{ES} > pK_A^{ES}$); (B) plot of $\log V$ (when $pK_B^{ES} < pK_A^{ES}$ and proton binding is therefore positively cooperative); (C) plot of $\log (V/K_m)$; (D) plot of pK_m. The alignment shows how pK values in ES and in E (or S) appear as changes of slope of 1 unit (see Dixon, 1953). In the example given, the form of the enzyme that exists between pK_A^E and pK_B^E has slightly more affinity for the substrate than the forms that exist outside this range. Hence, between pK_B^E and pK_B^E the substrate promotes protonation of the enzyme by binding mainly the protonated form. But above pK_B^{ES} the predominance of the deprotonated form of the free enzyme overcomes its lesser affinity for the substrate, and it is the main form ligated. After M. Dixon, *Biochem. J.* **55**, 161 (1953).

$$\log V = \log \tilde{V} \tag{33}$$

so the graph will have a slope of zero. Further increase in pH will give a region in which the dependence is governed by $\log (1 + K_B^{ES}/[H^+])$, followed by one where $K_B^{ES}/[H^+]$ is dominant and Eq. (31) simplifies to

$$\log V = \log \tilde{V} - \log \frac{K_B^{ES}}{[H^+]} = \log \tilde{V} + (pK_B^{ES} - pH) \tag{34}$$

which will give a slope of -1.

Thus, provided that $pK_B \gg pK_A$, a graph of $\log V$ against pH will be composed of linear regions of slopes $+1$, 0, and -1, which will be joined by curved sections where two of the terms in the second part of Eq. (31) contribute appreciably to the slope. From Eqs. (32)–(34) it can be seen that if the linear portions are extended they will intersect at the pK values. The curved portions will pass 0.3 unit (log 2) below the intersections (Fig. 8A). If, however, pK_B is below pK_A, then the horizontal portion is missing and the intersection is not at the pK values, but at their mean (Fig. 8B).

Similar behavior should be seen if log (V/K_m) is plotted against pH (see Eq. (27)), but in this case the intersection points will give the pK values for ionization of the free enzyme rather than of the enzyme-substrate complex (Fig. 8C). The behavior of graphs involving K_m is more complicated because this constant is affected by ionizations of both the free enzyme and the enzyme-substrate complex (Eq. (26)). Dixon (1953) has recommended that pK_m (i.e., $-\log K_m$) be plotted against pH. Since

$$pK_m = \log(V/K_m) - \log V \tag{35}$$

it follows that this plot represents a combination of the preceding two. The graph obtained should be composed of a number of sections with slopes of +1, 0, and −1, and each change of slope indicates a pK. Downward bends will correspond to ionizations of the free enzyme, since they will be given by the log V/K_m portion of Eq. (35), whereas upward bends will correspond to ionizations of the enzyme-substrate complex, since they will be given by the $-\log V$ portion. They are upward (i.e., they increase the slope by one unit) because $-\log V$ is here involved rather than downward for log V (Fig. 8C). A plot of pK_m is shown in Fig. 8D.

Since the plot of pK_m against pH contains all the information that can be obtained from the separate plots of log V and of log (V/K_m), one should be able to obtain all the pK values from it alone. In practice, however, pK values can be determined accurately only if they are well separated, so the presence of four pK values in the pK_m plot, compared with two in each of the others, may make it less satisfactory to use. As we will indicate later, however, there are cases in which the use of the pK_m plot in addition to the other two may help to resolve possible ambiguities.

The simple model being considered can be somewhat extended by supposing that two protonations or deprotonations are required to convert the form predominant over a given pH range into the active form. Then slopes of +2 or −2 may appear in the plots, although accurate determination of these requires sensitive assays capable of measuring activities that are small fractions of that at the pH optimum. This extension is just one example of the versatility of the logarithmic plot and the simplicity of the rules for interpreting it (Dixon, 1953).

4. Comparison of Graphical Methods

Three methods have been described above for obtaining ionization constants: (i) plots of $1/V$ and of K_m/V against [H^+] or 1/[H^+]; (ii) plots of V, V/K_m, and K_m (or its reciprocal) against pH; (iii) plots of the logarithms of these parameters against pH. All these methods are satisfactory if the ionization constants are well separated. The method of Malcolm Dixon (iii) is the most used and gives the clearest presentation of the results. Plots of kinetic constants against pH (method ii), especially for V and V/K_m which should show only two ionization constants each, have the advantage of the ease of applying the Alberty-Massey equation (Eq. (30)) and of fitting theoretical curves. It thus copes best when pK_B is close to pK_A or below it. The method (i) of plots against [H^+] or 1/[H^+] concentrates on a relatively narrow range of [H^+] around the ionization constants (there is little useful information outside the range 0.1–10*K* (Ainsworth, 1977). It has the advantage that it allows a more complete analysis of hydrogen ions as activators and inhibitors, using equations similar to those used in other aspects of kinetic studies (Segel, 1993; Cook and Cleland, 2007), especially by indicating the onset of complicating factors such as cooperativity or partial effects by curvature of the plots.

D. Complications

The mechanism shown in Eq. (22) oversimplifies real situations in several ways. In this section, we shall look at the assumptions it makes and examine the effects of dispensing with some of them.

1. The Presence of more than One Enzyme-Substrate Intermediate

The system in which two complexes of enzyme and substrate are envisaged (Eqs. (36)) gives a steady-state rate equation of the form shown in Eq. (24) where

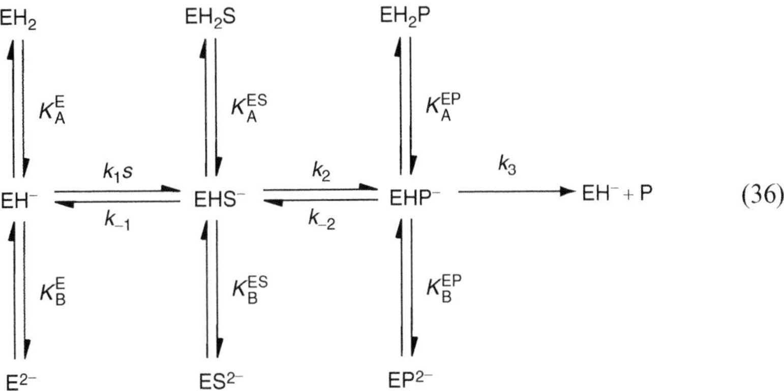

(36)

$$V = \frac{k_2 k_3 e}{(k_{-2}+k_3)(1+[\mathrm{H}^+]/K_\mathrm{A}^\mathrm{ES}+K_\mathrm{B}^\mathrm{ES}/[\mathrm{H}^+])+k_2(1+[\mathrm{H}^+]/K_\mathrm{A}^\mathrm{EP}+K_\mathrm{B}^\mathrm{EP}/[\mathrm{H}^+])} \tag{37}$$

and

$$V/K_\mathrm{m} = \frac{k_1 k_2 k_3 e}{(k_{-1}k_{-2}+k_{-1}k_3+k_2k_3)(1+[\mathrm{H}^+]/K_\mathrm{A}^\mathrm{E}+K_\mathrm{B}^\mathrm{E}/[\mathrm{H}^+])} \tag{38}$$

Equation (38) thus shows the usual pH dependence for V/K_m, which gives the ionization constants of the free enzyme. In fact, Eq. (37) is also simple in form, and the overall values it gives for $1/K_\mathrm{A}$ and for K_B are simply $(k_{-2}+k_3)/(k_2+k_{-2}+k_3)$

of the value of each of these for ES plus $k_2/(k_2 + k_{-2} + k_3)$ of the value of each for EP. It is thus a mean for the two complexes, weighted in the ratio of $(k_{-2} + k_3)$ to k_2 in favor of these for ES—that is, exactly in the ratio of the steady-state concentrations of EHS^- and EHP^-. It therefore does not matter how many enzyme-substrate complexes there are; the constants obtained from the pH dependence of V are average values weighted in favor of the predominant complex. If, further, equilibrium obtains between the complexes, then the predominance is governed by the equilibrium constant k_{-2}/k_2 instead of the steady-state ratio $(k_{-2} + k_3)/k_2$, since k_3 is then negligible in comparison with k_{-2}.

The scheme of Eq. (36) is much more plausible than that of Eq. (22), which is reasonable only when equilibrium is assumed between EH^- and EHS^-; otherwise Eq. (22) assumes that S dissociates slowly from EH^- but that P, which may be chemically similar to S, dissociates infinitely rapidly. It is therefore important that Eq. (36) gives similar kinetic relationships, although the makeup of the kinetic constants in terms of individual rate constants is changed.

2. More than One Form of the Enzyme Can Bind Substrate

Elaboration of the scheme of Eq. (22) into the mechanism of Eq. (39) by allowing differently protonated forms of the enzyme to combine with substrate leads to several alternative pathways by which the catalytically productive EHS^- complex can be formed. If it is assumed that the rate of breakdown of the EHS^- complex to yield products is slow relative to the dissociation steps, so that the latter remain at thermodynamic equilibrium, the situation does not differ from Eq. (22), because addition of pathways cannot affect an equilibrium, and thus Eq. (23) will apply to the mechanism shown in Eq. (39). If, however, steady-state conditions are

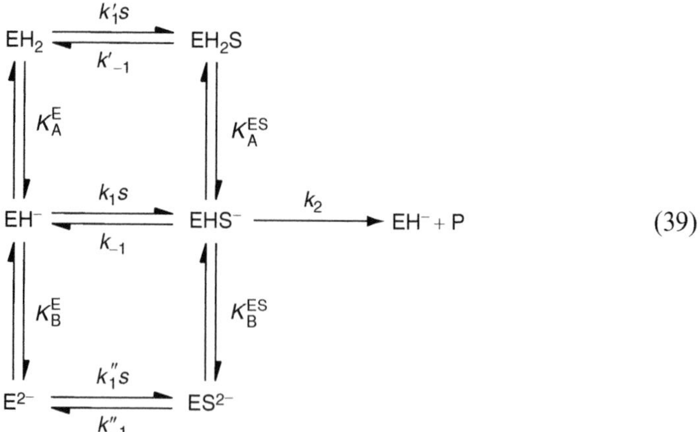

(39)

assumed to apply to the mechanism shown in Eq. (39), the resulting kinetic equation is extremely complicated (Alberty and Bloomfield, 1963; Kaplan and Laidler, 1967; Krupka and Laidler, 1960; Laidler, 1955; Ottolenghi, 1971; Peller and Alberty, 1963; Stewart and Lee, 1967) and contains terms in which the concentrations of the substrate and hydrogen ions are squared and cubed, respectively. This may lead to complex dependence of the initial velocity on the substrate and hydrogen ion concentrations, and generally the kinetic equation is too complicated to be of practical use. A number of simplifying conditions can result in equations similar to Eq. (23) (Laidler, 1955). One assumption, which was proposed by Alberty and his coworkers (Alberty and Bloomfield, 1963; Alberty and Massey, 1955; Peller and Alberty, 1963), is that the proton-transfer steps occur rapidly in relation to all the chemical steps, so that the proton-binding steps remain in thermodynamic equilibrium. Under these conditions the initial-rate equation becomes:

$$v = \frac{k_2 e/(1 + [\mathrm{H}^+]/K_\mathrm{A}^\mathrm{ES} + K_\mathrm{B}^\mathrm{ES}/[\mathrm{H}^+])}{1 + \{[k_2 + k_{-1}(1 + [\mathrm{H}^+]/K_\mathrm{A}^\mathrm{E} + K_\mathrm{B}^\mathrm{E}/[\mathrm{H}^+])]/k_1(1 + [\mathrm{H}^+]/K_\mathrm{A}^\mathrm{ES} + K_\mathrm{B}^\mathrm{ES}/[\mathrm{H}^+])\}1/s} \tag{40}$$

Because this equation differs from Eq. (23) only in the term representing K_m, V still depends solely on ionizations of the enzyme-substrate complex, so Eq. (25) still applies. The expression for V/K_m, however, shows four pK values and the plot of its logarithm against pH will show four downward turns and also two upward ones, which do not correspond to pK values. Two of the pK values are those of the free enzyme, but the other two are distorted by rate constants. It may be possible to tell which there are because the distorted ones are farther apart than the compensating upward turns by the same amount, namely log $[(k_2 + k_{-1})/k_{-1}]$ (Brocklehurst and Dixon, 1976). Hence if all six bends can be located the true pK values can be found, but the chances that all will be separated and within the range where the enzyme is both stable and appreciably active may not be great. As k_2 falls in comparison with k_{-1} the distorted pK values approach the compensating upward bends and therefore disappear, leaving Eq. (27) (Fig. 8C) to apply.

3. Change of Rate-Determining Step with pH

The above mechanism can show a change of rate-determining step with pH. The consequences of such a change are first examined in simple systems before returning to that of Eq. (39).

In a reaction of two consecutive steps, it is the ratio of the forward rate constant for the second reaction to the backward constant for the first reaction that determines which of the two is rate-limiting. Thus a reaction of the type will change its rate-determining step with pH. At low pH, the reactant is in its inert form AH^+ and the intermediate in its reactive form XH^+. Thus $k_2[\mathrm{XH}^+]$ will exceed k_{-1} [X] so

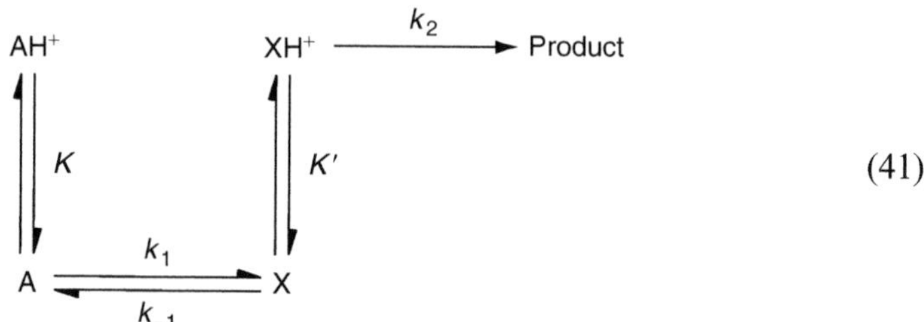

(41)

the first step will be rate-limiting. At high pH, the reverse will hold, all four species AH^+, A, X, and XH^+ will come to equilibrium, and the second step will be rate-determining. Reactions of this type, for which imine formation provides a nonenzymic model (Jencks, 1969), show a typical bell-shaped pH dependence. One of the two pK values that characterize the curve is the pK of the reactant, but the other is a "mirage"(Jencks, 1969) and is the same pK distorted by rate constants.(Dixon, 1973) The value of pK_B (compare Eq. (20)) cannot fall below p$K_A + 0.6$; that is, the width of the bell cannot be as narrow as that given by positively cooperative proton binding.

In demonstrating these statements, it was assumed (Dixon, 1973) that the total reactant concentration, that is, $[AH^+] + [A]$, was independent of pH, which is likely if it greatly exceeds the concentration of intermediate. In an enzymic reaction, however, no such assumption can be made. If only V is to be determined, then only species with bound substrate need be considered (Eq. (42)).

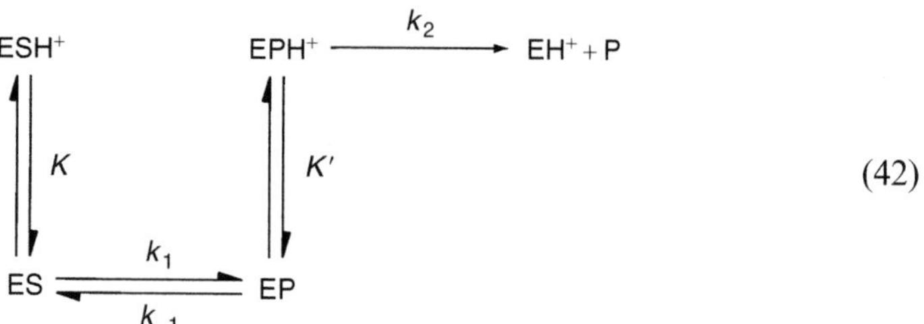

(42)

Steady-state treatment, considering $[ESH^+] + [ES] + [EP] + [EPH^+]$ constant, still gives an equation of the form of Eq. (25), with a bell-shaped curve of V against pH, but neither of the pK values that characterize it is undistorted by rate constants. It remains true, however, that $K_A \geq 4\ K_B$; that is, p$K_B \geq$ p$K_A + 0.6$.

The mechanism given in Eq. (39) can show a change in rate-determining step with pH (Brocklehurst and Dixon, 1976, 1977) because (provided $k_2 > k_{-1}$) the term $k_2\ [EHS^-]$ may exceed the rate of the reverse of the first step, that is,

$k'_{-1}[EH_2S] + k_{-1}[EHS^-] + k''_{-1}[ES^{2-}]$, at some pH values and be less than it at others. Renard and Fersht (1973) have analyzed a change of rate-determining step with pH. Demonstration of its occurrence normally requires a method for observing how the concentration of intermediates varies with pH and not merely the steady-state rate of reaction. In the case of dihydrofolate reductase, for example, it has been shown by stopped flow studies that the apparent pK value of 8.4 observed in steady-state kinetics results from a change in rate-determining step from product release at low pH to hydride transfer at pH values above 8.4 (Fierke *et al.*, 1987).

4. More than One Form of Enzyme-Substrate Complex Can Yield Products

So far we have assumed that the only ionization reactions that affect the breakdown of enzyme-substrate complex to products completely prevent this reaction. It is, however, possible that loss or gain of a proton near the active site may change the rate of the reaction. So we must modify the scheme of Eq. (39) by drawing another route to the product (Eq. (42a)).

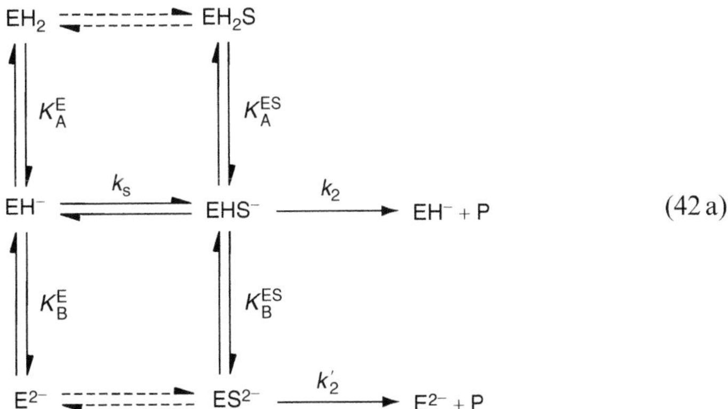

(42 a)

We will deal only with rapid equilibrium of substrate and hydrogen ion binding, because the full steady-state equations are complex. Hence we have assigned an equilibrium constant for binding of substrate to one form of the enzyme; whether or not the routes shown by dashed lines occur, their equilibrium constants are determined by K_s and the acid dissociation constants are marked. The resultant rate equation is

$$v = \frac{\{[k_2 + k'_2(K_B^{ES}/[H^+])]e\}/(1 + [H^+]/K_A^{ES} + K_B^{ES}/[H^+])}{1 + (\tilde{K}_m/s)[(1 + [H^+]/K_A^E + K_B^E/[H^+]/(1 + [H^+]/K_A^{ES} + K_B^{ES}/[H^+])]} \tag{43}$$

This predicts a simple dependence of K_m on pH (Eq. (26)), in accord with the simple assumption that binding is at equilibrium, and the dependence can give all four pK values. The pH dependence of V and of V/K_m is more complicated.

Putting K_2' to zero simplifies the scheme to that of Eq. (22) and simplifies Eq. (43) to Eq. (23). The possible types of pH dependence of V are shown in Fig. 9. It is possible that ES^{2-} may lose a further proton to render it completely incapable of reaction, as shown by the dashed portions of Fig. 9, but this may not occur within the pH range under study. Provided that the pK values are adequately separated, they can be found from the curves.

For the range over which $[H^+]/K_A^E$ and $[H^+]/K_A^{ES}$ are negligible, and likewise any further inactivating dissociation of ES^{2-} is negligible—that is, the range between the two plateaus of Fig. 9—hydrogen ions will appear to be nonessential activators if $k_2' < k_2$ or partial inhibitors if $k_2' > k_2$. Thus secondary plots of $1/V$ and of K_m/V (intercepts and slopes of plots of $1/v$ against $1/s$) against $[H^+]$ or $1/[H^+]$ will be hyperbolic. Linear secondary plots can be obtained by plotting different functions, as follows. Suppose that the plateau at higher pH is the higher one (Fig. 9A), that is, that $k_2' > k_2$, and that the value of V on the lower plateau is V', that is, $V' = k_2e$. Likewise let the height of the higher plateau be $\alpha V'$, so that $\alpha = k_2'/k_2$. From Eq. (43), considering only the range between the plateaus,

$$\frac{1}{V} + \frac{1 + K_B^{ES}/[H^+]}{V'(1 + \alpha K_B^{ES}/[H^+])} = \frac{[H^+] + K_B^{ES}}{V'([H^+] + \alpha K_B^{ES})} \quad (44)$$

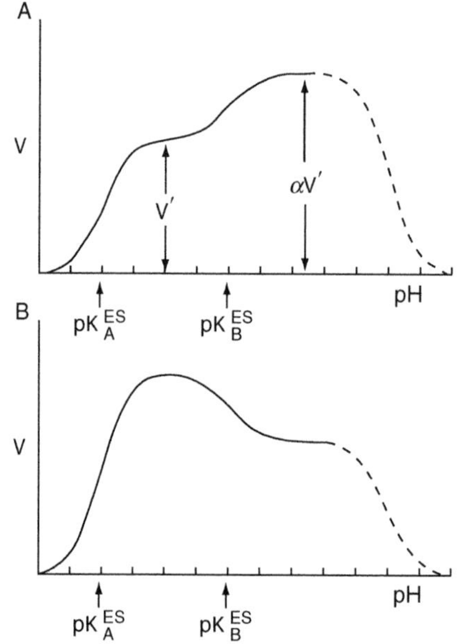

Fig. 9 pH Dependence of V when two protonic states of ES can yield products. Deprotonation of ES with a pK of pK_B^{ES} (A) increases and (B) diminishes the rate of the catalyzed reaction. Dashed lines indicate that a further deprotonation may prevent the reaction.

Hence

$$\frac{1}{V} - \frac{1}{V'} = \frac{[H^+] + K_B^{ES} - [H^+] - \alpha K_B^{ES}}{V'([H^+] + \alpha K_B^{ES})} = \frac{K_B^{ES}(1-\alpha)}{V'([H^+] + \alpha K_B^{ES})} \tag{45}$$

Taking reciprocals gives

$$1/(1/V - 1/V') = [H^+]V'/K_B^{ES}(1-\alpha) + V'\alpha/(1-\alpha) \tag{46}$$

Hence a plot of $1/(1/V - 1/V')$ against $[H^+]$ will be linear and will have a slope of $V'/K_B^{ES}(1-\alpha)$, an intercept on the vertical axis of $V'\alpha/(1-\alpha)$, and an intercept on the $[H^+]$ axis of $-K_B^{ES}$. Since α and V' are known from the heights of the plateaus, K_B^{ES} is determined from the results over the transition between them. Similar types of rearrangement can be used to obtain linear plots when $k_2' < k_2$ and also for plotting derivatives of V/K_m against $[H^+]$ or $1/[H^+]$. In plots of log V or log (V/K_m) against pH, this mechanism gives an extra bend as the velocity levels off to the finite value of V at high pH values. This bend does not correspond to a pK and will not be seen in a plot of pK_m against pH, and thus the use of this plot in addition to the other two may help to avoid a mistaken assignment of a pK value.

Clearly this mechanism can be extended to activity for more than two protonic states of the enzyme-substrate complex. Provided the pK values are separated enough so that plateaus exist, their values can be determined as above.

5. Abortive Complex Formation

An enzyme may be able to bind a substrate in unproductive ways. Fastrez and Fersht (1973) analyzed this in connection with the hydrolysis by chymotrypsin of Ac-Tyr-NH-Ph in which the phenyl group of the aniline residue bore various substituents. If the substrate binds with this phenyl ring in the site that should bind that of the tyrosine residue, no hydrolysis can occur. In the simplest case, the same protonic state of the enzyme binds substrate to form a productive complex ES and an unproductive one ES′ but cannot bind substrate in both ways at once, and both forms of binding are in equilibrium. The pK values of V/K_m remain those of the free enzyme, but those of V prove (Fastrez and Fersht, 1973) to be weighted average of all forms of enzyme-substrate complex whether productive or not. If abortive binding predominates, then the pK values seen in V are those of the abortive complex ES′. More complex situations involving abortive binding can easily be envisaged.

6. Comparison of Assumptions

In systems containing alternative pathways (see Eq. (39)) it was necessary to assume, in order to obtain manageable equations that all protonation steps were at thermodynamic equilibrium. This assumption has been made in most studies of the effects of pH on enzyme activity, but has been questioned (Laidler, 1955; Ottolenghi, 1971), and in a detailed criticism Knowles (1976) queried its validity, pointing out that dissociation of protons may be as slow as $10^4\ s^{-1}$ whereas other steps in the enzymic reaction may be faster. In some branches of the mechanism shown in Eq. (39) substrate binds to the enzyme immediately after a protonation step; thus, for equilibrium conditions to apply to these proton-binding steps alone, it is necessary for their dissociation to be much faster than the subsequent substrate-binding steps (see, e.g., Dalziel, 1969). Enzymes may, however, combine with their substrates at up to $5 \times 10^8\ M^{-1}\ s^{-1}$, so that rates of combination at natural substrate concentrations may be up to $10^6\ s^{-1}$, suggesting that equilibrium conditions for protonation may not always occur when the interaction with substrate is in steady state. Indeed, Knowles (1976) cited evidence that increasing the buffer concentration and thus accelerating equilibration of protons increases the activity of carbonic anhydrase (Tu and Silverman, 1975). In nonenzymic models, for example, catalysis of amide hydrolysis by a carboxyl group (Aldersley *et al.*, 1972), transfer of a proton can be rate-limiting. Cornish-Bowden (1976) also concluded that protonation reactions are unlikely to be slower than $10^4\ s^{-1}$ (unless compulsorily accompanied by conformational changes). He argued that the rates of individual steps in enzyme-catalyzed reactions are unlikely to be much faster than this, so that the equilibrium assumption may often be valid. We cannot assume, however, that this will always be so. Although an increase of rate on increasing buffer concentration gives evidence that a proton transfer is rate-limiting, absence of such an effect does not rule out a rate-limiting proton transfer catalyzed by groups in the enzyme, possibly at a site inaccessible to buffer components.

Cleland (1977) has called attention to the fact that bound substrate can hinder proton equilibration by an enzyme; thus the proton removed by fumarate hydratase from malate only exchanges relatively slowly with water while fumarate is on the enzyme (Hansen *et al.*, 1969). He has therefore considered the situation in which proton-transfer steps in enzyme-substrate complexes are slow, even though those in the free enzyme are considered to be in equilibrium. The rate equations given by such systems contain terms in which the hydrogen ion concentration is squared, and these can lead to waves in Dixon plots as well as to displacement of pK values.

Much more general cases have been considered by many authors (Alberty and Bloomfield, 1963; Kaplan and Laidler, 1967; Krupka and Laidler, 1960; Laidler, 1955; Ottolenghi, 1971; Peller and Alberty, 1963; Stewart and Lee, 1967), who have developed complex initial-rate equations to describe them and have considered how assumptions can simplify these equations. We have shown how even fairly probable departures from the simplest system represented in Eq. (22) lead to kinetic pK values that differ from the molecular values. Although we have indicated ways for finding molecular pK values in some specific cases, they apply when it is known that a particular complication is operative, and this may not be easy to discover.

E. Interpretation of the Results of pH Experiments

As pointed out already, the results of studies on the effects of pH on enzyme activity are difficult to interpret with any certainty. Even in the simplest system the pK values obtained are relatively complex molecular constants (Eqs. (16) and (17)) rather than simple group constants. It is often assumed without any arguments in favor that a simple mechanism is operative, although quite simple features often found in enzyme mechanisms may shift the kinetic pK values, and those observed may bear little relation to the molecular values in more complex mechanisms if the binding steps do not approach equilibrium. Despite these reservations, many such studies have yielded results that have been shown by other methods to be substantially correct in identifying the pK values of groups involved. Even if pK values are shifted, an approximate value for a pK may give an indication of the type of group involved in a process, and so can usefully supplement other methods.

The concept that a pK value corresponds to that of a single ionizable group in an enzyme is, however, misleading. Most proteins contain several groups of the same type, each of these can contribute to the observed pK if their ionization states affect one another. In such cases, the observed pK will be made up of fractional contributions from each of the groups. The pK determined will represent a single specific group only if it titrates in a different pH range from the other groups of the same type or if its affinity for H^+ is unaffected by their ionization state (Dixon *et al.*, 1991). In enzymes such as ribonucleases, where the activity depends on two histidine residues, one protonated and the other ionized, it is misleading to ask which residue has the lower pK but rather how the two identified groups partition between the two observed pK values. Similar considerations apply to the aspartyl peptidases where one carboxyl group must be ionized and another protonated; each would be expected to contribute some fraction of each of the observed pK values.

Possibly too much emphasis has been placed on the use of studies of pH dependence to identify specific ionizing groups. Other features of the responses of enzymes to pH changes may be informative about their mechanisms. Changes in the reaction mechanism of an enzyme as the pH is changed might yield valuable information on the role of ionizing groups in the reaction, and variation of the kinetic parameter V/K_m with the nature of the substrate may provide further information on the nature of the substrate-binding site in the enzyme (Shindler and Tipton, 1977).

F. pH-Independence of K_m

Haldane (1930) pointed out that the constancy of K_m over a pH range in which V showed a typical bell-shaped curve, observed for yeast invertase, might be explained by equality of K_m and K_s. Thus Eq. (40), given by the mechanism of Eq. (39), simplifies to give pH independence of K_m only if $k_2 \ll k_{-1}$, $K_A^E = K_A^{ES}$, and $K_B^E = K_B^{ES}$, and then K_m simplifies to k_{-1}/k_1, that is, K_s. Cornish-Bowden (1976, 1976) has extended the analysis and also concluded that the constancy of K_m required that $K_A^E = K_A^{ES}$ and $K_B^E = K_B^{ES}$ and that the rate constants for binding

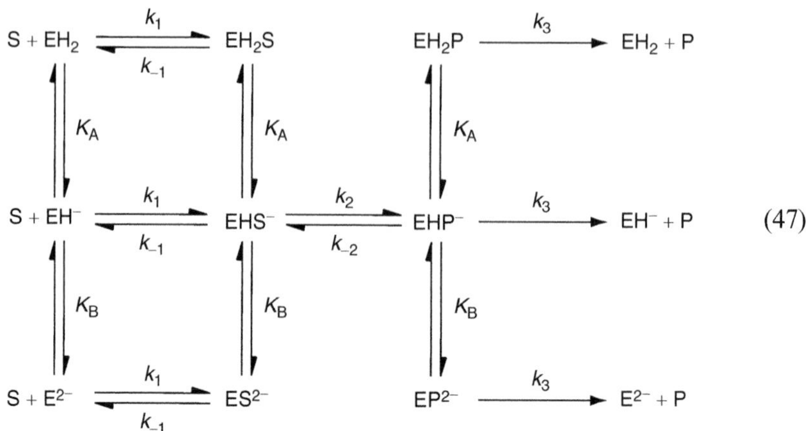

(47)

and dissociation of substrate and product must be unaffected by the protonation and deprotonation that inactivate the enzyme-substrate complex, so that the reaction can be represented as shown in Eq. (47).

Cornish-Bowden showed, however, that this mechanism gives constancy of K_m even if substrate binding does not approach equilibrium, that is, when the condition $k_2 \ll k_{-1}$ does not hold, provided that then $k_{-1} = k_3$, that is, that substrate and product dissociate equally rapidly, which may be quite likely if they are chemically similar. If this is the cause of constancy of K_m, it again proves to simplify to k_{-1}/k_1, so the pH independence gives the valuable information that $K_m = K_s$. It is only when $k_{-1} \gg k_2$, however, that the pK values derived from the pH dependence of V are the true values of pK_A and pK_B; if K_m is independent of pH because of equality of k_{-1} and k_3, then the observed pK values are shifted further apart, each moved from pK_A or pK_B by log $[(k_2 + k_{-2} + k_3)/k_3]$.

The assumptions $K_A^E = K_A^{ES}$ and $K_B^E = K_B^{ES}$ can give constancy of K_m without equality of K_m and K_s in a simpler system, such as that represented by Eqs. (22) and (23). Such systems are, however, somewhat unrealistic with these assumptions, because it would be surprising for substrate binding to have no effect on protonation equilibria and yet to be completely inhibited by these protonations and deprotonations. Hence constancy of K_m fairly reliably establishes that $K_m = K_s$, although, in view of the restrictions needed to give it, it is likely to be a much rarer phenomenon than equality of K_m and K_s.

G. Identification of Amino Acid Residues from Their Ionizations

1. pK Values

The table shows the normal ranges of pK values that have been found for the ionizing groups of amino acid residues in proteins. Comparison of the values with those obtained from studies of the effects of pH on enzyme activity may suggest the groups involved in different stages of the enzymic reaction. Such suggestions are only tentative for two reasons. The first is that, as already noted, the kinetic p*K* values may not be true p*K* values. The second is that the p*K* values of groups may be greatly affected by their environments, whereas the values given in the table are those for groups in an aqueous environment. A hydrophobic environment will stabilize the un-ionized forms of groups. This will be the protonated form for a "neutral acid group" (e.g., –COOH) so that its p*K* is raised, and the deprotonated form of a "cationic acid group" (one that bears a positive charge when protonated, e.g., $-NH_3^+$), whose p*K* will be lowered. Similarly, proximity to another charged group can greatly affect a p*K*, a positive charge lowering it and a negative charge increasing it. An extension of the effect of charge is the possibility of hydrogen bonding; a proton may be lost more easily because its binding site may hydrogen bond to another bound proton and be thereby stabilized, or hydrogen bonding to a neighboring group in addition to binding to its primary site of attachment may help to retain a proton. It is therefore not surprising that several groups in proteins are known to have p*K* values far outside the ranges shown. Perhaps the most extreme is the lysine residue of acetoacetate decarboxylase that forms an imine with the substrate. There is good evidence (Schmidt and Westheimer, 1971) that it has a p*K* of 5.9. The p*K* of 8.8 of a terminal amino group of δ-chymotrypsin proves to be made up of a p*K* of 7.9 in the conformation favored at high pH, in which the group is exposed to the medium (i.e., a normal value), and a p*K* of 10.0 in the conformation favored at low pH, in which it is adjacent to a negative charge inside the protein molecule (Fersht, 1971). Further, a carboxyl group in lysozyme is displaced in p*K* from 5.2 to 6.0 by ionization of a neighboring aspartic residue and to above 8 by combination with a long-chain substrate (Parsons and Raftery, 1972). In the case of the cysteine-peptidase papain, the p*K* of an aspartate carboxyl is lowered to 2.8 by interactions resulting from a reaction of a cysteine-histidine$^+$ pair, a minor species in which the imidazole group is protonated as a result of dissociation of the cysteine SH group, and the aspartate (Noble *et al.*, 2000). Clearly in such cases the simple question of which specific group gives rise to an observed p*K* value can be over simplification.

Typical pK Values for Side-Chain Ionizing Groups[a]

		ΔH of dissociation	
Residue	**p*K***	kJ/mol	kcal/mol
Asp, Glu	4	0	0
His	6	30	7
Cys	9	30	7
Lys	10	45	12
Tyr	10	30	7

[a]The p*K* of a terminal carboxyl is about 3.5 and that of a terminal ammonium is about 8; these values are lower than those of the corresponding groups in side chains because of the electron-withdrawing effect of the adjacent peptide bond. Other displacing factors are discussed in the text; because they include nearby charges, the values shown above do not apply to the free amino acids.

2. Heats of Ionization

Examination of the table shows that even without displacement of pK values the identification of groups may be difficult because the ranges of values for the different groups overlap. Determination of the heat of dissociation of a group from the temperature dependence of its pK may allow them to be distinguished. For a dissociation reaction (Eq. (48))

$$\Delta H = RT^2 \mathrm{d}(\ln K)/\mathrm{d}T = -R\mathrm{d}(\ln K)/\mathrm{d}(1/T) = 2.303R\mathrm{d}(\mathrm{p}K)/\mathrm{d}(1/T) \quad (48)$$

where ΔH is the enthalpy change on dissociation, R the gas constant, and T the absolute temperature. Thus a graph of pK against $1/T$ should have a slope of $\Delta H/(2.303R)$. Values of typical enthalpies of dissociation are shown in the table. Although they may be helpful when pK values are similar, they are not reliable for distinguishing groups whose pK values have been displaced by their environment, because this displacement may change the ΔH (Knowles, 1976).

As an approximation we may consider that a pK is displaced by a change in its ΔH of ionization. The change in ΔG^0 is $2.303RT$, that is, 6 kJ/mol, for each unit the pK is displaced, so we might expect ΔH to be changed by the same amount. Displacement of the pK of a typical amino group from 10 to 7 would therefore change its ΔH from 45 to 27 kJ/mol; similarly, displacement of the pK of a typical carboxyl group from 4 to 7 would change its ΔH of dissociation from 0 to 18 kJ/mol. The difference in ΔH that remains is rather small on which to attribute a pK of 7 to one of these groups rather than the other. Further, it can only be an approximation that the changes in ΔG and ΔH are equal. An ionization may be accompanied by a change in the ordering of water molecules. Factors that displace the pK may easily alter the degree of this change. At biological temperatures (i.e., not far from the freezing point of water), this will have little effect on ΔG, but a large one on ΔH (compensated in ΔG by an opposite one in $-T\Delta S$).

3. Effect of Dielectric Constant

An approach that allows neutral acid and cationic acid groups to be distinguished is the effect of alteration of the dielectric constant of the medium on the observed pK. A difficulty is how to define the pH of a medium other than water, and two ways round this have been found.

The more rigorous method (Findlay *et al.*, 1962) was used for studying the pK values of pancreatic ribonuclease. The pH is defined as the pH that the buffer used would give in water, that is, as $\mathrm{p}K_{\mathrm{buffer}} + \log$ ([buffer base]/[buffer acid]). This may be determined by measuring the pH of the solution before addition of organic solvent (Cleland, 1977). Then the effect of addition of dioxane or formamide is considered. A cationic acid group ($-\mathrm{GH}^+$ and $-\mathrm{G}$) would react with a neutral acid buffer (HB and B^-) as follows:

$$-\mathrm{GH}^+ + \mathrm{B}^- \rightleftharpoons -\mathrm{G} + \mathrm{HB} \quad (49)$$

Since lowering the dielectric constant favors uncharged species it will displace this reaction to the right. Hence at a fixed pH, that is, at a fixed ratio $[B^-]/[HB]$, a greater fraction of the group will be deprotonated, so that its pK is diminished. In cationic acid buffers, however, the reaction becomes

$$-GH^+ + B \rightleftharpoons -G + HB^+ \tag{50}$$

so the total charge on both sides is the same, and the equilibrium is unaffected by addition of organic solvent. Likewise the pK of a neutral acid group is raised in cationic acid buffers and unchanged in neutral acid buffers. Hence a study of the effect of addition of organic solvent on the pK of the group in the two types of buffer assigns the group to its class.

A more rapid method, which appears to work just as satisfactorily, is to calibrate a glass electrode in water and use it directly in the mixture of water and organic solvent. At least when dioxane is the solvent added, the electrode appears to measure correctly the activity of hydrogen ions. The two reactions to be considered, for cationic acid groups and neutral acid groups, respectively, are

$$-GH^+ \rightleftharpoons -G + H^+ \tag{51}$$

$$-GH \rightleftharpoons -G^- + H^+ \tag{52}$$

The charge is unchanged in the first reaction, and organic solvent proves to have little effect on the measured pK. In the second reaction, however, organic solvent inhibits dissociation, since this reaction produces charged species; thus addition of dioxane to water to 80% concentration progressively raises the pK of formic acid from 3.72 to 6.11 (Inagami and Sturtevant, 1960).

Two warnings should be given on the use of this method. One is that the organic solvent may change the conformation and ionizing properties of the enzyme under study. The other is that the molecular pK observed may have groups of different types contributing to it. Thus the carboxylate-imidazole system of chymotrypsin has a pK of 7 for protonation of form (I) to form (II) (Eq. (53)).

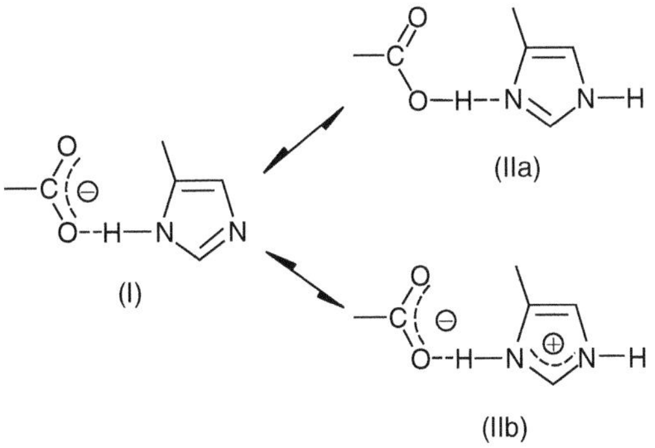

(I) ⇌ (IIa) ; (I) ⇌ (IIb) (53)

Since form (II) has an infrared spectrum like that of an undissociated carboxylic acid (Koeppe and Stroud, 1976), and the 2-proton of the imidazole has a nuclear magnetic resonance spectrum like that of unprotonated imidazole (Hunkapiller *et al.*, 1973), there is strong evidence (with some qualification (Dixon, 1976) that form (IIa) contributes to form (II) much more than does form (IIb). In both these forms the proton is added to the side of the histidine that is accessible to water, and the net charge of the complex is that of a neutral acid. Nevertheless, the test with organic solvent gives the answer of a cationic acid (Inagami and Sturtevant, 1960). Clearly the system is very different from that of Eqs. (51) and (52), where all the charged and uncharged forms are surrounded by water, since the carboxyl group and one side of the imidazole ring are buried.

4. Direct Titration of Groups

The identification of a specific ionizing group from its pK can be greatly strengthened if it can be shown that the group actually titrates with the determined pK. The complete titration curve of a protein with acid or alkali is usually too complex to be interpreted, because many groups are present and their titration curves overlap. The pK values of some specific groups may be found from the differences in proton binding or release between the native enzyme and enzyme in which a specific residue has been replaced by site-directed mutagenesis or chemically modified to prevent its ionization. Both these approaches have been used with lysozyme, for example. For example, to determine the pK values of the carboxyl groups one aspartic residue (Asp-52) in lysozyme by comparing titration curves before and after its esterification with Et_3O^+ ion (Parsons and Raftery, 1972), whereas the titration behavior of both Asp-52 and Glu-35 were studied after replacing each with by its corresponding amide by site-directed mutagenesis (Hashimoto *et al.*, 1996). Similarly the effects of changing Asp-52 to Ala on the pK value of Glu-35 have been studied in that enzyme (Malcolm *et al.*, 1989).

The titration of some specific classes of residues in proteins can be followed by measuring a property that changes when the residues ionize. Perhaps the best known of these approaches uses the change in ultraviolet absorbance of tyrosine residues when they ionize, and a similar method can be used for cysteine residues (Dyson *et al.*, 1997). The ionization of carboxyl groups can be followed by the accompanying shift in an infrared absorption band from about 1710 to about 1570 cm^{-1} (Koeppe and Stroud, 1976; Susi*et al.*, 1959; Timasheff and Rupley, 1972). Because water absorbs strongly near these bands, it is necessary to carry out these studies in ^2H_2O. Proton magnetic resonance in ^2H_2O may be used to observe the titration of histidine residues, because the resonance of the proton on C-2 of the imidazole ring is sensitive to the ionization state of the ring and is shifted out of the envelope of the numerous aliphatic protons of the protein (Hunkapiller *et al.*, 1973; Robillard and Schulman, 1972; Markley, 1975). Ionization of tyrosine residues can be similarly followed (Karplus *et al.*, 1973). ^{13}C-Nuclear magnetic resonance has been used to titrate lysine and aspartic residues (Lindman *et al.*, 2006).

These methods are not quite as direct as might appear. A difference titration between a protein with one group blocked and the unmodified protein can be interpreted simply only if blocking the group does not affect the titration of any other ionizing groups in the protein. Parsons and Raftery (1972) dealt ingeniously with one change in another group (see the Section IV). Similarly, the blocking of all groups except the one under study assumes that the modification does not affect this group. The methods based on specific properties of groups will give precise molecular pK values and an estimate of the contribution of the group to a molecular pK. Often all that is wanted is such an estimate, as the group may contribute overwhelmingly [$K_x \gg K_y$ in Eq. (1)]. But the precise contribution cannot be determined without assuming that the change of property on ionization is completely independent of environment (Dixon, 1976). Thus, in the procedures just cited for showing that form (IIa) contributes more than (IIb) in Eq. (53), the properties of carboxyl and carboxylate cannot be assumed to be exactly the same with N or HN groups of imidazole as close as they are in model compounds; the interaction that so much affects their pK is likely to affect their other properties.

5. Effects of pH on Enzyme Inhibition

A simple competitive inhibitor will bind to an enzyme in a rapid equilibrium, and thus the inhibitor constant will be the simple dissociation constant for the EI complex and some of the possible complications involved in consideration of the effects of substrate can be neglected. For example, for the system

$$\begin{array}{ccccccc} EH_2I & \overset{K_i'}{\rightleftharpoons} & EH_2 & & & & \\ \updownarrow K_A^{EI} & & \updownarrow K_A^{E} & & & & \\ EHI^- & \underset{I}{\overset{K_i}{\rightleftharpoons}} & EH^- & \overset{S}{\rightleftharpoons} & EHS^- & \longrightarrow & \text{etc.} \end{array} \tag{54}$$

the pH dependence of K_i will be given by:

$$K_i = \tilde{K}_i \frac{1 + [H^+]/K_A^E}{1 + [H^+]/K_A^{EI}} \tag{55}$$

and thus it should be possible to determine the pK values from a graph of pK_i against pH (Dixon, 1953).

The situation with simple uncompetitive inhibitors should also be relatively easy to analyze, giving the ionization constants for the ES and ESI complexes, but mixed and noncompetitive inhibition will be more complicated because of the alternative forms of the enzyme to which the inhibitor can bind.

6. Effects of pH on Individual Steps of an Enzyme Reaction

If it is possible to study the effects of pH on one step in an enzyme reaction by using rapid reaction techniques it should be possible accurately to determine the pK values of the groups controlling this step. An example of such an approach concerns the hydrolysis of 4-methyl umbelliferyl phosphate by alkaline phosphatase, where the reaction proceeds by way of a phosphoenzyme intermediate. Stopped-flow studies on the effects of pH on the reaction showed that the rate constant for the formation of the phosphoenzyme was relatively insensitive to pH in the range 5.3–7.5, whereas that for breakdown of the intermediate to yield phosphate increased with increasing pH, resulting in a change in the rate-limiting step of the reaction at about pH 6.5 (Halford and Schlesinger, 1974). Whereas the effects of pH on the steady-state kinetics of an enzyme would be expected to yield pK values only for the initial substrate-binding step and the rate-determining step of the reaction, rapid reaction measurements may yield pK values for intermediate steps in the reaction pathway. Such studies have, for example, also shown that substrate binding can result in changes in the pK values of amino acid residues involved in the catalytic process (Elfström and Widersten, 2006; Stines-Chaumeil *et al.*, 2006), proton release or uptake and conformational changes that do not affect the steady-state parameters (Caccuri *et al.*, 1998; Jentoft *et al.*, 1997).

7. Effects of pH on Chemical Modification of an Enzyme

This method is designed to identify the pK value of a specific group at the active site of an enzyme by measuring the effect of pH on the rate at which it reacts with a highly specific irreversible inhibitor (Schmidt and Westheimer, 1971). If the inhibitor (X) reacts with the enzyme in a single irreversible step

$$\begin{array}{l} \mathrm{EH} \\ \Big\updownarrow K_A \\ \mathrm{E}^- \xrightarrow{\tilde{k}_0 x} \mathrm{E}^- - \mathrm{X} \end{array} \tag{56}$$

the observed rate constant (k_0) will be given by

$$k_0 = \frac{\tilde{k}_0 x}{1 + [\mathrm{H}^+]/K_\mathrm{A}} \tag{57}$$

where x represents the concentration of X. Thus the value of pK_A can be found by determining the effect of pH on k_0. The scheme shown in Eq. (56) can easily be expanded to allow for two ionizations of the free enzyme.

If x is much greater than the initial enzyme concentration ($[\mathrm{E}]_0$) pseudo-first-order conditions apply, and thus

$$[\mathrm{E}] = [\mathrm{E}]_0 - [\mathrm{E} - \mathrm{X}] = [\mathrm{E}]_0 \mathrm{e}^{-k_0' t} \tag{58}$$

where [E] is enzyme concentration at any given time, t, and the pseudo-first-order rate constant (k'_0) is given by

$$k'_0 = k_0 x \tag{59}$$

Equation (58) can be written as

$$\ln \frac{[E]_0}{[E]} = k'_0 t \tag{60}$$

and thus k'_0 can be determined from a plot of log $[E]_0/[E]$ against time.

If the inhibitor first forms a reversible complex with the enzyme according to the mechanism

$$\begin{array}{l} \text{EH} \\ \Big\Updownarrow K_A \\ \text{E}^- \underset{k_{-1}}{\overset{k_1 x}{\rightleftharpoons}} \text{EX}^- \xrightarrow{\tilde{k}_0} \text{E}^- - \text{X} \end{array} \tag{61}$$

the rate equation becomes

$$k_0 = \frac{\tilde{k}_0}{1 + (\tilde{K}_m/x)(1 + [H^+]/K_A)} \tag{62}$$

where $\tilde{K}_m = (k_{-1} + k_2)/k_1$.

Systems obeying this mechanism can be distinguished from those obeying that shown in Eq. (56) because Eq. (57) predicts that k_0 will be a linear function of x whereas Eq. (62) predicts that the dependence will be hyperbolic (Kitz and Wilson, 1962). Thus K_A can be evaluated graphically by using the reciprocal form of Eq. (62).

$$\frac{1}{k_0} = \frac{\tilde{K}_m}{\tilde{k}_0}\left(1 + \frac{[H^+]}{K_A}\right)\frac{1}{x} + \frac{1}{\tilde{k}_0} \tag{63}$$

Thus a graph of $1/k_0$ against $1/x$ at a series of concentrations of H^+ will yield a family of straight lines that intersect on the $1/k_0$ axis at a value corresponding to $1/\tilde{k}_0$, and replots of the slopes of these lines against the concentration of H^+ will extend to cut the baseline at $-K_A$. An alternative procedure, suggested by Schmidt and Westheimer (1971), is to use such a low inhibitor concentration that Eq. (62) simplifies to

$$k_0 = \frac{\tilde{k}_0}{(K_m/x)(1 + [H^+]/K_A)} \tag{64}$$

when K_A can be determined directly from the effect of pH on k_0. If the intermediate enzyme-inhibitor complex is able to ionize

$$\begin{array}{ccccc} \text{EH} & & \text{EXH} & & \\ \updownarrow K_A & & \updownarrow K_B & & \\ \text{E}^- & \underset{k_{-1}}{\overset{k_1}{\rightleftharpoons}} & \text{EX}^- & \xrightarrow{\tilde{k}_0} & \text{E}^- - \text{X} \end{array} \tag{65}$$

the rate equation becomes

$$k_0 = \frac{\tilde{k}_0}{(\tilde{K}_m/x)(1 + [H^+]/K_A) + 1 + [H^+]/K_B} \tag{66}$$

Thus plots of $1/k_0$ against $1/x$ at a series of different H^+ concentrations will yield a family of lines that intersect to the left of the vertical axis (mixed inhibition). The values of K_A and K_B can be found from replots of the slopes and vertical axis intercepts against $[H^+]$.

If the reasonable assumption is made that the protonated form of the enzyme can also bind the inhibitor, then the system becomes

$$\begin{array}{ccccc} \text{EH} & \underset{k_{-2}}{\overset{k_2x}{\rightleftharpoons}} & \text{EHX} & & \\ \updownarrow K_A & & \updownarrow K'_A & & \\ \text{E}^- & \underset{k_{-1}}{\overset{k_1x}{\rightleftharpoons}} & \text{EX}^- & \xrightarrow{k_0} & \text{E}^- - \text{X} \end{array} \tag{67}$$

As in the case in which more than one form of the enzyme could bind substrate (see Eqs. (39) and (40)), steady-state treatment of this system gives a complex equation (Eq. (68)).

$$k_0 = \frac{\tilde{k}_0}{1 + [H^+]/K'_A} \Bigg/ \left[1 + \frac{(K_m + k_{-2}[H^+]/k_1K'_A)(1 + [H^+]/K_A)}{(1 + k_2[H^+]/k_1K_A)(1 + [H^+]/K'_A)x}\right] \tag{68}$$

This equation does not predict a simple dependence of k_0 upon pH and cannot be used to determine either of the pK values. If, however, it is assumed that all the binding processes remain in equilibrium, that is, that all proton-transfer steps are very fast and $k_0 \ll k_{-1,}$, the equation simplifies to Eq. (66) with K_m replaced by K_s (k_{-1}/k_1). If it is assumed that the proton-transfer reactions are fast but $k_0 \gg k_{-1,}$, the equation becomes (cf. Eq. (40))

$$k_0 = \frac{\tilde{k}_0}{1 + [H^+]/K'_A} \Bigg/ \left[1 + \frac{k_2 + k_{-1}(1 + [H^+]/K_A)}{K_1(1 + [H^+]/K'_A)}\frac{1}{x}\right] \tag{69}$$

which would allow pK'_A to be determined but results in a displacement of pK_A from the true value. It can be argued (Brocklehurst and Dixon, 1977) that k_1 is likely to exceed $10^6\ M^{-1}\ s^{-1}$ when the inhibitor is a substrate analog, so that observed values of much under this for the pH-independent value that k_0 approaches at its pH optimum imply that $k_0 < k_{-1}$ so that K_A may be determined. Clearly it is possible to devise more complicated reaction schemes, which can result in displacement of both pK values. Thus, as with studies on the effects of pH on the kinetic parameters of the overall enzyme-catalyzed reaction, this method will give true pK values only if certain simplifying assumptions are made about the kinetic mechanism operative.

8. The Competitive Labeling Method

Whereas the previous method required a reagent with a high specificity in reacting with a single residue of the enzyme, the competitive labeling method (Kaplan *et al.*, 1971) uses a reagent of low specificity that may react with all amino acid residues of a given type. An electrophilic reagent is radioactively labeled and allowed to react with a mixture of the enzyme and a standard nucleophile. So little reagent is used that the proportions of the nucleophiles of different reactivities that are present do not change in the course of the reaction, since only a small fraction of each is used up. The reagent concentration therefore falls to zero with pseudo-first-order kinetics. The fraction of any group modified is thus proportional to the second-order rate constant for its reaction with the reagent. After this reaction is complete, an excess of unlabeled reagent is added to complete the conversion of each nucleophile present into its modified form. The fraction modified in the first reaction is now represented by the degree of labeling of the modified form.

The enzyme is digested enzymically, its peptides are separated, and the specific radioactivity of each of the modified residues is determined. If the rate constant for modification of the standard nucleophile is known, the rate constant can be calculated for the reaction of each group with the reagent, from the ratio of labeling of the modified group to that of the modified standard nucleophile.

The procedure is repeated at a number of different pH values, and correction is made for the titration of the standard nucleophile. Hence the apparent rate constant for reaction of each group can be plotted against pH, to give both the pK of the group and the rate constant for modification of its unprotonated form.

The method was first used for the acetylation of amino groups with acetic anhydride (Kaplan *et al.*, 1971) and the pK values of several of these groups in elastase were found. The method assumes that the degree of binding of the reagent to specific sites is small, that is, that the reaction is first order in reagent. This can be checked by the constancy of relative labeling as the concentration of the reagent is diminished. It also assumes that no pH-dependent structural transition occurs in the enzyme over the operative pH range. In fact, however, Kaplan *et al.* (1971) found a marked increase in the rate constant with pH just above the range.

Fortunately, the rate of increase with pH had fallen enough (because of passing the p*K*) before the onset of the second increase (attributed to unfolding of the enzyme molecule), to allow p*K* values to be assigned. When, however, Cruikshank and Kaplan (1975) applied the method to the modification of histidine residues in chymotrypsin by 1-fluoro-2,4-dinitrobenzene, a fall of reactivity followed the rise as the pH was raised. They attributed the first slowing in the rate of increase to approach of a p*K* and the later fall to a change of folding, but it is not clear that the phenomena were well enough separated to allow great reliance on the p*K* values found.

Although the degree of binding must be small, and the fact that it is small can be checked, binding forces may make important contributions to the rate constants. Thus the conclusion that two histidine residues of chymotrypsin, found to be more reactive than expected from their p*K* values, are specifically activated by chemical means (Cruikshank and Kaplan, 1975) may be unjustified, since preferential binding of the hydrophobic reagent to a protein surface (or parts of it) may be all that is needed to explain the enhanced reactivity.

These two difficulties in interpretation illustrate some limitations of this powerful method.

II. Limitations of the Methods

The kinetic study of the effects of hydrogen ions as activators or inhibitors of enzyme-catalyzed reactions can yield useful information on the kinetic mechanism followed. As in all kinetic studies it is not possible to conclude that a given kinetic mechanism operates, since several mechanisms can usually be devised that will fit the results. It is necessary to select the simplest mechanism that does so and to recognize that more complex models may have to be considered when further results are obtained.

The first problem in using kinetic results on p*K* values for identifying the roles of ionizing groups is that simplifying assumptions must be made about the kinetic mechanisms that apply. There is usually no satisfactory justification for these assumptions. Hence although kinetic results can provide a useful indication of the types of groups involved, they can do little more than this. The p*K* values found may not be ionizations at all; if they are, they may be distorted by rate constants, and they will be molecular values to which more than one group may contribute. Even true p*K* values of groups may be far displaced by the environment from the normal values of the groups concerned.

It is sometimes assumed that p*K* values that affect enzyme activity are those of groups near the active site of the molecule. The ionization of a distant group, however, like that of the terminal isoleucine of chymotrypsin (Fersht, 1971; Himoe *et al.*, 1967) may result in a reversible conformational change that converts the enzyme into an inactive form. The coupling of an ionization to a conformational change may, however, raise its temperature coefficient enough to reveal such coupling.

III. Practical Aspects

A. Effects of pH on the Stability of Enzymes

Many enzymes are irreversibly denatured at extreme pH values; unless this is allowed for, it may be mistaken for a reversible ionization. The stability is more easily investigated by preincubating samples of the enzyme at different pH values. The duration of the preincubation should be at least as long as the usual assay time. The enzyme is then readjusted to a pH at which it is known to be stable for assay. Possible results that could be obtained in such studies are shown in Fig. 10.

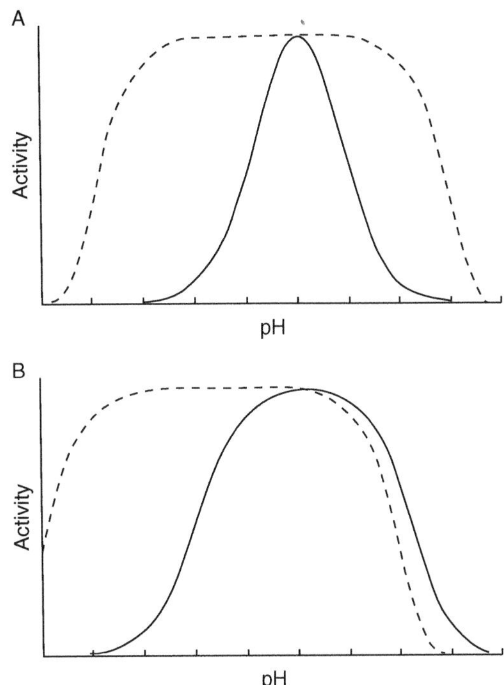

Fig. 10 Checks of enzyme stability. Solid lines represent assay at the pH indicated; dashed lines represent the results of assay at the optimum pH after incubation at the pH indicated for a period equal to the length of the assay. The result in (A) shows that the enzyme is almost fully active after preincubation at any pH at which it showed appreciable activity in the assay, so the assay can be taken to be valid in showing reversible effects. On the alkaline side of (B), however, the falloff in activity proves not to be a measure of reversible effects because the dashed line shows that the enzyme is irreversibly inactivated during assay. The form of irreversible inactivation shown by the dashed line assumes that the activity left at the end of the incubation is e^{-kt} of its value at the start, and that k is proportional to $[OH^-]$. The solid line on the alkaline side of (B) has been plotted on the assumption that there is no reversible loss of activity over the pH range. The amount of product produced over the course of the assay during which the enzyme is being inactivated as a fraction of what would have been formed if the enzyme had been stable is $(1 - e^{-kt})/k$. This can be substantial even when most of the enzyme is inactive by the end of the assay.

In Fig. 10A, the enzyme is stable for the required period over the relevant range of pH, so the effects of pH on activity can be interpreted in terms of fully reversible ionizations. In Fig. 10B, there is an irreversible loss of activity on the alkaline side of the optimum, so that results in this region should not be attributed to reversible ionizations. Although the amount of active enzyme left at the end of the preincubation falls rather sharply with increasing pH after 50% inactivation in the example shown in Fig. 10B (where it is assumed that the rate of inactivation is proportional to $[OH^-]$), the total activity shown over the assay period (Fig. 10B, solid line) follows a curve like that of a reversible titration and could easily be mistaken for one unless the stability check is made.

In the experimental test just outlined it is important that the conditions for preincubation should resemble those used in the assay itself. Enzyme concentrations should be comparable, because many enzymes become less stable on dilution. The ionic strength should also be similar. It may be convenient to work at such a high value of ionic strength that changes in buffers and substrate concentrations make a negligible difference to its total value. Alternatively, the ionic strength should be adjusted to the same value in each experiment, or the effect of ionic strength on enzyme activity and stability may be separately determined.

B. Effects of pH on Substrates

1. Stability of Substrate

Although it is generally recognized that enzymes may be unstable at extremes of pH, it is sometimes forgotten that the same may be true of substrates. Not only may their concentrations be diminished by breakdown, but their breakdown products are likely to be enzyme-inhibitors since they share some molecular features with the substrates. The pH stability of the substrate can be checked in much the same way as that of the enzyme. If one of the substrates is unstable at extremes of pH, a study of the time course of its breakdown should indicate whether it would still be possible to determine true initial rates near these extremes when the reaction is initiated by addition of that substrate. Examples of common substrates that are unstable at some pH values include NAD^+, which decomposes at high pH, and NADH, which decomposes at low pH (Lowry *et al.*, 1961).

2. Ionizations of Substrates

The simple interpretation of the effects of pH on enzyme activity assumed that the effects were due to ionizations of the free enzyme or of enzyme-substrate complexes and that the substrate did not ionize over the pH range considered. In fact every ionization attributed above to the free enzyme could equally well be due to free substrate. Similarly, ionizations of the enzyme-substrate complex may be due to ionization of a group on the bound substrate. The pK values of the substrates should be determined so that they may be recognized if they should appear in kinetic plots. It is best to titrate the substrate under the conditions used, since a literature value for its pK may apply to a different temperature or ionic strength.

Further, it may not be easy to avoid systematic errors associated with a particular electrode assembly, but these will not greatly matter if all pK values of a given study are self-consistent.

If the following mechanism applies

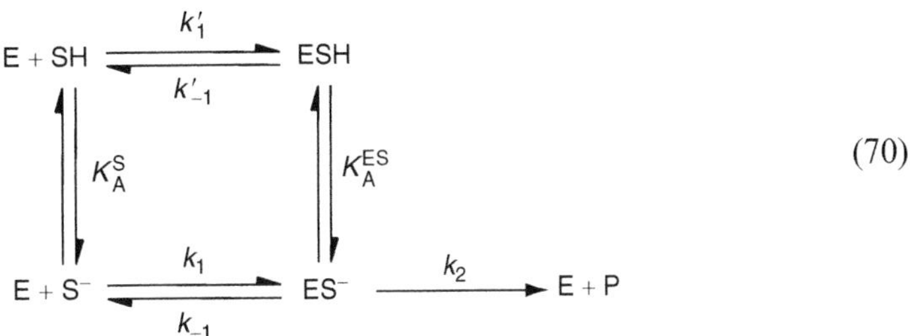

(70)

then this is essentially that of Eq. (39) except that the substrate ionization constant K_A^S replaces that of the free enzyme K_A^E. All the same conclusions apply; for example, Eq. (23) will apply with K^S for K^E if substrate binding is in equilibrium. In effect, the protonated form of the substrate is acting as a reversible inhibitor of the enzyme, since sequestration of some enzyme as ESH lowers the concentration of ES^-.

Enzymes often bind only one ionic form of the substrate; more strictly, binding greatly favors one form. Thus an enzyme acting on a phosphorylated substrate may bind only the dianion R–O–PO_3^{2-} (i.e., S^{2-}) and not R–O–PO(OH)–O^- (i.e., SH^-). The favored binding of the dianion lowers the pK of the bound form of substrate, possibly to below the pH range under study; indeed it may not be possible to obtain an appreciable concentration of bound SH^- at attainable values of the pH and substrate concentration.

Thus in the scheme

$$\begin{array}{ccccc} E + SH^- & \overset{K'}{\rightleftharpoons} & ESH^- & & \\ \updownarrow K_A & & \updownarrow K'_A & & \\ E + S^{2-} & \overset{K}{\rightleftharpoons} & ES^{2-} & \longrightarrow & E + P \end{array} \qquad (71)$$

$K'_A \gg K_A$ and $'K \gg K$. We can therefore ignore K'_A and K' as well as [ESH^-]. If the enzyme follows the simple scheme of Eq. (22), K_A^S will be seen in V/K_m

(Eq. (27)) but not in V (Eq. (25)), so the pK of the substrate will have disappeared from the enzyme-substrate complex, since this complex does not protonate appreciably over the pH range under study.

The complications of several overlapping protonation ranges are increased by substrate ionizations, so it is prudent, if possible, to use substrates that do not change their degree of ionization over the pH range under study. The important point, however, is to allow for substrate ionizations.

A further complication can occur if the true substrate for a reaction is the complex between the substrate and a metal ion. Ionization of the substrate may then alter its affinity for the metal ion, and this effect must be allowed for in calculating substrate concentrations. This problem has been considered by Fromm (1975), and details of a computer program for the calculation of the various species in solution have been published by Storer and Cornish-Bowden (1976).

3. Substrate Concentrations

An accurate determination of the effects of pH on an enzyme requires the determination of the apparent values of K_m and V at a series of different pH values. A number of workers have sought to diminish the amount of work involved by studying the effects of pH at only two substrate concentrations: very high and very low. By inspection of the Michaelis-Menten equation $v = Vs/(s + K_m)$ it can be seen that when $s \gg K_m$ then $v \rightarrow V$, so that variation of initial velocity with pH will correspond to the variation of V. Likewise at a very low substrate concentration when $s \ll K_m$ then $v \rightarrow Vs/K_m$, so that the variation in initial velocity will correspond to that of V/K_m. The disadvantage of using this method is that K_m would be expected to vary with pH so that the required condition that $s \gg K_m$ or that $s \ll K_m$ may hold for a fixed substrate concentration at some pH values but not at others. Hence the enzyme may not remain saturated over the whole pH range at the higher substrate concentration, and its degree of saturation may not remain negligible at the lower. Further, it may not be possible to use sufficiently high substrate concentrations for the condition $s \gg K_m$ because of insolubility or because of inhibition by high substrate concentrations. Similarly low substrate concentration may render measurements imprecise because of insensitivity of the assay. A full kinetic study would therefore involve less ambiguity and would give more accuracy than the use of only two substrate concentrations.

4. Enzyme Reaction Involving more than One Substrate

The theory developed in the preceding section has, for simplicity, dealt with an enzyme mechanism that involves only one substrate. The rate equations to give the effects of pH on the kinetics of reactions of two or more substrates may be readily derived if it is assumed that proton transfers occur so fast that they remain in equilibrium; a simplified procedure for deriving rate equations involving some steps at equilibrium has been developed by Cha (1968). Since a complete kinetic analysis of the effects of pH on such enzymes would take a long time, a high concentration of one of the substrates is often used so that the reaction may be treated by the equations that describe the single-substrate case. If this is done it is important to check that the concentration of the fixed substrate is sufficiently high to maintain saturation at all pH values. In the interpretation of the effects of pH on the kinetics of these reactions, it is also important to check that the kinetic mechanism does not change with pH; this necessitates a full kinetic analysis at the different pH values.

C. Effects of pH on the Assay Method

In all studies on the effects of pH on enzyme activity, it is important to ensure that the assay method used gives a true measure of the rate of the catalyzed reaction over the whole pH range studied. Complications that could occur include effects of pH on the activity of a coupling enzyme or on the absorbance of a product formed. If a stopped assay is used, that is, one in which the enzyme-catalyzed reaction must be stopped to allow determination of the amount of product formed, then a check must be made at each pH that the rate of reaction is constant over the time of incubation; a check at one pH is insufficient.

D. Buffers

There is a wide choice of buffers that may be used in studying enzymes (Good *et al.*, 1966; Dawson *et al.*, 1989). Lists of p*K* values from which buffers may be chosen are widely available in many standard reference works (see, e.g., CRC Handbook of Chemistry & Physics, 2006; Dawson *et al.*, 1989; The Merck Index, 2006) and a comprehensive list of p*K* values may be downloaded from several sources (e.g., Williams, 1967). The equation

$$\text{pH} = \text{p}K + \log[\text{buffer base}] - \log[\text{buffer acid}]$$

shows that both the acid and its conjugate base must be in high concentration if the pH is not to be much changed by a small addition or withdrawal of protons, which turns one into the other. Hence the pH must not be far from the p*K* of the buffer, else one of these concentrations will be small and therefore easily changed. The useful buffering range extends about a unit on either side of the p*K*. It will therefore be necessary to use several buffers to cover a wide pH range. Their ranges should overlap so that assays can be made at the same pH in different buffers to check for possible inhibitory effects of specific buffer components. Sometimes a mixture of buffers has been used to cover a wide pH range, but this could lead to a failure to recognize that one of the components was inhibitory.

Individual buffers may inhibit a particular enzyme in a number of ways. One component might be a specific inhibitor. If only one ionic form was inhibitory the enzyme activity would decline with the titration curve of the buffer. This could be detected by absence of such a decline over the same pH range with another buffer.

The inhibitor might be of any of the known types, competitive, truly noncompetitive, uncompetitive, and mixed. A less specific inhibition could be due to ligation by the buffer of an essential component of the reaction. Many buffer components are chelating agents and could thereby inhibit enzymes that require metal ions for activity. Since protons usually compete with metal ions for chelating agents, such an inhibitory effect could vary with the pH. Alternatively, a buffer component may appear to activate an enzyme if it removes an inhibitory metal ion. Lists of chelating ability (Sillén and Martell, 1964) are useful to check this, and Good and coworkers (Good *et al.*, 1966) have listed several buffers with little chelating ability. A buffer may also affect an enzyme by supplying an essential or nonessential activator. Thus, for example, several enzymes are activated by potassium ions, so a change from potassium phosphate to sodium phosphate buffer would appear inhibitory unless another potassium salt were present. An even less specific effect might be mediated by a change of ionic strength, which should be kept constant or at least within a range where it has been shown not to affect the enzyme. These considerations indicate that a preliminary survey of the effects of several buffers and of ionic strength on the enzyme under study is an important preliminary to the more detailed investigation of the effect of pH.

Just as the ionization constants of groups in proteins can vary with temperature (Eq. (48)), changes in temperature will affect the pK values of buffer components and hence the pH values of buffer solutions; it is important to allow for this if studies are made at different temperatures. The pK values of amines have particularly large temperature coefficients (see the table); thus Tris buffers, for example, fall in pH by 0.3 per 10° rise in temperature. A glass electrode calibrated at one temperature needs care in operating at others because its temperature coefficient will depend on details of the measurement procedure, according to whether the solution at the new temperature brings to its own temperature only the one half-cell consisting of the glass membrane and the solution and electrode inside it, or also the second half-cell consisting of the reference electrode.

If pH measurements are to be made with added organic solvents, one of the approaches indicated above under effects of dielectric constant must be used. It is important to specify exactly what has been done, because various (and probably equally valid) definitions of pH are used for media other than water.

A particular change of solvent that may need to be considered is the substitution of $^{2}H_2O$ for water. If the reading of a pH meter calibrated in water is used directly, then 0.4 has to be added to give the reading that the meter would show in water with the same concentration of strong acid, or in mixtures the amount $(0.3139\alpha + 0.0894\alpha^2)$ has to be added, where α is the atom fraction of deuterium, that is, $[^2H]/([^2H] + [^1H])$ (Pentz and Thornton, 1967). Since pK values are affected by substitution of deuterium, other systems (such as specifying the pH the buffer mixture would show in water, or using uncorrected meter readings) may be just as valid, provided that what is actually done is clearly specified.

Since dissociation of a buffer acid involves a change of charge, the pK is likely to be affected by the ionic strength and hence by the buffer concentration.

This effect is most marked with buffers (like phosphate) that contain multivalent species, both because the activity coefficients of such species are more sensitive to ionic strength than those of univalent species, and because multivalent species and their counterions contribute more to the ionic strength. Thus a buffer of 0.1 M NaH_2PO_4 and 0.1 M Na_2HPO_4 exhibits a pH of about 6.7 and this rises by about 0.2 on 10-fold dilution and 0.1 on a further 10-fold dilution.

A simple point too often overlooked is the importance of checking the pH of the reaction mixture rather than that of the added buffer, since other components may donate protons or combine with them. It may be possible to adjust all solutions to the pH of the assay system before they are mixed, but this may be inconvenient, especially if some of the components are not very stable under these conditions. It is also important to check the pH again after the reaction to ensure that the buffering has been effective.

IV. Some Examples of pH Studies

The effects of pH have been studied, in various degrees of detail, on many enzymes. Those given below are chosen to illustrate specific points made earlier in this article.

A. Carnitine Acetyltransferase

This enzyme (EC 2.3.1.7), which is responsible for maintaining equilibrium in the transfer of acetyl groups between carnitine and CoA (Pearson and Tubbs, 1967), shows fairly straightforward kinetics. Initial rate and product inhibition studies are consistent with a random-order mechanism in which substrate binding is at, or close to, thermodynamic equilibrium (Chase and Tubbs, 1966), and this conclusion has been supported by estimates of dissociation constants for the complexes of the enzyme with individual substrates and products obtained in studies of the optical rotatory dispersion of the enzyme (Tipton and Chase, 1969). Further, the binding site for the acetyl group is so located with respect to those for carnitine and CoA that a bromoacetyl group on bound carnitine alkylates the thiol group of bound CoA (Chase and Tubbs, 1969).

Chase (1967) studied the effect of pH on the activity of this enzyme over the range 6-9. He took the precautions mentioned above, such as checking several buffers and measuring substrate pK values. The results with V proved to be simple: it was constant over the range studied. Hence there is no group in the enzyme-substrate complex with a pK within or close to this range whose ionization has appreciable effect on the catalyzed reaction. The plots of pK_m against pH for both carnitine and acetylcarnitine showed unit slope at low pH and zero slope at high pH, with the bend at pH 7.1–7.2. Since these substrates have no pK near this value there must be a group in the enzyme whose deprotonated form is required for binding (although we should note the ever-present kinetic ambiguity: we only know the molecular pK, so the group may need to be protonated provided that some other group predominantly protonated over the range studied needs simultaneously to be deprotonated). In view of the likelihood that the trimethylammonium group of these substrates is bound to a negatively charged group in the protein, the finding that protonation prevents binding is not surprising. The group evidently has the pK of 7.1-7.2 in the free enzyme and cannot be protonated above pH 6 in the enzyme-carnitine complex.

The plots of pK_m against pH for CoA and acetyl-CoA were more complex, and again similar to each other. The pK_m again rose with pH (with unit slope) at low pH and showed a bend to zero slope with a pK of 6.4. This corresponded to the pK of the phosphate group of CoA, as determined by titration. Evidently the dianion was the form preferentially bound. Chase checked this elegantly by measuring the pK_m for dephospho-CoA, which gave the same value of V as CoA. With it there was no change in pK_m from pH 6 up to above pH 7.

It may seem strange that removal of one negative charge from the dianionic form of the phosphate group of CoA prevents binding, but removal of both in preparing dephospho-CoA does not. It should, however, be realized that the slope of unity from pH 6 upward with CoA only implies that the dianion is the predominant form bound. If stability had allowed measurements to be continued below pH 6, a further bend in the curve would be expected when the predominance of monoanion in solution would overcome its lesser affinity for the enzyme; in other words, the enzyme-substrate complex would exhibit a pK. We do not expect this pK to be very low, since the dianion of CoA is favored over dephospho-CoA in binding by only about 40-fold.

As the pH is further raised, the pK_m for CoA or acetyl-CoA turns down with a pK of 7.85 and up again to zero slope with a pK of 8.25. Hence these pK values exist in the enzyme and enzyme-substrate complex, respectively. Since that in the enzyme-substrate complex is not seen in V, it does not appreciably affect the reaction catalyzed. Evidently some group in the enzyme of pK 7.85 needs to be deprotonated for optimal binding, but the smallness of the shift of pK shows that binding is only 2.5-fold worse if it is deprotonated. When dephospho-CoA was used, the turndown at 7.85 was similar, diminishing the already low affinity of this substrate. Measurements were not continued to check whether the horizontal slope was regained at pH 8.25 and above.

Bound bromoacetylcarnitine alkylates a histidine residue of the enzyme provided no CoA is bound (because its thiol group would be preferentially alkylated) (Chase and Tubbs, 1970). There is therefore an ionizing group close to the active site, and it shows nucleophilic reactivity at pH 7.2. It is not possible to attribute any of the kinetic pK values to it; possibly its pK is below the range studied.

B. Fumarate Hydratase

This enzyme (fumarase, EC 4.2.1.2) might appear simpler than carnitine acetyltransferase because it has only a single substrate (malate) in one direction and water is one of the two substrates in the other direction. In fact, however, it shows far more complexities. Cleland (1977) has reviewed its action and included the effects of pH. Studies on the effect of pH can be complicated by the fact that a number of small anions, including many that are common buffer components, can activate the enzyme (Rose, 1998).

Its pH dependence has been studied in detail by Alberty and his coworkers (Alberty and Massey, 1955; Brant *et al.*, 1963). The pH dependence of V/K_m gave pK values of 5.8 and 7.1, which on simple theory would be those of the free enzyme. Wingler and Alberty (1960), however, had already found that several dicarboxylic acids were competitive inhibitors of the enzyme and that their binding was diminished by protonation (pK 6.9). The shift of pK from these values for the free enzyme to those of the enzyme-inhibitor complex varied, as is to be expected, with the nature of the inhibitor. Thus inhibitor binding provides correct pK values for the free enzyme, which V/K_m did not.

The discrepancy might be explained by the rapid action of the enzyme; it approaches the diffusion-controlled limit (Brant *et al.*, 1963), so substrate binding cannot be in equilibrium. Although this could shift the pK values seen in V/K_m from those of the free enzyme, the shift observed is not that predicted by Eq. (40). As discussed earlier, this mechanism would give two pK values for the free enzyme plus two distorted values; if all forms of enzyme combined with substrate at the same rate ($k_1 = k_1' = k_1''$) only the two distorted pK values would be seen but the shift would be symmetrical, with pK_A lowered as much as pK_B is raised. Since this is not found, there must be some other complexity, which can, in fact, be shown to be that protonation of the enzyme-substrate complex is relatively slow. Exchange of ^{2}H or ^{3}H between water and malate can, at high substrate concentrations, be markedly slower than ^{14}C exchange between malate and fumarate (Hansen *et al.*, 1969). It thus appears that the H^+ ion removed from malate to form fumarate can remain bound to the enzyme while the fumarate molecule dissociates and is replaced, so the reversible protonation of the enzyme-substrate complex is relatively slow.

Cleland (1977) has pointed out that the molecular pK values for the free enzyme of 6.3 and 6.9 are those that would be expected for groups of identical pK of 6.6 if there is no interaction between the groups. This is a useful warning against allotting these pK values to different groups. Usually negative interaction of closely situated groups would be expected because of electrostatic interactions, but the closeness of the molecular pK values shows that it is absent here.

The pK values shown in V in the direction from malate, that is, those attributable to the enzyme-malate complex, are 6.4 and 9.1 (Brant *et al.*, 1963). Those in the reverse direction are 4.9 and 7.0. Those of 4.9 and 9.1 have high temperature coefficients, and those of 7.0 and 6.4 low ones. Cleland (1977) has the reasonable interpretation that a histidine residue (pK 9.1) has to be protonated and a carboxylate (pK 6.4) deprotonated to attack malate. It is reasonable that in the enzyme-substrate complex the pK of the imidazole is raised from that of the free enzyme because its protonated form is stabilized by interaction with the hydroxyl group of malate; that is, the proton is ready to assist the hydroxyl to leave. However, in the predominant enzyme-substrate complex

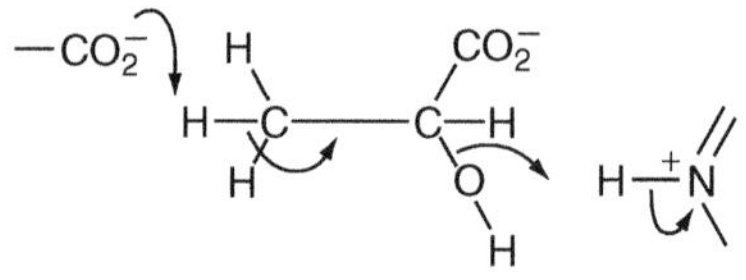

the H^+ ion that the carboxylate will accept is presumably still a carbon-bound proton of the substrate and so is incapable of stabilizing the carboxylate form of the group. Hence this pK is not much changed from that of the free enzyme. Likewise in the fumarate complex

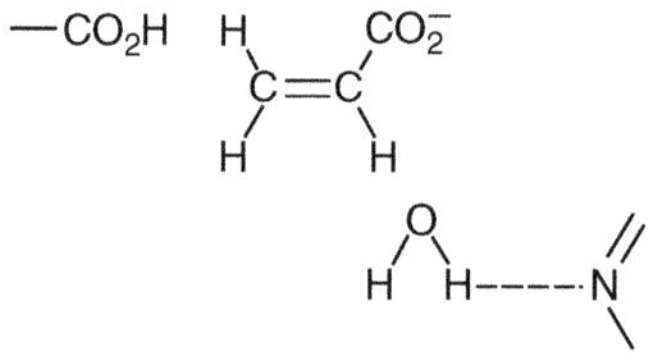

it is the imidazole group that contributes most to the lower pK (4.9); its deprotonated form is stabilized by interaction with the water molecule that can attack fumarate, and again the carboxyl group, which requires to be protonated, is little changed in pK.

This mechanism looks reasonable, but in view of the serious drawbacks to ascribing pK values on the basis of ΔH values (because shifts in pK are likely to shift ΔH) it must be taken as tentative. The study illustrates both the dangers of inferring much from pK values alone and also the way they can usefully supplement other studies.

C. Lysozyme

Much of the study of this enzyme (EC 3.2.1.17) has been made against the background of its molecular structure determined by X-ray crystallography and that of its complexes with inhibitors (Blake *et al.*, 1967). A mechanism of action was proposed with this structure (Blake *et al.*, 1967; Vernon, 1967), according to which the carboxyl group of a glutamic residue (Glu-35) protonates the R–O^- leaving group of the substrate (thus assisting its departure) and the carboxylate form of an aspartic residue (Asp-52) stabilizes the oxonium ion formed (Eq. (72)).

$$-CO_2^- \quad \text{Asp-52} \quad C\text{—}OR \quad H\text{—}O\text{—}CO\text{— Glu-35} \longrightarrow -CO_2^- \quad {}^{+}O{=}C \quad H\text{—}O\text{—}R \quad {}^{-}O\text{—}CO\text{—} \tag{72}$$

Reverse of this process with water in place of R–OH completes the hydrolysis (Eq. (73)).

$$-CO_2^- \quad {}^{+}O{=}C \quad H\text{—}O\text{—}H \quad {}^{-}O\text{—}CO\text{—} \longrightarrow -CO_2^- \quad O\text{—}C\text{—}OH \quad H\text{—}O\text{—}CO\text{—} \tag{73}$$

Many variants of this mechanism have been discussed (Lowe, 1967; Dunn and Bruice, 1973), for example, that the carboxylate of Asp-52 forms a covalent bond to some degree with the $-O^+{=}CH-$ system.

It is clear that knowledge of the properties of the carboxyl groups of the aspartic and glutamic residues would contribute to the understanding of such a mechanism. This knowledge has been provided by an ingenious study by Raftery and colleagues. First, they treated lysozyme at pH 4.5 with the Et_3O^+ ion and isolated from the products a derivative in which one carboxyl group was esterified. Two effects contribute to the specificity of the modification: (1) at pH 4.5 more powerful nucleophiles in the protein are mainly protonated, and their higher basicity than that of carboxylate increases their protonation (and hence diminishes their reactivity) to a greater extent than it raises their intrinsic reactivity (Jencks, 1969); (2) the reagent, being a cation, has particularly high reactivity with anionic nucleophiles. The product proved (Parsons and Raftery, 1969) to have Asp-52 esterified; the ester group had low reactivity to nucleophiles (e.g., borohydride and hydrazine), and this may have assisted the isolation of a peptide with the ester bond intact, which allowed characterization of the product.

Parsons and Raftery (1972, 1969) then measured the difference between the titration curves of lysozyme and that of the esterified derivative. The difference fitted the sum of two titration curves of pK values 4.4 and 6.1 less one of pK 5.2, all three being of one proton per enzyme molecule, so the results could be interpreted as showing that free lysozyme had two molecular pK values, which were replaced in the ester by one of 5.2. A dibasic acid has the same titration curve as an equimolar mixture of two monobasic acids provided that its protonation is not positively cooperative. It follows from the analysis of Simms (1926) that if the pK

values of the equivalent monobasic acids are called pK^* ± log p (this defines pK^* and p), then the molecular pK values of the dibasic acid to which they are equivalent are pK^* ± log (p + 1/p) (Dixon, 1979). Hence the pK values of 4.4 and 6.1 need correction in this way to provide true molecular constants of the system, but they are so far apart that this correction is less than 0.01. It thus appears that Asp-52 contributes to the system of molecular pK values 4.4 and 6.1, and so does another group, whose pK is 5.2 when Asp-52 is esterified. This other group is convincingly identified as Glu-35, mainly because it is the only group that dissociates in the required pH range that is close enough to be likely to interact with Asp-52.

It is, as we have indicated, impossible to determine the group pK values that make up molecular values. Nevertheless argument by analogy is possible. Parsons and Raftery assumed that the pK of Glu-35 seen when Asp-52 was esterified would also be its pK when Asp-52 was free but protonated. This identity cannot be exact, but it provides the most reasonable estimate possible. Using it gives the pK of Glu-35 a value of 6.0 when Asp-52 is dissociated, almost the molecular pK of 6.1. This conclusion that the molecular pK of 6.1 is almost entirely due to Glu-35 holds even if 5.2 should not be a very accurate estimate of the pK of Glu-35 when Asp-52 is protonated; what matters is that the value is well below 6.1. Likewise the molecular pK of 4.4 is almost entirely attributable to the group pK of Asp-52 when Glu-35 is protonated, which also has a value of 4.4.

The fact that the titration difference fitted the theoretical curves over the pH range from 3 to 7 showed that no other groups that titrated over this range appreciably affected the pK values under study.

Parsons and Raftery (1972) then proceeded to similar difference titrations in the presence of inhibitors and substrate. They found that the binding of methyl 2-acetamido-2-deoxyglucoside raised the pK of 6.1 to one of 6.6. This agreed well with the effect of pH on the binding of this compound (Dahlquist and Raftery, 1968) as studied by logarithmic plots. Evidently the inhibitor is bound slightly more tightly when Glu-35 is protonated. The pK of 4.4 was not affected, and this is consistent with the absence of any such pK when pK_s for this ligand was plotted against pH. When, however, substrate of high molecular weight was added, the difference curve no longer fitted the sum of two one-site titrations minus a third; evidently other groups that titrated in the pH range used now affected the pK values of the enzyme-substrate complex. Nevertheless, Parsons and Raftery were able to show that the pK of Glu-35 was raised to above 8, and this agreed well with a kinetic pK previously observed in plots of V against pH.

Several other elegant points were made in this study, especially ones arising from the effect of ionic strength and temperature on the pK values reported, but even without them important conclusions emerge. The first is the possibility of assigning the molecular pK values of 4.4 and 6.1 in the free enzyme predominantly to Asp-52 and Glu-35, respectively. The slightly low value for Asp-52 proved to be largely due to the net positive charge of the molecule. Three factors that contribute to the high value for Glu-35 can be recognized. First, there is the rise to 5.2 despite its presence in a positively charged molecule; conceivably this is due to its somewhat hydrophobic environment. Second, there is the rise from 5.2 to 6.1 due to the negative charge on Asp-52. Third, there is the further rise to above 8 when a long-chain substrate is bound. Although the results of mutagenesis are often not straightforward to interpret, the replacement of the Asp-52 by Ala was shown to result in a significant reduction of activity and a decrease in the apparent p*K* of the glutamate residue to 4.7, closer to its intrinsic p*K*. The authors interpreted this as supporting the mechanism in which that Asp-52 facilitates catalysis through the electrostatic field provided by its dissociated form and by stabilizing the oxocarbonium ion intermediate (Hashimoto *et al.*, 1996).

In the postulated mechanism Asp-52 is required for its charge; a normal p*K* guarantees this, and its p*K* is fairly normal. If Glu-35, however, is to act as a general-acid catalyst and supply a proton, an appreciable fraction of it must be protonated, and the raised p*K* brings this about. The study as a whole gives excellent examples of the combination of techniques, from chemical modification to kinetics, to achieve the extremely difficult task of assigning p*K* values to particular residues, and using the results to understand something of enzymic action.

V. Future Prospects

A number of studies with variant mutated enzymes have shown that replacement of one or more amino acid residues can result in changes to the stability and/or the p*K* values of groups involved in catalysis. The challenge is to "reengineer" enzymes to optimize both their stabilities and their response to pH for specific, often biotechnological applications. The foregoing treatments should indicate that this is a far from straightforward operation but several groups have attempted to address the problems. There are two interlinked aspects, whether it is possible to predict the ionization states and p*K* values from knowledge of the structure of a protein and whether one can use such knowledge for the rational design of "optimized" enzymes.

As discussed earlier the p*K* values of specific amino acid residues can be significantly affected by their local environments within a protein with the hydrophicity, hydrogen bonding electrostatic interactions and accessibility to polar compounds, such as water, all being potentially involved. Approaches to predicting pH behavior have, in the main, tried to simplify the system. Many have argued that electrostatic factors at the protein surface are dominant, to the extent that other factors can be ignored. The approximation, formulated in 1934 by Kirkwood, considered the native protein as a sphere surrounded by a water shell with charges located at discrete sites close to its surface. This has since been extended to incorporate deviations from the simple sphere, and to include surface accessibility of ionizing groups, solvent and dielectric effects and salt. The complexity of this approach necessitated some simplifications and the reduced mean-field approach considers the p*K* of each group being exposed to a field created by the all the other ionizable groups. Even this necessitates reiterative computation for each group involved and frequently computer-intensive Monte Carlo methods are used (see Dimitrov and Crichton, 1997; Grochowski and Trylska, 2007; Lindman *et al.*, 2006; Nielsen and McCammon, 2003, for reviews of such approaches). These models have been steadily refined as our knowledge of protein tertiary structure and data from NMR have increased.

The problem facing the investigator is one of deciding which of the methods available is more reliable. This is not straightforward, since they each make different simplifications. A number of such calculations have indicated that the results obtained theoretically are in quite good agreement with the p*K* values determined experimentally, but there are also papers indicating that erroneous results may be obtained (Riccardi *et al.*, 2005). Furthermore, the simplification that buried groups in a protein may be neglected, because they do no do not readily ionize has been shown to be untenable since local water penetration may occur (Harms *et al.*, 2008). A further cautionary example is provided by the two cysteine peptidases papain and caricain, which have apparently identical crystallographic structures in their active site regions but show different responses to pH (Noble *et al.*, 2000; Salih *et al.*, 1987). Further improvements to these computational approach might be expected as the theoretical background and assumptions are adjusted to take account of such factors.

If one could accurately predict the p*K* values of residues in a protein it should be possible to predict the effects of individual mutations. Therefore, it should be possible to devise to rational approaches to obtaining predictable results from reengineering. There have been several attempts to do this (Hirata *et al.*, 2004; Tynan-Connolly and Nielsen, 2007), but clearly such work must take account of the fact that ionization affecting enzyme activity may be different from those affecting stability. Furthermore, as discussed earlier, the apparent p*K* values of groups involved in catalysis change during the progress of the reaction, it may not be adequate to reengineer a protein to optimize their p*K* values in the resting state. One may expect many further developments in this area, because of its potential importance.

Acknowledgment

We are grateful to Science Foundation Ireland for their support of part of this work.

References

Adams, E. Q. (1916). *J. Am. Chem. Soc.* **38,** 1503.
Ainsworth, S. (1977). "Steady-State Enzyme Kinetics" p. 160. Macmillan, New York.
Alberty, R. A., and Bloomfield, V. (1963). *J. Biol. Chem.* **238,** 2804.
Alberty, R. A., and Massey, V. (1955). *Biochim. Biophys. Acta* **13,** 347.

Aldersley, M. F., Kirby, A. J., and Lancaster, P. W. (1972). *Chem. Commun.* 570.
Blake, C. C. F., Johnson, L. N., Mair, G. A., North, A. C. T., Phillips, D. C., and Sarma, V. R. (1967). *Proc. R. Soc. Lond. Ser. B* **167,** 378.
Brant, D. A., Barnett, L. B., and Alberty, R. A. (1963). *J. Am. Chem. Soc.* **85,** 2204.
Brocklehurst, K., and Dixon, H. B. F. (1976). *Biochem. J.* **155,** 61.
Brocklehurst, K., and Dixon, H. B. F. (1977). *Biochem. J.* **167,** 859.
Caccuri, A. M., Lo Bello, M., Nuccetelli, M., Nicotra, M., Rossi, P., Antonini, G., Federici, G., and Ricci, G. (1998). *Biochemistry.* **37,** 3028.
Cha, S. (1968). *J. Biol. Chem.* **243,** 820.
Chase, J. F. A. (1967). *Biochem. J.* **104,** 503.
Chase, J. F. A., and Tubbs, P. K. (1966). *Biochem. J.* **99,** 32.
Chase, J. F. A., and Tubbs, P. K. (1969). *Biochem. J.* **111,** 225.
Chase, J. F. A., and Tubbs, P. K. (1970). *Biochem. J.* **116,** 713.
Cleland, W. W. (1977). *Adv. Enzymol.* **45,** 427.
Cook, P.F. and Cleland, W. W. (2007). Enzyme Kinetics and Mechanism. Garland Science.
Cornish-Bowden, A. (1976). "Principles of Enzyme Kinetics" p. 101. Butterworth, London.
Cornish-Bowden, A. (1976). *Biochem. J.* **153,** 445.
CRC Handbook of Chemistry & Physics (2006). 90th ed. (or any other recent edition). CRC Press Boca Raton, FL.
Cruikshank, W. H., and Kaplan, H. (1975). *Biochem. J.* **147,** 411.
Dahlquist, F. W., and Raftery, M. A. (1968). *Biochemistry* **7,** 3277.
Dalziel, K. (1969). *Biochem. J.* **114,** 547.
Dawson, R. M. C., Elliott, D. C., Elliott, W. H., and Jones, K. F. (1989). "Data for Biochemical Research," 3rd ed., Chapter 18. Oxford Univ. Press, London and New York.
Dimitrov, R. A., and Crichton, R. R. (1997). *Proteins* **27,** 57.
Dixon, M. (1953). *Biochem. J.* **55,** 161 see also.Dixon, M., and Webb, E. C. (1953). "Enzymes" 2nd. p. 116. Longmans, Green, New York.
Dixon, H. B. F. (1973). *Biochem. J.* **131,** 149.
Dixon, H. B. F. (1974). *Biochem. J.* **137,** 443.
Dixon, H. B. F. (1976). *Biochem. J.* **153,** 627.
Dixon, H. B. F. (1979). *Biochem. J.* **177,** 249.
Dixon, H. B. F., Clarke, S. D., Smith, G. A., and Carne, T. K. (1991). *Biochem. J.* **278,** 279.
Dixon, H. B. F., and Tipton, K. F. (1973). *Biochem. J.* **133,** 837.
Dunn, B., and Bruice, T. C. (1973). *Adv. Enzymol.* **37,** 1.
Dyson, H. J., Jeng, M. F., Tennant, L. L., Slaby, I., Lindell, M., Cui, D. S., Kuprin, S., and Holmgren, A. (1997). *Biochemistry* **36,** 2622.
Elfström, L. T., and Widersten, M. (2006). *Biochemistry* **45,** 205.
Fastrez, J., and Fersht, A. R. (1973). *Biochemistry* **12,** 1067.
Fersht, A. R. (1971). *Cold Spring Harbor Symp. Quant. Biol.* **36,** 71.
Fersht, A. R. (1977). "Enzyme Structure and Mechanism" Freeman, Reading.
Fierke, C. A., Johnson, K. A., and Benkovic, S. J. (1987). *Biochemistry* **26,** 4085.
Findlay, D., Mathias, A. P., and Rabin, B. R. (1962). *Biochem. J.* **85,** 139.
Fromm, H. J. (1975). "Initial Rate Enzyme Kinetics," p. 55. Springer-Verlag, Berlin and New York.
Good, N. E., Winget, G. D., Winter, W., Connolly, T. N., Izawa, S., and Singh, R. M. M. (1966). *Biochemistry* **5,** 467.
Grochowski, P., and Trylska, J. (2007). *Biopolymers* **89,** 93.
Haldane, J. B. S. (1930). "Enzymes." Longmans, London (reprinted 1965 by M.I.T. Press, Cambridge, Massachusetts).
Halford, S. E., and Schlesinger, M. J. (1974). *Biochem. J.* **141,** 845.
Hansen, J. N., Dinovo, E. C., and Boyer, P. D. (1969). *J. Biol. Chem.* **244,** 6270.
Harms, M. J., Schlessman, J. L., Chimenti, M. S., Sue, G. R., Damjanovic, A., and Garcia-Moreno, B. (2008). *Protein Sci.* **17,** 833.

Hashimoto, Y., Yamada, K., Motoshima, H., Omura, T., Yamada, H., Yasukochi, T., Miki, T., Ueda, T., and Imotol, T. (1996). *J. Biochem.* **119,** 145.
Himoe, A., Parks, P. C., and Hess, G. P. (1967). *J. Biol. Chem.* **242,** 919.
Hirata, A., Adachi, M., Utsumi, S., and Mikami, B. (2004). *Biochemistry* **43,** 12523.
Hunkapiller, M. W., Smallcombe, S. H., Whitaker, D. H., and Richards, J. H. (1973). *Biochemistry* **12,** 4732.
Inagami, T., and Sturtevant, J. M. (1960). *Biochim. Biophys. Acta.* **38,** 64.
Jencks, W. P. (1969). "Catalysis in Chemistry and Enzymology" McGraw-Hill, New York.
Jentoft, J. E., Neet, K. E., and Stuehr, J. E. (1997). *Biochemistry* **16,** 117.
Kaplan, H., and Laidler, K. J. (1967). *Can. J. Biochem.* **45,** 539.
Kaplan, H., Stevenson, K. J., and Hartley, B. S. (1971). *Biochem. J.* **124,** 289.
Karplus, S., Snyder, G. H., and Sykes, B. D. (1973). *Biochemistry* **12,** 1323.
Kitz, R., and Wilson, I. B. (1962). *J. Biol. Chem.* **237,** 3245.
Knowles, J. R. (1976). *Crit. Rev. Biochem.* **4,** 165.
Koeppe, R. E., II, and Stroud, R. M. (1976). *Biochemistry* **15,** 3450.
Krupka, R. M., and Laidler, K. J. (1960). *Trans. Faraday Soc.* **56,** 1467.
Laidler, K. J. (1955). *Trans. Faraday Soc.* **51,** 528.
Lindman, S., Linse, S., Mulder, F. A. A., and Andre, I. (2006). *Biochemistry* **45,** 13993.
Lowe, G. (1967). *Proc. R. Soc. Lond. Ser. B.* **167,** 431.
Lowry, O. H., Passonneau, J. V., and Rock, M. K. (1961). *J. Biol. Chem.* **236,** 2756.
Malcolm, B. A., Rosenberg, S., Corey, M. J., Allen, J. S., de Baetselier, A., and Kirsch, J. F. (1989). *Proc. Natl. Acad. Sci. USA.* **86,** 133.
Markley, J. L. (1975). *Acc. Chem. Res.* **8,** 70.
Michaelis, L., and Davidsohn, H. (1911). *Biochem. Z.* **35,** 386.
Nielsen, J. E., and McCammon, J. A. (2003). *Protein Sci.* **12,** 1894.
Noble, M. A., Gul, S., Verma, C. S., and Brocklehurst, K. (2000). *Biochem. J.* **351,** 723.
Ottolenghi, P. (1971). *Biochem. J.* **123,** 445.
Parsons, S. M., and Raftery, M. A. (1969). *Biochemistry* **8,** 4199.
Parsons, S. M., and Raftery, M. A. (1972). *Biochemistry* **11,** 1623.
Pearson, D. J., and Tubbs, P. K. (1967). *Biochem. J.* **105,** 953.
Peller, L., and Alberty, R. A. (1963). *J. Am. Chem. Soc.* **81,** 5709.
Pentz, L., and Thornton, E. R. (1967). *J. Am. Chem. Soc.* **89,** 6931.
Renard, M., and Fersht, A. R. (1973). *Biochemistry* **12,** 4713.
Riccardi, D., Schaefer, P., and Cui, Q. (2005). *J. Phys. Chem. B* 2005 **109,** 17715.
Robillard, G., and Schulman, R. G. (1972). *J. Mol. Biol.* **71,** 507.
Rose, I. A. (1998). *Biochemistry* **37,** 17651.
Salih, E., Malthouse, J. P. G., Kowlessur, D., Jarvis, M., O'Driscoll, M., and Brocklehurst, K. (1987). *Biochem. J.* **247,** 181.
Segel, I. H. (1993). Enzyme Kinetics: Behavior and Analysis of Rapid Equilibrium and Steady-State Enzyme. New Edition I Systems. John Wiley & Sons.
Schmidt, D. E., and Westheimer, F. H. (1971). *Biochemistry* **10,** 1249.
Shindler, J. S., and Tipton, K. F. (1977). *Biochem. J.* **167,** 479.
Sillén, L. G., and Martell, A. E. (1964). *Chem. Soc. Spec. Publ.* **1725,** 1971.
Simms, H. S. (1926). *J. Am. Chem. Soc.* **48,** 1239.
Stewart, J. A., and Lee, H. S. (1967). *J. Phys. Chem.* **71,** 3888.
Stines-Chaumeil, C., Talfournier, F., and Branlant, G. (2006). *Biochem. J.* **395,** 107.
Storer, A. C., and Cornish-Bowden, A. (1976). *Biochem. J.* **159,** 1.
Susi, H., Zell, T., and Timasheff, S. N. (1959). *Arch. Biochem. Biophys.* **85,** 437.
The Merck Index 14th ed. (or any other recent edition). Merck Publishing, Rahway NJ.

Timasheff, S. N., and Rupley, J. A. (1972). *Arch. Biochem. Biophys.* **150,** 318.
Tipton, K. F., and Chase, J. F. A. (1969). *Biochem. J.* **115,** 517.
Tu, C. K., and Silverman, D. N. (1975). *J. Am. Chem. Soc.* **97,** 5935.
Tynan-Connolly, B. M., and Nielsen, J. E. (2007). *Protein Sci.* **16,** 239.
Vernon, C. A. (1967). *Proc. R. Soc. Lond. Ser. B* **167,** 389.
Waley, S. G. (1953). *Biochim. Biophys. Acta* **10,** 27.
Williams, R. (1967). pK data. http://nsdl.org/resource/2200/20061121124513540T.
Wingler, P. W., and Alberty, R. A. (1960). *J. Am. Chem. Soc.* **82,** 5482.

CHAPTER 7

Temperature Effects in Enzyme Kinetics

Keith J. Laidler*,✠ and Branko F. Peterman*

*Department of Chemistry
University of Ottawa, Ottawa
Ontario KIN 9B4, Canada

If an enzyme-catalyzed reaction is studied over a range of temperature, the overall rate passes through a maximum. The temperature at which the rate is a maximum is known as the optimum temperature and was at one time thought to be characteristic of the enzyme system; it is now known, however, to be an ill-defined quantity and to vary with concentrations and with other factors, such as pH.

The explanation of this behavior, first given by Tammann (1895), is that changing the temperature affects two independent processes, the catalyzed reaction itself and the thermal inactivation of the enzyme. In the lower temperature range, up to 30 °C or so for a typical enzyme, inactivation is very slow and has no appreciable effect on the rate of the catalyzed reaction; the overall rate therefore increases with rise in temperature, as with ordinary chemical reactions. At higher temperatures inactivation becomes more and more important, so that the concentration of active enzyme falls during the course of reaction. An essential feature of the explanation is that the temperature coefficient of the rate of inactivation must be greater than that of the rate of the catalyzed reaction; in the low-temperature range the rate of inactivation is negligible compared with the rate of the catalyzed reaction, whereas in the high-temperature range it is much higher.

The influence of temperature on the inactivation process is considered later; first, we consider enzyme-catalyzed reactions, and treat not only the temperature

✠ Deceased

CONTEMPORARY ENZYME KINETICS AND MECHANISM
Reprinted from *Methods in Enzymology*, Volume 63 (Academic Press, 1979).

DOI: 10.1016/B978-0-12-378608-1.00007-4

coefficients of these reactions but also certain other matters (e.g., entropies of activation), information about which is provided by the temperature studies.

The rate law for an enzyme-catalyzed reaction involves at least three kinetic constants, each of which has its own temperature dependence. The simple Michaelis-Menten mechanism is

$$E + S \underset{k_{-1}}{\overset{k_1}{\rightleftharpoons}} ES \xrightarrow{k_2} E + X$$

and the steady-state rate equation is then

$$v = \frac{k_2[E]_0[S]}{(k_{-1} + k_2)/k_1 + [S]} \quad (1)$$

where $[E]_0$ and [S] represent the concentrations of enzyme and substrate, respectively. Each of the three constants k_1, k_{-1}, and k_2 will obey the Arrhenius law to a good approximation, and there are three activation energies, E_1, E_{-1}, and E_2, for the individual steps. The Arrhenius law will not necessarily apply to the rate v, although it will do so in some special cases. Sometimes the law applies to quite complicated processes that certainly involve several rate constants (Laidler, 1972).

The Arrhenius law can be written as

$$k = Ae^{-E/RT} \quad (2)$$

where R is the gas constant, T the absolute temperature, E the energy of activation, and A the frequency factor. In terms of activated-complex theory, the equation, in the case of reactions in solution, can be written as (Laidler, 1969)

$$k = \frac{kT}{h} e^{-\Delta G^{\ddagger}/RT} = \frac{kT}{h} e^{\Delta S^{\ddagger}/R} e^{-E/RT} \quad (3)$$

Here k is the Boltzmann constant and h is Planck's constant. The quantity $\Delta G^{\ddagger}$ is the Gibbs energy of activation (formerly known as the free energy of activation) and is the change in Gibbs energy as the activated complex is formed from the reactants. The quantity $\Delta S^{\ddagger}$ is the entropy of activation, which is the corresponding change in entropy. Energies of activation are calculated from plots of ln k against $1/T$, and the frequency factors and entropies of activation are calculated from the value of E and from k at any temperature.

It follows from what has been said that a detailed study of the temperature dependence of an enzyme-catalyzed reaction requires that the individual constants be separated. Care must also be taken with the control of pH, since temperature affects the degree of ionization.

I. Separation of Rate Constants

A number of procedures have been used to separate rate constants in enzyme mechanisms. We shall mention these only briefly, with particular reference to the scheme $E + S \underset{k_{-1}}{\overset{k_1}{\rightleftharpoons}} ES \xrightarrow[\searrow X]{k_2} ES' \xrightarrow{k_3} E + Y$ which applies to many enzyme systems. We shall refer to this as the *double-intermediate* mechanism.

1. Transient-phase studies, using stopped-flow and T-jump techniques, can lead to a separation of constants.

2. Measurement of concentrations of ES and ES′ during the steady state can provide information about the relative values of k_2 and k_3. For example, in the chymotrypsin-catalyzed hydrolysis of *p*-nitrophenyl trimethylacetate (Bender *et al.*, 1962), ES is large and ES′ small, so that $k_2 \gg k_3$. From this it follows that at high substrate concentrations, when $v = V_{max} = k_c [E]_0$, the catalytic constant k_c can be identified with k_3.

3. Studies with alternative substrates (Wilson and Cabib, 1956), such that ES′ is the same for all of them, can provide information about the relative magnitudes of k_2 and k_3.

4. Work with alternative nucleophiles (Hinberg and Laidler, 1972) can also help to separate k_2 and k_3.

II. Energies and Entropies of Activation for Single-Substrate Reactions

Relatively few investigations have provided values for k_2 and k_3 over a temperature range. Tables I and II list some kinetic parameters obtained in such studies.

In considering the magnitudes of the entropies of activation listed in Tables I and II, it is necessary to take into account the two types of effect, which it is convenient to refer to as solvent and structural effects (Laidler, 1955). By the former is meant the interaction between the solvent and the reaction system, an interaction that may change during the course of reaction. By structural effects is meant the possibility that the enzyme itself actually undergoes some reversible change in conformation during the process of reaction.

The possibility of such structural changes was first proposed (Laidler, 1951) in an attempt to interpret the effects of pressure on enzyme reactions. It was suggested that in some cases the enzyme molecule assumes a more open conformation when it forms a complex, with an increase in entropy, and that when it undergoes subsequent reactions there is a refolding of the enzyme molecule, with a loss of entropy. Such a picture seems reasonable since it brings the enzyme reactions into

Table I
Kinetic Parameters Relating to the Conversion of the Enzyme-Substrate Complex into the Second Intermediate and the First Product

Enzyme	Substrate	Temperature (°C)	pH	k_2 (s^{-1})	E_2 (kcal mol^{-1})	$\Delta S_2^{\ddagger}$ (cal K^{-1} mol^{-1})	References
Chymotrypsin	Benzoyl-L-tyrosylglycinamide	25.0	7.5	37	11.5	−19.8	Butler (1941)
	Benzoyl-L-tyrosinamide	25.0	7.5	0.625	14.6	−13.0	Kaufman *et al.* (1949)
Trypsin	Benzoyl-L-argininamide	25.0	7.8	270	14.9	−8.2	Butler (1941)

Table II
Kinetic Parameters Relating to the Conversion of the Second Intermediate into the Enzyme and the Second Product

Enzyme	Substrate	Temperature (°C)	pH	k_3 (s^{-1})	E_3 (kcal mol^{-1})	$\Delta S_3^{\ddagger}$ (cal K^{-1} mol^{-1})	References
Chymotrypsin	Methyl-L-β-phenyllactate	25.0	7.8	1.38	11.1	−23.4	Snoke and Neurath (1950)
	N-Acetyl-L-tyrosine ethyl ester	25.0	8.7	0.0683	10.9	−13.4	Bender *et al.* (1964)
	Benzoyl-L-tyrosine ethyl ester	25.0	7.8	78.0	9.2	−21.4	Kaufman *et al.* (1949)
	Benzoyl-L-phenylalanine ethyl ester	25.0	7.8	37.4	12.5	−11.0	Snoke and Neurath (1950)
	N-*trans*-Cinnamoylimidazole	25.0	8.7	0.0125	11.8	−29.6	Bender *et al.* (1964)
Alkaline phosphatase (*Escherichia coli*)	*p*-Nitrophenyl phosphate	25.0	8.5	28.0	9.4	−22.8	Lazdunski and Lazdunski (1966)
Papain	Furylacryloylimidazole	25.0	7.0	0.02	14.9	−22.7	Hinkle and Kirsch (1970)

line with what takes place during enzyme inactivation, in which it seems clear that unfolding occurs.

This explanation leads to the conclusion that the entropy of activation will be negative for reactions (2) and (3), and Tables I and II show that this prediction is correct in all cases.

Reference must also be made to another type of conformational change for which there is some evidence in certain enzyme systems. In the above discussion it has been assumed that bonds (perhaps hydrogen bonds) are broken, so that the unfolded molecule possesses more freedom of movement, and therefore more entropy, than the folded molecule. A process of a different kind could also occur during complex formation, and give rise to an entropy increase. When the fully extended (β) form of a protein molecule becomes converted into the less extended (α) form, without the formation or breaking of bonds, there is an increase of entropy; such a process may conveniently be referred to as a "rubberlike coiling," since a similar change occurs during the shortening of stretched rubber. A structural change of this type has been considered by Morales and Botts (1952) for the myosin-ATP system and is supported by the temperature studies of Ouellet *et al.* (1952), the solvent studies of Laidler and Ethier (1953), and the pressure studies of Laidler and Beardall (1955). It is supposed that myosin is normally maintained in an elongated form by charges on the protein molecule, and that when these charges are neutralized on complex formation the myosin contracts with an increase of entropy.

The interpretation of entropies of activation in terms of solvent effect is as follows. During the course of reaction there may be polarity changes that will result in either an increase or a decrease in solvent binding. If charges are formed during a reaction, for example, the solvent will be bound more firmly, and there will be a loss of entropy. If, on the other hand, charges are neutralized during the reaction, there will be a release of solvent molecules and a corresponding gain of entropy.

The values in Tables I and II for $\Delta S_2^{\ddagger}$ and $\Delta S_3^{\ddagger}$ are all negative, and could therefore be explained on the hypothesis that there are increases in polarity during the processes ES $\rightarrow$ ES′ + X and ES′ $\rightarrow$ E + Y. Such increases are not unreasonable; in reactions involving solvent molecules such as water, it is commonly found that there are increases in polarity when the activated complex is formed (Laidler and Chen, 1958; Laidler and Landskroener, 1956).

There are only a few cases for which the kinetic constant k_1 for the bimolecular interaction between enzyme and substrate has been obtained at more than one temperature. This was done for catalase and peroxidase, for which Chance has shown that the overall rate constants correspond to k_1 for the initial step. Using a rapid titration method, Bonnichsen *et al.* (1947) have determined rate constants at two temperatures. Their results for catalase are summarized in Table III. Rates at two temperatures do not, of course, permit a test of the Arrhenius law, but if the validity of the law is assumed, the figures in Table IV can be calculated. For horse

Table III
Kinetic Parameters for the Decomposition of Hydrogen Peroxide Catalyzed by Catalase[a]

Type of catalase	Temperature (°C)	$k \times 10^{-7}$ ($dm^3\ mol^{-1}\ s^{-1}$)	E_1 ($kcal\ mol^{-1}$)	$\Delta S_1^{\ddagger}$ ($cal\ K^{-1}\ mol^{-1}$)
Horse blood	25.5	3.50	1.9	−19.6
	23.5	3.50	1.5	−20.9
	22.0	3.54	1.8	−19.9
	23.5	3.50	1.8	−19.9
Horse liver	22.0	3.00	1.3	−21.9

[a]Data from Bonnichsen *et al.* (1947).

Table IV
Comparison of Activation Energies for Hydrogen Peroxide Decompositions

Catalyst	*E* (kcal)	References
Horse blood catalase	1.7	Bonnichsen *et al.* (1947)
Platinum	11.7	Bredig and von Berneck (1899)
Fe^{2+}	10.1	Baxendale *et al.* (1946)
I^-	13.5	Walton (1904)
No catalyst	17–18	Pana (1928), Williams (1928)

blood catalase, the average value of E is 1.7 kcal. This is an unusually low-activation energy for any reaction and, as shown in Table IV, is much lower than for the uncatalyzed hydrogen peroxide decomposition and for the reaction with other catalysts. The negative entropies of activation are consistent with an increase in polarity during reaction, or with some tightening of the enzyme structure. The results for the formation of the peroxide-hydrogen peroxide complex are quite similar. The activation energy is zero, and the entropy of activation about -28 cal K^{-1} mol^{-1}.

The steady-state treatment of the double-intermediate mechanism leads to the conclusion that the limiting rate at low substrate concentrations is

$$v_0 = k_0[E]_0[S] = \frac{k_1 k_2}{k_{-1} + k_2}[E]_0[S] \tag{4}$$

The second-order rate constant k_0, equal to $k_1k_2/(k_{-1} + k_2)$ is composite, and the Arrhenius law will not necessarily apply to it directly. There are, however, two special cases under which this composite constant will obey the law, as follows:

(a) When $k_2 \gg k_{-1}$. If this is the case k_{-1} may be neglected in comparison with k_2, and k_0 is equal to k_1. This means that the rate constant for the overall reaction at low substrate concentrations is simply that for the first step, the formation of the complex. If that is so, the Arrhenius law should apply to k_0, and the corresponding activation energy will be E_1, that for the initial complex formation.

(b) When $k_{-1} \gg k_2$. The constant k_0 is now equal to k_2k_1/k_{-1}. The ratio k_1/k_{-1} is simply the equilibrium constant for the complex formation:

$$\mathrm{E} + \mathrm{S} \underset{k_{-1}}{\overset{k_1}{\rightleftharpoons}} \mathrm{ES}$$

Its variation with temperature is given by

$$\frac{k_1}{k_{-1}} = \frac{A_1 e^{-E_1/RT}}{A_{-1} e^{-E_{-1}/RT}} = \frac{A_1}{A_{-1}} e^{-\Delta E/RT} \tag{5}$$

where A_1 and A_{-1} are temperature-independent frequency factors and ΔE is the increase in energy per mole for the change from E + S to ES. Since k_2 varies exponentially with temperature according to the Arrhenius law

$$k_2 = A_2 e^{-E_2/RT} \tag{6}$$

it follows that

$$\frac{k_1 k_2}{k_{-1}} = \frac{A_2 A_1}{A_{-1}} e^{-(E_2 + E_1 - E_{-1})RT} \tag{7}$$

$$\frac{k_1 k_2}{k_{-1}} = \frac{A_1 A_2}{A_{-1}} e^{-(E_2 + \Delta E)/RT} \tag{8}$$

In other words, the Arrhenius law should apply to this composite constant k_1k_2/k_{-1}, but the activation energy does not correspond to a single elementary step; it is the sum of the activation energy for the second step (ES $\rightarrow$ ES′ + X) and the total energy increase for the first step (E + S $\rightarrow$ ES).

These relationships are illustrated by the energy diagrams shown in Fig. 1. In case (a), which corresponds to $k_2 \gg k_{-1}$ and $k_0 = k_1$, the highest energy barrier over which the system must pass corresponds to $ES^{\ddagger}$, the activated complex for the reaction E + S $\rightarrow$ ES. The measured activation energy corresponding to k_0, the overall rate constant at low substrate concentrations, is thus E_1, corresponding to this initial step. In case (b), on the other hand, the highest energy barrier corresponds to $ES^{\ddagger\ddagger}$, the activated complex for the second step, ES $\rightarrow$ ES′ + X. In order to reach this level the system must first reach the level of ES, where the energy is ΔE higher than that of E + S, and then acquire an additional E_2; the activation energy corresponding to k_0 is thus $\Delta E + E_2$. In this case, the overall rate is in no way dependent on the rate of the initial step, so that the data can give no information about k_1 or E_1; these quantities can then be obtained only by the use of special techniques, such as those of transient-phase investigations.

When the situation corresponds to neither of these special cases the plots of $\log_{10} k_0$ against $1/T$ will not necessarily be linear, but energies of activation can still be calculated from the slope at any one temperature. The activation energy E_0 at any temperature is defined by

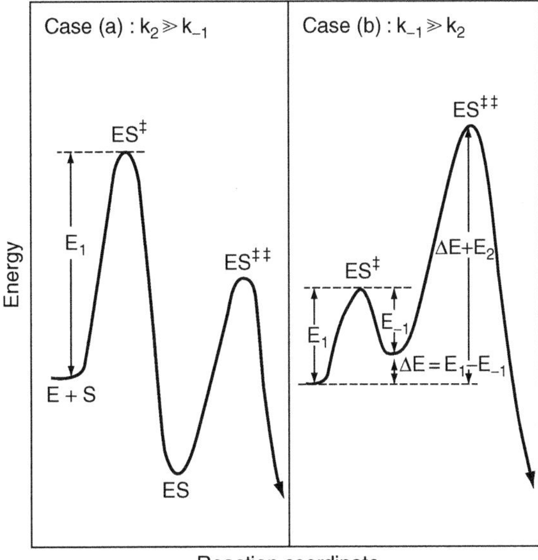

Fig. 1 Energy diagram for a reaction proceeding by the simple Michaelis-Menten mechanism, showing two extreme cases.

$$E_0 = RT^2 \frac{\mathrm{d} \ln k_0}{\mathrm{d}T} \tag{9}$$

By Eq. (4)

$$\ln k_0 = \ln k_1 + \ln k_2 - \ln(k_{-1} + k_2) \tag{10}$$

and therefore

$$\frac{E_0}{RT^2} = \frac{\mathrm{d} \ln k_0}{\mathrm{d}T} = \frac{\mathrm{d} \ln k_1}{\mathrm{d}T} + \frac{1}{k_2}\frac{\mathrm{d}k_2}{\mathrm{d}T} - \frac{1}{k_{-1} + k_2}\frac{\mathrm{d}(k_{-1} + k_2)}{\mathrm{d}T} \tag{11}$$

$$\frac{E_0}{RT^2} = \frac{\mathrm{d} \ln k_1}{\mathrm{d}T} + \frac{1}{k_2}\frac{\mathrm{d}k_2}{\mathrm{d}T} - \frac{1}{k_{-1} + k_2}\left(\frac{\mathrm{d}k_{-1}}{\mathrm{d}T} + \frac{\mathrm{d}k_2}{\mathrm{d}T}\right) \tag{12}$$

$$\frac{E_0}{RT^2} = \frac{\mathrm{d} \ln k_1}{\mathrm{d}T} + \frac{k_{-1}}{k_{-1} + k_2}\frac{\mathrm{d} \ln k_2}{\mathrm{d}T} - \frac{k_{-1}}{k_{-1} + k_2}\frac{\mathrm{d} \ln k_{-1}}{\mathrm{d}T} \tag{13}$$

The individual activation energies are defined by

$$E_1 = RT^2 \frac{\mathrm{d} \ln k_1}{\mathrm{d}T}, \quad E_{-1} = RT^2 \frac{\mathrm{d} \ln k_{-1}}{\mathrm{d}T}, \quad E_2 = RT^2 \frac{\mathrm{d} \ln k_2}{\mathrm{d}T} \tag{14}$$

and it follows that

$$E_0 = \frac{k_{-1}(E_1 + E_2 - E_{-1}) + k_2 E_1}{k_{-1} + k_2} \tag{15}$$

The overall activation energy E_0 is thus the weighted mean of the values E_1 and $E_1 + E_2 - E_{-1}$, the weighting factors being $k_2/(k_{-1} + k_2)$ and $k_{-1}/(k_{-1} + k_2)$.

Some values of E_0 and $\Delta S_0^{\ddagger}$, corresponding to k_0, are shown in Table V. The entropies of activation fall into two main groups. With pepsin, trypsin, and myosin the values are positive, whereas with chymotrypsin, carboxypeptidase, and urease they are negative. Light is thrown on these differences by consideration of the types of substrates that are hydrolyzed by these six enzymes. Chymotrypsin and urease invariably act upon electrically neutral molecules. Adenosine triphosphatase, on the other hand, is itself positively charged, and it acts upon a negatively charged substrate. Pepsin will act only upon substrates that contain a free $-CO_2H$ group, and at pH 4 this exists at least in part as a negatively charged group. Carboxypeptidase also acts only on substrates containing the $-CO_2H$ group, which will be ionized. The specificity requirement of trypsin is for a free $-NH_2$ group, and this will exist as the positively charged $-NH_3^+$ group. It is thus likely that with ATPase, pepsin, carboxypeptidase, and trypsin the interaction between enzyme and substrate involves a charge neutralization, whereas with chymotrypsin and urease there is no such neutralization.

Table V
Kinetic Values Relating to the Formation of the Enzyme-Substrate Complex

Enzyme	Substrate	Temperature (°C)	pH	k_0 (dm^3 mol^{-1} s^{-1})	E_0	$\Delta S_0^{\ddagger}$	References
Pepsin	Carbobenzoxy-L-glutamyl-L-tyrosine ethyl ester	31.6	4.0	0.57	23.1	14.1	Casey and Laidler (1950)
	Carbobenzoxy-L-glutamyl-L-tyrosine	31.6	4.0	0.79	20.2	2.6	Casey and Laidler (1950)
Trypsin	Chymotrypsinogen	19.6	7.5	2900	16.3	8.5	Butler (1941)
Chymotrypsin	Methyl hydrocinnamate	25.0	7.8	6.66	11.5	−23.2	Snoke and Neurath (1949, 1950)
	Methyl-DL-α-chloro-β-phenylpropionate	25.0	7.8	11.2	6.9	−33.0	Snoke and Neurath (1949, 1950)
	Methyl-D-β-phenyllactate	25.0	7.8	4.0	3.1	−47.2	Snoke and Neurath (1949, 1950)
	Methyl-L-β-phenyllactate	25.0	7.8	138	3.8	−38.5	Snoke and Neurath (1949, 1950)
	Benzoyl-L-tyrosine ethyl ester	25.0	7.8	19500	0.8	−38.5	Kaufman *et al.* (1949)
	Benzoyl-L-tyrosinamide	25.0	7.8	14.9	3.7	−43.0	Kaufman *et al.* (1949)
Carboxypeptidase	Carbobenzoxyglycyl-L-tryptophan	25.0	7.5	17444	9.9	−8.5	Snoke and Neurath (1949, 1950)
	Carbobenzoxyglycyl-L-phenylalanine	25.0	7.5	27900	9.6	−8.5	Snoke and Neurath (1949, 1950)
	Carbobenzoxyglycyl-L-leucine	25.0	7.5	3920	11.0	−7.5	Lumry *et al.* (1951)
Urease	Urea	20.8	7.1	5.0×10^6	6.8	−6.8	Wall and Laidler (1953)
Adenosine triphosphatase (myosin)	Adenosine triphosphate	25.0	7.0	8.2×10^6	21.0	44.0	Ouellet *et al.* (1952)

Reaction between an enzyme and an uncharged substrate involves certain electron shifts as a result of which the activated complex will be more polar than the reactants; there is therefore an increase in electro-striction and a corresponding negative entropy of activation. With ATPase, pepsin, and trypsin, on the other hand, this effect is presumably counteracted by the charge neutralization that occurs when the enzyme and substrate come together. This neutralization will lead to a release of water molecules, and there will be a corresponding increase of entropy. This explanation of the signs of the $\Delta S^{\ddagger}$ values derives some support from the general correlation between the sign of the entropy change and whether the substrate is charged or not. Carboxypeptidase, however, appears to be an exception, in that the substrates are charged but the entropies of activation are negative; a possible explanation is that the $-CO_2^-$ group on the substrate does not come into contact with a positive group on the enzyme, and is therefore not neutralized. The ionic strength effects for this enzyme tend to suggest that there is an approach of like (negative) charges.

Some of the entropy changes shown in Table V seem to be too large to be explained in terms of electrostatic effects alone, and structural effects must play a role. For chymotrypsin, where there is no charge neutralization, the entropies of activation are strongly negative, and the solvent studies of Barnard and Laidler (1952) suggest that the chymotrypsin unfolds during complex formation. Similarly, the solvent studies of Laidler and Ethier (1953) suggest that in the myosin system the large positive entropies of activation is partly due to electrostatic effects and partly to structural effects, the enzyme undergoing a contraction of the coiled helix when the substrate becomes attached.

For a further discussion of the solvent studies, in which mixed solvents are used, the reader is referred to Chapter 7 of the book by Laidler and Bunting (1973).

III. Reactions Involving More Than One Substrate

When there is more than one substrate the general scheme describing the one-substrate system has to be modified to include the binding and release steps of the additional substrate or substrates. In addition, some other interconversion steps must be considered.

Similar general principles apply to multisubstrate enzyme systems as to single-substrate systems. Quantities like V_{max} and K_m will not necessarily follow the Arrhenius law, since they are combinations of individual rate constants, but if certain rate constants predominate the law may be obeyed.

Little temperature-dependence work has been done on reactions involving more than one substrate. We will limit our discussion to some work on lactate dehydrogenase carried out by Borgmann *et al.* (1974, 1975, 1976) This enzyme catalyzes the overall reaction

$$E + NAD^+ + \text{lactate} \rightleftharpoons E + NADH + \text{pyruvate}$$

and was studied in both directions. The presteady-state and steady-state kinetics in the lactate → pyruvate direction (Borgmann *et al.*, 1974) are consistent with the four-step mechanism

$$E + A \underset{k_{-1}}{\overset{k_1[A]}{\rightleftharpoons}} EA \underset{k_{-2}}{\overset{k_2[B]}{\rightleftharpoons}} EAB \underset{k_{-3}[X]}{\overset{k_3}{\rightleftharpoons}} EY \underset{k_{-4}[Y]}{\overset{k_4}{\rightleftharpoons}} E + Y$$

where A and B are NAD^+ and lactate, respectively, and X and Y are pyruvate and NADH, respectively. It turned out that the above scheme was an oversimplification. Borgmann *et al.* (1976) showed that when presteady-state data in the forward direction are compared with similar data in the reverse reaction, it is necessary to include an additional step

$$\mathrm{EAB} \rightleftharpoons \mathrm{EXY}$$

We shall first discuss the results on the basis of the four-step model, and shall later consider the implications of adding the additional step.

The studies of the temperature dependence of the lactate dehydrogenase action (Borgmann *et al.*, 1974) reveal that the individual rate constants obey the Arrhenius law. The maximum steady-state velocities in the lactate → pyruvate direction are given by

$$\overrightarrow{V}_{\max} = \frac{k_3 k_4 [\mathrm{E}]_0}{k_3 + k_4} \tag{16}$$

and in the pyruvate → lactate direction the equation is

$$\overleftarrow{V}_{\max} = \frac{k_{-1} k_{-2} [\mathrm{E}]_0}{k_{-1} + k_{-2}} \tag{17}$$

Straight lines were obtained when $\log \overrightarrow{V}_{\max}$ and $\log \overleftarrow{V}_{\max}$ were plotted against $1/T$. This suggests that the maximum steady-state velocity in each direction is controlled by one rate constant. Further study showed that $\overrightarrow{V}_{\max}$ is equal to $k_4[\mathrm{E}]_0$. The Michaelis constants K_m with respect to lactate and pyruvate do not

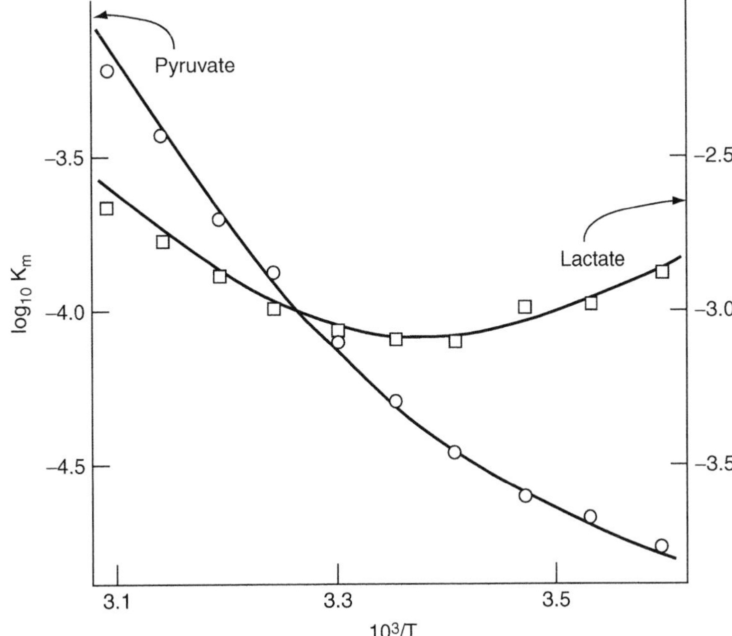

Fig. 2 Plots of log K_m against $1/T$ for the lactate dehydrogenase system, for the forward and reverse reactions. From Borgmann *et al.* (1974).

give straight lines when log K_m is plotted against $1/T$, as shown in Fig. 2, suggesting that the K_m's are controlled by more than one rate constant. For the methods of analysis of these curves, the reader is referred to the original paper (Borgmann *et al.*, 1974).

From the temperature-dependence measurements in the presteady-state and steady-state studies on this system, it was possible to construct the enthalpy, entropy, and Gibbs energy profiles shown in Fig. 3. The values for $\Delta H^{\ddagger}$, $\Delta S^{\ddagger}$, and $\Delta G^{\ddagger}$ are given in Table VI. The ternary complex EAB has a lower Gibbs energy than any other state, which means that the ternary complex is the most stable and predominant under standard conditions. The entropy profiles are of special interest, since they can provide insight into the state of the protein-substrate complex.

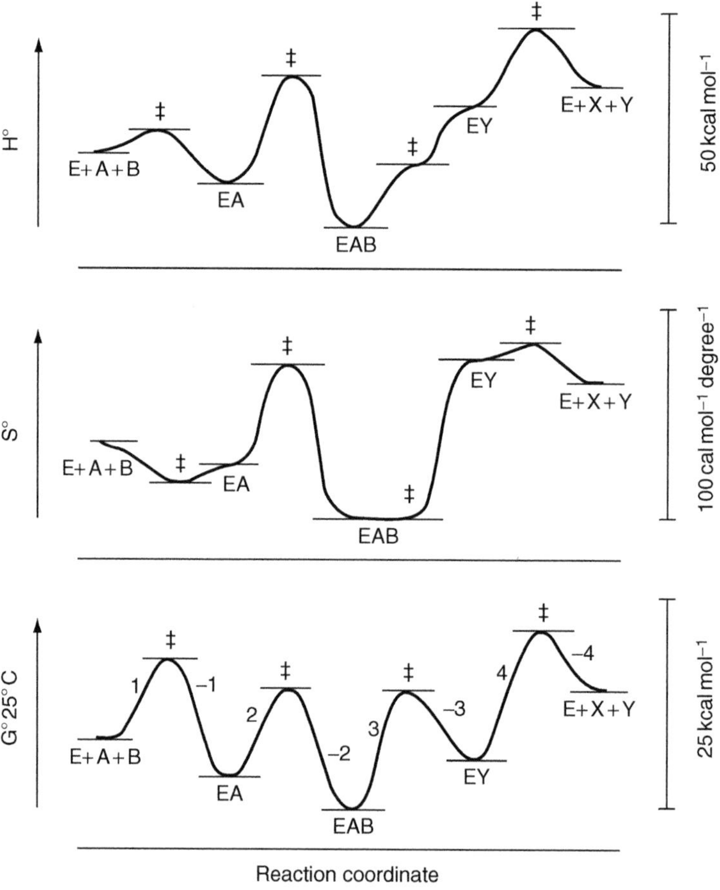

Fig. 3 Profiles showing the changes in standard enthalpy, entropy, and Gibbs energy for the lactate dehydrogenase system at 25 °C. This diagram is based on the results obtained in the lactate → pyruvate direction, which leads to the four-step mechanism. From Borgmann *et al.* (1974).

Table VI
Enthalpies, Entropies, and Gibbs Energies

	ΔH° or $\Delta H^\ddagger$ (kcal mol^{-1})	ΔS° or $\Delta S^\ddagger$ (cal K^{-1} mol^{-1})	ΔG° or $\Delta G^\ddagger$ at 25 °C (kcal mol^{-1})
k_1	4.1	−18.2	9.5
k_{-1}	12.5	−5.8	14.2
k_2	25.6	50.0	10.7
k_{-2}	36.2	74.2	14.1
k_3	13.9	−0.2	14.0
k_{-3}	−14.5	−74.8	7.9
k_4	17.5	7.6	15.2
k_{-4}	12.8	17.8	7.4
k_1/k_{-1}	−8.4	−12.4	−4.7
k_2/k_{-2}	−10.6	−24.2	−3.4
k_3/k_{-3}	28.4	74.6	6.1
k_4/k_{-4}	4.7	−10.2	7.8
K_{eq}	14.1	27.8	5.8

The low entropy for EAB suggests a more folded structure. The entropy profile also suggests that the binding of NAD^+ produces a more folded structure, and that the binding of NADH results in partial unfolding of the protein.

The studies of the presteady-state and the steady-state kinetics in the lactate → pyruvate direction did not provide any evidence for more than four steps. However, when the work was done on the reverse reaction (Borgmann *et al.*, 1976) it was found necessary to include the additional step EAB $\rightleftharpoons$ EXY, so that the mechanism is now

$$E + A \rightleftharpoons EA \rightleftharpoons EAB \rightleftharpoons EXY \rightleftharpoons EY \rightleftharpoons E + Y$$

The addition of this step affects almost all the thermodynamic parameters and the energy profiles. How the insertion of a step affects the enthalpy profile can be seen in Fig. 4. Comparison of Fig. 4A with B shows that interpretation of lactate-pyruvate data in terms of four-step model misplaces the barrier that is really between EXY and EY, placing it between EA and EAB. The use of the incomplete model does not affect the heights of the barriers and intermediates in the enthalpy profile, but it misplaces the barriers.

It is of interest to see what relationship exists between the adaptation of an animal to its thermal environment and the thermodynamic and kinetic parameters for the enzyme reactions. Ectothermic animals, such as fish, conform to environmental temperatures; endothermic animals, such as most mammals, maintain a relatively constant body temperature, which is more or less independent of the ambient temperature. Since ectothermic animals normally operate at lower temperatures than endothermic animals it might be expected that the kinetic parameters would favor more rapid reaction at lower temperatures, but this seems not to be the case. Low *et al.* (1973) compared enzymes from endothermic and ectothermic animals, but their work was confined to high substrate concentrations

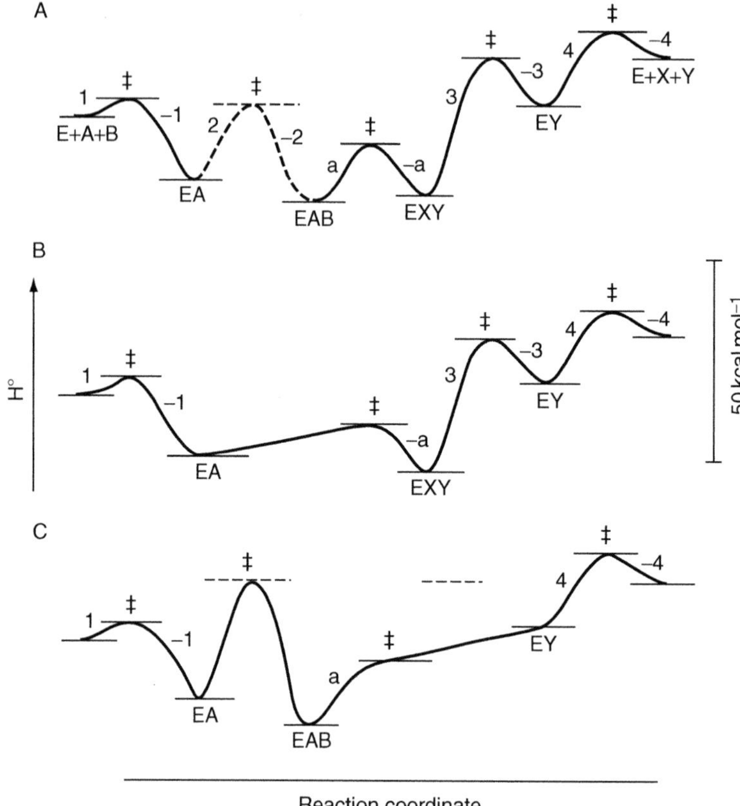

Fig. 4 Enthalpy profiles for the five-step mechanism (A) and the four-step mechanism (B and C). Profile B is based on the data for the pyruvate → lactate reaction, profile C for the reverse reaction. From Borgmann *et al.* (1976).

and therefore concerned only with V_{max} values; under physiological conditions the substrate concentrations are always too low to saturate the enzymes (Hochachka and Somero, 1973).

Borgmann *et al.* (1975) compared the individual thermodynamic and kinetic parameters for lactate dehydrogenase from beef heart and beef muscle (endothermic animals) and from flounder muscle (ectothermic). The comparison was based on the four-step model which, although incomplete, still provides a valuable characterization of the trends brought about by thermal adaptation. Figure 5 shows the Gibbs energy profiles for the three enzymes, and Fig. 6 compares the enthalpy and entropy profiles. It is particularly significant that there are greater differences between the thermodynamic parameters for the stable complexes AE, EAB, and EY than for the activated complexes. The changes affecting the stable and activated complexes can be interpreted in terms of the numbers of weak bonds (e.g., hydrogen bonds) formed or broken during the formation of the complexes

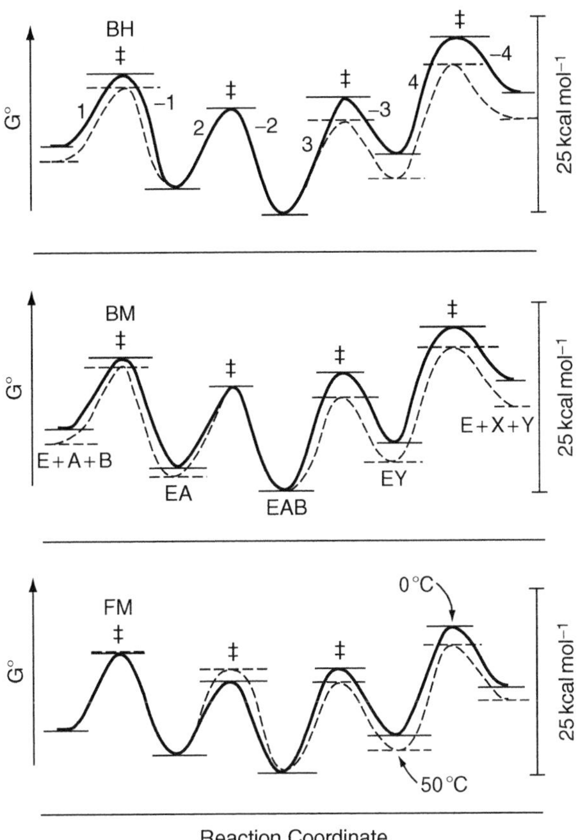

Fig. 5 Gibbs energy profiles for three forms of lactate dehydrogenase at 0 °C (boldface line) and 50 °C (dashed line). The three types of enzyme are from beef heart (BH), beef muscle (BM), and flounder muscle (FM). From Borgmann *et al.* (1975).

(Borgmann *et al.*, 1975). The main conclusion is that the number of weak bonds is greater for the beef heart muscle enzyme and smaller for the flounder muscle enzyme. This is consistent with the suggestion of Low *et al.* (1973) that endothermic enzymes have, through adaptability and selectivity, developed higher structural stability and can therefore more effectively resist higher temperatures.

Another interesting comparison is related to overall rates under physiological conditions, when pyruvate concentrations are about 0.05 mM. At this concentration the rate of the pyruvate → lactate conversion with beef heart enzyme goes through a maximum at about 35 °C, which is close to body temperature (Borgmann *et al.*, 1974). The reason for this maximum is related to the temperature variation of K_m; at lower temperatures the enzyme is almost completely saturated with substrate, but at higher temperatures the substrate concentration is well below the K_m, and the rate falls off. With the flounder muscle enzyme, on the other hand,

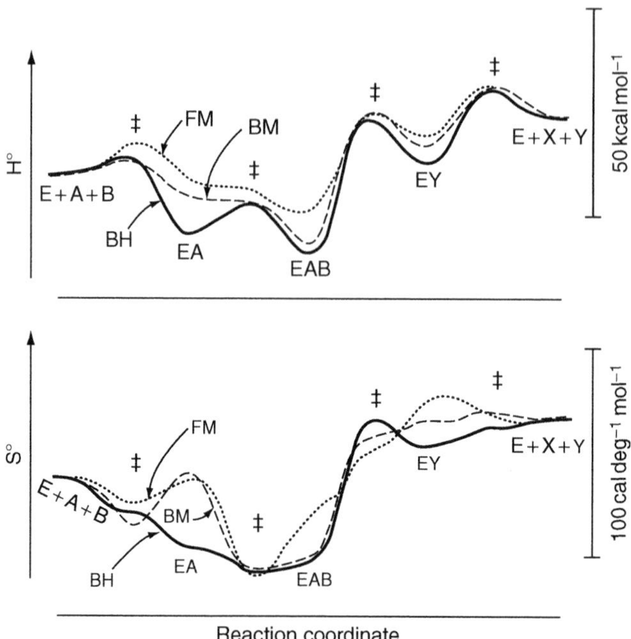

Fig. 6 Enthalpy and entropy profiles, superimposed for the three forms of lactate dehydrogenase (compare Fig. 5). From Borgmann *et al.* (1975).

the maximum temperature under physiological conditions is lower; this is significant in view of the lower temperature of the fish.

IV. Diffusional Effects in Enzyme Systems

Certain enzyme processes, such as the binding of enzymes to substrates, sometimes occur very rapidly. When this is the case, the question arises whether the rates are influenced by the rates of diffusion of the reacting substances. This has an important bearing on temperature coefficients and will now be briefly considered.

If the rate of the chemical interaction between two substances A and B is very much greater than the rate of diffusion, the rate of the process will be equal to the rate with which A and B diffuse together and the rate constant is then

$$k_D = 4\pi(D_A + D_B)d_{AB} \tag{18}$$

Here D_A and D_B are the diffusion coefficients of A and B and d_{AB} is the distance between the centers of A and B when reaction occurs. If the reactant molecules A and B are large compared with the solvent molecules (which is the case with enzyme reactions), Stokes's law will apply to a good approximation, and if d_{AB} is taken to be the sum of the molecular radii r_A and r_B, Eq. (18) becomes

$$k_D = \frac{2kT}{3\eta}\frac{(r_A + r_B)^2}{r_A r_B} \tag{19}$$

where k is the Boltzmann constant and η is the viscosity of the medium. For aqueous solutions at 25 °C, and with typical values for the radii, this expression leads to

$$k_D = 7.0 \times 10^9 \text{dm}^3\text{mol}^{-1}\text{s}^{-1}$$

for the rate constant of a fully diffusion-controlled reaction. This value is modified if electrostatic effects are involved (Alberty and Hammes, 1958); equal and opposite unit charges lead to an increase by a factor of about 4.

If the rate of the chemical interaction is not very much larger than the rate of diffusion there will be only partial diffusion control. If the rate constant for the chemical interaction is k_{chem}, the overall rate constant is given by

$$k = \frac{k_D k_{chem}}{k_D + k_{chem}} \tag{20}$$

If $k_{chem} \gg k_D$, the rate constant is k_D, as previously discussed; if $k_D \gg k_{chem}$, the rate constant is k_{chem}. The latter situation applies to most chemical reactions. Figure 7 shows a plot of k against k_{chem}, with k_D taken as 7.0×10^9 dm^3 mol^{-1} s^{-1}, the value given by Eq. (19) for water at 25 °C. When $k_{chem} = 7.0 \times 10^9$ dm^3 mol^{-1} s^{-1}

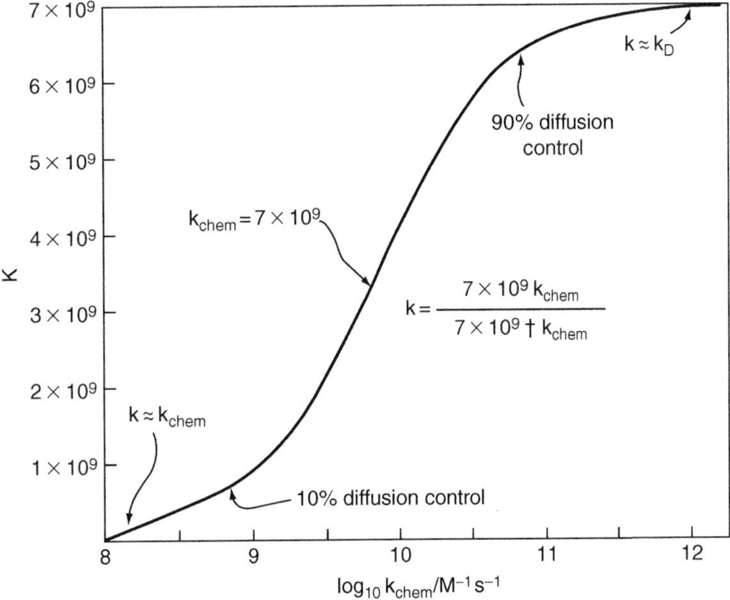

Fig. 7 Plots of the overall rate constant k against the rate constant k_{chem} for the chemical interaction, on the basis of Eq. (20). The rate constant for the diffusion process, k_D, is taken to be 7.0×10^9 dmRaster: F7 ^{3} mol^{-1} s^{-1}, a typical value for water at 25 °C.

the value of k is 3.5×10^9, one half of the value for full diffusion control, and one can say that there is 50% diffusion control. We reach 90% diffusion control only when k_{chem} is 6.3×10^{10} dm^3 mol^{-1} s^{-1}.

Inspection of the data given in Tables I–VI shows that in no case is the rate constant large enough for there to be any significant amount of diffusion control. The highest rate constant recorded in the tables is for horse blood catalase (Table III), for which $k_1 = 3.54 \times 10^7$ dm^3 mol^{-1} s^{-1} at 22 °C. This corresponds to a diffusional influence of less than 1%. This conclusion is consistent with the very low-activation energies shown in Table III; they are much lower than the values (4-5 kcal) characteristic of diffusion in water.

Although there is no evidence of diffusion control in the examples given in Tables I–VI, there have been some cases reported where the enzyme reaction occurs at diffusion-controlled rate (Alberty and Hammes, 1958; Jonsson and Wennerstrom, 1978). With immobilized enzymes, there may be substantial diffusional control, as discussed in Chapter 9 of Vol. 64.

V. Enzyme Inactivation

We have noted that the temperature optima found in studies of enzyme-catalyzed reactions are due to the inactivation of the enzyme at higher temperatures. The kinetics of enzyme inactivations, which are protein denaturations arising from conformational changes, have been investigated extensively, and a number of temperature studies have been made. Space does not permit a detailed treatment of the topic; the interested reader is referred to Chapter 13 of the book by Laidler and Bunting (1973), where references to other articles and reviews are given.

It is necessary to consider both temperature and pH effects on enzyme inactivations, since the two effects are closely related. It is usually found that rates of inactivation pass through a minimum as the pH is varied, and the behavior is frequently quite complex. A satisfactory treatment of pH and temperature effects has been given by Levy and Benaglia (1950) on the basis of the various ionized states, P_1, P_2, P_3, etc., in which a protein can occur. Suppose that P_1 is the form in which the protein exists in the most acid solutions, and that P_2 is P_1 which has lost a proton, etc:

$$\begin{array}{ccccccccc} P_1 & \underset{}{\overset{K_1}{\rightleftharpoons}} & P_2 & \overset{K_2}{\rightleftharpoons} & P_3 & \overset{K_3}{\rightleftharpoons} & P_4 & \overset{K_4}{\rightleftharpoons} & P_5 \\ \downarrow k_1 & & \downarrow k_2 & & \downarrow k_3 & & \downarrow k_4 & & \downarrow k_5 \\ D & & D & & D & & D & & D \end{array}$$

Each form can go into a denatured form D with a rate constant k_1, k_2, etc. If one of the intermediate rate constants, such as k_3, is smaller than the others, the overall rate of denaturation will pass through a minimum at the pH at which the corresponding form (e.g., P_3) is predominant.

A satisfactory treatment of temperature effects obviously requires that the rate constants k_1, k_2, etc., and the dissociation constants K_1, K_2, etc., are separated, on the basis of a pH study, and that their values are determined over a range of temperatures. This has been done in only a few cases, the most complete study being for the denaturation of ricin (Levy and Benaglia, 1950). On the basis of the above scheme, the form P_4 is the one that undergoes deactivation most slowly, the kinetic parameters for the process being, at 65 °C:

$$\Delta G^{\ddagger} = 38.9 \text{ kcal mol}^{-1}, \quad \Delta H^{\ddagger} = 53 \text{ kcal mol}^{-1}, \quad \Delta S^{\ddagger} = 42 \text{ cal K}^{-1}\text{mol}^{-1}$$

The form P_3 undergoes more rapid denaturation, with

$$\Delta G^{\ddagger} = 32.3 \text{ kcal mol}^{-1}, \quad \Delta H^{\ddagger} = 89 \text{ kcal mol}^{-1}, \quad \Delta S^{\ddagger} = 169 \text{ cal K}^{-1}\text{mol}^{-1}$$

It is of interest that the more rapid rate for P_3 is associated with a much larger entropy of activation, and that the enthalpy of activation is higher. The remaining forms, P_1, P_2, P_3, etc., undergo denaturation even more rapidly, but the kinetic parameters cannot be determined from the data available. For the process $P_3 \rightarrow P_4$ the thermodynamic parameters, at 65 °C, are

$$\Delta G = 11.7 \text{ kcal mol}^{-1}, \quad \Delta H = 30 \text{ kcal mol}^{-1}, \quad \Delta S = 54 \text{ cal K}^{-1}\text{mol}^{-1}$$

Of special interest are the very large enthalpy and entropy values associated with these processes. Further examples are given in Table VII, which shows the energies and entropies of activation obtained in studies at a fixed pH. If a process has a very high energy of activation and occurs at an appreciable speed at ordinary temperature, it must also have a large positive entropy of activation. It was noted at the beginning of this chapter that the occurrence of a temperature optimum requires that the activation energy for the inactivation process must be greater than that for the enzyme-substrate reaction.

The interpretation of these large enthalpy and entropy changes requires a detailed consideration of individual cases. We will here comment on only a few examples; further discussion is to be found in Chapter 13 of Laidler and Bunting's book (Laidler and Bunting, 1973). The process $P_3 \rightarrow P_4 + H^+$ in the ricin system involves an entropy increase of 54 cal K^{-1} mol^{-1} at 65 °C. The splitting-off of a proton can follow two patterns, of which the following are examples:

$$RNH_3^+ \rightleftharpoons RNH_2 + H^+$$
$$ROH \rightleftharpoons RO^- + H^+$$

Table VII
Energies and Entropies of Activation for Enzyme Inactivations

Enzyme	pH	Energy of activation (kcal mol^{-1})	Entropy of activation (cal K^{-1} mol^{-1})	References
Pancreatic lipase	6.0	46.0	68.2	McGillivray (1930)
Trypsin	6.5	40.8	44.7	Pace (1930)
ATPase	7.0	70.0	150.0	Ouellet *et al.* (1952)

In the first example the proton leaves a positively charged group, and there is no change in the number of charges when the process occurs. There will be no large change in entropy arising from electrostatic effects, and this conclusion is supported by experimental values obtained for simple ionizations of this kind (e.g., for $NH_4^+ \rightarrow NH_3 + H^+, \Delta S^\circ = -0.5$ cal $K^{-1}mol^{-1}$ at 25 °C). In the second example, two charged species are created when ionization occurs, and negative entropy changes will therefore arise from electrostatic causes and are found experimentally for simple systems (e.g., for $CH_3COOH \rightarrow CH_3COO^- + H^+, \Delta S = -22.0$ cal K^{-1} mol^{-1} at 25 °C. It follows that the value of 54 cal K^{-1} mol^{-1} cannot be interpreted in terms of electrostatic effects, and must be due to a very profound structural difference between P_3 and P_4.

The entropy of activation of 42 cal K^{-1} mol^{-1} observed for the denaturation of the P_4 form of ricin must be explained by a large structural change when P_4 becomes the activated-complex $P_4^\ddagger$. Inactivations usually involve a large overall entropy increase, because the denatured protein has a more open and disordered structure. Since the activated-complex $P_4^\ddagger$ is on its way toward becoming the denatured protein, it is not surprising that it also has a higher entropy than P_4. The same explanation applies to the values shown in Table VII, and it appears that the positive entropies of activation, and corresponding high energies of activation, are generally associated with enzyme inactivations, as a result of the loosening of the enzyme structure.

References

Alberty, R. A., and Hammes, G. G. (1958). *J. Phys. Chem.* **62,** 154.
Barnard, M. L., and Laidler, K. J. (1952). *J. Am. Chem. Soc.* **74,** 6099.
Baxendale, J. H., Evans, M. G., and Park, G. S. (1946). *Trans. Faraday Soc.* **42,** 155.
Bender, M. L., Schonbaum, G. R., and Zerner, B. (1962). *J. Am. Chem. Soc.* **84,** 2562.
Bender, M. L., Kézdy, F. J., and Gunter, C. R. (1964). *J. Am. Chem. Soc.* **86,** 3714.
Bonnichsen, R. K., Chance, B., and Theorell, H. (1947). *Acta Chem. Scand.* **1,** 685.
Borgmann, U., Moon, T. W., and Laidler, K. J. (1974). *Biochemistry* **13,** 5152.
Borgmann, U., Laidler, K. J., and Moon, T. W. (1975). *Can. J. Biochem.* **53,** 1196.
Borgmann, U., Laidler, K. J., and Moon, T. W. (1976). *Can. J. Biochem.* **54,** 915.
Bredig, G., and von Berneck, R. M. (1899). *Z. Phys. Chem. Stoechiomn. Verwandschaftslehre* **31,** 258.
Butler, J. A. V. (1941). *J. Am. Chem. Soc.* **63,** 2971.
Casey, E. J., and Laidler, K. J. (1950). *J. Am. Chem. Soc.* **72,** 2159.
Hinberg, I., and Laidler, K. J. (1972). *Can. J. Biochem.* **50,** 1334.
Hinkle, P. M., and Kirsch, J. F. (1970). *Biochemistry* **9,** 4633.
Hochachka, P. W., and Somero, G. N. (1973). "Strategies of Biochemical Adaptation," p. 11. Saunders, Philadelphia.
Jonsson, B., and Wennerstrom, H. (1978). *Biophys. Chem.* **7,** 285.
Kaufman, S., Neurath, H., and Schwert, G. W. (1949). *J. Biol. Chem.* **177,** 792.
Laidler, K. J. (1951). *Arch. Biochem.* **30,** 226.
Laidler, K. J. (1955). *Discuss. Faraday Soc.* **20,** 83.
Laidler, K. J. (1969). "Theories of Chemical Reaction Rates," McGraw-Hill, New York.
Laidler, K. J. (1972). *J. Chem. Educ.* **49,** 343.
Laidler, K. J., and Beardall, A. J. (1955). *Arch. Biochem. Biophys.* **55,** 138.

Laidler, K. J., and Bunting, P. S. (1973). "The Chemical Kinetics of Enzyme Action," Oxford University Press (Clarendon), London and New York.
Laidler, K. J., and Chen, D. T. Y. (1958). *Trans. Faraday Soc.* **54,** 1026.
Laidler, K. J., and Ethier, M. (1953). *Arch. Biochem. Biophys.* **43,** 338.
Laidler, K. J., and Landskroener, P. A. (1956). *Trans. Faraday Soc.* **52,** 200.
Lazdunski, C., and Lazdunski, M. (1966). *Biochim. Biophys. Acta* **113,** 551.
Levy, M., and Benaglia, A. E. (1950). *J. Biol. Chem.* **186,** 829.
Low, P. S., Bada, J. L., and Somaro, G. N. (1973). *Proc. Natl. Acad. Sci. USA* **70,** 430.
Lumry, R., Smith, E. I., and Glantz, R. R. (1951). *J. Am. Chem. Soc.* **73,** 4330.
McGillivray, I. H. (1930). *Biochem. J.* **24,** 891.
Morales, M. F., and Botts, J. (1952). *Arch. Biochem. Biophys.* **37,** 283.
Ouellet, L., Laidler, K. J., and Morales, M. F. (1952). *Arch. Biochem. Biophys.* **39,** 37.
Pace, J. (1930). *Biochem. J.* **24,** 606.
Pana, C. (1928). *Trans. Faraday Soc.* **24,** 486.
Snoke, J. E., and Neurath, H. (1949). *Arch. Biochem.* **21,** 351.
Snoke, J. E., and Neurath, H. (1950). *J. Biol. Chem.* **182,** 577.
Tammann, G. (1895). *Z. Phys. Chem. Stoechiom. Verwandschaftslehre* **18,** 426.
Wall, M. C., and Laidler, K. J. (1953). *Arch. Biochem. Biophys.* **43,** 299.
Walton, J. H. (1904). *Z. Phys. Chem. Stoechiom. Verwandschaftslehre* **47,** 185.
Williams, J. (1928). *Trans. Faraday Soc.* **24,** 245.
Wilson, I. B., and Cabib, E. (1956). *J. Am. Chem. Soc.* **78,** 202.

CHAPTER 8

Site-Directed Mutagenesis: A Tool for Studying Enzyme Catalysis

Bryce V. Plapp
Department of Biochemistry
Carver College of Medicine, The University of Iowa
Iowa City, IA 52242-1109, USA

I. Update

The use of this tool has become more convenient (with commercial kits and efficient DNA sequencing) and more common in recent years. Directed mutagenesis is applied to study amino acid residues that may participate directly in enzyme catalysis and to explore substrate specificities, structure-function relationships that arise from evolutionary studies, and the roles of protein dynamics in catalysis. Because protein enzymes are large and cooperative, and the effects of substitutions

DOI: 10.1016/B978-0-12-378608-1.00008-6

of amino acids are usually not additive, it is a continuing challenge to evaluate quantitatively the functional roles of a particular amino acid (Kraut *et al.*, 2003). Effects on structure, binding, catalysis, and dynamics can occur because of interactions of local or global features of the structure. Comprehensive structural and kinetic studies are still required to understand the manifold effects of each mutation. For instance, substitution of the presumed catalytic histidine (His-51) in the proton relay system of liver alcohol dehydrogenase affects the kinetics of coenzyme binding and proton and hydride transfers, but without perturbing the protein structure (Kovaleva and Plapp, 2005; LeBrun and Plapp, 1999; LeBrun *et al.*, 2004). Other substitutions greatly affect conformational changes, as demonstrated by structural studies, and thereby alter pH dependencies and kinetic constants (Ramaswamy *et al.*, 1999; Rubach *et al.*, 2001). Enzymes with substitutions that unmask the hydrogen transfer step (hydrogen transfer becoming predominantly rate-limiting) are useful for studying the hydrogen tunneling and the roles of protein dynamics in catalysis (Bahnson and Klinman, 1995; Rubach *et al.*, 2001). Engineering enzymes for use in biotechnology is an attractive goal, and even though some success is obtained by rational redesign and specific mutations, it appears that residues distal from the substrate-binding sites can substantially affect activity (Fan and Plapp, 1999; Ganzhorn *et al.*, 1987; Park and Plapp, 1992). The unidentified residues may slightly alter the structures at the active site or affect protein dynamics. Rather than site-directed mutagenesis, random mutagenesis and directed evolution may be more useful for producing specific enzymes. Nevertheless, for studying enzyme catalysis, the principles and procedures developed in the article remain useful guidelines today.

II. Introduction

Enzymes catalyze reactions by binding and orienting substrates in an active site formed with many amino acid residues that participate in the chemical transformation. Three-dimensional structures determined by X-ray crystallography show that enzymes bind substrates with a combination of ionic and hydrogen bonds and van der Waals interactions. The specificity of enzymes is due to multiple interactions that define the size and shape of the active site. The enzyme catalyzes the reaction by affecting solvation and the electrostatic environment and by facilitating proton transfers and covalent chemistry at the reaction center. Understanding the principles of enzyme action should allow us to design specific catalysts.

This chapter considers the purposes for using site-directed mutagenesis, some of the precautions for such studies, and an overall approach that is designed to provide maximal information. Some studies will be selected to illustrate what can be learned about the marvels of enzymes and to show why some conclusions are uncertain. Other articles have described the strategies and details for the techniques of site-directed mutagenesis and recombinant DNA technology (Methods in Enzymology Vols. 68, 100, 101, 153, 154, 155, 216, and 218). Results with some enzymes have been reviewed previously (Fersht, 1987; Gerlt, 1987; Hartman and Harpel, 1993; Miles, 1991; Silverman, this volume; Tsai and Yan, 1990).

III. Purposes

The ability to substitute selected amino acid residues by site-directed mutagenesis and to express large amounts of enzyme with recombinant DNA technology enables a powerful approach for testing hypotheses for catalysis derived from knowledge of enzyme structures, kinetics, chemical modifications, and bioorganic chemistry. What can we learn from site-directed mutagenesis? A primary goal has been to identify which amino acid residues in the active site are "essential," or critical, for catalysis. These usually include residues that appear to participate in acid-base catalysis or interact directly with the substrates. The residues may be conserved in homologous enzymes from different species or be identified by chemical modification or by inspection of the three-dimensional structure. An implicit goal of such studies is an estimation of the relative importance of a residue for catalysis. However, the change in activity cannot be the only criterion for assessing involvement of a residue, and circular arguments arise. If we are attempting to determine the magnitude of the contribution to catalysis, we cannot use the change in activity to judge whether the residue contributes. Some investigators seem to expect that substitution of an essential residue should totally inactivate the enzyme, but a 10-fold change in activity could represent the actual contribution. What we have here is not a failure to communicate, but rather a crisis in comprehension of catalysis.

A second purpose is to understand the evolution of catalytic efficiency, which requires chemical catalysis and specificity of binding. Bringing the substrates together and utilizing the binding energy provide the major contribution to catalysis (Jencks, 1975). Thus, residues that make van der Waals contacts or hydrogen bonds can be just as important for catalysis as reactive residues that participate in the chemistry (Fersht, 1987). Substituting amino acid residues permits the creative scientist to attempt to reproduce the evolution of catalytic function (Evolution of Catalytic Function, 1987). For a modest and practical goal, the origins of the different substrate specificities can be determined. Today, remodeling of substrate specificity is still difficult and unpredictable (Bone and Agard, 1991; Hedstrom *et al.*, 1991). Additional purposes have included studies on structural features that control conformational changes, allosteric effects, and protein dynamics.

IV. Precautions

Site-directed mutagenesis studies raise as many questions as they answer. There are some "successes," that is, results consistent with the predictions, and many surprises. Interpretation of results is difficult because many amino acid side chains contact the substrates, and it is not readily possible to assign a specific role to a particular group. A side-chain functional group involved in acid-base catalysis also participates in binding substrates and in the topography of the active site.

Some amino acid residues identified as being essential because chemical modification "inactivated" the enzymes have now been substituted with other amino acid residues without loss of activity (Profy and Schimmel, 1986; Teng *et al.*, 1993). This is not a real surprise, as it was recognized that attaching a large substituent to a reactive residue in the active site could simply block substrate binding. Furthermore, substitutions distant from the active site may decrease activity indirectly. Fromm pointed out one example. Modification of Lys-140 with pyridoxal 5′-phosphate or substitution with Ile inactivated *Escherichia coli* adenylosuccinate synthase, as did modification of Arg-147 with phenylglyoxal or substitution with Leu (Dong and Fromm, 1990; Dong *et al.*, 1990). Substrates protected against the chemical modifications, and it was concluded that these residues were at the active site. Nevertheless, a three-dimensional structure suggests that these residues are at the interface between monomers and far from the active site, which was identified by homology to p21^{ras} (Poland *et al.*, 1993). At this time, the modification results are not explained. Other mutagenesis studies on the consensus phosphate-binding site gave results consistent with expectations (Liu *et al.*, 1992).

As compared to chemical modifications, where residual activity below a few percent could be attributed to unmodified enzyme, site-directed mutations have lowered the limits so that activities reduced by 6 or more orders of magnitude can be meaningful in careful studies. If enzymes accelerate reactions by 9–12 orders of magnitude, the residual activities should be determined if possible. The criterion that an amino acid residue is essential if its substitution totally inactivates is obsolete. Furthermore, the magnitude of the change in activity should not be the only criterion used to judge the significance of the residue for catalysis. There are many amino acid residues at the active site (see, e.g., Fig. 1), and it is reasonable to suppose that each residue participates in catalysis. If each of ten residues contributed a factor of 10 and the effects were additive, the enzymatic rate enhancement could be explained. Thus, site-directed substitutions could allow estimations of each contribution, but this analysis can be confounded by the cooperative nature of catalysis.

Removal of a critical functional group typically produces an impaired enzyme, whose activity reflects the new constellation of residues. For instance, if a group involved in acid-base catalysis is substituted, the pH dependence of the modified enzyme will reflect the ionization of the remaining residues. Because active sites are different, the substitution of the same residue, for example, histidine, in different enzymes will naturally lead to effects of various magnitudes. The "context" complicates the estimation of catalytic contributions for particular interactions (Matthews, 1993). For instance, changing His-47 to Arg in three different alcohol dehydrogenases increased affinity for the coenzymes from 2- to 20-fold (Light *et al.*, 1992). Nevertheless, the results from sufficient studies should lead to a consensus. Eventually, computational biochemistry should be able to account for the interaction energies. Site-directed mutagenesis is one way to perturb the energies.

The definition of the active site is also being reconsidered. Once it included only the amino acid residues in contact with the substrates, but now we appreciate that

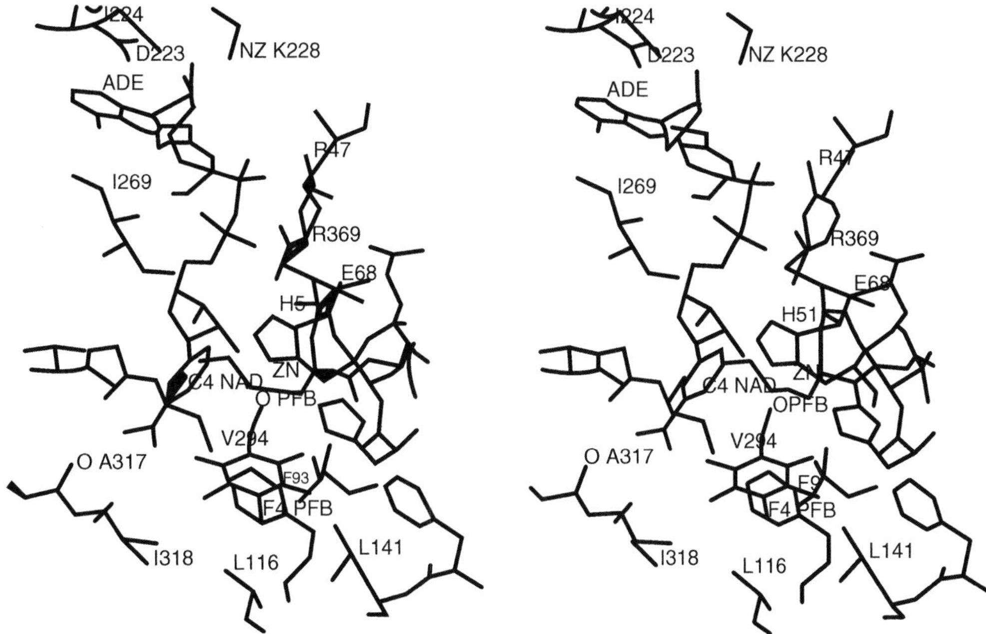

Fig. 1 Three-dimensional structure of horse liver alcohol dehydrogenase complexed with NAD^+ and 2,3,4,5,6-pentafluorobenzyl alcohol. The structure has been determined at 2.1 Å resolution with an *R* value of 18.3%. Some of the many interactions with the coenzyme and substrate analog are illustrated. The adenine ring is sandwiched between the side chains of Ile-224 and Ile-269; the adenosine ribose hydroxyl groups form hydrogen bonds with the carboxyl group of Asp-223 and the ε-amino group of Lys-228. The pyrophosphate moiety is neutralized by the guanidino groups of Arg-47 and Arg-369. The nicotinamide ribose sits on the side chain of Val-294, and its hydroxyl groups form hydrogen bonds with Ser-48, His-51, and Ile-269. The nicotinamide ring is hydrogen bonded to main-chain atoms of residues 317, 319, and 292. The catalytic zinc is tightly bound to the enzyme by Cys-46, Cys-174, and His-67. Near the zinc is the carboxyl group of Glu-68, which interacts with Arg-369. The hydroxyl group of the alcohol binds the zinc and is hydrogen bonded to the hydroxyl group of Ser-48, which is linked through the 2′-hydroxyl of the nicotinamide ribose to the imidazole of His-51. This system could shuttle a proton from the buried alcohol to His-51. The benzene ring of the alcohol is in a hydrophobic barrel that includes the side chains of Leu-57, Phe-93, Leu-116, Leu-141, Val-294, and Ile-318. Although the pentafluorobenzyl alcohol is not oxidized by the enzyme because of electron withdrawal by the fluorine atoms, the *pro-R* hydrogen is in a position that appears poised for direct transfer to C-4 of the nicotinamide ring. From Ramaswamy *et al.* (1994).

interactions from nearby residues modulate activity. Indeed, the whole protein structure is involved in catalysis, and distant changes can indirectly affect activity by long-range electrostatic effects or repositioning of residues near the reaction center. Substitutions of residues in mobile loops that form the active site during substrate-induced conformational changes also can significantly affect activity (Li *et al.*, 1992; Pan *et al.*, 1993). Proteins are dynamic molecules that are sensitive to global effects.

Distinguishing between direct effects of a substitution on catalysis and indirect effects due to structural changes cannot be resolved easily. Every substitution affects at least the local structure of the protein. It has been suggested that substitution of a residue at the active site decreases activity because of large changes in structure (Gerlt, 1987; Hibler *et al.*, 1987). However, substitutions of surface residues distant from the active site probably have relatively small effects on catalysis, as there are many substitutions (e.g., in homologous enzymes) that do not affect activity. X-ray crystallography of mutant forms of various proteins shows that the effects of local structural changes on nearby residues diminish with distance (Bone and Agard, 1991; Kavanaugh *et al.*, 1992; Matthews, 1993; Xue *et al.*, 1993). Proteins are structurally flexible ("plastic") and can accommodate changes readily, even with substitutions in the interior core (Anderson *et al.*, 1993; Bone *et al.*, 1989; Roper *et al.*, 1992; Wilson *et al.*, 1992). How small changes at the active site affect catalysis is the critical question.

The kinetic mechanisms of enzymes have several steps that are energetically balanced for optimal efficiency. Each amino acid residue contributes to some extent to the rate of each step. Thus, a site-directed substitution can change the rates differentially, and the rate of one step can be affected by various substitutions. In some cases, the effects of substitution of important residues may not be apparent from a standard assay with a fixed concentration of substrate (Shi *et al.*, 1993). This means that modified enzymes must be thoroughly studied to determine the rates of each step. Each mutant enzyme should become the subject of the various studies done on the wild-type enzyme.

V. Strategy

A. Design

A three-dimensional structure is invaluable for selecting amino acid residues to replace. At least a structure could be modeled on the basis of a homolog. Protein modeling programs such as O are useful for obtaining typical side-chain rotamers and interaction distances (Jones *et al.*, 1991; Langone, 1991). Comparison of amino acid sequences should also provide targets of conserved residues that presumably have similar functions or varied residues that could explain differences in specificity (Ganzhorn *et al.*, 1987). If a three-dimensional structure is not known, sequence alignments and chemical modification studies provide a secondary rationale for selection, and the interpretation of results is more speculative. "Alanine-scanning" mutagenesis, for instance, where all charged residues are systematically substituted with Ala, can identify residues that appear to be important and allow more detailed studies (Cunningham and Wells, 1989; Gibbs and Zoller, 1991).

The purpose of the experiment influences the choice of amino acid residues to use for substitution. There is considerable flexibility in the choice, as surface residues that make up the active site can often be changed without affecting protein stability significantly (Matthews, 1993). Consideration of the substitutions that have been accepted in the evolution of homologous proteins suggests that some interchanges are relatively conservative (Table I). However, any change has the potential to affect structure and catalysis. Each site-directed substitution provides a new enzyme that will be compared to the wild-type enzyme, and it is the differences between the contributions of two residues that will be evaluated.

For studies on the roles of functional side chains in the mechanism, nearly isosteric substitutions will minimize structural changes. Thus, His can be changed to Asn or Gln, Cys to Ser, Thr to Val, Asp to Asn, Glu to Gln, and vice versa. However, these changes can alter the pattern of hydrogen-bonding interactions, as was found for rat trypsin when Asp-102 was changed to Asn (Sprang *et al.*, 1987). The charge can be maintained with interchanges of Arg, Lys, or His and Glu with Asp. However, the accompanying change of size means that the contacts will be altered, which becomes the experimental variable. Removing the side chain by substituting with Ala is a large change in size and a more drastic test of the role of a residue since structural effects may be larger. Increasing the size, for instance, Asp to Glu, or Lys to Arg, risks the generation of steric interference, which is energetically more serious than creating a void. For tests of flexibility of the peptide backbone, which can be important when conformational changes are critical for activity, substitution of Pro or Gly with Ala, or another residue with Gly or Pro, will loosen or stiffen, respectively, the structure (Matthews, 1987). Substitutions of residues in conformational hinges or loops may give measurable changes in activity (Ahrweiler and Frieden, 1991; Bullerjahn and Freisheim, 1992; Mas *et al.*, 1988). Another approach is to use random or saturation mutagenesis, which eliminates bias in the choice of substitutions and produces a large number of enzymes that may have diverse activities (Climie *et al.*, 1990; Estell *et al.*, 1985; Hampsey *et al.*, 1986; Krebs and Fierke, 1993; Wente and Schachman, 1991).

Table I
Relatively Frequent Accepted Point Mutations[a]

Cys → Ser	Ser → Ala, Thr, Asn, Gly, Pro
Thr → Ser, Ala, Val	Pro → Ala, Ser
Ala → Ser, Gly, Thr, Pro	Gly → Ala, Ser
Asn → Asp, Ser, His, Lys, Gln, Glu	Asp → Glu, Asn, Gln, His, Gly
Glu → Asp, Gln, Asn, His	Gln → Glu, His, Asp, Asn, Lys, Arg
His → Asn, Gln, Asp, Glu, Arg	Arg → Lys, His, Trp, Gln
Lys → Arg, Gln, Asn	Met → Leu, Ile, Val
Ile → Val, Leu, Met	Leu → Met, Ile, Val, Phe
Val → Ile, Leu, Met	Phe → Tyr, Leu, Ile
Tyr → Phe, His, Trp	Trp → Phe, Tyr

[a]The relative frequency is adapted from the mutation data matrix derived from comparison of homologous sequences given by Dayhoff *et al.* (1983).

B. Integrity

Once the mutations are made, the integrity of the product should be established. The complete sequence of the coding region of the modified DNA should be verified in order to minimize errors from adventitious mutations arising from mutagenesis, subcloning, or biological artifacts. Because sequencing results and experimental missteps and failures are not usually reported, the origins of published discrepancies are difficult to trace. After the proteins are expressed and purified (Methods in Enzymology Vols. 22, 34, and 104), some criteria of homogeneity should be applied, such as chromatography and electrophoresis in two different systems. Confirmation of the molecular weight by sodium dodecyl sulfate-polyacrylamide gel electrophoresis or gel filtration, with comparison to the wild-type protein, is often sufficient. When charged residues are changed, evidence of the altered properties should be seen from ion-exchange chromatography or electrophoresis. Although rarely reported, peptide mapping and amino acid sequencing can add considerable confidence to the identification of the mutated enzyme and to the elimination of artifacts. When the proteins are expressed in heterologous systems, the processing may be different, which could also alter enzymatic characteristics. Special care is required to ensure that wild-type, another mutant, or host enzymes do not contaminate the desired enzyme. This can happen in laboratories where different enzymes are expressed in microbes and sterile practices are insufficient. Because the mutations are unique biological materials, investigators should store stocks of the mutated plasmids or transformed microbes to be able to provide them to others who wish to replicate the published work.

The structural integrity of the protein should be studied. Simple procedures, such as denaturation by heat, pH, or storage, can be useful. Spectral measurements of UV absorption, fluorescence, or circular dichroism (CD) may reveal some relevant differences (Ganzhorn and Plapp, 1988). In most cases, if the enzyme can be isolated by the usual methods, the structure is similar to that of the wild type. Instability is an indication of deleterious changes. Some practitioners insist that the three-dimensional structure must be determined by X-ray crystallography or nuclear magnetic resonance (NMR), but this is often not possible and does not resolve the fundamental issue of distinguishing between direct and indirect effects. As noted earlier, mutations cause local structural changes, and these can be observed. However, after a structure of the enzyme complexed with substrates, inhibitors, or transition state analogs is determined, one must still ask if the structure is relevant to one of the complexes in the catalytic pathway. This is not a question of whether the structure of enzyme in crystalline form is the same as that in solution, since there is little evidence to suggest that they are different. Rather, it is an issue because an enzymatic reaction has many steps, with side chains changing position and states of protonation or covalent chemistry, and substrates may bind differently at the multiple steps of the reaction.

For multisubunit enzymes, where the active site is formed from two subunits, structural and cooperative effects can be assessed by complementation studies in which two forms of inactive enzyme are rehybridized (Larimer *et al.*, 1987; Wente and Schachman, 1987). If activity is regained to the statistically expected level, it is reasonable to conclude that the structures of the subunits in the inactive enzymes are not grossly distorted. However, the regain of activity in heterodimers of the inactive carboxymethylated derivatives of the monomeric pancreatic ribonuclease indicates that structural rearrangements can be productive (Crestfield *et al.*, 1963). Furthermore, an altered subunit can inactivate a wild-type subunit in an oligomeric enzyme, as in a "dominant negative" mutant (Tsirka and Coffino, 1992).

C. Kinetics

Enzyme kinetics and mechanistic studies, using the whole arsenal of techniques (Methods in Enzymology Vols. 63, 64, and 87), are required to determine the changes in the altered enzymes. Initially, the activity of the enzyme in a standard assay and Michaelis-Menten parameters are determined. The K_m and V_{max} values for one of the substrates may be estimated with other substrates at a fixed "saturating" level. However, more complete studies are necessary if the kinetics studies are to be meaningful. The V_{max} values are underestimated if the concentrations of all substrates are not saturating or if inhibitory levels of one of the substrates are used. The K_m values determined at a fixed level of other substrates can be larger or smaller than the true values, and they should not be interpreted as measures of affinity in the absence of mechanistic studies. A complete kinetic characterization can define the mechanism, true turnover numbers, K_m values, catalytic efficiencies (V/K_m), and dissociation constants for some substrates and inhibitors. For multisubstrate enzymes, this means that initial velocity studies in both the forward and reverse reactions should be performed by varying in a systematic way the concentrations of all substrates. For a bisubstrate mechanism, a 5×5 array of values, determined in duplicate, can provide good estimates of the kinetic constants. Fitting the appropriate equations to the data provides confidence in the precision of the results and permits comparisons with the wild-type enzyme (Cleland, 1979). The necessary computer programs are readily available.

For the interpretation of the kinetic data, the enzyme concentration must be determined. The protein concentration can be estimated by using amino acid analysis, dry weight, or UV absorption with a calculated extinction coefficient. Colorimetric procedures must be standardized with the wild-type enzyme, even if serum albumin is used as a working standard. Without proper standardization, maximum velocities and the stoichiometry of ligand binding can be seriously in error. As a control for enzyme purity and functionality, the concentration of active sites should be titrated with a ligand that binds tightly and gives a measurable signal. Although a protein may appear to be homogeneous by electrophoresis or chromatography, active sites may be incapable of binding an exogenous ligand because of chemically altered side chains, metal loss or poisoning, or tightly bound impurities. Such sites may be inactive or have altered kinetic properties. The concentration of enzyme is used to calculate turnover numbers for the forward and reverse reactions (V/E_t or k_{cat}), which have units of reciprocal seconds (s^{-1}). For some mechanisms, combinations of turnover numbers and kinetic constants can provide estimates of the rates of binding and dissociation of substrates for some steps in the reactions.

Establishing the kinetic mechanism for a mutated enzyme is accomplished by using the techniques described in this series. Ideally, each mutated enzyme should be studied in as much detail as the wild-type enzyme. Product and dead-end inhibition studies are useful, as various mechanisms can be distinguished, and the inhibition constants give estimates of the affinities of ligands for the enzyme. Inhibitors that resemble the substrates are useful for examining changes in different parts of the active site. Substrate specificities may have changed. Kinetic isotope effects and isotope exchange experiments can help establish which steps are rate limiting for the new enzyme. Studies on the pH dependence of kinetic parameters can identify changes in groups involved in acid-base catalysis. Indeed, without pH studies and identification of rate-limiting steps, there is insufficient evidence to support conclusions of the involvement of ionizable groups in the mechanism.

Enzymologists describe a mechanism by identifying all of the intermediates and assigning rate constants for each of the steps. As steady-state kinetics analysis does not provide all of the relevant information, transient (presteady-state) kinetics is an important complementary tool. Isomerizations of transitory and central complexes can be detected and characterized when high concentrations (e.g., 1–10 μM) of enzymes and complexes are used. Kinetic simulation, in which the differential equations for the enzymatic reaction are numerically integrated and progress curves for various reactions are fitted, has become an important and accessible tool (Frieden, 1993, 1994; Zimmerle and Frieden, 1989). The analysis of site-directed mutants reaches new heights when the effect on each step of the reaction can be determined.

VI. Illustrations

A. Interpretation of Steady-State Kinetic Parameters

Horse liver and yeast alcohol dehydrogenases are used as primary examples since a variety of mutations have been made, and the kinetics have been extensively studied. The three-dimensional structure of the horse liver EE isoenzyme complexed with NAD^+ and a substrate analog has been determined to high resolution (Fig. 1). Inspection of the structure and studies of the mechanism have suggested that His-51 participates in acid-base catalysis, but it is apparent that many other amino acid residues participate in binding the substrates. Amino acid sequences of more than 47 members of the family have been determined (Sun and Plapp, 1992), and the yeast enzyme is similar enough to the liver enzyme to justify building models that represent the active site. Thus, the enzymes offer many opportunities to explore structure-function relationships and to test the roles of different residues in catalysis (Plapp *et al.*, 1990).

The kinetic mechanisms for the enzymes were established as Ordered Bi Bi (Wratten and Cleland, 1963), which can be written as follows:

$$E \underset{k_{-1}}{\overset{k_1\ NAD^+}{\rightleftharpoons}} E{\cdot}NAD^+ \underset{k_{-2}}{\overset{k_2 Alc}{\rightleftharpoons}} E{\cdot}NAD^+{\cdot}Alc \underset{k_{-3}}{\overset{k_3}{\rightleftharpoons}} E{\cdot}NADH{\cdot}Ald \underset{k_{-4}Ald}{\overset{k_4}{\rightleftharpoons}} E{\cdot}NADH \underset{k_{-5}NADH}{\overset{k_5}{\rightleftharpoons}} E \qquad (1)$$

The complete steady-state rate equation for the mechanism has 10 kinetic constants, which can be estimated from initial velocity and product inhibition studies with both substrates in each direction. The reaction in the forward direction in the absence of products is described by the Sequential Bi rate equation: $v = V_1AB/(K_{ia}K_b + K_aB + K_bA + AB)$, where A and B represent the concentrations of the substrates NAD^+ and alcohol, K_a and K_b are the Michaelis constants for these substrates, K_{ia} is the inhibition or dissociation constant for NAD^+, and V_1 is the maximum velocity (proportional to turnover number). A similar equation applies for the reverse reaction, where P and Q represent aldehyde and NADH. These kinetic constants can be estimated by varying the concentrations of both substrates in a systematic way or from product inhibition studies (Dalziel, 1963; Wratten and Cleland, 1963).

A variety of site-directed substitutions have been made in these enzymes, and Table II summarizes kinetic data for some of them (Fan and Plapp, 1995; Gould, 1988; Gould and Plapp, 1990; Kim, 1994; Park and Plapp, 1992).[1] The substitutions were made in the coenzyme- and substrate-binding sites, in the environment of the zinc, and in the acid-base system, as shown in Fig. 1.

The kinetic constants provide fundamental information about the mechanism and will be used to illustrate how the data may be interpreted for the various mutants. It is helpful to know the definitions of the kinetic constants in terms of the rate constants given in Eq. (1). Analogous expressions apply to the reverse reactions, and Haldane relationships relate these kinetic constants to the equilibrium constant for the overall reaction.

$$K_a = k_3k_4k_5/k_1(k_4k_5 + k_3k_4 + k_{-3}k_5)$$
$$K_b = k_5(k_{-2}k_4 + k_{-2}k_{-3} + k_3k_4)/k_2(k_4k_5 + k_3k_4 + k_3k_5 + k_{-3}k_5)$$
$$V_1 = k_3k_4k_5/(k_4k_5 + k_3k_4 + k_3k_5 + k_{-3}k_5)$$
$$K_{ia} = k_{-1}/k_1$$
$$V_1/k_a = k_1$$
$$V_1K_{ia}/K_a = k_{-1}$$
$$V_1/K_b = k_2k_3k_4/(k_{-2}k_4 + k_{-2}k_{-3} + k_3k_4)$$
$$V_1/K_{ia}K_b = k_1k_2k_3k_4/k_{-1}(k_{-2}k_4 + k_{-2}k_{-3} + k_3k_4)$$
$$K_i, CF_3CH_2OH = k_{-2}/k_2$$
$$K_{eq} = V_1K_pK_{iq}/V_2K_bK_{ia} = k_1k_2k_3k_4k_5/k_{-1}k_{-2}k_{-3}k_{-4}k_{-5}$$

[1] Mutations are conveniently designated with the single-letter abbreviations for the amino acids, wild-type first and mutant last, separated by the residue number.

Table II
Kinetic Constants for Liver and Yeast Alcohol Dehydrogenases[a]

	*Eq*ADH-E				*Sc*ADH1		
Constant	Wild type	D115Δ	F93A	I269S	Wild type	H51Q	E68Q
K_a (μM)	3.9	11	3.3	1000	160	96	410
K_b (mM)	0.35	36	0.21	11	21	18	41
K_p (mM)	0.40	87	41	11	0.74	15	56
K_q (μM)	5.8	24	1.9	570	95	150	160
K_{ia} (μM)	27	29	44	9500	950	460	3500
K_{iq} (μM)	0.50	0.91	0.036	180	31	6	29
V_1 (s^{-1})	3.5	7.5	0.35	90	360	27	9.9
V_2 (s^{-1})	47	240	20	1500	1800	2800	730
V_1/K_b (mM^{-1} s^{-1})	10	0.21	1.7	8.4	21	1.5	0.24
$V_1/K_{ia}K_b$ (mM^{-2} s^{-1})	370	7.2	38	0.88	22	3.3	0.068
V_2/K_p (mM^{-1} s^{-1})	120	2.8	0.49	140	2400	190	13
$V_2/K_{iq}K_p$ (μM^{-2} s^{-1})	0.24	0.0030	0.013	0.00078	0.080	0.032	0.00045
K_i, CF_3CH_2OH (μM)	8.4	630	53	7.8	2500	33,000	25,000

[a]Kinetic constants were determined in 33 mM sodium phosphate buffer, pH 8.0, at 25 °C for the horse liver enzymes, and in 83 mM sodium phosphate, 40 mM KCl buffer, pH 7.3, 30 °C for the yeast enzymes. NAD^+ and ethanol or NADH and acetaldehyde were the substrates. *Eq*ADH-E is the horse (*Equus*) liver EE isoenzyme, mutant forms of which include D115Δ with Asp-115 deleted (Park and Plapp, 1992), F93A (Kim, 1994), and I269S (Fan and Plapp, 1995); *Sc*ADH1 is the yeast (*Saccharomyces cerevisiae*) cytoplasmic isoenzyme I (Gould and Plapp, 1990), mutants of which include H51Q (Gould, 1988) and E68Q (Ganzhorn and Plapp, 1988).

Inspection of the data in Table II shows that the K_m values for coenzyme are not necessarily the same as the K_d values ($K_a \neq K_{ia}$ and $K_q \neq K_{iq}$ in most cases). Furthermore, changes in a K_m value due to mutation are not in general directly proportional to changes in a K_d value. The K_m for alcohol (K_b) would approximate the K_d (k_{-2}/k_2) only with certain values of rate constants, so that a better measure of affinity of the enzymes for alcohol is given by the K_i value for the competitive inhibitor trifluoroethanol. Comparison of the changes in K_b and K_i show that there can be 10-fold changes in one of these values and not in the other (F93A, I269S, H51Q enzymes). It is apparent that changes in K_m values should not be described as changes in affinity.

For the Ordered Bi Bi mechanism, V_1, or k_{cat} for the forward reaction, is controlled by several rate constants for the conversion of the central $ENAD^+$· alcohol complex to products. For the F93A enzyme, V_1 is decreased 10-fold relative to the wild-type enzyme. This change could conceivably result from an effect on the rate of hydrogen transfer, but in this case it is due to a slower release of the product NADH ($k_5 = V_2K_{iq}/K_q = 0.37$ s^{-1}). The V_1 for the I269S enzyme is increased by 25-fold, owing to an increased rate of release of NADH ($k_5 = 470$ s^{-1}), whereas the rate of binding of NADH ($k_{-5} = V_2/K_q = 2.6 \times 10^6$ M^{-1} s^{-1}) is decreased threefold. The V_2 for the I269S enzyme is increased 32-fold, because NAD^+ dissociates faster. The rate constants for binding were confirmed by transient kinetic studies using stopped-flow techniques.

B. Catalytic Efficiency

The increases in turnover numbers observed with some mutations raise the issues of catalytic efficiency and evolution of catalytic perfection. Can site-directed mutations or biological selection procedures produce "more active" enzymes? For analyzing the results of such studies, the ratio of constants, V/K_m or k_{cat}/K_m, is the important measure. When the concentration of substrate is low relative to the K_m value, the Michaelis-Menten equation reduces to $v = V[A]/K_m$ and describes the rate when the reaction is first order in substrate. The logarithm of this parameter is proportional to the energy change in attaining the transition state, and the logarithmic scale in figures is most informative for comparing a variety of substrates or mutant enzymes.

For alcohol dehydrogenase, V_1/K_b is the bimolecular rate constant describing the rate of reaction of the alcohol with the enzyme-NAD^+ complex to produce aldehyde and enzyme-NADH complex, including steps 2, 3, and 4 in Eq. (1). It is interesting that the liver and yeast enzymes have about the same catalytic efficiencies for ethanol even though these enzymes have very different Michaelis constants and turnover numbers. All of the mutant enzymes in Table II have decreased efficiencies, except for the I269S enzyme. The deletion of Asp-115 in the substrate-binding pocket of the E isoenzyme (D115Δ) decreases activity on ethanol but confers activity on steroids (Park and Plapp, 1992). The F93A substitution removes a phenyl group that interacts with ethanol. The H51Q change removes the catalytic base that facilitates loss of the proton from the hydroxyl group of the alcohol. In contrast, the I269S substitution affects the binding of the adenine ring of coenzyme, but not the catalytic efficiency with ethanol. Thus, many of the kinetic results are consistent with the proposed roles of the amino acid residues in reactions with ethanol. The decreased efficiency of the E68Q enzyme is not explained in the same way, however, since this residue is near the catalytic zinc and on the opposite side of the binding site for substrate. The E68Q substitution seems to affect the electrostatic environment of the zinc and indirectly affect catalysis. Although the changes in kinetic parameters may be consistent with simple interpretations, the ultimate explanation may be more complicated.

There are a few examples where catalytic efficiency is increased by mutation. Changing Thr-51 to Cys or Pro in tyrosyl-tRNA synthetase (tyrosyl-tRNA ligase) increased the efficiency of formation of the tyrosyl-adenylate complex 15- to 36-fold but decreased the overall k_{cat} and catalytic efficiency (Avis and Fersht, 1993; Ho and Fersht, 1986). Substitution of Asp-153 with Ala in an alkaline phosphatase led to a sixfold increase in k_{cat} and a 14-fold increase in k_{cat}/K_m for *p*-nitrophenyl phosphate (Matlin *et al.*, 1992). The increased turnover is due to faster release of the product inorganic phosphate, but the increased efficiency is more difficult to explain as Asp-153 does not seem to have a direct role in catalysis. Another mutation, K328A, also activates the enzyme in the presence of Tris buffer, which can stimulate transphosphorylation (Xu and Kantrowitz, 1991). The D101S substitution increased the efficiency fivefold of an alkaline phosphatase on *p*-nitrophenyl phosphate by altering the interactions with an Arg and decreasing affinity for phosphate (Chen *et al.*, 1992). Insertion of three amino acid residues into maize aldolase increased efficiency with fructose 1,6-bisphosphate by about threefold (Berthiaume *et al.*, 1993).

The magnitude of V/K depends on the substrate used and may be largest with the physiological reactant. It is a challenge to improve activity with the best substrate of the enzyme or to use protein engineering to make an enzyme more active on some substrate than the native enzyme is with its best substrate. There are some examples of mutations that increase activity with poor substrates, as would be expected if the specificity is reengineered. For instance, the Q102R substitution converted the *Bacillus stearothermophilus* lactate dehydrogenase to malate dehydrogenase (Wilks *et al.*, 1988). Changing Asp-179 to Asn in deoxycytidylate hydroxymethylase decreased k_{cat}/K_m for dCMP by 10^4-fold and increased activity on dUMP 60-fold. However, dCMP reacts 1.8×10^5 times faster than dUMP with native enzyme, and the efficiency of the D179N enzyme for dUMP is still 3000-fold less than the efficiency of native enzyme with dCMP (Graves *et al.*, 1992). The T48S and W93A substitutions in the alcohol-binding site of yeast alcohol dehydrogenase increased activity on octanol 16-fold, yielding 2.7-fold more activity than the native enzyme has with its best substrate, ethanol (Green *et al.*, 1993). Deletion of Asp-115 and substitutions of three nearby residues in the substrate-binding pocket of horse liver alcohol dehydrogenase E isoenzyme decreased V/K_b for ethanol 25-fold, but increased activity on hexanol by 2.7-fold. The changes generated activity on steroids that is about sixfold higher than the activity of natural horse S isoenzyme on steroids (Park and Plapp, 1992). Thus, some modest improvements in catalytic efficiency are possible.

The V/K_m parameter can be applied for reactions with one substrate, or for the second substrate as in the Ordered Bi reaction of alcohol dehydrogenase, but the evaluation of catalytic efficiency should be extended when there is more than one substrate. Using the kinetic principle that efficiency is the activity at low, limiting substrate concentrations, the catalytic efficiency for both substrates in the Sequential Bi reaction is expressed by the term $V_1/K_{ia}K_b$. This expression describes the termolecular reaction of enzyme with both substrates and is proportional to the free energy of activation of the overall reaction (Avis and Fersht, 1993).

A few studies have determined overall catalytic efficiencies. The results in Table II show that all of the mutations of alcohol dehydrogenase decreased catalytic efficiencies for the reactions in both directions. Studies on changing the specificity for nicotinamide coenzymes provide further examples. The specificity of NAD-dependent dehydrogenases for coenzyme appears to be controlled by an aspartic acid that interacts with the adenosine ribose hydroxyl groups (Fig. 1). Changing Asp-223 to Gly in the NAD-dependent yeast alcohol dehydrogenase made the enzyme almost equally reactive with NAD^+ or $NADP^+$, but reduced overall catalytic efficiency ($V_1/K_{ia}K_b$) by about 1000-fold (Fan *et al.*, 1991). When the corresponding Asp (residue 37) was changed to Ile in rat NADH-preferring dihydropteridine reductase, a complete kinetic study showed that selectivity for

NADPH was improved by 170-fold, but the wild-type enzyme was still 1300-fold more reactive with NADH than the D37I enzyme was with NADPH (Grimshaw *et al.*, 1992). These results show that the conserved Asp is not the only residue that determines specificity for coenzymes. Two studies show that multiple changes designed on the basis of homologous enzymes can reverse coenzyme specificity. Changing seven amino acid residues in the NADPH-preferring *E. coli* glutathione reductase inverted the preference for coenzyme (k_{cat}/K_a) by 18,000-fold, but the mutant was 30-fold less reactive with NADH than wild-type enzyme was with NADPH (Scrutton and Berry, 1990). Conversely, changing seven amino acid residues in the NAD-dependent *E. coli* dihydrolipoamide dehydrogenase produced an enzyme that was 1.3-fold more reactive with $NADP^+$ than the wild-type enzyme was with NAD^+ (Bocanegra *et al.*, 1993). Overall catalytic efficiencies and dissociation constants for the coenzymes should be determined for these multiply substituted enzymes.

Another complete kinetic study, on chloramphenicol acetyltransferase, showed that changing Thr-174 to Ala decreased k_{cat} by twofold, k_{cat}/K_m by fourfold, K_d by 12-fold, and $k_{cat}/K_{ia}K_b$ (which is ϕ_{AB}) by 42-fold (Lewendon and Shaw, 1993). Similar numbers were obtained for the Val-174 and Ile-174 enzymes. From these studies and inspection of the three-dimensional structure, it was concluded that a water molecule that formed a hydrogen-bonded bridge between Thr-174 and the tetrahedral intermediate stabilized the transition state of the reaction.

C. Transient Kinetics and Estimation of Individual Rate Constants

Steady-state kinetic analysis is useful to establish a mechanism and to indicate which steps are affected by mutations. It is conveniently applied when amounts of enzyme are limiting. Nevertheless, as shown above, V/K_m or catalytic efficiency parameters are collections of rate constants. When large amounts of enzyme are available, transient kinetics can be used to estimate magnitudes of individual rate constants (Johnson, 1992). Moreover, advances in computer simulation make it feasible to estimate the rate constants from progress curves (Frieden, 1994). With good data from a few experiments, a mechanism can be defined and rate constants estimated with small errors.

Kinetic simulation has been used to analyze a complete mechanism for liver alcohol dehydrogenase acting on benzyl alcohol (Shearer *et al.*, 1993), and some results with a mutant enzyme are given in Fig. 2 and Table III. The L57F substitution is in the substrate-binding pocket and has small effects on coenzyme binding and on the steady-state kinetic constants with ethanol and acetaldehyde. In contrast, rate constants for binding and reaction of benzyl alcohol and benzaldehyde are significantly affected. It is interesting that the rate constants for hydrogen transfer in the ternary complex (step 3) are increased two- to fivefold. This enzyme was found to have the greatest evidence for hydrogen tunneling, which is masked in the native enzyme because of kinetic complexity (Bahnson and Klinman, this volume; Bahnson *et al.*, 1993). Decreasing the size of the

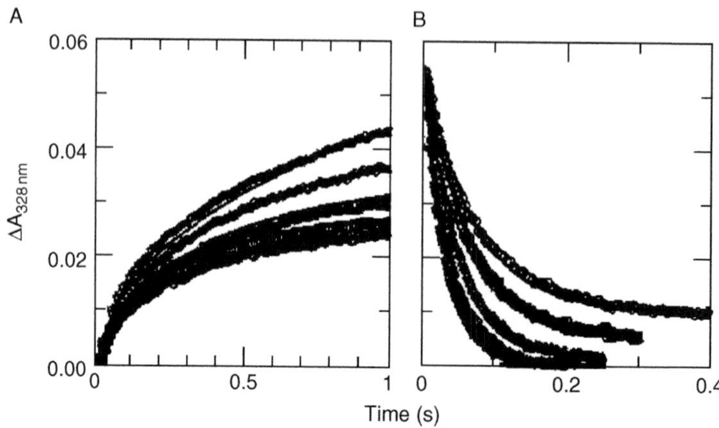

Fig. 2 Transient reactions of horse liver alcohol dehydrogenase, L57F enzyme. (A) The oxidation of benzyl alcohol (8–25 μM from lowest to highest curve) with 2 mM NAD^+ and 9 μM enzyme was determined with a BioLogic SFM3 stopped-flow instrument with a 1-cm light path. (B) The single turnover reduction of benzaldehyde (9.6–50 μM for highest to lowest curve) with 10 μM NADH and 7 μM enzyme was followed. The data points are shown, and the lines are simulated fits to all of the data together using KINSIM and FITSIM with the mechanism given in Eq. (1) and the rate constants given in Table III. (Kim, 1994)

Table III
Rate Constants for Reactions of Liver Alcohol Dehydrogenases with NAD^+ and Benzyl Alcohol or NADH and Benzaldehyde[a]

Constant	Native[b]	L57F[c]
k_1 (μM^{-1} s^{-1})	1.2	6.4
k_{-1} (s^{-1})	90	270
k_2 (μM^{-1} s^{-1})	3.7	35
k_{-2}(s^{-1})	58	110
k_3 (s^{-1})	38	60
k_{-3} (s^{-1})	310	1700
k_4 (s^{-1})	66	160
k_{-4} (μM^{-1} s^{-1})	0.83	4.2
k_5 (s^{-1})	5.5	6.4
k_{-5} (μM^{-1} s^{-1})	11	9.9

[a]The rate constants are defined in Eq. (1). The reactions were studied in 33 mM sodium phosphate buffer, pH 8.0, at 25 °C. Coenzyme binding to free enzyme (steps 1 and 5) was determined in the absence of substrates. Rate constants for steps 2, 3, and 4 were estimated using KINSIM and FITSIM with the family of progress curves for the transient reactions, such as shown in Fig. 2. Errors were usually less than 10% of the values.

[b]Sekhar and Plapp (1990).

[c](Kim 1994).

substrate-binding pocket causes substrates to dissociate more rapidly (k_{-2} and k_4) and exposes the tunneling to observation.

Rate constants for a complete kinetic mechanism of *E. coli* dihydrofolate reductase were also analyzed with KINSIM (Fierke *et al.*, 1987; Penner and Frieden, 1987). This was the beginning of detailed studies on catalytic energetics and the effects of mutations (Benkovic *et al.*, 1988). Substitutions of the conserved Leu-54 in the dihydrofolate-binding site of *E. coli* dihydrofolate reductase with Ile, Gly, and Asn decreased the hydride transfer rate by a constant factor of about 30-fold and increased dissociation constants for dihydrofolate by different factors. Thus, the contributions of residue 54 to binding and catalysis are separable (Murphy and Benkovic, 1989). Further study of a variety of mutated enzymes would be useful for developing our understanding of hydrogen transfer and substrate specificity.

Kinetic simulation has also been applied to mutants of bacterial luciferase, which has a complex mechanism described by 16 rate constants. Substitutions of Cys-106 were shown to decrease the stability of the hydroperoxy flavin intermediate most significantly (Abu-Soud *et al.*, 1993).

D. Acid-Base Catalysis

A central focus in studying enzymatic reactions is the evaluation of the contribution of residues to proton transfer steps. Imidazole, carboxyl, sulfhydryl, phenol, and amino groups are likely candidates. After substitution with a group that cannot be protonated in the physiological pH range, activity may decrease and the pH profiles should change. The magnitude of activity change is not a sufficient criterion to judge the participation of the residue, since (1) it is the magnitude we wish to determine, and (2) activity can decrease because of local structural effects. If the mutated enzyme has activity, it is important to study the mechanism of the enzyme and to determine at least the pH dependence for the steps using acid-base catalysis. Because enzymatic reactions are usually not limited by a single step in the mechanism and the pH-dependent step may be masked by kinetic complexity (Cleland, 1977), the analysis may require additional studies, for instance, with isotopes or NMR spectroscopy. Furthermore, removal of one ionizing group may expose the effects of other ionizable groups that can less directly affect catalysis and change the pH dependencies.

1. Altered pH Dependencies

The involvement of an amino acid residue in proton transfer steps is often inferred from a pH dependence study that shows evidence of p*K* values for free or complexed enzyme. Substitution of an involved residue should change the pH profile of the logarithm of *V* or *V*/*K* against the pH to a slope of 1.0 if hydroxide or hydronium were the catalyst or to a slope of zero if the reaction became pH independent. However, these expectations are not usually realized.

For alcohol dehydrogenases, substitution of His-51 with Gln illustrates some different results (Fig. 3). His-51 is part of a hydrogen-bonded system that includes the water or alcohol ligated to the catalytic zinc and the hydroxyl groups of Ser-48 (or Thr in yeast) and the nicotinamide ribose. The system could shuttle a proton from the buried alcohol to solvent. The S48A or T48A substitutions block the system and greatly reduce activity (Gould, 1988; Kim, 1994; Plapp *et al.*, 1990). The pH dependencies observed with these enzymes could originate from His-51, the water that is ligated to the catalytic zinc, or other groups. The pH dependence of V/K_b provides pK values for the E·NAD$^+$ complex. Significant substrate deuterium isotope effects indicated that the chemical reactions were predominantly rate determining. For the horse enzyme, the native enzyme exhibits a bell-shaped pH dependence with pK values of 6.7 and 9.0, whereas the H51Q enzyme has only a single pK of 8.6. The wild-type yeast enzyme has a wavelike pH dependence with finite activity at low and high pH with a pK value of 7.7 (Gould and Plapp, 1990). In contrast, the H51Q enzyme has a linear pH dependence with a slope of 0.45, and no certain pK values. The straight line can be fitted to an expression for a mechanism with pK values of 6.8 and 8.7, but the errors of ±0.2 are not acceptable. For both enzymes, catalytic efficiency at pH 7.3 is reduced by a factor of 10, which

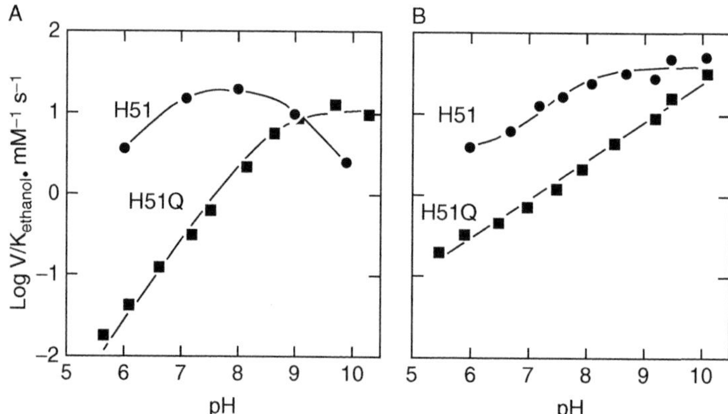

Fig. 3 Dependencies on pH for horse liver (A) and yeast (B) alcohol dehydrogenases and the H51Q mutants. Catalytic efficiencies for ethanol were determined with saturating concentrations of NAD$^+$. The data were fitted to the appropriate equations (Cleland, 1979), where the K terms refer to acid dissociation constants and k_x is a pH-independent rate. (A) Natural horse liver enzyme data (•) were determined by Dalziel (1963) at 23.5 °C and fitted to the BELL equation, $k = k_b/(1 + K_1/[H^+] + [H^+]/K_2)$. The data for the H51Q enzyme (■) were determined by Park (1991) at 25 °C using 10 mM $Na_4P_2O_7$ buffers adjusted to the desired pH and an ionic strength of 0.1 with H_3PO_4 or sodium phosphate for the pH range from 5.5 to 9.5 and 10 mM $Na_4P_2O_7$ and 5 mM sodium carbonate at the higher pH values. Data were fitted to the equation $k = k_b/(1 + K_1/[H^+])$. (B) Natural yeast enzyme data (•) were determined in the pyrophosphate buffer system at 30 °C (Gould and Plapp, 1990) and fitted to the equation $k = (k_a + k_bK_1/[H^+])/(1 + K_1/[H^+])$. Data for the H51Q enzyme (■) were determined by Gould (1988) using the pyrophosphate buffer system and fitted to the equation for a straight line, although the data could also be described by equations with five parameters.

could be significant physiologically. (Because His-51 is replaced by Thr or Tyr in alcohol dehydrogenase from some species, the residue is not required for activity.) This study does not directly measure the contribution of His-51 to catalysis, however, since the Gln isosterically replaces His and can form a hydrogen bond to the 2′-hydroxyl group of the nicotinamide ribose and block proton release. Substitution with another amino acid could change activity by a different magnitude. We conclude that His-51 contributes to catalysis and affects the pH dependence. Nevertheless, the assignment of a p*K* value to His-51 is problematic because it is part of the interacting system (with the zinc-water) that must be described by microscopic p*K* values.

Acid catalysis in *E. coli* dihydrofolate reductase was studied by substitution of Asp-27 with Asn or Ser (Howell *et al.*, 1986). Crystallography of the mutant enzymes complexed with methotrexate showed only minor changes in structure except for the positions of some water molecules at the site of substitution. Activities (V/K_m for dihydrofolate) of the two mutant enzymes were reduced about 4 orders of magnitude. The bell-shaped pH dependence of wild-type enzyme, with p*K* values of 5 and 8, was converted to amazingly linear dependencies with slopes between 0.55 and 0.75 describing increasing activity with pH decreasing from 8.7 to 4.3. The pH dependence was attributed to protonation of the substrate, with a p*K* of 3.5, but the linearity was not interpreted. Significant substrate deuterium isotope effects showed that hydrogen transfer steps were at least partially rate limiting in catalysis. It was concluded that Asp-27 participates in protonation of the substrate, but it is interesting that the proton appears to be shuttled to its destination on N-5 of the substrate through a hydrogen-bonded system that includes two water molecules and O-4 of the substrate (McTigue *et al.*, 1992). Interpretation of the effects of the D27N substitution may be complicated by the kinetic evidence that the interconversion between two conformational states of the enzyme is also affected (Appleman *et al.*, 1990). Furthermore, other changes of charge at the active site of the mouse enzyme with the L22R, W24R, and F31R substitutions also alter the pH profiles (Thillet *et al.*, 1988).

Substitution of Asp-102 with Asn in rat trypsin also altered the pH dependencies (Craik *et al.*, 1987). The pH profiles for k_{cat} and k_{cat}/K_m with wild-type enzyme show that activity is maximal above a p*K* of 6.8, but the D102N enzyme shows a complex dependence for k_{cat}/K_m with p*K* values of 5.4 and 9.9 and evidence for hydroxide catalysis on k_{cat}. Catalytic efficiency at pH 7.1 was decreased 11,000-fold for hydrolysis of a peptide analog and for reaction with diisopropyl fluorophosphate, which reacts with Ser-195. However, reactivity of His-57 with the active-site-directed reagent α-*N*-tosyl-L-lysylchloromethane was hardly affected. His-57 may be in the wrong tautomeric state to activate Ser-195, as suggested by X-ray crystallography of the D102N enzyme (Sprang *et al.*, 1987).

The effects of changing the charge balance in the active site were studied in muscle lactate dehydrogenase (Cortes *et al.*, 1992). Asn-140, which is a hydrogen bond donor to the pyruvate carbonyl oxygen, was changed to Asp. His-195 acts as an acid by donating a proton to the same keto group. Catalytic efficiency (V/K_m for pyruvate) was decreased at least 2 orders of magnitude at pH 7, and the pK value of 7, attributed to His-195, was abolished. Modeling of the resulting pH dependence, with an apparent pK of about 4.5, suggested that the pK of His-195 was shifted to about 10 and that only the form of the enzyme with both His and Asp protonated was active.

2. Elusive Bases

The combination of kinetics, specific substitutions, X-ray crystallography, and NMR spectroscopy provides different lines of evidence in support of catalytic roles for particular amino acid residues. However, such studies may lead to differing conclusions and may fail to identify "the base."

Ribulose-bisphosphate carboxylase/oxygenase from *Rhodospirillum rubrum* has been extensively studied with site-directed mutagenesis, and the results have been related to three-dimensional structures and chemical modification studies (Hartman and Harpel, 1993). The enzyme is attractive for study because the base-catalyzed enolization and carboxylation reactions can be studied as partial reactions. Each of seven substitutions decreased carboxylase activity by more than 10-fold, and at least four of these had much higher rates for enolization as compared to carboxylase activities, providing some evidence for separate catalytic roles. Hartman and Harpel suggest that substitution of a base involved in deprotonation of C-3 of ribulose bisphosphate should reduce activity by at least 6 orders of magnitude, as was found for proton transfers in glyceraldehyde-3-phosphate dehydrogenase, aspartate aminotransferase, Δ^5-3-ketosteroid isomerase, and mandelate racemase. By this criterion, only Lys-166 could be considered to be the base in ribulose-bisphosphate carboxylase, but according to a three-dimensional structure, the amino group is 6.5 Å away from the C-3 hydrogen of ribulose bisphosphate and poorly oriented. Substitutions of closer residues, His-321 and Ser-368, decreased k_{cat}/K_m for ribulose bisphosphate to about 1%, but enolization rates were about 10% of those of the wild type (Harpel *et al.*, 1991). Thus, identification of the base remains elusive. Analysis of a structure for the activated *Synechococcus* enzyme complexed with CO_2, Mg^{2+}, and 2-carboxyarabinitol 1,5-bisphosphate (analog of an intermediate) has suggested that the carbamate oxygen on Lys-191 (201 in *Synechococcus*) acts as the base and that His-287 (294) facilitates hydration of the carbonyl group formed at C-3 (Newman and Gutteridge, 1993). Alternative interpretations of the three-dimensional structures are possible, and it is difficult to predict the effects of substitutions of the potential bases.

In a similar enolization, catalyzed by pig heart aconitase, Ser-642 is proposed to act as a base (Zheng *et al.*, 1992). The S642A substitution decreased V_m by almost 5 orders of magnitude. X-ray crystallography of the enzyme complexed with the substrate isocitrate shows that Arg-452 can stabilize the serine alkoxide, which apparently forms on transfer of the proton to the Fe–OH of the [4Fe–4S] cluster (Lauble *et al.*, 1992). The mechanism for this proton transfer is not clear, in contrast to the situation in chymotrypsin, where Ser-195 is activated by His-57 in the catalytic triad. In one of the first "site-directed mutations," conversion of Ser-195 to dehydroalanine in chymotrypsin decreased activity by at least 4 orders of magnitude, and the S221A mutation in the catalytic triad of subtilisin reduced catalytic efficiency by almost 6 orders of magnitude (Carter and Wells, 1988; Weiner *et al.*, 1966)

The role of tyrosine residues 248 and 198 in pancreatic carboxypeptidase was tested by replacing them with Phe and determining the pH dependencies of k_{cat} and k_{cat}/K_m for a variety of substrates (Gardell *et al.*, 1987; Hilvert *et al.*, 1986). Because both the native and mutant enzymes showed similar bell-shaped pH dependencies with pK values of about 6 and 10, it appears that the tyrosine residues do not mediate acid catalysis. Certainly the pK values near 10 for the mutant enzymes are not due to these tyrosine residues. The catalytic efficiencies decreased with the substitutions, attributed to local structural effects. Although the work is exemplary, the results do not prove that a tyrosine does not participate in acid-base catalysis. If the proton transfer steps were not partially rate limiting and the pH dependencies were due to binding or conformational isomerizations, the role of tyrosine could be masked. The origins of the observed pH dependencies still need to be explained.

Substitution of Tyr with Phe could permit a water to compensate for the phenolic hydroxyl group. This possibility was tested by X-ray crystallography of Y52F and Y73F mutants of phospholipase A_2 (Thunnissen *et al.*, 1992). The tyrosines are involved in a hydrogen-bonded system at the reaction center. The substitutions did not significantly affect activity, and a void, rather than a water site, was created.

Inspection of the structure of *Lactobacillus casei* thymidylate synthase suggested that His-199 was the residue most likely to serve as a base in removing the hydrogen from C-5 of 2′-deoxyuridylate. The His is highly conserved in various species. Substitution of the corresponding His-147 in the *E. coli* enzyme to make the Gly, Asn, Gln, Val, or Leu enzymes decreased k_{cat} and k_{cat}/K_m by 10-fold (Dev *et al.*, 1969). The wild-type and the Gly, Asn, and Gln enzymes gave pH dependencies with a pK of about 5.7. Substitutions with Asp or Glu decreased k_{cat} by 10^3-fold, and with Arg or Lys by 10^4. Similar changes in activity were observed for the *L. casei* enzyme, and it was concluded that the results rule out a role for the His as an "essential" base, "even though there are no current alternative candidate residues to serve this function" (Climie *et al.*, 1990). An alternative conclusion would be that the contribution of His-147 to catalysis is about 10-fold.

Aspartate transcarbamoylase (aspartate carbamoyltransferase) from *E. coli* has been studied by a variety of techniques. X-ray crystallography of the complex with the bisubstrate analog *N*-(phosphonacetyl)-L-aspartate shows that His-134 and Arg-105 form hydrogen bonds to the carbonyl oxygen of the ligand, that Arg-54 binds OP-2 of the phosphate, and that Lys-84 is near the NH (Ke *et al.*, 1988). In the catalytic subunit, the H134A, K84R, R105A, and R54A substitutions decreased V/K_m for aspartate by 59, 580, 7900, and 60,000-fold, respectively (Stevens *et al.*, 1991). The pH dependence of wild-type enzyme shows that V and V/K_m for aspartate is maximal above a pK of about 7.2, and this pK could be due to His-134 (Léger and Hervé, 1988). However, the H134N enzyme, which has V/K_m

reduced by 75-fold, retains a pK of 6.3 (Xi *et al.*, 1990). Furthermore, NMR studies with (Teng *et al.*, 1993) C-enriched enzyme suggest that His-134 does not titrate in the pH range from 6.2 to 8.5 and that the imidazole ring is neutral, from which it was concluded that His-134 does not participate in acid or base catalysis (Kleanthous *et al.*, 1988). Instead, as supported by ^{31}P NMR studies on the native and R54A enzymes and by the failure of site-directed mutagenesis studies "to identify a residue in the active site that could act as a general base to accept the proton from aspartate," it is suggested that the phosphate of carbamoyl phosphate accepts the proton from the α-amino group of aspartate (Stebbins *et al.*, 1992). However, modeling of the reactive complex based on the complexes with inhibitors is uncertain, and the mechanism for the deprotonation of the aspartate ammonium group is not explained. Perhaps all four basic residues at the reaction center are involved in the dynamics of proton transfer.

3. Novel Catalysis

It is usually assumed that unprotonated imidazole is a base. Studies on triosephosphate isomerase lead to the possibility that neutral imidazole acts as an acid. A three-dimensional structure of the yeast enzyme complexed with phosphoglycolohydroxamate, an analog of the enediolate intermediate, provides a detailed picture of the active site, consistent with Glu-165 acting as a base and His-95 as an acid (Davenport *et al.*, 1991). The H95Q and H95N enzymes are about 200- and 3500-fold less reactive in both directions of reaction (k_{cat}/K_m), the E165D enzyme is about 300-fold less reactive, and the E165G or E165A enzymes are 10^6-fold less active than native enzyme (Blacklow and Knowles, 1990; Knowles, 1991; Nickbarg *et al.*, 1988). The H95Q enzyme should most closely mimic the hydrogen-bonding interactions of native enzyme and introduce the least structural perturbation. The NMR results show that His-95 is uncharged between pH 5 and 9.9 and suggest that a neutral imidazole participates in the transfer of the proton between O-1 and O-2 of the substrate (Lodi and Knowles, 1991). The structure and computer simulations suggest that other residues also participate in catalysis (Bash *et al.*, 1991; Davenport *et al.*, 1991).

The catalytic function of a base removed from an enzyme can be restored by added amines. Transaminases have a multifunctional lysine residue that forms an internal aldimine with pyridoxal 5′-phosphate and acts as a base in catalyzing a 1,3-prototropic shift in the conversion of the external aldimine to the ketimine. Substitution of Lys-258 with Ala in *E. coli* aspartate aminotransferase reduced activity by 6-8 orders of magnitude, but the loss of the ε-amino group permitted a Brønsted analysis of the proton transfer with exogenous amines. Ammonia could restore activity equivalent to 2% of the rate for the wild-type enzyme acting on cysteine sulfinate (Toney and Kirsch, 1992). In the K258A enzyme, the rate-determining step is abstraction by the exogenous amine of the Cα proton from the amino acid. The enzyme-bound aldimine is 2×10^5-fold more reactive with general base catalysts than a model aldimine in the imidazole-catalyzed transamination.

In contrast, substitution of the comparable Lys-245 with Gln in *Bacillus* D-amino-acid transaminase, which reduced k_{cat} to 1.5% and k_{cat}/K_m for amino acids or keto acid by 10^4- or 10^5-fold, was not rescued by exogenous amines, although ethanolamine stimulated the transaldimination (Bhatia *et al.*, 1993; Futaki *et al.*, 1990). The residual activity in the "attenuated" enzyme was suggested to result from a compensating role by Lys-267 (Yoshimura *et al.*, 1992). In this enzyme, the spectral changes on complexing substrates are so slow that the transients could be studied with ordinary spectrophotometers. Lys-229 binds the pyridoxal phosphate in *E. coli* serine hydroxymethyltransferase and was substituted with Gln (Schirch *et al.*, 1993). The observation that rates of partial reactions were similar to those for the wild-type enzyme led to the conclusion that the ϵ-amino group is not the base that removes the α-hydrogen of the amino acid substrate. Given the uncertainties in identifying the roles of particular residues in the native or mutant enzymes, the results from "rescue" experiments should be interpreted cautiously.

Activity has been restored to other enzymes that had functional groups removed. Guanidine derivatives increased by about 100-fold the activity (k_{cat}/K_m) of R127A carboxypeptidase on some substrates to produce about 1% of the activity of native enzyme (Phillips *et al.*, 1992). Because the reaction becomes bireactant, the concentrations of substrate and activator were varied so that correct K_m and k_{cat} values and dissociation constants for the guanidino compounds were determined. Such experiments test the importance of precise positioning of a group that stabilizes the transition state. The observation in this study that the R127A enzyme had 1% of the normal activity on one substrate and that guanidine derivatives did not restore activity limits the quantitative and structural interpretations.

The H51Q form of human liver alcohol dehydrogenase β_1 has about sixfold lower V/K_b values for ethanol at pH 7 than the natural enzyme, and activity is restored by 200 mM glycylglycine (Ehrig *et al.*, 1991). A Brønsted plot showed that buffering components affect activity, a cautionary note for interpretation of results of replacing groups that participate in acid-base catalysis.

E. Additivity

A review of the results of combining multiple mutations in a protein suggests that the effects on free energy of reactions are often additive (Wells, 1990). This can be expected when the mutations are distant from one another and do not grossly affect the protein structure (Carter *et al.*, 1984). When the groups interact with one another or when the mutations change the reaction mechanism or rate-limiting steps, effects may not be additive.

The large effects (4 or 5 orders of magnitude) of the Y14F and D38N mutations in a Δ^5-3-ketosteroid isomerase appeared to be additive for k_{cat} or k_{cat}/K_m, and the two residues appear to account for the total enzymatic rate acceleration, acting as acid-base catalysts for the rate-limiting enolization of the substrate (Kuliopulos *et al.*, 1990). The possible effects of mutations on individual steps in the mechanism were analyzed to show how different kinetic results can be interpreted.

In contrast, the studies on the Ser, His, and Asp residues in the catalytic triad of subtilisin are of interest because the k_{cat}/K_m values for the single, double, and triple substitutions with Ala all decreased activity by about 5 or 6 orders of magnitude (Carter and Wells, 1988). The enzyme hydrolyzes a peptide substrate at a rate about 9 orders of magnitude faster than the uncatalyzed rate. The lack of additivity apparently arises because the residues interact to facilitate nucleophilic attack on the carbonyl carbon of the peptide substrate. The results emphasize the difficulty of determining contributions to catalysis of a particular residue. Moreover, it was found that substitution of Asn-155, which is thought to stabilize the oxyanion tetrahedral intermediate, with Gly increased the activity of the S221A enzyme by a factor of 10 (Carter and Wells, 1990). This result led to the conclusion that mutagenesis results may exaggerate the importance of catalytic groups. A similar lack of additivity (named "antagonistic") was observed for the effects on V_m of the single and double D21E and R87G mutations in staphylococcal nuclease (Weber *et al.*, 1990). However, additive or synergistic effects were seen for binding of a substrate or substrate analog.

An extreme example of nonadditivity is found in the restoration of activity of the inactive W191F mutant of cytochrome-*c* peroxidase by a second mutation, H175Q. Residue 175 is the proximal ligand to the heme and interacts with residue 191 (Choudhury *et al.*, 1992). Further studies on multiple mutations in other enzymes are of interest for distinguishing between additivity and cooperativity in enzymatic catalysis.

VII. Concluding Remarks

In combination with other methods, site-directed mutagenesis is a useful tool for enzymologists. As with other experimental approaches, the procedures must be applied carefully and the results interpreted cautiously. The examples discussed here outline some uncertainties in such investigations. The original literature should be read critically for details. Further studies should continue to illuminate different catalytic mechanisms and resolve ambiguities in current understanding. Each enzyme offers a different challenge, and generalizations about the contributions of various amino acid residues to catalysis must await more comprehensive results. Enzymes are cooperative macromolecules, and the functions of individual residues must be integrated with the structural and dynamic aspects of catalysis.

Acknowledgments

I thank coworkers Robert M. Gould, David W. Green, Andrew D. Hershey, Axel J. Ganzhorn, Darla Ann Kratzer, Doo-Hong Park, Hong-Wei Sun, James A. Lorenzen, Tobias Jacobi, Edda Warth, Keehyuk Kim, Fan Fan, Susan K. Souhrada, and Suresh Pal, who developed and applied site-directed mutagenesis for the study of alcohol dehydrogenases, and the National Institute on Alcohol Abuse and Alcoholism, the National Science Foundation, and The University of Iowa Center for Biocatalysis and Bioprocessing for support of our work.

References

Abu-Soud, H. M., Clark, A. C., Francisco, W. A., Baldwin, T. O., and Raushel, F. M. (1993). *J. Biol. Chem.* **268,** 7699.

Ahrweiler, P. M., and Frieden, C. (1991). *Biochemistry* **30,** 7801.

Anderson, D. E., Hurley, J. H., Nicholson, H., Baase, W. A., and Matthews, B. W. (1993). *Protein Sci.* **2,** 1285.

Appleman, J. R., Howell, E. E., Kraut, J., and Blakley, R. L. (1990). *J. Biol. Chem.* **265,** 5579.

Avis, J. M., and Fersht, A. R. (1993). *Biochemistry* **32,** 5321.

Bahnson, B. J., and Klinman, J. P. (1995). *Methods Enzymol.* **249,** 374.

Bahnson, B. J., Park, D. H., Kim, K., Plapp, B. V., and Klinman, J. P. (1993). *Biochemistry* **32,** 5503.

Bash, P. A., Field, M. J., Davenport, R. C., Petsko, G. A., Ringe, D., and Karplus, M. (1991). *Biochemistry* **30,** 5826.

Benkovic, S. J., Fierke, C. A., and Naylor, A. M. (1988). *Science* **239,** 1105.

Berthiaume, L., Tolan, D. R., and Sygusch, J. (1993). *J. Biol. Chem.* **268,** 10826.

Bhatia, M. B., Futaki, S., Ueno, H., Manning, J. M., Ringe, D., Yoshimura, T., and Soda, K. (1993). *J. Biol. Chem.* **268,** 6932.

Blacklow, S. C., and Knowles, J. R. (1990). *Biochemistry* **29,** 4099.

Bocanegra, J. A., Scrutton, N. S., and Perham, R. N. (1993). *Biochemistry* **32,** 2737.

Bone, R., and Agard, D. A. (1991). *Methods Enzymol.* **202,** 643.

Bone, R., Silen, J. L., and Agard, D. A. (1989). *Nature (London)* **339,** 191.

Bullerjahn, A. M. E., and Freisheim, J. H. (1992). *J. Biol. Chem.* **267,** 864.

Carter, P. J., Winter, G., Wilkinson, A. J., and Fersht, A. R. (1984). *Cell (Cambridge, Mass.)* **38,** 835.

Carter, P., and Wells, J. A. (1988). *Nature (London)* **332,** 564.

Carter, P., and Wells, J. A. (1990). *Proteins Struct. Funct. Genet.* **7,** 335.

Chen, L., Neidhart, D., Kohlbrenner, W. M., Mandecki, W., Bell, S., Sowadski, J., and Abad-Zapatero, C. (1992). *Protein Eng.* **5,** 605.

Choudhury, K., Sundaramoorthy, M., Mauro, J. M., and Poulos, T. L. (1992). *J. Biol. Chem.* **267,** 25656.

Cleland, W. W. (1977). *Adv. Enzymol.* **45,** 273.

Cleland, W. W. (1979). *Methods Enzymol.* **63,** 103.

Climie, S., Ruiz-Perez, L., Gonzalez-Pacanowska, D., Prapunwattana, P., Cho, S. W., Stroud, R., and Santi, D. V. (1990). *J. Biol. Chem.* **265,** 18776.

Cortes, A., Emery, D. C., Halsall, D. J., Jackson, R. M., Clarke, A. R., and Holbrook, J. J. (1992). *Protein Sci.* **1,** 892.

Craik, C. S., Roczniak, S., Largman, C., and Rutter, W. J. (1987). *Science* **237,** 909.

Crestfield, A. M., Stein, W. H., and Moore, S. (1963). *J. Biol. Chem.* **238,** 2421.

Cunningham, B. C., and Wells, J. A. (1989). *Science* **244,** 1081.

Dalziel, K. (1963). *J. Biol. Chem.* **238,** 2850.

Davenport, R. C., Bash, P. A., Seaton, B. A., Karplus, M., Petsko, G. A., and Ringe, D. (1991). *Biochemistry* **30,** 5821.

Dayhoff, M. O., Barker, W. C., and Hunt, L. T. (1983). *Methods Enzymol.* **91,** 524.

Dev, I. K., Yates, B. B., Atashi, J., and Dallas, W. S. (1969). *J. Biol. Chem.* **264,** 19132.

Dong, Q., and Fromm, H. J. (1990). *J. Biol. Chem.* **265,** 6235.

Dong, Q., Liu, F., Myers, A. M., and Fromm, H. J. (1990). *J. Biol. Chem.* **266,** 12228.

Ehrig, T., Hurley, T. D., Edenberg, H. J., and Bosron, W. F. (1991). *Biochemistry* **30,** 1062.

Estell, D. A., Graycar, T. P., and Wells, J. A. (1985). *J. Biol. Chem.* **260,** 6518.

Evolution of Catalytic Function (1987). *Cold Spring Harbor Symp. Quant. Biol.* **52.**

Fan, F., Lorenzen, J. A., and Plapp, B. V. (1991). *Biochemistry* **30,** 6397.

Fan, F., and Plapp, B. V. (1995). *Biochemistry* **34,** 4709–4713.

Fan, F., and Plapp, B. V. (1999). *Arch. Biochem. Biophys.* **367,** 240–249.

Fersht, A. R. (1987). *Biochemistry* **26,** 8031.

Fierke, C. A., Johnson, K. A., and Benkovic, S. J. (1987). *Biochemistry* **26,** 4085.
Frieden, C. (1993). *Trends Biochem. Sci.* **18,** 58.
Frieden, C. (1994). *Methods Enzymol.* **240,** 311-322.
Futaki, S., Ueno, H., Martinez del Pozo, A., Pospischil, M. A., Manning, J. M., Ringe, D., Stoddard, B., Tanizawa, K., Yoshimura, T., and Soda, K. (1990). *J. Biol. Chem.* **265,** 22306.
Ganzhorn, A. J., Green, D. W., Hershey, A. D., Gould, R. M., and Plapp, B. V. (1987). *J. Biol. Chem.* **262,** 3754.
Ganzhorn, A. J., and Plapp, B. V. (1988). *J. Biol. Chem.* **263,** 5446.
Gardell, S. J., Hilvert, D., Barnett, J., Kaiser, E. T., and Rutter, W. J. (1987). *J. Biol. Chem.* **262,** 576.
Gerlt, J. A. (1987). *Chem. Rev.* **87,** 1079.
Gibbs, C. S., and Zoller, M. J. (1991). *J. Biol. Chem.* **266,** 8923.
Gould, R. M. (1988). Ph.D. Thesis. The University of Iowa, Iowa City.
Gould, R. M., and Plapp, B. V. (1990). *Biochemistry* **29,** 5463.
Graves, K. L., Butler, M. M., and Hardy, L. W. (1992). *Biochemistry* **31,** 10315.
Green, D. W., Sun, H. W., and Plapp, B. V. (1993). *J. Biol. Chem.* **268,** 7792.
Grimshaw, C. E., Matthews, D. A., Varughese, K. I., Skinner, M., Xuong, N. H., Bray, T., Hoch, J., and Whiteley, J. M. (1992). *J. Biol. Chem.* **267,** 15334.
Hampsey, D. M., Das, G., and Sherman, F. (1986). *J. Biol. Chem.* **261,** 3259.
Harpel, M. R., Larimer, F. W., and Hartman, F. C. (1991). *J. Biol. Chem.* **266,** 24734.
Hartman, F. C., and Harpel, M. R. (1993). *Adv. Enzymol.* **67,** 1.
Hedstrom, L., Graf, L., Stewart, C. B., Rutter, W. J., and Phillips, M. A. (1991). *Methods Enzymol.* **202,** 671.
Hibler, D. W., Stolowich, N. J., Reynolds, M. A., Gerlt, J. A., Wilde, J. A., and Bolton, P. H. (1987). *Biochemistry* **26,** 6278.
Hilvert, D., Gardell, S. J., Rutter, W. J., and Kaiser, E. T. (1986). *J. Am. Chem. Soc.* **108,** 5298.
Ho, C. K., and Fersht, A. R. (1986). *Biochemistry* **25,** 1891.
Howell, E. E., Villafranca, J. E., Warren, M. S., Oatley, S. J., and Kraut, J. (1986). *Science* **231,** 1123.
Jencks, W. P. (1975). *Adv. Enzymol.* **43,** 219.
Johnson, K. A. (1992). *In* "The Enzymes" (P. D. Boyer, ed.), 3rd edn., Vol. 20, p. 1. Academic Press, New York.
Jones, T. A., Zou, J. Y., Cowan, S. W., and Kjeldgaard, M. (1991). *Acta Crystallogr. B* **A47,** 110.
Kavanaugh, J. S., Rogers, P. H., Case, D. A., and Arnone, A. (1992). *Biochemistry* **31,** 4111.
Ke, H., Lipscomb, W. N., Cho, Y., and Honzatko, R. B. (1988). *J. Mol. Biol.* **204,** 725.
Kim, K. (1994). Ph.D. Thesis. The University of Iowa, Iowa City.
Kleanthous, C., Wemmer, D. E., and Schachman, H. K. (1988). *J. Biol. Chem.* **263,** 13062.
Knowles, J. R. (1991). *Nature (London)* **350,** 121.
Kovaleva, E. G., and Plapp, B. V. (2005). *Biochemistry* **44,** 12797.
Kraut, D. A., Carroll, K. S., and Herschlag, D. (2003). *Annu. Rev. Biochem.* **72,** 517–573.
Krebs, J. F., and Fierke, C. A. (1993). *J. Biol. Chem.* **268,** 948.
Kuliopulos, A., Talalay, P., and Mildvan, A. S. (1990). *Biochemistry* **29,** 10271.
Langone, J. J. (ed.) *Methods Enzymol.,* Vol. 202 (1991).
Larimer, F. W., Lee, E. H., Mural, R. J., Soper, T. S., and Hartman, F. C. (1987). *J. Biol. Chem.* **262,** 15327.
Lauble, H., Kennedy, M. C., Beinert, H., and Stout, C. D. (1992). *Biochemistry* **31,** 2735.
LeBrun, L. A., and Plapp, B. V. (1999). *Biochemistry* **38,** 12387.
LeBrun, L. A., Park, D., Ramaswamy, S., and Plapp, B. V. (2004). *Biochemistry* **43,** 3014.
Léger, D., and Hervé, G. (1988). *Biochemistry* **27,** 4293.
Lewendon, A., and Shaw, W. V. (1993). *J. Biol. Chem.* **268,** 20997.
Light, D. R., Dennis, M. S., Forsythe, I. J., Liu, C. C., Green, D. W., Kratzer, D. A., and Plapp, B. V. (1992). *J. Biol. Chem.* **267,** 12592.
Li, L., Falzone, C. J., Wright, P. E., and Benkovic, S. J. (1992). *Biochemistry* **31,** 7826.
Liu, F., Dong, Q., and Fromm, H. J. (1992). *J. Biol. Chem.* **267,** 2388.

Lodi, P. J., and Knowles, J. R. (1991). *Biochemistry* **30,** 6948.
Mas, M. T., Bailey, J. M., and Resplandor, Z. E. (1988). *Biochemistry* **27,** 1168.
Matlin, A. R., Kendall, D. A., Carano, K. S., Banzon, J. A., Klecka, S. B., and Solomon, N. M. (1992). *Biochemistry* **31,** 8196.
Matthews, B. W. (1987). *Biochemistry* **26,** 6885.
Matthews, B. W. (1993). *Annu. Rev. Biochem.* **62,** 139.
McTigue, M. A., Davies, J. F., II, Kaufman, B. T., and Kraut, J. (1992). *Biochemistry* **31,** 7264.
Miles, E. W. (1991). *Adv. Enzymol.* **64,** 93.
Murphy, D. J., and Benkovic, S. J. (1989). *Biochemistry* **28,** 3025.
Newman, J., and Gutteridge, S. (1993). *J. Biol. Chem.* **268,** 25876.
Nickbarg, E. B., Davenport, R. C., Petsko, G. A., and Knowles, J. R. (1988). *Biochemistry* **27,** 5948.
Pan, Q. W., Tanase, S., Fukumoto, Y., Nagashima, F., Rhee, S., Rogers, P. H., Arnone, A., and Morino, Y. (1993). *J. Biol. Chem.* **268,** 24758.
Park, D.-H. (1991). Ph.D. Thesis. The University of Iowa, Iowa City.
Park, D. H., and Plapp, B. V. (1992). *J. Biol. Chem.* **267,** 5527.
Penner, M. H., and Frieden, C. (1987). *J. Biol. Chem.* **262,** 15908.
Phillips, M. A., Hedstrom, L., and Rutter, W. J. (1992). *Protein Sci.* **1,** 517.
Plapp, B. V., Ganzhorn, A. J., Gould, R. M., Green, D. W., Jacobi, T., Warth, E., and Kratzer, D. A. (1990). *In* "Enzymology and Molecular Biology of Carbonyl Metabolism 3" (H. Weiner, B. Wermuth and D. W. Crabb, eds.), p. 241. Plenum, New York.
Poland, B. W., Silva, M. M., Serra, M. A., Cho, Y., Kim, K. H., Harris, E. M. S., and Honzatko, R. B. (1993). *J. Biol. Chem.* **268,** 25334.
Profy, A. T., and Schimmel, P. (1986). *J. Biol. Chem.* **261,** 15474.
Ramaswamy, S., Eklund, H., and Plapp, B. V. (1994). *Biochemistry* **33,** 5230.
Ramaswamy, S., Park, D. H., and Plapp, B. V. (1999). *Biochemistry* **38,** 13951.
Roper, D. I., Moreton, K. M., Wigley, D. B., and Holbrook, J. J. (1992). *Protein Eng.* **5,** 611.
Rubach, J. K., Ramaswamy, S., and Plapp, B. V. (2001). *Biochemistry* **40,** 12686.
Schirch, D., Fratte, S. D., Iurescia, S., Angelaccio, S., Contestabile, R., Bossa, F., and Schirch, V. (1993). *J. Biol. Chem.* **268,** 23132.
Scrutton, N. S., Berry, A., and Perham, R. N. (1990). *Nature (London)* **343,** 38.
Sekhar, V. C., and Plapp, B. V. (1990). *Biochemistry* **29,** 4289.
Shearer, G. L., Kim, K., Lee, K. M., Wang, C. K., and Plapp, B. V. (1993). *Biochemistry* **32,** 11186.
Shi, Z., Byeon, I. J. L., Jiang, R. T., and Tsai, M. D. (1993). *Biochemistry* **32,** 6450.
Silverman, D. N. (1995). *Methods Enzymol.* **249,** 479.
Sprang, S., Standing, T., Fletterick, R. J., Stroud, R. M., Finer-Moore, J., Xuong, N. H., Hamlin, R., Rutter, W. J., and Craik, C. S. (1987). *Science* **237,** 905.
Stebbins, J. W., Robertson, D. E., Roberts, M. F., Stevens, R. C., Lipscomb, W. N., and Kantrowitz, E. R. (1992). *Protein Sci.* **1,** 1435.
Stevens, R. C., Chook, Y. M., Cho, C. Y., Lipscomb, W. N., and Kantrowitz, E. R. (1991). *Protein Eng.* **4,** 391.
Sun, H. W., and Plapp, B. V. (1992). *J. Mol. Evol.* **34,** 522.
Teng, H., Segura, E., and Grubmeyer, C. (1993). *J. Biol. Chem.* **268,** 14182.
Thillet, J., Absil, J., Stone, S. R., and Pictet, R. (1988). *J. Biol. Chem.* **263,** 12500.
Thunnissen, M. M. G. M., Franken, P. A., de Haas, G. H., Drenth, J., Kalk, K. H., Verheij, H. M., and Dijkstra, B. W. (1992). *Protein Eng.* **5,** 597.
Toney, M. D., and Kirsch, J. F. (1992). *Protein Sci.* **1,** 107.
Tsai, M. D., and Yan, H. (1990). *Biochemistry* **30,** 6806.
Tsirka, S., and Coffino, P. (1992). *J. Biol. Chem.* **267,** 23057.
Weber, D. J., Serpersu, E. H., Shortle, D., and Mildvan, A. S. (1990). *Biochemistry* **29,** 8632.
Weiner, H., Jr, White, W. N., Hoare, D. G., and Koshland, D. E. (1966). *J. Am. Chem. Soc.* **88,** 3851.
Wells, J. A. (1990). *Biochemistry* **29,** 8509.
Wente, S. R., and Schachman, H. K. (1987). *Proc. Natl. Acad. Sci. USA* **84,** 31.

Wente, S. R., and Schachman, H. K. (1991). *J. Biol. Chem.* **266,** 20833.
Wilks, H. M., Hart, K. W., Feeney, R., Dunn, C. R., Muirhead, H., Chia, W. N., Barstow, D. A., Atkinson, T., Clarke, A. R., and Holbrook, J. J. (1988). *Science* **242,** 1541.
Wilson, K. P., Malcolm, B. A., and Matthews, B. W. (1992). *J. Biol. Chem.* **267,** 10842.
Wratten, C. C., and Cleland, W. W. (1963). *Biochemistry* **2,** 935.
Xi, X. G., Van Vliet, F., Ladjimi, M. M., Cunin, R., and Hervé, G. (1990). *Biochemistry* **29,** 8491.
Xue, Y., Liljas, A., Jonsson, B. H., and Lindskog, S. (1993). *Proteins Struct. Funct. Genet.* **17,** 93.
Xu, X., and Kantrowitz, E. R. (1991). *Biochemistry* **30,** 7789.
Yoshimura, T., Bhatia, M. B., Manning, J. M., Ringe, D., and Soda, K. (1992). *Biochemistry* **31,** 11748.
Zheng, L., Kennedy, M. C., Beinert, H., and Zalkin, H. (1992). *J. Biol. Chem.* **267,** 7895.
Zimmerle, C. T., and Frieden, C. (1989). *Biochem. J.* **258,** 381.

CHAPTER 9

Cooperativity in Enzyme Function: Equilibrium and Kinetic Aspects

Kenneth E. Neet

Department of Biochemistry and Molecular Biology
Chicago Medical School
Rosalind Franklin University of Medicine and Science

DOI: 10.1016/B978-0-12-378608-1.00009-8

I. Update

Advances in the study of cooperative systems in the past 10–15 years have not been in the analysis of general models or the application of new mathematical models, but rather in the realization that each enzyme system is unique unto itself. In most cases, some refinement to the general models is required to understand the details of the cooperative mechanisms. Therefore, detailed kinetic, binding, and structural studies have been applied to many enzymes with useful and interesting results. Phenomenological models (MWC, KNF) have given way to detailed structural models with microscopic rate constants determined by newer technology. In this update, I will mention three systems that have been more thoroughly studied and highlight the methods that have allowed such refinement.

Before these examples are discussed, however, one literature survey is worth noting that has documented the frequent occurrence of cooperativity and the distribution among types (Koshland and Hamadani, 2002). For nearly 600 enzymes consistent with *i*nduced, *i*ntramolecular conformational changes from essentially *i*dentical sites (termed i^3) over a 13 year literature period, approximately 55% showed positive cooperativity, 35% had negative cooperativity, and 10% revealed both negative and positive cooperativity in one or more substrates. The authors conclude (Koshland and Hamadani, 2002) that "the sensitivity capabilities... have about equal evolutionary value with a slight evolutionary advantage to positive cooperativity."

Intensive, continued studies on the cooperativity displayed by hemoglobin using NMR, spin labels, mutations, and partially ligated intermediates (reviewed in Ackers and Holt, 2006; Koshland and Hamadani, 2002) have led to a view in which the $\alpha_1\beta_1$ dimers act as units that switch states when 2 oxygens are bound (an extension of Fig. 5). Intradimer coupling produces autonomously functioning, yet energetically coupled, $\alpha\beta$ dimers within the tetramer. A related description appears to be the tertiary two-state (TTS) model which focuses on rebinding of carbon monoxide and allows for an equilibrium between high and low affinity tertiary conformations within R and T states (Eaton *et al.*, 2007). A multistep, detailed story is now available for hemoglobin, but perhaps not yet the final chapter. For example, dynamics of the protein structure have recently been compellingly proposed to contribute to the allosteric behavior of hemoglobin (Yonetani and Laberge, 2008). Other reaction pathway calculations suggest that the *rate* of tertiary structural changes, relative to the rate of the quaternary T to R transition, may be important in developing cooperativity in hemoglobin (Cui and Karplus, 2008).

A story with similar complexity holds for the kinetic cooperativity of the monomeric enzyme glucokinase. While the kinetics were originally interpreted on the basis of two catalytically active conformations of the enzyme to generate kinetic cooperativity in the monomer (Fig. 4), sophisticated rapid kinetic methodology has permitted the description of multiple conformations and at least four different relaxation times that influence activity (Antoine *et al.*, 2009). Thus the concepts of ligand-induced slow transitions (LISTs) or mnemonics have to be extended to include such multiple states. Structural analysis has allowed postulation that one of the slow steps is the movement of a C-terminal α-helix past or through a loop near the active site; experimental manipulation of this helix has confirmed its importance for cooperativity and activity (Larion and Miller, 2009).

Studies on glucokinase and hemoglobin have revealed another, more general and important, similarity. Rare amino acid point substitutions in glucokinase have been shown to give rise to variants with decreased cooperativity and affinity that cause type 2 diabetes (MODY2) or increased cooperativity and affinity that produce persistent hyperinsulinemic hypoglycemia of infancy (PHHI) in the human population (Pal and Miller, 2009). This outcome is reminiscent of the long-known hemoglobinopathies (e.g., polycythemia) that arise due to point mutations in hemoglobin that affect the cooperativity and/or affinity of oxygen binding. So the oft-asked question "what is the purpose of cooperativity?" can be answered for these two proteins in the sense that without allostery, bad things happen, that is, disease.

Cooperativity, both negative and positive, in ligand binding to cellular receptors has been noted for years without much mechanistic clarity. The epidermal growth factor (EGF) receptor had been known to demonstrate positive linkage between EGF binding and dimerization of the EGFR but, surprisingly, with a resultant negative cooperativity of steady-state binding of EGF to EGFR. A series of recent papers (Macdonald and Pike, 2008; Macdonald-Obermann and Pike, 2009; Saffarian *et al.*, 2007; Yang *et al.*, 2009) from the Pike laboratory have illuminated the complex equilibria in the EGFR system. These elegant studies used controllable expression of precise amounts of EGFR in membrane, selected coexpression of EGFR mutants (e.g., kinase dead), fusion EGFRs with luciferase fragment complementation, global analysis of extensive binding data, coupled to knowledge of the EFGR ECD structure to provide probably the best current description of cooperative binding in a receptor system. In the first step, ligand binding to naïve, unphosphorylated EGF receptors is positively linked to dimer formation and conformational change(s) (Yang *et al.*, 2009). This linkage is then lost after receptor tyrosine kinase activation and receptor autophosphorylation. Nevertheless, both phosphorylated and unphosphorylated EGF receptors exhibit negative cooperativity with EGF binding to one dimer site decreasing the affinity for the next. Furthermore, these studies determined the K_d for the monomer-dimer equilibrium in intact (transfected) cells and could explain changes in ligand-binding properties with a model including negative cooperativity in a dimerizing system (Macdonald and Pike, 2008) (see also Sections V.E and IX.A). Thus, cooperativity is mechanistically distinct from dimerization linkage. The intracellular juxtamembrane domain is required for both linkage and cooperativity, thereby providing "inside-out signaling" in this system (Macdonald-Obermann and Pike, 2009). Quantitative live cell fluorescence intensity distribution analysis supports the framework for these interpretations (Saffarian *et al.*, 2007) and promises extension of the understanding of cooperativity in the EGF system to truly physiological situations. Indeed, the past 15 years have been an exciting era for the regulatory enzymologist—the excitement continues.

II. Introduction

Cooperativity in protein or enzyme systems commonly refers to ligand binding with the type of positive cooperativity that gives rise to the sigmoid curves of oxygen binding to hemoglobin. However, the concept and the practical applications are much more encompassing than that single interpretation. Cooperativity in its broadest sense includes protein folding-unfolding reactions (e.g., denaturation, helix-coil transitions), macromolecular assembly (e.g., tubulin, sickling of hemoglobin S), binding of proteins as ligands to DNA (e.g., transcriptional factors, single-stranded DNA binding proteins), and binding of proteins as ligands to membrane receptors (e.g., hormones, growth factors). Cooperativity, in general, is any process in which the initial event (hydrogen bonding, protein-protein interaction, ligand binding) affects subsequent similar events, in the cases cited via communication through intra- or intermolecular protein interactions. Protein structures are uniquely suited for the ability to communicate the information needed for cooperativity through conformational changes and intersubunit binding affinities. This chapter mainly deals with substrate and ligand binding to enzymes but also briefly discusses the extension to proteins as ligands binding to other macromolecular systems. The current discussion of cooperativity, which is an update of an earlier chapter in this series (Neet, 1980), describes the basics of enzyme cooperativity, emphasizes advances since the early 1980s, and utilizes examples of several systems to describe how cooperativity is currently studied and analyzed. The reader is referred to previous reviews (Neet, 1980; Neet and Ainslie, 1980) for historical perspectives, more details of kinetic analysis of some systems, and a description of mechanisms not described here.

Cooperativity in enzyme systems is initially phenomenological, that is, it usually begins with the observation of non-Michaelis-Menten (nonhyperbolic) kinetic assays and later develops into a mechanistic explanation in the particular enzyme. Thus, a variety of both equilibrium (e.g., site-site models, oligomer association) and kinetic (e.g., slow transition, random order) mechanisms can give essentially identical velocity (v) versus substrate concentration ([S]) plots. The form of the cooperative curves and their fundamental analyses are basic to understanding the physiological relevance and the underlying mechanism of the cooperativity. The shape of positive, negative, and noncooperative curves is shown in Fig. 1 in six standard plotting formats (discussed in Section III). The investigator must eliminate potential sources of artifacts leading to the appearance of cooperativity such as nonrelevant adsorption of substrate at low concentration, impure enzymes, complex formation (e.g., MgATP), and substrate inhibition (discussed by Cornish-Bowden and Cardenas, 1987).

Homotropic cooperativity is the effect of the interaction of a ligand on subsequent binding or reaction of the same ligand (e.g., the first molecule of O_2 binding to hemoglobin affects the binding of the next three O_2 molecules, giving rise to the now classic sigmoidal binding curve). Homotropic cooperativity can be either positive or negative. In positive cooperativity the initial binding makes it easier for subsequent molecules to interact (e.g., O_2 enhances the binding of the next O_2 to hemoglobin). With positive cooperativity in its most general sense, the logical intermediate does not accumulate, that is, the initial and the final states are the most predominant at equilibrium. In negative cooperativity, the initial interaction makes it more difficult for subsequent molecules to bind or interact [e.g., NAD^+ binding to glyceraldehyde phosphate dehydrogenase (Conway and Koshland, 1968)]. In the extreme, negative cooperativity can give rise to half-the-sites reactivity (e.g., three molecules of carbamoyl phosphate (Suter and Rosenbusch, 1976) binding to the aspartate transcarbamylase (ATCase; aspartate carbamoyltransferase) hexamer).

Heterotropic cooperativity is the effect of a different ligand (or the same ligand acting at a quite different site on the protein) on the binding or reaction of the first molecule. Negative heterotropic effector, inhibitor, and antagonist are synonyms for the same type of ligand, producing a decrease in binding or reaction at the site (catalytic) of interest. Similarly, positive heterotropic effector is synonymous with activator, producing an increase in binding or reaction at the initial site. Heterotropic responses are the classic allosteric effectors, at least when the action is at a site different from the active site. Several examples of isosteric regulation have also been observed which produce similar kinetic effects (see below).

Kinetic effects in enzyme catalysis can alter existing cooperativity. The most comprehensive extension of cooperative binding to kinetic analysis has been developed by Ricard and coworkers (Ricard, 1985) as "structural kinetics." Alternatively, purely kinetic mechanisms can give rise to phenomenological "cooperativity" that appears only in kinetic assays and, thus, can be assessed by comparison of equilibrium binding and kinetic measurements with the same ligand. Hysteresis or LISTs produced by enzyme isomerizations that are on the same order of magnitude as the catalytic cycle can produce nonhyperbolic kinetics (Ainslie *et al.*, 1972; Neet and Ainslie, 1980; Ricard *et al.*, 1974). A steady-state random addition of substrates, with the appropriate combination of rate constants, may produce the appearance of cooperativity in enzyme catalysis (Ferdinand, 1966). These kinetic consequences are also discussed

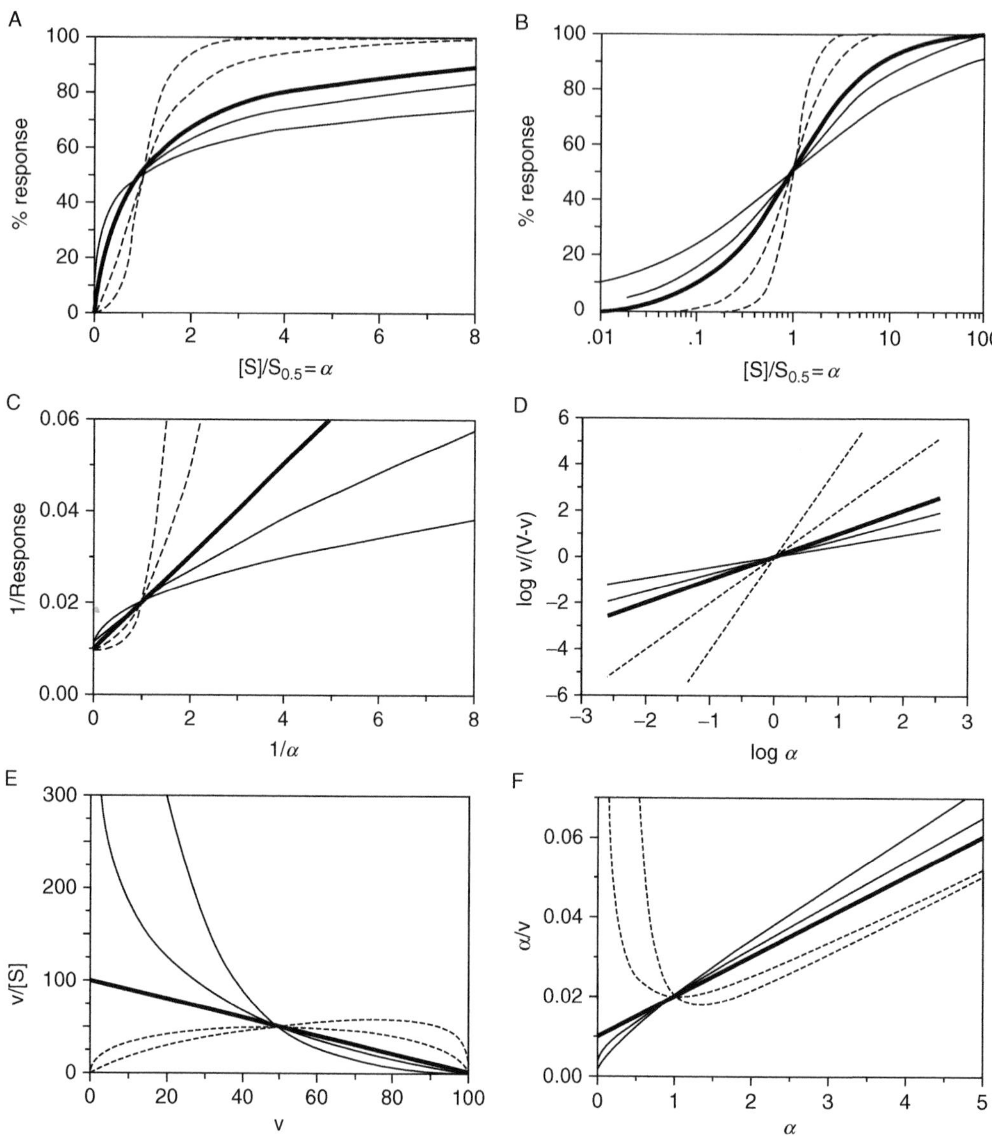

Fig. 1 Comparison of six ways of plotting kinetic or binding data to analyze the cooperativity. In each graph the noncooperative (Michaelis-Menten) case is the heavy line ($n_H = 1$), the two positively cooperative cases are dashed lines ($n_H = 4, 2$), and the two negatively cooperative cases are thin solid lines ($n_H = 0.75, 0.5$). Note that heterogeneous binding sites would appear the same as "negative cooperativity" in each of the plots. The % RESPONSE in (A) and (B) could represent enzymatic velocity or moles of bound ligand. The response is given as v in (C)-(F). The values of [S] are normalized to the $S_{0.5}$ value (α) for comparison purposes and would represent free ligand concentration in binding experiments.

in this chapter. Finally, certain structural features and thermodynamic considerations of cooperative enzymes are emerging and are presented.

III. Graphical Representation and Evaluation of Extent of Cooperativity

Six standard ways of graphing kinetic or binding data have been developed (Fig. 1) based mainly on rearrangements of the Michaelis-Menten, multisite binding isotherm, or Hill equations. Examples of different degrees of both positive and negative cooperativity are shown in Fig. 1 to contrast to the noncooperative (Michaelis-Menten) case.

The straight-up plot of response versus ligand concentration (Fig. 1A) gives rise to the "sigmoidal" curve of positive cooperativity but does not readily distinguish the hyperbola from the near-hyperbola of negative cooperativity. The log-response curve (Fig. 1B), commonly used in pharmacological studies, makes all common binding patterns sigmoidal. This plot provides an easy way to assess cooperativity visually from the slope of the graph or to calculate quickly the R_s, that is, the ratio of ligand concentration at 90% response to ligand concentration at 10% response ($R_s = 81$ for no cooperativity; $R_s < 81$ for positive cooperativity; $R_s > 81$ for negative cooperativity). The value of R_s is easily related to the Hill coefficient ($R_s = 81^{1/n_H}$).

The double-reciprocal plot (Lineweaver-Burke) (Fig. 1C) is standardly used for enzyme kinetics, and its merits and faults for obtaining good estimates of V_{max} and K_m have been extensively discussed (Rudolph and Fromm, 1979). The various forms of cooperativity are clearly distinct in the double-reciprocal plot by their deviations from linearity. Note that the asymptote to the positive cooperativity curve at low concentrations would intersect the Y axis at a negative value, distinguishing it from substrate inhibition, which would have a positive Y intercept.

The Hill plot (Fig. 1D) linearizes all common binding patterns (over the range of 10-90% saturation) and provides the value of the Hill coefficient, n_H, from the slope of the graph. The n_H is the standard and most common means of evaluating or comparing cooperativity ($n_H = 1$ for no cooperativity; $n_H < 1$ for negative cooperativity; $n_H > 1$ for positive cooperativity). Deviations from linearity in the central region of the Hill plot are indicative of mixed cooperativity or more complex binding patterns (Cornish-Bowden and Koshland, 1975). The value for V_{max} must be estimated to calculate the values for the Y axis in the Hill plot. For positive cooperativity, this estimate is easily obtained from the log-response or the double-reciprocal plot. For negative cooperativity, the estimate of V_{max} is more difficult because of the slow asymptotic approach (in the v versus [S], the log-response, or the Scatchard plot) or the steep curvature at high concentrations (in the double-reciprocal plot). Fortunately, the minimum slope of the Hill plot in this case is not very sensitive to the value of the V_{max}. However, the shape of the Hill plot does change with wrong choices of V_{max}. Incorrect choice of a V_{max} that is 75 or 85% of the true value produces a curved Hill plot with a linear segment below the $S_{0.5}$, whose slope is about 10% higher than the true value, then breaks sharply upward with a slope approaching infinity.

The Scatchard-Eadie plot (or Hofstee-Augustinsson when rotated 90°) of v/[S] versus v (Fig. 1E) is commonly used in ligand-binding analyses and has proponents for enzyme kinetics. The advantage of using the v/[S] versus v plot is that the data (or the curvature) are spread more evenly with retention of the distinction between the different forms of cooperativity. The position of the maximum in the positive cooperativity curves is easily related to the Hill coefficient; thus, $v/V_{max} = (1 - 1/n_H)$ at the optimum in the v/[S] curve.

Finally, the Hanes plot of [S]/v versus [S] (Fig. 1F) has similar characteristics to the Scatchard plot but is infrequently utilized. Note that, again, the [S] position of the minimum in the [S]/v curve is related to the degree of positive cooperativity.

Unfortunately, true negative cooperativity gives rise to the same shape of velocity or binding curve as the curve from a mixture of enzymes or binding sites within an enzyme (Cornish-Bowden and Cardenas, 1987). Thus, the curves termed "negative cooperativity" in Fig. 1 are indistinguishable from a preexisting heterogeneity of binding sites. The shape of the curve is a result of nonidentical binding after ligand binding and does not depend on whether the protein initially had identical (negative cooperativity) or nonidentical (heterogeneous) sites. Therefore, demonstration of the purity of a protein preparation, including the possibility of partially denatured components, is essential for the study of negative cooperativity. Because ligand binding or kinetics cannot discriminate between these two cases, attempts to resolve such issues normally depend on physical methods to demonstrate or eliminate preexisting asymmetry or heterogeneity of ligand-binding sites.

Although several alternate means of assessing the degree of cooperativity have been proposed (see Neet, 1980, for a comparison), the Hill coefficient remains the most generally used and conveys the most immediate, conceptual meaning to practicing biochemists. The n_H for positive cooperativity is equal to or less than the number of ligand-binding sites, n; how different n_H is from n is a qualitative estimate of the strength of the site-site interactions. The n_H for negative cooperativity is not indicative of, nor related to, the number of sites. The important consideration in evaluating cooperativity is to obtain good data over a wide range of ligand concentration that is appropriately chosen for the graphical means of assessing the data. The use of several graphical methods is recommended to extract the most information, but such plots should only be used for visualization and/or presentation. An unbiased computer fitting algorithm of the raw data (v and [S]) to the equation for the appropriate model is the final and best means to judge the type and extent of cooperativity and obtain relevant parameters.

IV. Estimation of Parameters by Curve Fitting

The estimation of cooperativity parameters and the evaluation of models by data fitting have been greatly facilitated with the general availability of computers, digital data acquisition, and appropriate software for analysis. The graph has become a means of presentation and "eyeballing" results, rather than a means of obtaining parameter values. An extensive discussion of statistical analysis of enzyme kinetic data and nonlinear least squares fitting procedures has been previously presented in this series (Cleland, 1979; Mannervik, 1982; Neet, 1980) and is not elaborated on here. Those discussions tend to deal with Michaelian enzymes, but the approaches are still generally valid.

The appropriate cooperativity equation for the model to be tested should be chosen from those discussed below and the data fitted with a nonlinear regression procedure. The Marquardt algorithm is most commonly used. Even simple estimations of the Hill coefficient are more easily done with the Hill equation in a nonlinear fitting program that estimates V_{max} as well as K_m and n_H. Numerous programs for both MacIntosh and IBM-compatible computers are now commercially available that can handle higher order rate (or binding) equations for cooperativity and exponential functions for hysteretic transient analyses either intrinsic to the application or by writing the desired equation into the program. Typical commercial programs include LIGAND (Munson, 1983) (Lunden Software, Biosoft), ENZFITTER (Sigma, St Louis, MO), GRAFIT (Erithacus Software Ltd.), ULTRAFIT (Biosoft), EZFIT (Noggle, 1993) (Prentice Hall), JMP (SAS Institute, Inc.), KALEIDOGRAPH (Synergy Software), and MATHEMATICA (Wolfram Research, Inc.). Simulations of data with chosen parameters can also be done with many of the programs and can be helpful in the design of experiments, for example, to determine the best concentration range for a ligand in the presence of several other ligands. No software, however, can substitute for precise data spread over an adequate concentration range and application of good common sense.

V. Mechanisms for Enzyme Cooperativity

Both the concerted (Monod *et al.*, 1965) model and the sequential (Koshland *et al.*, 1966) model are well described in biochemistry textbooks such as Berg *et al.* (2007), Lehninger *et al.* (2009), or Mathews and van Holde (1999). This section emphasizes the differences in the models, the basis for the resultant cooperativity, and the means for distinguishing the models. A later section describes studies that have chosen one model as the most appropriate for a particular system.

A. General Binding Model of Adair

$$n\overline{Y} = \frac{\Psi_1[\mathrm{S}] + 2\Psi_2[\mathrm{S}]^2 + 3\Psi_3[\mathrm{S}]^3 + 4\Psi_4[\mathrm{S}]^4}{1 + \Psi_1[\mathrm{S}] + \Psi_2[\mathrm{S}]^2 + \Psi_3[\mathrm{S}]^3 + \Psi_4[\mathrm{S}]^4} \tag{1}$$

The general Adair (1925) binding equation [Eq. (1)] defines the ligand saturation function, $\overline{Y}$, in terms of the number of binding sites n (usually subunits) and n coefficients (Ψ_i), where $\Psi_i = \Pi_{j=1}^{i} K_j$. The Adair coefficients are readily related to the simple stoichiometric, K_i, or intrinsic, K_i', ligand association constants (see Table I for the tetramer case).

Stoichiometric constants are obtained by counting all liganded species with a certain stoichiometry, namely, $K_i = [\mathrm{PS}_i]/[\mathrm{S}][\mathrm{PS}_{i-1}]$, whereas intrinsic constants are treated as if each liganded subunit were measured in isolation and therefore take into account the probability of binding to individual sites. An alternative definition of binding focuses on individual, discrete sites of binding on the oligomer and results in different, but related, equations utilizing these site binding constants (Klotz, 1974). The statistical relationship (Neet, 1980) between stoichiometric, K_i, and intrinsic K_i', constants for the ith site, is given by Eq. (2). In the case of a noninteracting oligomer, successive intrinsic constants will be identical.

$$K_i = \frac{(n+1-i)}{i} K_i' = \frac{\#\,\text{free sites on oligomer before binding}}{\#\,\text{occupied sites after binding}} K_i' \tag{2}$$

Intrinsic binding constants are frequently utilized instead of stoichiometric constants when the derivation of a binding (or rate) equation is made by considering the distribution of forms at equilibrium (Ricard and Cornish-Bowden, 1987; Segal, 1975). In this case, the saturation function may be presented in a slightly different form, where the numerical coefficients are the binomial expansion of $(n-1)$ in the numerator and n in the denominator [Eq. (3)] and $\Psi_i' = \Pi_{j=1}^{i} K_j'$.

$$\overline{Y} = \frac{\Psi_1'[\mathrm{S}] + 3\Psi_2'[\mathrm{S}]^2 + 3\Psi_3'[\mathrm{S}]^3 + \Psi_4'[\mathrm{S}]^4}{1 + 4\Psi_1'[\mathrm{S}] + 6\Psi_2'[\mathrm{S}]^2 + 4\Psi_3'[\mathrm{S}]^3 + \Psi_4'[\mathrm{S}]^4} \tag{3}$$

These constants, K_i or K_i', are phenomenological (or observed) constants and do not directly relate to molecular properties of the protein such as conformational changes or subunit interactions. However, they serve a useful purpose for initially fitting primary binding data and for comparison of molecular models.

B. Concerted, Symmetric Model of Monod-Wyman-Changeux

The concerted or Monod-Wyman-Changeux (MWC) model (Monod *et al.*, 1965) provides the simplest explanation for cooperativity in ligand binding to enzymes and has been widely applied to many enzymes. This model occupies a

Table I
Coefficients of Adair Equation for Cooperative Tetramer Models

			KNF		
	Adair	Intrinsic	Tetrahedral	Square	MWC concerted
Ψ_1	K_1	$4K_1'$	$4K_{AB}^3K_SK_T$	$4K_{AB}^2K_SK_T$	$\frac{4(1+Lc)}{(1+L)K_R}$
Ψ_2	K_1K_2	$6K_1'K_2'$	$6K_{AB}^4K_{BB}K_S^2K_T^2$	$(2K_{AB}^4+4K_{BB}K_{AB}^2)K_S^2K_T^2$	$\frac{6(1+Lc^2)}{(1+L)K_R^2}$
Ψ_3	$K_1K_2K_3$	$4K_1'K_2'K_3'$	$4K_{AB}^3K_{BB}^3K_S^3K_T^3$	$4K_{AB}^2K_{BB}^2K_S^3K_T^3$	$\frac{4(1+Lc^3)}{(1+L)K_R^3}$
Ψ_4	$K_1K_2K_3K_4$	$K_1'K_2'K_3'K_4'$	$K_{BB}^6K_S^4K_T^4$	$K_{BB}^4K_S^4K_T^4$	$\frac{(1+Lc^4)}{(1+L)K_R^4}$
$\frac{\Psi_1}{4}$		K_1'	$K_{AB}^3K_SK_T$	$K_{AB}^2K_SK_T$	$\frac{(1+Lc)}{(1+Lc)K_R}$
$\frac{3\Psi_2}{2\Psi_1}$		K_2'	$K_{AB}K_{BB}K_SK_T$	$(K_{AB}^2+2K_{BB})K_SK_T$	$\frac{(1+Lc^2)}{(1+Lc)K_R}$
$\frac{2\Psi_3}{3\Psi_2}$		K_3'	$\frac{K_{BB}^2K_SK_T}{(K_{AB})}$	$\frac{K_{BB}^2K_SK_T}{(K_{AB}^2+2K_{BB})}$	$\frac{(1+Lc^3)}{(1+Lc^2)K_R}$
$\frac{4\Psi_4}{\Psi_3}$		K_4'	$\frac{K_{BB}^3K_SK_T}{K_{AB}^3}$	$\frac{K_{BB}^2K_SK_T}{K_{AB}^2}$	$\frac{(1+Lc^4)}{(1+Lc^3)K_R}$
		K_2'/K_1'	$\frac{K_{BB}}{K_{AB}^2}$	$\frac{(K_{AB}^2+2K_{BB})}{K_{AB}^2}$	$\frac{(1+Lc^2)(1+L)}{(1+Lc)^2}$
		K_3'/K_2'	$\frac{K_{BB}}{K_{AB}^2}$	$\frac{K_{BB}^2}{(K_{AB}^2+2K_{BB})^2}$	$\frac{(1+Lc^3)(1+Lc)}{(1+Lc^2)^2}$
		K_4'/K_3'	$\frac{K_{BB}}{K_{AB}^2}$	$\frac{(K_{AB}^2+2K_{BB})}{K_{AB}^2}$	$\frac{(1+Lc^4)(1+Lc^2)}{(1+Lc^3)^2}$

historical place in enzymology, since it was the first general explanation for cooperativity and served as the basis for defining K systems (those effectors that primarily alter K_m or $S_{0.5}$), V systems (those effectors that primarily alter V_{max}), homotropic and heterotropic interactions, and the linkage between these interactions. The main advantages of the MWC model are that it is conceptually straightforward and requires few parameters to fit the experimental data. The main disadvantage is that the MWC model cannot account for negative cooperativity in equilibrium binding (or rapid equilibrium kinetic) systems.

The MWC model (Fig. 2) is based on a concerted change of an oligomeric structure (shown here as a tetramer) from the T-state (taut) to the thermodynamically more favored R-state (relaxed). The MWC model is "concerted" because symmetry is maintained as all subunits (or protomers) undergo the same conformational change concomitantly. The oligomeric conformational equilibrium is shifted by binding of substrate or activator with higher affinity to the R-state.

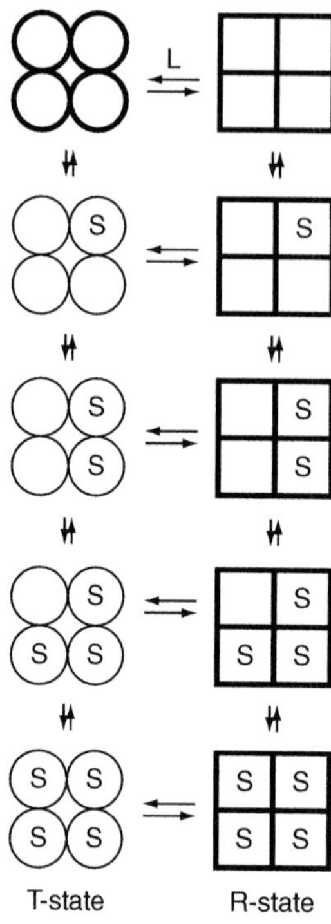

Fig. 2 Concerted Monod-Wyman-Changeux (MWC) model for cooperativity of a tetramer. The circles (T) and squares (R) represent the two symmetrical conformations of the tetramer. Ligand (S) binding shifts the equilibrium to the energetically favored R-state. The forms in heavy lines represent the original exclusive binding MWC model, and the full figure represents the complete nonexclusive binding model.

The cooperativity arises because the shift in the number of high-affinity R-subunit conformations (squares in Fig. 2) is greater than the amount of ligand binding to trigger the shift, providing nonoccupied high-affinity sites to which subsequent ligand can bind. The original proposal for the MWC model (Monod *et al.*, 1965) presented exclusive binding of substrate (shown in bold, Fig. 2) but was subsequently extended to nonexclusive binding (Rubin and Changeux, 1966) (Fig. 2, all forms). Because of the linkage between equilibria, a negative heterotropic effector is predicted to increase the homotropic cooperativity in addition to raising the apparent $S_{0.5}$ value; a positive heterotropic effector is predicted to lower or eliminate the homotropic cooperativity with a lower apparent $S_{0.5}$. These relationships are frequently found with enzymes, for example, ATCase.

The cooperativity derived from the nonexclusive binding MWC model can be described by the number of binding sites, n, and three parameters: the equilibrium constant between the two unliganded oligomer conformations, L; and the intrinsic dissociation constants for ligand dissociation from each conformer, K_R and K_T [Eq. (4)]. In Eq. (4), $L = [T_0]/[R_0]$; $K_R = [S][R]/[SR]$; $K_T = [S][T]/[ST]$; and

$c = K_R/K_T$. The parameter α is simply a ligand normalization term equal to $[S]/K_R$. The concentrations of ligand, R-subunit, T-subunit, ligand bound to R-subunit, and ligand bound to T-subunit are given by [S], [R], [T], [SR], and [ST], respectively. Also shown in the right-hand side of Eq. (4) is the equivalent relationship using terms similar to those for the KNF model (see below) with $c = K_A/K_B$, $\alpha = [S]K_B$, $K_T = 1/K_A$, and $K_R = 1/K_B$.

$$\overline{Y} = \frac{Lc\alpha(1+c\alpha)^{n-1} + \alpha(1+\alpha)^{n-1}}{L(1+c\alpha)^n + (1+\alpha)^n} = \frac{LK_A[S](1+K_A[S])^{n-1} + K_B[S](1+K_B[S])^{n-1}}{L(1+K_A[S])^n + (1+K_B[S])^n} \tag{4}$$

The degree of cooperativity in the MWC model is a function of the equilibrium between the R- and T-states, L, and the differential binding of ligand to the two states, c. The nature of the cooperativity and the lack of intermediates in a positively cooperative situation are clearly seen when typical concentrations of intermediate forms are examined. The conformational (or state) saturation, $\overline{R}$, occurs at lower ligand concentration than the ligand saturation, $\overline{Y}$, and is a characteristic structural feature of the MWC model. The concentration of enzyme forms with intermediate degrees of bound ligand, $\overline{Y}_1$, $\overline{Y}_2$, and $\overline{Y}_3$, never become large.

C. Sequential Model of Koshland-Nemethy-Filmer

The sequential or Koshland-Nemethy-Filmer (KNF) model (Koshland *et al.*, 1966) provides a more general explanation for cooperativity but, as a consequence, has more molecular parameters. The model is termed a sequential mechanism because of the sequential changes in subunit conformation within an oligomer. The advantages of the KNF model are that it can adequately describe negative (as well as positive) cooperativity of binding, which is frequently seen with enzymes, and can account for activators or inhibitors that do not produce the predicted results of the MWC model. The main disadvantage is that since more parameters are initially available for fitting (see below), more extensive data may be required to attain the same certainty in the final values. No independent means of assessing some of the molecular constants have been developed.

The KNF model (Koshland *et al.*, 1966) is based on ligand-induced conformational changes that consequently change subunit-subunit interactions in an oligomeric structure without (necessarily) affecting the conformation of the neighboring subunit (Fig. 3). The interaction constant (or implicitly the affinity) between the A-conformation (circles) and the B-conformation (squares), K_{AB}, would be different than the A-A interaction, K_{AA}, or the B-B interaction, K_{BB}. The cooperativity arises because of the alteration of the energetics of the subsequent ligand-induced conformational change of the subunit from A to B. A different number of K_{AA}, K_{BB}, and K_{AB} interactions modify the isomerization of the neighboring subunit in the complex. Because K_{AB} can be either greater or less than K_{BB}, the next ligand

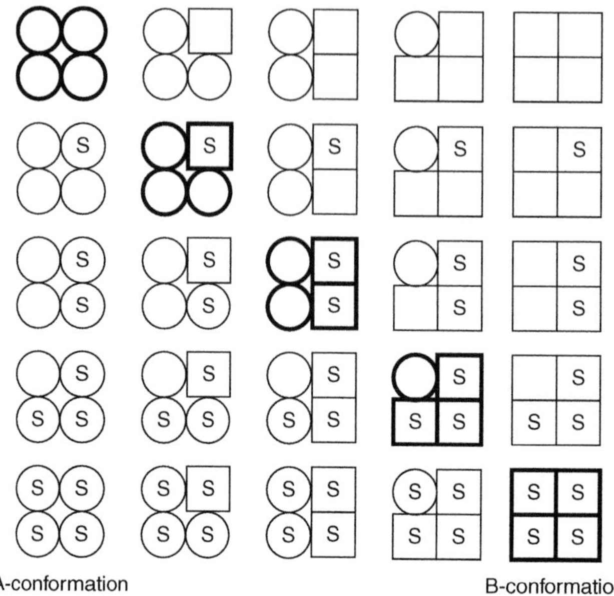

Fig. 3 Sequential Koshland-Nemethy-Filmer (KNF) model for cooperativity of a tetramer. The circles (A) and squares (B) represent two conformations of the subunit. Ligand (S) binds preferentially to the B-conformation and shifts the equilibria to the right. The diagonal forms in heavy lines represent the strict, ligand-induced KNF model with only unligated circles and ligated squares. The "square" geometry is implied in the drawing but actually depends on how the interactions between subunits are mathematically defined. The full figure is the most general model with nonexclusive binding to both A and B conformers and unligated hybrids containing A and B conformers. The MWC model is contained within the first and fifth columns that have no hybrids (cf. Fig. 2).

molecule can either bind more easily ($K_{BB} > K_{AB}$, positive cooperativity) or more difficultly ($K_{BB} < K_{AB}$, negative cooperativity), and either form of cooperativity can be generated. The original proposal of the sequential model was the most restrictive with an exact correspondence (Fig. 3, bold forms in diagonal sequence) between the bound ligand and each new subunit conformation, B. Generalization to include unliganded hybrid species and nonexclusive binding to A and B forms (Conway and Koshland, 1968; Haber and Koshland, 1967) and to allosteric modifiers (Kirtley and Koshland, 1967) was subsequently made (Fig. 3, lighter forms).

The cooperativity derived from the restricted KNF model can be quantitatively described by n and three molecular parameters: the relative interaction constant between the oligomer conformations in their quaternary context, K_{AB} and K_{BB}, and the constant for binding of ligand to the A-conformation with the concomitant isomerization, K_tK_s. Here, $K_{AB} = [AB][A]/[B][AA]$, $K_{BB} = [BB][A]^2/[AA][B]^2$, $K_t = [B]/[A]$, and $K_s = [SB]/[S][B]$. Because the interaction constants are defined relative to the A-A interaction ($K_{AA} = [AA]/[A]^2$) and K_{AA} is taken to be equal to unity, the actual parameters simply represent subunit affinities or association constants for formation of the AB or BB dimeric components: $K_{AB} = [AB]/[B][A]$ and $K_{BB} = [BB]/[B]^2$. The concentrations of ligand, A-subunit, B-subunit, ligand bound to A-subunit, and ligand bound to B-subunit are given by [S], [A], [B], [SA], [SB], respectively. The product of the equilibrium constant for the isolated protomer conformational transition and the intrinsic association constant for ligand with the B-conformer are never separated in the binding expression of the KNF model and thus represent a single fittable term, K_tK_s.

The KNF molecular constants (K_{AB}, K_{BB}, and K_tK_S) are determined from the Adair Ψ_i values fit to the data, based on n and the molecular geometry of the oligomer (Koshland *et al.*, 1966). The "geometry" enters ($n > 2$) because interactions may exist or not exist with each other subunit. The Ψ parameters for the square and tetrahedral tetramer cases are given in Table I. In the original paper (Koshland *et al.*, 1966), the "concerted" case was also considered in which the interactions of the oligomer were so tightly coupled that all subunits underwent the conformational transition simultaneously, conceptually the same as that of the MWC model.

Cooperativity is dependent on the relative subunit interactions, as can be seen by the ratio of successive intrinsic constants for the tetrahedral model which is simply equal to K_{BB}/K_{AB}^2. Either positive or negative cooperativity can be generated with the restricted KNF model with the appropriate relationship of K_{AB} to K_{BB} (Fig. 3). Again, only low concentrations of intermediate liganded forms, $\overline{Y}_1$, $\overline{Y}_2$, and $\overline{Y}_3$ occur in the positively cooperative case when concentrations of intermediate forms are examined. Conversely, in a negatively cooperative case the intermediates accumulate, since full binding of ligand becomes more difficult. For either cooperativity, the conformational saturation, $\overline{R}$, of the restricted KNF model occurs concomitantly with ligand saturation, $\overline{Y}$.

D. Comparison of Monod-Wyman-Changeux and Koshland-Nemethy-Filmer Models

The concept of symmetry (the concerted transition) in the MWC model is seen by some as a simplifying assumption tantamount to a physical law; others view it as an unnecessary restriction on the behavior of proteins. Some view the KNF model as conceptually difficult, whereas others readily grasp the implicit thermodynamic interactions and linkages. Regardless of the subjective responses to the two models, certain objective evaluations may be made. Putative differences in the two models based on a preexisting equilibrium compared to a ligand-induced fit are no longer considered to be relevant (Neet, 1980; Ricard, 1985). The same thermodynamic model applies for the two concepts with simply different occupation of various molecular populations (Neet, 1980).

Initially, quite different relationships of the state, $\overline{R}$, and binding, $\overline{Y}$, functions were predicted for the MWC and KNF models. In principle, a plot of $\overline{Y}$ versus $\overline{R}$ should give a straight line with a slope of unity for the restricted KNF model and a curvilinear plot with an initial slope larger than unity for the MWC model (cf. Figs 2 and 3). Nonexclusive binding in both models and unliganded hybrids in the KNF model, however, diminish the effectiveness of using this distinction to differentiate the models. In fact, as pointed out by several authors (Eigen, 1968; Hammes and Wu, 1971; Whitehead, 1970), the two models are simply extremes of the most general model, depicted here by all forms in Fig. 3, with the restricted KNF being the diagonal and the MWC being the first and fifth columns without any hybrid intermediates.

Three molecular fitting parameters arise in both the nonexclusive MWC model (c, L, and K_R) and the KNF model ($K_t K_S$, K_{BB}, and K_{AB}). Although for the latter model the number of Adair constants used to initially fit the primary data is equal to n, the relation between constants restricts values that Ψ_i can assume and still be consistent with the model. For example, the sequential tetrameric (square or tetrahedral) KNF model requires $\Psi_1^2\Psi_4 = \Psi_3^2$, whereas the concerted MWC model has the relation among Ψ values shown in Eq. (5) that demonstrates the lack of negative cooperativity since Ψ'_n/Ψ'_{n-1} must always be greater than unity. The relationship of these coefficients in the binding equation can, in principle, be used to discriminate between models (Cornish-Bowden and Koshland, 1970; Ricard and Cornish-Bowden, 1987) since the two models predict slightly different relative values (Table I). This experimental comparison requires extremely precise and extensive data and is tantamount to distinguishing between models based on the shape of very similar saturation curves (Cornish-Bowden and Koshland, 1970). For this reason, comparison of Ψ ratios, or saturation curves, is not an efficient nor widely used procedure.

$$\frac{\Psi'_n}{\Psi'_{n-1}} = \frac{(1 + Lc^n)}{(1 + Lc^{n-1})K_R} \tag{5}$$

General equations for the effect of activators and inhibitors, nonexclusive binding, the relationship of the KNF or MWC parameters to the intrinsic constants for a dimer, differential catalytic rates in the R and T or A and B forms, differential catalytic rates with two substrates, and the influence of products on differential catalytic rates have previously been presented and discussed for both the MWC and KNF models (Neet, 1980). Most attempts to determine whether the MWC or KNF model best applies to a specific enzyme have been based on various detailed physical assessments of structural or energetic changes on partial ligand binding, rather than on binding curves or simple state functions. Hemoglobin and ATCase are discussed in detail in Section VI to demonstrate how these methods may be used.

E. Oligomer Association-Dissociation Models

If, on binding ligand, the change in subunit affinity (K_{BB} and K_{AB} in the KNF model) is sufficiently great or if the initial equilibrium involves a dissociated species (R or T in the MWC model), then cooperativity can be directly due to the association or dissociation reaction itself (Frieden, 1967; Kurganov, 1967; Nichol *et al.*, 1967). Simply, the ligand binds preferentially either to the associated or the dissociated form. Systems that favor binding to the associated species or ones that favor the dissociated species can equally well occur, but they imply different molecular mechanisms of protein-ligand interaction (Neet, 1980). Either exclusive or nonexclusive binding to the two association states can occur. If the substrate favors dissociation of an oligomer, lower enzyme concentrations will enhance activity but will depress cooperativity. The converse relationship also holds; maximum cooperativity is generated at intermediate enzyme concentrations. The equations (presented elsewhere (Frieden, 1967; Kurganov, 1967, 1982; Neet, 1980; Nichol *et al.*, 1967) and extensively discussed (Kurganov, 1982; Levitzki and Schlessinger, 1974; Nichol and Winzor, 1976) are similar in form to those of the MWC or KNF models but with the addition of a term dependent on the protein concentration.

Interestingly, both positive and negative cooperativity can occur with this mechanism either because of the possibility of differing numbers of ligand bound per subunit in the monomer and oligomer state or because of the maintenance of two active forms, monomer and oligomer, at certain protein concentrations. The critical test for this mechanism is the dependence of the homotropic cooperativity (e.g., the n_H) on the concentration; the specific activity of the enzyme will also vary with its concentration. The cooperativity, in most cases, should also occur in equilibrium binding studies, since no kinetic terms are directly involved. Estimates of the K_D for the association reaction can be calculated by fitting the appropriate equations. Similarly, the effects of heterotropic effectors can occur as a shift of the monomer-oligomer equilibrium toward a more active or a less active association state. This type of allosterism has been discussed adequately in other places (Kurganov, 1982; Neet, 1980) and will not be belabored. Numerous enzymes show changes in oligomerization state on substrate binding (Kurganov, 1982; Neet, 1979), and several appear to generate cooperativity by this mechanism; muscle 6-phosphofructo-1-kinase is a good example in which the subunit interactions are weakened in the tetrameric state and dissociation to dimers and monomers occurs (Lee *et al.*, 1988). Assessing the applicability of this mechanism in physiological situations depends on accurately determining the concentration of the particular enzyme in its native environment. The EGF receptor shows this effect in a membrane (see update in front).

F. Kinetic Considerations of Site-Site Cooperativity Models

Both the MWC and the KNF models of cooperativity are based on thermodynamic considerations and strictly apply only to equilibrium binding measurements. The most rigorous utilization of these models, therefore, has been with such systems as O_2 binding to hemoglobin (Ackers and Hazzard, 1993) or with ATCase where substrate binding measurements have been made in parallel with kinetics (Newell *et al.*, 1989). Nevertheless, application to kinetic assays is common because of the simplicity of the experiments and the fundamentalism of the models. The assumption is frequently made that rapid equilibrium binding of substrates would reduce the kinetic measurements to the equivalent of thermodynamic equilibrium situations, that is, a simple proportionality exists between substrate binding and reaction rate. However, this simple relation is not generally adequate, nor supported by theoretical considerations (Ricard, 1985). Alternatively, attempts have been made to extend the equilibrium models to incorporate rate constants for each of the E·S complexes in either cooperativity model (Goldbeter, 1976; Neet, 1980; Paulus and DeRiel, 1975; Ricard *et al.*, 1974). The problems of such simplified considerations have been pointed out by Ricard and coworkers (Ricard, 1985; Ricard and Cornish-Bowden, 1987) who have developed a general theory of "structural kinetics."

Structural kinetics (Ricard and Noat, 1982, 1984, 1985, 1986) attempts to apply principles derived from the study of enzyme mechanisms to the effects on subunits in a cooperative, oligomeric enzyme. In other words, the kinetic properties of allosteric enzymes are derived from those of an ideal, isolated subunit utilizing the thermodynamics of structural changes. The catalytic principles (Fersht *et al.*, 1986; Jencks, 1975) include transition state complementarity, induced fit, strain, release, and evolution. The fundamental principles of structural kinetics partition the free energy of conversion of substrate to product into three terms, (i) The "intrinsic energy component" is equivalent to the ΔG^* that would occur if an idealized free subunit, that is, with no subunit interactions, were to catalyze the same reaction. (ii) The "protomer arrangement energy" (α) describes the contribution to the free energy of catalysis of the energy of interaction of the subunits on association together with no conformational change of each subunit. (iii) The "quaternary constraint contribution" (σ) accounts for the contribution of inter-subunit strain on catalysis owing to a conformational change affecting "strain" in the catalytic site. The overall free energy of catalysis by the oligomer is simply the sum of these three energy terms, or, correspondingly, the phenomenological rate constant is the product of the intrinsic rate constant of the subunit and the equilibrium constants for subunit association and subunit conformational change. These three energy terms are essentially the same as those of the KNF model (Koshland *et al.*, 1966) to describe quaternary effects on substrate binding, but here applied to catalysis.

Three additional postulates are introduced into structural kinetics that simplify the rate equation. (i) In each catalytic transition state the quaternary constraints are relieved, similar to the relief of intrasubunit strain in theories of catalysis. (ii) The minimum number of conformations is assumed (as in the MWC and KNF binding models), that is, each free subunit can exist in only two conformations, the unliganded or the liganded (with substrate or product). (iii) The conformation of the liganded transition state complex is the same, or energetically indistinguishable, for all transition states. The three postulates effectively take into account the generally accepted belief that maximum catalytic efficiency occurs when the enzyme conformation is complementary to the transition state of the reaction. These reasonable assumptions allow straightforward derivation of rate expressions containing conceptual terms. The full structural rate equation for the dimer is given in Eq. (6):

$$\frac{v}{[\mathrm{E}]_0} = \frac{2\overline{k}^{*}\sigma_{\mathrm{AA}}\overline{K}^{*}[\mathrm{S}] + 2\overline{k}^{*}\sigma_{\mathrm{AA}}\frac{\alpha_{\mathrm{AA}}}{\alpha_{\mathrm{AB}}}\overline{K}^{*2}[\mathrm{S}]^2}{1 + 2\frac{\sigma_{\mathrm{AA}}}{\sigma_{\mathrm{AB}}}\frac{\alpha_{\mathrm{AA}}}{\alpha_{\mathrm{AB}}}\overline{K}^{*}[\mathrm{S}] + \frac{\sigma_{\mathrm{AA}}}{\sigma_{\mathrm{BB}}}\frac{\alpha_{\mathrm{AA}}}{\alpha_{\mathrm{BB}}}\overline{K}^{*2}[\mathrm{S}]^2} \quad (6)$$

where α can be considered a dissociation constant for AA or AB; σ is an isomerization constant, and σ^*, k^*, and K^* are the apparent intrinsic isomerization, catalytic, and reciprocal Michaelis constant, respectively, of the isolated subunit. The observed catalytic rate constant for a subunit, k, is equal to $k^*\ \sigma_{\mathrm{AB}}(\alpha_{\mathrm{AB}}/\alpha_{\mathrm{AA}})$. The constants $1/\alpha_{\mathrm{AA}}$, $\alpha_{\mathrm{AA}}/\alpha_{\mathrm{AB}}$, $\alpha_{\mathrm{AA}}/\alpha_{\mathrm{BB}}$, and σ^* of the Ricard structural kinetic model are analogous to K_{AA}, K_{AB}, K_{BB}, and K_{t}, respectively, of the KNF model. When Eq. (6) is compared to the KNF formulation (the dimer equivalent of Eq. (1) and Table I), both have the same form but with a different arrangement and relationship between terms.

Two limiting cases of structural kinetics are revealing. If $\sigma_i = 1$, then the system is "loosely" coupled with no intersubunit interactions other than association and no propagation of substrate-induced conformational changes. Cooperativity is determined for a dimer by the term $\alpha_{\mathrm{AB}}^2/\alpha_{\mathrm{AA}}\alpha_{\mathrm{BB}}$ with different dependencies on this term for binding and kinetics (compare the analogous term for the KNF model given above). However, both binding cooperativity and kinetic cooperativity have the same sign (i.e., catalysis enhances any preexisting binding cooperativity). On the other hand, if $\sigma_i \neq 1$ then the dimer is "tightly" coupled. If quaternary constraints are relieved in the unliganded and fully liganded states ($\sigma_{\mathrm{AA}} = \sigma_{\mathrm{BB}} = 1$), then the conformation of the idealized monomer is the same as those of the oligomer; if the constraints are not relieved, then distinct conformations can occur in the oligomer. Binding cooperativity for the tightly coupled enzyme depends on the term $\alpha_{\mathrm{AB}}^2\sigma_{\mathrm{AB}}^2/\alpha_{\mathrm{AA}}\alpha_{\mathrm{BB}}\sigma_{\mathrm{AA}}\sigma_{\mathrm{BB}}$, but the kinetic cooperativity also depends on the ratio $\sigma_{\mathrm{AA}}/\sigma_{\mathrm{AB}}$. Therefore, kinetic cooperativity does not have to have the same sign as binding cooperativity, and "inversion" of observed cooperativity may occur. If subunits of the oligomer are so tightly coupled that conformational changes are propagated to the entire molecule, then a concerted molecular change occurs, analogous to the MWC binding model. In the absence of binding cooperativity, kinetic cooperativity for this model can only be positive (Ricard, 1985) as was originally reported for kinetic extensions to the MWC model (Goldbeter, 1976; Paulus and DeRiel, 1975). The intermediate cases of $\sigma \neq 1$ are more difficult to generalize. Tight and loose coupling situations for a dimer and for a tetramer with "square" or "tetrahedral" geometry have been analyzed (Giudici-Orticoni *et al.*, 1990; Ricard *et al.*, 1990) and reviewed (Ricard, 1985; Ricard and Cornish-Bowden, 1987).

Application of the theory to tetrameric chloroplast fructose bisphosphatase has utilized precise rate data to discriminate between models (Giudici-Orticoni *et al.*, 1990); the tetrahedral structural kinetic model fits the data well, whereas Adair, MWC, or KNF binding models do not. Quaternary constraints are not relieved in the unliganded and fully liganded states (only in the transition states) of fructose bisphosphatase, and free energies associated with individual subunit interaction steps have been calculated (Giudici-Orticoni *et al.*, 1990).

The reader is referred to two reviews (Ricard, 1985; Ricard and Cornish-Bowden, 1987) for further discussion of the analysis by structural kinetics. The frequent finding that catalytic sites have shared contributions from neighboring subunits (see below) may influence application of the theory of structural kinetics. At a minimum, the concept of the idealized, isolated subunit that can catalyze the reaction with maximum efficiency would not hold. Future detailed interpretations of cooperative, allosteric enzymes will require the extension, refinement, and application of the concepts of structural kinetics to enzymes with known catalytic mechanisms and crystallographic structures. The concepts of the structural kinetics model have not been widely adopted, probably because of their difficulty and the desire to approach each enzyme case individually.

G. Kinetic Cooperativity and Cooperativity in Monomeric Enzymes

Several models have been proposed for the generation of cooperativity by kinetic mechanisms, rather than by equilibrium effects of site-site interactions discussed in the previous paragraphs. With some enzymes, such as AMP deaminase, a thorough analysis of the kinetic mechanism has shown that the allosteric and cooperative kinetics can be attributed to direct effects on catalytic turnover of individual active sites in the tetramer without invoking binding interactions (Merkler and Schramm, 1990). Other models based on the type of kinetic mechanism (Ferdinand, 1966; Pettersson, 1986a,b; Wells *et al.*, 1976) and on a slow conformational change, or hysteresis (Frieden, 1970), of the enzyme (Ainslie *et al.*, 1972; Ricard *et al.*, 1974, 1977) are more prevalent in the literature and are discussed here. The kinetic cooperativity observed in these cases will have an appearance similar to that arising from other mechanisms and cannot be distinguished simply from appearance of the steady-state velocity plots; however, the curves from kinetic mechanisms do tend to be less symmetrical around the $S_{0.5}$ than binding mechanisms. Kinetic models of these types are dependent on catalytic turnover; therefore, cooperativity will not be observed for the same substrate in equilibrium binding measurements. This marked distinction is the most diagnostic feature of these mechanisms. Single-site mechanisms which purport to generate cooperativity of equilibrium binding invariably neglect thermodynamic constraints, microscopic reversibility, or detailed balance and are, therefore, invalid. Kinetic mechanisms are maintained at nonequilibrium by the conversion of substrate to product. Because interactions between multiple active sites are not required in kinetic mechanisms, cooperativity can be produced in monomeric enzymes

(Cornish-Bowden and Cardenas, 1987) or in oligomers with independent sites (Cornish-Bowden and Cardenas, 1987; Neet, 1980; Neet and Ainslie, 1980). These kinetic mechanisms that generate cooperativity should not be confused with the effect of kinetic constants on cooperativity generated by site binding mechanisms (see above and Goldbeter, 1976; Paulus and DeRiel, 1975; Ricard *et al.*, 1974).

1. Steady-State Random Kinetic Mechanism

When an enzyme catalyzes a reaction with two (or more) substrates (A and B) by a steady-state (nonequilibrium) random mechanism, the potential exists for non-Michaelian concentration dependence (Ferdinand, 1966; Pettersson, 1986a; Wells *et al.*, 1976; Whitehead, 1970). Second-order terms for each substrate occur in the general rate equation (Segal, 1975). No cooperativity will occur if the kinetic mechanism is rapid equilibrium, which reduces to an equilibrium binding situation, or if the reaction is ordered with one substrate binding exclusively before the other(s). Kinetic cooperativity may occur in the true steady-state random mechanism because the relationship among the individual rate constants determines the flux through each alternate pathway. At low substrate concentrations the reaction may proceed through one limb (e.g., A binds before B), and at higher substrate concentrations the other limb (i.e., B binds before A) may predominate. This shift in flux can produce nonhyperbolic kinetics of either the positively or negatively cooperative type. The degree of cooperativity of each substrate will depend on the concentration of the other substrate and will diminish either at very low or at very high concentrations of the fixed substrate, since the reaction flux will then proceed through a single pathway. Simulated analysis has indicated (Wells *et al.*, 1976) that "positive cooperativity" of one substrate will be associated only with noncooperative or substrate inhibition kinetics of the other substrate, whereas "negative cooperativity" (sometimes called substrate activation) with both substrates can occur. Not all steady-state random mechanisms will generate this type of cooperativity because a special relationship must exist among the various rate constants. The requirements and details of this mechanism have been discussed (Cornish-Bowden and Cardenas, 1987; Neet, 1980; Pettersson, 1986a; Wells *et al.*, 1976).

The steady-state random mechanism probably generates deviations from Michealian behavior with many enzymes (Hill *et al.*, 1977). The extent of the deviation will frequently be low (Cleland, 1970; Gulbinsky and Cleland, 1968), that is, producing Hill coefficients that are not far from unity. No clear examples have been reported of significant cooperativity from this mechanism that contributes to physiological behavior of an enzyme, although several possibilities have been suggested (Cornish-Bowden and Wong, 1978; Ivanetich *et al.*, 1990; Pettersson, 1986a,b; Trost and Pupillo, 1993) and in some cases contested (Cornish-Bowden and Storer, 1986; Ricard and Noat, 1985; Ricard *et al.*, 1986). Nevertheless, if cooperative enzyme kinetics are observed with a multisubstrate reaction, the first consideration should be that of a steady-state random mechanism until the exact kinetic mechanism is worked out. In a closely related situation, relevant substrate inhibition of *Escherichia coli* phosphofructokinase has been demonstrated to be due to a steady-state random mechanism by construction of a mutant that converts the enzyme to rapid equilibrium kinetics and eliminates MgATP inhibition (Zheng and Kemp, 1992).

2. Ligand-Induced Slow Transition (Hysteretic, Mnemonic) Mechanism

Enzyme hysteresis (Frieden, 1970), in the form of a substrate-induced slow conformational change, slow association-dissociation, or slow binding, has been observed on a standard assay time scale of minutes with many enzymes (Ainslie *et al.*, 1972; Frieden, 1970, 1979; Kurganov, 1982; Neet, 1980). These slow transitions can have the form of either bursts or lags, depending on whether the activity decreases or increases with time, respectively. The assay transient can be characterized by the initial velocity, v_i, the steady-state velocity, v_{ss}, and the relaxation time for the transient, τ [Eq. (7)]:

$$P = v_{ss}t - (v_{ss} - v_i)(1 - e^{-t/\tau})\tau \tag{7}$$

where P is the product concentration at time t. The hysteretic or slow transition can be readily observable in standard assays if τ is on the order of seconds to minutes; rapid mixing techniques may be required if τ is between 10 ms and 1 s.

Of vital importance when working with enzymes that show a hysteretic response is to eliminate possible artifacts. All enzymes will show nonlinear progress curves if substrate depletion or production inhibition occurs. Calculation of final substrate and product concentrations, measurement of product inhibition constants, readdition of substrate to the steady-state portion of the assay, and addition of product in the initial assay are effective means of checking for such possibilities. Other simple considerations such as temperature change, dilution, or inactivation of the enzyme can also lead to nonlinearity and should be evaluated. Last, the utilization of different assay systems, different preincubation conditions, and alternate substrates can help validate the presence of true enzymatic hysteresis.

If the slow transition is on the same order of magnitude (milliseconds) as the substrate binding and the catalytic cycle, then cooperativity of v_{ss} can be generated (Ainslie *et al.*, 1972; Rabin, 1967; Ricard *et al.*, 1974; Whitehead, 1970), provided that the appropriate relationship among all rate constants exists (Ainslie *et al.*, 1972; Neet and Ainslie, 1980; Ricard *et al.*, 1974). Either bursts or lags can be associated with either positive or negative cooperativity in v_{ss}. This type of cooperativity mechanism has been termed a LIST (Ainslie *et al.*, 1972), a mnemonic mechanism (Ricard *et al.*, 1974, 1977), or, more recently, rate-dependent recycling of free enzyme conformational states (Rose *et al.*, 1993). These concepts emphasize the relatively slow isomerization between two enzyme forms under nonequilibrium conditions. These mechanisms are sometimes considered to involve "temporal" cooperativity in contrast to the "spatial" cooperativity of site-site mechanisms. Thus, a hysteretic response in an enzyme can give rise to kinetic cooperativity in a single-site, monomeric enzyme (Ainslie *et al.*, 1972; Cornish-Bowden and

Cardenas, 1987). Alternatively, the inherent cooperativity of a site-site mechanism of an oligomeric enzyme can be altered because of the additional power of substrate added to the rate equation.

In the LIST mechanism, the initial steady-state velocity, v_i, can be measured after the pre-steady-state adjustments of substrate binding (this volume, Neet, 1980) occur but before the slow enzyme isomerization. Multiple turnovers occur during v_i (i.e., it is not a stoichiometric burst). The v_i, extrapolated to zero time, should either be hyperbolic, if the enzyme starts in a single form, or give the appearance of negative cooperativity (owing to heterogeneity), if the enzyme initially exists in two kinetically significant forms; in other words, no true temporal cooperative interactions exist in that "frozen" initial form. The rate of the transient, τ^{-1}, will be a second-order term in substrate concentration with complex combinations of rate constants (Ainslie *et al.*, 1972; Frieden, 1970) and, in principle, can be used to help evaluate parameters of the mechanism. However, unrealistically good data are required to determine the concentration dependency of τ with enough accuracy to define the parameters well. Graphical methods can be used to assess the v_i, v_{ss}, and τ^2 (Frieden, 1970), but a more rigorous procedure is to utilize a nonlinear fitting algorithm to Eq. (7) and obtain unbiased estimates of the parameters with associated standard errors (Neet *et al.*, 1990).

The cooperativity from the LIST mechanism occurs in v_{ss} after a linear steady state is reached. A second-order rate equation for the steady-state velocity is generated because of substrate addition steps to two distinct enzyme forms (Fig. 4). Conceptually, cooperativity occurs because the E conformer (Fig. 4, front upper left) can partition between formation of EA ($v = k_1$[A]) or isomerization to E′ ($v = k_{t1}$). At low [A], the isomerization will dominate, and v_{ss} will be mainly due to the cycle favored by the E $\rightleftharpoons$ E′ distribution. At high [A], E or E′ will

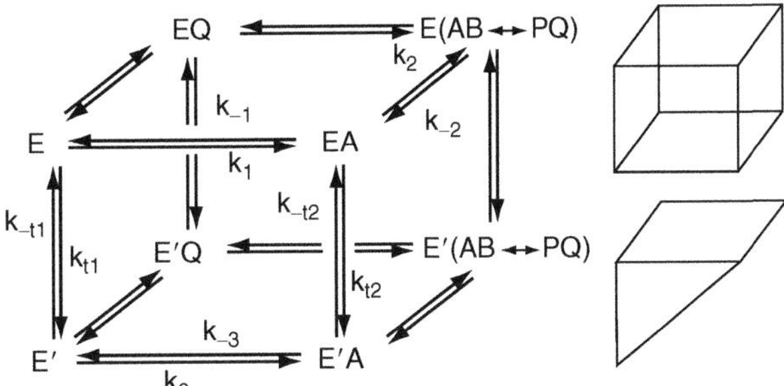

Fig. 4 Ligand-induced slow transition (LIST) model for a two-substrate monomeric enzyme with ordered addition of substrates. E and E′ represent two conformations of the enzyme with the vertical steps (k_{t1}, k_{t2}) being the isomerization. The heavy arrows and bold forms represent the mnemonic mechanism contained within the LIST model with one (upper) catalytic cycle and $k_{t2} \gg k_3[A]$ so that E′ + A → A on the diagonal and the E′ A intermediate is not kinetically significant. The small diagrams at right distinguish the full LIST (upper) and the mnemonic (lower) pathways. For glucokinase: A, glucose; B, MgATP; P, MgADP; and Q, glucose 6-phosphate.

be trapped in the form from which product is released and that same cycle will predominate. Either positive or negative cooperativity can occur with either substrate. The cooperativity of v_{ss} for substrate A will depend on the concentration of substrate B, since the kinetic partitioning of EA (Fig. 4, front upper right) between EAB ($v = k_2$[B]) and E′A ($v = k_{t2}$) depends on the second substrate. At low [B], EA⇌E′A should be nearly able to equilibrate and thus no cooperativity will be found; at high [B], EA or E′ A will tend to form EAB or E′ AB rapidly, and cooperativity of A in v_{ss} can occur. Thus, the three conditions for the generation of v_{ss} cooperativity are as follows: (a) the rate constants for isomerization must be of the same magnitude as k_{cat}, (b) substrate binding cannot be at rapid equilibrium (i.e., k_{cat} cannot be slow), and (c) the flux through the two catalytic cycles must change with changing substrate concentration. Procedures to identify the key catalytic pathways in the mechanism and the requirements among the rate constants that contribute to the latter condition have been discussed (Neet and Ainslie, 1980; Whitehead, 1976) Similar considerations for a single-substrate enzyme have been made (Ainslie *et al.*, 1972; Neet and Ainslie, 1980).

The mnemonic mechanism (Ricard *et al.*, 1974, 1977) emphasizes the "memory" that the enzyme conformation retains of the substrate after product conversion and release. This mechanism can be considered a special case of the LIST mechanism (Cornish-Bowden and Cardenas, 1987; Neet and Ainslie, 1980; Ricard and Cornish-Bowden, 1987); an even more general mechanism with six explicit conformations has been presented (Ricard, 1978). The mnemonic simplification allows relatively straightforward predictions in many cases. The mnemonic mechanism for a two-substrate enzyme can be described by the heavy lines in Fig. 4 and the condition that A binding to E′ leads directly to EA (no E′ A). Cooperativity is due to the lack of relaxation of E (less stable) back to E′ (more stable) at higher concentrations of A (i.e., a slow isomerization transition). At low [A] the pathway proceeds primarily through E′ + A → EA → E(AB⇌PQ), whereas at high concentration the first substrate binds before the isomerization can occur and the pathway is predominantly through E + A → EA → E(AB⇌PQ). Either negative or positive cooperativity of A (as measured by the Hill coefficient) is enhanced by increasing concentrations of B.

A major difference between the two kinetic mechanisms is that only one catalytic cycle occurs in the mnemonic mechanism whereas the LIST mechanism has two catalytic cycles (i.e., both E and E′ are catalytic active). For a single-substrate enzyme, the two mechanisms are virtually indistinguishable. With two substrates, ordered binding is implied in the mnemonic mechanism, and, therefore, cooperativity can only be generated with the first substrate. The LIST mechanism can produce cooperativity with either substrate and can accommodate random binding of two (or more) substrates. Product inhibition can help discriminate between the mechanisms. The second product released in the mnemonic (ordered, one catalytic cycle) mechanism is predicted to increase negative cooperativity or decrease positive cooperativity of the first substrate while showing linear slope inhibition (Meunier *et al.*, 1974). In the LIST mechanism predictions of product inhibition effects on cooperativity cannot be made, but product slope inhibition can itself be cooperative (or nonlinear) as in a simple Ordered Bi Bi mechanism (Neet and Ainslie, 1980). A complete set of rate constants has not been experimentally obtained for either the LIST or the mnemonic mechanisms for any enzyme, primarily owing to the inability to make adequate kinetic measurements of each individual step.

The mnemonic mechanism has been extensively applied to the interpretation of the negatively cooperative kinetics of glucose phosphorylation found with wheat germ hexokinase L_I, a monomeric enzyme. The predictions of initial velocity kinetics, product inhibition, and effects of substrates on conformation of the enzyme are consistent with the mnemonic mechanism (Buc *et al.*, 1977; Meunier *et al.*, 1974, 1979; Ricard *et al.*, 1977). Other mechanisms for kinetic cooperativity of wheat germ hexokinase have been proposed (Pettersson, 1986b) and rebutted (Ricard and Cornish-Bowden, 1987; Ricard and Noat, 1985).

Relatively few monomeric enzymes have been shown to have cooperativity (Cornish-Bowden and Cardenas, 1987), and few oligomeric enzymes have convincingly been proved to utilize a LIST or mnemonic mechanism as part of their cooperativity. Newer approaches analyzing reaction flux by the isotope counterflow method may prove useful in establishing the presence of alternate conformational pathways (Gregoriou *et al.*, 1981; Rose *et al.*, 1993). Cooperativity and slow transitions in enzymes that have been reasonably well characterized include rat liver glucokinase (Pollard-Knight and Cornish-Bowden, 1982) (discussed in detail below), wheat germ hexokinase (Meunier *et al.*, 1974), octopine dehydrogenase (Monneuse-Doublet *et al.*, 1978), and fumarase (fumarate hydratase) (Rose *et al.*, 1993). Furthermore, although many enzymes have hysteretic properties (Ainslie *et al.*, 1972; Frieden, 1970, 1979; Kurganov, 1982; Neet and Ainslie, 1980), the contribution of the hysteresis, itself, to a physiological function (e.g., see glucose-6-phosphatase, Nelson-Rossow *et al.*, 1993) has yet to be proved in most cases. Thus, the concepts of hysteresis and kinetic cooperativity are currently an intriguing aspect of enzyme kinetics and mechanism, but their importance, frequency of occurrence, and contribution to biological function remain speculative, for the most part. Emerging techniques for the future (see below) may help in such evaluations. See, for example, the update on glucokinase at the beginning of this chapter.

3. Structural Basis of Slow Conformational Changes

Since the proposal in 1980 of four structural types of conformational changes that could be slow enough to account for hysteretic mechanisms (Neet, 1980), little progress has been made in proving or refining these suggestions. Possible three-dimensional changes proposed were *cis-trans* isomerization of proline residues, cleft closure by domain movement, segmental movement of polypeptide chains, and sliding of α helices past or around are another. Each has the potential for a high activation energy. The crystallographic description of more complex and intricate conformational changes provide additional possibilities, but little can be said at the moment about the kinetics of such slow changes.

A slow cleft closure with glucokinase is highly probable, based on the glucose-induced domain movement of the homologous yeast hexokinase (Bennett and Steitz, 1978), but no direct data have yet been presented. The hysteresis of *E. coli* dihydrofolate reductase (Penner and Frieden, 1985) may also be due to slow domain movement, but attempts to modify the $NADP^+$/dihydrofolate-dependent slow isomerization by mutagenic substitution of several amino acid residues in the hinge region have not yet been successful (Ahrweiler and Frieden, 1991). Proline isomerization certainly is involved in slow steps in protein folding (Wood *et al.*, 1988), but no ligand-induced isomerization of proline has been documented. Standard X-ray crystallographic methods used to define the structural basis of allosteric transitions are not capable of direct extension to time-dependent changes. Laue diffraction methods, now being used to study time-dependent changes during catalysis (Johnson, 1992; Stoddard *et al.*, 1991), could be utilized in future studies to describe slow (minutes) conformational changes of allosteric, hysteretic enzymes at high resolution. Solution nuclear magnetic resonance (NMR) methods, because of the time required for data collection subsequent to ligand addition, will probably remain too slow for the study of ligand-induced slow conformational changes.

VI. Experimental Evaluation of Models

A. Concerted Versus Sequential Models: Classic Case of Oxygen Binding to Hemoglobin

Hemoglobin is the prototypical cooperative allosteric protein that has been studied with numerous techniques since the R- and T-states were first described by crystallography (Perutz, 1970, 1989). A simple concerted transition between the two states can be explained in detail by the triggering of the Fe^{2+} movement in the heme ring being transmitted over 25–40 Å to the other hemes by shifts in the tertiary and quaternary structure (Perutz, 1989). The $\alpha_1\beta_1$ dimer rotates about 15° relative to the other dimer with changes in the ionic and hydrogen bonds between the FG bend (residues 94–102) and the C helix (residues 40–45) at the $\alpha_1\beta_2$ and $\alpha_2\beta_1$ contacts.

More recent data, however, indicate that the simple concerted (MWC) model is not sufficient to explain completely the energetics of the reaction and that intermediate microstates occur. Precise O_2 binding data at protein concentrations in which the hemoglobin heterotetramer dissociates to $\alpha\beta$ dimers and measurement of subunit affinities by gel filtration from Ackers laboratory (Ackers and Hazzard, 1993; Ackers *et al.*, 1992) have allowed calculation of the free energy for individual oxygen-binding steps. Other studies with nonlabile ligands, allosteric modifiers, and hemoglobin mutants have provided further details of the subunit interactions during ligand binding. The clearest scenario, at least with cyanide binding, is that a symmetry switch occurs in each $\alpha\beta$ dimer when one of the subunits is ligated (Ackers and Hazzard, 1993). The cooperative free energy (i.e., a measure of the free energy altering dimer-dimer interactions on ligation) is propagated to the other subunit of the $\alpha\beta$ dimeric half-tetramer as about 3 kcal/mol in two steps per tetramer (Fig. 5). The full T-R transition of the quaternary structure of the tetramer only occurs when at least one subunit of each half-tetramer is ligated.

A careful study of the energetics of 60 mutant or modified hemoglobins confirmed that the $\alpha_1\beta_2$ interface was uniquely coupled to the cooperativity (Turner

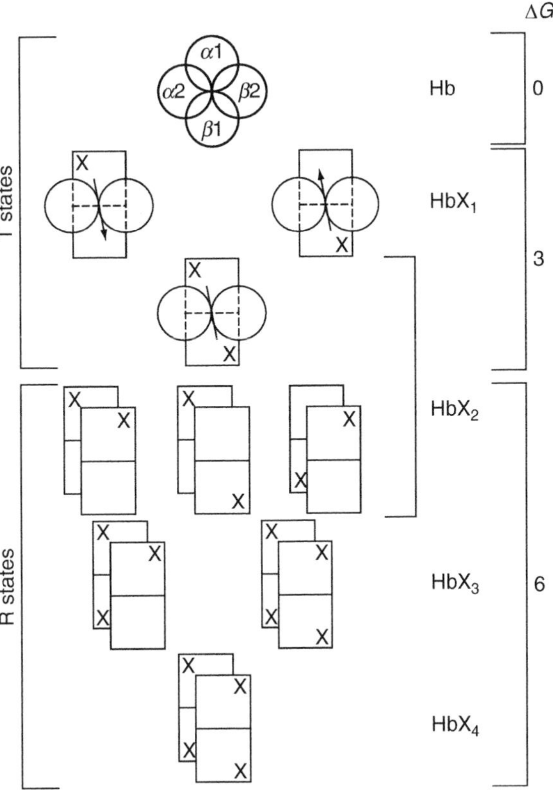

Fig. 5 Model for ligand binding to the substates of cyanomethemoglobin. The 10 possible ligation substates of hemoglobin are shown with the $(\alpha_1\beta_1)(\alpha_2\beta_2)$ composition of the dimer of dimers indicated in the top, unligated line. Oxygenation is shown increasing vertically downward, with the stoichiometry indicated. When ligand binds to either α or β subunit, the effect is transmitted across the dimer interface, indicated by the arrow, and changes the conformation of the dimer from circle (T-state) to square (R-state). The T (four substates) and R (six substates) macroquaternary states are bracketed. The cooperative free energy, ΔG_c, at each stage is indicated at right. Quarternary switching from the T- to the R-state, indicated by rotation of the $\alpha_2\beta_2$ dimer relative to the $\alpha_1\beta_1$ dimer, occurs when at least one subunit of each $\alpha\beta$ dimer is ligated. Adapted from Ackers and Hazzard (1993) and Ackers *et al.* (1992).

et al., 1992). Detailed proton NMR studies support the conclusion that intermediate states exist between R and T (see review by Ho, 1992). The resulting mechanism is neither purely MWC nor sequential (KNF) but encompasses aspects of each within the dimer of dimer context. This conclusion is reminiscent of early pseudosymmetrical dimer models for hemoglobin and enzymes (Guidotti, 1967; Viratelle and Seydoux, 1975). The description of this mechanism for hemoglobin is facilitated by utilization of the definition of molecular constants proposed for the KNF model. Furthermore, the microstate model may also explain macroscopic deviations from the MWC model [e.g., a 10-fold difference in K_T in the presence of the effector 2,3-bisphosphoglycerate (Johnson, 1992; Perutz, 1989)]. However, a similar detailed analysis of the heterotropic effects on hemoglobin is needed to test the model further. See (Yonetani and Tsuneshige 2003) and Ciaccio *et al.* (2008).

B. Concerted Versus Sequential Models: Case of Aspartate Transcarbamylase

Aspartate transcarbamylase (ATCase; EC 2.1.3.2, aspartate carbamolytransferase) from *E. coli* is the most extensively studied allosteric enzyme. Many different experimental approaches have been utilized over the years to test the theoretical models of cooperativity with ATCase: How much concerted symmetry is retained in the molecular allosteric mechanism? This section briefly reviews the molecular structure of the enzyme and the allosteric transitions involved, but it mainly focuses on methods that have been utilized to understand the conformational changes and provides a critical view of the proposed models as applied to ATCase. The enzyme plays a central role in the regulation of the pyrimidine pathway in bacteria; its structure and allosteric regulation are described in many introductory biochemistry texts. More extensive reviews and interpretations in terms of allosteric models and structure have appeared (Allewell, 1989; Johnson, 1992; Kantrowitz and Lipscomb, 1988, 1990; Schachman, 1988). Determinations of the crystal structure of the T-state, the T-state with CTP bound, the R-state with *N*-phosphonacetyl-L-aspartate (PALA) bound, and the R-state with phosphonoacetamide plus malonate bound (Gouaux and Lipscomb, 1990; Ke *et al.*, 1984; Krause *et al.*, 1987; Stevens *et al.*, 1990) have been extremely useful in interpreting kinetic and mutational studies.

When ATCase is fit to the MWC model (Krause *et al.*, 1987), the equilibrium constant (L_0) between the R- and T-states is 250 (or an energetic difference of 3.3 kcal/mol); this value is about 2.5 kcal/mol in the presence of ATP and 4.2 kcal/mol with CTP. The ratio for substrate dissociation constants in the two states, c, for PALA is 0.03 and that for Asp is 0.05. The Hill coefficient is 1.8–2.0 in the absence of effectors, 2.3 in the presence of CTP, and 1.4 in the presence of ATP. No significant kinetic contribution occurs, since similar values have been measured both by equilibrium binding studies with PALA in the absence of carbamyl phosphate (Newell *et al.*, 1989) (i.e., a purely thermodynamic situation) and by the kinetics of catalysis of the carbamoylation of aspartate (Howlett *et al.*, 1977).

1. Structural Model

The enzyme ATCase (310 kDa) exists as a dimer of catalytic trimers (3 × 33 kDa) that are held together by three dimeric (2 × 17 kDa) regulatory subunits. The dihedral molecular symmetry axes of the heterododecamer allow description of interactions among the catalytic polypeptides (C1-C2-C3, C4-C5-C6), between the regulatory peptides (R1-R2, etc.), and between the regulatory and catalytic chains (C1-R1, etc.).

Significant changes in the tertiary structure of the subunits occur on binding the bisubstrate analog PALA and the consequent transition from the T- to the R-state. The aspartate moiety of the substrate binds to Arg-167, Arg-229, and Gln-231 in the aspartate domain and to Lys-84′ (from the neighboring C-polypeptide chain in the trimer), changing several intrachain ionic interactions and triggering a 2 Å closure (residues 50–55) between the Asp domain and the carbamoyl phosphate domain with a 5 Å movement of the 70–75 sequence culminating in a 10 Å movement of Lys-84′. The movement of Arg-229 displaces the sequences around 230–245 (the 240s loop) by 8 Å and disrupts interaction with the other catalytic trimer (the C1-C4 interface) by eliminating the E239(C1)-Y165, K164(C4) bonding that only occurs in the T-state. Major changes in quaternary structure result. A screw movement occurs between the two catalytic trimers as a result of a "differential gear" effect (Perutz, 1989) between R- and C-subunit interactions (Fig. 6). This movement results in a 12 Å separation along the threefold axis and a net rotation of the two C-trimers of about 10° relative to one another. The C1-R1-R6-C6 (and symmetry related) contacts are maintained as the R1-R6 dimer rotates

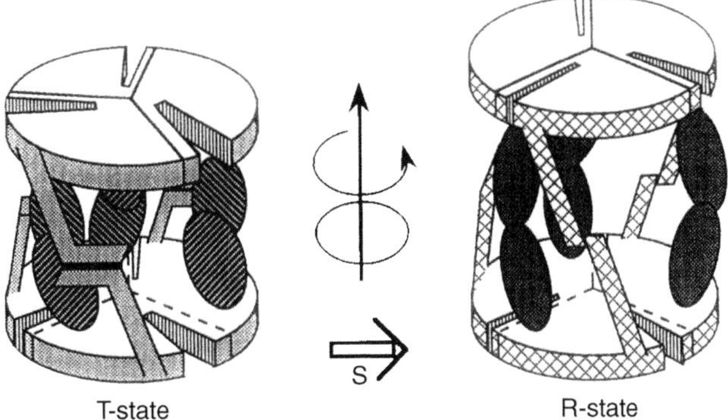

Fig. 6 Model of the allosteric cooperative transition of ATCase as determined from crystal structures. The two disks represent the two catalytic subunits containing three polypeptide chains each. The darker ellipsoids represent the regulatory dimers. The interaction of the 240s loop between the two catalytic subunits is shown by the vertical extensions of the cross-hatched or shaded volume with a more extensive interaction in the T-state (heavy bar). The concerted transition from the T- to the R-state causes an expansion of 15 Å along the vertical axis and a rotation of 15° of the upper catalytic subunit with respect to the lower; the resultant conformational change closes the active site cleft to promote interactions with the substrate in the R-state. Adapted from (Kantrowitz and Lipscomb 1990).

about 15° around the local twofold symmetry axis. The T-state has C1-R1, C1-R4, and C1-C4 (and corresponding symmetry related) interactions intact, whereas the R-state has only C1-R1 interactions. Thus, the cleft closure around the substrate displaces an adjacent (240s) loop, weakens the C1-C4 interaction, releases the "inhibition" by the regulatory subunits (C1-R4), and allows the other C-chains to relax into the higher affinity (Asp or PALA), closed domain conformation.

Hydrodynamic and structural studies in solution have proved sensitive and useful in support of the gross structural changes seen in the crystallographic studies. The R-state has been found to be more swollen or larger than the T-state by sedimentation velocity (Howlett *et al.*, 1977), low-angle X-ray scattering (Hervë *et al.*, 1985; Moody *et al.*, 1979), and analytical gel-exclusion chromatography (Bromberg *et al.*, 1990). Changes on the order of 3–6% in molecular size have been observed subsequent to substrate binding. Such methods should be useful with other enzymes that undergo relatively large quaternary structural changes. Direct comparisons of $\overline{R}$ to $\overline{Y}$ have not been completely satisfactory for distinguishing allosteric mechanisms for ATCase (Blackburn and Schachman, 1977; McClintock and Markus, 1969; Neet, 1980).

2. Homotropic Cooperativity

Not only is the MWC model useful for interpreting kinetic data for ATCase, but several lines of evidence strongly support the details of the concerted mechanism for the positive cooperativity seen with aspartate or PALA. The binding of one molecule of PALA can produce about five new R-state catalytic sites in the hexamer under the correct conditions, as measured by arsenolysis of carbamoyl aspartate (Foote and Schachman, 1985). This observation is an elegant and straightforward demonstration of one of the tenets of the MWC model. Spectroscopic studies of hybrid molecules containing nitrated catalytic polypeptides showed that unliganded chains change conformation (Yang and Schachman, 1980), consistent with the symmetry requisite of the MWC model. Boundary spreading of sedimentation velocity patterns detected no intermediate forms between the different hydrodynamic sizes of the R- and T-states in the presence of a 1.5 molar ratio of PALA to ATCase (Werner and Schachman, 1989; Werner *et al.*, 1989). The changes in sedimentation velocity of the wild-type (Howlett *et al.*, 1977) and mutant (Eisenstein *et al.*, 1990) enzymes parallel the activity changes as judged by the correspondence of the calculated [R] to [T] ratio. Linkage between heterotropic and homotropic ligands has been observed in many instances, for example, with mutations[1] such as cE239Q (Ladjimi and Kantrowitz, 1988), rN111A (Eisenstein *et al.*, 1989), rK143A (Eisenstein *et al.*, 1990), cD236A (Newton and Kantrowitz, 1990), or cD162A (Newton *et al.*, 1992) that effect both types of cooperativity (but see exceptions below), suggesting that the same equilibria are fundamental to both processes. These results are consistent in that cD236 and

[1] The letter c or r denotes a mutant in the catalytic or regulatory chains, respectively.

rK143 interact across the R1-C4 interface in the T-state (Gouaux and Lipscomb, 1990). The double mutant cK164E/E239K (and cK164E only) that lost both heterotropic and homotropic effects was shown to have sedimentation properties consistent with the R-state (Newell and Schachman, 1990).

Finally, the molecular model discussed above is so dependent on quaternary structural changes via the screw shift of the catalytic trimers through the three analogous C1-R1-R6-C6 connections that sequential conformational changes with part T-like and part R-like hybrid states is almost inconceivable; the reorientation of subunits with respect to one another precludes nonsymmetrical intermediates. A sequential mechanism using the ATCase R and T structures would be like trying to sit on a tripod stool with one leg 20% longer than the other two. Thus, structural and kinetic data are in good agreement with respect to the most appropriate allosteric model for homotropic cooperativity in ATCase.

3. Heterotropic Cooperativity

The situation is less clear with the heterotropic effectors, which bind at a distance of some 60 Å from the active site, partly because the structural changes have only relatively recently been described at high resolution. The uncoupling of heterotropic effects from homotropic effects (i.e., nucleotide activation or inhibition with no aspartate homotropic cooperativity) has been taken as evidence that the simplest MWC model does not strictly hold. Several situations have been found in which this linkage is lost, namely, with the alternate substrates L-alanosine (Baillon *et al.*, 1985) and cysteine sulfinate (Foote *et al.*, 1985), the mutants rY77F (Van Vliet *et al.*, 1991), cK83Q (Robey *et al.*, 1986), cE50Q (Ladjimi *et al.*, 1988), and C-terminal deletion of two residues of the regulatory chain (pAR5-related mutations) (Xi *et al.*, 1990, 1991). However, these uncoupling observations must be interpreted cautiously because of the different ways in which homotropic cooperativity can be lost. If sigmoid binding with a mutant is lost because $L_0 \approx 1$, the potential still exists for an effector-induced transition to a full R- or T-state with activation or inhibition. For the alternate substrates, Schachman (1988) has used an extended MWC model with $c = 1$ and $V_R > V_T$ to explain the apparent uncoupling with alternate substrates.

Evidence for a global conformational change produced by ATP or CTP was obtained by sedimentation velocity analysis of the rK143A mutant that had nearly equal proportions of R- and T-state enzyme ($L_0 = 2.7$) (Eisenstein *et al.*, 1990), consistent with the effectors simply shifting the R-T equilibrium. Detailed catalytic mechanism studies, including ^{13}C kinetic isotope effects, have supported the conclusion that only one active form—the R-state—is present in the presence of nucleotides (Parmentier *et al.*, 1992). On the other hand, crystallographic studies of wild-type ATCase suggest that the C-trimers move by only 0.4 (CTP) or 0.5 Å (ATP) along the threefold axis in the presence of nucleotide (Stevens *et al.*, 1990), at least in the crystal environment. An intriguing suggestion has been made (Glackin *et al.*, 1989), based on electrostatic calculations, that chains of ionizable groups link subunit interfaces and binding sites, thus providing a means of transmission of allosteric information independent of conformational changes *per se*. If this were the case, the circle and squares of Fig. 2 would represent "states" of the protein and not actual conformations.

The inhibitor CTP [as well as the substrate carbamoyl phosphate (Suter and Rosenbusch, 1976)] has been shown to have "negative cooperativity" of binding (Matsumoto and Hammes, 1973; Suter and Rosenbusch, 1977), and asymmetry is present in CTP-ligated ATCase crystals (Kim *et al.*, 1987). Although solution studies cannot distinguish (see above) between preexisting heterogeneity of the structure (inherent) and negative cooperativity (induced), a high-resolution (2.5 Å) structure that enabled visualization of the N terminus of the R-chains has demonstrated, within the limits of the atomic structural analysis, that part if not all of the CTP binding heterogeneity is due to negatively cooperative effects propagated through the R1-R6 interface to the Lys-60 that directly interacts with CTP (Kosman *et al.*, 1993). The nucleotide perturbation model for heterotropic interactions proposed by Stevens and Lipscomb (1992) on the basis of the small effector-induced conformational changes in the crystal structure (Stevens *et al.*, 1990) provides a detailed explanation for transmittal of the effector signal across the C1-R1-R6-C6 interfaces. This rationale for heterotropic interactions has some elements of induced fit by ATP and CTP with states attained that are neither R nor T, but asymmetrical hybrids are not implicit in the mechanism. The presence of negative cooperativity with CTP and the implications of the nucleotide perturbation model suggest that the simplest MWC model is not sufficient to explain entirely the heterotropic effects.

4. Summary

The concerted model of Monod-Wyman-Changeux satisfies most of the data relevant to the homotropic interactions of ATCase and eliminates the necessity of invoking a sequential model, that is, symmetry appears to be conserved as originally proposed on theoretical grounds. However, the implications of the heterotropic effector interactions are that more than two states must exist to account for the structural changes that occur on ATP or CTP binding, perhaps multiple T-like substates. Whether these states retain molecular symmetry has not yet been conclusively proved. Sequential changes of subunit conformation, or significant concerted conformational changes (Conway and Koshland, 1968), along the reaction coordinate for effector binding could account both for the "global" effects on nucleotide binding (Eisenstein *et al.*, 1990; Kosman *et al.*, 1993) and for the negative cooperativity of the CTP binding (Matsumoto and Hammes, 1973; Suter and Rosenbusch, 1977). Recent studies using low-angle X-ray scattering, fluorescence, mutations, and substrate analogs have provided further details of the structural changes (Cardia *et al.*, 2008).

C. Use of Mutants to Distinguish Models

Mutational analysis (Plapp, 1995) has been very useful with ATCase, hemoglobin, and other proteins in testing the importance or the role of individual amino acid residues, particularly in contact regions that transmit effects between subunits or domains. However, assignment of interaction surfaces from mutational data without a three-dimensional structure is clearly dangerous because of the transmission of substitution effects through the flexibility of a protein molecule. Mutations have provided a range of molecules with varying allosteric properties (e.g., altered L_0, n_H, c, heterotropic linkage) that probe the MWC versus the KNF models when used in conjunction with other methods. Study of the energetics of hemoglobin mutants has been extremely enlightening (Turner *et al.*, 1992). Hybrid ATCase (Eisenstein *et al.*, 1992) and hemoglobin (Ackers *et al.*, 1992) molecules containing varying mixtures of wild-type polypeptide chains with mutant or modified chains have elucidated other details; such methods will probably see more extensive usage in the future. Nevertheless, it is clear that mutants do not provide a simple, definitive conclusion regarding models by themselves. Spectroscopic, binding, catalytic, thermodynamic, crystallographic, and hydrodynamic studies as well as cleverly designed experiments have been needed to provide the kind of understanding that currently exists with ATCase and hemoglobin.

D. Allosteric Structures and Model Testing

The study of crystal structures leads to a detailed description of a limited number of conformations, usually termed R and T as in the MWC model, but may impose certain symmetry constraints. A comparison of the structural changes in the R- and T-states of five well-characterized proteins, namely, hemoglobin, ATCase, phosphofructokinase, glycogen phosphorylase, and fructose bisphosphatase, has been made (Johnson, 1992; Johnson and Barford, 1990; Liang *et al.*, 1993; Lipscomb, 1991; Perutz, 1989). The transmission of homotropic and heterotropic regulatory information over molecular distances has been described in each case by the movement of polypeptide backbone and side chains. As expected, the interactions across the subunit interfaces are very important. Common features are as follows: a minimal, but critical, movement at the active site (as little as 0.5 Å in hemoglobin); the rotation of subunits with respect to one another by 7–17 Å around a cyclic axis; the alteration of salt bonds and hydrogen bonds at the subunit interface in the T-R transition; only two apparent modes of docking of the subunits, implying a concerted mechanism; and a more highly constrained and extensive subunit interface in the T-state. Variable features include the following: translation along the subunit rotational axis may occur (ATCase); the involvement of each type of tertiary structure at the subunit interface, that is, α helices (hemoglobin, fructose bisphosphatase), β strands (phosphofructokinase), unordered loops (ATCase), and both α and β structures (phosphorylase); and the occurrence of the substrate binding directly at the interface to facilitate communication (only in phosphofructokinase and fructose bisphosphatase).

The study of the kinetics and structural properties in solution add to the critical tests of cooperative models that can be applied. Hemoglobin and ATCase come close to fitting the MWC model but vary from it when sufficient solution data are obtained (discussed above). The same may be true for phosphofructokinase, which fits the MWC model at one level but also has elements of association-dissociation in the mammalian enzyme (Lee *et al.*, 1988) and kinetic effects in the bacterial enzyme (Deville-Bonne *et al.*, 1991; Zheng and Kemp, 1992). Other well-studied enzymes, such as glycogen phosphorylase (Johnson, 1992; Johnson *et al.*, 1988), pyruvate kinase (Consler *et al.*, 1992), and fructose-1,6-bisphosphatase (Liang *et al.*, 1993), have for the moment been fit to the MWC model but are still being tested. When enough types of data are obtained, each individual enzyme appears to have its own peculiarity, and the general models serve as a starting framework, rather than a final solution. See update at the beginning of this chapter.

E. Kinetic Cooperativity in Monomeric Enzymes: Case of Glucokinase

Vertebrate glucokinase (hexokinase D or IV, EC2.7.1.2) is essential for the uptake of glucose by the liver and is involved in the regulation of blood glucose levels by the liver (Hue, 1981) and pancreas (Matschinsky, 1990). Positive cooperativity with glucose (Niemeyer *et al.*, 1975; Storer and Cornish-Bowden, 1976) ($n_H = 1.5$) is thought to be important for the enhanced responsiveness of the enzyme to changes in blood glucose levels (Hue, 1981; Matschinsky, 1990). A portion of noninsulin-dependent diabetes mellitus of the young, NIDDM type II, has been shown to be due to a genetic defect in glucokinase (Stoffel *et al.*, 1992; Vionnet *et al.*, 1992).

Because glucokinase is a monomer (52 kDa) under all conditions with no evidence of multiple sites or cooperativity of binding (Cardenas *et al.*, 1978; Holroyde *et al.*, 1976), the cooperativity cannot be generated by site-site interactions. Evidence has been presented that the kinetic mechanism is essentially ordered (Gregoriou *et al.*, 1981; Pollard-Knight and Cornish-Bowden, 1982) and that a random mechanism (Pettersson, 1986a) or abortive complex formation (Pettersson, 1986b) is unlikely to explain the cooperativity (Cornish-Bowden and Storer, 1986). A glucose-induced slow conformational change, in the form of a mnemonic mechanism (Pollard-Knight and Cornish-Bowden, 1982; Storer and Cornish-Bowden, 1977) and subsequently a LIST mechanism (Cardenas *et al.*, 1984, 1985; Neet *et al.*, 1988), has been proposed as the cause of the kinetic cooperativity. The predicted relationship with the second substrate occurs with glucokinase since the cooperativity decreases from an n_H of 1.5–1.0 as the MgATP concentration is decreased (Cardenas *et al.*, 1979; Storer and Cornish-Bowden, 1976). Glycerol added to assays depresses cooperativity (Cardenas *et al.*, 1985; Pollard-Knight *et al.*, 1985) and decreases the rate of the molecular transition sufficiently so that a slow transient has been observed in reaction assays (Neet *et al.*, 1990). Consistent with effects on an enzyme isomerization, a burst was observed under these conditions if the glucokinase was preincubated in high glucose and a lag occurred if the enzyme was initially in low glucose. The steady-state velocities after the transient showed the cooperativity and other features of the linear glucokinase assay without glycerol (Neet *et al.*, 1990). The LIST mechanism is also consistent with the effects of the inhibitors *N*-acetylglucosamine (Cardenas *et al.*, 1979, 1984; Neet *et al.*, 1990) and palmitoyl-CoA (Tippett and Neet, 1982) on cooperativity (Neet *et al.*, 1988), the requirement for two catalytic cycles (Neet *et al.*, 1988, 1990), and the observed negative cooperativity with MgATP (Neet *et al.*, 1990) or glucose (Cardenas *et al.*, 1979; Neet *et al.*, 1990) under certain conditions. Furthermore, a glucose-induced conformational change in the glucokinase molecule has been observed by protein intrinsic fluorescence measurements (Cardenas *et al.*, 1985; Lin and Neet, 1990) and shows a slow transition in glycerol similar to that in the kinetic assay under similar conditions (Lin and Neet, 1990). The half-times for the kinetic assay transient and the fluorescent conformational change are both between 0.5 and 5 min, have similar dependencies on glucose concentration, and provide strong evidence that the basis for the glucokinase cooperativity is a LIST (Lin and Neet, 1990). The current model is shown in Fig. 4 with ordered addition of glucose (A) before MgATP (B); the vertical lines represent the enzyme isomerizations. (See update in front.)

VII. Thermodynamic Analysis of Cooperativity, Linkage, and Multiple Ligand Binding

A. Thermodynamic Analysis of Cooperativity from Graphs

Hill plots of positive cooperativity can also be used to assess the energy of interaction between the first and the last ligand to bind to the protein, if the data are extensive enough. Typical Hill plots have a linear segment with a slope of one at each extreme (i.e., very low and very high ligand saturation). These regions represent the noninteractive binding of the first molecule to the unliganded protein and the last molecule to the last empty site. The perpendicular distance between the asymptotes is equal to the total free energy of interaction and is, therefore, a measure of the difference in free energy of binding of the first and last molecules (Neet, 1980; Ricard, 1985; Whitehead, 1970; Wyman, 1964). This difference in free energy is the energy invested in cooperative interactions within the oligomer.

B. Concept and Utility of Linkage Relationships

Heterotropic effects of binding are related to homotropic effects by the thermodynamic linkage principles (Wyman, 1964). For example, if an inhibitor increases the K_d^S for a substrate, then the reciprocal effect must occur and the substrate would increase the K_d^I for the inhibitor. This concept is a direct consequence of the principle of detailed balance of reactions (or thermodynamic "boxes"). Consider a protein that binds n_1 molecules of X_1 and n_2 molecules of X_2. The simplest expression to describe the relationship of the binding isotherms for each ligand, Y_1 and Y_2, is by the differential equation that relates the variation in binding of X_1 and X_2. It can be shown (Wyman, 1964) that

$$n_1\left(\frac{\partial \overline{Y}_1}{\partial \ln[X_2]}\right)_{[X_1]} = n_2\left(\frac{\partial \overline{Y}_2}{\partial \ln[X_1]}\right)_{[X_2]} \tag{8}$$

If there is no effect of X_1 on X_2, then the partial differentials are equal to zero, no interaction occurs, and there would be no effect of X_2 on X_1 either.

For the Bohr effect of hemoglobin the difference between the free energy of oxygenating a mole of hemoglobin at acidic or alkaline pH is exactly equal to the difference between the free energy of protonating a mole of hemoglobin and a mole of oxyhemoglobin (Wyman, 1964). Because this principle is thermodynamic in nature, it is independent of a particular model and is quite useful in designing experiments and testing the consistency of data. However, the linkage principle applies rigorously only to equilibrium situations, that is, kinetic parameters for a reaction may not follow linkage considerations owing to the unique contribution of kinetic constants. Furthermore, reciprocal effects arising from linkage on the cooperativity, n_H, of binding are predictable, but they are usually model dependent (see cooperativity models above).

In a practical sense, it is often useful to know if two different ligands affect an enzyme by binding to the same enzyme form (mutually exclusive binding) or whether they bind to different sites and have additive or synergistic effects. Exclusive binding is usually interpreted that both ligands bind to the same site, although this structural constraint is not required to explain the experimental result. The most rigorous procedure is to measure binding of one ligand in the presence of the other and directly look for displacement. The interaction of multiple inhibitors (or activators), however, and whether they are inhibiting (or activating) nonexclusively or exclusively can also be assessed by appropriate design of experiments. If v_0/v_i is plotted against ($[I_1]$ plus $[I_2]$), a straight line is obtained for mutually exclusive binding; a curve or deviation from linearity supports concurrent binding to the same enzyme form (Yagi and Ozawa, 1960). An alternate, more rigorous graphical method (Yonetani and Theorell, 1964) is to measure v_0/v_i as a function of $[I_1]$ at constant $[I_2]$ and plot the result as a function of $[I_1]$. In this case, a series of parallel lines is obtained for exclusive binding and straight lines with different slopes occur for nonexclusive, interacting binding. A complete discussion of these methods has been provided (Yonetani, 1982). A more general, but rigorous, method that is not limited to Michaelis-Menten enzymes or to specific mechanisms has been introduced (Chou and Talalay, 1977, 1981) and is based on the linearity of simple graphs of $\log[(f_i)^{-1}-1]$ versus $[I_1 + I_2]$, where f_i is the fractional inhibition. This procedure is straightforward but appears to have been underutilized.

VIII. Ligand-Binding Sites and Intramolecular Communication Distance: Kinetic Consequences

Two major findings that were not entirely anticipated in early concepts of regulatory enzymes have evolved from structural studies of allosteric proteins. Binding sites for ligands, either substrates or effectors, have frequently been found to lie on interfaces between two subunits of both allosteric and other oligomeric enzymes, rather than being contained within a single subunit. Distances between binding sites can vary from large distances (e.g., 65 Å), to vanishingly small (i.e., overlapping or isosteric sites) in the case of some covalent modifications. Neither of these observations could have been, nor can be, readily inferred from kinetic or other solution studies.

A. Intersubunit Ligand-Binding Sites

The placing of ligand-binding sites on intersubunit interfaces provides for an exquisite means of communicating cooperative or allosteric effects between subunits in an oligomeric enzyme. Examples include 2,3-bisphos-phoglycerate with hemoglobin (Perutz, 1970, 1989), glutathione with glutathione reductase (Schulz *et al.*, 1978), the glutamine analog methionine sulfoximine with glutamine synthetase (glutamate-ammonia ligase) (Almassy *et al.*, 1986; Yamashita *et al.*, 1989), fructose 6-phosphate at the active site and ADP at the effector site of phosphofructokinase (Rypniewski and Evans, 1989; Shirakihara and Evans, 1988), fructose 1,6-bisphosphate and fructose 2,6-bisphosphate with fructose-1,6-bisphosphatase (Ke *et al.*, 1989; Liang *et al.*, 1992), and the bisubstrate analog PALA with aspartate transcarbamylase (Ke *et al.*, 1984; Krause *et al.*, 1987). Such shared sites almost by necessity have to be responsive to changes in subunit interactions. With many more crystal structures now available, many more examples of interfacial binding have been found. Whether the high incidence of interfacial binding would effect how one views the validity of the general allosteric models will be left to the reader to consider—pros and cons for either MWC or KNF models can be debated vigorously. The lack of activity or loss of cooperativity of a dissociated enzyme can certainly be more easily understood if the active site is shared by neighboring subunits. In this case, of course, kinetics must be done on the intact oligomeric enzyme. Nevertheless, the formalism of the analysis in terms of kinetic mechanism or allosteric model does not change.

Comparison of kinetics and binding constants between oligomer and subunits can give insight into the makeup of binding sites (Dautry-Varsat and Garel, 1981; Newell *et al.*, 1989). However, the subunits must be dissociated gently enough to maintain a folded structure, and elaborate proof is needed for "nativeness" of any inactive subunit that results. Proteolytic fragments and kinetic analysis of refolding of the bifunctional aspartokinase-homoserine dehydrogenase I have shown that the kinase activity occurs in the monomeric species but that the dehydrogenase activity is generated only in the dimeric intermediate (Dautry-Varsat and Garel, 1981; Vaucheret *et al.*, 1987). Cooperativity due to subunit association mechanisms (see Section VII) could also be readily understood with intersubunit substrate binding sites. In the final analysis, crystal or NMR structures are required for definitive answers regarding the subunit contributions of binding sites.

The concept that a distinct subunit would likely contain the allosteric effector binding sites, suggested from the early example of the separable regulatory and catalytic subunits of ATCase, has not held up well as a general finding; relatively few other similar examples currently exist, such as the cAMP-dependent protein kinase (Taylor, 1989) and calmodulin in phosphorylase kinase (Pickett-Giles and Walsh, 1987). In most cases a single polypeptide chain contains both the allosteric and catalytic site, as in phosphofructokinase (Shirakihara and Evans, 1988), pyruvate kinase (Engstrom *et al.*, 1987), and homoserine dehydrogenase (Dautry-Varsat and Garel, 1981).

In many other instances, binding sites lie on the interface of two domains of the same polypeptide, that is, a crevice or cleft, as foretold by early crystal studies of lysozyme and yeast hexokinase. Thus, evolution has apparently found it easy to form a functional binding site by juxtaposition of two rather separate subsites from distant domains of a protein. Ligand-induced movement of two domains or alteration of the spatial relationship of two subunits also appears to be a relatively common way by which a protein can achieve significant effects on catalysis or regulation through conformational changes.

B. Interaction Distances

Distances between binding sites can be measured, in principle, in solution by such techniques as NMR and fluorescence. Many examples exist in which these distances are in reasonable agreement with those subsequently determined from crystal structures. Transmittal of regulatory information over long distances is exemplified by ATCase and glycogen phosphorylase with distances of 60 Å or more (Table II). The movement of polypeptide backbone and amino acid residues has been described in detail in several cases. Communication of regulatory information over shorter distances also occurs in fructose-1,6-bisphosphatase and isocitrate dehydrogenase (Table II).

A recent demonstration has been made that isosteric regulatory effects can occur, that is, regulation at the same site rather than another (allosteric) site (Dean and Koshland, 1990). Phosphorylation of isocitrate dehydrogenase introduces a negative charge from the phosphate at the residue, Ser-113, that would normally hydrogen bond to the negatively charged γ-carboxyl of isocitrate; substrate repulsion occurs directly without a conformational change (Hurley *et al.*, 1990). Identical types of kinetic responses may be observed with isosteric regulation as with allosteric effects in which the phosphorylation site is more distant (e.g., 12 Å away in glycogen synthase), and significant conformational changes are

Table II
Atomic Distances in Selected Allosteric Proteins

Enzyme (protein)	Effect	Site(s)	Distance (Å)	References
Hemoglobin	Homotropic	O_2	25–40	Perutz (1970)
Aspartate transcarbamylase	Homotropic	Aspartate	40–50	Ke *et al.* (1984) and Krause *et al.* (1987)
Aspartate transcarbamylase	Heterotropic	ATP/CTP to aspartate	60	
Phosphofructokinase	Homotropic	Fru-6-P	45–55	Shirakihara and Evans (1988)
Phosphofructokinase	Heterotropic	ADP to Fru-6-P	20–25	
Glutamine synthetase[a]	Homotropic	Gln to Gln	45–52	Yamashita *et al.* (1989) and Almassy *et al.* (1986)
Glutamine synthetase	Heterotropic	Covalent adenylyl to Gln	22	
Glycogen phosphorylase	Homotropic	Glucoside to glucoside	66	Johnson (1992), Johnson *et al.* (1988), and Sprang *et al.* (1988)
Glycogen phosphorylase	Heterotropic	AMP to glucoside	32	
Glycogen phosphorylase	Heterotropic	Covalent P to AMP	12	
Fructose-1, 6-bisphosphatase	Heterotropic	AMP to Fru-1,6-P_2	28	Liang *et al.* (1992, 1993) and Ke *et al.* (1989)
Fructose-1, 6-bisphosphatase	Heterotropic	Fru-2,6-P_2 to Fru-1,6-P_2	0	
Isocitrate dehydrogenase	Heterotropic	Covalent P to isocitrate	0	(Hurley *et al.* 1990)

[a]Glutamate-ammonia ligase.

required to transmit the regulatory information. Although these examples concern regulation by covalent modification, similar effects could occur with regulation by noncovalent modifiers (i.e., by apparent "allosteric" effectors). Thus, more fuel is provided for long-standing kinetic controversies. Crystallographic data indicate that fructose 2,6-bisphosphate, a negative effector of fructose-1,6-bisphosphatase, binds to the fructose 1,6-bisphosphate (active) site (Ke *et al.*, 1989; Liang *et al.*, 1992, 1993); in this case, of course, the effector and substrate have similar structures. Whether product inhibition of mammalian brain hexokinase I by glucose 6-phosphate occurs at an allosteric site or at the active site has been discussed for 30 years (Magnani *et al.*, 1992). Inhibition at the active site (an isosteric effect), as proposed by Solheim and Fromm (Solheim and Fromm, 1981) from kinetic arguments, is probably correct (but see Baijal and Wilson, 1992; Jarori *et al.*, 1990) since site-specific mutagenesis (Smith and Wilson, 1992) and a truncated recombinant hexokinase (Magnani *et al.*, 1992) show that the glucose 6-phosphate binding site is in the same C-terminal domain as the active site, rather than being on a distinct domain. Confirmed by X-ray crystal structure analysis in Aleshin *et al.* (2000). Furthermore, NMR data indicate the glucose 6-phosphate binds to hexokinase I within 3–19 Å of the glucose site (White and Wilson, 1987). Interpretation of similar types of purely kinetic data in terms of the underlying structure must be made with caution.

IX. Cooperativity in Other Macromolecular Systems

A. Ligand Binding to Membrane Receptors

Numerous receptor systems have demonstrable cooperativity, either positive or negative. Because the macromolecular receptor is binding a ligand (frequently a hormone or growth factor that is itself a peptide or protein), the mathematical analysis of binding to the receptor is similar to that described above for enzymes (Munson, 1983). Analysis of cellular response data is not analogous to that of enzyme kinetic data, however, since the signal transduction by coupling and amplification complicates any rigorous interpretation of cooperativity of the responses. The experimental methods for obtaining binding data are different since the receptor will be either in the membrane of a whole cell, in a membrane fragment, or solubilized in detergent. The reader is referred to Vols. 109, 146, and 147 in Methods in Enzymology for a detailed discussion of receptor techniques. A significant difference when dealing with receptors on cells in culture is that dynamic internalization of the ligand-receptor complex occurs and usually a steady-state binding situation, rather than a true equilibrium, must be assumed (Wiley, 1981). Furthermore, the fluidity of the membrane allows dynamic alterations in receptor-receptor interactions (e.g., dimerization) that can influence cooperativity (DeLean and Rodbard, 1979; DeLisi and Chabay, 1979; Jacobs and Cuatrecasas, 1976), much as in the enzyme oligomerization model discussed above.

Phenomenological (Adair-type) models are initially applied to describe the binding of proteins to membrane receptors. From graphs of the binding data (analogous to those in Fig. 1), the possible presence of cooperativity can be deduced. The MWC model is frequently utilized to explain both homotropic and heterotropic effects, and R- and T-states of receptors are commonly invoked (Changeux *et al.*, 1967). A good example is the nicotinic acetylcholine receptor (Galzi *et al.*, 1991), which is an ion channel regulated by the effect of agonists, competitive antagonists, and noncompetitive inhibitors at several allosteric sites. Multiple, presumably pseudosymmetric, states of the heterologous pentameric acetylcholine receptor are postulated to exist (Lena and Changeux, 1993). Whether this model will withstand further testing, when conformational changes and molecular interactions are described at high resolution, remains to be seen.

An additional complexity of membrane receptors is the frequent heterogeneity of receptors in a membrane that gives rise to biphasic binding curves. This heterogeneity may be due to pharmacological subtypes (isoreceptors) (e.g., neuronal dopamine receptors), cross talk with related receptors (e.g., the nerve growth factor family of neurotrophin receptors), functional diversity (e.g., interleukin 2 receptors), or environmental variation (e.g., insulin receptors). Because the biphasic curves have the same appearance as "negative cooperativity" (see earlier discussion), much controversy has arisen about the cause of the biphasic curves. One interesting test with membrane receptors has been to compare simple dissociation rates from dilution of the ligand-receptor complex to the apparent dissociation rate of labeled ligand in the presence of an excess of unlabeled ligand (De Meyts *et al.*, 1976; Jacobs and Cuatrecasas, 1976). If negative cooperativity exists, the increased receptor occupancy in the presence of excess ligand should increase the rate of dissociation of the bound radioactive ligand (assuming the association rate is constant), whereas the rate for independent heterogeneous receptors should be unaffected by the cold chase. Although some success was reported with this method, others have found it not to be a useful discriminator (de Vries *et al.*, 1991; DeLisi and Chabay, 1979). Further use of this interesting approach is warranted.

B. Proteins Binding to DNA

An active area of research is the study of nucleic-binding proteins, many of which show positive cooperativity of binding. Well-studied examples include Gene-32 protein of T4 bacteriophage, the single-stranded DNA-binding protein of *E. coli*, and numerous transcriptional factors. Methodology often utilizes labeled oligonucleotides and gel-shift assays, so the analysis of data can be somewhat different than that with enzymes and low molecular weight ligands. Because the protein-protein cooperativity can be influenced by the presence of multiple binding sites within the DNA template, a different model (Kowalczykowski *et al.*, 1986) has been applied to this system. The equilibrium binding of a protein to an infinite lattice, such as DNA, can be described by three thermodynamic parameters: n, the binding site size of the protein for the DNA (number of bases "covered"); K, the intrinsic binding constant for a nucleotide sequence; and ω, the cooperativity term. The cooperativity parameter, ω, is the relative affinity for a protein to bind adjacent to an already bound protein ligand as opposed to an isolated DNA site and can be greater than 1 (positive cooperativity) or less than 1 (negative cooperativity). The binding equations, methods to measure the parameters with either varying ligand or lattice concentration, and a rigorous interpretation of the results have been presented (Kowalczykowski *et al.*, 1986).

The steroid receptor complexes also bind specific DNA sequences with positive cooperativity. The molecular structures of the glucocorticoid (Luisi *et al.*, 1991) and the estrogen (Schwabe *et al.*, 1993) receptor oligonucleotide complexes have been determined by X-ray crystallography and provide a molecular model to explain the cooperativity. The monomeric glucocorticoid or estrogen DNA-binding domain dimerizes on binding to DNA. The apparent specificity (and affinity) of the glucocorticoid DNA-binding domain depends on the correct spacing (3 bp) between the two hexameric half-sites. The cooperativity (ω or dimerization) on DNA binding is due to the ordering of a region of the binding domain that is disordered in solution and makes positive protein-protein contacts on the correctly spaced palindromic response element (Luisi *et al.*, 1991; Schwabe *et al.*, 1993). The steroid receptors have the most molecular detail and the most specific proposal of any cooperative system other than soluble enzyme systems.

C. Macromolecular Assembly

Self-assembling, homomeric proteins frequently show (positive) cooperativity, that is, no (or few) intermediates exist between the monomer state and the high polymeric state. Examples include the classic case of hemoglobin S (which leads to sickling), tubulin, and other cytoskeletal proteins. Analysis of these equilibria differs from that of ligand-binding systems because the change in the protein-protein association constant is the criterion for cooperativity. Methods employed usually involve a measure of the size distribution of the system by sedimentation equilibrium, viscosity, gel filtration, and so on. The exchange of labeled monomers into the polymer can also be utilized to estimate relative affinities.

X. Future Directions

Considering the fact that the field of allosteric enzymes is three decades old, there is surprisingly poor understanding of the details of most regulatory enzymes. The two examples discussed, hemoglobin and ATCase, have a nearly complete molecular description but still lack complete agreement on the ultimate mechanism. Others (e.g., phosphofructokinase, glycogen phosphorylase, isocitrate dehydrogenase) have atomic coordinates and a molecular hypothesis that needs further testing. Finally, less well-characterized proteins (e.g., glucokinase, wheat germ hexokinase, chloroplast fructose bisphosphatase) have kinetic models but lack information at the molecular structure level. The examples discussed in this chapter quite clearly show the necessity of having one or more three-dimensional coordinate structures from X-ray crystallography in order to interpret allosteric enzymes. However, it is equally clear that such a structure is not in itself sufficient and that extensive solution studies of conformation, structural kinetics, and energetics are required for a thorough understanding of the enzyme and to substantiate an appropriate model.

Physical techniques that can characterize protein conformations and interactions in solution are making tremendous progress in application to allosteric mechanisms. Such techniques not only are able to reveal hybrid or asymmetrical intermediates that crystallography may not, but can also provide information on the dynamics of the allosteric transition. The most visible technique of NMR (and the one with the most information content next to X-ray crystallography) is already making significant contributions to smaller proteins and will undoubtedly continue to be important to allosteric systems as the study of larger proteins becomes possible. Sedimentation and gel filtration techniques that have made such important contributions to ATCase and hemoglobin may not be broadly applicable to other systems. However, microcalorimetry [both differential scanning (DSC) and isothermal titration] and quasielastic light scattering (QEL) may provide similar types of detailed information on the energetics of protein-protein interactions between subunits. Raman spectroscopy and fluorescence relaxation may complement steady-state fluorescence as a means of studying the kinetics of (undefined) conformational changes. Fourier transform infrared spectroscopy (FTIR) may supplant circular dichroism as a means for determining secondary structure in solution and provide more dynamic details of allosteric transitions. A limiting amount of protein for such demanding studies has become much less of a problem with the advent of recombinant DNA methodology. Indeed, the molecular biology tools of protein overexpression and site-specific mutagenesis will continue to be of tremendous benefit to the practicing enzymologist. Application of these physical methods to membrane receptor systems will remain more difficult.

Important questions for the enzymologist extend beyond the mechanistic approaches posed above into the classic arena of the physiological importance of the regulatory aspects of enzymes. What is really the importance of the allosteric site of a key enzyme for the regulation of a metabolic pathway? What would be the biological result of losing the positive cooperativity of an enzyme such as phosphofructokinase or glucokinase? Is the slow isomerization of a hysteretic enzyme essential for life? What is the role of negative cooperativity in physiological function? What would happen to the organism if an enzyme could not form tetramers? See the update at the beginning of this chapter for the answers to some of these questions.

Fortunately, the means to answer these questions appears to be at hand with a combination of molecular biology, cell biology, genetics, and biology to assist the enzymologist. Most enzyme mutants have only been studied mechanistically *in vitro*, but these same mutants can readily be studied in a physiological setting in mammalian cell culture or in transgenic animals. Homologous recombination of a mutant regulatory enzyme, deficient only in its allosteric binding site or in its substrate cooperativity, into a metabolically important cell (e.g., hepatocyte) or into a mouse embryo is technically possible. The former system would allow study of metabolic flux in a living cell with altered regulation; the latter system would determine the consequences of altered regulation of pathways for a living organism. Such regulation-deficient transgenic mice should be much more informative than "knockout" experiments, although somewhat more difficult to construct. The answer may well be that the animal adapts to the lack of regulation, much to the chagrin of the enzymologist. In any event, models of various metabolic disorders (including rather mild effects) may be provided with the production of such genetically altered animals. The next few decades may have as many advances and increase our knowledge as much as the past 15–30 years; it should be an equally exciting era for the regulatory enzymologist.

Acknowledgments

The author gratefully acknowledges support by grants from the National Institutes of Health and from the Juvenile Diabetes Foundation. I also thank the following colleagues who critically read portions of the manuscript: Drs Marilu Cardenas, Athel Cornish-Bowden, Antony M. Dean, Robert G. Kemp, Jacque Ricard, and Howard K. Schachman.

References

Ackers, G. K., Doyle, M. L., Myers, D., and Daugherty, M. A. (1992). *Science* **255,** 54.
Ackers, G. K., and Hazzard, J. H. (1993). *Trends Biochem. Sci.* **18,** 385.
Ackers, G. K., and Holt, J. M. (2006). *J. Biol. Chem.* **281,** 11441.
Adair, G. S. (1925). *J. Biol. Chem.* **63,** 529.
Ahrweiler, P. M., and Frieden, C. (1991). *Biochemistry* **30,** 7801.
Ainslie, G. R., Jr., Shill, J. P., and Neet, K. E. (1972). *J. Biol. Chem.* **247,** 7088.
Aleshin, A. E., Kirby, C., Liu, X., Bourenkov, G. P., Bartunik, H. D., Fromm, H. J., and Honzatko, R. B. (2000). *J. Mol. Biol.* **296,** 1001.
Allewell, N. M. (1989). *Annu. Rev. Biophys. Biophys. Chem.* **18,** 71.
Almassy, R. J., Janson, C. A., Hamlin, R., Xuong, N. H., and Eisenberg, D. (1986). *Nature (London)* **323,** 304.
Antoine, M., Boutin, J. A., and Ferry, G. (2009). *Biochemistry* **48,** 5466.
Baijal, M., and Wilson, J. E. (1992). *Arch. Biochem. Biophys.* **298,** 271.
Baillon, J., Tauc, P., and Hervë, G. (1985). *Biochemistry* **24,** 7182.
Bennett, W. S., and Steitz, T. A. (1978). *Proc. Natl. Acad. Sci. USA* **75,** 4848.
Berg, J. M., Tymoczko, J. L., and Stryer, L. (2007). "Biochemistry," 6th ed, Freeman, New York.
Blackburn, M. N., and Schachman, H. K. (1977). *J. Biol. Chem.* **16,** 5084.
Bromberg, S., Burz, D. S., and Allewell, N. M. (1990). *J. Biochem. Biophys. Methods* **20,** 143.
Buc, J., Ricard, J., and Meunier, J. C. (1977). *Eur. J. Biochem.* **80,** 593.
Cardenas, M. L., Rabajille, E., and Niemeyer, H. (1978). *Arch. Biochem. Biophys.* **190,** 142.
Cardenas, M. L., Rabajille, E., and Niemeyer, H. (1979). *Arch. Biol. Med. Exp.* **12,** 571.
Cardenas, M. L., Rabajille, E., and Niemeyer, H. (1984). *Eur. J. Biochem.* **145,** 163.
Cardenas, M. L., Rabajille, E., Trayer, I. P., and Niemeyer, H. (1985). *Arch. Biol. Med. Exp.* **18,** 273.
Cardia, J. P., Eldo, J., Xia, J., O'Day, E. M., Tsuruta, H., Gryncel, K. R., and Kantrowitz, E. R. (2008). *Proteins* **71,** 1088.
Changeux, J. P., Thiery, J. P., Tung, Y., and Kittel, C. (1967). *Proc. Natl. Acad. Sci. USA* **57,** 335.
Chou, T. C., and Talalay, P. (1981). *Eur. J. Biochem.* **115,** 207.
Chou, T. C., and Talalay, P. (1977). *J. Biol. Chem.* **252,** 6438.
Ciaccio, C., Coletta, A., De Sanctis, G., Marini, S., and Coletta, M. (2008). *IUBMB Life* **60,** 112.
Cleland, W. W. (1970). *In* "The Enzymes" (P. D. Boyer, ed.), 3rd ed., Vol. 2, p. 1. Academic Press, New York.
Cleland, W. W. (1979). *Methods Enzymol.* **63,** p.103.
Consler, T. G., Jennewein, M. J., Cai, G. Z., and Lee, J. C. (1992). *Biochemistry* **31,** 7870.
Conway, A., and Koshland, D. E., Jr. (1968). *Biochemistry* **7,** 4011.
Cornish-Bowden, A., and Cardenas, M. L. (1987). *J. Theor. Biol.* **124,** 1.
Cornish-Bowden, A., and Koshland, D. E., Jr. (1970). *Biochemistry* **9,** 3325.

Cornish-Bowden, A., and Koshland, D. E., Jr. (1975). *J. Mol. Biol.* **95,** 201.
Cornish-Bowden, A., and Storer, A. C. (1986). *Biochem. J.* **240,** 293.
Cornish-Bowden, A., and Wong, J. T. (1978). *Biochem. J.* **175,** 969.
Cui, Q., and Karplus, M. (2008). *Protein Sci.* **17,** 1295.
Dautry-Varsat, A., and Garel, J. R. (1981). *Biochemistry* **20,** 1396.
Dean, A. M., and Koshland, D. E., Jr. (1990). *Science* **249,** 1044.
DeLean, A., and Rodbard, D. (1979). *In* "The Receptors" (R. D. O'Brien, ed.), p. 143. Plenum, New York.
DeLisi, C., and Chabay, R. (1979). *Cell Biophys.* **1,** 117.
De Meyts, P., Bianco, A. R., and Roth, J. R. (1976). *J. Biol. Chem.* **251,** 1877.
Deville-Bonne, D., Fourgain, F., and Garel, J. R. (1991). *Biochemistry* **30,** 5750.
de Vries, C. P., van Haeften, T. W., and van der Veen, E. A. (1991). *Endocr. Res.* **17,** 331.
Eaton, W. A., Henry, E. R., Hofrichter, J., Bettati, S., Viappiani, C., and Mozzarelli, A. (2007). *IUBMB Life* **59,** 586.
Eigen, M. (1968). *Q. Rev. Biophys.* **1,** 3.
Eisenstein, E., Han, M. S., Woo, T. S., Ritchey, J. M., Gibbons, I., Yang, Y. R., and Schachman, H. K. (1992). *J. Biol. Chem.* **267,** 22148.
Eisenstein, E., Markby, D. W., and Schachman, H. K. (1990). *Biochemistry* **29,** 3724.
Eisenstein, E., Markby, D. W., and Schachman, H. K. (1989). *Proc. Natl. Acad. Sci. USA* **86,** 3094.
Engstrom, L., Ekman, P., Humble, E., and Zetterqvist, O. (1987). *In* "The Enzymes" (P. D. Boyer, ed.), 3rd ed., Vol. 18, p. 47. Academic Press, New York.
Ferdinand, W. (1966). *Biochem. J.* **98,** 278.
Fersht, A. R., Leatherbarrow, R. J., and Wells, T. N. C. (1986). *Trends Biochem. Sci.* **11,** 321.
Foote, J., Lauritzen, A. M., and Lipscomb, W. N. (1985). *J. Biol. Chem.* **260,** 9624.
Foote, J., and Schachman, H. K. (1985). *J. Mol. Biol.* **186,** 175.
Frieden, C. (1979). *Annu. Rev. Biochem.* **48,** 471.
Frieden, C. (1967). *J. Biol. Chem.* **242,** 4045.
Frieden, C. (1970). *J. Biol. Chem.* **245,** 5788.
Galzi, J. L., Revah, F., Bessis, A., and Changeux, J. P. (1991). *Annu. Rev. Pharmacol. Toxicol.* **31,** 37.
Giudici-Orticoni, M. T., Buc, J., Bidaud, M., and Ricard, J. (1990). *Eur. J. Biochem.* **194,** 483.
Giudici-Orticoni, M. T., Buc, J., and Ricard, J. (1990). *Eur. J. Biochem.* **194,** 475.
Glackin, M. P., McCarthy, M. P., Mallikarachchi, D., Matthew, J. B., and Allewell, N. M. (1989). *Proteins* **5,** 66.
Goldbeter, A. (1976). *Biophys. Chem.* **4,** 159.
Gouaux, J. E., and Lipscomb, W. N. (1990). *Biochemistry* **29,** 389.
Gregoriou, M., Trayer, I. P., and Cornish-Bowden, A. (1981). *Biochemistry* **20,** 499.
Guidotti, G. (1967). *J. Biol. Chem.* **242,** 3704.
Gulbinsky, J. S., and Cleland, W. W. (1968). *Biochemistry* **7,** 566.
Haber, J. E., and Koshland, D. E., Jr. (1967). *Proc. Natl. Acad. Sci. USA* **58,** 2087.
Hammes, G. G., and Wu, C. W. (1971). *Science* **172,** 1205.
Hervë, G., Moody, M. F., Tauc, P., Vachette, P., and Jones, P. T. (1985). *J. Mol. Biol.* **185,** 189.
Hill, C. M., Waight, R. D., and Bardsley, G. (1977). *Mol. Cell. Biochem.* **15,** 173.
Ho, C. (1992). *Adv. Protein Chem.* **43,** 153.
Holroyde, M. J., Allen, M. B., Storer, A. C., Warsy, A. S., Chesher, J. M. E., Trayer, I. P., Cornish-Bowden, A., and Walker, D. G. (1976). *Biochem. J.* **153,** 363.
Howlett, G. J., Blackburn, M. N., Compton, J. G., and Schachman, H. K. (1977). *Biochemistry* **16,** 5091.
Hue, L. (1981). *Adv. Enzymol.* **52,** 247.
Hurley, J. H., Dean, A. M., and Koshland, D. E., Jr. (1990). *Science* **249,** 1012.
Ivanetich, K. M., Goold, R. D., and Sikakana, C. N. (1990). *Biochem. Pharmacol.* **39,** 1999.
Jacobs, S., and Cuatrecasas, P. (1976). *Biochim. Biophys. Acta* **433,** 482.
Jarori, G. K., Iyer, S. B., Kasturi, S. R., and Kenkare, U. W. (1990). *Eur. J. Biochem.* **188,** 9.

Jencks, W. P. (1975). *Adv. Enzymol.* **43,** 219.
Johnson, L. N. (1992). *In* "Receptor Subunits and Complexes" (A. Burgen and E. A. Barnard, eds.), p. 39. Cambridge University Press, Cambridge.
Johnson, L. N. (1992). *Protein. Sci.* **1,** 1237.
Johnson, L. N., and Barford, D. (1990). *J. Biol. Chem.* **265,** 2409.
Johnson, L. N., Hajdu, J., Acharya, K. R., Stuart, D. I., McLaughlin, P. J., Oikonomakos, N. G., and Barford, D. (1988). *In* "Allosteric Proteins" (G. Herve, ed.), p. 81. CRC Press, Boca Raton, Florida.
Kantrowitz, E. R., and Lipscomb, W. N. (1988). *Science* **241,** 669.
Kantrowitz, E. R., and Lipscomb, W. N. (1990). *Trends Biochem. Sci.* **15,** 53.
Ke, H. M., Honzatko, R. B., and Lipscomb, W. N. (1984). *Proc. Natl. Acad. Sci. USA* **81,** 4037.
Ke, H., Thorpe, C. M., Seaton, B. A., Marcus, F., and Lipscomb, W. N. (1989). *Proc. Natl. Acad. Sci. USA* **86,** 1475.
Kim, K. H., Pan, Z., Honzatko, R. B., Ke, H. M., and Lipscomb, W. N. (1987). *J. Mol. Biol.* **196,** 853.
Kirtley, M. E., and Koshland, D. E., Jr. (1967). *J. Biol. Chem.* **242,** 4192.
Klotz, I. M. (1974). *Acc. Chem. Res.* **7,** 162.
Koshland, D. E., Jr., and Hamadani, K. (2002). *J. Biol. Chem.* **277,** 46841.
Koshland, D. E., Jr., Nemethy, G., and Filmer, P. (1966). *Biochemistry* **5,** 365.
Kosman, R. P., Gouaux, J. E., and Lipscomb, W. N. (1993). *Proteins* **15,** 147.
Kowalczykowski, S. C., Paul, L. S., Lonberg, N., Newport, J. W., McSwiggen, J. A., and von Hippel, P. H. (1986). *Biochemistry* **25,** 1226.
Krause, K. L., Volz, K. W., and Lipscomb, W. N. (1987). *J. Mol. Biol.* **193,** 527.
Kurganov, B. I. (1982). "Allosteric Enzymes: Kinetic Behaviour." Wiley, New York.
Kurganov, B. I. (1967). *Khim. Tekhnol. Polim.* **11,** 140.
Ladjimi, M. M., and Kantrowitz, E. R. (1988). *Biochemistry* **27,** 276.
Ladjimi, M. M., Middleton, S. A., Kelleher, K. S., and Kantrowitz, E. R. (1988). *Biochemistry* **27,** 268.
Larion, M., and Miller, B. G. (2009). *Biochemistry* **48,** 6157.
Lee, J. C., Hesterberg, L. K., Luther, M. A., and Cai, G. Z. (1988). *In* "Allosteric Enzymes" (G. Herve, ed.), **231**CRC Press, Boca Raton, Florida.
Lehninger, A. L., Nelson, D. L., and Cox, M. M. (2009). "Principles of Biochemistry" 5th ed., Freeman, New York.
Lena, C., and Changeux, J. P. (1993). *Trends Neurosci.* **16,** 181.
Levitzki, A., and Schlessinger, J. (1974). *Biochemistry* **13,** 5214.
Liang, J. Y., Huang, S., Zhang, Y., Ke, H., and Lipscomb, W. N. (1992). *Proc. Natl. Acad. Sci. USA* **89,** 2404.
Liang, J. Y., Zhang, Y., Huang, S., and Lipscomb, W. N. (1993). *Proc. Natl. Acad. Sci. USA* **90,** 2132.
Lin, S. X., and Neet, K. E. (1990). *J. Biol. Chem.* **265,** 9670.
Lipscomb, W. N. (1991). *Chemtracts* **2,** 1.
Luisi, B. F., Xu, W. X., Otwinowski, Z., Freedman, L. P., Yamamoto, K. R., and Sigler, P. B. (1991). *Nature (London)* **352,** 497.
Macdonald, J. L., and Pike, L. J. (2008). *Proc. Natl. Acad. Sci. USA* **105,** 112.
Macdonald-Obermann, J. L., and Pike, L. J. (2009). *J. Biol. Chem.* **284,** 13570.
Magnani, M., Bianchi, M., Casabianca, A., Stocchi, V., Daniele, A., Altruda, F., Ferrone, M., and Silengo, L. (1992). *Biochem. J.* **285,** 193.
Mannervik, B. (1982). *Methods Enzymol.* **87,** 370.
Mathews, C. K., and van Holde, K. E. (1999). "Biochemistry," 3rd ed, Benjamin/Cummings, Redwood City, California.
Matschinsky, F. M. (1990). *Diabetes* **39,** 647.
Matsumoto, S., and Hammes, G. G. (1973). *Biochemistry* **12,** 1388.
McClintock, D. K., and Markus, G. (1969). *J. Biol. Chem.* **244,** 36.
Merkler, D. J., and Schramm, V. L. (1990). *J. Biol. Chem.* **265,** 4420.

Meunier, J. C., Buc, J., Navarro, A., and Ricard, J. (1974). *Eur. J. Biochem.* **49,** 209.
Meunier, J. C., Buc, J., and Ricard, J. (1979). *Eur. J. Biochem.* **97,** 573.
Monneuse-Doublet, M. O., Olomucki, A., and Buc, J. (1978). *Eur. J. Biochem.* **84,** 441.
Monod, J., Wyman, J., and Changeux, J. P. (1965). *J. Mol. Biol.* **12,** 88.
Moody, M. F., Vachette, P., and Foote, A. M. (1979). *J. Mol. Biol.* **133,** 517.
Munson, P. J. (1983). *Methods Enzymol.* **92,** 543.
Neet, K. E. (1979). *Bull. Mol. Biol. Med.* **4,** 101.
Neet, K. E. (1980). *Methods Enzymol.* **64,** 139.
Neet, K. E., and Ainslie, G. R., Jr. (1980). *Methods Enzymol.* **64,** 192.
Neet, K. E., Keenan, R. P., and Tippett, P. S. (1990). *Biochemistry* **29,** 770.
Neet, K. E., Tippett, P. S., and Keenan, R. P. (1988). *In* "Enzyme Dynamics and Regulation" (P. B. Chock, C. Y. Huang, C. L. Tsou and J. H. Wang, eds.), p. 28. Springer-Verlag, New York.
Nelson-Rossow, K. L., Sukalski, K. A., and Nordlie, R. C. (1993). *Biochim. Biophys. Acta* **1163,** 297.
Newell, J. O., Markby, D. W., and Schachman, H. K. (1989). *J. Biol. Chem.* **264,** 2476.
Newell, J. O., and Schachman, H. K. (1990). *Biophys. Chem.* **37,** 183.
Newton, C. J., and Kantrowitz, E. R. (1990). *Proc. Natl. Acad. Sci. USA* **87,** 2309.
Newton, C. J., Stevens, R. C., and Kantrowitz, E. R. (1992). *Biochemistry* **31,** 3026.
Nichol, L. W., Jackson, W. J. H., and Winzor, D. J. (1967). *Biochemistry* **6,** 2449.
Nichol, L. W., and Winzor, D. J. (1976). *Biochemistry* **15,** 3015.
Niemeyer, H., Cardenas, M. L., Rabajille, E., Ureta, T., Clark-Turri, L., and Penaranda, J. (1975). *Enzyme* **20,** 321.
Noggle, J. H. (1993). "Practical Curve Fitting and Data Analysis." Prentice-Hall, Englewood Cliffs, NJ.
Pal, P., and Miller, B. G. (2009). *Biochemistry* **48,** 814–816.
Parmentier, L. E., O'Leary, M. H., Schachman, H. K., and Cleland, W. W. (1992). *Biochemistry* **31,** 6570.
Paulus, H., and DeRiel, J. K. (1975). *J. Mol. Biol.* **97,** 667.
Penner, M. H., and Frieden, C. (1985). *J. Biol. Chem.* **260,** 5366.
Perutz, M. F. (1970). *Nature (London)* **228,** 726.
Perutz, M. F. (1989). *Q. Rev. Biophys.* **22,** 139.
Pettersson, G. (1986a). *Biochem. J.* **233,** 347.
Pettersson, G. (1986b). *Eur. J. Biochem.* **154,** 167.
Pickett-Giles, C. A., and Walsh, D. A. (1987). *In* "The Enzymes" (P. D. Boyer, ed.), 3rd ed., **17,** p. 396. Academic Press, New York.
Plapp, B. V. (1995). *Methods Enzymol.* **249,** 91.
Pollard-Knight, D., Connolly, B. A., Cornish-Bowden, A., and Trayer, I. P. (1985). *Arch. Biochem. Biophys.* **237,** 328.
Pollard-Knight, D., and Cornish-Bowden, A. (1982). *Mol. Cell. Biochem.* **44,** 71.
Rabin, B. R. (1967). *Biochem. J.* **102,** 22c.
Ricard, J. (1978). *Biochem. J.* **175,** 779.
Ricard, J. (1985). *In* "Organized Multienzyme Systems" (G. R. Welch, ed.), p. 177. Academic Press, New York.
Ricard, J., and Cornish-Bowden, A. (1987). *Eur. J. Biochem.* **166,** 255.
Ricard, J., Giudici-Orticoni, M. T., and Buc, J. (1990). *Eur. J. Biochem.* **194,** 463.
Ricard, J., Meunier, J. C., and Buc, J. (1974). *Eur. J. Biochem.* **49,** 195.
Ricard, J., Meunier, J. C., and Buc, J. (1977). *Eur. J. Biochem.* **80,** 581.
Ricard, J., Mouttet, C., and Nari, J. (1974). *Eur. J. Biochem.* **41,** 479.
Ricard, J., and Noat, G. (1985). *Eur. J. Biochem.* **152,** 557.
Ricard, J., and Noat, G. (1984). *J. Theor. Biol.* **111,** 737.
Ricard, J., and Noat, G. (1985). *J. Theor. Biol.* **117,** 633.
Ricard, J., and Noat, G. (1986). *J. Theor. Biol.* **123,** 431.
Ricard, J., and Noat, G. (1982). *J. Theor. Biol.* **96,** 347.

Ricard, J., Soulie, J. M., and Bidaud, M. (1986). *Eur. J. Biochem.* **159,** 247.

Robey, E. A., Wente, S. R., Markby, D. W., Flint, A., Yang, Y. R., and Schachman, H. K. (1986). *Proc. Natl. Acad. Sci. USA* **83,** 5934.

Rose, I. A., Warms, J. V. B., and Yuan, R. G. (1993). *Biochemistry* **32,** 8504.

Rubin, M. M., and Changeux, J. P. (1966). *J. Mol. Biol.* **21,** 265.

Rudolph, F. B., and Fromm, H. J. (1979). *Methods Enzymol.* **63,** 138.

Rypniewski, W. R., and Evans, P. R. (1989). *J. Mol. Biol.* **207,** 805.

Saffarian, S., Li, Y., Elson, E. L., and Pike, L. J. (2007). *Biophys J.* **93,** 1021.

Schachman, H. K. (1988). *J. Biol. Chem.* **263,** 18583.

Schulz, G. E., Schirmer, R. H., Sachsenheimer, W., and Pai, E. F. (1978). *Nature (London)* **273,** 120.

Schwabe, J. W. R., Chapman, L., Finch, J. T., and Rhodes, D. (1993). *Cell (Cambridge, Mass.)* **75,** 567.

Segal, I. H. (1975). "Enzyme Kinetics." Wiley (Interscience), New York.

Shirakihara, Y., and Evans, P. R. (1988). *J. Mol. Biol.* **204,** 973.

Smith, A. D., and Wilson, J. E. (1992). *Arch. Biochem. Biophys.* **292,** 165.

Solheim, L. P., and Fromm, H. J. (1981). *Arch. Biochem. Biophys.* **211,** 92.

Sprang, S. R., Acharya, K. R., Goldsmith, E. J., Stuart, D. I., Varvill, K., Fletterick, R. J., Madsen, N. B., and Johnson, L. N. (1988). *Nature (London)* **336,** 215.

Stevens, R. C., Gouaux, J. E., and Lipscomb, W. N. (1990). *Biochemistry* **29,** 7691.

Stevens, R. C., and Lipscomb, W. N. (1992). *Proc. Natl. Acad. Sci. USA* **89,** 5281.

Stoddard, B. L., Koenigs, P., Porter, N., Petratos, K., Petsko, G. A., and Ringe, D. (1991). *Proc. Natl. Acad. Sci. USA* **88,** 5503.

Stoffel, M., Froguel, P.h., Takeda, J., Zouali, H., Vionnet, N., Nishi, S., Weber, I. T., Harrison, R. W., Pilkis, S. J., Lesage, S., Vaxillaire, M., Velho, G., *et al.* (1992). *Proc. Natl. Acad. Sci. USA* **89,** 7698.

Storer, A. C., and Cornish-Bowden, A. (1976). *Biochem. J.* 159.

Storer, A. C., and Cornish-Bowden, A. (1977). *Biochem. J.* **165,** 61.

Suter, P., and Rosenbusch, J. P. (1976). *J. Biol. Chem.* **251,** 5986.

Suter, P., and Rosenbusch, J. P. (1977). *J. Biol. Chem.* **252,** 8136.

Taylor, S. S. (1989). *J. Biol. Chem.* **264,** 8443.

Tippett, P. S., and Neet, K. E. (1982). *J. Biol. Chem.* **257,** 12846.

Trost, P., and Pupillo, P. (1993). *Arch. Biochem. Biophys.* **306,** 76.

Turner, G. J., Galacteros, F., Doyle, M. L., Hedlund, B., Pettigrew, D. W., Turner, B. W., Smith, F. R., Moo-Penn, W., Rucknagel, D. L., and Ackers, G. K. (1992). *Proteins* **14,** 333.

Van Vliet, F., Xi, X. G., de Staercke, C., de Wannemaeker, B., Jacobs, A., Cherfils, J., Ladjimi, M. M., Hervë, G., and Cunin, R. (1991). *Proc. Natl. Acad. Sci. USA* **88,** 9180.

Vaucheret, H., Signon, L., LeBras, G., and Garel, J. R. (1987). *Biochemistry* **26,** 2785.

Vionnet, N., Stoffel, M., Takeda, J., Yasuda, K., Bell, G. I., Zouali, H., Lesage, S., Velho, G., Iris, F., and Passa, P. (1992). *Nature (London)* **356,** 721.

Viratelle, O. M., and Seydoux, F. J. (1975). *J. Mol. Biol.* **92,** 193.

Wells, B. D., Stewart, T. A., and Fisher, J. R. (1976). *J. Theor. Biol.* **60,** 209.

Werner, W. E., Cann, J. R., and Schachman, H. K. (1989). *J. Mol. Biol.* **206,** 221.

Werner, W. E., and Schachman, H. K. (1989). *J. Mol. Biol.* **206,** 221.

Whitehead, E. (1970). *Prog. Biophys. Mol. Biol.* **21,** 321.

Whitehead, E. P. (1976). *Biochem. J.* **159,** 449.

White, T. K., and Wilson, J. E. (1987). *Arch. Biochem. Biophys.* **259,** 402.

Wiley, H. S., and Cunningham, D. D. (1981). *Cell (Cambridge, Mass.)* **25,** 433.

Wood, L. C., White, T. B., Ramdas, L., and Nall, B. T. (1988). *Biochemistry* **27,** 8562.

Wyman, J. J. (1964). *Adv. Protein Chem.* **19,** 223.

Xi, X. G., Van Vliet, F., Ladjimi, M. M., de Wannemaeker, B., de Staercke, C., Glansdorff, N., Piërard, A., Cunin, R., and Hervë, G. (1991). *J. Mol. Biol.* **220,** 789.

Xi, X. G., Van Vliet, F., Ladjimi, M. M., de Wannemaeker, B., de Staercke, C., Piërard, A., Glansdorff, N., Hervë, G., and Cunin, R. (1990). *J. Mol. Biol.* **216,** 375.

Yagi, K., and Ozawa, T. (1960). *Biochim. Biophys. Acta* **42,** 381.

Yamashita, M. M., Almassy, R. J., Janson, C. A., Cascio, D., and Eisenberg, D. (1989). *J. Biol. Chem.* **264,** 17681.
Yang, K. S., Ilagan, M. X., Piwnica-Worms, D., and Pike, L. J. (2009). *J. Biol. Chem.* **284,** 7474.
Yang, Y. R., and Schachman, H. K. (1980). *Proc. Natl. Acad. Sci. USA* **77,** 6187.
Yonetani, T. (1982). *Methods Enzymol.* **87,** 500.
Yonetani, T., and Laberge, M. (2008). *Biochim. Biophys. Acta* **1784,** 1146.
Yonetani, T., and Theorell, H. (1964). *Arch. Biochem. Biophys.* **106,** 243.
Yonetani, T., and Tsuneshige, A. (2003). *C. R. Biol.* **326,** 523.
Zheng, R. L., and Kemp, R. G. (1992). *J. Biol. Chem.* **267,** 23640.

PART II

Inhibitors as Probes of Enzyme Catalysis

CHAPTER 10

Reversible Enzyme Inhibitors as Mechanistic Probes

Herbert J. Fromm
Department of Biochemistry and Biophysics
Iowa State University
Ames, Iowa 50011

I. Introduction

A number of experimental protocols have been employed by kineticists over the years in order to choose between bireactant ordered and random mechanisms. Unfortunately, most suffer from technical limitations. Among these, the most popular methods are the use of the Haldane relationship, product inhibition, and isotope exchange. The first method, which compares kinetically determined parameters with the apparent equilibrium constant for a reaction, requires precise initial rate data, which are frequently difficult to obtain. As Fromm and Nelson (1962) pointed out, interpretation of results from product inhibition studies is complicated by the formation of abortive complexes of enzyme, substrate, and product. Finally, isotope exchange cannot be used with systems, such as fructose1,6-bisphosphatase, in which the equilibrium constant is extremely large.

CONTEMPORARY ENZYME KINETICS AND MECHANISM
Reprinted from *Methods in Enzymology*, Volume 249 (Academic Press, 1995).

DOI: 10.1016/B978-0-12-378608-1.00010-4

Although competitive inhibitors have been used extensively for a variety of reasons in enzyme experiments, their value as tools for making a choice of kinetic mechanism from among possible alternatives was not realized until 1962, when Fromm and Zewe (1962) suggested that competitive inhibitors of substrates could be used to differentiate between random and ordered mechanisms. Furthermore, in the latter case, a determination of the substrate binding order could be made from such experiments. This protocol is quite likely the simplest approach for differentiating between Ordered and Random Bi Bi possibilities.[1] In addition, it has the advantage of permitting the kineticist to come to definitive conclusions from studies of reactions in a single direction only. Its obvious limitation involves the requirement that a competitive inhibitor be available for each substrate. However, when other initial rate data are available, a good deal of information concerning the kinetic mechanism may be provided from experiments with only one-substrate analog even for Bi and Ter reactant systems.

In this discussion it is assumed that the competitive inhibitor, which is usually a substrate analog, when bound to the enzyme, will not permit product formation to occur. These inhibitors are then dead-end inhibitors as contrasted with "partial" competitive inhibitors, which when associated with the enzyme allow formation of product either at a reduced or accelerated rate (Fromm, 1975a). In addition, it will be assumed that enzyme-inhibitor complex formation occurs rapidly relative to other steps in the reaction pathway.

II. Theory

A. One-Substrate Systems

Before describing how competitive inhibitors are used to make a choice of mechanism in the case of multisubstrate system, some discussion is warranted concerning reversible dead-end inhibition for one-substrate reactions. Scheme 1 illustrates a typical Uni Uni mechanism and now a linear competitive inhibitor enters into the reaction mechanism.

By definition, a competitive inhibitor competes with the substrate for the same site on the enzyme. Identical kinetic results are obtained, however, if the inhibitor and substrate compete for different sites, but where binding is mutually exclusive. Although there is usually a structural similarity between the substrate and the competitive inhibitor, this is not always the case. Finally, competitive inhibition is reversed when the enzyme is saturated with substrate.

The derivation of the rate equation for a competitive inhibitor can be done algebraically or by using any of the more sophisticated procedures described elsewhere (Fromm, 1979; Huang, 1979). Derivation by the former method is as follows.

[1] The nomenclature of Cleland will be used throughout this chapter. See Cleland (1963).

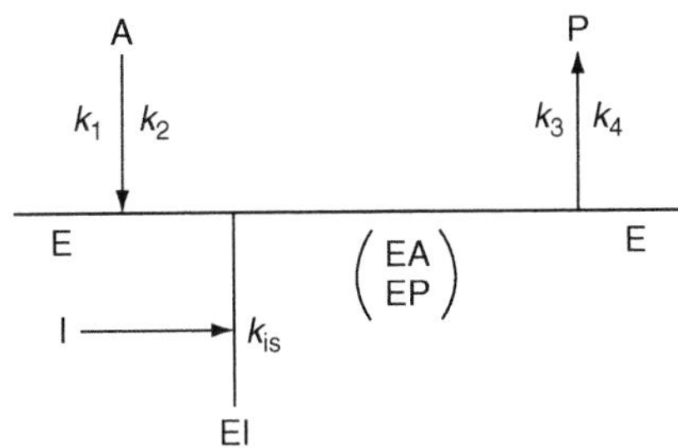

Scheme 1

The velocity expression is $\upsilon = \kappa_3 \begin{pmatrix} \text{EA} \\ \text{EP} \end{pmatrix}$, and the conservation of enzyme equation is $\text{E}_0 = \text{E} + \begin{pmatrix} \text{EA} \\ \text{EP} \end{pmatrix} + \text{EI}$, when E_0 is total enzyme. The dissociation constant for the enzyme-inhibitor complex is taken as $K_{is} = (\text{E})(\text{I})/(\text{EI})$. Thus,

$$\text{E}_0 = \text{E} + \begin{pmatrix} \text{EA} \\ \text{EP} \end{pmatrix} + \frac{(\text{E})(\text{I})}{K_{is}} = (\text{E})\left(1 + \frac{\text{I}}{K_{is}}\right) + \begin{pmatrix} \text{EA} \\ \text{EP} \end{pmatrix} \tag{1}$$

From the expression $\text{E} = \left(\frac{K_a}{\text{A}}\right)\begin{pmatrix} \text{EA} \\ \text{EP} \end{pmatrix}$, and the equation for initial velocity,

$$\upsilon = \frac{V_1}{1 + (K_a/\text{A})(1 + \text{I}/K_{is})} \tag{2}$$

where K_a is the Michaelis constant, $(k_2 + k_3)/k_1$, and V_1 is the maximal velocity, $k_3\text{E}_0$.

It can be seen from Eq. (1) that the free enzyme component in the conservation of enzyme equation is multiplied by the factor $(1 + \text{I}/K_{is})$. A shorthand method for including the effect of a reversible dead-end inhibitor in the rate equation is simply to first identify the enzyme species in the noninhibited rate equation that reacts with the inhibitor, then multiply that enzyme form by the proper factor. This will perhaps be somewhat clearer when the rate equation for noncompetitive inhibition is discussed.

Figure 1 illustrates a double-reciprocal plot for an enzyme system in the presence and in the absence of a competitive inhibitor. The rationale for this particular plot comes by taking the reciprocal of both sides of Eq. (2):

$$\frac{1}{\upsilon} = \frac{1}{V_1}\left[1 + \frac{K_a}{\text{A}}\left(1 + \frac{\text{I}}{K_{is}}\right)\right] \tag{3}$$

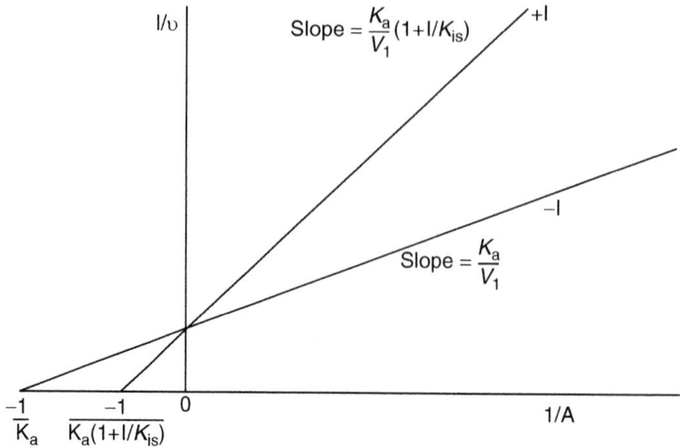

Fig. 1 Plot of 1/initial velocity (υ) versus 1/substrate concentration (A) in the presence and absence of a linear competitive inhibitor (I).

The inhibition described in Eq. (3) is referred to as linear competitive, because a replot of slopes versus I gives a straight line. It will be seen later that competitive inhibition need not be linear.

A number of procedures are currently in vogue for linearizing initial rate plots for inhibitors. The method used most is that illustrated in Fig. 1. Another graphing method, the Dixon plot, is frequently used; however, it suffers from very serious inherent limitations (Fromm, 1975b; Purich and Fromm, 1972a) and its use is not recommended.

Scheme 2 describes linear noncompetitive enzyme inhibition. In this model the substrate and the inhibitor are not mutually exclusive binding ligands, and

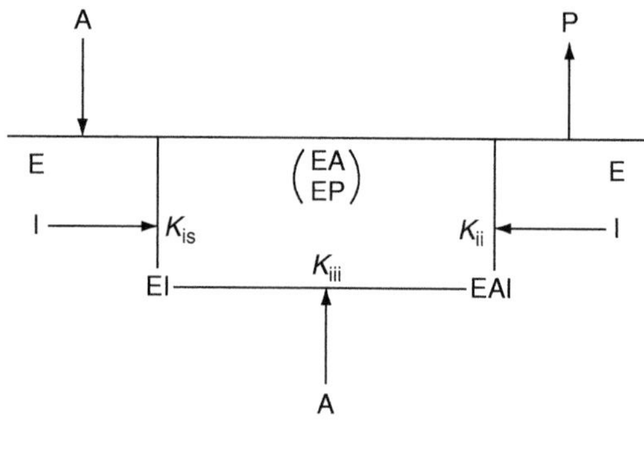

Scheme 2

complexes of enzyme, inhibitor, and substrate are thus possible. The rate equation for linear noncompetitive inhibition, in double-reciprocal form, is

$$\frac{1}{\upsilon} = \frac{1}{V_1}\left[\left(1 + \frac{\mathrm{I}}{K_{\mathrm{ii}}}\right) + \frac{K_{\mathrm{ia}}}{\mathrm{A}}\left(1 + \frac{\mathrm{I}}{K_{\mathrm{is}}}\right)\right] \tag{4}$$

where K_{ia} is the dissociation constant of the EA complex. In the derivation of Eq. (4), it is assumed that the inhibitory complexes are dead-end and that all steps shown in Scheme 2 are in rapid equilibrium relative to the breakdown of the central complex to form product.

In the derivation of Eq. (4), the enzyme forms containing inhibitor must be added to the conservation of enzyme equation. Thus, $E_0 = E + EA + EI + EAI$. From the interactions shown in Scheme 2, it can be seen that EI arises from reaction of the inhibitor and E, whereas EAI, may arise from interaction of the inhibitor and EA. In deriving Eq. (4), it is then necessary merely to multiply the factor $(1 + I/K_{ii})$ by the EA term of the uninhibited rate equation and the factor $(1 + I/K_{is})$ by the E term of the uninhibited rate expression. The very same result will be obtained if the K_{iii} step, rather than the K_{ii} pathway, is used. This is true because the four dissociation constants are not independent, but are related by the following expression: $K_{is}K_{iii} = K_{ia}K_{ii}$.

A typical plot of $1/\upsilon$ versus 1/A in the presence and in the absence of the noncompetitive inhibitor is shown in Fig. 2. It can be seen from the graph that the lines converge in the second quadrant. Convergence of the curves may also

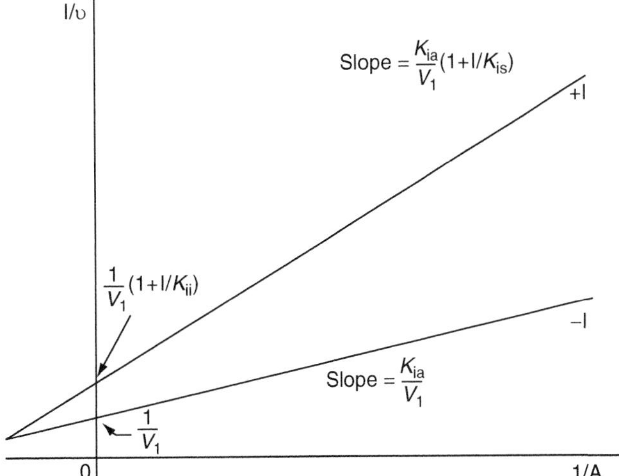

Fig. 2 Plot of 1/initial velocity (υ) versus 1/substrate concentration (A) in the presence and in the absence of a linear noncompetitive inhibitor (I).

occur in the third quadrant or on the abscissa, depending on the relationship between the dissociation constants for the enzyme-inhibitor complexes. The *x, y* coordinates of the intersection point for noncompetitive inhibition are $-K_{is}/K_{ia}K_{ii}$, $(1/V_1)(1 - K_{is}/K_{ii})$. If $K_{is} = K_{ii}$, that is, the binding of the inhibitor to the enzyme is not affected by the presence of the substrate on the enzyme, the curves will intersect on the 1/A axis. When $K_{is} < K_{ii}$, intersection will be above the abscissa, whereas when $K_{is} > K_{ii}$, intersection of the curves will be in the third quadrant.

It can be seen from Eq. (4) that replots of slopes and intercepts against inhibitor concentration will give linear lines. This type of noncompetitive inhibition is more formally referred to as S-linear (slope) and I-linear (intercept) noncompetitive.

Another type of reversible dead-end enzyme inhibition that will be useful for later discussion is linear uncompetitive inhibition. Scheme 3 depicts the pathway that illustrates this phenomenon. The rate expression for linear uncompetitive inhibition is shown in Eq. (5), where K_i is the dissociation constant for the EAI complex:

$$\frac{1}{\upsilon} = \frac{1}{V_1}\left[\left(1 + \frac{\mathrm{I}}{K_i}\right) + \frac{K_a}{\mathrm{A}}\right] \tag{5}$$

Figure 3 describes the results of a plot of 1/υ versus 1/A in the presence and absence of an uncompetitive inhibitor. It can be seen that linear uncompetitive inhibition gives rise to a family of parallel lines. It is often difficult to determine whether the lines in inhibition experiments are really parallel. Rather sophisticated computer programs are available that both test and fit *weighted* kinetic data in an attempt to address this problem (Fromm, 1975c). A discussion of this point is not presented here; however, it may be helpful to point out that uncompetitive inhibition may be graphed as a Hanes plot (i.e., A/υ versus A). This plot is illustrated in Fig. 4 and has the advantage, relative to the double-reciprocal plot, that the inhibited and uninhibited lines must converge at a common point on the A/υ axis. This approach does not eliminate the necessity of model fitting and testing,

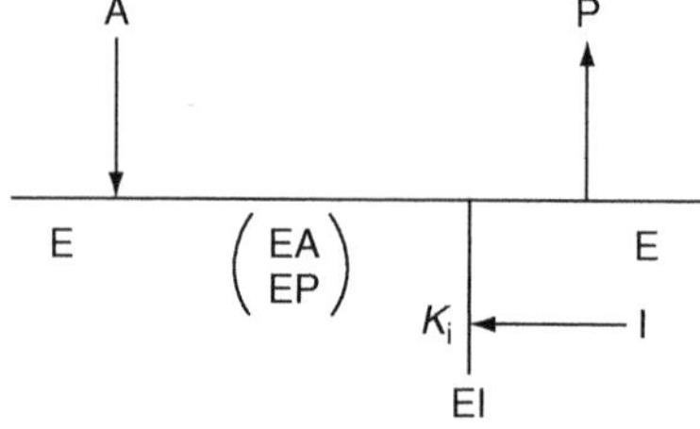

Scheme 3

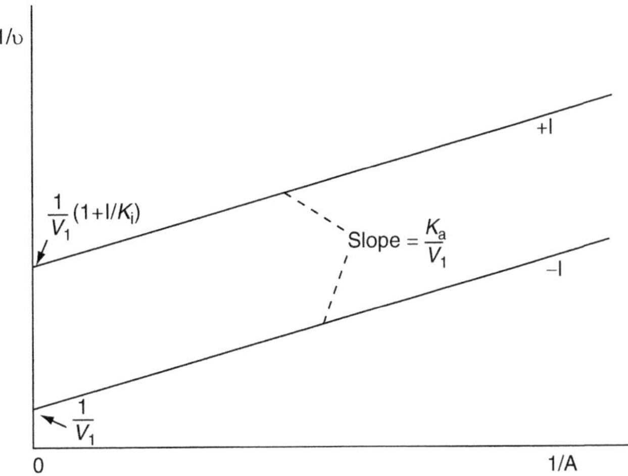

Fig. 3 Plot of 1/initial velocity (υ) versus 1/substrate concentration (A) in the presence and in the absence of a linear noncompetitive inhibitor (I).

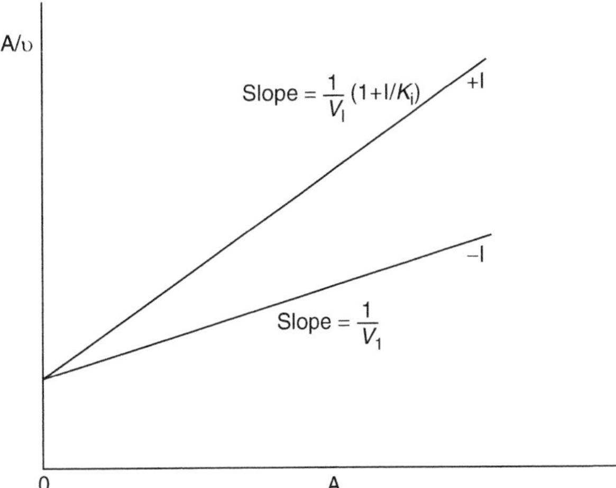

Fig. 4 Plot of substrate concentration (A) divided by initial velocity (υ) versus substrate concentration (A) in the presence and in the absence of a linear uncompetitive inhibitor (I).

but it does permit the investigator to make a preliminary judgment more easily on the nature of the inhibition.

The last type of reversible dead-end inhibition to be considered in conjunction with unireactant systems is nonlinear inhibition. Nonlinear enzyme inhibition may be obtained from replots of primary double-reciprocal plots as a result of multiple dead-end inhibition, substrate and product inhibition, partial inhibition, and allostery. This discussion is limited to multiple dead-end inhibition.

When an inhibitor adds reversibly to different enzyme forms (e.g., Scheme 2), inhibition is linear; however, when there is multiple inhibitor binding to a single enzyme form, or to enzyme forms that are sequentially connected, replots of either slopes or intercepts against inhibitor may be nonlinear. If the following equilibrium is added to the simple Uni Uni mechanism of Scheme 1, $EI + I = EI_2$, K_{ii}, the rate equation obtained is

$$\frac{1}{v} = \frac{1}{V_1}\left[1 + \frac{K_{ia}}{A}\left(1 + \frac{I}{K_{is}} + \frac{I^2}{K_{is}K_{ii}}\right)\right] \tag{6}$$

Equation (6) describes parabolic competitive inhibition. The slope of Eq. (6) is

$$\text{Slope} = \frac{K_{ia}}{V_1}\left(1 + \frac{I}{K_{is}} + \frac{I^2}{K_{is}K_{ii}}\right) \tag{7}$$

and a plot of slope as a function of I will give rise to a parabola. It is also possible to obtain parabolic uncompetitive inhibition. In this case only the intercept will be affected by inhibitor. In the event that parabolic noncompetitive inhibition is encountered, either the slope or the intercept, or both, may contain inhibitor terms of greater than first degree.

Nonlinear noncompetitive inhibition may affect slopes and intercepts; in this case it is called S-parabolic I-parabolic noncompetitive inhibition. If the replots of intercept against I are linear, whereas the slope replot is parabolic, the inhibition is called S-parabolic I-linear noncompetitive inhibition.

Parabolic inhibition will also result from interactions of the following type:

$$E + I = EI, K_{is}; \quad EI + A = EIA, K_{ii}; \quad EIA + I = EI_2A, K_{iii} \tag{8}$$

Slope and intercept replots versus inhibitor may be of a more complicated nature. Cleland has referred to some of these as 2/1, 3/2, etc., functions. Equation (9) illustrates an example of a $S - 2/1$ function, in which a second-order polynomial is divided by a first-order polynomial:

$$\text{Slope} = \frac{K_{ia}(1 + aI + bI^2)}{V_1(1 + cI)} \tag{9}$$

It may be difficult to distinguish a plot of slope versus inhibitor concentration for Eq. (9) from linear replots.

B. Two-Substrate Systems

Let us now consider the case of a random mechanism to determine how the dead-end competitive inhibitor affects the kinetics of the system. The rapid-equilibrium random pathway of enzyme and substrate interaction (Random Bi Bi) is illustrated by Fromm (1979). In the case of a competitive inhibitor for substrate A, the inhibitor, I, would participate at every step in the kinetic mechanism in which the substrate A normally reacts. Thus, the following interactions of enzyme with inhibitor might be expected:

$$\mathrm{E} + \mathrm{I} = \mathrm{EI}, K_{\mathrm{i}}; \quad \mathrm{EB} + \mathrm{I} = \mathrm{EIB}, K_{\mathrm{ii}}; \quad \mathrm{EI} + \mathrm{B} = \mathrm{EIB}, K_{\mathrm{iii}} \tag{10}$$

When the expressions EI and EIB are added to the conservation of enzyme expression, and the rate equation derived for the effect of the competitive inhibitor of substrate A, the following relationship is obtained:

$$\frac{1}{\upsilon} = \frac{1}{V_1} + \frac{K_{\mathrm{a}}}{V_1(\mathrm{A})}\left(1 + \frac{\mathrm{I}}{K_{\mathrm{ii}}}\right) + \frac{K_{\mathrm{b}}}{V_1(\mathrm{B})} + \frac{K_{\mathrm{ia}}K_{\mathrm{b}}}{V_1(\mathrm{A})(\mathrm{B})}\left(1 + \frac{I}{K_{\mathrm{i}}}\right) \tag{11}$$

where K_{b} is the Michaelis constant for substrate B.

When double-reciprocal plots of $1/\upsilon$ as a function of 1/A are made at different fixed concentrations of inhibitor, only the slope term of the rate expression is altered; that is

$$\mathrm{Slope} = \frac{K_{\mathrm{a}}}{V_1(\mathrm{A})}\left(1 + \frac{\mathrm{I}}{K_{\mathrm{ii}}}\right) + \frac{K_{\mathrm{ia}}K_{\mathrm{b}}}{V_1(\mathrm{B})}\left(1 + \frac{\mathrm{I}}{K_{\mathrm{i}}}\right) \tag{12}$$

On the other hand, when B is the variable substrate, double-reciprocal plots at different fixed levels of inhibitor will exhibit increases in both slopes and intercepts:

$$\mathrm{Intercept} = \frac{1}{V_1}\left[1 + \frac{K_{\mathrm{a}}}{\mathrm{A}}\left(1 + \frac{\mathrm{I}}{K_{\mathrm{ii}}}\right)\right] \tag{13}$$

$$\mathrm{Slope} = \frac{1}{V_1}\left[K_{\mathrm{b}} + \frac{K_{\mathrm{ia}}K_{\mathrm{b}}}{\mathrm{A}}\left(1 + \frac{\mathrm{I}}{K_1}\right)\right] \tag{14}$$

Equation (11) predicts then that a dead-end competitive inhibitor for substrate A of the Random Bi Bi mechanism is a noncompetitive inhibitor of substrate B.

If now a dead-end competitive inhibitor for substrate B is used, the following interactions are to be expected:

$$\mathrm{E} + \mathrm{I} = \mathrm{EI}, K_{\mathrm{i}}; \quad \mathrm{EA} + \mathrm{I} = \mathrm{EAI}, K_{\mathrm{ii}}; \quad \mathrm{EI} + \mathrm{A} = \mathrm{EAI}, K_{\mathrm{iii}} \tag{15}$$

The rate equation for the effect of a dead-end competitive inhibitor of substrate B is described by Eq. (16):

$$\frac{1}{v} = \frac{1}{V_1} + \frac{K_a}{V_1(A)} + \frac{K_b}{V_1(B)}\left(1 + \frac{I}{K_{ii}}\right) + \frac{K_{ia}K_b}{V_1(A)(B)}\left(1 + \frac{I}{K_i}\right) \tag{16}$$

It can be seen from Eq. (16) that a dead-end competitive inhibitor of substrate B will show noncompetitive inhibition relative to substrate A. In summary, then, for the rapid-equilibrium Random Bi Bi mechanism, a competitive inhibitor for either substrate will act as a noncompetitive inhibitor for the other substrate. These observations are consistent with the symmetry inherent in the random mechanism. Similar inhibition patterns are to be expected for the rapid-equilibrium Random Bi Uni mechanism.

Very few Random Bi Bi mechanisms are truly rapid-equilibrium random in both directions; however, this condition will be approximated in the "slow direction." When steady-state conditions prevail, that is, when the interconversion of the ternary complexes is not slow relative to other steps of the kinetic mechanism, it may be supposed that the initial rate plots in double-reciprocal form would not be linear. This is to be expected because of the second-degree substrate terms generated under steady-state conditions; however, Schwert (1954) suggested that the deviation from linearity might be too subtle to discern. A similar point was also made by Cleland and Wratten (1969) and Rudolph and Fromm (1971) concluded from computer simulations of the steady-state Random Bi Bi mechanism proposed for yeast hexokinase that the kinetics approximate the limiting equilibrium assumption. These workers also found that the competitive inhibition patterns proposed for the rapid-equilibrium case would be indistinguishable from the situation in which steady-state conditions prevail.

In the case of the Ordered Bi Bi mechanism, competitive dead-end inhibitors of the first substrate to add to the enzyme give inhibition patterns relative to the other substrate that are distinctively different from the pattern obtained when a competitive dead-end inhibitor of the second substrate is employed. It is this very point that permits the kineticist to make a choice between Random and Ordered Bi Bi mechanisms and allows one to identify the first and second substrates to add in the ordered mechanism.

In the case of the Ordered Bi Bi mechanism, substrate A adds only to free enyme (Fromm, 1979). By analogy, the competitive dead-end inhibitor should add only to this enzyme form. In addition, it is assumed that the conformation of the enzyme has been distorted enough by the inhibitor so as to preclude addition of substrate B to the enzyme-inhibitor complex.

The competitive dead-end inhibitor for substrate A may react as follows with the enzyme:

$$E + I = EI, \quad K_i = \frac{(E)(I)}{EI} \tag{17}$$

If the conservation of enzyme equation for the Ordered Bi Bi mechanism is modified to account for the additional complex, EI, the initial rate expression is

$$\frac{1}{v} = \frac{1}{V_1} + \frac{K_a}{V_1(\mathrm{A})}\left(1 + \frac{\mathrm{I}}{K_i}\right) + \frac{K_b}{V_1(\mathrm{B})} + \frac{K_{ia}K_b}{V_1(\mathrm{A})(\mathrm{B})}\left(1 + \frac{\mathrm{I}}{K_i}\right) \tag{18}$$

Equation (18) predicts that the competitive inhibitor for substrate A, the first substrate to add in the ordered mechanism, will be noncompetitive relative to substrate B. On the other hand, for this mechanism, a dead-end competitive inhibitor for substrate B would be expected not to react with free enzyme, but rather with the EA binary complex. This interaction may be described by the following relationship:

$$\mathrm{EA} + \mathrm{I} = \mathrm{EAI}, \quad K_i = \frac{(\mathrm{EA})(\mathrm{I})}{\mathrm{EAI}} \tag{19}$$

The kinetic expression obtained when this effect is included in the Ordered Bi Bi mechanism is

$$\frac{1}{v} = \frac{1}{V_1} + \frac{K_a}{V_1(\mathrm{A})} + \frac{K_b}{V_1(\mathrm{B})}\left(1 + \frac{\mathrm{I}}{K_i}\right) + \frac{K_{ia}K_b}{V_1(\mathrm{A})(\mathrm{B})} \tag{20}$$

It is quite clear that a dead-end competitive inhibitor for the second substrate will yield uncompetitive inhibition relative to substrate A. This unique inhibition pattern allows a distinction to be made between ordered and random bireactant kinetic mechanisms, and it permits determination of the substrate binding order in the former case. These points are summarized in Table I.

C. Three-Substrate Systems

The kinetic mechanisms for enzymes that utilize three substrates may be divided into Ping Pong and Sequential categories. It is then possible to make a choice of mechanism from among these terreactant systems using dead-end competitive inhibitors for the substrates. The kinetic mechanisms and expected inhibition patterns are illustrated in Table II.

III. Practical Aspects

Experimentally, it is very important that the fixed, or nonvaried, substrate be held at a subsaturating level, preferably in the region of the Michaelis constant. If, for example, when considering Eq. (11), the concentration of substrate A is

Table I
Use of Dead-End Competitive Inhibitors for Determining Bireactant Kinetic Mechanisms

Mechanism	Competitive inhibitor for substrate	1/A plot	1/B plot
Random Bi Bi, Ordered Bi Bi-subsite	A	C[b]	N[c]
mechanism[a] and Random Bi Uni	B	N	C
Ordered Bi Bi	A	C	N[d]
and Theorell-Chance	B	U[e]	C
Ping Pong Bi Bi	A	C	U
	B	U	C

[a]See Fromm (1979).
[b]C refers to a double-reciprocal plot that shows competitive inhibition.
[c]N refers to a double-reciprocal plot that shows noncompetitive inhibition.
[d]In the ordered mechanism convergence may be on, above, or below the abscissa; however, the point of intersection with the inhibitor must have the same ordinate as a family of curves in which the other substrate is substituted for the inhibitor.
[e]U refers to a double-reciprocal plot that shows uncompetitive inhibition.

held very high when B is the variable substrate, it is possible that the intercept increases to be expected in the presence of inhibitor may not be discernible, and the inhibition may appear to be competitive with respect to either substrate. It is important to note that, when replots of slopes and intercepts are made as a function of inhibitor concentration for the type of inhibition illustrated, the plots will be linear.

In studies in which dead-end competitive inhibitors are employed, it is often useful to evaluate the various inhibition constants. This can be done in a number of ways, and a few of the methods that may be used will be illustrated.

It is possible to evaluate either K_i or K_{ii} in Eqs. (11) and (16) from secondary plots of slopes and intercepts versus inhibitor concentration. It can be seen from Eq. (14) that a plot of slope versus I will give a replot in which the slope of the secondary ploy is

$$\text{Slope} = \frac{K_{ia}K_b}{V_1 K_1(\text{A})} \tag{21}$$

The value of K_i may also be evaluated by determining the intersection point of the secondary plot on the abscissa, that is, where slope = 0. In this case,

$$\text{I} = -K_i\left(1 + \frac{\text{A}}{K_{ia}}\right) \tag{22}$$

Table II
Competitive Inhibition Patterns for Various Three-Substrate Mechanisms[a]

Mechanism[b]	Competitive inhibitor for substrate	1/A plot	1/B plot	1/C plot
B 2a and 2b:	A	C[c]	N[d,e]	N[f]
Ordered Ter	B	U[g]	C	N[h]
Ter and Ordered Ter Bi	C	U	U	C
B 3: Random Ter	A	C	N	N
Ter and Ter Bi[m]	B	N	C	N
	C	N	N	C
B 4: Random AB[m]	A	C	N	C[i]
	B	N	C	C[j]
	C	U	U	C
B 5: Random BC[m]	A	C	N	N
	B	U	C	N
	C	U	N	C
B 6: Random AC[m]	A	C	N	N
	B	N	C	N
	C	N	N	C
D 1: Hexa Uni Ping Pong	A	C	U	U
	B	U	C	U
	C	U	U	C
D 2: Ordered Bi Uni	A	C	N[k]	U
Uni Bi Ping Pong	B	U	C	U
	C	U	U	C
D 3: Ordered Uni	A	C	U	U
Uni Bi Bi Ping Pong	B	U	C	N[l]
	C	U	U	C
D 4: Random Bi Uni	A	C	N	U
Bi Ping Pong[n]	B	N	C	U
	C	U	U	C
D 5 and D 6: Random	A	C	U	U
Uni Uni Bi Bi	B	U	C	N
Ping pong	C	U	N	C

[a]The various interactions of the competitive inhibitors are presented elsewhere along with the inhibited rate equations (Fromm, 1975d).

[b]The numbers refer to the mechanisms listed by Fromm (1979).

[c]C refers to a double-reciprocal plot that shows competitive inhibition.

[d]N refers to a double-reciprocal plot that shows noncompetitive inhibition.

[e]If EI reacts with B to form EIB, the plots would be nonlinear.

[f]If EIB reacts with C to form EIBC, the plots would be nonlinear.

[g]U refers to a double-reciprocal plot that shows uncompetitive inhibition.

[h]If EAI reacts with C to form EAIC, the plot would be nonlinear.

[i]If EIB reacts with C to form EIBC, the plot would be noncompetitive.

[j]If EIA reacts with C to form EIAC, the plot would be noncompetitive.

[k]If EI reacts with B to form EIB, the plot would be nonlinear.

[l]If FI reacts with C to form FIC, the plot would be nonlinear.

[m]It is assumed that all steps equilibrate rapidly relative to the breakdown of the productive quaternary complex to form products.

[n]It is assumed that all steps equilibrate rapidly relative to the breakdown of the productive ternary complex to form products.

The advantage of using Eq. (22) rather than Eq. (21) is that it is not necessary to evaluate V_1. Presumably, data for A and K_{ia} will be in hand.

The value for K_{ii} can be determined with a knowledge of A and K_a by evaluating the point of intersection on the abscissa of data from Eq. (13). In this case, where the intercept equals 0, in the replot,

$$\mathrm{I} = -K_{ii}\left(1 + \frac{\mathrm{A}}{K_a}\right) \tag{23}$$

Methods similar to those described for the Random Bi Bi mechanism can be used to determine the dissociation constants in the case of the Ordered Bi Bi mechanism. For example, K_i can be evaluated from either a slope or intercept versus inhibitor replot using Eq. (18), in which B is the variable substrate. It is of interest to note that the inhibition constant must be the same for this mechanism regardless of whether the determination is made from the slope or intercept. This may or may not be true for the Random pathway, depending on whether $K_i = K_{ii}$. It will be possible to determine K_i in Eq. (20) readily, using the methods already described.

IV. Example

The use of dead-end competitive inhibitors for choosing between the Random and Ordered Bi Bi mechanisms has been employed with many enzyme systems. The basic protocol involves segregation of mechanisms into either the Ping Pong or Sequential class from initial rate experiments. After the Sequential nature of the system has been established, the dead-end competitive inhibitors may be used to establish whether the kinetic mechanism is Random or Ordered.

Two examples can be used to illustrate this point. Fromm and Zewe (1962) reported that yeast hexokinase is Sequential when they demonstrated that double-reciprocal plots of 1/υ versus 1/[$MgATP^{2-}$] at different fixed concentrations of glucose converged to the left of the axis of ordinates. From the same data, they observed that, when 1/υ was plotted as a function of 1/[glucose] at different fixed concentrations of $MgATP^{2-}$, the resulting family of curves also converged to the left of the 1/υ axis. In addition, both sets of primary plots intersected on the abscissa. These investigators also demonstrated that AMP, a competitive dead-end inhibitor for MgATP2-, was a noncompetitive inhibitor with respect to glucose. From these experiments, it was concluded that the kinetic mechanism for yeast hexokinase was either Random Bi Bi or Ordered Bi Bi with $MgATP^{2-}$ as the initial substrate to add to the enzyme. If glucose were to add to hexokinase before $MgATP^{2-}$ in an Ordered Bi Bi mechanism, AMP inhibition would have been uncompetitive with respect to glucose.

The same investigators employed oxalate, a dead-end competitive inhibitor of L-lactate, to help establish the kinetic mechanism of the muscle lactate dehydrogenase reaction (Zewe and Fromm, 1965). They observed that oxalate was uncompetitive with respect to NAD^+ and concluded from these findings, and other studies, that the kinetic mechanism was Iso Ordered Bi Bi with the nucleotide substrates adding to the enzyme first.

Lindstad *et al.* (1992) studied the kinetic mechanism of sheep liver sorbitol dehydrogenase using a variety of kinetic techniques including the Haldane relationship (Haldane, 1930), the Dalziel relationship (Dalziel, 1957), isotope effects, and a competitive inhibitor for sorbitol, dithiothreitol. Both the Haldane relationship and the Dalziel relationship require highly accurate data in order to obtain precise kinetic parameters. In addition, as has been pointed out, no Dalziel relation exists in the case of the rapid-equilibrium Random Bi Bi mechanism, and, therefore, this mechanism cannot be excluded using Dalziel relationships (Dalziel, 1957).

The initial rate data obtained by Lindstad *et al.* (1992) apparently were not fit to an entire kinetic model; for example, in the case of Eq. (11), the data must be fit to the entire family of lines described by Eq. (11) rather than to a single line. This is obviously the case by Lindstad *et al.* (1992) and often leads to the calculation of erroneous kinetic parameters. It is precisely these parameters that are used in the Haldane and Dalziel relationships to make a choice of mechanism. Fortuitously, the data obtained by Lindstad *et al.* (1992) gave excellent agreement for the limiting Sequential Ordered Bi Bi case, the Theorell-Chance mechanism. In addition, the double-reciprocal plots in the forward direction intersect above the 1/substrate axis (Fig. 1) and below the 1/substrate axis in the reverse direction (Fig. 2). As pointed out by Lueck *et al.* (1973) these observations are also consistent with the Theorell-Chance mechanism.

A strong argument for the Ordered Bi Bi mechanism was obtained by Lindstad *et al.* (1992) using dithiothreitol as a competitive inhibitor for sorbitol.

Inhibition relative to NAD^+ was found to be uncompetitive. This unique inhibition pattern, namely, competitive-uncompetitive, shows that the kinetic mechanism is ordered and that sorbitol adds after NAD^+ to sorbitol dehydrogenase.

Dead-end competitive inhibitors have been used to study the kinetic mechanism of bireactant systems in which the inhibitor exhibits multiple binding to the enzyme (Tabatabai and Graves, 1978). Tabatabai and Graves (1978) established that phosphorylase kinase has a sequential mechanism. They observed that adenyl–5′-yl(β,γ-methylene) diphosphate was a linear competitive inhibitor for ATP and a linear noncompetitive inhgibitor relative to a tetradecapeptide, the other substrate. On the other hand, a heptapeptide was found to be nonlinear competitive (S-parabolic) with respect to the tetradecapeptide and nonlinear noncompetitive (I-parabolic, S-parabolic) relative to ATP. From these findings, it was concluded that the kinetic mechanism for phosphorylase kinase was Random Bi Bi. The rationalization of the nonlinear inhibition data was as follows. From the linear inhibition data provided by the ATP analog, the mechanism was either Random Bi Bi or Ordered Bi Bi with ATP as the first substrate to add to the enzyme. In the case of the ordered mechanism with multiple binding of the heptapeptide, if A and B are ATP and tetradecapeptide, respectively, then

$$\mathrm{EA} + \mathrm{I} = \mathrm{EAI}, K_{\mathrm{i}}; \quad \mathrm{EAI} + \mathrm{I} = \mathrm{EAI}_2, K_{\mathrm{ii}} \tag{24}$$

and thus,

$$\frac{1}{v} = \frac{1}{V_1} + \left[1 + \frac{K_{\mathrm{a}}}{\mathrm{A}} + \frac{K_{\mathrm{b}}}{\mathrm{B}}\left(1 + \frac{\mathrm{I}}{K_{\mathrm{I}}} + \frac{\mathrm{I}^2}{K_{\mathrm{i}}K_{\mathrm{ii}}}\right) + \frac{K_{\mathrm{ia}}K_{\mathrm{b}}}{(\mathrm{A})(\mathrm{B})}\right] \tag{25}$$

It was possible to exclude the ordered mechanism for phosphorylase kinase from consideration, because the inhibition pattern relative to ATP was not nonlinear uncompetitive (I-parabolic).

In the case of the Random Bi Bi mechanism, multiple binding by the inhibitor would be expected to occur as follows:

$$\mathrm{E} + \mathrm{I} = \mathrm{EI}, K_{\mathrm{i}}; \quad \mathrm{EI} + \mathrm{I} = \mathrm{EI}_2, K_{\mathrm{ii}}; \quad \mathrm{EA} + \mathrm{I} = \mathrm{EAI}, K_{\mathrm{iii}} \tag{26}$$

and thus,

$$\mathrm{EAI} + \mathrm{I} = \mathrm{EAI}_2, K_{\mathrm{iv}}; \quad \mathrm{EI} + \mathrm{A} = \mathrm{EAI}, K_{\mathrm{v}}; \quad \mathrm{EI}_2 + \mathrm{A} = \mathrm{EAI}_2, K_{\mathrm{vi}}$$

$$\frac{1}{v} = \frac{1}{V_1} + \left[1 + \frac{K_{\mathrm{a}}}{\mathrm{A}} + \frac{K_{\mathrm{b}}}{\mathrm{B}}\left(1 + \frac{\mathrm{I}}{K_{\mathrm{iii}}} + \frac{\mathrm{I}^2}{K_{\mathrm{iii}}K_{\mathrm{iv}}}\right) + \frac{K_{\mathrm{ia}}K_{\mathrm{b}}}{(\mathrm{A})(\mathrm{B})}\left(1 + \frac{\mathrm{I}}{K_{\mathrm{i}}} + \frac{\mathrm{I}^2}{K_{\mathrm{iii}}K_{\mathrm{iv}}}\right)\right] \tag{27}$$

The initial rate results were consistent with Eq. (27), and Tabatabai and Graves (1978) concluded that the kinetic mechanism for phosphorylase kinase is rapid-equilibrium Random Bi Bi.

Wolfenden (1972) and Lienhard (1973) have outlined how transition-state or multisubstrate (sometimes referred to as geometric) analogs may be used to provide information on the chemical events that occur during enzymatic catalysis. If it were possible to design an inactive compound that resembles the transition state, this analog would be expected to bind very tightly to the enzyme. In theory, a good deal of binding energy when enzyme and substrate interact is utilized to alter the conformation of both the enzyme and substrate so that proper geometric orientation for catalysis is provided between enzyme and substrate. Therefore, some of this binding energy is conserved because the multisubstrate analog more closely resembles the transition state than does the substrate. It certainly does not follow that, if the inhibition constant is lower than the dissociation constant for enzyme and substrate, the inhibitor is a transition-state analog. There are many examples in the literature where competitive inhibitors bind more strongly to enzymes than do substrates and yet are clearly not transition-state analogs.

These suggestions may be formalized by considering the following two reactions:

$$\mathrm{E} + \mathrm{A} = \mathrm{EA}', \quad K_1 = 10^{-7}\,\mathrm{M}$$
$$\mathrm{EA}' = \mathrm{EA}, \quad K_2 = 10^4$$

for the overall reaction

$$\mathrm{E} + \mathrm{A} = \mathrm{EA}, \quad K_{\mathrm{ia}} = 10^{-3}\,\mathrm{M}$$

The first reaction represents the thermodynamically favorable process of enzyme-substrate binding. The second reaction may be taken to be the enzyme-induced distortion of both the substrate and enzyme leading to the transition state.

From the perspective of kinetics, the use of geometric analogs may not be clear-cut in the case of multisubstrate systems. The analog should bind free enzyme, and, in theory, for a two-substrate system, the analog and substrate should not be able to bind to the enzyme simultaneously. This situation is different to check experimentally because it is not easy to determine whether, for example, 50% of the enzyme has substrate bound and the other 50% of the enzyme is associated with both substrate and analog or analog alone. For example, the kinetic pattern for a random mechanism where the analog can bind free enzyme and the EB complex will be the same as that for an ordered mechanism in which the analog binds exclusively to free enzyme.

It becomes fairly clear when considering the effect of multisubstrate or geometric analogs on the kinetics of bireactant enzyme systems that only in the case of the rapid-equilibrium Random Bi Bi mechanism may one obtain unequivocal results, and then only under certain circumstances. Consider, for example, the interaction of the analog and enzyme in an Ordered Bi Bi mechanism. The inhibitor will bind enzyme and will not permit the addition of the second substrate. Thus, the analog will act like any other competitive inhibitor of substrate A for this mechanism (see Table I); that is, it will be a noncompetitive inhibitor of substrate B. Therefore, the inhibition patterns provided by geometric analogs are identical to those to be expected for dead-end competitive inhibitors of the first substrate of the Ordered Bi Bi mechanism.

In the case of the Random Bi Bi mechanism, multisubstrate analogs may indeed give unique inhibition patterns, and this observation has been used to provide support for the Random Bi Bi mechanism for muscle adenylate kinase (Lienhard and Secemski, 1973; Purich and Fromm, 1972b). The multisubstrate analog used to test this theory with adenylate kinase was P^1,P^4-di(adenosine-5′) tetraphosphate (AP_4A) (Purich and Fromm, 1972b). It was subsequently shown that AP_5A binds even more strongly to the enzyme than does AP_4A (Lienhard and Secemski, 1973).

When considering the Random Bi Bi mechanism, the geometric analog should bind exclusively to free enzyme. This binding should effectively preclude binding of substrates A and B, and thus only the E term of the rate equation will be affected by the analog I. The rate expression is therefore

$$\frac{1}{v} = \frac{1}{V_1} + \frac{K_a}{V_1(\mathrm{A})} + \frac{K_b}{V_1(\mathrm{B})} + \frac{K_{ia}K_b}{V_1(\mathrm{A})(\mathrm{B})}\left(1 + \frac{\mathrm{I}}{K_i}\right) \tag{28}$$

where K_I is the dissociation constant of the enzyme-multisubstrate inhibitor complex.

Equation (28) indicates that the multisubstrate analog will function as a competitive inhibitor for both substrates. This effect is unique to the Random mechanism and suggests that the inhibitor bridges both substrate-binding pockets.

The physiologically important regulator fructose 2,6-bisphosphate has been used to establish the kinetic mechanism of the mammalian fructose-1,6-bisphosphatase reaction (Ganson and Fromm, 1982). It was well established that the inhibitor is competitive relative to the substrate, fructose 1,6-bisphosphate (Van Schaftigen and Hers, 1981). When fructose 2,6-bisphosphate was used as an inhibitor for the reverse reaction, it was found to be competitive relative to both of the reverse reaction substrates, orthophosphate and fructose 6-phosphate (Ganson and Fromm, 1982). These observations are fully consistent with Eq. (28) and the Random Bi Bi mechanism previously proposed for the phosphatase (Stone and Fromm, 1980).

If the inhibitor binds only at one-substrate site in either the Random or the Ordered Bi Bi cases, or if, for the latter mechanism, the compound does resemble the transition state and substrate B does not add, inhibition patterns will be competitive and noncompetitive relative to the two substrates. Thus, it will not be possible to differentiate between the two mechanisms based on these inhibition patterns, nor will be possible to determine whether the inhibitor is really a transition-state analog in the Ordered mechanism. In the case of the Random pathway, the enzyme-inhibitor complex will permit binding of one-substrate and the enzyme-substrate complex will allow analog to bind. In summary, then, only the unique inhibition pattern illustrated by Eq. (28) allows one to use multisubstrate analogs to differentiate unambiguously between kinetic mechanisms.

Frieden (1976) presented a rapid-equilibrium Ordered Bi Bi subsite mechanism (Fromm, 1979) that cannot be differentiated from the rapid-equilibrium Random Bi Bi mechanism from studies of product inhibition, substrate analog inhibition, isotope exchange at equilibrium, and the Haldane relationship. Multisubstrate analogs can be used to make this differentiation, along with other kinetic procedures (Fromm, 1976).

In the case of Frieden's mechanism (Frieden, 1976), the transition-state analog should bind at both the active site and the subsite for B, that is, the EB site. This will result in modification of a number of terms in the rate equation. The dead-end complexes to be expected are EI_1, EI_2, EI_1I_2, and EAI_2, where I_1 and I_2 respresent binding of the transition-state analog at the active and subsites, respectively.

The resulting rate expression is described by Eq. (29), where K_i, K_{ii}, K_{iii}, and K_{iv} represent dissociation constants for EI_1, EI_2, EI_1I_2, and EAI_2 complexes, respectively:

$$\frac{1}{v} = \frac{1}{V_1}\left[1 + \frac{K_{ia}K_b}{K_{ib}(A)} + \frac{K_b}{(B)}\left(1 + \frac{I}{K_{iv}}\right) + \frac{K_{ia}K_b}{(A)(B)}\left(1 + \frac{I}{K_i} + \frac{I}{K_{ii}} + \frac{I^2}{K_iK_{iii}}\right)\right] \quad (29)$$

Equation (29) indicates that the multisubstrate analog acts like a competitive inhibitor relative to substrate B and as a noncompetitive inhibitor with respect to substrate A. In addition, replots of the slopes from the primary plots versus inhibitor concentration will yield a concave-up parabola. This analysis indicates that a differentiation can be made between the Ordered Bi Bi subsite and the Random Bi Bi mechanisms.

V. Limitations

When considering competitive substrate inhibitors, the possibility is automatically excluded that the inhibitor may bind to an enzyme-product complex. In the case of the rapid-equilibrium Random Bi Bi mechanism, a competitive inhibitor for substrate B could in theory bind the EQ complex; however, this complex occurs after the rate-limiting step and is not part of the kinetic equation. Similarly, although this binary complex is kinetically important in the Ordered Bi Bi case; if an EQI complex did form, inhibition would be noncompetitive rather than competitive relative to substrate B. Under these conditions, the approach would not be a viable technique, and another inhibitor should be sought.

Huang (1977) suggested that it is not possible to differentiate between Ordered and Random Bi Bi mechanisms if in the Ordered case an inhibitor for B binds free enzyme and the EA complex to form complexes EI and EAI. This will be true only when all steps in the kinetic mechanism equilibrate rapidly relative to the breakdown of the productive ternary complex to form products. This mechanism is easily separated from those under discussion here by its unique double-reciprocal plot patterns in the absence of inhibitors (Fromm, 1975e).

It should be pointed out that the dead-end competitive inhibitors cannot be used to differentiate between normal and Iso mechanisms. Nor can they be used to make a choice as to whether ternary complexes are kinetically important in Ordered mechanisms.

The competitive substrate inhibitors cited above have been referred to as "dead-end" inhibits (Fromm and Zewe, 1962). The question arises as to what happens if the inhibitions are not of the dead-end type, that is, if the enzyme-inhibitor complexes of the ordered mechanism act in a manner similar to those analogous complexes in the random mechanism. This possibility was considered by Hanson and Fromm (1965). If in the ordered mechanism the EI complex permits substrate

B to add, the additional reaction would be

$$\mathrm{EI} + \mathrm{B} = \mathrm{EIB}, K_{\mathrm{ii}} \tag{30}$$

and Eq.(18) would be modified as shown in Eq. (31):

$$\frac{1}{v} = \frac{1}{V_1} + \frac{K_a}{V_1(\mathrm{A})}\left[1 + \frac{\mathrm{I}}{K_i} + \frac{(\mathrm{I})(\mathrm{B})}{K_i K_{ii}} + \frac{K_{ia}K_b(\mathrm{I})}{K_a K_i K_{ii}}\right] + \frac{K_b}{V_1(\mathrm{B})} + \frac{K_{ia}K_b}{V_1(\mathrm{A})(\mathrm{B})}\left(1 + \frac{\mathrm{I}}{K_i}\right) \tag{31}$$

Inhibition relative to substrate A would of course be competitive; however, a $1/v$ versus $1/\mathrm{B}$ plot would show concave-up hyperbolic inhibition. This effect is obviously readily distinguishable from the case in which a dead-end binary complex is formed.

VI. Concluding Remarks

The use of competitive substrate inhibitors for studying kinetic mechanism has received wide attention. If one is interested in determining the kinetic mechanism for an enzyme, this protocol should be used immediately after a determination is made as to whether the initial rate kinetics are Sequential or Ping Pong. Theoretically, using competitive substrate inhibitors to study the sequence of enzyme and substrate binding is far less ambiguous than either substrate or product inhibition or isotope exchange experiments. The protocol is no more complicated than initial velocity experiments in which inhibitors are not used.

The procedure does suffer from certain limitations, as do most kinetic methods. For example, it will not permit one to differentiate between Iso and conventional mechanisms, nor can it be used to choose between Theorell-Change mechanisms and analogous mechanisms involving long-lived central complexes. However, on balance, using dead-end competitive inhibitors is a powerful tool for studying enzyme kinetics.

Acknowledgments

This work was supported by research grants from the National Institutes of Health (NS 10546) and the National Science Foundation (MCD-9218763).

References

Cleland, W. W. (1963). *Biochim. Biophys. Acta* **67,** 104.

Cleland, W. W., and Wratten, C. C. (1969). "The Mechanism of Action of Dehydrogenases," p. 103. University Press of Kentucky, Lexington.

Dalziel, K. (1957). *Acta Chem. Scand.* **11,** 1706.
Frieden, S. C. (1976). *Biochem. Biophys. Res. Commun.* **68,** 914.
Fromm, H. J. (1975a). "Initial Rate Enzyme Kinetics," p. 86. Springer-Verlag, Berlin and New York.
Fromm, H. J. (1975b). "Initial Rate Enzyme Kinetics," p. 99. Springer-Verlag, Berlin and New York.
Fromm, H. J. (1975c). "Initial Rate Enzyme Kinetics," p. 63. Springer-Verlag, Berlin and New York.
Fromm, H. J. (1975d). "Initial Rate Enzyme Kinetics," p. 102. Springer-Verlag, Berlin and New York.
Fromm, H. J. (1975e). "Initial Rate Enzyme Kinetics," p. 37. Springer-Verlag, Berlin and New York.
Fromm, H. J. (1976). *Biochem. Biophys. Res. Commun.* **72,** 55.
Fromm, H. J. (1979). *Methods Enzymol.* **63,** 42.
Fromm, H. J. (1979). *Methods Enzymol.* **63,** 84.
Fromm, H. J., and Nelson, D. (1962). *J. Biol. Chem.* **237,** 215.
Fromm, H. J., and Zewe, V. (1962). *J. Biol. Chem.* **237,** 3027.
Ganson, N. J., and Fromm, H. J. (1982). *Biochem. Biophys. Res. Commun.* **108,** 233.
Haldane, J. B. S. (1930). "Enzymes." Longmans, Green.
Hanson, T. L., and Fromm, H. J. (1965). *J. Biol. Chem.* **240,** 4133.
Huang, C. Y. (1977). *Arch. Biochem. Biochem. Biophys.* **184,** 488.
Huang, C. Y. (1979). *Methods Enzymol.* **63,** 54.
Lienhard, G. E. (1973). *Science* **180,** 149.
Lienhard, G. E., and Secemski, I. I. (1973). *J. Biol. Chem.* **248,** 1121.
Lindstad, R. I., Hermansen, L. F., and KcKinley-McKee, J. S. (1992). *Eur. J. Biochem.* **210,** 641.
Lueck, J. D., Ellison, W. R., and Fromm, H. J. (1973). *FEBS Lett.* **30,** 321.
Purich, D. L., and Fromm, H. J. (1972a). *Biochim. Biophys. Acta* **268,** 1.
Purich, D. L., and Fromm, H. H. (1972b). *Biochim. Biophys. Acta* **276,** 563.
Rudolph, F. B., and Fromm, H. J. (1971). *J. Biol. Chem.* **246,** 6611.
Schwert, G. W. (1954). *Fed. Proc. Fed. Am. Soc. Exp. Biol.* **13,** 971Abstr..
Stone, S. R., and Fromm, H. J. (1980). *J. Biol. Chem.* **255,** 3454.
Tabatabai, L., and Graves, D. J. (1978). *J. Biol. Chem.* **253,** 2196.
Van Schaftigen, E., and Hers, H. G. (1981). *Proc. Natl. Acad. Sci. USA* **78,** 2861.
Wolfenden, R. (1972). *Acc. Chem. Res.* **5,** 10.
Zewe, V., and Fromm, H. J. (1965). *Biochemistry* **4,** 782.

CHAPTER 11

Application of Affinity Labeling for Studying Structure and Function of Enzymes

Bryce V. Plapp

Department of Biochemistry
Carver College of Medicine
The University of Iowa
Iowa City, IA 52242-1109, USA

DOI: 10.1016/B978-0-12-378608-1.00011-6

I. Update

Affinity-labeling reagents resemble enzyme substrates and thereby react more specifically and rapidly with their targets. They are useful for identifying amino acid residues that are important for catalytic activity and for studying enzyme mechanisms. In recent years, the application of such reagents has been complemented by site-directed mutagenesis that is used to substitute specific amino acid residues and is especially useful for enzymes for which three-dimensional structures are known. Chemical modification remains useful for exploring enzymes for which specific expression systems are not available and structures are not known. Moreover, according to a recent survey, 24 of the 71 enzymes that are targets for therapeutic drugs are covalently modified by the drugs (Robertson, 2005). These mechanism-based inactivators can be more specific and longer lasting than reversible inhibitors. Examples of promising drug leads are the carbamates that inactivate fatty acid amide hydrolase (Ahn *et al.*, 2007, 2008). The continued development of such drugs may bring to fruition the vision of Baker (1967) for the rational design of efficacious drugs. The principles and procedures developed in the accompanying chapter and in later chapters (Silverman, 1995; Szedlacsek and Duggleby, 1995) are relevant for evaluating potential drugs and for investigating enzymes.

II. Introduction

Affinity labeling is a popular method that may be used to determine topography of active sites, to elucidate enzyme mechanisms, and to provide new, rationally designed drugs. But despite the promise of the approach and the investment of much empirical work, our great expectations have not yet been fully realized. This is due, in part, to our lack of knowledge about the specific enzymes we wish to modify and to our ignorance of fundamental principles that explain the reactions of proteins with small molecules. But it is also partly due to the limited goals of some researchers who only wish to prove that they have prepared an active-site-directed reagent (or to invent a new type of inactivation) when the ultimate goal is to learn something about the enzyme. As William H. Stein would ask when discussing research results presented to him, "What did you learn about Nature?"

There have been many reviews on affinity labeling, including a volume in this series (Jacoby and Wilchek, 1977). Here, we will discuss some of the uses of active-site-directed reagents and consider the design and evaluation of such reagents for the benefit of those investigators who may wish to use this approach in a critical manner. Since "suicide" or "k_{cat}" inactivators have been recently reviewed, we will not discuss them even though study of their chemical transformations also leads to mechanistic information (Seiler *et al.*, 1978; Walsh, 1977). "Photoaffinity" labeling has also been reviewed (Bayley and Knowles, 1977; Benisek *et al.*, 1982; Chowdhry and Westheimer, 1979).

III. Uses of Affinity Labeling

A. Identification of Essential Amino Acid Residues

In the characterization of an enzyme, it is interesting to try to identify amino acid residues that are at the active site and possibly involved in catalytic activity. An investigator may attempt to design and synthesize a new active-site-directed reagent. In my judgment, a more rewarding approach is to evaluate simple chemical reagents first. This avoids the delay of perhaps 1 man-year for the synthesis of a well-designed active-site-directed reagent and the disappointment if the compound is inactive. Furthermore, the methodology for using simple reagents—for example, bromoacetic acid—is well established (Means and Feeney, 1971; Hirs, 1967). Finally, such studies are useful for determining what kinds of functional groups to place on an active-site-directed reagent and are essential for determining whether the reagent reacts in a facilitated manner, as it should.

Of course, it usually turns out that a simple reagent modifies more than one amino acid residue of the enzyme during complete inactivation, and the investigator may consider several possible explanations: (a) only one residue is "essential" for catalysis, (b) modification of several residues in the active site interferes with substrate binding, or (c) modification of several residues denatures the enzyme. (In this context, an "essential" residue is one that cannot be modified without "completely" inactivating the enzyme—for example, to $<1\%$ residual activity. This is probably still more activity than the enzyme would have if a residue directly involved in catalysis were modified. Thus, alcohol dehydrogenase carboxymethylated on Cys-46 with about 2% residual activity (Reynolds and McKinley-McKee, 1975), or liver alcohol dehydrogenase with a phosphopyridoxyl group on Lys-228 and 10% activity (Sogin and Plapp, 1975), or ribonuclease carboxymethylated on His-12 with about 5% activity (Machuga and Klapper, 1977) are partially active and apparently not modified on "essential" residues, whereas chymotrypsin methylated on His-57 with about 0.05% activity (Henderson, 1971) may be said to be modified on an essential residue.) In order to distinguish among the possibilities, "differential labeling" (Phillips, 1977) may be used, where the active site is first protected with the bound ligand while the enzyme is modified with one reagent, and then the ligand is removed and the enzyme is modified at the active site with another reagent.

In favorable cases, a simple reagent may modify just one amino acid residue, if the environment of the residue makes it especially reactive or if the "simple" reagent binds and reacts like an active-site-directed reagent. For instance, the reactions of liver alcohol dehydrogenase (Reynolds *et al.*, 1970) or pancreatic ribonuclease (Plapp, 1973) with iodoacetate seem to fit the criteria normally required for affinity labeling, apparently because of ionic interactions with the enzyme.

Whenever a reagent, by whatever experimental method, is found to give essentially complete inactivation with nearly stoichiometric modification, the modified amino acid residue should be identified by protein sequence analysis. Amino acid analysis is not sufficient to conclude that a particular amino acid is involved in binding some part of a ligand. When the three-dimensional structure of the enzyme is determined, assignment of a function and comparison of results from solution and crystal studies become possible if the amino acid is firmly identified. Without the identification, one can only speculate.

Active-site-directed reagents are useful in identifying amino acid residues in active sites, and many examples are compiled in Table I.

B. Topography of Active Site

With respect to the goal of "mapping" an active site, it is clear that X-ray crystallography provides more detailed (and higher resolution) three-dimensional information about more amino acid residues than affinity labeling ever will. The problem with an active-site-directed reagent is that one does not know for sure how it binds to the enzyme while it is modifying a residue; affinity labeling is inherently a "low-resolution" approach. Thus, an enzymologist may find that trying to crystallize an enzyme will be more rewarding than trying to make specific reagents. Although it has been argued that the structure in the crystal is "static" and not necessarily the same as the "dynamic" structure in solution, I am not aware of any evidence from chemical modification studies that establishes a different structure in solution. Of course enzyme structures are dynamic, but other tools, such as NMR, are more appropriate for describing such dynamics.

Nevertheless, when the three-dimensional structure of the enzyme is not known, mapping of active sites is useful. It is interesting, in fact, that in 1963 it was possible to describe the structure of the active site of chymotrypsin by use of three different affinity-labeling agents that placed functional groups in different regions of the substrate binding pocket (Baker, 1967). As shown in Fig. 1, Met-192, Ser-195, and His-57 could be identified. A little model building produced a picture of the active site that fits the three-dimensional structure as determined later by X-ray crystallography of chymotrypsin and some of its inhibited forms (Blow, 1971; Segal *et al.*, 1971).

In contrast, the careful studies of Cuatrecasas and coworkers on the affinity labeling of staphylococcal nuclease gave a picture of the active site that was quite different from the picture determined independently by X-ray crystallography (Fig. 2). Although the affinity labeling allowed the identification of several amino acid residues that are near the active site, only one of these (Tyr-85) is in contact with the parent inhibitor, deoxythymidine 3′, 5′-diphosphate, as determined in the crystal. Interestingly, two residues in the active site (Lys-84 and Tyr-113) did not appear to react. Furthermore, two labeled residues (Lys-24 and Trp-140) were so far from the "active site" that it was possible that the reagents reacted in a subsite. More importantly, it was concluded from the labeling studies that Tyr-85 was near the 5′-phosphate and possibly involved in catalytic action. But the crystallography places that residue near the 3′-phosphate and Tyr-113 near the probable cleavage site on the 5′-phosphate. It may be concluded that the flexibility of the reagents or the protein led to many of the apparent

Table I
Identification of Amino Acid Residues in Enzymes Modified by Affinity Labeling

1. Chymotrypsin, bovine pancreas
   a. Diisopropylphosphorofluoridate: Ser-195 (Oosterbaan *et al.*, 1958)
      pH 7, 25 °C, 45 M^{-1} s^{-1} (Fahrney and Gold, 1963)
   b. Tosyl-L-phenylalanylchloromethane: His-57 (Ong *et al.*, 1965)
      pH 7, 25 °C, 7.7 M^{-1} s^{-1} (Shaw and Ruscica, 1971)
   c. *p*-Nitrophenylbromoacetyl-α-aminoisobutyrate: Met-192 (Lawson and Schramm, 1965)
   d. Photolysis of diazoacetyl-Ser-195 led to *O*-carboxymethyl-Tyr-146 of another molecule of enzyme (Hexter and Westheimer, 1971)
2. Trypsin, bovine pancreas
   a. α-*N*-Tosyl-L-lysylchloromethane: His-46 (Shaw and Glover, 1970)
      pH 7, 25 °C $K_I = 0.21$ mM, $k_3 = 2.6 \times 10^{-3}$ s^{-1}
   b. *p*-Guanidinophenacyl bromide: Ser-183 (Schroeder and Shaw, 1971)
      pH 7.1, 25 °C, $K_I = 1.63$ mM, $k_3 = 1.3 \times 10^{-3}$ s^{-1}
3. Trypsinogen, bovine pancreas
   Diisopropylphosphorofluoridate: Ser-183 (Morgan *et al.*, 1972)
   pH 7, 25 °C, $k_3/K_I = 6.8 \times 10^{-4}$ M^{-1} s^{-1}
   (Note: trypsin is inactivated with $k = 5.0$ M^{-1} s^{-1})
4. Thrombin
   α-*N*-Tosyl-L-lysylchloromethane: His-43 (Glover and Shaw, 1971)
   pH 7.5, $K_I = 2.3$ mM, $k_3 = 9 \times 10^{-4}$ s^{-1}
5. Thermolysin
   N-Chloroacetyl-d,L-*N*-hydroxyleucine methyl ester: Glu-143 (Rasnick and Powers, 1978)
   pH 7.2, 25 °C, $K_I = 7.5$ mM, $k_3 = 7.5 \times 10^{-3}$ s^{-1}
6. Carboxypeptidase-A_{γ}^{Leu}, bovine pancreas
   N-Bromoacetyl-*N*-methyl-l-phenylalanine: Glu-270 (Hass and Neurath, 1971)
   (Side reactions: α-*N*-Asp-1, His-13)
   pH 7.5, 25 °C, $K_I = 4.8$ mM, $k_3 = 3.2 \times 10^{-3}$ s^{-1}
7. Carboxypeptidase B, bovine pancreas
   a. α-*N*-Bromoacetyl-d-arginine
      Thr-Phe-Glu*-Leu-Arg-Asp-Lys-Gly-Arg-Tyr-Gly-Phe (Kimmel and Plummer, 1972)
      (Homologous to Glu-270 in carboxypeptidase A)
   b. 4-(Bromoacetamido)butylguanidine
      Thr-Ile-Tyr*-Pro-Ala-Ser-Gly-Gly-Ser-Asp-Asp-Trp (Plummer, 1969)
      (Homologous to Tyr-248 in carboxypeptidase A)
      pH 8, Inactivation 15 times faster than with bromoacetamide
8. d-Alanine carboxypeptidase, *Bacillus subtilis*
   Penicillin G: Ser-36 (Leu-Pro-Ile-Ala-Ser*-Met) (Waxman and Strominger, 1980)
   (Homologous to Ser-44 in β-lactamase)
   pH 7.5, 25 °C, K_I large, $k_3/K_I = 440$ M^{-1} s^{-1} (Blumberg and Strominger, 1971)
9. β-Lactamase, *B. cereus*
   6β-Bromopenicillanic acid—"suicide substrate"
   Phe-Ala-Phe-Ala-Ser*-Thr-Tyr-Lys (Ser-44 = Ser-70) (Cohen and Pratt, 1980; Knott-Hunziker *et al.*,1979; Loosemore *et al.*, 1980)
   pH 7.5, 30 °C, $K_I > 2$ mM, $k_3/K_I = 1.8 \times 10^4$ M^{-1} s^{-1}
10. Staphylococcal nuclease (Cuatrecasas and Wilchek, 1977)
    Deoxythymidine derivatives, BrAcNH = bromoacetamido
    Ph = *p*-phenyl, P = phosphate or phosphodiester

I. 3′-BrAcNHPhP, 5′-P	1 mol/mol incorporated
$K_i \approx 1$ μM, $k_3/K_I = 0.3$ M^{-1} s^{-1}	0.36 Lys-48, 0.40 Lys-49
pH 9.4	0.15 Tyr-115

(*continued*)

Table I (*continued*)

II. 5′-BrAcNHPhP	2 mol/mol incorporated
$K_i \approx 2$ mM	0.8 Tyr-85, but enzyme active with higher K_m
III. 3′-BrAcNHPhP	3 mol/mol incorporated including Lys-24
$K_i \approx 20$ mM, bound to subsite?	(or Met-26?)
IV. 3′-N_2PhP, 5′-P	0.9 Tyr-115
V. 5′-N_2PhP	Tyr-85
VI. 3′-N_2PhP	0.5 His-46
bound to subsite	0.5 Trp-140 (not in active site)

11. Ribonuclease, bovine pancreas
   a. 5′-(*p*-Diazophenylphosphoryl)uridine 2′(3′)-phosphate: Tyr-73 (Gorecki *et al.*, 1971)
   Not at active site
   b. 2′(3′)-*O*-Bromoacetyluridine: His-12 (Pincus *et al.*, 1975)
   pH 5.5, 30 °C, $K_I = 87$ mM, $k_3 = 3.5 \times 10^{-3}$ s^{-1}
12. Carbonic anhydrase, erythrocyte (Cybulsky *et al.*, 1973)
   Bromoacetazolamide: human type C, His-64
   Bovine type B, His-64
   N-Bromoacetylacetazolamide: human type B, His-67
13. Lysozyme, chicken egg white
   2′,3′-Epoxypropyl *β*-glycoside of di(*N*-acetyl)-d-glucosamine: Asp-52 (Eshdat *et al.*, 1973)
14. Sucrase-isomaltase, rabbit small intestine (Braun *et al.*, 1977; Quaroni and Semenza, 1976)
   1-D-1,2-anhydro-*myo*-inositol (active isomer of conduritol-*β*-epoxide)
   Sucrase: Ile-Asp*-Met-Asn-Gln-Pro-Asn-Ser-Ser
   pH 6.8, 37 °C, 0.027 M^{-1} s^{-1}
   Isomaltase: Ile-Asp*-Met
   pH 6.8, 37 °C, 0.21 M^{-1} s^{-1}
15. *β*-Galactosidase, *Escherichia coli*
   β-D-Galactopyranosylmethyl-*p*-nitrophenyltriazene: Met-500 (Fowler *et al.*, 1978; Sinnott and Smith, 1978)
   pH 7, 25 °C, $K_I = 0.48$ mM, $k_3 = 9.8 \times 10^{-3}$ s^{-1}
16. Galactosyltransferase, bovine colostrums
   Uridine 5′-diphosphate cleaved by periodate and irreversibly attached to enzyme with borohydride reduction
   Ser-Gly-Lys* (Powell and Brew, 1976)
17. Triose phosphate isomerase, rabbit muscle
   3-Chloroacetolphosphate: Glu-165 (Hartman, 1971)
   pH 6.5, 2 °C, 2.3 × 10^3 M^{-1} s^{-1}
18. Glucosephosphate isomerase, rabbit muscle
   N-Bromoacetylethanolamine phosphate
   Val-Leu-His*-Ala-Glu-Asn-Val-Asp (Gibson *et al.*, 1980)
   pH 8, 37 °C, $K_I = 0.056$ mM, $k_3 = 1.8 \times 10^{-3}$ s^{-1}
19. Phosphoglycerate mutase, rabbit muscle
   N-Bromoacetylethanolamine phosphate
   (Trp,Lys,Cys*,Asp,Ser,Glu_2,Gly,Ala,Leu,Phe_2) (Hartman and Norton, 1976)
   pH 7, room temperature, $K_I = 0.32$ mM, $k_3 = 6.8 \times 10^{-3}$ s^{-1}
20. Ribulosebisphosphate carboxylase, spinach
   N-Bromoacetylethanolamine phosphate (Schloss *et al.*, 1978)
   Tyr-Gly-Arg-Pro-Leu-Gly-Cys*-Thr-Ile-Lys*-Pro-Lys
   Trp-Ser-Pro-Glu-Leu-Ala-Ala-Ala-Cys*-Glu-Val-Trp-Lys
   pH 8, 30 °C, +5 mM Mg^{2+}, $K_I = 3.0$ mM, $k_3 = 4.8 \times 10^{-4}$ s^{-1} Lys
   pH 8, 30 °C, no Mg^{2+}, $K_I = 0.8$ mM, $k_2 = 8.4 \times 10^{-5}$ s^{-1} 2 Cys
21. Aldolase, rabbit muscle
   N-Bromoacetylethanolamine phosphate (Hartman and Brown, 1976)

(*continued*)

Table I (*continued*)

pH 8.5, room temperature, $K_I = 0.76$ mM, $k_3 = 1.4 \times 10^{-4}$ s^{-1}, Lys-146
pH 6.5, room temperature, $K_I = 0.87$ mM, $k_3 = 3.3 \times 10^{-5}$ s^{-1}, His-359

22. 2-Keto-3-deoxygluconate-6-phosphate aldolase, *Pseudomonas putida*
   Bromopyruvate: Glu-56 (Meloche *et al.*, 1978; Suzuki and Wood, 1980)
   pH 6, 24.5 °C, $K_I = 1$ mM, $k_3 = 0.011$ s^{-1}
23. Aspartate aminotransferase, pig heart cytoplasmic and mitochondrial
   β-Chloro-L-alanine (suicide inactivator): Lys-258 (Morino *et al.*, 1974)
   (Same lysine as binds pyridoxal phosphate)
   pH 7.4, 18 °C, 3.2 M formate, $K_I = 0.2$ M, $k_3 = 0.33$ s^{-1}
24. Aspartate aminotransferase, pig heart cytoplasmic only
   4′-*N*-(2,4-dinitro-5-fluorophenyl)pyridoxamine-5′-phosphate: Lys-258 (Riva *et al.*, 1980)
25. Tryptophan synthase, β2 subunit, *E. coli*
   Bromoacetylpyridoxamine phosphate: Cys-230 (Miles and Higgins, 1980; Higgins *et al.*, 1980)
   (Or Cys-62 if Cys-230 is first blocked)
   pH 8, 37 °C, $K_I = 0.6$ mM, $k_3 = 0.0023$ s^{-1}
26. Formylglycinamide ribonucleotide amidotransferase, *Salmonella typhimurium*
   l-azaserine: Ala-Leu-Gly-Val-Cys* (French *et al.*, 1963)
27. Anthranilate synthetase component II, *Serratia marcescens*
   l-(αS,5S)-αAmino-3-chloro-4,5-dihydro-5-isoxazoleacetic acid: Cys-83 (Tso *et al.*, 1980)
   pH 5.3, room temperature, 0.5 mM chorismate, $K_I = 0.14$ mM, $k_3 = 0.052$ s^{-1}
28. Thymidylate synthetase, *Lactobacillus casei*
   5-Fluorodeoxyuridine 5′-phosphate
   Ala-Leu-Pro-Pro-Cys*-His-Thr-Leu-Tyr (Bellisario *et al.*, 1976)
   Labeling by iodoacetate in presence of 5, 10-methylene tetrahydrofolate and FdUMP protects CySH
29. L-Isoleucine tRNA ligase, *E. coli*
   l-Isoleucylbromomethane
   Ile-Glu-Ser-Met-Val-Ala-Asp-Arg-Pro-Asn-Trp-Cys*-Ile-Ser-Arg (Rainey *et al.*, 1976)
   pH 7.5, 27 °C, $K_d = 0.7$ mM, $k_3 = 1.8 \times 10^{-3}$ s^{-1}
30. Phosphofructokinase, sheep heart
   p-Fluorosulfonylbenzoyl-5′-adenosine
   Asn-Phe-Ala-Thr-Lys*-Met-Gly-Ala-Lys, allosteric site (Weng *et al.*, 1980)
31. Phosphofructokinase, rabbit muscle
   N^6-(6-Bromoacetamidohexyl)-AMP-PCP: Cys*-Lys-Asp-Phe-Arg (Nagata *et al.*, 1979), ATP inhibitory site
32. Mitochondrial F_1-ATPase, bovine
   p-Fluorosulfonylbenzoyl-5′-adenosine: β subunit
   Ile-Met-Asp-Pro-Asn-Ile-Val-Gly-Ser-Glu-His-Tyr*-Asp-Val-Ala-Arg (Esch and Allison, 1978, 1979)
   pH 7, 20% glycerol, $k_3/[I] = 0.09$ M^{-1} s^{-1}
33. cAMP-dependent protein kinase II, porcine skeletal muscle
   p-Fluorosulfonylbenzoyl-5′-adenosine: catalytic subunit
   Glx-Ile(Asp$_2$, Met, Thr,Ser,Glu,Gly,Leu,Phe,His,Lys*) (Zoller and Taylor, 1979)
   pH 7, 37 °C, 10% glycerol, $K_I = 57$ μM, $k_3 = 6 \times 10^{-4}$ s^{-1}
34. cAMP-dependent protein kinase II, porcine heart
   8-Azidoadenosine 3′:5′-monophosphate: regulatory subunit
   Lys-Arg-Asn-Ile-Ser-His-Tyr*-Glu-Glu-Gln-Leu-Val-Lys-Met (Kerlavage and Taylor, 1980)
   Photoaffinity labeling, K_d about 70 nM, 0.5 mol/mol
35. Δ^5-3-Ketosteroid isomerase, *Pseudomonas testosterone*
   3-Oxo-4-estren-17β-yl acetate
   Photoinactivation: Asp-38 → Ala-38 (Ogez *et al.*, 1977)

(*continued*)

Table I (*continued*)

36. Glutamate dehydrogenase, bovine liver
    Iodoacetyldiethylstilbestrol: Cys-89, allosteric site (Michel *et al.*, 1978)
        pH 7.6, 37 °C, 1 mM NADH, $K_I = 10\ \mu M$, $k_3 = 0.0021\ s^{-1}$
37. Aspartate-β-semialdehyde dehydrogenase
    l-2-Amino-4-oxo-5-chloropentanoic acid: Phe-Val-Gly-Gly-Asp-His*-Thr-Val-Ser (Biellmann *et al.*, 1980)
        pH 7.2, 0 °C, $K_I = 0.2$ mM, $k_3 = 5.1 \times 10^{-3}\ s^{-1}$
38. Alcohol dehydrogenase
    a. Nicotinamide-5-bromoacetyl-4-methylimidazole dinucleotide
        Horse liver: Cys-174 (Zn ligand in nicotinamide site!) (Woenckhaus and Jeck, 1971)
            pH 6.5, 37 °C, $K_I = 0.7$ mM, $k_3 = 1.6 \times 10^{-3}\ s^{-1}$
        Yeast: Cys-43 (Zn ligand homologous to Cys-46 in horse enzyme, in nicotinamide site) (Jörnvall *et al.*, 1975)
    b. 4-(3-Bromoacetylpyridino)butyl diphosphoadenosine (Woenckhaus *et al.*, 1979)
        Horse liver: Cys-46
            pH 6.2, 25 °C, $K_I = 1$ mM, $k_3 = 1.7 \times 10^{-3}\ s^{-1}$
        Yeast: Cys-43
            pH 6.6, 25 °C, $K_I = 4$ mM, $k_3 = 2.8 \times 10^{-2}\ s^{-1}$

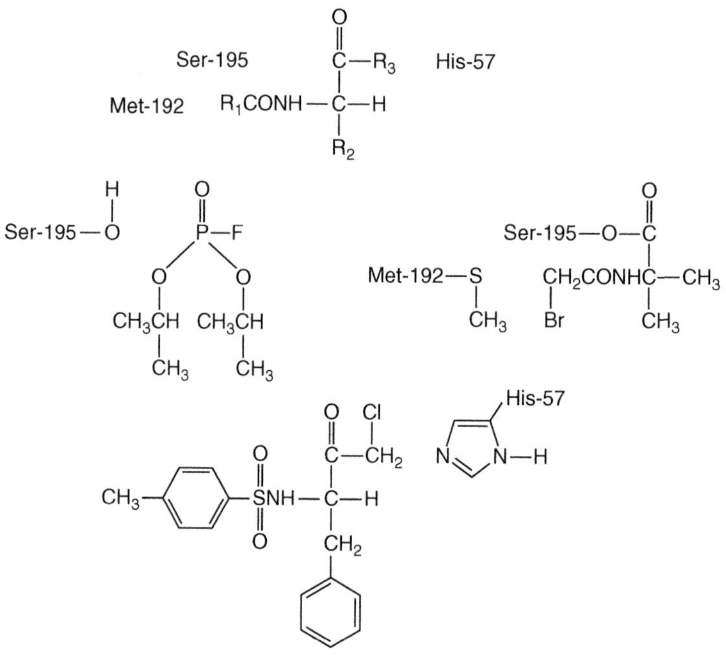

Fig. 1 Topography of active site of chymotrypsin as determined by sequence analysis of protein labeled with the indicated reagents (see Table I, item 1, for references).

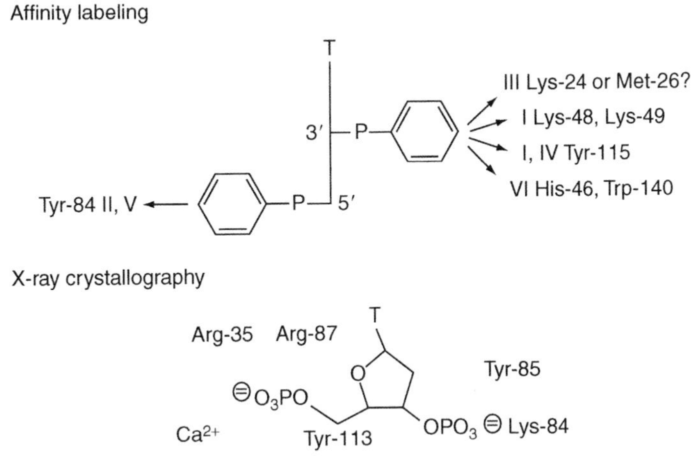

Affinity labeling

T
3′
P
III Lys-24 or Met-26?
I Lys-48, Lys-49
I, IV Tyr-115
VI His-46, Trp-140
Tyr-84 II, V
P
5′

X-ray crystallography

T
Arg-35 Arg-87
O
Tyr-85
$^{\ominus}O_3PO$
Ca^{2+}
Tyr-113
$OPO_3{}^{\ominus}$ Lys-84

Fig. 2 Topography of active site of staphylococcal nuclease as derived by affinity-labeling studies (see Table I, item 10, for reagents and references) or by X-ray crystallography (Arnone *et al.*, 1971).

differences in active site topography, but the lesson to be learned is that affinity-labeling results need to be interpreted cautiously.

An extensive series of steroid affinity labels have also been used to study the topography of the active site of human placental 17β-dehydrogenase (Chin *et al.*, 1980; Pons *et al.*, 1976), and 3α,20β-hydroxysteroid dehydrogenase (Benisek *et al.*, 1982), and a number of different amino acid residues have been modified. Since the reactive functional group is bromoacetyl, which has the flexibility to move several angstroms, the exact positioning of the amino acids is difficult (Warren and Mueller, 1977). It will be interesting to determine the structures by X-ray crystallography.

C. Investigating Catalytic Mechanisms

The identification of "essential" amino acid residues opens the way for speculation about the roles of the amino acid side chains in the binding of substrates and in the catalytic mechanism. Unfortunately, modification of a side chain in the active site with a bulky (or even a small) group can inactivate for steric reasons, and thus it is difficult to determine if the modified group directly participates in catalysis. Despite the ambiguity, affinity labeling can yield some information.

Once the reactive amino acids have been identified, the active-site-directed reagent can be used to determine how environmental factors affect reactivity. Thus, the pH dependence of modification can be used to give pK values for the reactive amino acid, or at least for the system of which it is a part. (The ionization of another nonmodifiable residue can affect the binding of the reagent, the state of the residue to be modified, or the structure of the enzyme.) For instance, the alkylation of His-57 in chymotrypsin by tosyl-L-phenylalanylchloromethane depends on a basic group with a pK of 6.8 and an acidic group with a pK of 8.9. The pH dependencies for catalytic action show similar pK values, and it was concluded that His-57 has the pK of 6.8 (Kézdy *et al.*, 1967b). But note that acylation of Ser-195 by phenylmethanesulfonyl fluoride also depends on a group with a pK of 7 (Gold and Fahrney, 1964).

The use of very well-designed active-site-directed reagents that closely resemble a substrate, bind tightly to the enzyme, and react with considerable facilitation may also lead one to attempt to position the modifiable residue near a part of the substrate and to assign a catalytic role. The first compelling example of this was the use of tosyl-L-phenylalanylchloromethane to modify His-57 (Ong *et al.*, 1965). One could then imagine that the imidazole group could participate as an acid-base catalyst for the scission of the amide bond of a substrate. The identification of Ser-195 as the site of acylation by diisopropylphosphorofluoridate (Oosterbaan *et al.*, 1958) or phenylmethanesulfonyl fluoride (Gold, 1965) is less compelling because of the uncertainty of the mode of binding of these simpler reagents and the possibility that the acyl group might have been transferred from another group (e.g., the histidine) that was initially acylated.

This discussion illustrates the value of using an active-site-directed reagent that is isosteric with a substrate. A reagent that closely resembles a substrate should produce more facilitation of the reaction and give one more confidence in assigning a catalytic function to the modifiable residue. Tosyl-L-phenylalanylchloromethane is almost isosteric with a substrate. If we assume that one of the structures on the catalytic pathway is **A** shown below (the imidazole is about to accept a proton from a serine hydroxyl), then **B** might be the ground-state structure leading to attack by the imidazole on the $-CH_2Cl$.

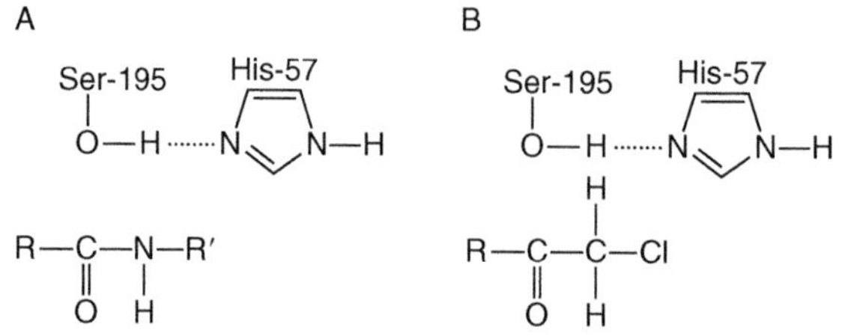

A

Ser-195 — O—H ······ N (His-57 imidazole) N—H

R—C(=O)—N(H)—R′

B

Ser-195 — O—H ······ N (His-57 imidazole) N—H

R—C(=O)—C(H)(H)—Cl

However, it appears to me that the imidazole might be as much as 2 Å away (the diameter of a hydrogen atom) from its optimum position for alkylation. Perhaps the flexibility of the enzyme or reagent is sufficiently high so that reactivity is close to maximal. But then one wonders why the serine hydroxyl is not alkylated significantly. A possible explanation is that the reagent forms a hemiketal with the serine before it alkylates the histidine, as suggested on the basis of high-resolution X-ray crystallography of subtilisin inactivated with halomethyl ketones (Poulos *et al.*, 1976). Nevertheless, benzyloxycarbonyl-L-phenylalanyl-L-alanyl-chloromethane reacts with a cysteine (Cys-25) in papain rather than with the histidine (Drenth *et al.*, 1976). Of course, a sulfhydryl is more reactive than a hydroxyl group, but tosyl-L-phenylalanylchloromethane reacts with papain 100 times faster than it reacts with chymotrypsin (Bender and Brubacher, 1966) or 2000 times faster than it reacts with cysteine (Whitaker and Perez-Villaseñor, 1968). Other chloromethylketones, tosyl-L-lysylchloromethane (Whitaker and Perez-Villaseñor, 1968), and benzyloxycarbonyl-L-phenylalanylchloromethane (Leary *et al.*, 1977) react even faster with papain. Furthermore, trypsin reacts with *p*-guanidino-phenacyl bromide to form an ether linkage with the active-site serine (Schroeder and Shaw, 1971). Other examples have been summarized previously (Shaw, 1967). Thus, one must conclude that chloromethylketone reagents are not completely specific. It would be interesting to design other functional groups for these active-site-directed reagents that are closer to being isosteric. In this regard, it is impressive that peptidyl diazomethylketones are highly specific and greatly facilitated (10^{10}-fold) in their reactions with thiol proteinases (Green and Shaw, 1981).

It may be noted that suicide, or mechanism-based, inactivators probably must be close to being isosteric with a substrate, since the enzyme presumably acts on the reagent to convert it to a reactive species. Even these species often seem to have a poor shape or be the width of one hydrogen atom too far away, however.

To reiterate, if we wish to understand enzyme mechanisms, in particular how enzymes recognize and bind substrates, the design of active-site-directed reagents should be relatively sophisticated.

Affinity labeling can also be used to analyze the "inherent reactivity" of amino acid side chains, which could be a source of catalytic power for enzymes. Of course, one must separate the inherent reactivity from the reactivity due to the "Circe" effect (Jencks, 1975). This will be discussed later in Section V.

IV. Considerations in the Design of Active-Site-Directed Reagents

A. Specificity

What characteristics allow a reagent to react specifically and rapidly? The affinity group is very important in this respect, and one should examine the structures and affinities of substrates and inhibitors to find out (a) groups that are essential for the binding, (b) tolerance for adding or removing groups, and (c) groups that might be on the enzyme and could be modified.

It is often assumed that the affinity group should resemble a substrate, product, or inhibitor. But remember that "resemblance" is in the eye of the biochemist, whereas the enzyme apparently uses more sophisticated criteria to determine how well it binds a ligand. (We cannot easily predict how well an enzyme will bind some structure that we design.) Therefore, extensive empirical studies are required to determine the size, shape, and functionality of the best ligands. Baker (1967) emphasized that studies on "bulk tolerance" are prerequisites for successful design. But we should also look for the minimum structure that is required for good binding. Often, one can learn much about the active site by examining space-filling models of compounds that are known to bind to the enzyme.

In evaluating binding, it is necessary to do proper kinetic studies to be able to decide which part of the active site a ligand binds to and what the dissociation constant is. It is not sufficient to determine an I_{50} (concentration of inhibitor giving 50% inhibition in a particular assay). The type of inhibition against one of the substrates should be determined; if it is competitive, the inhibitor probably binds to the same site as the substrate and the K_i is the dissociation constant. If the inhibition is noncompetitive or uncompetitive, different sites may be involved and the apparent K_i usually should be corrected—for example, for the effect of nonsaturation by the nonvaried substrate(s) or for the simplification of the kinetic equation (Segel, 1975). With such kinetic constants in hand, it is possible to rationalize binding affinity in terms of structure and to begin to design an appropriate reagent.

For extensive series of compounds, Hansch's correlation analysis is required. This approach has been applied to the monumental results from Baker's laboratory and shows, among other things, that related enzymes may have similar binding sites and that the specificity of the binding site must be explored with carefully chosen substituents (Silipo and Hansch, 1976; Yoshimoto and Hansch, 1976).

Eventually, an affinity group should be found that binds to the enzyme with a dissociation constant ≤ 1 mM (1 μM for some photoaffinity reagents), since the tighter the compound binds, the more facilitated the reaction should be. If the dissociation constant is 10 or 100 mM, the reagent may be quite nonspecific because it binds by simple ionic or hydrophobic interactions at various sites on the enzyme.

The importance of the structure of the affinity group for obtaining specific modification directed toward particular amino acid residues was illustrated with the studies in Figs. 1 and 2. Furthermore, as shown in Table II, the location of the functional group on the affinity portion of the reagent is critical for determining whether a facilitated reaction is obtained. Although the affinity group for reagent **2** binds tightly, the compound does not inactivate acetylcholinesterase any faster than **1**, which has no "affinity group." Compounds **3** and **4** have more suitable designs and react in a moderately facilitated manner. In contrast, **5** and **6** do not react measurably. Just how these compounds bind into the active site is not yet known. It is clear that simply attaching a functional group somewhere onto the affinity group is not good enough. If the group is attached in the wrong location, the compound could still bind to the enzyme, but then it could protect the enzyme against a bimolecular attack by another molecule of reagent. As discussed later, this mechanism of reaction can also result in saturation kinetics and can deceive the investigator about the nature of the reaction.

An impressive example of the role of the affinity group in providing rate-enhancement facilitation and selectivity is the series of studies by Kettner and Shaw, illustrated in Table III. By varying the peptide structure, reagents that are selective enough to be used *in vivo* may be obtained.

Table II
Role of Reagent Structure in Determining Rate of Inactivation of Acetylcholinesterase[a]

Structure no.	Reagent	K_I (mM)	k_3 (s^{-1})	k_3/K_I (M^{-1} s^{-1})	K_I, inhibitor (mM)
1	CH_3SO_2F	Large	–	2.5	–
2	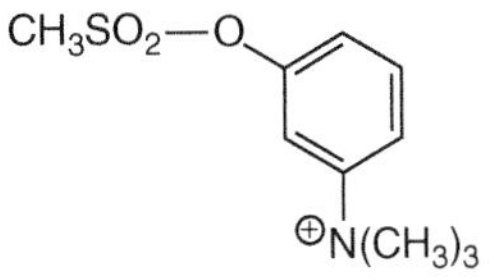 CH_3SO_2—O–(phenyl, meta-$^{\oplus}N(CH_3)_3$)	0.4	8.3×10^{-4}	2	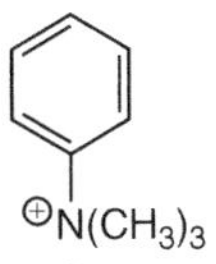 phenyl–$^{\oplus}N(CH_3)_3$ 0.053
3	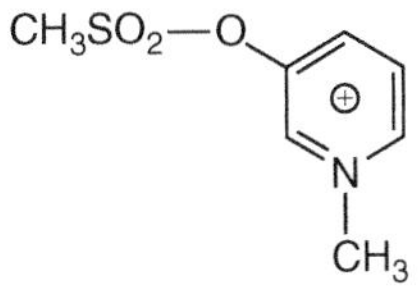 CH_3SO_2—O–(N-CH_3 pyridinium, 3-position)	0.1	5×10^{-3}	50	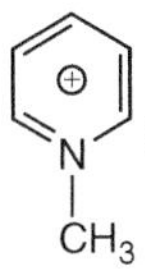 N-CH_3 pyridinium 0.11
4	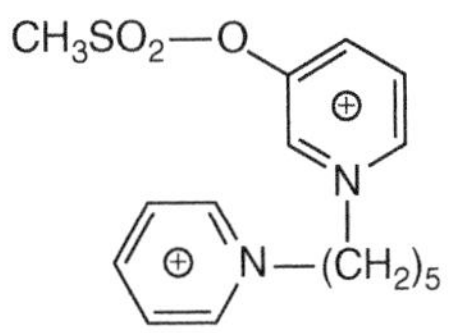 CH_3SO_2—O–(pyridinium, 3-position)–N—$(CH_2)_5$–N-pyridinium	0.02	5.5×10^{-4}	27	–
5	CH_3SO_2—$OCH_2CH_2\overset{\oplus}{N}(CH_3)_3$		No inactivation		$^{\oplus}N(CH_3)_4$ 1.2
6	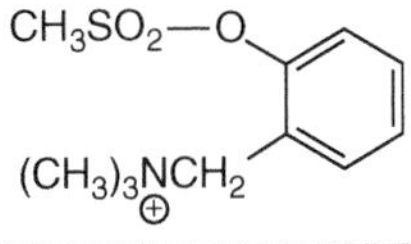 CH_3SO_2—O–(phenyl, ortho-$CH_2\overset{\oplus}{N}(CH_3)_3$)		No inactivation		

[a]Rates of inactivation are taken from Kitz and Wilson (1962). Inhibition constants are from Kitz and Wilson (1963), who also showed that quaternary ammonium salts can stimulate the rate of inactivation by CH_3SO_2F by up to 33-fold, and the stimulation has subsequently been studied by others (Belleau and DiTullio, 1970; Pavlič,1973).

B. Covalent Chemistry

In choosing the functional group to place onto the affinity group, the knowledge obtained from the use of simple reagents on the enzyme is valuable. A great variety of alkylating, acylating, photolabile groups, and other groups have been used as can be seen by perusal of the book on affinity labeling (Jacoby and Wilchek, 1977).

Depending on the purpose of the reagent, one could choose either an "exo" or "endo" type of affinity labeling (Cory *et al.*, 1977). When Baker (1967) originally chose to use "exo" affinity labeling, he postulated that enzymes from different sources (e.g., normal and cancerous tissues) would have the same amino acid residues at the active site, but might differ in nonessential residues outside the active site. However, even quite large and sophisticated exo-site reagents may inactivate a variety of homologous enzymes (Robinson *et al.*, 1980). Furthermore,

Table III
Rate-Enhancement Specificity and Selectivity of Peptidyl Chloromethyl Ketones,[a] P-Arg-CH_2Cl

	$10^{-4} \times k_{obs}/[I]$ (M^{-1} min^{-1}, 25 °C, pH 7.0)			
P	Thrombin	Plasma kallikrein	Plasmin	Urokinase
Val-Ile-Pro-	73	2.2	0.35	0.18
Val-Pro-	54	2.9	0.35	0.54
Ile-Pro-	42	2.0	0.31	0.39
Dns-Glu-Gly-	26	140	28	4.2
Phe-Ala-	8	0.86	0.09	0.35
Ile-Leu-	5.2	8.9	0.36	0.014
Glu-Gly-	1.9	16	1.3	20
Pro-Gly-	1.2	3.3	0.091	0.79
Ac-Gly-Gly-	0.74	1.4	0.053	2.6
Ala-Phe-	0.17	440	14	0.0059
Pro-Phe-	0.12	150	3.7	0.0015

[a]The estimated bimolecular rate constants for inactivation, $k_{obs}/[I]$, were taken from several publications by Kettner and Shaw (1977, 1978, 1979a). The most reactive inactivator of thrombin is D-Phe-L-Pro-L-Arg-CH_2Cl, which has a $k_{obs}/[I]$ value of 6.8×10^8 M^{-1} min^{-1} (Kettner and Shaw, 1979b).

we now know that enzymes catalyzing the same reactions (e.g., serine proteases or NAD-dependent dehydrogenases) may have very similar tertiary structures, even with several differences in the amino acids within the active site. These differences frequently can be exploited by varying the chemistry and location of the functional group. Thus, the impetus for using "exo" affinity labeling is somewhat reduced. Nevertheless, we have used such reagents in attempting to modify the only alkylatable residue in the substrate binding pocket of liver alcohol dehydrogenase. A methionine residue located 14 Å from the catalytic zinc ion was the target, and we found that reagents of just the right length, size, and shape would inactivate the enzyme in a facilitated manner. The reagents that were too short, too long, or too rigid reacted with less facilitation and presumably with less specificity (Chen and Plapp, 1978). One difficulty with using very long and flexible groups onto which is attached the reactive functional group is that one does not necessarily known how these groups are bound by the enzyme. The flexibility reduces the resolution with which one can map the active site.

"Endo" alkylators should be utilized if one wishes to identify amino acid residues that may participate in catalysis and also if one hopes to obtain catalysis by the enzyme of the chemical reaction. "Endo" alkylators should be isosteric, if possible, with a natural substrate, and the functional group could replace some portion of the normal substrate. Suicide inactivators, or enzyme-facilitated or enzyme-catalyzed modifications, are examples of this type of "endo" modification.

To illustrate the difficulties that one may encounter in trying to use isosteric reagents, the work by Woenckhaus and coworkers on liver alcohol dehydrogenase may be cited (Table I, item 38). NAD analogs in which either the nicotinamide ring

Table IV
Effect of Leaving Group on Reactivity[a]

X–CH_2COOH (Plapp, 1973)				Cbz-Phe-CH_2X (Larsen and Shaw, 1976)	
X	NBP[b] (rel. rate)	RNase (M^{-1} s^{-1})	DNase (M^{-1} s^{-1})	X	Chymotrypsin (M^{-1} s^{-1})
I	52	0.050	0.014	Mesyl	1800
Br	32	0.085	0.116	Br	790
Tosyl	6.3	0.0083	0.00036	Tosyl	7400
Cl	0.63	0.0028	0.0085	Cl	69
Condition		pH 5.5, 37 °C	pH 7.2, 25 °C, + 4 mM $CuCl_2$		pH 6.8, 25 °C

[a]The rates of reaction are the pseudobimolecular rate constants, k_3/K_I; see Section V.A.
[b]Relative reactivities with 4-(*p*-nitrobenzyl)pyridine in 75% 2-methoxyethanol at pH 4.2 and 37 °C, using the procedure of Baker and Jordaan (1965). The numbers are expressed as the change in absorbance at 570 nm (1-cm path) per min divided by the final molarity of the reagent in the reaction mixture. For comparison, other compounds were $BrCH_2CH_2OH$, 0.3; CH_3I, 45. An extensive series of aziridines and related compounds have been studied previously (Bardos *et al.*, 1965, and references therein).

or the adenine ring contain the reactive functional groups each modify cysteine residues that are ligands to zinc in the nicotinamide-binding pocket. Obviously, the enzyme finds it difficult to distinguish the two ends of the NAD analogs.

The chemistry of the leaving group is important in designing reagents. We found that variation of the leaving group (Cl, Br, I, tosyl) did not affect the relative facilitation of carboxymethylation of pancreatic ribonuclease as compared to reaction with a model compound, whereas with pancreatic deoxyribonuclease, smaller leaving groups produced more facilitation (Table IV). We attribute this to an interaction of deoxyribonuclease with the leaving group itself (Plapp, 1973). With tosyl-L-phenylalanylchloromethane analogs, the sulfonate esters appear to have enhanced reactivities with chymotrypsin (Table IV).

One of the reasons for studying the reaction of ribonuclease with tosylglycolate (carboxymethyltosylate) was that we thought that the tosyl group would bind into the active site, but then be displaced when the carboxymethyl group reacted with the enzyme. In other words, the affinity group is itself the leaving group. This experimental design was also used by Nakagawa and Bender (1970) to methylate His-57 in chymotrypsin with methyl benzenesulfonate and by White and Branchini (1977) to ethylate leuciferase with another sulfonate ester. In these cases, the changes in activity of the enzyme are not due to the irreversible binding of the affinity group into the active site; rather, they result from modification by very small substituents. Such reagents could also be applied to other enzymes in order to determine whether the amino acid residue that is modified is really involved in catalytic activity or is just so close to the active site that its modification interferes with substrate binding.

V. Evaluation of Active-Site-Directed Reagents

After designing and synthesizing an affinity-labeling reagent, we should determine whether it is active-site-directed and collect the data required to learn about the enzyme structure and function. Accordingly, the following experimental criteria for affinity labeling will be discussed, in order to explain why certain questions should be asked and how to use the answers.

A. Kinetics

Do the kinetics of inactivation of the enzyme by the reagent show saturation behavior, as predicted by the mechanism of affinity labeling? The kinetics should be studied under conditions that maintain the activity of the enzyme in the absence of reagent. Usually, the concentration of reagent should exceed the concentration of the enzyme, so that pseudo-first-order kinetics may be observed. Note that it is insufficient to simply report the molar ratio of the reagent to the enzyme because (a) it is the concentration of the reagent that determines the rate of inactivation when reagent exceeds enzyme concentration and (b) it is often difficult to determine the molarity of the enzyme. If the kinetics of inactivation are pseudo-first-order, then the concentration of reagent should be varied and the pseudo-first-order rate constants for inactivation calculated, so that one can determine the kinetic constants that characterize the reaction of an active-site-directed reagent. If the kinetics of inactivation are not pseudo-first-order, the investigator may have to distinguish among a variety of possible explanations: reagent instability, partial activity of modified enzyme, etc.

Kitz and Wilson (1962) derived an equation that is based on the appropriate assumption that the enzyme is a reagent, as contrasted to the less rigorous assumption of steady state (Cornish-Bowden, 1979). (The steady-state concentration of the reversible enzyme-activator complex does change as a function of time.) Thus, for affinity labeling,

$$\mathrm{E} + \mathrm{I} \underset{k_2}{\overset{k_1}{\rightleftharpoons}} \mathrm{E{\cdot}I} \overset{k_3}{\rightarrow} \mathrm{E} - \mathrm{X}, \quad K_\mathrm{I} = \frac{k_2}{k_1},$$

it is assumed that formation of the reversible complex ($\mathrm{E \cdot I}$) is in rapid equilibrium compared to the formation of the irreversibly inactivated enzyme ($\mathrm{E} - \mathrm{X}$); that $[\mathrm{I}] \gg [\mathrm{E}]$; and that when the reaction mixture is assayed for enzymatic activity, E and $\mathrm{E \cdot I}$ produce full activity while $\mathrm{E} - \mathrm{X}$ is inactive. Then

$$\frac{\mathrm{d}[\mathrm{E}-\mathrm{X}]}{\mathrm{d}t} = k_3[\mathrm{E{\cdot}I}], \quad K_\mathrm{I} = \frac{[\mathrm{E}][\mathrm{I}]}{[\mathrm{E{\cdot}I}]}$$

$$[\mathrm{E}]_\mathrm{t} = [\mathrm{E}] + [\mathrm{E{\cdot}I}] + [\mathrm{E}-\mathrm{X}] = \frac{[\mathrm{E{\cdot}I}]K_\mathrm{I}}{[\mathrm{I}]} + [\mathrm{E{\cdot}I}] + [\mathrm{E}-\mathrm{X}]$$

$$\frac{\mathrm{d}[\mathrm{E}-\mathrm{X}]}{\mathrm{d}t} = \frac{k_3([\mathrm{E}]_\mathrm{t} - [\mathrm{E}-\mathrm{X}])}{1 + K_\mathrm{I}/[\mathrm{I}]}$$

or

$$\frac{\mathrm{d}[\mathrm{E}-\mathrm{X}]}{[\mathrm{E}]_{\mathrm{t}}-[\mathrm{EX}]}=\frac{k_3\mathrm{d}t}{1+K_{\mathrm{I}}/[\mathrm{I}]}$$

Integrating,

$$\ln([\mathrm{E}]_{\mathrm{t}}-[\mathrm{E}-\mathrm{X}])=-\frac{k_3t}{1+K_{\mathrm{I}}/[\mathrm{I}]}+\ln[\mathrm{E}]_{\mathrm{t}}$$

If the logarithm of enzyme activity $[\mathrm{E}]_{\mathrm{t}}-[\mathrm{E}-\mathrm{X}]=[\mathrm{E}]+[\mathrm{E}\cdot\mathrm{I}]$ is plotted against time (most conveniently on semilog paper), the observed first-order rate constant, k_{obs}, may be calculated from the slope or by $k_{\mathrm{obs}}=0.693/t_{1/2}$, and is related to the desired constants by Eq. (1):

$$k_{\mathrm{obs}}=\frac{k_3[\mathrm{I}]}{K_{\mathrm{I}}+[\mathrm{I}]} \tag{1}$$

Since this equation predicts hyperbolic saturation kinetics, a plot of $1/k_{\mathrm{obs}}$ against $1/[\mathrm{I}]$ allows a graphical estimation of k_3 and K_{I}. For better estimates, one can use the programs of Cleland (1979), which provide the values and their standard errors. Note that when $[\mathrm{I}]\ll K_{\mathrm{I}}$, k_{obs} equals the pseudobimolecular rate constant, k_3/K_{I}, which has units of $\mathrm{M}^{-1}\,\mathrm{s}^{-1}$ and is used to compare reactivities of various reagents—for example, as a measure of the extent of facilitation obtained with active-site-directed reagents as compared to other reagents.

We usually assume that an active-site-directed reagent should give saturation kinetics, but note that the reaction of human serum cholinesterase with diisopropylfluorophosphate did not give saturation kinetics, whereas another organophosphate, malaoxon, did (Main, 1964). Saturation kinetics may not arise if the rate of dissociation (k_2) for the complex and the rate of the unimolecular reaction (k_3) are relatively rapid. This makes it difficult to use concentrations of reagent approaching K_{I}, since the rate of reaction gets too fast to measure. If saturation kinetics are not observed and k_3 and K_{I} cannot be determined accurately, the apparent bimolecular rate constant, k_3/K_{I}, can be calculated from the slope of a plot of k_{obs} against [I].

The K_{I} usually is taken as a measure of the affinity of the reagent for the enzyme, and it is gratifying when the K_{I} agrees with the inhibition constant determined for the reagent from competitive inhibition kinetics. However, the K_{I} determined from the inactivation kinetics may be larger than that observed from competitive inhibition kinetics if the reagent binds in a less favorable way when it undergoes the chemical reaction than when it binds as a reversible inhibitor.

On the other hand, if the K_{I} values agree, it may be because one molecule of active-site-directed reagent binds to and protects the active site against bimolecular reaction by a second molecule of the reagent.

$$E + I \xrightarrow{k_b} E - X$$

$$E + I \overset{K_I}{\rightleftharpoons} E{\cdot}I$$

Saturation kinetics are also observed, as the equation for this "self-protection mechanism" (Baker, 1967) is kinetically equivalent to Eq. (1):

$$k_{obs} = \frac{k_b K_I[I]}{K_I + [I]} \tag{2}$$

This mechanism can be rendered unlikely if the following criteria are met.

B. Reactivity

How much faster does the reagent react with the enzyme than does a simple reagent that does not have the affinity group? If the affinity label had a bromoacetyl function, bromoacetate or bromoacetamide could be used as the simple reagent for determination of the bimolecular rate of inactivation. If the pseudobimolecular rate constant, k_3/K_I, for reaction of the enzyme with the active-site-directed reagent is considerably faster than the bimolecular rate constant for the simple reagent, it is reasonable to conclude that true affinity labeling is occurring. On the other hand, if the affinity reagent reacts more slowly than the simple reagent, one cannot conclude that self-protection is occurring since the (usually large) affinity reagent could have reduced reactivity because of electronic or steric effects. In doing this experiment, a good control is to test also a mixture of two compounds that separately represent the affinity and functional groups of the active-site-directed reagent, for instance, bromoacetate and a substrate or substrate analog. The binding of a substrate may induce a conformational change that exposes an amino acid side chain for reaction. In a similar way, reaction of a reagent that resembles one substrate of the enzyme may be stimulated by the presence of one or more of the other substrates of the enzymes. For instance, inactivation of lactate dehydrogenase by bromopyruvate, which is actually a substrate, was stimulated fivefold by NAD (Berghäuser *et al.*, 1971).

Another criterion for true affinity labeling is that active-site-directed reagents with the functional group in different positions on the affinity group should have different rates of inactivation, as reflected in k_3/K_I. If the functional group of a reagent is not juxtaposed to react with an amino acid side chain while the affinity group is bound, that reagent should not react with facilitation, but could still react by the bimolecular mechanism at a rate that may approach the rate with a simple reagent, or could react more slowly if the affinity group hinders access of the reagent to the site of reaction.

Interpreting these experiments requires some information about the relative reactivities of the various reagents with an amino acid side chain incorporating a functional group into a more complex structure could conceivably alter its reactivity because of electronic or steric effects. The model nucleophile, 4-(*p*-nitrobenzyl) pyridine, which reacts to form a colored alkylated product, is useful for this purpose (see Table IV). It would be better, of course, to compare the rate constants for reactions of the reagent with the same type of group that reacts in the protein. Table V presents selected data on the reactivities of simple reagents with functional groups of amino acids and proteins. These data illustrate the range of possible reactivities and show that the microscopic environment of the protein can depress or considerably enhance the rate of reaction of simple reagents. This fact makes it difficult to eliminate the self-protection mechanism if the rates of reaction of simple reagents with the enzyme are not determined.

C. Inactivation

Does the active-site-directed reagent completely inactivate the enzyme? If an essential residue is being modified, one expects no residual activity. However, many investigators fail to follow a reaction after enzyme activity is reduced to less than 10% of the initial activity, probably because the assays are inconvenient. Moreover, the kinetics may start to deviate from first order, and the investigator chooses not to be concerned with this complication. Critical information is lost thereby which could be obtained by simply starting a reaction mixture with a concentration of enzyme that is 10 or 100 times higher than is normally used and making dilutions in order to follow the reaction from 0% to more than 99% inactivation. If the reaction begins to slow down substantially, more reagents can be added as a check for reagent decomposition. (One can also analyze for reagent, of course, or determine its rate of decomposition in the reaction medium.) If the enzyme cannot be completely inactivated, it may be because (a) some impurity or decomposed reagent binds tightly to the active site and protects against reagent, or (b) the modified enzyme has residual activity, or (c) the enzyme preparation is heterogeneous, containing some unmodifiable isoenzyme, or (d) other reasons.

As a start toward distinguishing among these possibilities, the enzyme can be isolated from the reaction mixture and retreated with reagent, or its kinetics of action on substrates can be studied. If (b) or (c) hold, the kinetics may be significantly different than those for native enzyme. For instance, chymotrypsin alkylated on Met-192 with bromoacetyl-α-aminoisobutyrate has 20% of the residual activity of native chymotrypsin in a standard assay, but its V_{max} is increased by 1.4-fold (accompanied by a 10-fold increased K_m) with acetyltyrosine ethyl ester as substrate (Lawson and Schramm, 1965). On other substrates, k_{cat} is increased as much as eightfold (Kézdy *et al.*, 1967a). D-Amino acid oxidase treated with the suicide substrate D-propargylglycine also appears to have residual activity, and its K_m values for amino acids appear to be differentially altered (Marcotte and Walsh, 1978).

Table V
Selected Data on Reactivities of Simple Reagents with Model Compounds and Enzymes[a]

1. Cysteine sulfhydryl
   a. $ClCH_2CONH_2$
      Gly-Cys-Gly, 30 °C, $\bar{k} = 0.27M^{-1}s^{-1}$, p*K* 9.0 (Lindley, 1962)
      Ficin, 30.1 °C, $\bar{k} = 16M^{-1}s^{-1}$, p*K* 8.3 (Whitaker and Lee, 1972)
      Papain, 30.5 °C, $\bar{k} = 6.2M^{-1}s^{-1}$, p*K* 8.5 (Chaiken and Smith, 1969)
   b. ICH_2CONH_2
      Glutathione, 25 °C, pH 11.2 (mercaptide), $k_0 = 27\ M^{-1}\ s^{-1}$ (Halász and Polgár, 1976)
      Thiolsubtilisin, 25 °C, pH 7 (ion-pair), $k_0 = 7.2\ M^{-1}\ s^{-1}$ (Halász and Polgár, 1976)
      Papain, pH 5.5 (ion-pair), $k_0 = 14\ M^{-1}\ s^{-1}$ (Halász and Polgár, 1976)
      Papain, pH 10 (mercaptide), $k_0 = 976\ M^{-1}\ s^{-1}$ (Halász and Polgár, 1976)
      Liver alcohol dehydrogenase, 25 °C, pH 7.2, 0.021 $M^{-1}\ s^{-1}$ (Evans and Rabin, 1968)
      Yeast alcohol dehydrogenase, 25 °C, pH 7.6, $k_0 = 0.43\ M^{-1}\ s^{-1}$ (Whitehead and Rabin, 1964)
      Yeast hexokinase B, 35 °C, $\bar{k} = 14M^{-1}s^{-1}$, p*K* 10 (Jones *et al.*, 1975)
      Glyceraldehyde-3-phosphate dehydrogenase 25 °C, $\bar{k} = 280M^{-1}s^{-1}$, p*K* = 8.2 (Polgár, 1975)
   c. $BrCH_2CONH_2$
      Liver alcohol dehydrogenase, 25 °C, pH 8, $k_0 = 0.027\ M^{-1}\ s^{-1}$ (Fries *et al.*, 1975)
      Yeast alcohol dehydrogenase, 25 °C, pH 7.9, $k_0 = 0.37\ M^{-1}\ s^{-1}$ (Plapp *et al.*, 1968)
   d. CH_3I (Halász and Polgár, 1976)
      Glutathione, 25 °C, pH 11.2 (mercaptide), $k_0 = 0.92\ M^{-1}\ s^{-1}$
      Thiolsubtilisin, 25 °C, pH 7 (ion-pair), $k_0 = 4.2\ M^{-1}\ s^{-1}$
      Papain, 25 °C, pH 5.5 (ion-pair), $k_0 = 0.028\ M^{-1}\ s^{-1}$
      Papain, 25 °C, pH 10 (mercaptide), $k_0 = 0.6\ M^{-1}\ s^{-1}$
   e. *N*-Ethylmaleimide
      Glutathione, 25 °C, pH 6.5, $k_0 = 263\ M^{-1}\ s^{-1}$ (Evans *et al.*, 1981)
      Papain, 25 °C, pH 6.4, $k_0 = 2.5\ M^{-1}\ s^{-1}$ (Evans *et al.*, 1981)
      Yeast alcohol dehydrogenase, 20 °C, pH 7.0, $k_0 = 0.22\ M^{-1}\ s^{-1}$ (Heitz *et al.*, 1968)
   f. Acrylonitrile
      Glutathione, 30 °C, $\bar{k} = 0.59M^{-1}s^{-1}$, p*K* = 8.6 (Friedman *et al.*, 1965)
      Bovine serum albumin (reduced), 30 °C, pH 7, 6 M urea, $k_0 = 0.01\ M^{-1}\ s^{-1}$ (Cavins and Friedman, 1968)
2. Methionine thioether
   a. ICH_2COOH
      Pancreatic ribonuclease, 25 °C, pH 3.5, 8 M urea, $k_0 = 7.2 \times 10^{-4}\ M^{-1}\ s^{-1}$ (Link and Stark, 1968)
   b. CH_3I (Link and Stark, 1968)
      α-*N*-Acetylmethionine, 25 °C, pH 3.0, $k_0 = 5 \times 10^{-4}\ M^{-1}\ s^{-1}$
      Pancreatic ribonuclease, 25 °C, pH 3.3, $k_0 = 8 \times 10^{-4}\ M^{-1}\ s^{-1}$
   c. $BrCH_2CONH–C_6H_5$ (Bittner and Gerig, 1970)
      α-*N*-Acetylmethionine, 25 °C, pH 6.0, 10% ethanol, $k_0 = 1.3 \times 10^{-3}\ M^{-1}\ s^{-1}$
      α-Chymotrypsin-Met-192, 27.5 °C, pH 6.0, 10% ethanol, $k_0 = 0.26\ M^{-1}\ s^{-1}$
3. Histidine imidazole
   a. $BrCH_2COOH$ (Lennette and Plapp, 1979)
      Histidine hydantoin (*N*-1 + *N*-3), 25 °C, pH 7.72, $k_0 = 5.9 \times 10^{-5}\ M^{-1}\ s^{-1}$, (p*K* = 6.4)
      Pancreatic ribonuclease (His-12 + His-119), 25 °C, $\bar{k} = 2.6 \times 10^{-2}M^{-1}s^{-1}$, pH optimum = 5.5, p*K* values = 4.7, 6.3
   b. $BrCH_2CH_2COOH$ (Heinrikson *et al.*, 1965)
      Histidine, 25 °C, pH 5.5, $2.3 \times 10^{-6}\ M^{-1}\ s^{-1}$
      Pancreatic ribonuclease-His-119, 25 °C, pH 5.5, $k_0 = 6.3 \times 10^{-4}\ M^{-1}\ s^{-1}$
   c. $BrCH_2COCOOH$ (Heinrikson *et al.*, 1965)
      Pancreatic ribonuclease-His-119, 25 °C, pH 5.5, $k_0 = 9.1 \times 10^{-2}\ M^{-1}\ s^{-1}$
   d. ICH_2CONH_2

(*continued*)

Table V (*continued*)

Histidine, 25 °C, pH 5.3, $k_0 = 1.2 \times 10^{-6}$ M^{-1} s^{-1} (Fruchter and Crestfield, 1967)
Pancreatic ribonuclease, 25 °C, pH 5.3 (pH optimum) $k_0 = 1.1 \times 10^{-4}$ M^{-1} s^{-1} (Fruchter and Crestfield, 1967)
Pancreatic trypsin, 25 °C, pH 7.0, $k_0 = 5.3 \times 10^{-6}$ M^{-1} s^{-1} (pK = 6.7) (Inagami and Hatano, 1969)
e. 1-Fluoro-2,4-dinitrobenzene (Cruickshank and Kaplan, 1972)
α-*N*-Acetylhistidine, 20 °C, $\bar{k} = 7.4 \times 10^{-4} M^{-1} s^{-1}$, p$K$ 7.2
α-Chymotrypsin, 20 °C, $\bar{k} = 7.5 \times 10^{-3} M^{-1} s^{-1}$, pK 6.8
4. Lysine, ε-amino
a. $BrCH_2COOH$
Pancreatic ribonuclease-Lys-41, 25 °C, pH 8.5, $k_0 = 2.6 \times 10^{-3}$ M^{-1} s^{-1} (Heinrikson, 1966)
b. 1-Fluoro-2,4-dinitrobenzene (Murdock *et al.*, 1966)
Gly-Lys, 15 °C, $\bar{k} = 0.22 M^{-1} s^{-1}$, p$K$ = 10.1
Ribonuclease-Lys-41, 15 °C, $\bar{k} = 0.44 M^{-1} s^{-1}$, p$K$ = 8.9
c. Acrylonitrile
ε-Aminocaproic acid, 30 °C, $\bar{k} = 1.96 \times 10^{-2} M^{-1} s^{-1}$, p$K$ 10.6 (Friedman *et al.*, 1965)
5. Peptidyl, α-amino
a. 1-Fluoro-2,4-dinitrobenzene
Gly-Gly, 30 °C, 6% dioxane, pH 10, $k_0 = 0.37$ M^{-1} s^{-1} (Gerig and Reinheimer, 1975)
Valyl terminal of *Streptomyces griseus* trypsin, 20 °C, $\bar{k} = 6.7 \times 10^{-3} M^{-1} s^{-1}$, p$K$ 8.1 (Duggleby and Kaplan, 1975)
b. *N*-Ethylmaleimide
H_2N-Val-Leu-Ser ..., 25 °C, pH 7.4, $k_0 = 6 \times 10^{-3}$ M^{-1} s^{-1} (Smyth *et al.*, 1964)
c. Acrylonitrile
Tetraglycine, 30 °C, $\bar{k} = 5.7 \times 10^{-3} M^{-1} s^{-1}$, p$K$ = 7.6 (Friedman *et al.*, 1965)
6. Serine hydroxyl
CH_3SO_2F: chymotrypsin-Ser-195, 25 °C, pH 7, $k_0 = 0.021$ M^{-1} s^{-1} (Gold, 1967)

[a]Since the unprotonated forms of the amino acid side chains usually react much faster than the protonated forms, the pH dependence of the rate constant should be measured. Where this was done, the pH-independent rate constant k is reported, together with the apparent pK value. In other cases, the observed rate constant k_0 is given, along with pH values. In general, reactions were studied in buffers of about 0.1 ionic strength but the original papers should be consulted for specific details and other examples.

D. Protection

Do substrates, products, or reversible inhibitors protect against inactivation? Such protection is the usual evidence that reaction occurs at the active site, although another possible explanation is that the "active-site-directed" reagent reacts at an "allosteric" site. The protective agent (L) should give competitive inhibition with an inhibition constant of K_L against inactivation by the reagent, according to the following equation:

$$k_{obs} = \frac{k_3[\mathrm{I}]}{K_\mathrm{I}(1 + [\mathrm{L}]/K_\mathrm{L}) + [\mathrm{I}]}$$

Therefore, one should do an experiment with varied concentrations of I and L, in order to evaluate k_3, K_I, and K_L by the usual procedures used in enzyme kinetics

(Todhunter, 1979). If the inhibition pattern is not competitive or if K_L is not of about the same magnitude as the constant determined by other methods (equilibrium dialysis, inhibition of enzyme activity), inactivation by the reagent may not be due to reaction at the active site. It should be apparent from this discussion that just using one (high) concentration of a substrate to test for protection is poor experimental design; enzymes may have, in addition to the active site, low-affinity sites where substrates or inhibitors can bind.

E. Specificity

Does modification of one amino acid residue correlate with complete inactivation? This is the ultimate criterion for the specificity of labeling and the efficacy of the active-site-directed reagent. Usually, the investigator varies the extent of inactivation by treatment for varied times of reaction or with varied concentrations of inactivator, isolates the enzyme from the reaction mixture, and determines the incorporation of reagent. Typically, a plot of residual activity against incorporation deviates from linearity (1:1 stoichiometry) and shows that some secondary sites are reacting before the active site is completely modified. In some unusual cases, modification of one site appears to inactivate two sites ("half-of-the-sites" reactivity) (Bernhard and MacQuarrie, 1973; Levitzki, 1974; Stallcup and Koshland, 1973).

After finding the reagent or conditions that give specific labeling, the amino acid residues that are modified should be identified. Amino acid analysis can often be used to determine which amino acid is modified, either because of loss of an amino acid or because of conversion to a derivative. Studies with model compounds are essential for identification and have been carried out with many reagents. Amino acid sequence analysis of labeled peptides is then required. If the three-dimensional structure has been determined, X-ray crystallography can be used to identify the modified sites (e.g., Walder *et al.*, 1980). Sequence analysis is often difficult, but the investigator should try to determine the extent of labeling of each of the various sites modified and the recovery of label throughout the purification of labeled peptides. Note that it was the careful analysis of the products of carboxymethylation of pancreatic ribonuclease that led to the concept that two histidine residues were involved in catalytic activity of that enzyme (Crestfield *et al.*, 1963). Quantitative analyses on the carboxymethylated derivatives of *Streptococcus faecium* dihydrofolate reductase also allowed identification of three methionine residues that could be essential when about 1.5 methionines reacted in a differential labeling experiment (Gleisner and Blakley, 1975). The three methionines modified—residues 28, 36, and 50—correspond to Leu-27, Val-35, and Phe-49 in the homologous sequence of the *Lactobacillus casei* enzyme, for which X-ray crystallography identifies Leu-27 and Phe-49 as part of the binding site for the inhibitor methotrexate (Matthews *et al.*, 1978). The reaction of human serum prealbumin with *N*-bromoacetyl-L-thyroxine also revealed several sites of modification, giving a rough map of the binding site even while showing how nonspecific affinity labels can be

(Cheng *et al.*, 1977). Since it is useful to know which secondary sites react, for new insights into structure-function relationships, investigators should try to characterize the products of reaction as completely as possible, rather than to presume that modification is absolutely specific. Even tosyl-L-phenylalanylchloromethane, which alkylates predominately His-57 in chymotrypsin, may modify 0.3 residues of Met-192 per molecule (Stevenson and Smillie, 1968).

Although photoaffinity labeling appears, in principle, to be a good method to label many different residues in the active site, the low yields of each product make sequence analysis difficult. A possible complication of photoaffinity "labeling" is that the reagent may not become attached to the protein, and it could become difficult to locate the site of reaction. Ogez *et al.* (1977) found that Asp-38 in Δ^5-3-ketosteroid isomerase is converted to an alanine upon photoinactivation in the presence of 3-oxo-4-estren-17β-yl acetate. A review of other photolabeling studies suggests that this complication may occur in other cases (Benisek *et al.*, 1982). On the other hand, if the enzyme can be inactivated without attaching a sterically perturbing group, conclusions about the functional role of the modifiable amino acid may be more definitive. Not many sites of labeling by "suicide" reagents have been identified, but there appear to be two problems with these reagents. The activated reagent may react with a coenzyme at the active site rather than with an amino acid residue, or the reagent may diffuse away from the active site and react with other sites on the same, or other, proteins (Walsh, 1977).

VI. Facilitation of Reaction Achieved with Active-Site-Directed Reagents

How much faster do affinity-labeling agents react with an enzyme than do simple reagents without the affinity groups? That is, what is the ratio of the pseudobimolecular rate constant for the reaction of the specific reagent (k_3/K_I) to the bimolecular rate constant for the simple reagent (k_b)? We assume that large values of this ratio will yield more specific reaction, and thus this ratio should be determined as part of the characterization of the reagent. Unfortunately, many investigators just report that a simple reagent did not inactivate under the same conditions employed for the specific reagent. Furthermore, there is too little information available on the rates of reaction of simple reagents with compounds that model amino acid side chains (Table V) to serve as a basis for comparison.

How much facilitation could be expected for an active-site-directed reagent? If it is assumed that such reagents bind as do normal substrates, the extent of facilitation could approach the amount of catalytic power exerted by the enzyme as a result of binding of the reactive groups in stereochemically correct juxtaposition. The magnitude of this catalytic factor has been estimated in a variety of ways, depending on the presumed origin of catalysis "propinquity" (Bruice and Benkovic, 1966), "proximity and orientation" (Koshland, 1962), or "Circe effect" (Jencks,

Table VI
Effect of Reagent Structure on Magnitude of Facilitation of Reaction of Chymotrypsin[a]

R–Phenylalanyl–CH_2–X		k (M^{-1} s^{-1})	
R–	–X	Chymotrypsin	N^{α}-Acetylhistidine
H–	–Cl	0.041	
N^{α}-Formyl–	–Cl	1.35	
N^{α}-Tosyl–	–Cl	7.7	
N^{α}-Benzyloxycarbonyl–	–Cl	69	4.5×10^{-5}
N^{α}-Formyl–	–Br	26	
N^{α}-Benzyloxycarbonyl–	–Br	790	1.7×10^{-4}

[a]At pH 7, 25 °C; chymotrypsin in 5–10% methanol, N^{α}-acetylhistidine in 80% methanol (Shaw and Ruscica, 1971).

1975). In the most favorable cases, we expect the facilitation ratio to be at least 10^3. Of course, the facilitation obtained can be much less if the reagent is poorly designed and binds so that the reactive groups are not properly positioned for reaction. As discussed earlier, there would be no facilitation if the reagent bound only in a nonproductive mode. Clearly, the structure of the reagent is critical for achieving facilitation. This can be demonstrated by considering the results of Shaw *et al.* on serine proteases. As shown in Table VI, the magnitude of facilitation varies from about 10^3 to about 10^6 with a homologous series of chloromethyl ketones. The best reagent is 1700 times more reactive than the poorest with chymotrypsin. Now, these ratios are calculated with reference to the rate of reaction with α-*N*-acetylhistidine; another point of reference should be the rate of reaction of something like CH_3COCH_2Cl with chymotrypsin. This number would allow one to estimate the "inherent" reactivity of His-57. The acylation of Ser-195 in chymotrypsin by sulfonyl fluorides is facilitated 10^4-fold by the addition of a phenyl ring to methanesulfonyl fluoride (Fahrney and Gold, 1963; Gold, 1967). More facilitation would be expected with more specific reagents.

This discussion raises an old issue in protein chemistry: Is the reaction of an amino acid residue with a reagent facilitated because of binding of reagent to the enzyme (according to the classical theory of affinity labeling) or because the amino acid is "hyperreactive," due to its microenvironment? It is well known that simple reagents can react faster by an order of magnitude or more with certain groups on a protein than with other groups of the same type on the protein or on free amino acids. In one such case, fluorodinitrobenzene reacts selectively with Lys-41 in RNase A because the pK value of the ε-amino group is depressed about 1.4 units, although the pH-independent rate constant is similar to that for a normal primary amine (Table V, item 4b). In a contrasting case, the pH-independent rate constant for reaction of fluorodinitrobenzene with His-57 in chymotrypsin is 10 times that for α-*N*-acetylhistidine, apparently because the carboxylate of Asp-102 increases the nucleophilicity of His-57 (Table V, item 3e). It is apparent from these examples that one cannot determine whether a reagent is "active-site-directed" simply by comparing the rates of reaction of the enzyme and a model amino acid.

In order to differentiate between possible mechanisms of facilitation (i.e., affinity labeling or hyperreactivity), we have explored the use of a simple active-site-directed reagent, bromoacetate, which reacts about 10^3 times faster with histidine residues 12 and 119 in ribonuclease than with a model histidine compound (Lennette and Plapp, 1979) and binds with a dissociation constant of about 23 mM at pH 5.5 and 37 °C before reaction with the enzyme. Stark *et al.* (1961) explained the facilitation of the reaction by postulating that the protonated imidazole nitrogen of one histidine attracted the carboxylate of the haloacetate, orienting the reagent for nucleophilic attack by the unprotonated nitrogen of the other imidazole. This mechanism is consistent with the three-dimensional structure of the enzyme (Richards and Wyckoff, 1971). Jencks (1975) explained the facilitation by saying that the interaction of ribonuclease and the haloacetate serves to decrease the free rotation and translation of the reagent. These explanations are equivalent to saying that the haloacetate is an affinity label and imply that the binding of reagent results in a more favorable entropy of reaction. An alternative explanation is that the environments of the histidines make them hyperreactive, which would imply a more favorable enthalpy of reaction. (In our definition, a hyperreactive group has enhanced inherent reactivity due to electronic effects and would be more reactive even with a reagent that did not bind as an affinity label.)

Based on these ideas, we proposed that transition-state analysis could be used at least to characterize and possibly to differentiate between the mechanisms of facilitation (Lennette and Plapp, 1979). If the entropy of activation (ΔS^*) was more favorable, affinity labeling is implied; a more favorable enthalpy of activation (ΔH^*) implies hyperreactivity. In order to have a valid basis of reference, the activation parameters for the reaction of bromoacetate with ribonuclease were compared to the parameters for the reaction of bromoacetate with histidine hydantoin, which is a good model of a histidine residue in a protein. Since the imidazole ring has two nitrogens that can react, the rates of reaction of each nitrogen were determined. The surprising finding was that the enhanced reactivities of histidine 12 and 119 in ribonuclease are due almost entirely to more favorable magnitudes of ΔH^*. The reactivity of His-12 could be increased because of hydrogen bonding to the carbonyl oxygen of Thr-45, and the reactivity of His-119 because of contact with the carboxylate of Asp-121. If the two imidazoles have increased nucleophilicities, they should also be hyperreactive with a simple alkylating agent such as iodoacetamide. Although His-12 does react readily with iodoacetamide, His-119 does not (Fruchter and Crestfield, 1967). Thus, we must assume that the carboxylate group of the haloacetate is involved in the reaction mechanism, and we proposed that the binding of bromoacetate stabilizes the enzyme in a particular conformation, with His-119 fixed in a position that confers upon it hyperreactivity, as shown in the following scheme:

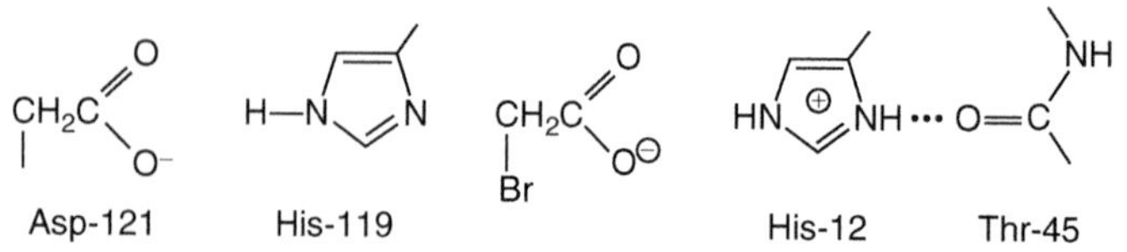

Asp-121 His-119 His-12 Thr-45

Of course, this explanation implies that binding energy is used to orient the reagent, and thus we could expect changes in ΔS^*. However, differential solvation effects on the carboxymethylation of histidine hydantoin as compared to ribonuclease could result in compensation between the changes in ΔH^* and ΔS^*. Complications of this sort can limit the utility of activation parameters. Nevertheless, we believe such data are required to provide an experimental basis for theories being developed to explain enzyme catalytic power.

VII. Concluding Remarks

It is clear that affinity labeling has been useful for studying enzymes, even though considerable effort is required to design, synthesize, and evaluate such reagents. Unfortunately, many of the reagents that have been prepared have not been demonstrated to be significantly more reactive or specific than simpler chemical reagents. This has led to the publication of many incomplete studies. Progress in this area would be facilitated if investigators would carefully assess their objectives and cautiously apply the approach as one of several parallel courses of study. It is especially important that investigators be more critical of their results and perform more of the experiments required to evaluate the candidate active-site-directed reagents.

Despite the work involved and the difficulties that have been encountered, the applications of affinity-labeling reagents are potentially extremely valuable. In addition to their uses for studying enzyme structure and function, they offer one of the best avenues for the rational development of specific chemotherapeutic agents. Unfortunately, so little is known about the detailed interactions of small molecules with large molecules that extensive empirical studies are required before effective compounds can be prepared. In this context, study of enzymes can serve as models for more complicated regulatory systems. Furthermore, affinity reagents are being applied in ever more sophisticated ways to study receptor systems. Thus, I believe that significant efforts in the development of affinity-labeling reagents are justified and will be rewarding.

Acknowledgments

This work was supported by research grant AA00279 from the National Institute on Alcohol Abuse and Alcoholism. The editorial assistance of Rosemary K. Plapp was appreciated.

References

Ahn, K., Johnson, D. S., Fitzgerald, L. R., Liimatta, M., Arendse, A., Stevenson, T., Lund, E. T., Nugent, R. A., Nomanbhoy, T. K., Alexander, J. P., and Cravatt, B. F. (2007). *Biochemistry* **46,** 13019.

Ahn, K., McKinney, M. K., and Cravatt, B. F. (2008). *Chem. Rev.* **108,** 1687–1707.

Arnone, A., Bier, C. J., Cotton, F. A., Day, V. W., Hazen, E. E., Jr., Richardson, D. C., Richardson, J. S., and Yonath, A. (1971). *J. Biol. Chem.* **246,** 2302.

Baker, B. R. (1967). "Design of Active-Site Directed Irreversible Enzyme Inhibitors." Wiley, New York.

Baker, B. R., and Jordaan, J. H. (1965). *J. Heterocycl. Chem.* **2,** 21.

Bardos, T. J., Datta-Gupta, N., Hebborn, P., and Triggle, D. J. (1965). *J. Med. Chem.* **8,** 167.

Bayley, H., and Knowles, J. R. (1977). *Methods Enzymol.* **46,** 69.

Belleau, B., and DiTullio, V. (1970). *J. Am. Chem. Soc.* **92,** 6320.

Bellisario, R. L., Maley, G. F., Galivan, J. H., and Maley, F. (1976). *Proc. Natl. Acad. Sci. USA* **73,** 1848.

Bender, M. L., and Brubacher, L. J. (1966). *J. Am. Chem. Soc.* **88,** 5880.

Benisek, W. F., Ogez, J. R., and Smith, S. B. (1982). *Adv. Chem. Ser.* **198,** 267.

Berghäuser, J., Falderbaum, I., and Woenckhaus, C. (1971). *Hoppe-Seyler's Z. Physiol. Chem.* **352,** 52.

Bernhard, S. A., and MacQuarrie, R. A. (1973). *J. Mol. Biol.* **74,** 73.

Biellmann, J. F., Eid, P., Hirth, C., and Jörnvall, H. (1980). *Eur. J. Biochem.* **104,** 59.

Bittner, E. W., and Gerig, J. T. (1970). *J. Am. Chem. Soc.* **92,** 2114.

Blow, D. M. (1971). *In* "The Enzymes" (P. D. Boyer, ed.), Vol. 3, 3rd edn. p. 185. Academic Press, New York.

Blumberg, P. M., and Strominger, J. L. (1971). *Proc. Natl. Acad. Sci. USA* **68,** 2814.

Braun, H., Legler, G., Deshusses, J., and Semenza, G. (1977). *Biochim. Biophys. Acta* **483,** 135.

Bruice, T. C., and Benkovic, S. J. (1966)."Bioorganic Mechanisms," Vol. 1, Benjamin, New York.

Cavins, J. F., and Friedman, M. (1968). *J. Biol. Chem.* **243,** 3357.

Chaiken, I. M., and Smith, E. L. (1969). *J. Biol. Chem.* **244,** 5087.

Chen, W. S., and Plapp, B. V. (1978). *Biochemistry* **17,** 4916.

Cheng, S. Y., Wilchek, M., Cahnmann, H. J., and Robbins, J. (1977). *J. Biol. Chem.* **252,** 6076.

Chin, C. C., Asmar, P., and Warren, J. C. (1980). *J. Biol. Chem.* **255,** 3660.

Chowdhry, V., and Westheimer, F. H. (1979). *Annu. Rev. Biochem.* **48,** 293.

Cleland, W. W. (1979). *Methods Enzymol.* **63.** Article [6].

Cohen, S. A., and Pratt, R. F. (1980). *Biochemistry* **19,** 3996.

Cornish-Bowden, A. (1979). *Eur. J. Biochem.* **93,** 383.

Cory, M., Andrews, J. M., and Bing, D. H. (1977). *Methods Enzymol.* **46,** 115.

Crestfield, A. M., Stein, W. H., and Moore, S. (1963). *J. Biol. Chem.* **238,** 2413.

Cruickshank, W. H., and Kaplan, H. (1972). *Biochem. J.* **130,** 1125.

Cuatrecasas, P., and Wilchek, M. (1977). *Methods Enzymol.* **46,** 358.

Cybulsky, D. L., Kandel, S. I., Kandel, M., and Gornall, A. G. (1973). *J. Biol. Chem.* **248,** 3411.

Drenth, J., Kalk, K. H., and Swen, H. M. (1976). *Biochemistry* **15,** 3731.

Duggleby, R. G., and Kaplan, H. (1975). *Biochemistry* **14,** 5168.

Esch, F. S., and Allison, W. S. (1978). *J. Biol. Chem.* **253,** 6100.

Esch, F. S., and Allison, W. S. (1979). *J. Biol. Chem.* **254,** 10740.

Eshdat, Y., McKelvy, J. F., and Sharon, N. (1973). *J. Biol. Chem.* **248,** 5892.

Evans, N., and Rabin, B. R. (1968). *Eur. J. Biochem.* **4,** 548.

Evans, B. L. B., Knopp, J. A., and Horton, H. R. (1981). *Arch. Biochem. Biophys.* **206,** 362.

Fahrney, D. E., and Gold, A. M. (1963). *J. Am. Chem. Soc.* **85,** 997.

Fowler, A. V., Zabin, I., Sinnott, M. L., and Smith, P. J. (1978). *J. Biol. Chem.* **253,** 5283.

French, T. C., Dawid, I. B., and Buchanan, J. M. (1963). *J. Biol. Chem.* **238,** 2186.

Friedman, M., Cavins, J. F., and Wall, J. S. (1965). *J. Am. Chem. Soc.* **87,** 3672.

Fries, R. W., Bohlken, D. P., Blakley, R. T., and Plapp, B. V. (1975). *Biochemistry* **14,** 5233.
Fruchter, R. G., and Crestfield, A. M. (1967). *J. Biol. Chem.* **242,** 5807.
Gerig, J. T., and Reinheimer, J. D. (1975). *J. Am. Chem. Soc.* **97,** 168.
Gibson, D. R., Gracy, R. W., and Hartman, F. C. (1980). *J. Biol. Chem.* **255,** 9369.
Gleisner, J. M., and Blakley, R. L. (1975). *Eur. J. Biochem.* **55,** 141.
Glover, G., and Shaw, E. (1971). *J. Biol. Chem.* **246,** 4594.
Gold, A. M. (1967). *Methods Enzymol.* **11,** 706.
Gold, A. M. (1965). *Biochemistry* **4,** 897.
Gold, A. M., and Fahrney, D. (1964). *Biochemistry* **3,** 783.
Gorecki, M., Wilchek, M., and Patchornik, A. (1971). *Biochim. Biophys. Acta* **229,** 590.
Green, G. D. J., and Shaw, E. (1981). *J. Biol. Chem.* **256,** 1923.
Halász, P., and Polgár, L. (1976). *Eur. J. Biochem.* **71,** 563, 571.
Hartman, F. C. (1971). *Biochemistry* **10,** 146.
Hartman, F. C. (1977). *Methods Enzymol.* **46,** 130.
Hartman, F. C., and Brown, J. P. (1976). *J. Biol. Chem.* **251,** 3057.
Hartman, F. C., and Norton, I. L. (1976). *J. Biol. Chem.* **251,** 4565.
Hass, G. M., and Neurath, H. (1971). *Biochemistry* **10,** 3535, 3541.
Heinrikson, R. L. (1966). *J. Biol. Chem.* **241,** 1393.
Heinrikson, R. L., Stein, W. H., Crestfield, A. M., and Moore, S. (1965). *J. Biol. Chem.* **240,** 2921.
Heitz, J. R., Anderson, C. D., and Anderson, B. M. (1968). *Arch. Biochem. Biophys.* **127,** 627.
Henderson, R. (1971). *Biochem. J.* **124,** 13.
Hexter, C. S., and Westheimer, F. H. (1971). *J. Biol. Chem.* **246,** 3928.
Higgins, W., Miles, E. W., and Fairwell, T. (1980). *J. Biol. Chem.* **255,** 512.
Hirs, C. H. W., Ed. (1967). *Methods Enzymol.* Vol. 11.
Inagami, T., and Hatano, H. (1969). *J. Biol. Chem.* **244,** 1176.
Jacoby, W. B., and Wilchek, M., Eds. (1977). *Methods Enzymol.* Vol. 46.
Jencks, W. P. (1975). *Adv. Enzymol.* **43,** 219.
Jones, J. G., Otieno, S., Barnard, E. A., and Bhargava, A. K. (1975). *Biochemistry* **14,** 2396.
Jörnvall, H., Woenckhaus, C., and Johnscher, G. (1975). *Eur. J. Biochem.* **53,** 71.
Kerlavage, A. R., and Taylor, S. S. (1980). *J. Biol. Chem.* **255,** 8483.
Kettner, C., and Shaw, E. (1977). *In* "Chemistry and Biology of Thrombin" (R. L. Lundblad, K. G. Mann and J. W. Fenton, eds.), p. 129. Ann Arbor Sci. Publ., Ann Arbor, MI.
Kettner, C., and Shaw, E. (1978). *Biochemistry* **17,** 4778.
Kettner, C., and Shaw, E. (1979). *Biochim. Biophys. Acta* **569,** 31.
Kettner, C., and Shaw, E. (1979). *Thromb. Res.* **14,** 969.
Kézdy, F. J., Feder, J., and Bender, M. L. (1967). *J. Am. Chem. Soc.* **89,** 1009.
Kézdy, F. J., Thomson, A., and Bender, M. L. (1967). *J. Am. Chem. Soc.* **89,** 1004.
Kimmel, M. T., and Plummer, T. H., Jr. (1972). *J. Biol. Chem.* **247,** 7864.
Kitz, R., and Wilson, I. B. (1962). *J. Biol. Chem.* **237,** 3245.
Kitz, R., and Wilson, I. B. (1963). *J. Biol. Chem.* **238,** 745.
Knott-Hunziker, V., Waley, S. G., Orlek, B. S., and Sammes, P. G. (1979). *FEBS Lett.* **99,** 59.
Koshland, D. E., Jr. (1962). *J. Theor. Biol.* **2,** 75.
Larsen, D., and Shaw, E. (1976). *J. Med. Chem.* **19,** 1284.
Lawson, W. B., and Schramm, H. J. (1965). *Biochemistry* **4,** 377.
Leary, R., Larsen, D., Watanabe, H., and Shaw, E. (1977). *Biochemistry* **16,** 5857.
Lennette, E. P., and Plapp, B. V. (1979). *Biochemistry* **18,** 3933, 3938.
Levitzki, A. (1974). *J. Mol. Biol.* **90,** 451.
Lindley, H. (1962). *Biochem. J.* **82,** 418.
Link, T. P., and Stark, G. R. (1968). *J. Biol. Chem.* **243,** 1082.
Loosemore, M. J., Cohen, S. A., and Pratt, R. F. (1980). *Biochemistry* **19,** 3990.
Machuga, E., and Klapper, M. H. (1977). *Biochim. Biophys. Acta* **481,** 526.
Main, A. R. (1964). *Science* **144,** 992.

Marcotte, P., and Walsh, C. (1978). *Biochemistry* **17,** 2864.

Matthews, D. A., Alden, R. A., Bolin, J. T., Filman, D. J., Freer, S. T., Hamlin, R., Hol, W. G. J., Kisliuk, R. L., Pastore, E. J., Plante, L. T., Xuong, N., and Kraut, J. (1978). *J. Biol. Chem.* **253,** 6946.

Means, G. E., and Feeney, R. E. (1971). "Chemical Modification of Proteins." Holden-Day, San Francisco, CA.

Meloche, H. P., Monti, C. T., and Hogue-Angeletti, R. A. (1978). *Biochem. Biophys. Res. Commun.* **84,** 589.

Michel, F., Pons, M., Descomps, B., and Crastes de Paulet, A. (1978). *Eur. J. Biochem.* **84,** 267.

Miles, E. W., and Higgins, W. (1980). *Biochem. Biophys. Res. Commun.* **93,** 1152.

Morgan, P. H., Robinson, N. C., Walsh, K. A., and Neurath, H. (1972). *Proc. Natl. Acad. Sci. USA* **69,** 3312.

Morino, Y., Osman, A. M., and Okamoto, M. (1974). *J. Biol. Chem.* **249,** 6684.

Murdock, A. L., Grist, K. L., and Hirs, C. H. W. (1966). *Arch. Biochem. Biophys* **114,** 375.

Nagata, K., Suzuki, K., and Imahori, K. (1979). *J. Biochem. (Tokyo)* **86,** 1179.

Nakagawa, Y., and Bender, M. L. (1970). *Biochemistry* **9,** 259.

Ogez, J. R., Tivol, W. F., and Benisek, W. F. (1977). *J. Biol. Chem.* **252,** 6151.

Ong, E. B., Shaw, E., and Schoellmann, G. (1965). *J. Biol. Chem.* **240,** 694.

Oosterbaan, R. A., Kunst, P., van Rotterdam, J., and Cohen, J. A. (1958). *Biochim. Biophys. Acta* **27,** 549, 556.

Pavlič, M. R. (1973). *Biochim. Biophys. Acta* **327,** 393.

Phillips, A. T. (1977). *Methods Enzymol.* **46,** 59.

Pincus, M., Thi, L. L., and Carty, R. P. (1975). *Biochemistry* **14,** 3653.

Plapp, B. V. (1973). *J. Biol. Chem.* **248,** 4896.

Plapp, B. V., Woenckhaus, C., and Pfleiderer, G. (1968). *Arch. Biochem. Biophys.* **128,** 360.

Plummer, T. H., Jr. (1969). *J. Biol. Chem.* **244,** 5246.

Polgár, L. (1975). *Eur. J. Biochem.* **51,** 63.

Pons, M., Nicolas, J. C., Boussioux, A. M., Descomps, B., and Crastes de Paulet, A. (1976). *Eur. J. Biochem.* **68,** 385.

Poulos, T. L., Alden, R. A., Freer, S. T., Birktoft, J. J., and Kraut, J. (1976). *J. Biol. Chem.* **251,** 1097.

Powell, J. T., and Brew, K. (1976). *Biochemistry* **15,** 3499.

Quaroni, A., and Semenza, G. (1976). *J. Biol. Chem.* **251,** 3250.

Rainey, P., Holler, E., and Kula, M. R. (1976). *Eur. J. Biochem.* **63,** 419.

Rasnick, D., and Powers, J. C. (1978). *Biochemistry* **17,** 4363.

Reynolds, C. H., and McKinley-McKee, J. S. (1975). *Arch. Biochem. Biophys.* **168,** 145.

Reynolds, C. H., Morris, D. L., and McKinley-McKee, J. S. (1970). *Eur. J. Biochem.* **14,** 14.

Richards, F. M., and Wyckoff, H. W. (1971). *In* "The Enzymes" (P. D. Boyer, ed.), 3rd edn. Vol. 4, p. 647. Academic Press, New York.

Riva, F., Carotti, D., Barra, D., Giartosio, A., and Turano, C. (1980). *J. Biol. Chem.* **255,** 9230.

Robertson, J. G. (2005). *Biochemistry* **44,** 5561.

Robinson, D. J., Furie, B., Furie, B. C., and Bing, D. H. (1980). *J. Biol. Chem.* **255,** 2014.

Schloss, J. V., Stringer, C. D., and Hartman, F. C. (1978). *J. Biol. Chem.* **253,** 5707.

Schroeder, D. D., and Shaw, E. (1971). *Arch. Biochem. Biophys.* **142,** 340.

Segel, I. H. (1975). "Enzyme Kinetics." Wiley, New York.

Segal, D. M., Powers, J. C., Cohen, G. H., Davies, D. R., and Wilcox, P. E. (1971). *Biochemistry* **10,** 3728.

Seiler, N., Jung, M. J., and Koch-Weser, J. (1978). "Enzyme-Activated Irreversible Inhibitors." Elsevier, Amsterdam.

Shaw, E. (1967). *Methods Enzymol.* **11,** 677.

Shaw, E., and Glover, G. (1970). *Arch. Biochem. Biophys.* **139,** 298.

Shaw, E., and Ruscica, J. (1971). *Arch. Biochem. Biophys.* **145,** 484.

Silipo, C., and Hansch, C. (1976). *J. Med. Chem.* **19,** 62.

Silverman, R. B. (1995). *Methods Enzymol.* **249,** 240.

Sinnott, M. L., and Smith, P. J. (1978). *Biochem. J.* **175,** 525.
Smyth, D. G., Blumenfeld, O. O., and Konigsberg, W. (1964). *Biochem. J.* **91,** 589.
Sogin, D. C., and Plapp, B. V. (1975). *J. Biol. Chem.* **250,** 205.
Stallcup, W. B., and Koshland, D. E., Jr. (1973). *J. Mol. Biol.* **80** 4163, 77.
Stark, G. R., Stein, W. H., and Moore, S. (1961). *J. Biol. Chem.* **236,** 436.
Stevenson, K. J., and Smillie, L. B. (1968). *Can. J. Biochem.* **46,** 1357.
Suzuki, N., and Wood, W. A. (1980). *J. Biol. Chem.* **255,** 3427.
Szedlacsek, S. E., and Duggleby, R. G. (1995). *Methods Enzymol.* **249,** 144.
Todhunter, J. A. (1979). *Methods Enzymol.* **63,** 383.
Tso, J. Y., Bower, S. G., and Zalkin, H. (1980). *J. Biol. Chem.* **255,** 6734.
Walder, J. A., Walder, R. Y., and Arnone, A. (1980). *J. Mol. Biol.* **141,** 195.
Walsh, C. (1977). *Horiz. Biochem. Biophys.* **3,** 36.
Warren, J. C., and Mueller, J. R. (1977). *Methods Enzymol.* **46,** 447.
Waxman, D. J., and Strominger, J. L. (1980). *J. Biol. Chem.* **255,** 3964.
Weng, L., Heinrikson, R. L., and Mansour, T. E. (1980). *J. Biol. Chem.* **255,** 1492.
Whitaker, J. R., and Lee, L. S. (1972). *Arch. Biochem. Biophys.* **148,** 208.
Whitaker, J. R., and Perez-Villaseñor, J. (1968). *Arch. Biochem. Biophys.* **124,** 70.
White, E. H., and Branchini, B. R. (1977). *Methods Enzymol.* **46,** 537.
Whitehead, E. P., and Rabin, B. R. (1964). *Biochem. J.* **90,** 532.
Woenckhaus, C., and Jeck, R. (1971). *Hoppe-Seyler's Z. Physiol. Chem.* **352,** 1417.
Woenckhaus, C., Jeck, R., and Jörnvall, H. (1979). *Eur. J. Biochem.* **93,** 65.
Yoshimoto, M., and Hansch, C. (1976). *J. Med. Chem.* **19,** 71.
Zoller, M. J., and Taylor, S. S. (1979). *J. Biol. Chem.* **254,** 8363.

CHAPTER 12

Mechanism-Based Enzyme Inactivators

Richard B. Silverman

Department of Chemistry
Northwestern University
Evanston, Illinois 60208

CONTEMPORARY ENZYME KINETICS AND MECHANISM
Reprinted from *Methods in Enzymology*, Volume 249 (Academic Press, 1995).

DOI: 10.1016/B978-0-12-378608-1.00012-8

I. Terminology

An enzyme inactivator, in general, is a compound that produces irreversible inhibition of the enzyme; that is, it irreversibly prevents the enzyme from catalyzing its reaction (Silverman, 1992a). Irreversible in this context, however, does not necessarily mean that the enzyme activity never returns, only that the enzyme becomes dysfunctional for an extended (but unspecified) period of time. A compound that leads to formation of a covalent bond to the enzyme often destroys enzyme activity indefinitely; however, there are many cases where the covalent bond, that forms, is reversible, and enzyme activity slowly returns. There are other cases of irreversible inhibition in which no covalent bond forms at all, but the compound binds so tightly to the active site of the enzyme that k_{off}, the rate constant for release of the compound from the enzyme, is exceedingly small. This, in effect, produces irreversible inhibition. Yet another type of irreversible inhibitor is one that is converted by the enzyme to a product that binds very tightly to the enzyme and, therefore, has a very slow off rate.

The name mechanism-based enzyme inactivator (MBEI) conjures up the idea of an enzyme inactivator that depends on the mechanism of the targeted enzyme. In the broadest sense of the term, any of the inactivators described above that utilize the enzyme mechanism could be classified as MBEIs. In fact, Pratt (1992) and Krantz (1992) have suggested this general classification for mechanism-based enzyme inhibitors. However, there is a more narrow definition of MBEI that I have invoked previously (Escribano *et al.*, 1989; Garcia-Canovas *et al.*, 1989; Silverman, 1988a; Varon *et al.*, 1990) and which is used here. In this chapter, a MBEI is defined as an unreactive compound whose structure resembles that of either the substrate or product of the target enzyme, and which undergoes a catalytic transformation by the enzyme to a species that, prior to release form the active site, inactivates the enzyme. By this definition the ultimate cause for inactivation can result from any of the above-cited covalent or noncovalent mechanisms, provided a catalytic step was required to convert the inactivator to the inactivating species and the species responsible for inactivation did not leave the active site prior to inactivation.

The inactivators as defined above can be differentiated from other types of enzyme inactivators, such as affinity-labeling agents (ALAs), transition state analogs (TSAs), and slow, tight-binding inhibitors (STBIs). ALAs are generally reactive (electrophilic) compounds that alkylate or acylate enzyme nucleophiles. Often more than one enzyme nucleophile reacts with these compounds, and if a crude system containing more than one enzyme is being used, reactions with multiple enzymes can occur. The closer the structure of the ALA resembles that of the substrate for the target enzyme, the greater the specificity that will be attained. A less reactive variation of ALAs, termed quiescent ALAs, has been described (Krantz, 1992). These inactivators contain an electrophilic carbon that is stable in solution in the presence of external nucleophiles but which can react with an enzyme active site nucleophile involved in covalent catalysis.

TSAs depend on the property of enzymes to bind substrates most tightly at the transition states of reactions. A TSA is a compound whose structure resembles that of the hypothetical transition state (or intermediate) structure. Because of this structural similarity to the transition state structure, the enzyme binds these compounds exceedingly tightly, leading to very small k_{off} values.

STBIs have relatively slow rates of binding (i.e., the k_{on} values are relatively small and, as a result, time-dependent inhibition is observed), and then they form complexes that are exceedingly stable (i.e., the k_{off} rate constants are even smaller). The causes for the slowness and tightness of binding of these inhibitors can be a conformational change in the enzyme during binding, a change in the protonation state of the enzyme, displacement of a water molecule at the active site, or reversible, covalent bond formation. Although these are often noncovalent inactivators, their affinities for the enzyme are so great that sometimes stoichiometric amounts of inactivator are sufficient to inactivate the enzyme.

A MBEI, by the definition used here, is a compound that is transformed by the catalytic machinery of the enzyme into a species that acts as an ALA, a TSA, or a tight-binding inhibitor (either covalent or noncovalent) prior to release from the enzyme.

II. Basic Kinetics of Enzyme Inactivators

The different types of enzyme inactivators can be differentiated by their kinetic schemes. An ALA has a kinetic mechanism as shown in Scheme 1 Generally, the k_{on} and k_{off} rates are very fast, thereby establishing the equilibrium with E·ALA leads to the covalent enzyme adduct depicted as E-ALA′. Because of the covalent nature of this type of inactivator, dialysis or gel filtration would not restore enzyme activity.

$$\mathrm{E+ALA} \underset{k_{off}}{\overset{k_{on}}{\rightleftharpoons}} \mathrm{E{\cdot}ALA} \xrightarrow{k_2} \mathrm{E\text{-}ALA'}$$

Scheme 1

A TSA generally is a tight-binding, noncovalent inactivator which displays the kinetic scheme shown in Scheme 2. In this case, k_{on} is generally rapid, but k_{off} is slow. Because initial binding is rapid, inactivation occurs immediately, and there is

$$\mathrm{E+TSA} \underset{k_{off}}{\overset{k_{on}}{\rightleftharpoons}} \mathrm{E{\cdot}TSA}$$

Scheme 2

no time dependence to the inactivation. Also, dialysis or gel filtration would restore the enzyme activity, because of the noncovalent nature of the interaction.

The STBI has a relatively small k_{on} and an even smaller k_{off}, as shown in Scheme 3. These types of inactivators have been known to be both noncovalent and covalent in nature. Because of the slowness of the binding, a time-dependent loss of enzyme activity is observed. Dialysis or gel filtration can restore enzyme activity, but it may require longer dialysis times or a longer gel filtration column.

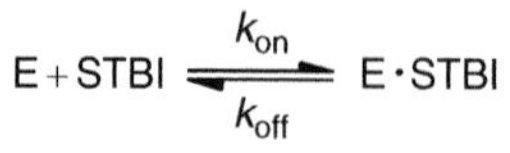

$$\mathrm{E+STBI} \underset{k_{off}}{\overset{k_{on}}{\rightleftharpoons}} \mathrm{E \cdot STBI}$$

Scheme 3

A MBEI requires a step to convert the compound to the inactivating species (k_2), as depicted in Scheme 4. This step, which is generally responsible for the observed time dependence of the enzyme inactivation, usually is irreversible and forms a new complex (E·MBEI′) which can have three fates: (1) if MBEI′ is not reactive, but forms a tight complex with the enzyme, then the inactivation may be the result of a noncovalent tight-binding complex (E · MBEI′); (2) if MBEI′ is a reactive species,[1] then a nucleophilic, electrophilic, or radical reaction with enzyme may ensue (k_4) to give the covalent complex E-MBEI″; and (3) the species generated could be released from the enzyme as a product (k_3). Most often these inactivators result in covalent bond formation with the enzyme, and therefore, dialysis or gel filtration does not restore enzyme activity.

$$\mathrm{E+MBEI} \underset{k_{off}}{\overset{k_{on}}{\rightleftharpoons}} \mathrm{E \bullet MBEI} \overset{k_2}{\rightleftharpoons} \mathrm{E \bullet MBEI'} \xrightarrow{k_4} \mathrm{E \bullet MBEI''}$$

$$\mathrm{E \bullet MBEI'} \xrightarrow{k_3} \mathrm{E + MBEI'}$$

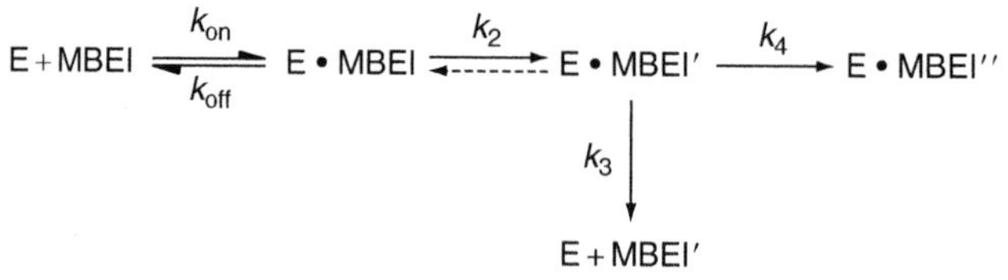

Scheme 4

The two principal kinetic constants that are useful in describing MBEIs are k_{inact} and K_I. Based on Scheme 4, k_{inact} is a complex mixture of k_2, k_3, and k_4 (Eq. (1)), and K_I is a complex mixture of k_{on}, k_{off}, k_2, k_3, and k_4 (Eq. (2)) (Waley, 1980, 1985):

[1] The reactive species generated may be like an affinity-labeling agent, in which case it could undergo a reaction with an active site nucleophile; or it could be radical in nature, in which case it could react with an enzyme radical; or it could be a nucleophile, in which case it could react with an electrophilic species, such as an oxidized cofactor on the enzyme.

$$k_{inact} = \frac{k_2 k_4}{k_2 + k_3 + k_4} \quad (1)$$

$$K_I = \left(\frac{k_{off} + k_2}{k_{on}}\right)\left(\frac{k_3 + k_4}{k_2 + k_3 + k_4}\right) \quad (2)$$

Only if k_2 is rate determining ($k_2 \ll k_4$) and k_3 is 0 (or nearly 0), k_{inact} will be equal to k_2 ($k_{inact} = k_2$). Values given for k_{inact} are assumed to represent the inactivation rate constants at infinite concentrations of inactivator.

An expression can be derived (Jung and Metcalf, 1975) that relates the enzyme concentration (E), the inactivator concentration (I), k_{inact}, and K_I (Eq. (3)). The half-life for inactivation ($t_{1/2}$), then, is described by Eq. (4). Extrapolation to infinite inactivator concentrations reduces Eq. (4) to $t_{1/2} = \ln 2/k_{inact}$:

$$\frac{\partial \ln E}{\partial t} = \frac{k_{inact} I}{K_I + I} \quad (3)$$

$$t_{1/2} = \frac{\ln 2}{k_{inact}} + \frac{\ln 2 K_I}{k_{inact} I} \quad (4)$$

These are the basic equations needed for typical MBEIs; more complex discussion of kinetics related to MBEIs can be found elsewhere (Escribano *et al.*, 1989; Funaki *et al.*, 1991; Garcia-Canovas *et al.*, 1989; Kuo and Jordan, 1983; Silverman, 1988b; Varon *et al.*, 1990). The determination of these kinetic constants is described in Section V.

The ratio of product release to inactivation is termed the partition ratio and represents the efficiency of the MBEI. When inactivation is the result of the formation of an E-MBEI″ adduct, the partition ratio is described by k_3/k_4. The partition ratio depends on the rate of diffusion of MBEI′ from the active site, its reactivity, and the proximity of an appropriate nucleophile, radical, or electrophile on the enzyme for covalent bound formation. It does not depend on the initial inactivator concentration. There are some cases where the partition ratio has been shown to be 0, that is, every turnover of inactivator produces inactivated enzyme (Silverman and Invergo, 1986).

For substrates and reversible competitive inhibitors, when k_{on} and k_{off} are very large, then the K_m and K_i represent the dissociation constants for the breakdown of the E · S and E · I complexes, respectively. The K_i value is obtained experimentally by determining the effect on the rate of conversion of subsaturating concentrations of substrate to product on addition of a constant amount of an inhibitor. The K_I value, a term used for MBEIs, however, is obtained experimentally (see Section V) by determining the effect on the rate of inactivation by a change in the inactivator concentration. When k_{on} and k_{off} are very large and k_2 is rate determining, then K_i (for initial binding of the inactivator to the enzyme) and K_I should be the same for the same compound under identical conditions.

When k_4 becomes partially rate determining, then $K_I \geq K_i$. Just as the K_m represents the concentration of substrate that gives one-half the maximal velocity, the K_I represents the concentration of the inactivator that produces one-half the maximal rate of inactivation.

III. Uses of Mechanism-Based Enzyme Inactivators

The two principal areas where MBEIs have been most useful are in the study of enzyme mechanism and in the design of new potential drugs.

A. Study of Enzyme Mechanisms

The reason that MBEIs are so important to the study of enzyme mechanisms is because they are really nothing more than modified substrates for the target enzymes. Once inside the active site of the enzymes, though, they are converted to products that inactivate the enzyme. However, the mechanisms by which these compounds are converted to the species that inactivate the enzyme proceed, at least initially, by the catalytic mechanisms for normal substrates of the enzymes. Therefore, whatever information can be obtained regarding the inactivation mechanism is directly related to the catalytic mechanism of the enzyme. One approach for the use of mechanism-based inactivators to elucidate enzyme mechanisms would involve formulating a hypothesis regarding the catalytic mechanism of the enzyme, then designing a compound that would require that mechanism to convert it to the activated form. If inactivation occurs, it supports the hypothetical mechanism. Of course, it does not prove the mechanism, because there may be other mechanisms that were not considered that also would lead to enzyme inactivation. An alternative approach arises when a time-dependent inactivator is serendipitously discovered, and then mechanisms are contemplated that might lead this compound to inactivate the enzyme.

B. Drug Design

Most enzyme inhibitor drugs (EID) are noncovalent, reversible inhibitors. The basis for their effectiveness is the tightness of their binding to the enzyme (i.e., the stability of E · EID and a small k_{off}), so that they compete with the binding of substrates (S) for the enzyme (Scheme 5). The efficacy of the drug continues as long as the enzyme is complexed with the drug (E · EID). Because the enzyme concentration is low and fixed, the equilibrium between E + EID and E · EID will depend on the concentration of EID and S. When the concentration of EID diminishes because of metabolism, the concentration of E · EID diminishes and the concentration of E · S increases. To maintain the pharmacological effect of the drug, administration of the drug several times a day, then, becomes necessary. An effective MBEI, however, could form a covalent bond to the enzyme, thereby

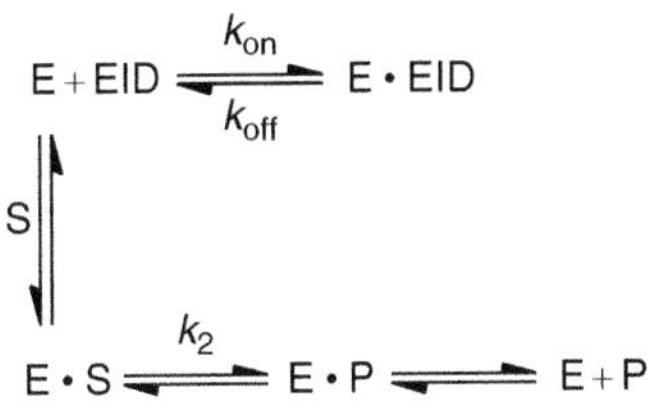

$$\begin{array}{ccc} E + EID & \underset{k_{off}}{\overset{k_{on}}{\rightleftharpoons}} & E \cdot EID \\ S \updownarrow & & \\ E \cdot S & \overset{k_2}{\rightleftharpoons} E \cdot P \rightleftharpoons & E + P \end{array}$$

Scheme 5

preventing the dissociation of the inactivator from the enzyme. This would mean that frequent administration of the drug would not be necessary. Inactivation of an enzyme, however, induces gene-encoded synthesis of that enzyme, but this could take hours to days before sufficient newly synthesized enzyme is present.

Although ALAs could also form a covalent bond to the enzyme, the reactivity of these compounds renders them generally unappealing for drug use because of the possibility that they could react with multiple enzymes or other biomolecules, thereby leading to toxicity with multiple enzymes or other biomolecules, thereby leading to toxicity and side effects. MBEIs, however, are unreactive compounds, so nonspecific reactions with other biomolecules would not be a problem. Only enzymes that are capable of catalyzing the conversion of these compounds to the form that inactivates the enzyme, and also have an appropriately positioned active site group to form a covalent bond, would be susceptible to inactivation. With a sufficiently clever design, it should be possible to minimize (to one?) the number of enzymes that would be affected. Provided the partition ratio is zero or a small number, in which case potentially toxic product release would not be important, then a MBEI could have the desirable drug properties of specificity and low toxicity.

IV. Criteria for Mechanism-Based Enzyme Inactivators

The following experimental criteria have been established to characterize MBEIs. The protocols to support these criteria are described in Section V.

A. Time Dependence of Inactivation

Following a rapid equilibrium between the enzyme and the MBEI to give the E · MBEI complex (see Scheme 4), there is a slower reaction that ensues to convert the inactivator to the form that actually inactivates the enzyme (k_2). This produces a time-dependent loss of enzyme activity in which the pH-inactivation rate profile should be consistent with the pH-rate profile for substrate conversion to product.

B. Saturation

Formation of the E · MBEI complex occurs rapidly, and the rate of inactivation is proportional to added inactivator until sufficient inactivator is added to saturate all of the enzyme molecules. Then there is no further increase in rate with additional inactivator, that is, saturation kinetics are observed.

C. Substrate Protection

MBEIs act as modified substrates for the target enzymes and bind to the active site. Therefore, addition of a substrate or competitive reversible inhibitor slows down the rate of enzyme inactivation. This is referred to as substrate protection of the enzyme.

D. Irreversibility

Most cases of mechanism-based inactivation result in the formation of covalent irreversible adducts. Consequently, dialysis or gel filtration does not restore enzyme activity. Some alternate substrates for enzymes give products that form tight, noncovalent complexes or weak covalent complexes with the enzyme. These enzyme product complexes may dissociate during dialysis or gel filtration. It is not clear at what point in the continuum of enzyme-product complex stabilities that an alternate substrate should be reclassified as a MBEI.

E. Inactivator Stoichiometry

When a radioactively labeled inactivator is used, it is possible to measure the number if inactivator molecules attached per enzyme molecule. Because MBEIs require the catalytic machinery at the active site of the enzyme to convert them to the form that inactivates the enzyme, at most one inactivator molecule should be attached per enzyme active site. In the case of multimeric enzymes, not all of the subunits have functional active sites. Sometimes only one-half of the subunits are catalytically active, and, therefore, only one-half as much inactivator is incorporated per enzyme inactivated. This "half-sites" reactivity may occur because of negative cooperativity after one-half of the subunits are labeled.

F. Involvement of Catalytic Step

One of the most important aspects of MBEIs is that they require the enzyme to convert them to the species that actually inactivates the enzyme. Therefore, it must be demonstrated that there is an enzyme-catalyzed reaction on the compound that is responsible for the activity of the inactivator.

G. Inactivation Prior to Release of Active Species

Another requirement for MBEIs is that the species which is formed by enzyme catalysis inactivates the enzyme prior to release from the active site of the enzyme. It must be shown that a product which is released does not return to the enzyme to cause inactivation. Particularly if the activated product is highly electrophilic, release from the enzyme would be equivalent to the generation of an ALA. A species that inactivates the enzyme after release may produce inactivation by attachment to a site other than the active site, or the compound may have undergone a rearrangement prior to its return to enzyme, which may confuse the conclusion about the inactivation mechanism.

V. Experimental Protocols for Mechanism-Based Enzyme Inactivators

Each of the above-mentioned criteria for MBEIs can be tested experimentally as described in this section.

A. Time Dependence of Inactivation

The experiment to determine time dependence of inactivation is as follows. All of the enzyme assay components except substrate are equilibrated at the desired temperature (generally 25 or 37 °C) in a preincubation tube. It is generally preferable to use a buffer at or near the pH optimum for the enzyme, because the maximum rate of inactivation should be observed at the pH optimum of the enzyme. Sufficient enzyme for 6–10 assays is added to give a solution of *at least* 50 times the concentration necessary for a standard assay of enzyme activity. When investigating a new inactivator, a series of arbitrary inactivator concentrations (e.g., 10 and 100 μM, and 1 and 10 mM final concentrations in the preincubation tube) are tried in separate preincubation tubes to determine the appropriate concentration range to monitor time-dependent inactivation. Sometimes it is more convenient or appropriate to add the enzyme last to initiate the reaction, particularly if the enzyme is somewhat unstable. These experiments are much quicker if a continuous enzyme assay (such as a spectrophotometric assay) is possible rather than a single time-point assay. With a continuous assay the relative rates can be observed immediately, and if complete inactivation has occurred at an early time point, the assay can be discontinued and a lower concentration of inactivator tried. Also, if no change in enzyme activity has occurred after an hour or so, the experiment can be aborted and a higher concentration of inactivator used. With a single-point assay, the rates generally are not known until the entire experiment has been completed, Of course, in a single-point assay, the time selected to make the assay measurement must be in the linear part of the assay rate.

As soon as all of the components of the preincubation tube are added, a stopwatch is begun to mark that time as time zero. Once the solution has been mixed, an aliquot is removed (preferably within 15 s of inactivator addition) and is diluted at least 50-fold into the enzyme assay mixture containing the substrate at saturating concentration and preequilibrated at the desired temperature (it is most convenient to use 25 °C so that everything is at about room temperature); then the assay is carried out. This 50-fold dilution is why the enzyme in the preincubation tube must be at least 50 times, more concentrated than is required for the assay. Even greater dilution would be desirable, because the purpose of the dilution is to quench the inactivation reaction by rapidly decreasing the inactivator concentration and rapidly increasing the substrate concentration so that the substrate is able to occupy the active sites of uninactivated enzyme molecules. If a 100-fold dilution is used, then 100 times more concentrated enzyme is used in the preincubation tube than is needed to monitor an assay. However, it should be noted that when the enzyme concentration is too concentrated in the preincubation tube, rates of inactivation decrease, possibly because of the increased viscosity or protein-protein interactions.

Periodically (seconds to hours, depending on the observed inactivation rate) identical aliquots are removed from the preincubation tube, diluted as above, and assayed. With the continuous assay, it is easy to determine when to take aliquots, because the change in enzyme activity is observed immediately. Additional time points need to be taken with a single-point assay because the rates generally are not known until after the entire experiment is completed. Once a convenient concentration range is determined (one in which complete inactivation occurs within a couple of hours or less), the inactivation experiment is repeated with a series of concentrations (at least five different concentrations, but more is better) in that range spanning about an order of magnitude of concentrations. At least five time points, over at least two half-lives, should be taken; more time points will give more accurate data. Loss of enzyme activity should be monitored until no activity remains to determine if there is a change in kinetics with time. When a crude enzyme system is used or when a single-point assay is used, it is best to do duplicate or triplicate runs per concentration and average the results.

All inactivation experiments are monitored relative to a control sample which is done exactly as the experiment except without inactivator added (an equal volume of buffer is added instead). The enzyme activity in this sample is set to 100% at each time point. This control takes into account the normal loss of enzyme activity under the conditions of the experiment not as a result of the presence of the inactivator.

Once the data have been collected, the logarithm of the percentage of enzyme activity remaining (as measured by whatever means the enzyme activity is determined) relative to the uninactivated control is plotted against the time of preincubation (Fig. 1). Increasing inactivator concentrations should produce increased rates of inactivation. Often these plots exhibit pseudo-first-order kinetics because the inactivator is added in a large excess relative to the enzyme concentration.

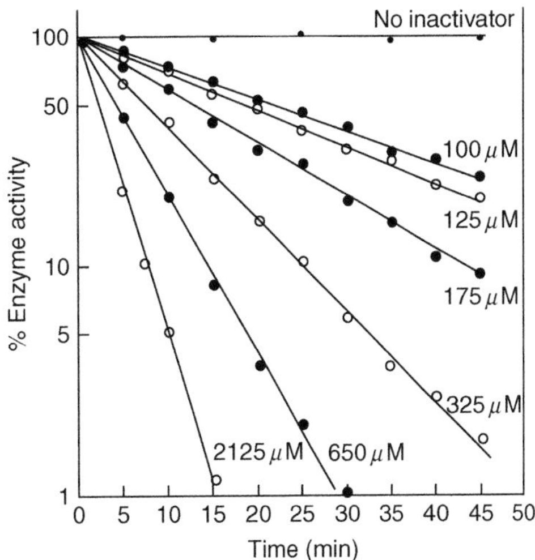

Fig. 1 Time- and concentration-dependent inactivation of an enzyme by a MBEI. The inactivator concentrations for each line are given.

However, this is the ideal case. Equations have been derived (Funaki *et al.*, 1991) showing that pseudo-first-order kinetics are followed only when the ratio of the initial concentration of the mechanism-based inactivator to that of the enzyme is greater than the partition ratio. When the inactivator is very potent (i.e., has a low K_I and a large k_{inact}), it may not be possible to measure the inactivation rates fast enough unless the inactivator concentration is lowered to a point approaching the concentration of the enzyme; in that case, pseudo-first-order kinetics would not be observed. To slow down the inactivation rate, the use of nonideal conditions, such as lowering the temperature or changing the pH of the buffer from that needed to get optimum enzyme activity, can be used. However, not only will the rate change, but all kinetic constants (e.g., K_I and k_{inact}; see Section V.B) also will be altered. Because of the effects of temperature, pH, ionic strength, and buffer, kinetic constant comparisons cannot be made between inactivators measured under different conditions.

Another common problem in doing these experiments arises when the partition ratio for the inactivator is very high and the concentration of the inactivator diminishes during the time course of the inactivation experiment. This may lead to nonpseudo-first-order kinetics (Fig. 2). If this is suspected, then some type of analytical determination of the inactivator concentration during the time period of the experiment should be carried out, such as high-performance liquid chromatography (HPLC) analysis. If inactivator consumption is a problem, then the early time points would be more dependable than the later ones.

Nonpseudo-first-order kinetics also would be observed if a product generated is a good inhibitor of the enzyme. As the concentration of the metabolite increases, more inhibition occurs, and this prevents the inactivator from being effective.

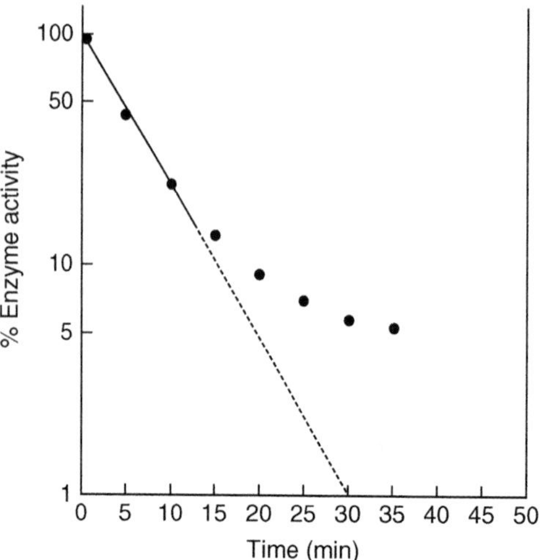

Fig. 2 Nonpseudo-first-order inactivation as a result of consumption of inactivator.

This problem is more prevalent with compounds having high partition ratios. In these cases, the earlier time points and higher inactivator concentrations give more reliable data.

In the two cases above of nonpseudo-first-order kinetics, there is an upward deviation to the loss of enzyme activity (i.e., the rate decreases with time) relative to pseudo-first-order kinetics. In some cases, the opposite may be observed, namely, there is an increase in the rate of inactivation with time (Fig. 3). The rate of increase of enzyme inactivation does not have to be as pronounced as is shown in Fig. 3. In general, this result suggests that the initial compound incubated with the enzyme is converted to another compound, which is the actual inactivator of the enzyme. As the concentration of the actual inactivator species increases, the rate of inactivation increases, until it reaches the maximum saturation rate. If a fresh aliquot of enzyme is then added to that same solution, which already is saturated with the actual inactivator species, then there will be no lag period to inactivation, and the maximum inactivation rate will be observed immediately.

There are several possible sources for the formation of the actual inactivator. It could be the result of a nonenzymatic conversion of the presumed inactivator into either a reactive species or a new mechanism-based inactivator. Alternatively, the presumed inactivator may be a promechanism-based inactivator; that is, the enzyme may convert the compound to a MBEI that is released into solution, and this new compound is the actual inactivator of the enzyme. Alternatively, the enzyme may convert the compound to a moderately reactive species, which escapes the active site, then as it builds in concentration, returns to inactivate the enzyme. As described below (see Section V.G), a test for the generation of a reactive species is to add an electrophile-trapping agent (such as a thiol) and see if this prevents

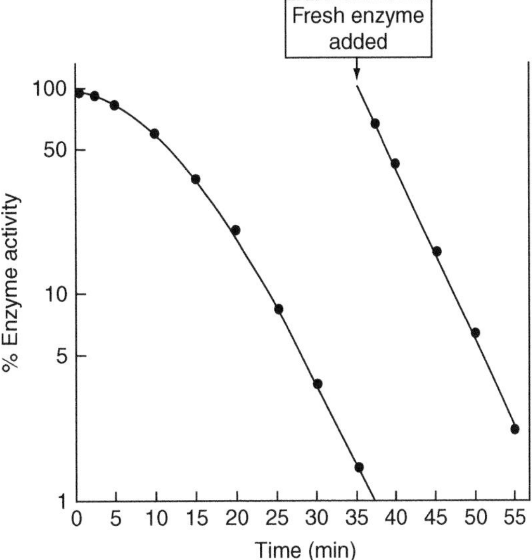

Fig. 3 Nonpseudo-first-order inactivation as a result of conversion of a presumed inactivator to the actual inactivating species.

inactivation. In any case, if the actual inactivator species is released from the enzyme, the compound is not classified as a MBEI.

Finally, there are times when biphasic kinetic are observed. This can arise when two or more inactivation mechanisms are occurring simultaneously, if the inactivated enzyme adduct is not stable and the breakdown of this adduct is rate determining, if there is negative cooperativity and attachment of the inactivator to one subunit causes an adjoining subunit to be less active, and if there is heterogeneity of subunit composition that results in nonequivalent binding to the subunits.

B. Saturation and Determination of K_I and k_{inact}

Saturation is demonstrated by carrying out the time-dependent experiment described above at several (at least five) different concentrations at a constant enzyme concentration. If little or no change in inactivation rate is observed with increasing concentration of inactivator, it suggests that the initial inactivator concentration selected is near saturation. In this case, lower concentrations should be used to produce sufficiently large differences in the rates of inactivation so that results are not ambiguous. The half-life for inactivation ($t_{1/2}$) at each inactivator concentration is plotted against 1/[inactivator concentration] (known as a Kitz and Wilson plot; Kitz and Wilson, 1962). If inactivation proceeds with saturation, then a plot like that shown in Fig. 4 will be observed. Saturation is indicated by the

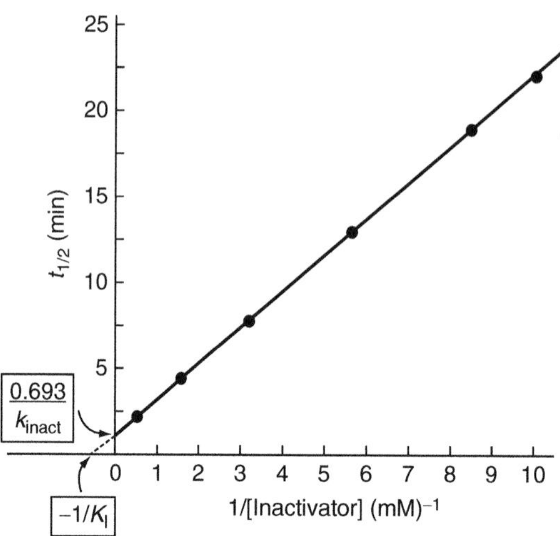

Fig. 4 Kitz and Wilson (1962) replot of the half-life of enzyme inactivation as a function of the reciprocal of the mechanism-based inactivator concentration.

intersection of the experimental line at the positive y-axis. When that occurs, then there is a finite rate of inactivation at inln $2/k_{inact}$,finite inactivator concentration. The point of intersection at the y-axis is, from which the value of k_{inact} can be determined. The extrapolated negative x-axis intercept is $-1/K_I$. From Fig. 4, the k_{inact} is 0.46 min^{-1} and the K_I is 1.48 mM.

There are several kinetics and graphics programs that can be used to obtain these results instead of determining them manually. Inactivator concentrations should be selected so that the data points are not bunched at one end of the experimental line. This allows the points to be given equal weighting in the analysis. Therefore, choose concentrations based on 1/[I], not [I]. If K_I is low and k_{inact} is large, all concentrations of the inactivator used may give the same rate of inactivation. This indicates that saturation has already been reached. If the inactivator concentration is lowered, nonlinear kinetics may result because the inactivator concentration may approach that of the enzyme. Carrying out the experiment at nonideal conditions, for example, at a lower temperature or at a nonideal pH, may permit subsaturating concentrations of inactivator to be used. Typically, the values for k_{inact} are quite small (on the order of 10^{-3}-1 min^{-1}) relative to values of k_{cat} for substrates (10–10^4 min^{-1}).

When the experimental line intersects at the origin, then saturation is not being observed (Fig. 5). This indicates that a bimolecular reaction is occurring and that k_{inact} is fast relative to formation of the E · MBEI complex. When this is observed, alteration of the experimental conditions, for example, lowering the temperature (to slow down k_{inact}) or changing the pH, may give saturation kinetics.

Intersection of the experimental line at the negative y-axis has no physical meaning (a negative rate of inactivation). When this occurs, more concentrations

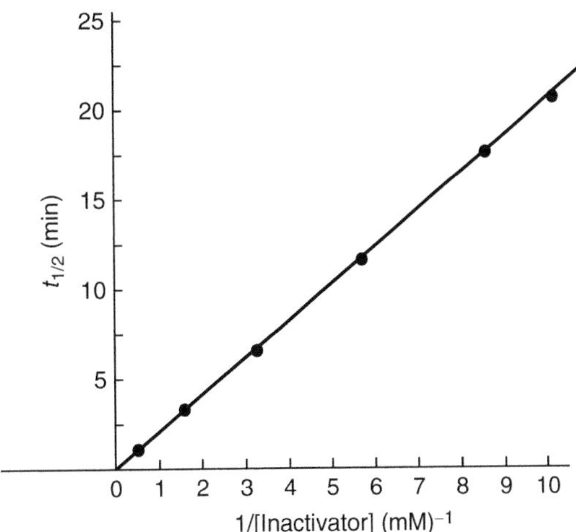

Fig. 5 Mechanism-based inactivation without saturation.

should be used in different concentration ranges. If increasing the inactivator concentration leads to an upward deviation of the experimental line (large half-lives, i.e., slower rates, at higher inactivation concentrations), then the rate of inactivation may be a function of $[I]^2$, that is, there may be two binding sites or two inactivator molecules involved in inactivation. Kinetics for two-site inactivation has been described (Kuo and Jordan, 1983).

C. Substrate (or Competitive Inhibitor) Protection

To show that inactivation is occurring at the active site of the enzyme, the inactivation experiment described by Fig. 1 is repeated under identical conditions in the presence and absence of a known substrate for the enzyme. Because both the MBEI and the substrate must bind at the active site, the presence of the substrate will prevent the binding of the inactivator, and, therefore, the rate of inactivation will decrease compared with the rate in the absence of substrate. It is best to do the experiment with several different substrate concentrations as is shown in Fig. 6. A competitive inhibitor of the enzyme also can be used to protect the enzyme from the inactivator.

D. Irreversibility

After inactivation is complete, the enzyme is dialyzed against several changes of buffer at 4 °C or subjected to gel filtration at 4 °C to remove any reversibly bound inactivator molecules. A noninactivated enzyme control should be run simultaneously and carried through the same operations as the inactivated enzyme.

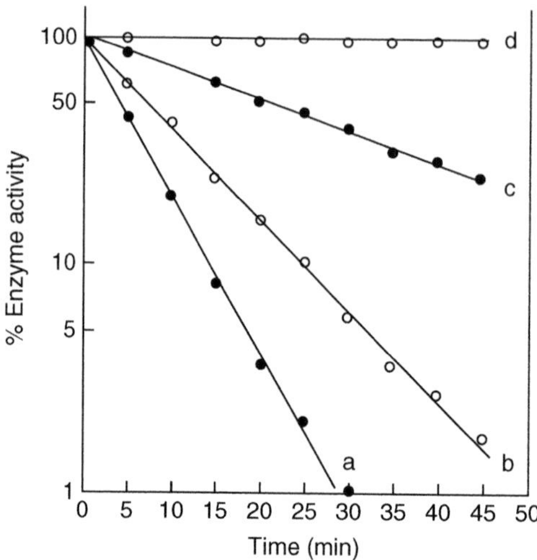

Fig. 6 Substrate protection from inactivation: (A) 650 μM inactivator with no substrate; (B) 650 μM inactivator with 100 μM substrate; (C) 650 μM inactivator with 500 μM substrate; (D) 650 μM inactivator with 2 mM substrate or no inactivator.

This control is required to show that under the conditions of the experiment, the enzyme would still be active if it were not treated with the inactivator, and the control enzyme activity is set to 100% for comparison. If the inactivated enzyme complex is marginally stable, dialysis or gel filtration at room temperature instead of at 4 °C may result in partial or complete return of enzyme activity. If more than one adduct were formed, possibly only one of the adducts would decompose at room temperature with concomitant reactivation of that portion of the enzyme activity. Extended dialysis at room temperature may release more adduct. Arrhenius activation energies and free energies can be estimated if the rates of reactivation can be determined at various temperatures. Dialysis at higher or lower pH also may yield useful information about the adduct stability and structure.

E. Inactivator Stoichiometry

Often the most difficult or time-consuming part of the characterization of a mechanism-based inactivator is the chemical synthesis and purification of a high specific activity radioactively labeled analog. As a rule of thumb, a specific activity of 0.5 mCi/mmol [1.1×10^6 disintegrations/min (dpm)/μmol] is satisfactory for a ^{14}C-labeled compound, but higher specific activities (usually more easily done by tritium incorporation, but the tritium must be in a nonexchangeable position) will permit the use of less enzyme. Once the labeled inactivator has been obtained, it is incubated with the enzyme. If aliquots are removed during inactivation and each assayed for enzyme activity remaining and for radioactivity attached to the protein

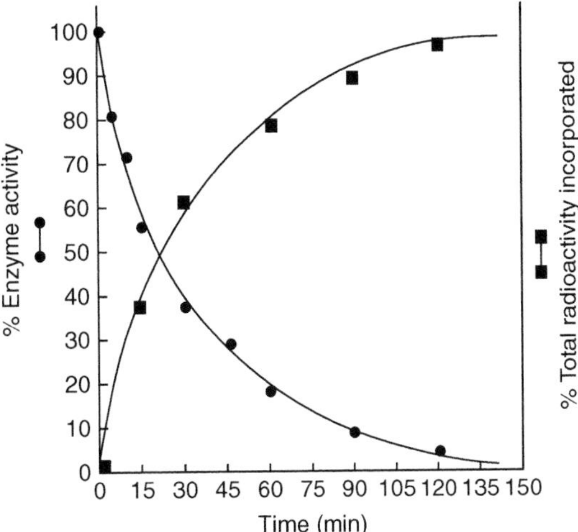

Fig. 7 Correspondence of the time-dependent inactivation of an enzyme with the incorporation of radioactivity from a radiolabeled mechanism-based inactivator. Periodically, aliquots are removed and assayed for remaining enzyme activity and for radioactivity covalently bound to the enzyme.

(either by acid precipitation or gel filtration followed by scintillation counting), there should be a correlation of loss of enzyme activity with gain of radioactivity; at 50% loss of enzyme activity there should be approximately 50% incorporation of total radioactivity (Fig. 7). When no activity remains, exhaustive dialysis or gel filtration is carried out to remove all of the unbound inactivator (a small volume dialysis can be carried out after several large volume dialyses to determine if additional radioactivity is being released). A protein assay and radioactivity determination are made, and from the specific activity of the inactivator the stoichiometry of inactivation (equivalents of compound bound per enzyme molecule) can be calculated.

This experiment is most effective if homogeneous enzyme is used, so that a direct protein and radioactivity determination can be made. However, gel electrophoresis of a labeled enzyme mixture can be done, the bands eluted, and protein and radioactivity determinations made on the eluted protein.

If some enzyme activity remains after dialysis, then only the radioactivity bound per inactivated enzyme should be calculated (multiply the total protein by the fraction inactivated). Generally, for a mechanism-based inactivator a 1:1 stoichiometry of binding is observed (1.0 ± 0.3 is usually considered to be 1 equivalent, given the errors in all of these determinations). If greater than 1:1 stoichiometry is observed, it may indicate that incomplete dialysis or gel filtration was done or nonspecific covalent reactions took place, possibly as a result of an activated form of the inactivator being released during inactivation. An electrophile-trapping agent, such as 2-mercaptoethanol or dithiothreitol, would be useful to have present during inactivation in order to prevent potent released electrophiles from labeling the periphery of the enzyme. Sometimes excess inactivator is tightly but reversibly bound to the protein. If the covalent adduct is stable, the enzyme can be denatured to release unstable adducts or excess reversibly bound radioactivity. Denaturants such as 8 M urea or 1% sodium dodecyl sulfate are very gentle means of denaturation, but trichloroacetic acid precipitation or heat denaturation also can be effective. When less than 1:1 stoichiometry is observed, it may indicate one or more of the following: (1) some of the enzyme used was inactive (possibly denatured) prior to treatment with the inactivator; (2) more than one adduct formed, but at least one of the adducts was unstable to dialysis or gel filtration; (3) in the case of multimeric enzymes, some kind of negative cooperativity was important. When a 0.5:1 stoichiometry is observed, often it indicates that half-site reactivity is occurring.

F. Involvement of Catalytic Step

Experiments to test the involvement of a catalytic step have the most variability of any of the experiments for MBEIs because they depend on the specific reaction catalyzed by the particular enzyme. If an oxidation or reduction of a cofactor is involved, then this may be observed spectrophotometrically. If a C–H bond is cleaved, then either deuteration or tritiation of that hydrogen can be used; a deuterium isotope effect or release of tritium, respectively, can be monitored. If part of the inactivator is cleaved during inactivation, radioactive labeling of that part of the compound and demonstration that it is released during inactivation would be an appropriate experiment. The important aspect is to show that the enzyme is involved in the activation process.

G. Inactivation Prior to Release of Activated Species

There are several experiments that can be carried out to test whether inactivation occurs before or after an activated species is released from the enzyme. If inactivation occurs as a result of a species that is released, then the rate of inactivation may increase with time as the species increases in concentration outside of the enzyme. In this case, pseudo-first-order kinetics would not be evident; rather, there would be an increase in the rate of inactivation over time (see Fig. 3 in Section V.A). However, because the reactive species also may react with the buffer, buffer quenching may compete with the enzyme for the released species.

Another experiment to test this phenomenon is first to measure the rate of inactivation of the enzyme, then add a fresh aliquot of enzyme. If the inactivating species is building in concentration, then the rate of inactivation of the second aliquot of enzyme will be as fast as the maximum rate of inactivation with the first aliquot or faster than that for the first aliquot, if saturation had not yet been achieved with the first aliquot (see Fig. 3 and note that the lag period does not have to be as long as shown). With a MBEI, there should be no difference in inactivation rates no matter how many successive aliquots are added (unless the concentration of the inactivator is depleted).

A third test is to have a trapping agent present in the preincubation solution to react with any electrophilic or radical species generated and released by the enzyme. If inactivation occurs subsequent to release of an activated species, then the trapping agent would prevent that species from inactivating the enzyme. In this case, there would be a considerable decrease in the inactivation rate in the presence of the trapping agent. In fact, it may totally prevent inactivation from occurring. The presence of the trapping agent should have no effect on the inactivation rate of a MBEI. Because thiols are both excellent nucleophiles and radical traps, 2-mercaptoethanol or dithiothreitol are commonly used in this experiment. It is also useful to use a bulky thiol, such as reduced glutathione, to minimize the possibility of the thiol entering the active site and trapping the reactive species prior to its release from the active site.

H. Determination of Partition Ratio

There are three common methods for the determination of the partition ratio. When a product is released, the rate constant that relates to the formation of a product is k_{cat}, the catalytic rate constant, determined from Michaelis-Menten kinetics. The rate constant that relates to inactivation is k_{incat}, determined as described above. The ratio k_{cat}/k_{incat}, then, defines the partition ratio. If k_{cat} can be measured easily, then this is the simplest procedure to follow. When a potent MBEI is used or if the partition ratio is small, inactivation may occur too rapidly for accurate k_{cat} measurements to be made.

Another method of partition ratio determination requires the synthesis of a radioactively labeled analog of the inactivator and is only effective for covalent inactivation. After complete inactivation occurs, the small molecules are separated from the enzyme by gel filtration, ultrafiltration, protein precipitation, or microdialysis. The latter technique involves suspension of the labeled enzyme solution contained in narrow dialysis tubing into a test tube containing a minimum volume of buffer (a dialysis clip at the top will hold it suspended; the bottom of the tubing is knotted). A flea stir bar is added to the test tube, and the buffer (3–5 ml) is stirred while dialysis takes place. The dialyzate is saved, a fresh aliquot of buffer is added, and dialysis is continued. After the second microdialysis is completed, the combined dialyzates are retained for further analyses of the small molecules. These two dialyses remove a large percentage of the radioactive small molecules. If, for example, 300 μl of enzyme solution is dialyzed twice in 3 ml each of buffer, then greater than 99% of the small molecules will be separated from the enzyme solution. The enzyme solution is then exhaustively dialyzed to remove the traces of remaining unbound radioactivity, and the stoichiometry of labeling is determined by protein and radioactivity assays. Because a large excess of inactivator is generally used for inactivation, the small molecules must be chromatographed (ion-exchange or reversed-phase silica gel chromatography) to separate the excess inactivator from the products. The amount of radiolabeled metabolites produced per radiolabeled enzyme can be determined to give the partition ratio directly.

Whereas standard gel filtration techniques dilute the small molecule fraction considerably and require more time to set up than microdialysis, a modified version of gel filtration, described by Penefsky (1977, 1979) is fast and results in much less dilution than the standard method. Dilution is not a problem with ultrafiltration, but protein has a tendency to adhere to the filtration membrane. A fast and convenient microultrafiltration tool, however, is the Centricon centrifugal microconcentrator[2] which allows recovery of both filtrate and protein without dilution (generally, though, buffer is added to the concentrated protein to remove the last of the small molecules, so there may be some dilution).

A third approach for the determination of the partition ratio is to titrate the enzyme with the inactivator; this measures the number of inactivator molecules required to inactivate the enzyme completely. Increasing concentrations of inactivator are added to a fixed amount of enzyme, and each is incubated until no more enzyme activity is observed. After dialysis or gel filtration, a plot is constructed of the enzyme activity remaining versus the ratio of amount of inactivator per enzyme active sites (Fig. 8). If the enzyme solutions of all of the samples are normalized to a constant protein concentration after dialysis or gel filtration, then direct comparisons can be made between samples.

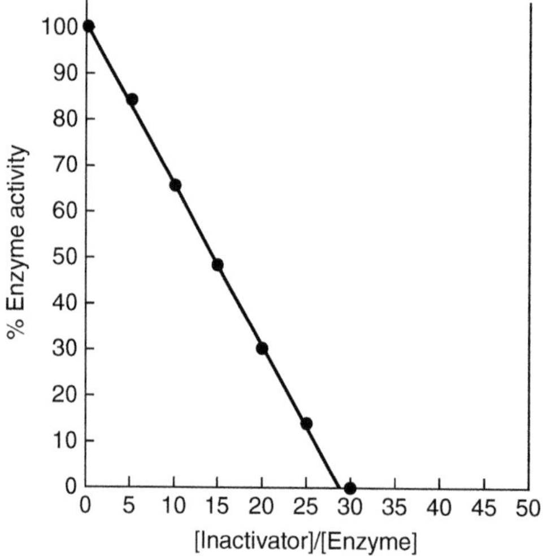

Fig. 8 Titration of an enzyme with a MBEI. The loss of enzyme activity is measured as a function of the ratio of the inactivator to enzyme concentration.

[2] Registered trademark of the Amicon Division of W. R. Grace & Co., Beverly, MA 01915-1065; Tel.: 1-800-343-0696.

In the ideal case, a straight line will be obtained from 100% to 0% enzyme activity remaining. The intercept with the x-axis gives the number of inactivator molecules required to inactivate each enzyme molecule (the turnover number). This number includes the one molecule of inactivator required to inactivate the enzyme; consequently, the partition ratio is the turnover number minus one (assuming there is a 1:1 stoichiometry of inactivator and enzyme). Often the higher ratio data points deviate from linearity because of product inhibition or product protection of the enzyme from further inactivation (Fig. 9). In this case, the linear portion (the lower inactivator concentrations) should be used for extrapolation to the turnover number. If the inactivated enzyme solution is not dialyzed or gel filtered prior to the determination of the enzyme activity remaining, and if a product formed is a potent reversible inhibitor, then each assay will show an artificially low enzyme activity, resulting in a falsely low partition ratio. This method is the least reliable because there are several ways in which falsely high and falsely low partition ratios can be obtained.

The point in the inactivation mechanism where product formation branches from inactivation can be determined when C–H bond cleavage is involved (Fitzpatrick and Villafranca, 1986; Fitzpatrick *et al.*, 1985). In this case, the deuterium isotope effect on the partition ratio, on the rate of product formation, and on the rate of inactivation should be measured. If there is no isotope effect on the partition ratio or on V_{max}, but the same isotope effect on k_{inact}/K_I and V_{max}/K_I, then inactivation is occurring from a species in which C–H bond cleavage has already occurred, and both product formation and inactivation occur from a common intermediate. If there is no isotope effect on the partition ratio, but a different isotope effect on V_{max} than on k_{inact}, the product formation and

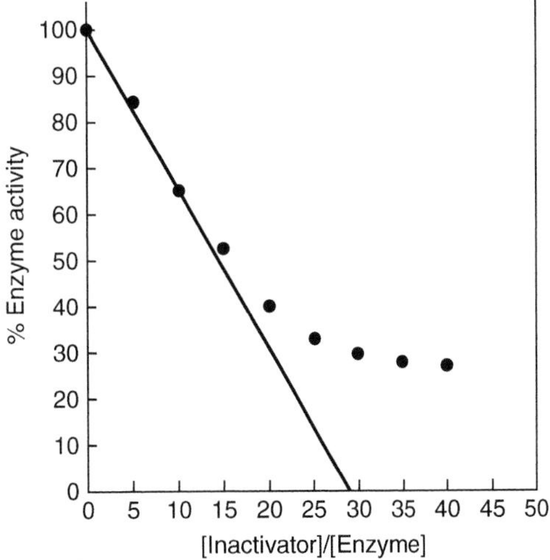

Fig. 9 Titration of an enzyme with a MBEI in which there is product inhibition.

inactivation occur from different species, and both pathways involve C–H bond cleavage. If there is a normal deuterium isotope effect on the partition ratio, an inverse isotope effect on k_{inact}/K_I, and a normal isotope effect on V_{max}, then partitioning is occurring at the point of C–H bond cleavage, and this bond breakage only is involved in product formation, not in inactivation.

VI. Applications

The remainder of this chapter deals with examples on the above-described methodologies for MBEIs and approaches that can be taken to elucidate inactivator mechanisms. The experiments described for the given enzymes are general approaches that can be applied to other enzymes as well. The examples used come from the author's laboratory. It is not the intent of the author to lead the reader to believe that the only or best examples are from the author's laboratory; these are just the most convenient and well known to the author.

A. Mechanism-Based Inactivation of γ-Aminobutyric Acid Aminotransferase

γ-Aminobutyric acid (GABA) aminotransferase is a pyridoxal 5′-phosphate (PLP)-dependent enzyme whose mechanism is shown in Scheme 6. 4-Amino-5-halopentanoic acids (**1**, X = F, Cl, Br; Scheme 7) are mechanism-based inactivators of GABA aminotranferase (Silverman and Levy, 1980). The standard experiments described above were carried out. Inactivation is time dependent, pH dependent, exhibits saturation, is blocked by the presence of substrate, and is not affected by the presence of the electrophile-trapping agent 2-mercaptoethanol. Dialysis of the inactivated enzyme does not result in return of enzyme activity. Inactivation by (*S*)-[*U*-^{14}C]-4-amino-5-chloropentanoic acid leads to incorporation of 0.85–1.25 equivalents of radioactivity per active site (Silverman and Levy, 1980, 1981). No radioactivity was released on 8 M urea denaturation, indicating that a stable covalent bond was formed. Concomitant with enzyme inactivation, there is a time-dependent change in the UV-visible spectrum of the PLP cofactor to a spectrum that resembles that of pyridoxamine 5′-phosphate (PMP) (Silverman and Levy, 1981). The spectral change indicates that inactivation requires the catalytic action of the enzyme, an essential characteristic of a MBEI.

Another experiment to demonstrate that inactivation was enzyme catalyzed was to show that (*S*)-[4-^{2}H]-4-amino-5-chloropentanoic acid inactivated GABA aminotransferase with a deuterium isotope effect of 6.7. The fact that inactivation occurred with an isotope effect indicates that C–H bond cleavage is a rate-determining step in inactivation. When 4-amino-5-fluoropentanoic acid (**1**, X = F) was the inactivator, it was shown that one fluoride ion was released for every radioactive molecule incorporated into the enzyme. On the basis of these initial results, the inactivation mechanism shown in Scheme 7 was proposed (Silverman and Levy, 1980, 1981).

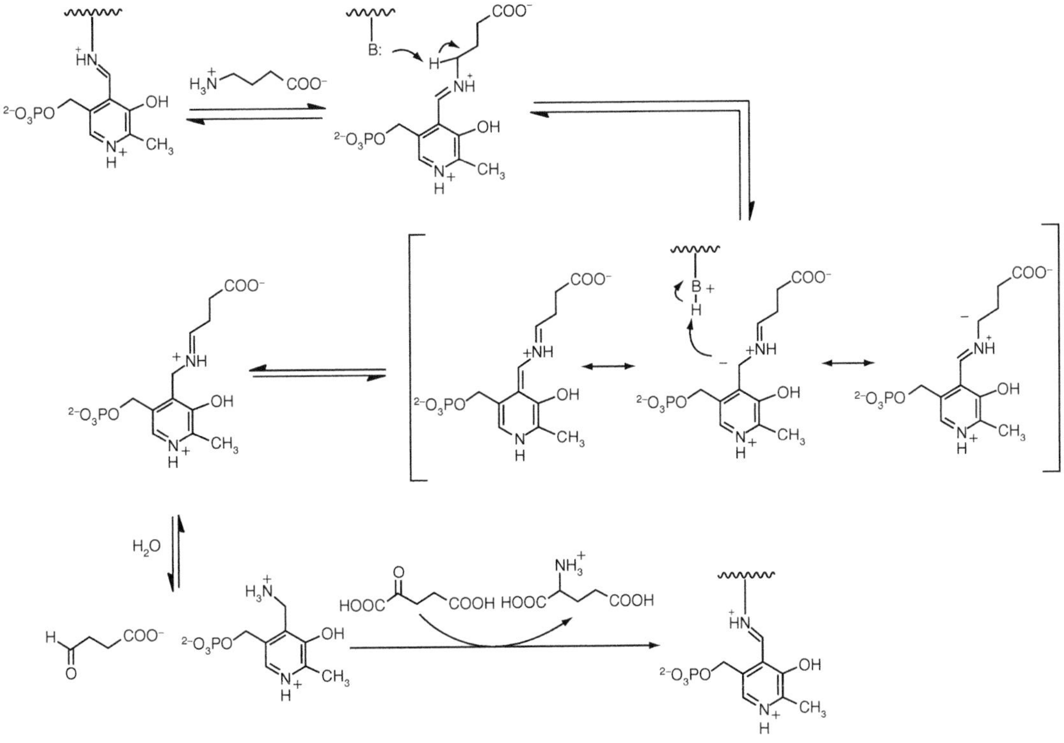

Scheme 6

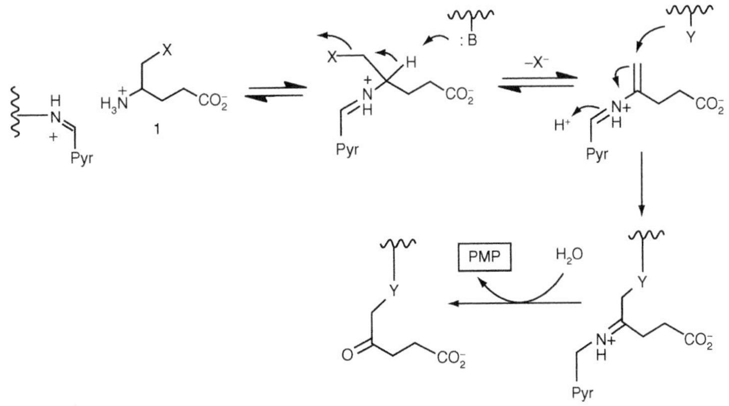

Scheme 7

The partition ratio for this MBEI was determined by several methods to be zero. First, complete inactivation of the enzyme by these inactivators occurs in the absence of α-ketoglutarate, the second substrate required to convert any PMP back to PLP (see Scheme 6). When α-ketoglutarate is added to the inactivated enzyme, no activity is regenerated. If any of the inactivator molecules are turned over without inactivation, then some of the enzyme would have been converted to the PMP form, and active enzyme would be generated on incubation with α-ketoglutarate. Furthermore, when α-[*U*-^{14}C]ketoglutarate was present during inactivation with 4-amino-5-fluoropentanoic acid, no [*U*-^{14}C]glutamate, the product of transamination of PMP with α-ketoglutarate, was detected. Second, only one fluoride ion is released per inactivation event. If multiple turnovers were occurring, then more than one fluoride ion would be released. Third, when (*S*)-[*U*-^{14}C]-4-amino-5-chloropentanoic acid is used to inactivate the enzyme, essentially no radioactive nonamines are generated, indicating that no products are produced during inactivation.

All of the above results are consistent with the mechanism shown in Scheme 7, but additional labeling experiments proved otherwise. Apo-GABA aminotransferase was reconstituted with [4-^{3}H]PLP, inactivated with (*S*)-4-amino-5-fluoropentanoic acid, then denatured. According to the mechanism shown in Scheme 7, [^{3}H] PMP should be released; however, essentially no [^{3}H]PMP was detected. Instead, the entire radioactivity was released from the enzyme as **3** (Scheme 8), indicating that a different inactivation mechanism via an enamine intermediate occurs. This mechanism is based on the earlier work of Metzler and coworkers (Likos *et al.*,

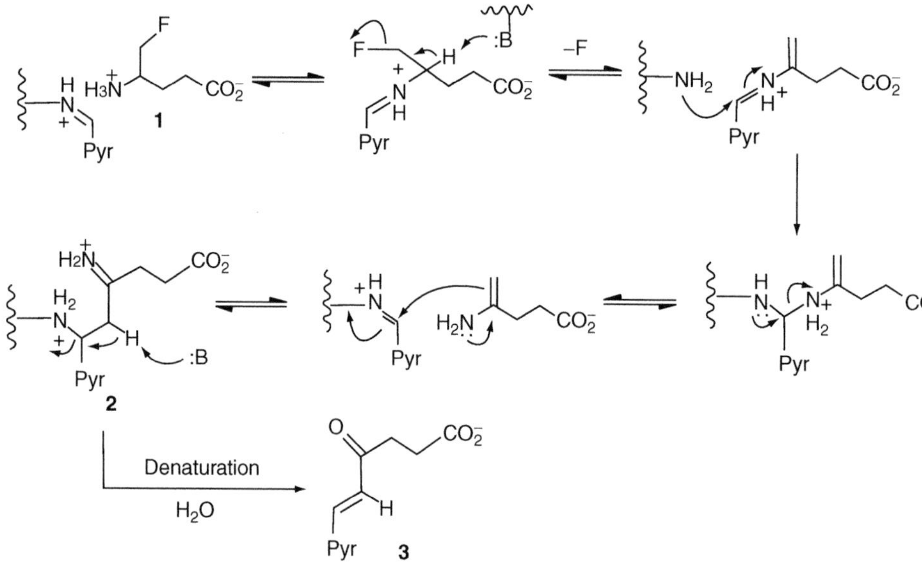

Scheme 8

1982; Ueno *et al.*, 1982). Compound **3** is derived from the covalent adduct **2** by β-elimination. Part of the initial confusion about the inactivation mechanism was derived from the UV-visible spectrum of the inactivated enzyme, which appeared to that of PMP. Adduct **2** is similar in structure to PMP and would be expected to have a similar UV-visible spectrum.

Another example of how radioactive labeling can be used to differentiate inactivation mechanisms was described for the inactivation of GABA aminotransferase by γ-ethynyl-GABA (**4**) (Burke and Silverman, 1991). At least nine different possible inactivation adducts, derived from three different inactivation mechanisms, can be imagined for this inactivation reaction (Fig. 10; see Burke and Silverman (1991) for these mechanisms, if interested). Four are PLP adducts (**5**, **6**, **9**, and **10**), two are PMP adducts (**7** and **8**), two are adducts derived from an enamine mechanism as described above for 4-amino-5-fluoropentanoic acid (**11** and **12**), and the other is derived from an enamine, but by a mechanism different from that described above. This mechanism does not affect the PLP cofactor at all (**13**).

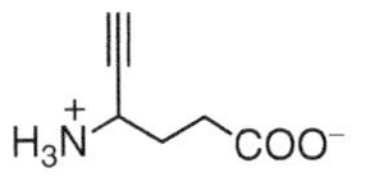

$H_3\overset{+}{N}$ COO⁻
4

These possibilities were differentiated with the use of two radioactive compounds, [4-^{3}H] PLP and γ-ethynyl[2-^{3}H]-GABA. As described above, apo-GABA aminotransferase was reconstituted with [4-^{3}H]PLP, and the fate of the

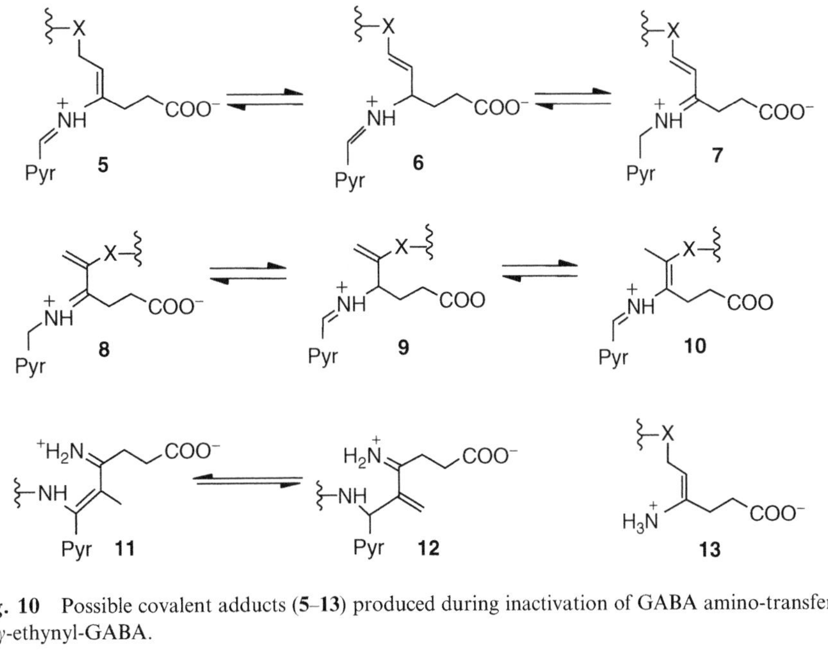

Fig. 10 Possible covalent adducts (**5**–**13**) produced during inactivation of GABA amino-transferase by γ-ethynyl-GABA.

PLP was determined after inactivation. More than 95% of the tritium was released on denaturation as [4-^{3}H]PLP; no ^3H_2O was detected. These results exclude the two PMP adducts that could have been formed and the two enamine adducts. The four PLP adducts (**5**, **6**, **9**, and **10**) and the other enamine-derived product (**13**) were differentiated with γ-ethynyl[2-^{3}H]-GABA. Inactivation of GABA aminotransferase with γ-ethynyl[2-^{3}H]-GABA followed by exhaustive dialysis led to the incorporation of 1.0 equivalent of inactivator bound to the enzyme. Denaturation of the labeled enzyme with 6 M urea, however, resulted in the release of about one-half of the radioactivity, unless the denaturation was carried out at pH 7.4 or pH 9.5 and 4 °C, in which case no radioactivity was released. This suggests that either there are two different adducts formed, one stable to denaturation and the other unstable, or that one adduct can hydrolyze to give two different products, one stable to denaturation and one unstable. If the labeled enzyme is treated with sodium borohydride prior to denaturation, then, even at room temperature, little or no radioactivity is released, indicating that the unstable product is probably attached to the enzyme in the form of a Schiff base. To be consistent with that result, the linkage to the enzyme (X in **5**–**10** and **13**) must be an amino group, presumably from a lysine residue. The unstable product (that released on denaturation) was isolated and identified by HPLC comparison to a synthesized standard compound as **14**, which could be derived from either **9** or **10** as shown in Scheme 9.

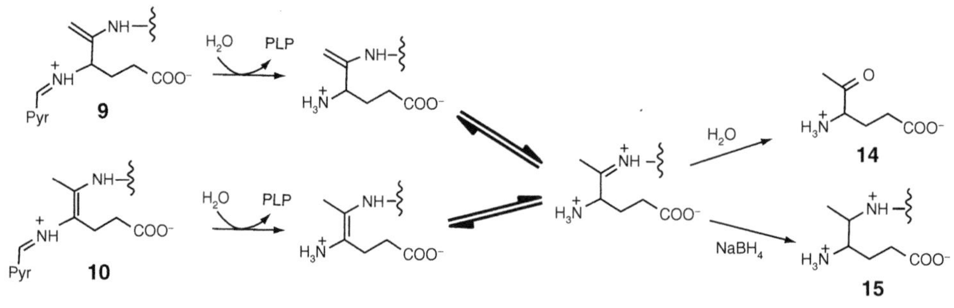

Scheme 9

The stable adduct also must be derived from the PLP adducts (**5**, **6**, **9**, and **10**). Consider the products of hydrolysis of these adducts (structures **15**–**18**, Scheme 10). For **16** or **17** to be candidates for the stable adduct, X cannot be NH; otherwise, they would be unstable. If X is another heteroatom (O or S), then they could be stable. Treatment of the stable adduct under conditions known to hydrolyze vinyl ethers (X = O) and vinyl sulfides (X = S) did not release the radioactivity from the enzyme; therefore, **16** and **17** are not the stable adduct. Adducts **15** and **18** (X = NH) were differentiated by reduction with sodium borohydride followed by oxidation with sodium periodate (Scheme 11). If the adduct has structure **15** (X = lysine), then reduction with sodium borohydride

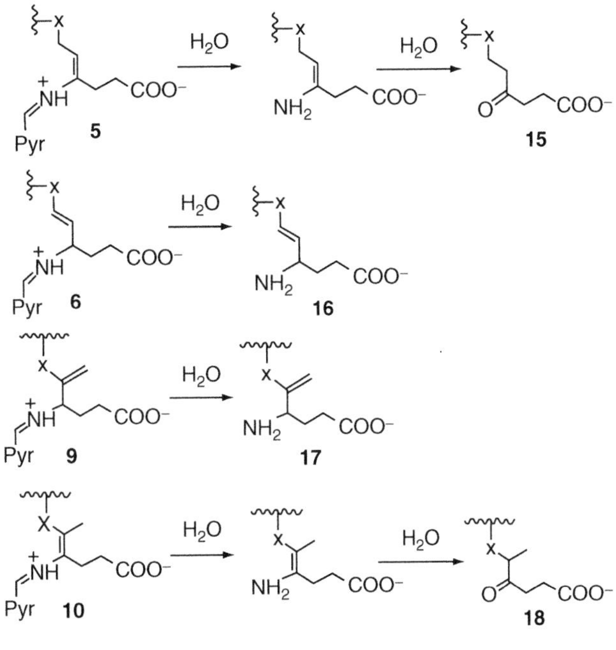

Scheme 10

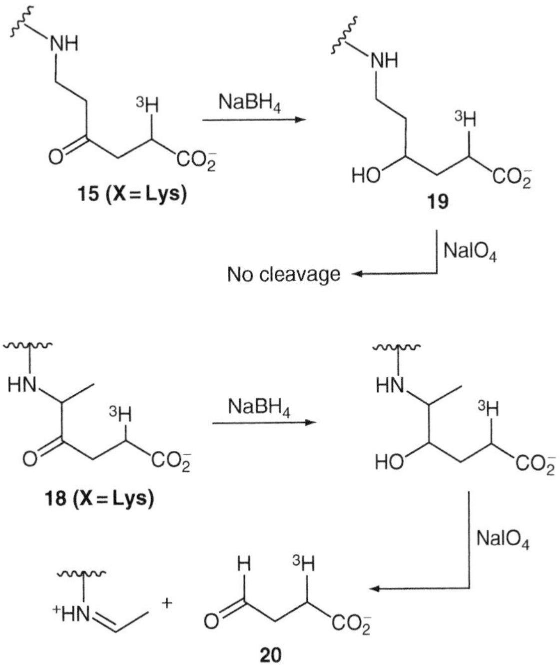

No cleavage

Scheme 11

would give adduct **19** which would be stable to treatment with sodium periodate; consequently, all of the tritium would remain attached to the enzyme. If the adduct has structure **18** (X = lysine), however, the sodium borohydride-reduced adduct would be oxidized by sodium periodate to give tritiated succinic semialdehyde (**20**). This experiment resulted in complete release of the tritium from the enzyme as succinic semialdehyde which was identified by HPLC. Therefore, the stable adduct has the structure **18** (X = lysine). Given that the unstable adduct could have come from **9** or **10** and the stable adduct is derived from **10**, it is most likely that both the unstable and the stable adduct are derived from **10** (Scheme 12). Hydrolysis of **10** gives **21** (Scheme 12), which can break down equally by pathways (a) and (b) to give the unstable adduct and release **14** or the stable adduct (**18**), respectively.

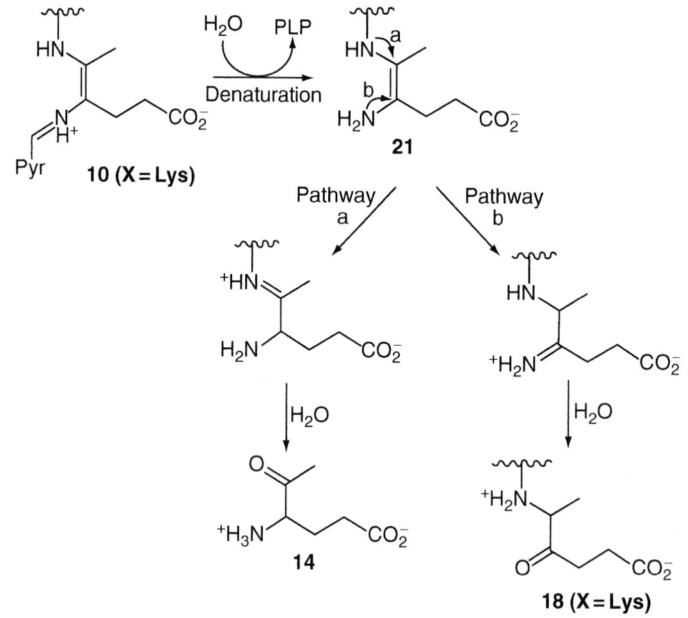

Scheme 12

To determine the partition ratio and to complete the metabolic profile of inactivation of GABA aminotransferase by γ-ethynyl-GABA, turnover products formed during inactivation were isolated. Incubation of GABA aminotransferase with γ-ethynyl-GABA in the presence of α-[5-^{14}C]ketoglutarate produced about 1 equivalent of [5-^{14}C]glutamate; consequently, in addition to inactivation there is one molecule of γ-ethynyl-GABA transaminated, presumably (but not identified) to 4-oxo-5-hexynoic acid (**22**, Scheme 13). Inactivation with γ-ethynyl[2-^{3}H]-GABA produced about 4 equivalents of radioactive nonamines, of which at least

2 equivalents were identified by HPLC to be 4-oxo-5-hexenoic acid (**23**, Scheme 13). Presumably one of these 4 equivalents of radioactive nonamines is derived from the transamination noted above to give **22**. About 8 equivalents of a radioactive amine, identified as **14**, the same product identified as coming from the unstable adduct, also were generated by incubation of GABA. Therefore, the turnover number is 13, and the partition ratio is 12 (12 product molecules generated for each inactivation event). An overall set of pathways to accommodate these data is shown in Scheme 13.

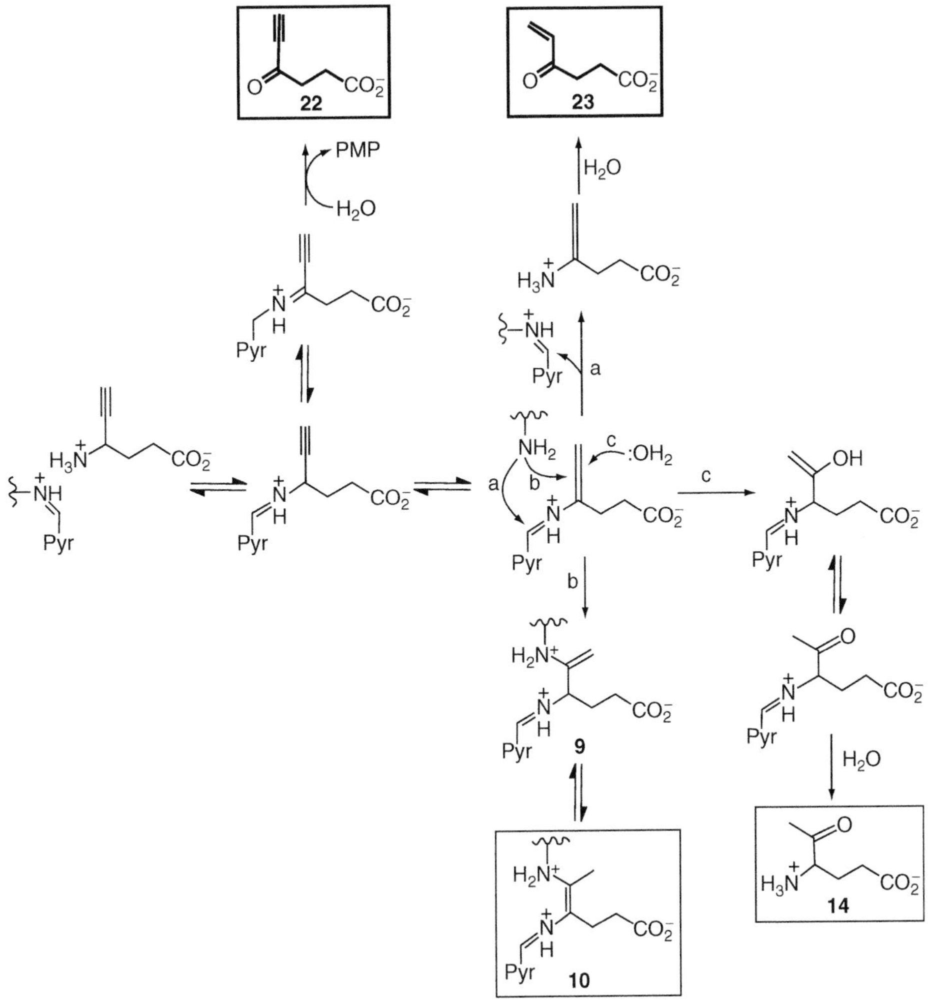

Scheme 13

B. Mechanism-Based Inactivation of Monoamine Oxidase

N-(1-Methylcyclopropyl)benzylamine (**24**) (Silverman and Hoffman, 1981; Silverman and Yamasaki, 1984) and 1-phenylcyclopropylamine (**25**) (Silverman and Zieske, 1985) are mechanism-based inactivators of the flavoenzyme monoamine oxidase (MAO). The currently accepted mechanism for this enzyme, mostly derived from studies with mechanism-based inactivators, is shown in Scheme 14 (Silverman, 1992; Silverman *et al.* 1993b). In addition to the criteria for mechanism-based inactivation described above for the inactivators of GABA aminotransferase, it was shown for **24** and **25** that there was no lag time for inactivation, and a second aliquot of MAO added to the inactivation solution with **24** was inactivated at the identical rate as was the first aliquot (Silverman and Yamasaki, 1984). To demonstrate that enzyme catalysis accompanies inactivation, the flavin absorption spectrum was recorded during and after inactivation by **24** and **25**. In both cases (Silverman and Yamasaki, 1984; Silverman and Zieske, 1985), the flavin was shown to be reduced during inactivation and to remain reduced after denaturation of the inactivated enzyme. This indicates that two electrons are transferred during inactivation and that the inactivator becomes attached to the flavin.

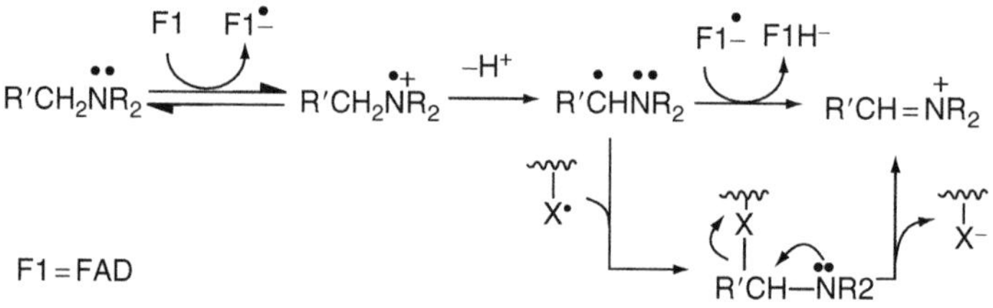

F1 = FAD

Scheme 14

24

25

1-Phenylcyclopropylamine (**25**, 1-PCPA) inactivates MAO via two different pathways; one pathway is irreversible and the other is reversible (Scheme 15). The observation that initially indicated that more than one pathway may be involved was that pseudo-first-order loss of enzyme activity occurred for the first two to three half-lives, then the rate of inactivation slowed. When a series of concentrations of 1-PCPA were used in ratios from 1 to 10 times the enzyme concentration, it was found that, over an extended period of time, the enzyme activity slowly returned but not completely (Fig. 11). This suggests that two adducts form, one that is irreversible and other that is reversible. The higher the ratio of inactivator concentration to enzyme, the less enzyme activity returned

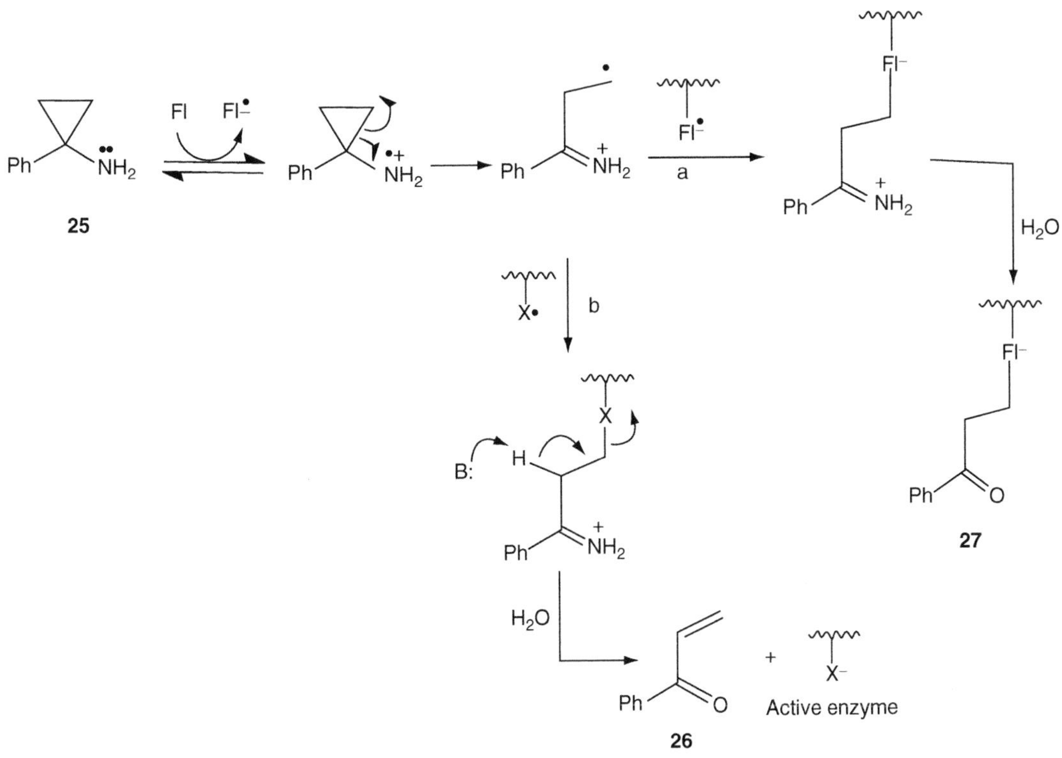

Scheme 15

(relative to a control that was not inactivated). When the final percentage of enzyme activity remaining was plotted against the inactivator/enzyme ratio (as shown in Fig. 8), a straight line was obtained that intersected the x-axis at a ratio of eight inactivator molecules required to inactivate the enzyme completely, indicating a partition ratio of seven (seven product molecules plus one covalent adduct). Inactivation of MAO with 1-[phenyl-^{14}C]**25** resulted in the covalent attachment of 1.2 ± 0.2 equivalents of ^{14}C in addition to producing 7 equivalents of radioactive acrylophenone (**26**). The mechanism shown in Scheme 15 accounts for both the attachment of 1 equivalent of radioactivity from 1-[phenyl-^{14}C]**25** and the formation of acrylophenone.

The structure of the irreversible adduct (**27**, after denaturation) was determined by carrying out three organic reactions on it (Scheme 16). Treatment of **27** with NaB^3H_4 resulted in the incorporation of 0.73 equivalent of tritium after subtracting out the noninactivated control, suggesting the presence of a ketone. Baeyer-Villiger oxidation of phenyl alkyl ketones with peroxytrifluoroacetic acid is known to give phenyl esters exclusively; saponification of phenyl esters give phenol. When this reaction was carried out on the protease-digested **27**, 87% of the theoretical

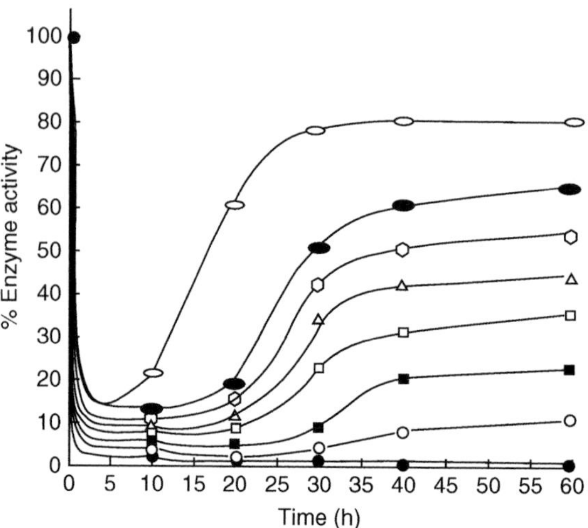

Fig. 11 Effect on enzyme activity by treatment of monoamine oxidase with varying concentrations of 1-phenylcyclopropylamine. The ratios of inactivator/enzyme used were as follows: open oval, 1 equivalent; closed oval, 2 equivalents; hexagon, 3 equivalents; triangle, 4 equivalents; open square, 5 equivalents; closed square, 6 equivalents; open circle, 7 equivalents; closed circle, 10 equivalents.

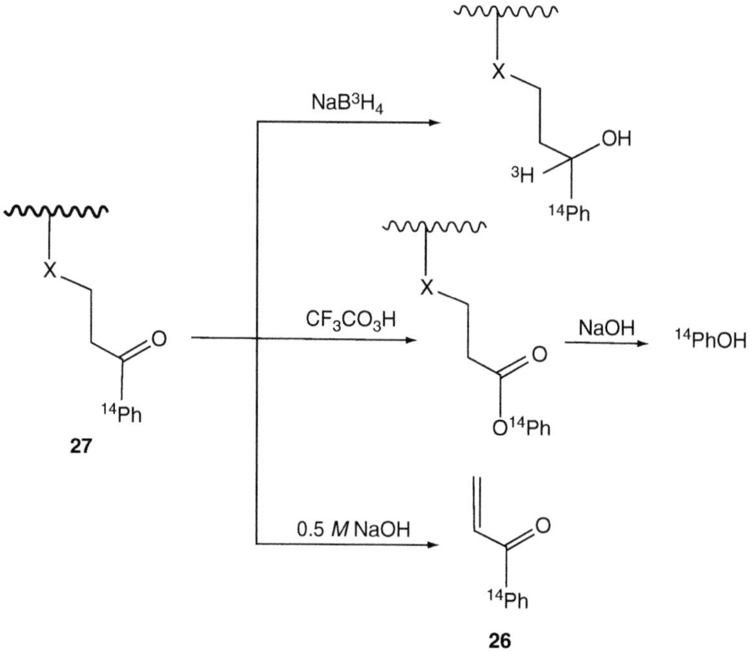

Scheme 16

amount of [^{14}C]phenol was released from the enzyme, consistent with the presence of a [^{14}C]phenyl, ketone in the adduct structure. Finally, hydroxide treatment resulted in the release of [^{14}C]acrylophenone, the expected product of a retro-Michael reaction on **27**. The identity of the X group in the irreversible adduct was determined in two ways. The absorption spectrum of the flavin is reduced on inactivation and remains reduced after denaturation, suggesting that the inactivator becomes attached to the reduced form of the flavin. Pronase digestion of the enzyme labeled with 1-[phenyl-^{14}C]**25** gives peptide fragments that contain the radioactivity from the inactivator and which have the absorption spectrum of the flavin.

The reversible adduct, which has a half-life of about 1 h, is formed seven times more readily than is the irreversible adduct. On denaturation of this adduct, the flavin spectrum becomes reoxidized, even though the radioactivity is still mostly bound to the enzyme. Therefore, this adduct is not attached to the flavin. The group on the enzyme to which the reversible adduct is attached was determined to be a cysteine residue (Silverman and Zieske, 1986a). Reduction of the adduct with sodium borohydride followed by treatment with Raney nickel gave *trans*-β-methylstyrene, the dehydration product of 1-phenyl-1-propanol, which is the expected reduction product of a (3-hydroxy-3-phenylpropyl) cysteine adduct (Scheme 17). This is consistent with the structure of the reversible adduct shown in Scheme 17; the instability of this reversible adduct could account for the fact that 7 equivalents of [^{14}C]acrylophenone are spontaneously produced from 1-[phenyl-^{14}C]**25** after complete irreversible inactivation.

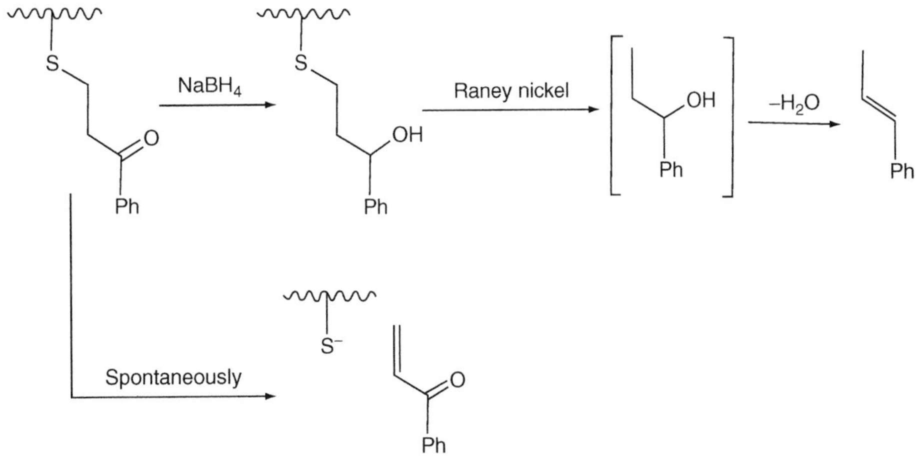

S
O
Ph
$NaBH_4$
S
OH
Ph
Raney nickel
OH
Ph
$-H_2O$
Ph
Spontaneously
S^-
O
Ph

Scheme 17

Further support for a reversible cysteine adduct came from 5,5′-dithiobis(2-nitrobenzoic acid) titration of the native enzyme and of the reversibly inactivated and sodium borohydride-reduced enzyme. In this experiment, it was shown that the native enzyme had one more cysteine than the inactivated enzyme, lending support to the hypothesis that the reversible adduct is a cysteine adduct. The difference in the stabilities of the cysteine and the flavin adducts could be in the leaving group ability of the group to which the inactivator is attached. Cysteine, the amino acid to which the reversible adduct is bound, would be a much better leaving group than the flavin cofactor, to which the irreversible adduct is attached. The rationale for the formation of a cysteine adduct is that a hydrogen atom transfer from an active site cysteine residue to the flavin semiquinone would generate a thiyl radical, which could undergo radical combination with the inactivator radical in competition with flavin semiquinone combination to the inactivator radical (Scheme 18).

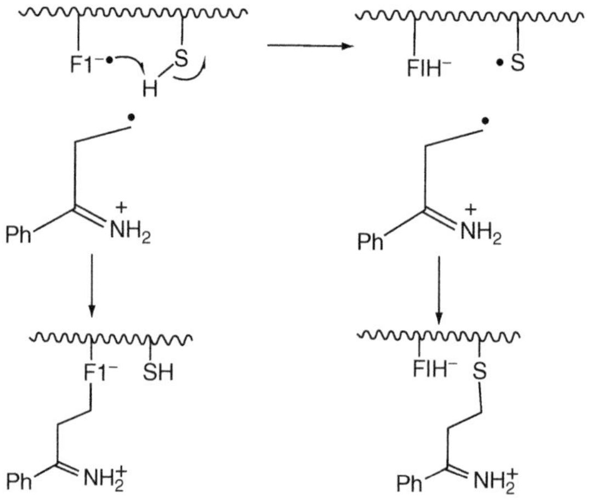

Scheme 18

To obtain further support for a radical mechanism, 1-phenylcyclobutylamine (**28**) was synthesized and its reaction with MAO studied (Scheme 19) (Silverman and Zieske, 1986b). If the mechanism proceeds by a one-electron oxidation mechanism, then, based on related nonenzymatic one-electron rearrangements (Menapace and Kuivila, 1964; Tanner and Rahimi, 1979; Wilt *et al.* 1966), **31** should be produced (pathway (a) in Scheme 19). Intermediate **29** is related to several intermediates known to undergo endocyclizations to intermediates related to **30** (Tanner and Rahimi, 1979). The alternative pathway from **29** is attachment to the flavin (pathway (b), Scheme 19), leading to enzyme inactivation. In fact, both occur. Incubation of MAO with **28** leads to a slow loss of enzyme activity. Analysis by HPLC of aliquots removed periodically showed that concomitant with the consumption of **28** is the formation of **31**. Therefore, MAO catalyzes the oxidation and rearrangement of **28** to give the product expected from chemical

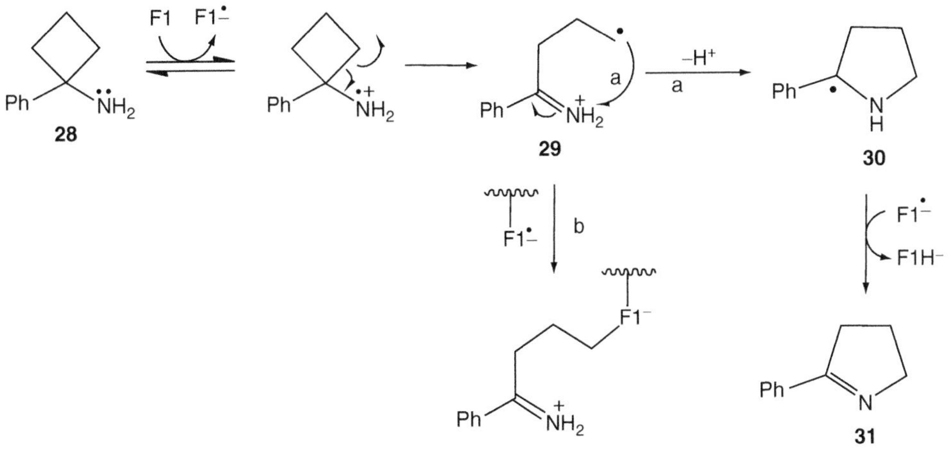

Scheme 19

studies known to proceed by one-electron pathways. Furthermore, if the MAO-catalyzed oxidation of **28** or of a related compound (**32**) is carried out in the presence of a nitrone spin-trapping agent, such as α-phenyl-*N*-*tert*-butylnitrone (**33**) in an electron paramagnetic resonance (EPR) spectrometer, the expected triplet of doublets centered at a g value of 2.006 (an organic radical) is observed (Yelekci *et al.*, 1989).

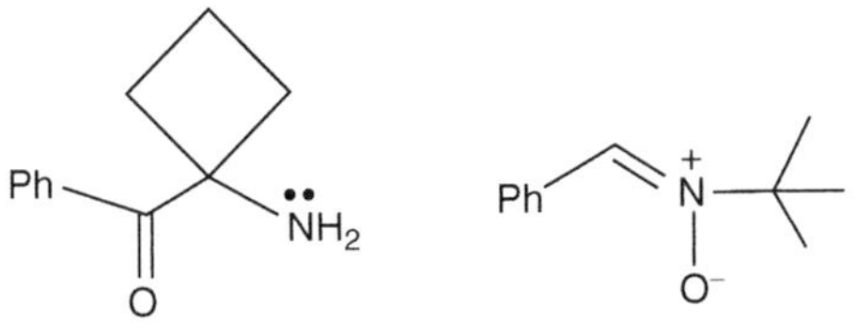

Mass spectrometry of deuterated (or other isotopically labeled) metabolites can be a useful tool for the elucidation of inactivator mechanisms. (Aminomethyl) trimethylsilane (**34**) was shown to be a mechanism-based inactivator of MAO (Banik and Silverman, 1990). Two mechanisms were imagined (Scheme 20) based on the enzyme mechanism and the known reactivity of silicon that is β to an electron-deficient center (Brumfield *et al.*, 1984; Lan *et al.*, 1984; Ohga *et al.*, 1984). Pathway (a) in Scheme 20 is the one based on the earlier work indicating that silicon β to an electron-deficient center is highly electrophilic (Brumfield *et al.*, 1984; Lan *et al.*, 1984; Ohga *et al.*, 1984). In this case, an active site nucleophile reacts with the silicon atom, which is β to an electron-deficient nitrogen atom as a result of one-electron transfer of **34** to the flavin, cleaving the carbon-silicon bond and eventually producing formaldehyde. Pathway (b) (Scheme 20) can also produce formaldehyde by a reaction of water with released (formyl)trimethylsilane (via what is referred to as a Brook rearrangement; Brook, 1974). If the inactivation

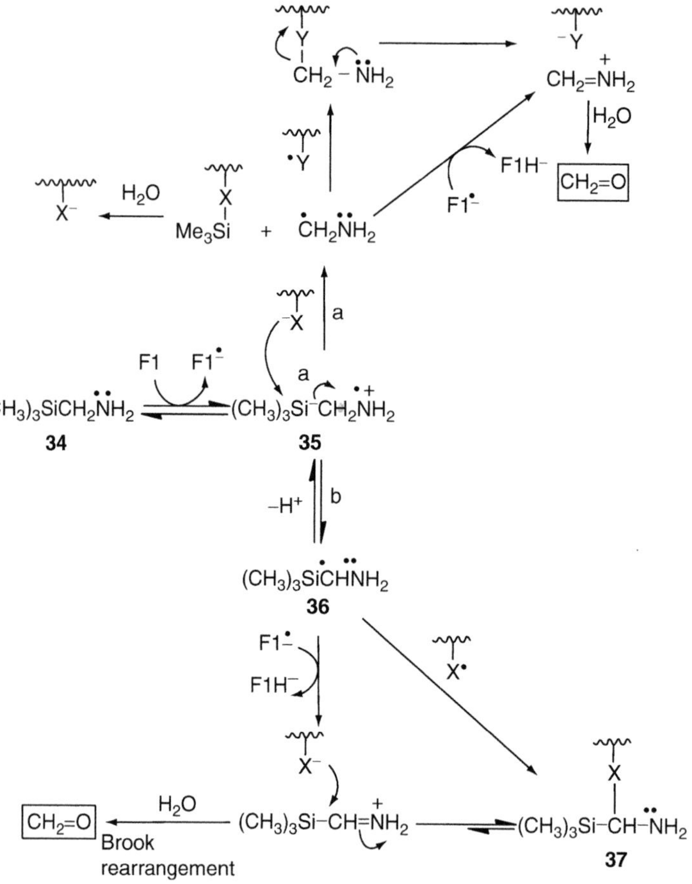

Scheme 20

is carried out with [1-$^2H_2$]**34** [$(CH_3)_3SiCD_2NH_2$], however, pathway (a) would produce dideuterioformaldehyde, but pathway (b) would produce monodeuterioformaldehyde because one deuterium would be removed from **35** to give **36**. Following inactivation of MAO with [1-$^2H_2$]**34**, treatment with 2,4-dinitrophenylhydrazine reagent to trap the formaldehyde(s) produced, extraction, and mass spectrometry, it was found that both dideuterioformaldehyde and monodeuterioformaldehyde 2,4-dinitrophenylhydrazones were formed in the ratio of about 1:3.5. This indicates that both pathways are relevant. Furthermore, there is a deuterium isotope effect on the rate of inactivation ($k^{H}_{inact}/k^{D}_{inact}$) of 2.3 with no effect on the K_I. This kinetic isotope effect is almost identical with the deuterium isotope effects for MAO-catalyzed oxidation of tyramine (Belleau and Moran, 1963) and dopamine (Yu *et al.*, 1986).

The observed deuterium isotope effect on inactivation indicates that C–H bond cleavage is a rate-determining step in inactivation, consistent with pathway (b) but not with pathway (a) (Scheme 20). Apparently, however, since dideuterioformaldehyde is generated, pathway (a) is a relevant pathway as well, but not relevant to enzyme inactivation. The converse experiment, namely, the inactivation of MAO with **34** in ^2H_2O, gave a mixture of undeuterated and monodeuterated formaldehyde, as expected for oxidation by pathways (a) and (b), respectively. To confirm this conclusion, MAO was inactivated with [1-^{3}H]**34**. According to pathway (a), no tritium would be incorporated into the enzyme; all would be released as [^{3}H] formaldehyde. Pathway (b), however, would lead to incorporation of tritium into the enzyme (**37**) and, as a result of the release of (formyl)trimethylsilane and subsequent Brook rearrangement, also to formation of [^{3}H]formaldehyde and in the incorporation of 1.2 equivalents of tritium into the enzyme. The stoichiometric incorporation of tritium into the enzyme is not consistent with pathway (a) but is consistent with pathway (b). Given the 1:3.5 ratio of dideuterioformaldehyde to monodeuterioformaldehyde noted above and the 1 equivalent of tritium incorporated during inactivation, it suggests that for every 5.5 molecules of inactivator oxidized by MAO, one molecule goes via pathway (a) without inactivation and 4.5 go via pathway (b), 3.5 of which are converted to formaldehyde and one becomes attached to the enzyme (**37**).

Isotopically labeled analogs have been used to clarify the pathway for inactivation of MAO by the anticonvulsant agent milacemide (**38a**) (Silverman *et al.*, 1993). Milacemide is both a substrate and inactivator of MAO. Two pathways for oxidation and inactivation were envisioned subsequent to initial electron transfer to the flavin (Scheme 21). Pathway (a) involves the removal of the α-proton from the pentyl chain, leading to both inactivation (pathway (c)) and metabolite formation (pathway (d)). Pathway (b) proceeds via removal of the acetamido methylene proton, leading to inactivation (pathway (e)) and metabolite formation (pathway (f)). To test which pathway(s) was reasonable, the corresponding dideuterated analogs **38b** and **38c** were prepared. Analog **38b** inactivated MAO with a negligible isotope effect (1.25) on k_{inact}/K_I, but **38c** exhibited a large isotope effect (4.55) on k_{inact}/K_I. As a substrate, **38b** showed no isotope effect (1.03) on k_{cat}/K_m, but **38c** had an isotope effect of 4.53. These results indicate that pathway (a) is the most reasonable one for both inactivation and oxidation of milacemide. The partition ratios for **38a-c** are similar (within a factor of 1.4), consistent with a mechanism in which partitioning between metabolite formation and inactivation occurs subsequent to the C–H bond cleavage step.

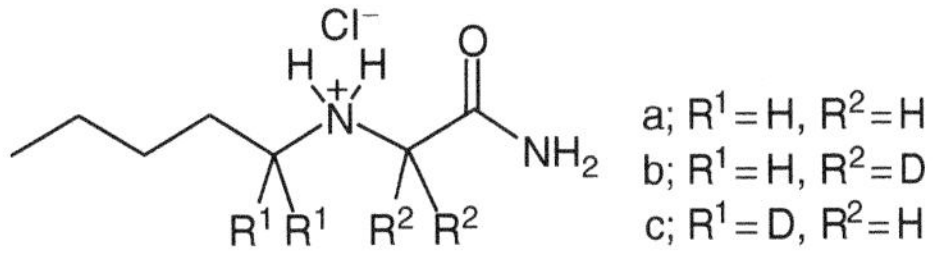

Cl⁻ H H N⁺ O NH$_2$ R^1 R^1 R^2 R^2

a; R^1 = H, R^2 = H
b; R^1 = H, R^2 = D
c; R^1 = D, R^2 = H

38

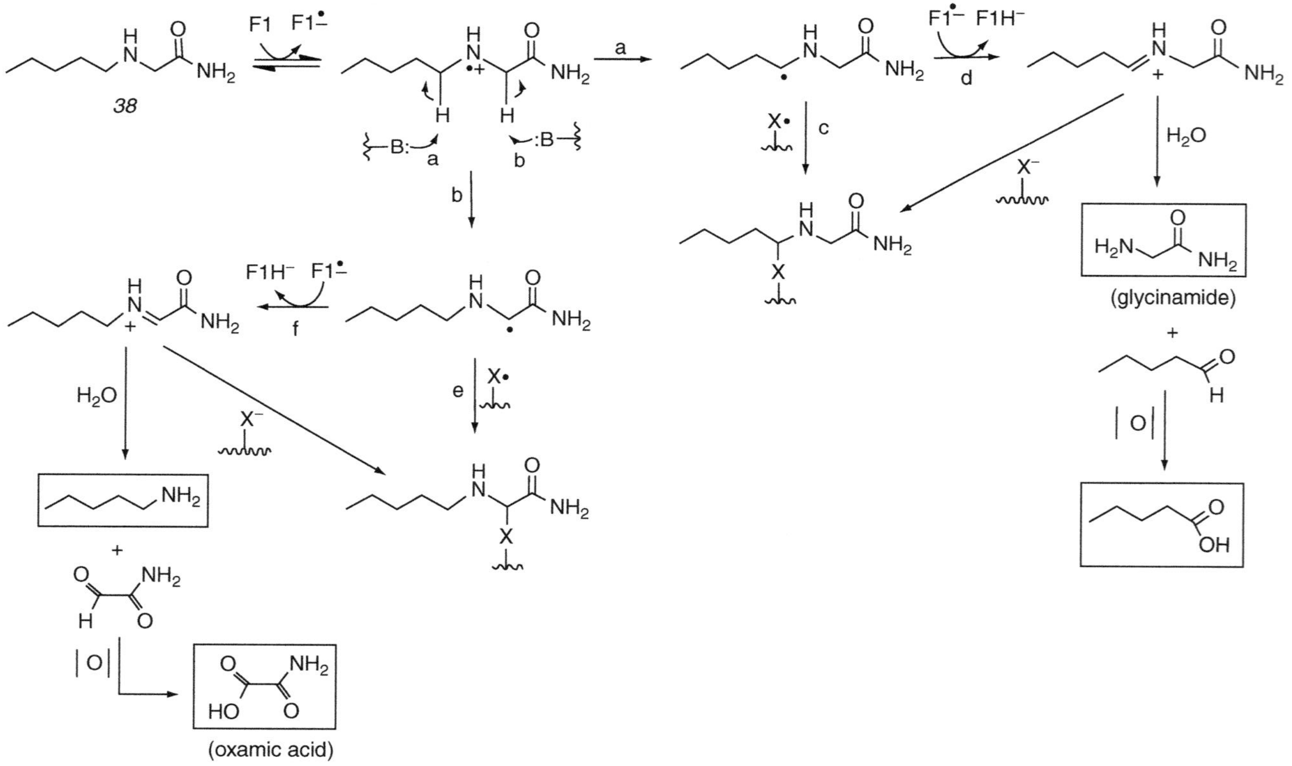

Scheme 21

A slightly different picture emerged, however, when two ^{14}C-labeled analogs were used (**39**) and the radiolabeled metabolites were isolated. Incubation of MAO with **39a** gave both [^{14}C]glycinamide (from pathway (a) involving oxidation of the pentyl side chain) and [^{14}C]oxamic acid (from pathway (b) involving oxidation of the acetamide methylene to give 2-oxoacetamide, which was nonenzymatically oxidized). Inactivation of MAO with **39b** produced both [^{14}C]pentanoic acid (from pathway (a) involving oxidation of the pentyl side chain to give pentanal, which was nonenzymatically oxidized) and [^{14}C]pentylamine (from pathway (b) involving oxidation of the acetamide methylene). These result indicate that oxidation does occur at both methylenes adjacent to the amine nitrogen, but the deuterium isotope effect results support inactivation as resulting from pathway (a).

39

a, * = ^{12}C; ‡ = ^{14}C b, * = ^{14}C; ‡ = ^{12}C

VII. Conclusion

Mechanism-based enzyme inactivation is a powerful tool for studies of enzyme mechanisms and mechanisms of enzyme inactivation by small molecules. Mechanistic hypotheses can be tested by appropriate molecular design, utilizing isotopically labeled analogs to permit the elucidation of structures of metabolites produced and to determine what portions of the mechanism-based inactivators become covalently attached to the target enzyme. This approach to studies of enzyme mechanisms is well suited for those who are geared more to the organic chemistry of enzyme-catalyzed reactions and who have insights into the chemical machinery of active sites of enzymes. The use of MBEIs is yet another of the very important methods in enzymology.

References

Banik, G. M., and Silverman, R. B. (1990). *J. Am. Chem. Soc.* **112,** 4499.
Belleau, B., and Moran, J. (1963). *Ann. N. Y. Acad. Sci.* **107,** 822.
Brook, A. G. (1974). *Acc. Chem. Res.* **7,** 77.
Brumfield, M. A., Quillen, S. L., Yoon, U. C., and Mariano, P. S. (1984). *J. Am. Chem. Soc.* **106,** 6855.
Burke, J. R., and Silverman, R. B. (1991). *J. Am. Chem. Soc.* **113,** 9329.
Escribano, J., Tudela, J., Garcia-Carmona, F., and Garcia-Conovas, F. (1989). *Biochem. J.* **262,** 597.
Fitzpatrick, P. F., and Villafranca, J. J. (1986). *J. Biol. Chem.* **261,** 4510.
Fitzpatrick, P. F., Flory, D. R., Jr., and Villafranca, J. J. (1985). *Biochemistry* **24,** 2108.
Funaki, T., Ichihara, S., Fukazawa, H., and Kuruma, I. (1991). *Biochim. Biophys. Acta* **1118,** 21.
Garcia-Canovas, F., Tudela, J., Varon, R., and Vazquez, A. M. (1989). *J. Enzyme Inhib.* **3,** 81.
Jung, M. J., and Metcalf, B. W. (1975). *Biochem. Biophys. Res. Commun.* **67,** 301.
Kitz, R., and Wilson, I. B. (1962). *J. Biol. Chem.* **237,** 3245.
Krantz, A. (1992). *Biomed. Chem. Lett.* **2,** 1327.

Kuo, D. J., and Jordan, F. (1983). *Biochemistry* **22,** 3735.
Lan, A. J. Y., Quillen, S. L., Heuckeroth, R. O., and Mariano, P. S. (1984). *J. Am. Chem. Soc.* **106,** 6439.
Likos, J. J., Ueno, H., Feldhaus, R. W., and Metzler, D. E. (1982). *Biochemistry* **21,** 4377.
Menapace, L. W., and Kuivila, H. G. (1964). *J. Am. Chem. Soc.* **86,** 3047.
Ohga, K., Yoon, Y. C., and Mariano, P. S. (1984). *J. Org. Chem.* **49,** 213.
Penefsky, H. S. (1977). *J. Biol. Chem.* **252,** 2891.
Penefsky, H. (1979). *Meth. Enzymol.* **56,** 527.
Pratt, R. F. (1992). *Biomed. Chem. Lett.* **2,** 1323.
Silverman, R. B. (1988a). "Mechanism-Based Enzyme Inactivation: Chemistry and Enzymology", Vols. 1 and 2. CRC Press, Boca Raton, FL.
Silverman, R. B. (1988b). "Mechanism-Based Enzyme Inactivation: Chemistry and Enzymology", Vol. 1, p. 12. CRC Press, Boca Raton, FL.
Silverman, R. B. (1992a). "The Organic Chemistry of Drug Design and Drug Action" Chap. 5. Academic Press, San Diego, CA.
Silverman, R. B. (1992b). *In* "Advances in Electron Transfer Chemistry" (P. S. Mariano, ed.), Vol. 2, p. 177. JAI Press, Greenwich, CT.
Silverman, R. B., and Hoffman, S. J. (1981). *Biochem. Biophys. Res. Commun.* **101,** 1396.
Silverman, R. B., and Invergo, B. J. (1986). *J. Am. Chem. Soc.* **25,** 6817.
Silverman, R. B., and Levy, M. A. (1980). *Biochem. Biophys. Res. Commun.* **95,** 250.
Silverman, R. B., and Levy, M. A. (1981). *Biochemistry* **20,** 1197.
Silverman, R. B., and Yamasaki, R. B. (1984). *Biochemistry* **23,** 1322.
Silverman, R. B., and Zieske, P. A. (1985). *Biochemistry* **24,** 2128.
Silverman, R. B., and Zieske, P. A. (1986a). *Biophys. Res. Commun* **135,** 154.
Silverman, R. B., and Zieske, P. A. (1986b). *Biochemistry* **25,** 341.
Silverman, R. B., Nishimura, K., and Lu, X. (1993a). *J. Am. Chem. Soc.* **115,** 4949.
Silverman, R. B., Zhou, J. P., and Eaton, P. E. (1993b). *J. Am. Chem. Soc.* **115,** 8841.
Tanner, D. D., and Rahimi, R. M. (1979). *J. Org. Chem.* **44,** 1674.
Ueno, H., Likos, J. H., and Metzler, D. E. (1982). *Biochemistry* **21,** 4387.
Varon, R., Garcia, M., Garcia-Canovas, F., and Tudela, J. (1990). *J. Mol. Catal.* **59,** 97.
Waley, S. G. (1980). *Biochem. J.* **185,** 771.
Waley, S. G. (1985). *Biochem. J.* **227,** 843.
Wilt, J. W., Maravetz, L. L., and Zawadzki, J. F. (1966). *J. Org. Chem.* **31,** 3018.
Yelekci, K., Lu, X., and Silverman, R. B. (1989). *J. Am. Chem. Soc.* **111,** 1138.
Yu, P. H., Bailey, B. A., Durden, D. A., and Boulton, A. A. (1986). *Biochem. Pharmacol.* **35,** 1027.

PART III

Detection of Enzyme Reaction Intermediates

CHAPTER 13

Transient Kinetic Approaches to Enzyme Mechanisms

Carol A. Fierke* and Gordon G. Hammes†

*Department of Biochemistry
College of Medicine
Duke University Medical Center
Durham, North Carolina 27710

†Department of Biochemistry
Duke University Medical Center
Durham, North Carolina 27710

I. Introduction

The elucidation of enzymatic mechanisms through kinetic investigations has been an important area of research for many years (Fersht, 1985; Hammes, 1982). Enzymes have a special fascination for many reasons: their role in physiological processes, their often unstable nature, their incredible efficiency as catalysts, and their unusual kinetic patterns. Initially, kinetic studies were carried out at very low enzyme concentrations relative to substrate concentrations. This was due to the difficulty in obtaining large quantities of high-purity enzymes and the

DOI: 10.1016/B978-0-12-378608-1.00013-X

relative simplicity of the kinetic equations for this situation. Under such conditions, all of the enzyme species can be assumed to be in a steady state so that any mechanism, regardless of complexity, can be described by a single rate equation. If initial rates are measured, with no products present, the rate measurements are usually easy to carry out and easy to interpret.

Steady-state kinetics can provide information about the overall reaction pathway for multiple substrate reactions, the specificity of the enzyme for specific substrate structures, and lower bounds for the specific rate constants in a mechanism (Cleland, 1977; Fersht, 1985; Hammes, 1982; Peller and Alberty, 1959). However, the kinetic parameters measured are complex functions of the specific rate constants so that individual rate constants rarely can be determined, and little information is obtained about mechanistic intermediates. If the pH dependencies of the steady-state kinetic parameters are measured, the nature of the ionizable groups of the enzymes that are important for catalysis can be inferred; in addition, the use of isotopes sometimes can shed light on the reaction intermediates.

To elucidate the details of an enzymatic mechanism and to measure rate constants for elementary steps, kinetic experiments must be carried out at high enzyme concentrations where enzyme-substrate species are significantly populated and can be directly detected. Because enzymes are very efficient catalysts, fast reaction techniques are often needed to study the transient kinetics. Such studies are vital for the establishment of detailed reaction mechanisms. The primary difficulties in carrying out transient kinetics are the special technologies and the large amounts of enzyme required. However, commercial stopped-flow and rapid quench equipment are now available, and cloning methods can be used to produce large amounts of most proteins. Therefore, the role of transient kinetics in understanding enzyme mechanisms should markedly expand in the near future.

This chapter is concerned with introducing transient kinetic studies to biochemists. First, the theoretical bases for analyzing the rate equations associated with typical mechanisms will be considered. Second, a brief description of the various types of stopped-flow, rapid quench, and temperature-jump equipment is given. Finally, some examples of transient kinetic studies are presented. Additional information can be found in other reviews (Bernasconi, 1986; Hammes, 1982; Hammes and Schimmel, 1970; Johnson, 1992).

II. Theory of Transient Kinetics

A complete treatment of all the kinetic complexities that may be encountered in carrying out transient studies obviously is not possible. Instead, the principles associated with the kinetic analysis of complex mechanisms are presented, with specific examples of commonly encountered situations. In analyzing a complex mechanism, the first question that must be asked is how many rate equations are required to describe the mechanism. The answer is deceptively simple: the number of independent rate equations is equal to the number of independent concentration variables. The number of independent concentration variables in turn is equal to the total number of concentration variables minus the number of conservation relationships that exist between the concentration variables. The conservation relationships are usually simple mass conservation, but they also include more subtle relationships such as those determined by steady-state situations, that is, the rate of change of a concentration with time may equal zero.

The experimental design of transient kinetic experiments is important since significant simplification in the kinetic analysis often can be accomplished. For example, whenever possible, reactions should be carried out under conditions where the observed kinetics are first order. This is usually done by making all concentrations large relative to the concentration of the species whose rate of change with time is being measured. Other potential simplifications are discussed throughout the theoretical analysis.

The simplest type of reaction that might be anticipated is an irreversible first-order reaction. For example, a conformational change of an enzyme from state E to state E′, characterized by a rate constant k_1, might be triggered by a change in pH:

$$\mathrm{E} \xrightarrow{k_1} \mathrm{E}' \tag{1}$$

This mechanism is obviously characterized by a single rate equation since there are two species and the total enzyme concentration, $[\mathrm{E}]_0$, is conserved:

$$[\mathrm{E}] + [\mathrm{E}'] = [\mathrm{E}]_0 \tag{2}$$

For this mechanism

$$\frac{-\mathrm{d}[\mathrm{E}]}{\mathrm{d}t} = k_1[\mathrm{E}] \tag{3}$$

or after integration

$$[\mathrm{E}] = [\mathrm{E}]_0 \mathrm{e}^{-k_1 t} \tag{4}$$

if $[\mathrm{E}] = [\mathrm{E}]_0$ at $t = 0$, where "t" is the time. Data analysis is simple as a plot of ln[E] versus t is linear with a slope of $-k_1$, or k_1 can be obtained by fitting the time course of E to Eq. (4) with a nonlinear least squares analysis. If this reaction is reversible,

$$\mathrm{E} \underset{k_{-1}}{\overset{k_1}{\rightleftharpoons}} \mathrm{E}' \tag{5}$$

the rate equation becomes

$$\frac{-\mathrm{d}[\mathrm{E}]}{\mathrm{d}t} = k_1[\mathrm{E}] - k_{-1}([\mathrm{E}]_0 - [\mathrm{E}]) \tag{6}$$

which can be integrated with the same boundary conditions as above to give

$$[\mathrm{E}] = [\mathrm{E}]_\mathrm{e} + ([\mathrm{E}]_0 - [\mathrm{E}]_\mathrm{e})\mathrm{e}^{-(k_1+k_{-1})t} \tag{7a}$$

or

$$\ln\left(\frac{[\mathrm{E}] - [\mathrm{E}]_\mathrm{e}}{[\mathrm{E}]_0 - [\mathrm{E}]_\mathrm{e}}\right) = -(k_1 + k_{-1})t \tag{7b}$$

where the subscript "e" denotes the equilibrium concentration. A plot of $\ln([E] - [E]_e)$ versus time is a straight line with a slope of $-(k_1 + k_{-1})$. If the equilibrium constant, $K_1 = [E']_e/[E]_e = k_1/k_{-1}$, is known, both rate constants can be calculated.

Concentrations are usually not measured directly. Instead, some property that is related to the concentration such as absorbance or fluorescence is measured. If the property, P_E, is a linear function of the concentration,

$$P_E = a[E] + b \tag{8}$$

where "a" and "b" are constants, then P_E can be substituted for [E] in the integrated rate equations. For example, the absorbance and fluorescence are directly proportional to the concentration.

As a second example, consider the combination of enzyme, E, and substrate, S, to form a complex, X:

$$E + S \underset{k_{-2}}{\overset{k_2}{\rightleftharpoons}} X \tag{9}$$

A single independent rate equation is required to describe this mechanism and can be written as

$$\frac{-d[E]}{dt} = k_2[E][S] - k_{-2}[X] \tag{10}$$

If the mass conservation relationships,

$$[E]_0 = [E] + [X] \tag{11}$$

$$[S]_0 = [S] + [X] \tag{12}$$

are utilized, the concentrations [S] and [X] can be expressed in terms of [E], $[E]_e$, and $[S]_0$, and Eq. (10) can be integrated. The result is very complex and rarely used (Hammes, 1978).

This integrated rate equation can be considerably simplified by adjusting the concentrations so that either E or S remains essentially constant during the experiment. In practice, this means that one of the concentrations must be about an order of magnitude greater than the other. For example, if $[S]_0 \gg [E]_0$, the integration of Eq. (10) gives Eq. (7a), except that $k_1 = k_2[S]_0$ and $k_{-1} = k_{-2}$. This is frequently called a "pseudo"-first-order rate equation where the observed rate constant, $k_{obs} = k_2[S]_0 + k_{-2}$. In this case, if $k_2[S]_0 + k_{-2}$ is determined at several concentrations of S_0, a plot of k_{obs} versus $[S]_0$ is linear with a slope of k_2 and an intercept of k_{-2}. The ratio of rate constants is equal to the equilibrium constant. Reduction of higher order kinetic equations to first-order rate equations by experimental design should be done whenever possible.

An alternative method of reducing higher order rate equations to first order is to carry out kinetic measurements near equilibrium, that is, chemical relaxation experiments (Bernasconi, 1976; Eigen and deMaeyer, 1963; Hammes, 1982).

Consider the combination of enzyme and substrate (Eq. (9)) near equilibrium. New concentration variables can be defined as

$$\begin{aligned}[E] &= [E]_e + \Delta E \\ [S] &= [S]_e + \Delta S \\ [X] &= [X]_e + \Delta X\end{aligned} \tag{13}$$

where "Δ" designates the deviation from equilibrium and $\Delta E = \Delta S = -\Delta X$ because of mass conservation. Equation (10) becomes

$$\frac{-d\Delta E}{dt} = \{k_2([E]_e + [S]_e) + k_{-2}\}\Delta E + k_2[E]_e[S]_e - k_{-2}[X]_e + k_2(\Delta E)^2 \tag{14}$$

Note that $k_2[E]_e[S]_e = k_{-2}[X]_e$ by definition, and near equilibrium $(\Delta E)^2$ can be neglected because it is small. Thus Eq. (14) becomes

$$\frac{-d\Delta E}{dt} = \frac{\Delta E}{\tau} \tag{15}$$

where $1/\tau = k_2([E]_e + [S]_e) + k_{-2}$ and "τ" is the chemical relaxation time. This example illustrates that near equilibrium all rate equations become first order. Equation (15) can be integrated to give

$$\Delta E = (\Delta E)_0 e^{-t/\tau} \tag{16}$$

so that a plot of ln ΔE versus t gives a straight line with a slope of $-1/\tau$. If τ is determined at a series of equilibrium concentrations, a plot of $1/\tau$ versus $([E]_e + [S]_e)$ has a slope of k_2 and an intercept of k_{-2} (Eq. (15)). If the equilibrium constant is not known, a value can be assumed to get preliminary values of k_2 and k_{-2}. Then the ratio of the rate constants obtained can be used to calculate a new equilibrium constant, and the process can be repeated until the ratio of rate constants equals the equilibrium constant used to calculate the equilibrium concentrations.

This linearization of the rate equations is typically valid up to 10% from equilibrium and can be used to analyze stopped-flow data, as well as those obtained from more conventional relaxation techniques. Typical experimental data are shown in Fig. 1, where the reciprocal relaxation time ($1/\tau = k_{obs}$) is plotted versus the sum of the equilibrium concentrations for the reaction of NADPH with fatty-acid synthase (Cognet *et al.*, 1983).

Thus far only single-step reaction mechanisms have been considered. However, reaction mechanisms of interest usually have multiple steps. A two-step mechanism is now considered in detail, with the methodology developed being applicable to more complex mechanisms.

Consider a mechanism whereby an enzyme and substrate combine to form a complex which undergoes a conformational change:

$$E + S \underset{k_{-1}}{\overset{k_1}{\rightleftharpoons}} X_1 \underset{k_{-2}}{\overset{k_2}{\rightleftharpoons}} X_2 \tag{17}$$

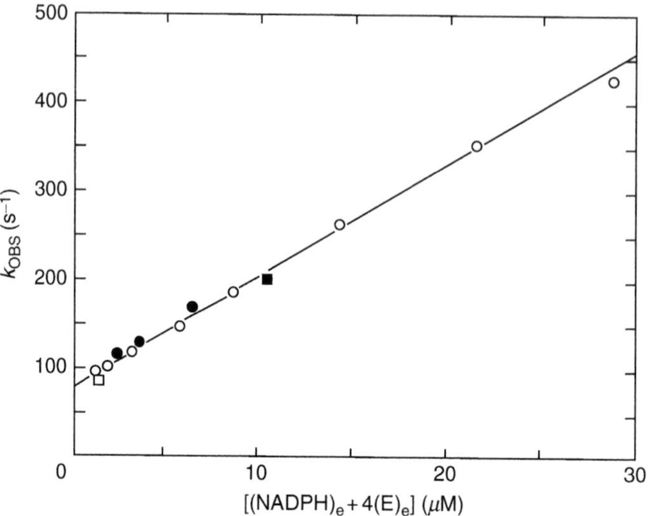

Fig. 1 Plot of the first-order rate constant for binding of NADPH to fatty-acid synthase, k_{obs}, versus the sum of the equilibrium concentrations of unbound NADPH and unoccupied NADPH binding sites on the enzyme, ($[NADPH]_e + 4[E]_e$); reprinted with permission from Cognet *et al.* (1983), copyright 1983 American Chemical Society.

Two independent rate equations are needed to describe this mechanism as four concentration variables and two conservation relationships exist:

$$[E]_0 = [E] + [X_1] + [X_2] \tag{18}$$

or

$$\Delta E + \Delta X_1 + \Delta X_2 = 0$$

and

$$[S]_0 = [S] + [X_1] + [X_2] \tag{19}$$

or

$$\Delta S + \Delta X_1 + \Delta X_2 = 0$$

The rate equations can be written as

$$\frac{-d[E]}{dt} = k_1[E][S] - k_{-1}[X_1] \tag{20}$$

$$\frac{-d[X_2]}{dt} = -k_2[X_1] + k_{-2}[X_2] \tag{21}$$

The same procedures previously utilized can be used to linearize Eqs. (20) and (21) in the neighborhood of equilibrium, namely, $[E] = [E]_e + \Delta E$, etc. The result is

$$\frac{-\mathrm{d}\Delta\mathrm{E}}{\mathrm{d}t} = \{k_1([\mathrm{E}]_e + [\mathrm{S}]_e) + k_{-1}\}\Delta\mathrm{E} + k_{-1}\Delta\mathrm{X}_2 \\ = a_{11}\Delta\mathrm{E} + a_{12}\Delta\mathrm{X}_2 \tag{22}$$

$$\frac{-\mathrm{d}\Delta\mathrm{X}_2}{\mathrm{d}t} = k_2\Delta\mathrm{E} + (k_{-2} + k_2)\Delta\mathrm{X}_2 \\ = a_{21}\Delta\mathrm{E} + a_{22}\Delta\mathrm{X}_2 \tag{23}$$

where the a_{ij} terms are defined by these equations. The integration of coupled first-order linear homogeneous differential equations is described in textbooks on differential equations, and only the results are presented here. On integration of Eqs. (22) and (23), the two concentration variables are given as a sum of two exponentials:

$$\Delta\mathrm{E} = A_1\mathrm{e}^{-t/\tau_1} + A_2\mathrm{e}^{-t/\tau_2} \tag{24}$$

$$\Delta X_2 = A_3\mathrm{e}^{-t/\tau_1} + A_4\mathrm{e}^{-t/\tau_2} \tag{25}$$

where the A_i values are constants and the τ_i are obtained by solving the determinant

$$\begin{vmatrix} a_{11} - 1/\tau & a_{12} \\ a_{21} & a_{22} - 1/\tau \end{vmatrix} = 0 \tag{26}$$

The τ_i are equivalent to reciprocal first-order rate constants and are functions of equilibrium concentrations and rate constants. The values of A_i are determined by the boundary conditions and are not explored further here. The determinant can be expanded to give

$$(1/\tau)^2 - (a_{11} + a_{22})(1/\tau) + a_{11}a_{22} - a_{12}a_{21} = 0 \tag{27}$$

This quadratic equation can be solved for the relaxation times:

$$\frac{1}{\tau_{1,2}} = \frac{a_{11} + a_{22}}{2}\left\{1 \pm \left[1 - \frac{4(a_{11}a_{22} - a_{12}a_{21})}{(a_{11} + a_{22})^2}\right]^{1/2}\right\} \tag{28}$$

where one relaxation time corresponds to the positive square root and the other to the negative square root.

These relationships can be directly related to experimental measurements: observing concentrations as a function of time should yield a plot of two exponentials characterized by the two relaxation times which can be obtained from the data, for example, by a nonlinear least squares analysis. The relaxation times are complex functions of equilibrium concentrations. However, the analysis of the relaxation times is usually not difficult. For example,

$$(1/\tau_1) + (1/\tau_2) = a_{11} + a_{22} = k_1([\mathrm{E}]_\mathrm{e} + [\mathrm{S}]_\mathrm{e}) + k_{-1} + k_2 + k_{-2} \tag{29}$$

$$(1/\tau_1)(1/\tau_2) = a_{11}a_{22} - a_{12}a_{21} = k_1(k_2 + k_{-2})([\mathrm{E}]_\mathrm{e} + [\mathrm{S}]_\mathrm{e}) + k_{-1}k_{-2} \tag{30}$$

Linear plots of the sum and product of the reciprocal relaxation times versus ($[\mathrm{E}]_\mathrm{e} + [\mathrm{S}]_\mathrm{e}$) can be used to obtain all four individual rate constants.

Often the bimolecular step is much more rapid than the conformational change. In fact, this can usually be made the case by working at sufficiently high concentrations of enzyme and/or substrate. In this case, $k_1([\mathrm{E}]_\mathrm{e} + [\mathrm{S}]_\mathrm{e}) + k_{-1} \gg (k_2 + k_{-2})$ (i.e., $a_{11} \gg a_{22}, a_{21}, a_{12}$), and the square root in Eq. (28) can be expanded as $(1 - \mathrm{X})^{1/2} \approx 1 - \mathrm{X}/2$ for $\mathrm{X} \ll 1$. This expansion for $\mathrm{X} = 4(a_{11}a_{22} - a_{12}a_{21})/(a_{11} + a_{22})^2$ gives

$$\frac{1}{\tau_{11}} = k_1([\mathrm{E}]_\mathrm{e} + [\mathrm{S}]_\mathrm{e}) + k_{-1} \tag{31}$$

$$\frac{1}{\tau_2} = k_{-2} + \frac{k_2}{1 + k_{-1}/\{k_1([\mathrm{E}]_\mathrm{e} + [\mathrm{S}]_\mathrm{e})\}} \tag{32}$$

The expression for $1/\tau_1$ is the same as for an isolated bimolecular reaction because that step essentially equilibrates before the conformational change occurs. The second relaxation time primarily characterizes the rate of the conformational change because the bimolecular step is essentially at equilibrium while the conformational change occurs. When the sum of the equilibrium concentrations ($[\mathrm{S}]_\mathrm{e} + [\mathrm{E}]_\mathrm{e}$) is large relative to the equilibrium dissociation constant for the first step (k_{-1}/k_1), then $1/\tau_2 = k_2 + k_{-2}$, the relaxation time for the isolated second step, because all of the enzyme is present as X_1 and X_2.

From the analysis above, it should be evident that in the neighborhood of equilibrium, all mechanisms can be described by a set of independent rate equations that are homogeneous first-order (linear) differential equations. They can be written symbolically as

$$\frac{-\mathrm{d}\Delta c_i}{\mathrm{d}t} = \sum_{j=1}^{n} a_{ij}\,\Delta c_j \tag{33}$$

where "n" is the number of independent concentration variables. The solution to this set of equations is a sum of n exponential terms:

$$\Delta c_i = \sum_{j=1}^{n} A_{ij}\,\mathrm{e}^{-t/\tau_j} \tag{34}$$

where the "τ_j" are relaxation times and the "A_{ij}" are constants determined by the boundary conditions. The relaxation times are functions of equilibrium concentrations and rate constants and can be calculated from the following determinant:

$$\begin{vmatrix} a_{11} - 1/\tau & \dots & a_{1n} \\ \vdots & & \vdots \\ a_{n1} & \dots & a_{nn} - 1/\tau \end{vmatrix} = 0 \tag{35}$$

Application of this analysis to experiments is quite feasible since multi-experimental curves can be deconvoluted readily with standard statistical computer packages. In practice, the deconvolution often can be accomplished experimentally by adjusting the concentrations and conditions such that the rate of the different steps is quite different. In a study of the interaction of erythro-β-hydroxyaspartate with aspartate aminotransferase, nine relaxation times associated with transamination were resolved ranging from microseconds to seconds (Hammes and Haslam, 1969). Although the amplitude constants, A_{ij}, contain useful information, they are difficult to calculate and are rarely utilized in kinetic analyses of experimental data.

An analysis of the mechanism in Eq. (17) far from equilibrium is now considered. The coupled rate equations (Eqs. (20) and (21)) cannot be solved analytically. Numerical integration can be carried out, but this is difficult to apply to experimental situations. Instead, the common practice is to make the coupled reactions "pseudo"-first order by having the substrate concentration much greater than the enzyme concentration, $[S]_0 \gg [E]_0$ (of course, $[E]_0 \gg [S]_0$ works equally well). In this case, Eqs. (20) and (21) become

$$\frac{-d[E]}{dt} = (k_1[S]_0 + k_{-1})[E] + k_{-1}[X_2] - k_{-1}[E]_0 \tag{36}$$

$$\frac{-dX_2}{dt} = k_2[E] + (k_{-2} + k_2)[X_2] - k_2[E]_0 \tag{37}$$

These equations are identical in form to Eqs. (22) and (23), except that $k_1[S]_0$ appears rather than $k_1([E]_e + [S]_e)$ and constant terms containing $[E]_0$ are present. The solution to Eqs. (36) and (37), which are nonhomogeneous analogs of Eqs. (22) and (23), is a sum of two time-dependent exponentials and a constant:

$$[E] = A_1 e^{-\lambda_1 t} + A_2 e^{-\lambda_2 t} + [E]_e \tag{38}$$

$$[X_2] = A_3 e^{-\lambda_1 t} + A_4 e^{-\lambda_2 t} + [X_2]_e \tag{39}$$

where the "λ_i" values are obtained by solving the same determinant as Eq. (26), except that $k_1[S]_0$ is substituted for $k_1([E]_e + [S]_e)$ and λ for $1/\tau$. The analysis of experimental data proceeds exactly as for the near-equilibrium situation.

A special case worth further consideration is the slower time constant when the bimolecular reaction equilibrates rapidly relative to the conformational change. In this case

$$\lambda_2 = k_{-2} + \frac{k_2}{1 + K_1/[S_0]} \tag{40}$$

with $K_1 = k_{-1}/k_1$. This is a general result for a binding equilibrium preceding a first-order reaction, namely, the "apparent" first-order rate constant is dependent on the ligand concentration. Measurement of the dependence of λ_2 on the ligand concentration permits determination of the equilibrium dissociation constant and the two first-order rate constants characterizing the second step.

If all rate equations can be made "pseudo"-first order, an analytical solution of the coupled differential equations can be obtained, analogous to the situation near equilibrium. The solutions are sums of time-dependent exponentials and constants. The number of exponential terms observed establishes the minimum member of independent rate equations characterizing the mechanism.

One additional mechanism is considered, namely, consecutive first-order reactions. This mechanism illustrates the commonly observed situation of an intermediate appearing and then disappearing. The simplest case is

$$\mathrm{A} \xrightarrow{k_1} \mathrm{B} \xrightarrow{k_2} \mathrm{C}. \tag{41}$$

If the experiment starts with an initial concentration of A equal to $[\mathrm{A}]_0$, then

$$[\mathrm{A}] = [\mathrm{A}]_0\, \mathrm{e}^{-k_1 t} \tag{42}$$

and

$$\begin{aligned} \frac{\mathrm{d}[\mathrm{B}]}{\mathrm{d}t} &= k_1[\mathrm{A}] - k_2[\mathrm{B}] \\ &= k_1[\mathrm{A}]_0\, \mathrm{e}^{-k_1 t} - k_2[\mathrm{B}] \end{aligned} \tag{43}$$

Integration of Eq. (43) for $k_1 \neq k_2$ gives

$$[\mathrm{B}] = \frac{[\mathrm{A}]_0 k_1}{k_2 - k_1}\left(\mathrm{e}^{-k_1 t} - \mathrm{e}^{-k_2 t}\right) \tag{44}$$

Equation (44) is not valid for $k_1 = k_2$ since B becomes indeterminant. If Eq. (43) is integrated for $k_1 = k_2$, then

$$[\mathrm{B}] = k_1[\mathrm{A}]_0 t\, \mathrm{e}^{-k_1 t} \tag{45}$$

The time dependencies of A and B completely describe this system since only two independent concentration variables exist. If the time dependence of C is desired, it can be obtained from the mass conservation relationship $[\mathrm{A}]_0 = [\mathrm{A}] + [\mathrm{B}] + [\mathrm{C}]$:

$$[\mathrm{C}] = [\mathrm{A}]_0\left(1 - \frac{k_2}{k_2 - k_1}\mathrm{e}^{-k_1 t} + \frac{k_1}{k_2 - k_1}\mathrm{e}^{-k_2 t}\right) \tag{46}$$

If $k_2 \gg k_1$, B decays to C more rapidly than it is formed, and the concentration of B is very low at all times. Under these conditions, direct observation of the intermediate is virtually impossible, and the concentration of C essentially increases exponentially with k_1 being the characteristic rate constant (see Fig. 2). For the other extreme case, $k_1 \gg k_2$, the intermediate rapidly accumulates until essentially all of A has been converted to B which then decays to C as a first-order process with rate constant k_2. The buildup and decay occur in different time domains and can be treated essentially as independent processes. Likewise, the time dependence of the appearance of C shows two phases; a "lag" at short times while the intermediate [B] is accumulating followed by an exponential decay of [B] to [C]. If k_1 and k_2 are comparable, the concentration of *B* goes through a

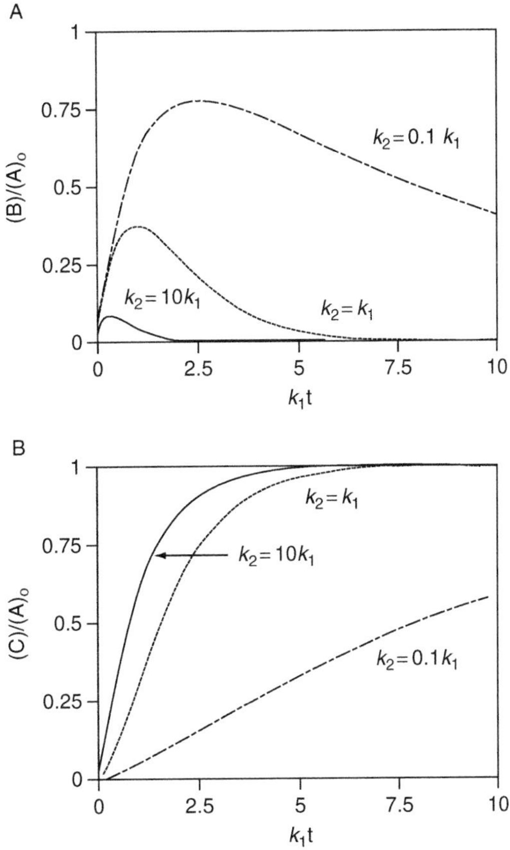

Fig. 2 Time dependence of the concentration ratio of either $[B]/[A]_0$ (A) or $[C]/[A]_0$ (B) for irreversible consecutive first-order reactions (Eqs. (41), (44), and (46)), where the ratio k_2/k_1 is 10 (—), 1 (. . .), or 0.1 (— - —).

well-defined maximum, and its time dependence requires consideration of both the rate of formation and the rate of decay. This behavior is shown graphically in Fig. 2.

The analysis of consecutive reactions can be applied directly to enzymatic processes as illustrated by consideration of a simple Michaelis-Menten mechanism:

$$\mathrm{E+S} \underset{k_{-1}}{\overset{k_1}{\rightleftharpoons}} \mathrm{X} \xrightarrow{k_2} \mathrm{E+P} \tag{47}$$

For this mechanism

$$\frac{\mathrm{d[X]}}{\mathrm{d}t} = k_1[\mathrm{S}][\mathrm{E}] - (k_2 + k_{-1})[\mathrm{X}] \tag{48}$$

If the initial substrate concentration, $[\mathrm{S}]_0$, is much greater than the total enzyme concentration, $[\mathrm{E}]_0$, then [S] can be assumed to remain constant at early times so that

$$\frac{\mathrm{d[X]}}{\mathrm{d}t} = k_1'[\mathrm{E}]_0 - (k_1' + k_{-1} + k_2)[\mathrm{X}] \tag{49}$$

where $k_1' = k_1[\mathrm{S}]$ and the conservation relationship, $[\mathrm{E}]_0 = [\mathrm{E}] + [\mathrm{X}]$, has been utilized. Equation (49) can be integrated to give

$$[\mathrm{X}] = \frac{k_1'[\mathrm{E}]_0}{k_1' + k_{-1} + k_2}[1 - \mathrm{e}^{-(k_1'+k_{-1}+k_2)t}] \tag{50}$$

Furthermore, the initial rate of product formation is given by

$$\frac{\mathrm{d[P]}}{\mathrm{d}t} = k_2[\mathrm{X}] \tag{51}$$

If Eq. (50) is inserted into Eq. (51) and the latter is integrated, the time dependence of product formation can be written as follows:

$$[\mathrm{P}] = \frac{k_1'k_2[\mathrm{E}]_0 t}{k_1' + k_{-1} + k_2} + \frac{k_1'k_2[\mathrm{E}]_0}{(k_1' + k_{-1} + k_2)^2}(\mathrm{e}^{-(k_1'+k_{-1}+k_2)t} - 1) \tag{52}$$

Note that at long times the second term becomes much smaller than the first, and the time dependence of [P] is given by the usual steady-state Michaelis-Menten equation. At very short times, if $k_1[\mathrm{S}] \gg (k_{-1} + k_2)$, there is a "lag" in the formation of product as the concentration of the intermediate, [X], increases to a steady-state concentration (see Fig. 3), and then the concentration of product increases linearly with time. However, if the concentration of both X and P are assayed

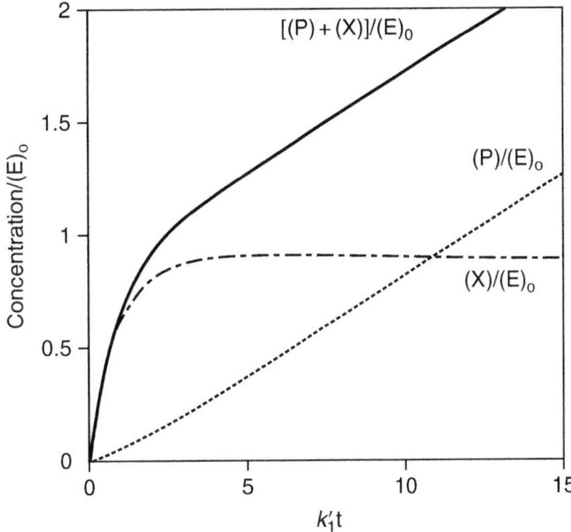

Fig. 3 Time dependence of the concentration ratios [P]/[E]$_0$ (...), [X]/[E]$_0$ (— - —), and [P] + [X]/[E]$_0$ (—) for a Michaelis-Menten mechanism (Eq. (47)), where the ratio $k_2/k_1' = 0.1$ and $k_{-1}/k_1' = 0.1$. These variations in concentration ratios are described by Eqs. (50), (52), and (53).

as product, as in the case where "X" is a species in which product is noncovalently bound to the enzyme, the observed product concentration, ([X] + [P]), is described by

$$[\mathrm{X}] + [\mathrm{P}] = \frac{k_1'[\mathrm{E}]_0}{k_1' + k_{-1} + k_2}\left[k_2 t + \frac{k_1' + k_{-1}}{k_1' + k_{-1} + k_2}\left(1 - \mathrm{e}^{-(k_1' + k_{-1} + k_2)t}\right)\right] \tag{53}$$

In this case when $k_1' \gg (k_2 + k_{-1})$, the observed product concentration initially increases rapidly with time with a maximum amplitude of [X] + [P]/[E]$_0$ = 1 and then becomes a linear function of time (see Fig. 3). Similarly, because Δ[S] = $-\Delta$([X] + [P]), the substrate concentration as a function of time displays a rapid, initial decrease followed by a linear decrease. The initial phase is called a "burst." Burst kinetics are observed for any mechanism in which product is formed in a rapid, first step followed by regeneration of active enzyme in a slower, second step as illustrated by chymotrypsin (Gutfreund and Sturtevant, 1956).

A well-known paradigm in chemical kinetics is that the number of possible mechanisms is limitless. Therefore, only some general procedures for carrying out analyses have been presented, with specific solutions to some relatively simple cases. Application of the principles developed to other mechanisms, although not always obvious, is possible.

III. Computer Simulation of Transient Kinetics

As discussed, the time dependence of substrate disappearance for simple kinetic mechanisms can be determined analytically by integration of the differential equations describing this scheme. Furthermore, two methods for simplifying these equations, relaxation kinetics and pseudo-first-order kinetics, allow explicit integration of the coupled differential equations for kinetic mechanisms with several steps. As the kinetic mechanisms become more complex, however, the differential equations describing the time dependence of the concentration terms become more difficult to integrate, and analytic solutions are impossible without simplification. For these reasons, it is often useful to integrate numerically the differential equations for complex reactions. Even though these numerical solutions are approximate, in practice they can be calculated as accurately as desired.

Advances in computing have allowed investigators to develop computer algorithms that will numerically integrate differential equations (Barshop *et al.*, 1983; Frieden, 1993; Hecht *et al.*, 1990; Holzhütter and Colosimo, 1990; Zimmerle and Frieden, 1989), providing rapid solutions to even complex reactions. The computer programs are readily available, flexible, easy to use, and run on a variety of microcomputers. Originally the programs simply allowed simulation of the time dependence of the concentration of substrate, intermediates, and products; however, some programs have been extended to allow the fitting of the reaction kinetics based on simulation (Holzhütter and Colosimo, 1990; Zimmerle and Frieden, 1989). This analysis provides an estimate of the confidence limits of each kinetic parameter.

Computer simulation of complex kinetic data has the following distinct advantages: (i) no simplifying assumptions are required to fit the data; (ii) the solutions must account for the concentration dependence of both the rate and the amplitude of a reaction; (iii) the simulation can explicitly fit data from a variety of experiments; and (iv) the simulations clearly demonstrate that a specific mechanism can account for the given data. However, a good fit between a simulated mechanism and the data does not prove that the mechanism is correct. Simulations are most useful for establishing a minimal mechanism required to fit the data and for quantitatively eliminating other possible mechanisms. To obtain the maximum information from numerical simulation, experiments should be designed to isolate specifically a few rate constants in order to provide narrow limits for their value in the simulation. Additional experiments can then build on this initial mechanism in an iterative fashion until all of the rate constants in the proposed mechanism are sufficiently constrained. This type of analysis has been essential for determination of the complete kinetic mechanism for a number of enzymes including dihydrofolate reductase (DHFR) (Fierke *et al.*, 1987; Penner and Frieden, 1987), alcohol dehydrogenase (Sekhar and Plapp, 1990), and 5-enolpyruvoylshikimate-3-phosphate synthase (EPSP synthase; 3-phosphoshikimate 1-carboxyvinyltransferase) (Anderson *et al.*, 1988).

IV. Rapid Mixing Methods

Stopped-flow instrumentation is useful for measuring rapid transient kinetics whenever there is an observable signal change that is easily related to concentration changes, including absorbance, fluorescence, light scattering, circular dichroism, nuclear magnetic resonance (NMR), and conductivity. The kinetics of enzyme-catalyzed reactions are determined most frequently by monitoring changes in absorbance or fluorescence. Instruments are equipped with a monochromator for determining the excitation wavelength and interference filters for the emission wavelength. A variety of band-pass and cutoff filters optimized for different wavelengths and purposes are available commercially (Johnson, 1986). There are several commercially available instruments designed for use with biological materials that combine high sensitivity with low sample volumes.

The design of a stopped-flow instrument is quite simple and has been discussed in detail in previous reviews (Johnson, 1992, 1986). The reaction is initiated by pushing two or more solutions (maintained in separate syringes) through a mixing chamber and into an observation cell. The flow of liquids is limited by a stop syringe (hence the name stopped-flow), and spectral data are collected by a computer after the flow has stopped. Stopped-flow enzyme kinetic data can be obtained under conditions of either excess substrate (presteady-state experiment) or excess enzyme (single turn-over experiment).

The maximum rate measurable by the stopped-flow technique is limited by the amount of time it takes to mix the reactants and to transport the solutions from the mixing chamber, where the reaction is initiated, to the observation point. This time, also called the "dead time" of the instrument, is dependent on the efficiency of mixing, the flow rate, and the distance between the mixer and the observation port. The dead time can be varied by changing the flow velocity; however, at high pressures the shear force through the mixer may damage some biological samples. In modern commercial instruments, the dead time is approximately 1–2 ms. This sets a practical upper limit of about 700 s^{-1} for first-order rate constants observable by this technique, since at 1 ms half of the reactants have been converted to products for this rate constant (Eq. (4)). For second-order (or higher) rate constants, the maximum rate constant observable is dependent on the concentration of reactants. For example, the observed pseudo-first-order rate constant for a simple association reaction (Eq. (9)) is dependent on the second-order association rate constant, the concentration of the reactant in excess, and the first-order dissociation rate constant (Eq. (7)).

Another limitation of stopped-flow spectroscopy is that the workable concentration of substrates is dictated by the extinction coefficients and quantum yields. The enzyme and substrate concentrations must be large enough to produce an observable signal, yet the total absorbance of the sample must be low enough so as to not interfere with the observed fluorescence or absorbance signal (via the inner filter effect or deviations from the Beer-Lambert law (Lakowicz, 1983).

Variation in the path length of the observation cell in modern instruments (0.2–2.5 cm) allows some flexibility in the design of these experiments. Additionally, averaging of several reaction traces significantly increases the ratio of signal to noise (S/N). Although in many cases a fluorescence signal is more sensitive, this measurement has the disadvantage that it is difficult to obtain stoichiometric amplitude data. The development of a rapid scanning diode array stopped-flow spectrophotometer allows observation of absorbance at multiple wavelengths at a rate of about 800 spectra s^{-1}. This technology can determine high-resolution difference spectra of transient intermediates and has been utilized for detecting and identifying reaction intermediates with pyridoxal 5′-phosphate-dependent enzymes (Brzović *et al.*, 1992; Phillips, 1991).

A second technique for obtaining nonequilibrium transient kinetic data is the quench-flow approach. This method has several advantages: an optical signal is not necessary for following the reaction; the chemical nature of intermediates and products can be probed directly by a variety of methods, including NMR spectroscopy (Anderson *et al.*, 1990; Barlow *et al.*, 1989); and the amplitude of the reaction is more straightforward to determine. However, it requires more time and material since each reaction measures only a single time point in a kinetic curve, rather than determining the entire kinetic transient, as in stopped-flow spectroscopy.

Several commercial quench-flow instruments are available for use with biological materials that combine gentle mixing methods with either a minimal sample volume (20–100 μl reaction^{-1}) (Johnson, 1992) for measurement of enzyme kinetics or a larger volume used for determination of protein folding kinetics (Roder, 1989). The design of a quench-flow instrument (Johnson, 1992) is reminiscent of the stopped-flow instrument. The reaction is started by pushing two or more solutions (maintained in separate syringes) through a mixing chamber. At this point the solution flows through a tube of defined length where it is allowed to react for various amounts of time before being mixed with a second solution, or quench, that stops the reaction. The reaction time is controlled by varying both the length of the reaction tube (i.e., the distance between the first and second mixing chambers) and the velocity of the liquid traveling through this tube. In this case, the dead time of the instrument ($\sim$2 ms) is determined by the minimum length of tubing needed to connect the two mixing chambers and the maximum flow velocity.

In chemical quench-flow experiments, the choice of quench reagents is critical. To observe the correct composition of intermediates, the reaction must be quenched quickly relative to the observed rate constants ($k_{\text{quench}} \gg 700\ s^{-1}$), and the materials must be stable under the conditions of the quench. In practical terms, this means that the quench usually involves denaturation of the enzyme by acid or base since proton transfer reactions are fast (Eigen and Hammes, 1963). However, metal chelators (such as EDTA) have been used to quench magnesium-catalyzed reactions, such as those catalyzed by ribozymes (Beebe and Fierke, 1994; Herschlag and Cech, 1990), where the substrates are unstable under both basic and acidic conditions. Verification that the same intermediates are obtained using different quench conditions is an essential control. For example, a tetrahedral intermediate formed in the EPSP synthase-catalyzed reaction of shikimic 3-phosphate with phosphoenol pyruvate is observed only when quenched by base (either triethylamine or 0.2 M NaOH) since it is unstable in acidic conditions (0.2 M HCl) (Anderson *et al.*, 1988).

A variant of the chemical-quench experiment, called a pulse-chase experiment, can be used to measure rate constants for formation of noncovalent enzyme intermediates (Rose, 1980). In this experiment, the enzyme is mixed with radioactive substrate under pseudo-first-order conditions, and the reaction is quenched at various times with a large excess ($\geq$10-fold) of either unlabeled substrate, product, or inhibitor. The quench stops additional reaction of the enzyme with radiolabeled substrate by either diluting the label or binding to and inhibiting the enzyme. However, noncovalent enzyme complexes are able to partition between dissociation (k_{-1}) and reaction to form products (k_2) under these conditions:

$$\mathrm{E} + \mathrm{S} \underset{k_{-1}}{\overset{k_1}{\rightleftharpoons}} \mathrm{E{\cdot}S} \xrightarrow{k_2} \mathrm{E} + \mathrm{P} \tag{54}$$

Finally, after allowing the reaction to occur for 5–10 half-times, the reaction is stopped with a chemical quench and the products analyzed. If substrate dissociation is faster than the chemical transformation, the kinetics of product formation should be identical to those observed in the chemical quench. On the other hand, if substrate dissociation is slow, additional product formation should be observed, reflecting the noncovalent enzyme complexes (such as E • S in Eq. (54)). The percentage of radiolabeled substrate observed as product is indicative of the ratio $k_2/(k_{-1} + k_2)$. The rate constant for formation of E • S can be calculated from the slope of a plot of k_{obs} versus concentration of substrate (or enzyme, depending on which is in excess).

V. Relaxation Methods

Rapid mixing methods have been predominantly used for transient kinetic studies of enzymes. However, these methods are inherently limited by how rapidly solutions can be mixed. This limitation is due to the hydrodynamics of mixing, and mixing times less than about 1 ms are difficult to achieve. Many of the elementary steps in enzymatic reactions require a shorter resolution time, which can be obtained with relaxation techniques (Bernasconi, 1976; Eigen and deMaeyer, 1963; Hammes and Schimmel, 1970). With relaxation methods, the reactants are mixed and allowed to come to equilibrium. If the equilibrium is shifted by rapidly changing an external parameter, such as the temperature, the rate of change of the concentrations of the reactants to new equilibrium values permits the kinetics of the reactions occurring to be studied.

The temperature-jump method has proved to be the most useful relaxation technique for studying enzymatic reactions (Rapid mixing methods also can be used as relaxation techniques through concentration jumps, pH jumps, etc.). The equilibrium constant, K, for a chemical reaction changes with temperature, T, at constant pressure, according to the relationship

$$\frac{d \ln K}{\mathrm{d}T} = \frac{\Delta H^\circ}{RT^2} \tag{55}$$

where "ΔH°" is the standard enthalpy change for the chemical reaction and R is the gas constant. The simplest way to change the temperature of an electrolyte solution rapidly is to discharge a high voltage across the solution. The energy discharged is $\frac{1}{2}CV^2$ where "C" is the capacitance used to store the high voltage and "V" is the voltage. For example, if an 0.1 μF capacitor is used to store 10,000 V, a temperature jump of 7.5 °C can be obtained in 0.15 ml. The time constant for the exponential discharge is $rC/2$ where "r" is the resistance of the solution (French and Hammes, 1969). Typically r is 100–200 ohms, which gives a discharge time constant of 5–10 μs. An inert electrolyte, such as 0.1 M KNO_3, is used to achieve such low resistances.

A schematic diagram of a temperature-jump apparatus is shown in Fig. 4. The practical problems associated with building such an apparatus have been discussed elsewhere and are not presented here (Bernasconi, 1976; Eigen and deMaeyer, 1963; French and Hammes, 1969). Although commercial equipment is not readily available, a temperature-jump apparatus can be built relatively easily. The typical time resolution is a few microseconds although submicroseconds are feasible.

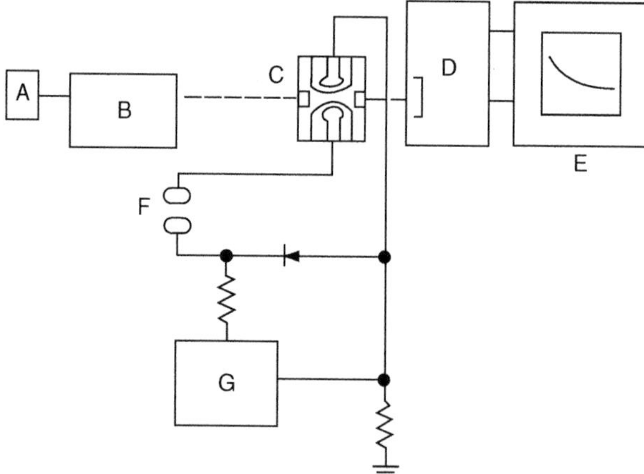

Fig. 4 Schematic diagram of a temperature-jump apparatus utilizing absorption spectrophotometry for the detection of concentration changes. (A), light source; (B), monochromator; (C), observation cell; (D), photomultiplier; (E), oscilloscope and/or computer; (F), spark gap; and (G), high voltage; from French and Hammes (1969), with permission.

Other methods have been used to obtain temperature jumps, such as dielectric heating and laser flashes, but have not proved as generally useful as a high-voltage discharge (Bernasconi, 1976; French and Hammes, 1969).

External parameters other than temperature can be used to perturb chemical equilibria (e.g., pressure and electric field) and have been extensively utilized in nonenzymatic systems (Bernasconi, 1976; Eigen and deMaeyer, 1963; Kustin, 1969). Chemical equilibria, however, are generally more sensitive to temperature changes than to pressure and electric field changes. Consequently, only temperature jumps have been extensively used to study enzyme mechanisms. In using this method, the enzymatic activity must be constantly monitored to be sure that the enzyme is not denatured by the temperature/electric field jumps.

Relaxation methods are useful only if appreciable concentrations of intermediates, reactants, and/or products are present at equilibrium. Obviously, if a reaction goes essentially to completion (i.e., all products), no equilibria are present to perturb. However, relaxation methods also can be used to perturb steady states. Thus, for example, reactants might be rapidly mixed and a temperature jump applied to the reactants and intermediates during a steady state, even if the reaction ultimately goes to completion. This method is applicable for steady states with time constants longer than about 10 ms, with the temperature-jump time resolution being a few microseconds. The methodology for carrying out such an experiment is somewhat difficult (Erman and Hammes, 1966a; French and Hammes, 1969), but a stopped-flow temperature-jump method has been used to study the ribonuclease reaction (Erman and Hammes, 1966b).

VI. Dihydrofolate Reductase

DHFR catalyzes the reduction of 7,8-dihydrofolate (H_2F) by NADPH to form 5,6,7,8-tetrahydrofolate (H_4F) (Benkovic *et al.*, 1988; Blakeley, 1985; Brown and Kraut, 1992). The enzyme maintains the intracellular pools of H_4F and derivatives, which are essential cofactors for the biosynthesis of thymidylate, purines, and several amino acids. Hence, DHFR is the target enzyme of both antitumor and antimicrobial drugs. Transient kinetics have been essential for investigation of the roles of amino acid residues in catalysis and inhibitor binding (Benkovic *et al.*, 1988; Taira *et al.*, 1987). Complete kinetic schemes have been described for DHFRs from a variety of sources, including *Escherichia coli* (Fierke *et al.*, 1987; Penner and Frieden, 1987), *Lactobacillus casei* (Andrews *et al.*, 1989), mouse (Thillet *et al.*, 1990), and human (Appleman *et al.*, 1989, 1990), and they show remarkable similarities. Salient features of the kinetics of *E. coli* DHFR are described here.

Determination of the association (k_1) and dissociation (k_{-1}) rate constants for substrate and product was particularly illuminating for the kinetic mechanism of DHFR. The rate of binding ligands to DHFR can be measured under conditions of excess ligand by following either the quenching of intrinsic enzyme fluorescence or fluorescence energy transfer between the enzyme and NADPH (Cayley *et al.*, 1981;

Dunn and King, 1980; Fierke *et al.*, 1987). In the formation of binary complexes of DHFR, two exponentials of equal amplitude are observed, consistent with a mechanism in which substrate (L_1, L_2) binds rapidly to only one of two enzyme conformers (E_1 and E_2) and interconversion between these conformers is slow:

$$\begin{array}{c} E_1 + L_1 \underset{k_{-1}}{\overset{k_1}{\rightleftharpoons}} E_1 \cdot L_1 + L_2 \underset{k'_{-1}}{\overset{k'_1}{\rightleftharpoons}} E_1 \cdot L_1 \cdot L_2 \\ k_{-2} \Downarrow\Uparrow k_2 \\ E_2 \end{array} \tag{56}$$

The fast exponential phase is primarily due to the rate of formation and dissociation of the binary complex, $E_1 \cdot L_1$, whereas the slow phase reflects the interconversion of E_1 and E_2. The association and dissociation rate constants can be determined from the slope and intercept, respectively, of a plot of the observed pseudo-first-order rate constants characterizing the ligand binding reaction versus ligand concentration (Eq. (7a)). Typical association and dissociation rate constants determined using this technique are listed in Table I (Fierke *et al.*, 1987).

A single, ligand-dependent exponential is observed in the formation of ternary complexes from binary complexes. The binding rate constants for formation of the

Table I
Association and Dissociation Rate Constants for Ligands Binding to *Escherichia coli* Dihydrofolate Reductase[a]

Ligand	Enzyme species	k_{on}[b,c] ($\mu M^{-1}\ s^{-1}$)	k_{off}[b,c] (s^{-1})	k_{off}[b,d] (s^{-1})
NADPH	E	20	3.5	3.6
	E · H_4F	8	85	85
	E · MTX[e]	20	≤2	–
H_2F	E	42	47	22
	E · TNADPH[e]	25	40	43
	E · $NADP^+$	26	10	7
$NADP^+$	E	13	300	290
	E · H_2F	5	60	50
	E · H_4F	–	–	200
H_4F	E	24	≤1	1.4
	E · $NADP^+$	28	5	2.4
	E · NADPH	2	10	12

[a]Measured at pH 6.0, 25 °C; reprinted with permission from Fierke *et al.* (1987), copyright 1987 American Chemical Society.

[b]Here, k_{on} is k_1 or k_1' and k_{off} is k_{-1} or k_{-1}' (Eq. (56)), depending on whether a binary or ternary complex is formed.

[c]The values of k_{on} and k_{off} are determined from the slope and intercept, respectively, of a plot of k_{obs} for the formation of E · L versus [L].

[d]Determined from mixing E · L_1 with excess competitive ligand [L_3] to form E · L_3 (Eq. (57)).

[e]MTX, Methotrexate; TNADPH, thio-NADPH.

reactive E · NADPH · H_2F ternary complex are estimated from E · I · S complexes where "I" is methotrexate for binding NADPH and either $NADP^+$ or thio-NADPH for binding H_2F (Table I). More direct measurements using a pulse-chase technique (Rose, 1980) would be useful since these inhibitors bind differently than substrate (Bystroff *et al.*, 1990; McTigue *et al.*, 1993).

Dissociation rate constants can also be measured from the fluorescence change observed when an enzyme-ligand complex (E · L_1) is mixed with a large excess of a second ligand (L_3) that competes for the binding site:

$$\mathrm{E{\cdot}L_1} \underset{k_{-1}}{\overset{k_1}{\rightleftharpoons}} \mathrm{E + L_1} \underset{k''_{-1}}{\overset{k''_1[\mathrm{L_3}]}{\rightleftharpoons}} \mathrm{E{\cdot}L_3} \tag{57}$$

To measure k_{-1} accurately, the effective rate constant for binding the second ligand ($k''_1[L_3]$) must be significantly faster than $k_1[L_1]$, k_{-1}, and k''_{-1}. Experiments of this type (see Table I) demonstrate the following (Fierke *et al.*, 1987): (1) association rate constants for the binding of all ligands are near the diffusion-controlled limit, about $2 \times 10^7\ M^{-1}\ s^{-1}$; (2) dissociation of tetrahydrofolate is the rate-limiting step for steady-state turnover; and (3) NADPH binding increases the rate constant for dissociation of H_4F, leading to the following preferred pathway of product dissociation:

$$\mathrm{E \cdot NADP^+ \cdot H_4F} \xrightarrow{\nearrow \mathrm{NADP^+}} \mathrm{E \cdot H_4F} \xrightarrow[\mathrm{NADPH}]{} \mathrm{E \cdot NADPH \cdot H_4F} \xrightarrow{\nearrow \mathrm{H_4F}} \mathrm{E \cdot NADPH} \tag{58}$$

At saturating substrate concentrations, the kinetic mechanism of DHFR can be simplified to two steps, namely, reduction of H_2F by NADPH to form products followed by dissociation of H_4F:

$$\mathrm{E \cdot NADPH \cdot H_2F} \xrightarrow{k_2} \mathrm{E \cdot NADP^+ \cdot H_4F} \xrightarrow[\mathrm{NADPH}]{\mathrm{NADP^+}} \mathrm{E \cdot NADPH \cdot H_4F} \xrightarrow{k_3} \mathrm{E \cdot NADPH + H_4F} \tag{59}$$

Equilibration between the E · $NADP^+$ · H_4F and E · NADPH · H_4F species is rapid at saturating NADPH. Because H_4F dissociation is slow ($k_2 \gg k_3$) (Fierke *et al.*, 1987), a burst of disappearance of NADPH may be observed during the first turnover by measuring either ultraviolet absorbance, fluorescence, or fluorescence energy transfer (Fig. 5). The fluorescence energy transfer experiments provide the highest signal-to-noise ratio, and therefore, the kinetic measurements were done using this technique. However, the absolute amplitude of the burst was determined from absorbance data. At low pH, the rate constant of the burst is about 950 s^{-1}, with an amplitude approaching 1 mol NADPH/mol DHFR. The large fluorescence energy transfer signal allows measurement of this fast rate constant even

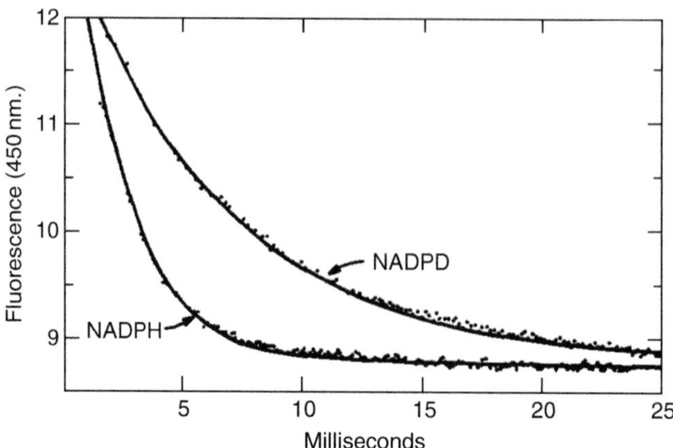

Fig. 5 Measurement of the presteady-state transient by stopped-flow fluorescence energy transfer for the reaction of DHFR with substrates. The enzyme was preincubated with either NADPH or [4′ (*R*)-^{2}H] NADPD, and the reaction was initiated by addition of H_2F. Final conditions were as follows: 15 μM DHFR, 125 μM NADPH(D), 100 μM H_2F, pH 6.5, and 25 °C. The data were fit by a single exponential decay followed by a linear, steady-state rate, $k_{cat} = 12\ s^{-1}$. Reprinted with permission from Fierke *et al.* (1987), copyright 1987 American Chemical Society.

though more than half of the signal disappears within the dead time of the instrument. This observed rate constant is independent of the substrate concentration, indicating that association is not the rate-limiting step. Furthermore, a kinetic deuterium isotope effect (k_{NADPH}/k_{NADPD}) of three on the rate constant of the burst indicates that this fluorescence signal directly reflects hydride transfer from NADPH to H_2F (Fig. 5). If the rate constant for hydride transfer is significantly faster than product dissociation, it can be determined directly by fitting the burst data to a single exponential followed by a linear rate. If not, simulation of the presteady-state data to more complicated models is required (Fierke *et al.*, 1987).

Both the rate constant and the amplitude of the presteady-state burst show a marked pH dependence (Fierke *et al.*, 1987). At low pH, the observed rate constant decreases, with a p*K* of 6.5 characterizing ionization of the ternary E · NADPH · H_2F complex, while the burst amplitude is invariant. These data indicate that the proton transfer component of the reaction is fast relative to hydride transfer. The p*K* of approximately 8 observed in the burst amplitude reflects the pH at which the observed rate constant for hydride transfer equals the rate constant for H_4F dissociation.

The mechanism of DHFR was completed by determining the overall equilibrium constant for the reaction, and by measuring the rate constant for hydride transfer from H_4F to $NADP^+$. In the *E. coli* enzyme, hydride transfer is the rate-limiting step for steady-state turnover in the reverse direction ($H_4F + NADP^+ \rightarrow H_2F + NADPH$) (Fierke *et al.*, 1987). However, in mouse DHFR dissociation of NADPH and/or H_2F is the rate-limiting step in this direction so that the rate constant for hydride transfer must be determined from presteady-state transients (Thillet *et al.*, 1990).

A necessary requirement for a kinetic mechanism derived using transient kinetic techniques is that it predicts the observed steady-state behavior. The proposed kinetic mechanism for *E. coli* DHFR predicts the observed steady-state kinetic parameters (k_{cat}, k_{cat}/K_m, and K_m) in both directions as well as the pH dependence and isotope effects (Fierke *et al.*, 1987). Furthermore, this model also predicts full time course kinetic curves (Fierke *et al.*, 1987; Penner and Frieden, 1987). Finally, this mechanism is essential for understanding the effects of single amino acid variants in DHFR and allows evaluation of the importance of a given amino acid for both binding ligands and catalyzing hydride transfer.

VII. Ribonuclease P

The ribonucleoprotein complex ribonuclease P (RNase P) catalyzes an essential step in tRNA maturation, namely, the cleavage of precursor-tRNA (pre-tRNA) to produce a mature 5′-terminus (Altman, 1990; Darr *et al.*, 1992; Pace and Smith, 1990). In diverse organisms, the enzyme is composed of two subunits; one is RNA (about 400 nucleotides) and the other is protein (about 120 amino acids). Although both are essential for *in vivo* activity, the RNA component from bacterial RNase P is sufficient to catalyze specific cleavage of pre-tRNA *in vitro* in the presence of high salt. Transient kinetics approaches have been useful for investigating the catalytic mechanism of RNase P (Beebe and Fierke, 1994; Reich *et al.*, 1988; Tallsjo and Kirsebom, 1993) as well as the *Tetrahymena* (Herschlag and Cech, 1990) and hammerhead ribozymes (Fedor and Uhlenbeck, 1992). A minimal kinetic description of a single turnover of RNase P RNA-catalyzed tRNA maturation includes (i) binding of pre-tRNA to RNase P; (ii) cleavage of the phosphodiester bond generating a 5′-product with a 3′-hydroxyl terminus and a 3′-tRNA product with a 5′-phosphate terminus (Altman, 1990; Darr *et al.*, 1992); and (iii) independent dissociation of both products.

The time course for hydrolysis of pre-tRNA catalyzed by the RNA component of RNase P (RNase P RNA) is characterized by two phases: a burst of tRNA product at short times followed by a linear increase in the concentration of tRNA (Beebe and Fierke, 1994; Reich *et al.*, 1988) (see Fig. 3). The data are consistent with a two-step mechanism in which hydrolysis occurs in the first step followed by regeneration of the active catalyst in the second step. The size of the pre-steady-state burst is dependent on the RNase P RNA concentration, with an amplitude of approximately 0.85 mol tRNA/mol RNase P RNA. This indicates that at least 85% of the enzyme molecules are catalytically competent and that the rate constant of the first step is at least 10-fold faster than that of the second step. Assuming two consecutive first-order (or pseudo-first-order) reactions, the amplitude and observed rate constant of the burst are dependent on both k_1 and k_2, as in Eq. (53).

To investigate whether product dissociation is the rate-limiting step for steady-state turnover catalyzed by RNase P RNA, the rate and equilibrium constants for product binding were measured. Bound and free product were resolved by gel shift or gel filtration centrifuge column chromatography (Beebe and Fierke, 1994; Hardt *et al.*, 1993). These experiments demonstrate the following: (i) association rate constants are quite large, about 5×10^6 M^{-1} s^{-1}; (ii) dissociation of tRNA is the rate-limiting step for steady-state turnover with a rate constant of 0.013 s^{-1}; and (iii) dissociation of the 5′-fragment is rapid.

The hydrolytic cleavage step catalyzed by the RNA component of RNase P can be isolated by measuring a single turnover in the presence of excess enzyme ([E]/[S] $\geq$ 5). Under these conditions, product dissociation is not observable since E · tRNA and tRNA are indistinguishable on the denaturing polyacrylamide gels used for assaying cleavage. The reaction is initiated by mixing RNase P RNA and pre-tRNA in a quench-flow instrument and is quenched by the addition of either EDTA or acid (Beebe and Fierke, 1994). At low concentrations of RNase P RNA, the appearance of tRNA product can be fit by a single first-order exponential with the rate constant linearly dependent on the concentration of RNase P RNA (Fig. 6A). However, at higher concentrations there is a lag in product formation, and the observed rate constant is independent of the concentration of RNase P RNA (Fig. 6B). The data of Fig. 6B are best explained by two consecutive first-order irreversible reactions (Eq. (46)), where the first step is association order of RNase P RNA and pre-tRNA to form a binary complex, $k_1 \approx 5 \times 10^6$ M^{-1} s^{-1}, and the second step is the chemical cleavage step, $k_2 = 6$ s^{-1}:

$$\mathrm{E} + \text{pre-tRNA} \xrightarrow{k_1} \mathrm{E}\cdot\text{pre-tRNA} \xrightarrow{k_2} \mathrm{E}\cdot\mathrm{tRNA}\cdot\mathrm{P} \tag{60}$$

Measurement of the effect of thiophosphate substitution at the cleavage site on k_2 would verify that it directly reflects hydrolysis (Herschlag *et al.*, 1991). Furthermore, at high RNase P RNA concentration more than 95% of the pre-tRNA is hydrolyzed in the first turnover, indicating that either the equilibrium for hydrolysis is very favorable or the dissociation rate constant of the 5′-fragment is significantly faster than religation. This sets an upper limit for the rate constant for religation.

The model predicts that at high RNase P RNA concentrations a significant amount of the E · pre-tRNA binary complex forms and does not exchange readily with ligands in solution. This can be tested by stopping the reaction with the addition of a high concentration of tRNA before the EDTA quench (Beebe and Fierke, 1994). In this case, any intermediate enzyme-substrate complexes which do not dissociate rapidly should react to form products before the reaction is quenched with EDTA; this is observed as additional product formation in the tRNA trap (Fig. 7). The tRNA trap data can be fit by a single exponential with a rate constant equal to k_1[RNase P RNA], where "k_1" is similar to the enzyme-dependent rate constant observed in the EDTA quench (Fig. 6A).

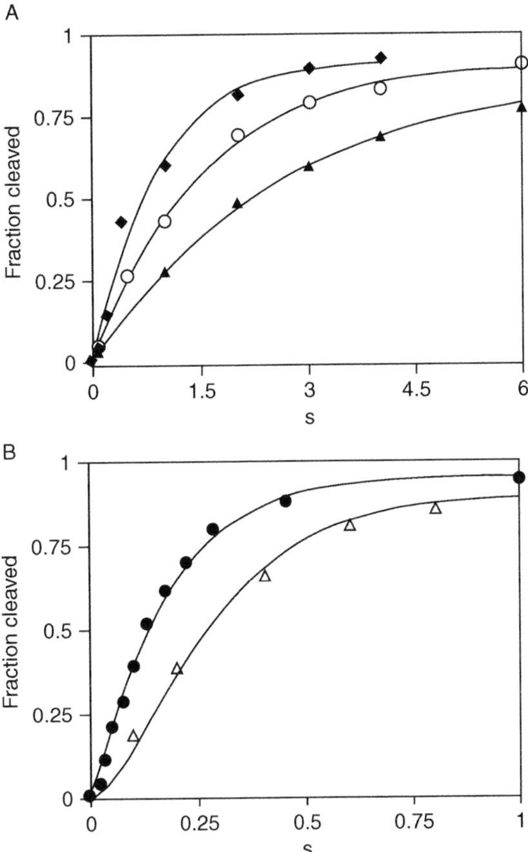

Fig. 6 Single turnover measurements of the hydrolysis of pre-tRNA^{Asp} catalyzed by the RNA component of *Bacillus subtilis* RNase P. Pre-tRNA^{Asp} (24 nM) was mixed with varying concentrations of excess RNAse P RNA in a quench-flow instrument at 37 °C in 100 mM $MgCl_2$, 800 mM NH_4Cl, 50 mM Tris, pH 8.0, 0.05% Nonidet P-40, and 0.1% sodium dodecyl sulfate. The reaction was quenched by a threefold dilution into 90 mM EDTA followed by addition of urea to 5 M. (A) At low concentrations of RNase P RNA [0.12 μM (▲), 0.24 μM (○), and 0.45 μM (◆)], the data are fit by a single exponential (Eq. (4)). (B) At high concentration of RNase P RNA [1.4 μM (Δ) and 19 μM (•)], the data are fit to a mechanism of two consecutive first-order reactions (Eq. (46)) with $k_1 = 6 \times 10^6$ M^{-1} s^{-1} [RNase P RNA]$_0$ and $k_2 = 6$ s^{-1}; reprinted with permission from Beebe and Fierke (1994), copyright 1994 American Chemical Society.

Furthermore, since at least 95% of the radiolabeled pre-tRNA is observed as labeled tRNA product in the tRNA trap experiment, hydrolysis of pre-tRNA is at least 20-fold faster than dissociation (Rose, 1980).

This set of experiments allowed development of a scheme which describes the kinetic behavior of a single turnover and simulates the steady-state turnover number, k_{cat}. However, it predicts that the steady-state kinetic parameter

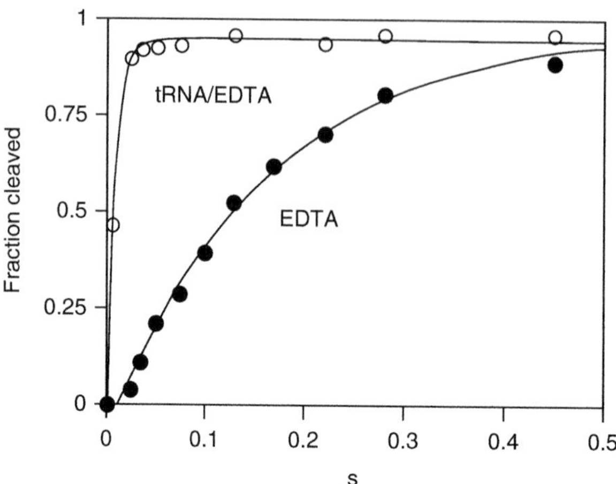

Fig. 7 Formation of an RNase P RNA • pre-tRNA^{Asp} binary complex assayed by a cold tRNA trap. Pre-tRNA^{Asp} (24 nM) was mixed with excess RNase P RNA (19 μM) in a quench-flow instrument at 37 °C in 100 mM $MgCl_2$, 800 mM NH_4Cl, 50 mM Tris, pH 8.0, 0.05% Nonidet P-40, and 0.1% sodium dodecyl sulfate. The reaction was quenched by a threefold dilution into either (•) 90 mM EDTA followed by 5 M urea or (○) 58 μM tRNA followed by addition of 90 mM EDTA and 5 M urea after 10 s. The cold tRNA trap data are fit by a single exponential (Eq. (4)), whereas the EDTA data are fit to a mechanism of two consecutive first-order reactions (Eq. (46)), $k_1 = 6 \times 10^6$ M^{-1} s^{-1} [RNase P RNA]$_0$ and $k_2 = 6$ s^{-1}; reprinted with permission from Beebe and Fierke (1994), copyright 1994 American Chemical Society.

measuring the second-order rate constant for hydrolysis of pre-tRNA catalyzed by RNase P RNA at low substrate concentration, k_{cat}/K_m, should be about 30-fold faster than the observed rate constant (Reich *et al.*, 1988). Because the single turnover experiments measure the reaction up to product formation, the only model consistent with all the data is that an additional enzyme species (E′) is produced during the first turnover that binds substrate significantly more slowly and does not equilibrate rapidly with the original conformer (E). This different enzyme species could vary in RNA conformation, bound metals, or bound products. This example clearly illustrates the utility of testing whether a kinetic mechanism determined from transient kinetic methods is sufficient to explain the observed steady-state behavior.

These experiments have determined a minimal kinetic scheme for the hydrolysis of pre-tRNA^{Asp} catalyzed by the RNA component of *Bacillus subtilis* RNase P under high salt conditions (Beebe and Fierke, 1994). This scheme will allow pinpointing of the role of metals, pH, and the protein component on each step of the reaction. Furthermore, it is an essential background for comprehending the effect of structural changes in both the substrate and enzyme on catalytic activity. Additional studies should lead to a greater understanding of the chemistry and mechanisms of rate acceleration utilized by ribozymes.

VIII. Fatty-Acid Synthase

Chicken liver fatty-acid synthase is a multifunctional enzyme that catalyzes the synthesis of palmitic acid according to the following overall reaction:

$$\begin{aligned}&\text{Acetyl-CoA} + 7\text{malonyl-CoA} + 14\text{NADPH} + 14\text{H}^+\\&\rightarrow \text{palmiticacid} + 8\text{CoA} + 14\text{NADP}^+ + 6\text{H}_2\text{O} + 7\text{CO}_2\end{aligned} \tag{61}$$

The enzyme has been extensively reviewed (Chang and Hammes, 1990; Wakil, 1989; Wakil *et al.*, 1983), and only some aspects of transient kinetic studies are discussed here. The enzyme consists of two identical polypeptide chains and two independent catalytic sites composed of two different polypeptide chains. During the catalysis, reaction intermediates are covalently bound to three different sites, a serine hydroxyl, 4′-phosphopantetheine, and a cysteine sulfhydryl. The sequence of reactions in a catalytic cycle is shown in Fig. 8.

The reaction is initiated by nucleophilic attack of serine on acetyl-CoA and formation of an acetyl oxyester; the acetyl moiety is then passed sequentially to 4′-phosphopantetheine and cysteine, where thioesters are formed. Subsequently the enzyme binds malonyl-CoA, which reacts with serine, and a malonyl residue is passed from serine to 4′-phosphopantetheine. The acetyl then condenses onto the malonyl, with the release of carbon dioxide, to form a ketothioester. The ketothioester is reduced to an alcohol by NADPH, then dehydrated, and the double bond is reduced by NADPH to give a saturated four-carbon thioester (Fig. 9). The saturated acid is then transferred to the cysteine from 4′-phosphopantetheine, malonyl is loaded onto the enzyme, and the reaction sequence is repeated until the thioester of palmitic acid is present on the 4′-phosphopantetheine. The thioester is cleaved by a thioesterase activity to release palmitic acid. The separate activities are acetyl/malonyl transacylase, β-ketoacyl synthase, β-ketoacyl reductase, dehydratase, enoyl reductase, and palmitoyl thioesterase.

Transient kinetic methods have been used to investigate the individual steps in the first catalytic cycle. The reaction of acetyl-CoA with the enzyme was studied with a quench-flow apparatus utilizing radioactive acetyl, with perchloric acid as the quench (Cognet and Hammes, 1983). The kinetics were first order because the acetyl-CoA concentration was much greater than that of the enzyme. The apparent first-order rate constant increased with increasing acetyl-CoA concentration, ultimately leveling off at high concentrations. This is indicative of a first-order reaction preceded by a binding equilibrium. Because this reaction is slightly reversible, the analysis of the kinetics also must take into account dissociation of the acetylated enzyme; however, this complication is not considered here. The deacetylation of the enzyme by reaction with CoA was studied independently

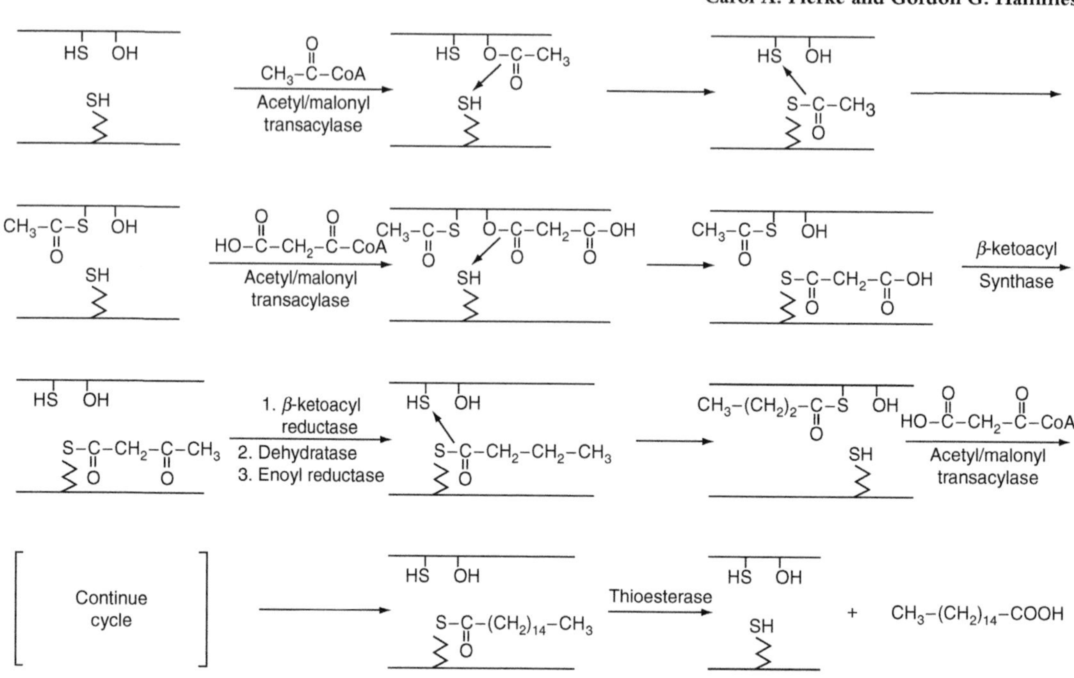

Fig. 8 Schematic representation of palmitic acid synthesis by chicken liver fatty-acid synthase. The symbol ~SH represents 4′-phosphopantetheine, -SH is a cysteine sulfhydryl, and -OH is a serine hydroxyl. The initial sequence of reactions, involving malonyl transacylase, β-ketoacyl synthase, β-ketoacyl reductase, dehydratase, and enoyl reductase, is repeated seven times to give enzyme-bound palmitic acid.

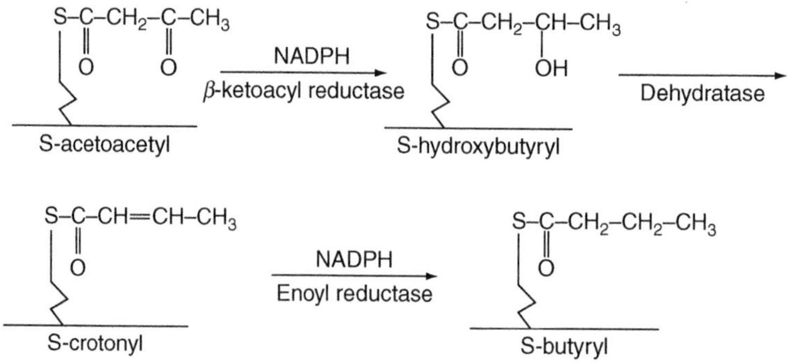

NADPH
β-ketoacyl reductase
Dehydratase
S-acetoacetyl
S-hydroxybutyryl
NADPH
Enoyl reductase
S-crotonyl
S-butyryl

Fig. 9 Reduction of enzyme-bound acetoacetate to butyrate by fatty-acid synthase. The cycle is repeated for each addition of malonate to the growing fatty-acid chain. Here "S" is part of 4′-phosphopantetheine that is covalently coupled to the enzyme.

with the quench-flow approach. The overall process was fit to the following mechanism:

$$\text{AcCoA} + \text{E} \underset{}{\overset{K_1}{\rightleftharpoons}} \text{E}\cdot\text{AcCoA} \underset{k_{-2}}{\overset{k_2}{\rightleftharpoons}} \text{EAc}\cdot\text{CoA} \underset{}{\overset{K_3}{\rightleftharpoons}} \text{EAc} + \text{CoA} \qquad (62)$$

At pH 7.0, 0.1 M potassium phosphate, 23 °C, the equilibrium dissociation constants, K_1 and K_3, are estimated to be 85 and 70 μM, respectively, and the rate constants, k_2 and k_{-2}, are 43 and 103 s^{-1}. Note that the rate constant for deacetylation of the enzyme is larger than the acetylation rate constant. However, the deacetylation does not interfere with the formation of palmitic acid under physiological conditions owing to the low concentration of CoA.

Although the observed acetylation kinetics are deceptively simple, the chemistry is actually complex because the acetyl group has three different binding sites on the enzyme. Therefore, the apparent first-order rate constant characterizes the rate of acetyl transfer from enzyme-bound CoA to three different sites and/or the rate of internal transfer between the sites. The rate of acetylation of individual sites can be determined by specifically blocking either the cysteine or cysteine and 4′-phosphopantetheine through chemical modification of the protein (Yuan and Hammes, 1985). The results obtained with both sulfhydryls blocked demonstrate that formation of the oxyester is relatively rapid, $k_2 = 150\ s^{-1}$. If the transfer between sites is strictly ordered (Fig. 8), then results with other modified enzymes indicate that intramolecular transfer of acetyl from serine to 4′-phosphopantetheine has a rate constant of less than 110 s^{-1} and that for transfer to cysteine is less than 43 s^{-1} (If binding to the three sites is random, these rate constants represent lower bounds for the transfer of acetyl from enzyme-bound CoA to each of the binding sites). Similar studies of the binding of malonyl to the enzyme have not been carried out owing to the expense of radioactive substrate and the complications of side reactions.

Kinetic investigation of the next step in the reaction sequence, namely, the condensation of acetyl and malonyl, is problematic because spectral changes specific to this reaction do not occur and isolation of the reaction product is difficult. However, the binding of NADPH to the enzyme is accompanied by a large enhancement in the fluorescence of NADPH. The kinetics of binding is consistent with a bimolecular reaction characterized by a second-order rate constant of $1.27 \times 10^7\ M^{-1}\ s^{-1}$ at pH 7.0, 25 °C (Cognet *et al.*1983). If the substrate concentrations, acetyl-CoA, malonyl-CoA, and NADPH, are large relative to the enzyme concentration, the enzyme will cycle through the entire reaction, and determination of specific rate constants is not possible. If the concentration of NADPH is much less than that of the enzyme, however, the reaction will stop with formation of the 3-hydroxybutyryl/crotonyl- or butyryl-enzyme intermediate (Fig. 9). All that remains to be done is to adjust the substrate reactions so that the kinetics for the formation of the acetoacetyl and 3-hydroxybutyryl/crotonyl intermediates are pseudo-first order.

If the concentrations of acetyl-CoA and malonyl-CoA are large relative to the enzyme concentration, the formation of acetoacetyl-enzyme, EAcAc, is first order with respect to the enzyme E:

$$\frac{d[\mathrm{EAcAc}]}{dt} = k_1[\mathrm{E}] \tag{63}$$

where "k_1" is a function of the concentrations of acetyl-CoA and malonyl-CoA owing to the binding equilibria of the substrates with the enzyme. Integration of Eq. (63) gives

$$[\mathrm{EAcAC}] = [\mathrm{E}]_0(1 - e^{-k_1 t}) \tag{64}$$

where "$[\mathrm{E}]_0$" is the enzyme concentration at $t = 0$.

If the formation of the acetoacetyl-enzyme is coupled to the subsequent β-ketoacyl reductase reaction under conditions where the enzyme concentration is much greater than that of NADPH, the reduction is first order with respect to NADPH:

$$\begin{aligned}\frac{-d[\mathrm{NADPH}]}{dt} &= k_2[\mathrm{EAcAC}][\mathrm{NADPH}] \\ &= k_2[\mathrm{E}]_0(1 - e^{-k_1 t})[\mathrm{NADPH}]\end{aligned} \tag{65}$$

Integration of Eq. (65) gives

$$[\mathrm{NADPH}] = [\mathrm{NADPH}]_0 \exp\{-k_2[\mathrm{E}]_0(t + (1/k_1)(e^{-k_1 t} - 1))\} \tag{66}$$

where "$[\mathrm{NADPH}]_0$" is the NADPH concentration at $t = 0$. Note that $k_2[\mathrm{E}]_0$ is a pseudo-first-order rate constant that is a function of the total enzyme concentration because of the rapid binding equilibrium between enzyme and NADPH. This analysis depends on several assumptions: (i) binding of acetyl-CoA, malonyl-CoA, and NADPH is rapid relative to the formation of acetoacetyl- and 3-hydroxybutyryl-enzyme; and (ii) further reaction of 3-hydroxybutyryl does not interfere with the observed kinetics. All these assumptions have been shown to be valid (Cognet *et al.*, 1983).

The time course for the reduction of NADPH should display a lag as acetoacetyl-enzyme is formed, followed by a first-order decay with rate constant $k_2[\mathrm{E}]_0$. A typical time course is shown in Fig. 10. If the binding equilibria are taken into account by extrapolating the pseudo-first-order rate constants to the approximate limits at high substrate concentrations, the rate constants for acetoacetyl- and 3-hydroxybutyryl-enzyme formation are 30.9 and 17.5 s^{-1}, respectively, at pH 7.0, 0.1 M potassium phosphate, 25 °C.

The reduction of acetoacetyl-enzyme can be independently studied because this intermediate can be directly formed by reaction of acetoacetyl-CoA with the

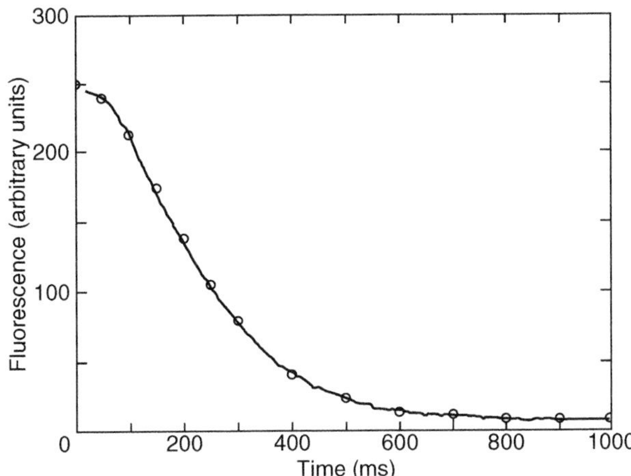

Fig. 10 Typical stopped-flow kinetic trace of the change in fluorescence following mixing of fatty-acid synthase (2.14 μM)-NADPH (0.75 μM) with acetyl-CoA (56 μM) and malonyl-CoA (98 μM) in 0.1 M potassium phosphate buffer—1 mM EDTA, pH 7.0, at 25 °C. The circles have been calculated with the best fit parameters $k_1 = 4.3\ s^{-1}$, $k_2 = 9.5\ s^{-1}$ (Eq. (66)); reprinted with permission from Cognet *et al.* (1983), copyright 1983 American Chemical Society.

enzyme (Cognet and Hammes, 1985). This intermediate can then be reacted directly with NADPH, with [NADPH] $\gg$ [E] so the reaction is pseudo-first order. The first-order rate constant obtained when the enzyme is saturated with NADPH is 20 s^{-1}, in good agreement with the result from the coupled reaction. Similarly, the reduction of the hydroxybutyryl-/crotonyl-enzyme by NADPH can be studied directly. The rate constant obtained is 36.6 s^{-1} (Cognet and Hammes, 1985). In both cases, the dependence of the pseudo-first-order rate constant on the concentration of NADPH was determined in order to obtain the first-order rate constants cited. Unfortunately no method has been devised to study the kinetics of the interconversion of the hydroxybutyryl and crotonyl intermediates. In fact, these intermediates are in equilibrium on the enzyme so that the reaction of NADPH is with an equilibrium mixture.

The studies described above have permitted delineation of most of the rate constants for the first catalytic cycle. However, the rate constants may change as the fatty-acid chain length of the covalently bound intermediate increases. An assessment of how the rate constants vary with chain length can be obtained by examining the distribution of chain lengths covalently bound during catalysis (Anderson and Hammes, 1985). Because this chapter is primarily concerned with transient kinetics, these studies are not described here. The results suggest that the rate constants for the condensation and reduction steps become larger as the chain length increases, by approximately a factor of two per cycle.

Finally, the turnover number for the thioesterase has been estimated as about 2 s^{-1} with palmitoyl-CoA as the substrate (Liu and Hammes, 1988).

Although the mechanism of action of fatty-acid synthase is quite complex, transient kinetics have been used to elucidate many of the elementary steps in the mechanism. These studies illustrate the experimental design and interpretation that are often needed for transient kinetic investigations.

IX. Conclusion

This chapter has presented a description of transient kinetic approaches to the study of enzyme mechanisms. In particular, transient kinetic analysis of three enzymes, DHFR, RNase P, and fatty-acid synthase, demonstrates that determination of the individual rate constants in an enzyme pathway provides novel insights into enzyme mechanisms, particularly if chemical transformation is not rate limiting for steady-state turnover. Transient kinetic analysis is an essential method for determination of enzyme mechanism because of the vast wealth of unique information it provides. Although this chapter could not cover all possible kinetic mechanisms, it provides a useful blueprint for the utilization of these methods to elucidate the molecular bases of enzyme catalysis.

Acknowledgments

We thank Teaster Baird, Michael Been, Shawn Zinnen, Jane Beebe, and Lora LeMosy for help with the preparation of this review and the National Institutes of Health (GM40602) and the Office of Naval Research (C.A.F.) for support of this work. C. A. F. received an American Heart Association Established Investigator Award and a David and Lucile Packard Foundation Fellowship in Science and Engineering.

References

Altman, S. (1990). *J. Biol. Chem.* **265,** 20053.

Anderson, V. E., and Hammes, G. G. (1985). *Biochemistry* **24,** 2147.

Anderson, K. S., Sikorski, J. A., and Johnson, K. A. (1988). *Biochemistry* **27,** 7395.

Anderson, K. S., Sammons, R. D., Leo, G. D. C., Sikorski, J. A., Benesi, A. J., and Johnson, K. A. (1990). *Biochemistry* **29,** 1460.

Andrews, J., Fierke, C. A., Birdsall, B., Ostler, G., Feeney, J., Roberts, G. C. K., and Benkovic, S. J. (1989). *Biochemistry* **28,** 5743.

Appleman, J. R., Beard, W. A., Delcamp, T. J., Prendergast, N. J., Freisheim, J. H., and Blakely, R. L. (1989). *J. Biol. Chem.* **264,** 2625.

Appleman, J. R., Beard, W. A., Delcamp, T. J., Prendergast, N. J., Freisheim, J. H., and Blakely, R. L. (1990). *J. Biol. Chem.* **265,** 2740.

Barlow, P. N., Appleyard, R. J., Wilson, B. J. O., and Evans, J. N. S. (1989). *Biochemistry* **28,** 7985.

Barshop, B. A., Wrenn, R. F., and Frieden, C. (1983). *Anal. Biochem.* **130,** 134.

Beebe, J. A., and Fierke, C. A. (1994). *Biochemistry* **33,** 10294.

Benkovic, S. J., Fierke, C. A., and Naylor, A. M. (1988). *Science* **239,** 1105.

Bernasconi, C. F. (1976). "Relaxation Kinetics," Academic Press, New York.

Bernasconi, C. F. (ed.) (1986). *In* "Investigation of Rates and Mechanisms of Reactions," Part 2 4th edn. Wiley (Interscience), New York.
Blakeley, R. L. (1985). *In* "Folates and Pterins" (R. L. Blakely and S. J. Benkovic, eds.), Vol. 2, p. 91. Wiley, New York.
Brown, R. A., and Kraut, J. (1992). *Faraday Discuss. Chem. Soc.* **93,** 217.
Brzović, P. S., Sawa, Y., Hyde, C. C., Miles, E. W., and Dunn, M. F. (1992). *J. Biol. Chem.* **267,** 13028.
Bystroff, C., Oakley, S. J., and Kraut, J. (1990). *Biochemistry* **29,** 3263.
Cayley, P. J., Dunn, S. M. J., and King, R. W. (1981). *Biochemistry* **20,** 874.
Chang, S. I., and Hammes, G. G. (1990). *Acc. Chem. Res.* **23,** 363.
Cleland, W. W. (1977). *Adv. Enzymol.* **45,** 273.
Cognet, J. A. H., and Hammes, G. G. (1983). *Biochemistry* **22,** 3002.
Cognet, J. A. H., and Hammes, G. G. (1985). *Biochemistry* **24,** 290.
Cognet, J. A. H., Cox, B. G., and Hammes, G. G. (1983). *Biochemistry* **22,** 6281.
Darr, S. C., Brown, J. W., and Pace, N. R. (1992). *Trends Biochem. Sci.* **17,** 178.
Dunn, S. M. J., and King, R. W. (1980). *Biochemistry* **19,** 766.
Eigen, M., and deMaeyer, L. (1963). *In* "Investigation of Rates and Mechanisms of Reactions" (S. L. Friess, E. S. Lewis and A. Weissberger, eds.), Part 2 2nd edn. p. 895. Wiley, New York.
Eigen, M., and Hammes, G. G. (1963). *Adv. Enzymol.* **25,** 1.
Erman, J. E., and Hammes, G. G. (1966). *Rev. Sci. Instrum.* **37,** 746.
Erman, J. E., and Hammes, G. G. (1966). *J. Am. Chem. Soc.* **88,** 5607.
Fedor, M. J., and Uhlenbeck, O. C. (1992). *Biochemistry* **31,** 12042.
Fersht, A. (1985). "Enzyme Structure and Mechanism," 2nd edn. Freeman, San Francisco.
Fierke, C. A., Johnson, K. A., and Benkovic, S. J. (1987). *Biochemistry* **26,** 4085.
French, T. C., and Hammes, G. G. (1969). *Methods Enzymol.* **16,** 3.
Frieden, C. (1993). *Trends Biochem. Sci.* **18,** 58.
Gutfreund, H., and Sturtevant, J. M. (1956). *Biochem. J.* **63,** 656.
Hammes, G. G. (1978). *In* "Principles of Chemical Kinetics," p. 72. Academic Press, New York.
Hammes, G. G. (1982). "Enzyme Catalysis and Regulation," Academic Press, New York.
Hammes, G. G., and Haslam, J. L. (1969). *Biochemistry* **8,** 1591.
Hammes, G. G., and Schimmel, P. R. (1970). *In* "Enzymes" (P. D. Boyer, ed.), 3rd edn.,Vol. 2, p. 67. Academic Press, New York.
Hardt, W. D., Schlegl, J., Erdmann, V. A., and Hartmann, R. K. (1993). *Nucleic Acids Res.* **21,** 3521.
Hecht, J. P., Nikonov, J. M., and Alonso, G. L. (1990). *Comput. Methods Prog. Biomed.* **33,** 13.
Herschlag, D., and Cech, T. R. (1990). *Biochemistry* **29,** 10159.
Herschlag, D., Piccirilli, J. A., and Cech, T. R. (1991). *Biochemistry* **30,** 4844.
Holzhütter, H. G., and Colosimo, A. (1990). *Comput. Appl. Biosci.* **6,** 23.
Johnson, K. A. (1992). *In* "Enzymes" (P. D. Boyer, ed.), 4th edn., Vol. 20, p. 1. Academic Press, New York.
Johnson, K. A. (1986). *Methods Enzymol.* **134,** 677.
Kustin, K. (Ed.). (1969). *Methods Enzymol.* **16.**
Lakowicz, J. R. (1983). *In* "Principles of Fluorescence Spectroscopy," p. 44. Plenum, New York.
Liu, W., and Hammes, G. G. (1988). Unpublished results.
McTigue, M. A., Davie, J. F., II, Kaufman, B. T., and Kraut, J. (1993). *Biochemistry* **32,** 6855.
Pace, N. R., and Smith, D. (1990). *J. Biol. Chem.* **265,** 3587.
Peller, L., and Alberty, R. A. (1959). *J. Am. Chem. Soc.* **81,** 5907.
Penner, M. H., and Frieden, C. (1987). *J. Biol. Chem.* **262,** 15908.
Phillips, R. S. (1991). *Biochemistry* **30,** 5927.
Reich, C., Olsen, G. J., Pace, B., and Pace, N. R. (1988). *Science* **239,** 178.
Roder, H. (1989). *Methods Enzymol.* **176,** 446.
Rose, I. A. (1980). *Methods Enzymol.* **64,** 47.
Sekhar, V. C., and Plapp, B. V. (1990). *Biochemistry* **29,** 4289.

Taira, K., Fierke, C. A., Chen, J. T., Johnson, K. A., and Benkovic, S. J. (1987). *Trends Biochem. Sci.* **12,** 275.
Tallsjo, A., and Kirsebom, L. A. (1993). *Nucleic Acids Res.* **21,** 51.
Thillet, J., Adams, J. A., and Benkovic, S. J. (1990). *Biochemistry* **29,** 5195.
Wakil, S. J. (1989). *Biochemistry* **28,** 4523.
Wakil, S. J., Stoops, J. K., and Joshi, V. C. (1983). *Annu. Rev. Biochem.* **52,** 537.
Yuan, Z. Y., and Hammes, G. G. (1985). *J. Biol. Chem.* **260,** 13532.
Zimmerle, C. T., and Frieden, C. (1989). *Biochem. J.* **258,** 381.

CHAPTER 14

Rapid Quench Kinetic Analysis of Polymerases, Adenosinetriphosphatases, and Enzyme Intermediates

Kenneth A. Johnson

Department of Chemistry and Biochemistry
Institute for Cell and Molecular Biology
University of Texas
Austin, TX 78735, USA

DOI: 10.1016/B978-0-12-378608-1.00014-1

I. Update

Methods for performing rapid quench kinetic studies to establish the rates of enzyme-catalyzed reactions have been refined in the past decade but remain largely as described in the original manuscript. For example, advances in computer-controlled servo motors have made stepper motors obsolete and have improved the performance of rapid quench instruments, but the basic methods remain the same. However, major advances have been achieved in the way experimental data are fit to a model using computer simulation (Johnson *et al.*, 2009a,b). Global data fitting based upon numerical integration of the rate equations affords the most rigorous data analysis because it bypasses the errors inherent in simplifying a reaction scheme so that the math becomes tractable, and because multiple experiments involving different starting conditions and various signals can all be fit simultaneously to a single, unifying model. Now, the results from rapid quench experiments and those obtained using stopped-flow methods and steady-state measurements can all be fit simultaneously, which, in turn, allows each data set to be interpreted more rigorously. For example, it has often been difficult to extract meaningful mechanistic information from fluorescence stopped-flow experiments because the origin of the fluorescence signal is not always known. Correlation of rates measured in the stopped-flow with rates of the chemical reaction observed by rapid quench methods has been difficult and has often relied upon attempts to overlay the two data sets with some arbitrary scaling factor (Johnson and Taylor, 1978). Global data fitting now allows both rapid quench and fluorescence stopped-flow data to be fit simultaneously and the global fitting affords a more robust assessment of errors in fitted parameters including fluorescence scaling factors. Fast computer simulation algorithms also provide direct visual feedback by showing how the fitted curves change in shape as parameters are varied in approaching the best fit, which is useful both before the fit in finding starting values for nonlinear regression analysis, and after the fit in exploring the extent to which parameters are constrained by the data. Thus, the major advance in rapid quench methods is due to our ability to now interpret the results more rigorously and to use the richness of the data to expand our ability to interpret other experiments to achieve an overall global fit to define a mechanism.

Rapid quench methods remain as the most reliable method to assess fidelity of DNA polymerases by measurement of k_{pol} and K_d governing nucleotide incorporation as outlined in this manuscript. However, based upon studies using a fluorescently labeled protein to monitor the conformational change following nucleotide binding, we now know that the conformational change is faster than chemistry (Tsai and Johnson, 2006). Nonetheless, measurements of k_{pol}/K_d define the specificity constant (k_{cat}/K_m) for sequential nucleotide incorporation during processive DNA synthesis, but the K_d measured by rapid quench methods is closer to a K_m measurement, not a true K_d for nucleotide binding.

II. Introduction

Transient-state kinetic methods allow definition of the sequence of reactions occurring at the active site of an enzyme including steps following substrate binding and leading up to product release. It is precisely these reaction steps that define the elementary steps accounting for the high fidelity of DNA polymerization or establishing the pathway for coupling of ATP hydrolysis to energy transduction by molecular motors. In each case, analysis of the reactions occurring at the active site of the enzyme by single turnover kinetic methods has allowed definition of the reaction sequence (Johnson, 1986, 1992, 1993; Kati *et al.*, 1992). On the other hand, steady-state kinetic analysis only establishes the order of substrate binding and the order of product release; the steady-state kinetic parameters k_{cat} and k_{cat}/K_m only define the maximum rate of substrate to product conversion and a lower limit estimate of the rate of substrate binding, respectively. Steady-state kinetic analysis cannot address questions regarding the pathway of events occurring at the active site following substrate binding and prior to product release (Johnson, 1992).

This chapter summarizes the methods of kinetic analysis used to define the sequence and rates of reactions catalyzed by DNA polymerases and by force-transducing adenosinetriphosphatases (ATPases) and gives a brief review of the detection of enzyme intermediates. The methods are of general utility in studying enzyme reaction mechanisms and have led to the discovery of several new enzyme intermediates (Anderson and Johnson, 1990a; Anderson *et al.*, 1988a,b, 1991a,b). For a more detailed analysis of transient kinetic methods in general, the reader is referred to Chapter 13 of this volume and Johnson (1992).

III. Principles of Transient-State Kinetic Analysis

The rapid quench transient kinetic methods described in this chapter involve the rapid mixing of enzyme with substrate to initiate the reaction. The reaction is then terminated after a short time interval by mixing with a quenching agent, such as a denaturant, and the amount of product formed is then quantified. By analyzing the time course of product formation the reaction can then be analyzed. If the experiment is done over a sufficiently short time interval and at a high enough enzyme concentration, then the kinetics of a single enzyme turnover can be measured.

Basically, two types of experiments can be performed. With substrate in excess over enzyme, one may observe a rapid "burst" of product formation, followed by steady-state turnover. Analysis of the rate and amplitude of the burst provides information to define the rate of the chemical reaction occurring at the active site, while the linear steady-state phase defines the rate-limiting steps leading to product release. However, a burst of product formation is only seen if the release of product, or some other step after chemistry, is slower than catalysis. If chemistry or a preceding step is rate limiting, then there is no burst and the initial rate would then define the rate of the chemical reaction. In this case, the absence of a burst could serve to establish that chemistry (or a step preceding chemistry) is rate limiting in the steady state.

A single enzyme turnover can be observed if the experiment is performed with enzyme in excess over substrate. If the enzyme is at sufficiently high concentration so that substrate binding is fast, then the kinetics measure the rate of the chemical conversion of substrate to product at the active site. Alternatively, at lower concentrations, analysis of the enzyme concentration dependence of the observed rate defines the kinetics of reactions involved in substrate binding and any steps leading to the observed chemical reaction. A single turnover experiment is most useful in attempts to define new enzyme intermediates because the 100% conversion of substrate to product within a single pass through the enzyme-bound species provides the most sensitivity to allow observation of any potential intermediates (Anderson and Johnson, 1990a,b; Anderson *et al.*, 1988a, 1991a,b; Marquardt *et al.*, 1993; Brown *et al.*, 1994; Cho *et al.*, 1992; Johnson, 1992).

In the analysis of transient-state reaction kinetics, the enzyme is considered as a stoichiometric reactant rather than a trace catalyst. Therefore, the reactions occurring at the active site of the enzyme can be observed and quantified directly. Moreover, because the data define the kinetics of the reaction in terms of absolute concentration units that can be related to enzyme active-site concentrations, the pathway of events occurring at the active site can be established without the indirect inference typical of steady-state kinetic methods. It is important to stress the value of information provided by the amplitude of a chemical reaction in absolute concentration units. Such information is often just as valuable as measuring the rate of a reaction in order to establish the pathway (Anderson and Johnson, 1990a,b; Anderson *et al.*, 1988a; Johnson, 1992).

In this chapter, I summarize the logic and the methods of analysis used to establish the reaction pathways of polymerases and force-transducing ATPases. Previous studies on these two systems provide a recipe for kinetic analysis in terms of the order and type of kinetic experiments to perform to examine other enzymes in these classes. In addition, I briefly summarize the key elements of kinetic analysis leading to the identification of new enzyme intermediates.

IV. Chemical Quench-Flow Methods

Rapid chemical quench-flow methods are based on the rapid mixing of substrate with enzyme to initiate the reaction, followed at a specific time interval by the mixing with a quenching agent to terminate the reaction. The product formed in this time interval can then be quantified. In principle, these experiments are done by driving syringes first to mix enzyme with substrate such that a reaction occurs while the mixture is flowing through a reaction loop of tubing (see Fig. 1). The reaction is then terminated when the enzyme-substrate mixture is combined with the quenching agent from a third syringe. The time of reaction is determined by the time it takes the enzyme-substrate mixture to flow from the point of mixing, through the reaction loop, and to the point of quenching. The reaction time can be varied by changing the length of the reaction loop and the rate of flow. In practice, there are limits on the rate of flow between the maximum rate that can be achieved at reasonable drive pressures and the minimum flow rate necessary to maintain efficient mixing. Within a factor of 2 or 3 in flow rate, a given length of tubing can be used to obtain a range of reaction times.

The slowest rate of flow is limited by the conditions to maintain turbulence. Turbulent flow through a tube can be predicted by calculation of a Reynolds

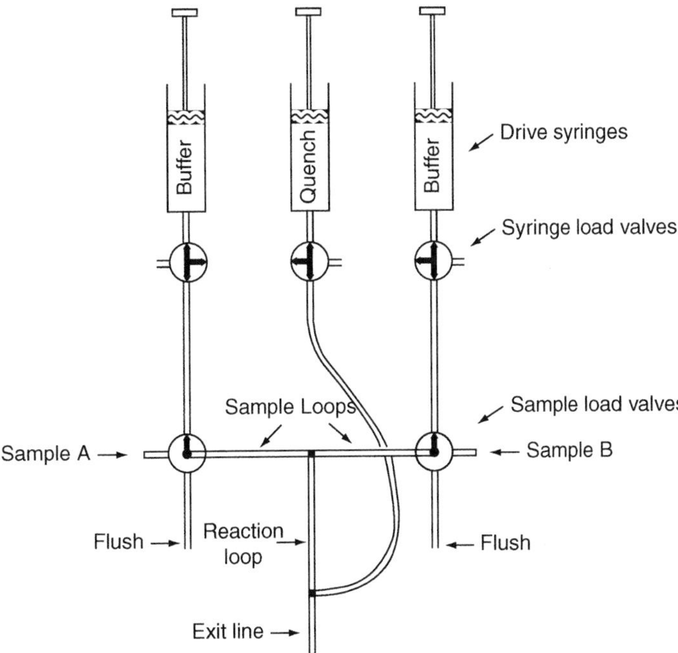

Fig. 1 Schematic of a pulsed-quench-flow apparatus. (Courtesy of KinTek Corporation, State College, PA.)

number, which is a dimensionless parameter dependent on the flow rate, V, the viscosity, ν, and the diameter of the tube, d: $R = Vd/\nu$. Turbulent flow occurs when the Reynolds number exceeds 2000 (Wiskind, 1964). Thus, for a 0.08-cm diameter tubing, and a viscosity of 0.01 poise (g/cm-s), a linear flow rate of 2.5 m/s (1.25 ml/s) must be maintained. Although there has been some debate as to the validity of this calculation, in practice an instrument designed within these limits succeeds in measuring the rate of known chemical reactions (Froehlich *et al.*, 1976; Johnson, 1986; Lymn and Taylor, 1970), and the failures of both private and commercial ventures in the manufacture of quench-flow instruments can be traced to a flow rate too slow to maintain turbulence by this criterion.

A. Quench-Flow Design

Most quench-flow instruments utilize a design based on loading enzyme and substrate into the drive syringes. This design is extremely wasteful of the precious reagents, however, because at the end of each drive the reaction loop still contains the two reactants. This solution, often 200 μl in volume, must be discarded in order to flush the system prior to collecting the next time point. An alternative design is shown in Fig. 1 (KinTek Corporation, State College, PA). In this instrument, the reactants are loaded into small sample loops, and then buffer in the drive syringes is used to force the reactants together, through the reaction loop to the point of mixing with the quenching agent, and then out into a collection tube. With this design, sample volumes as low as 15 μl can be obtained with nearly 100% sample recovery, making these experiments feasible for almost any enzyme system. An eight-way valve is used to select various reaction loop volumes, and a computer-controlled stepper motor is used to drive the reactants at a precise speed. By using eight different reaction loop volumes and with variation of the flow rate over a factor of about 2, reaction times ranging from 2 to 100 ms can be obtained. Longer reaction times are obtained in a push-pause-push mode, where the reactants are forced together, held in the reaction loop for a defined period of time, and then forced out to mix with the quench solution. Using this mode, reaction times from 100 ms to 100 s (or longer) can be obtained.

B. Choice of Quenching Agent

Most commonly employed quenching agents are 1 N HCl and 1 N NaOH (final concentrations). In some experiments with 5-enolpyruvoyl-shikimate 3-phosphate synthase (EPSP synthase; 3-phosphoshikimate 1-carboxyvinyltransferase), 1 N trifluoroacetic acid gave better results than HCl. After stopping the reaction with acid, the reaction mixtures can be neutralized by addition of a small volume of a high concentration of buffer and base (i.e., 4 M Tris, 2 M NaOH). In some cases, it is necessary to vortex the acid-quenched solutions with chloroform prior to neutralization to prevent renaturation of the enzyme (Anderson *et al.*, 1988a); that is, the enzyme must be killed twice to keep it from coming back to life. To first isolate the tetrahedral intermediate in EPSP synthase, neat triethylamine was used as a quenching agent, which when mixed 1:2 with sample gives an aqueous layer with approximately pH 12 (Anderson *et al.*, 1988b). In experiments with DNA polymerases, 0.3 M EDTA has often been used to chelate the Mg^{2+} to prevent further reaction without the complications involved with the use of acids. However, use of EDTA as a quenching agent relies upon the rapid release of Mg^{2+} from the enzyme to stop the reaction.

Regardless of the reagent used to quench the reaction, it is important to perform the appropriate control experiments involving mixing enzyme with quenching agent prior to adding substrate. This sample provides the most appropriate "blank" for measurement of background in the assay for product in addition to providing a necessary control for stopping the reaction.

C. Analysis of Reaction Products

Ultimately, the quality of the data is limited most by the methods used to separate substrate from product and quantify the amount of product formed. Methods have ranged from ion-exchange high-performance liquid chromatography (HPLC) (Anderson *et al.*, 1988a) to the use of charcoal columns with batch elution (Johnson, 1983) to isolate [^{32}P]phosphate from [γ-^{32}P]ATP. For studies on ATPase reactions, the best results are obtained using [α-^{32}P]ATP and resolving ADP from ATP by thin-layer chromatography (TLC) on poly(ethylenimine) (PEI) cellulose F plates (EM Separations, Gibbstown, NJ). Small aliquots (1-2 μl) are applied to the TLC plates. The plates can be washed with methanol to remove salts and then dried and developed with 0.6 M KH_2PO_4 at pH 3.4. ADP and ATP standards run in the outside lanes can be visualized with a hand-held fluorescent lamp so that appropriate areas of each lane of the plate can be cut and quantified by scintillation counting. Alternatively, the TLC plate can be quantified directly using a two-dimensional radioactivity detector, such as a phosphorimager.

The kinetics of DNA polymerization can be followed on a sequencing gel by first labeling the DNA primer on the 5′ end with ^{32}P using T4 polynucleotide kinase. The products of reaction are then separated by electrophoresis of a 16% (w/v) polyacrylamide gel containing 8 M urea. The gel can then be quantified using a phosphorimager. Alternatively, by first exposing and developing a sheet of X-ray film, the gel can be placed on the X-ray film and the appropriate bands cut with a razor blade and quantified by scintillation counting. In addition, methods have been developed to resolve and quantify fluorescently labeled oligonucleotides by capillary electrophoresis (Tsai and Johnson, 2006).

Regardless of the method, it is important to have an internal standard for the recovery of sample from the rapid quench-flow instrument. For HPLC, TLC, or gel analysis, the total radioactivity in a given sample or lane is used to normalize the data such that the analysis is dependent on determining the fraction of product formed relative to starting reactant concentration. For example, in the case of gel analysis for DNA polymerization, the ratio of the elongated DNA (i.e., 26-mer) product relative to total labeled DNA (25-mer plus 26-mer) establishes the fraction of the reaction that has occurred. This fraction times the concentration of DNA in the original reaction mixture gives the concentration of product formed.

D. Note on Utility of Concentrations Rather than Moles of Product

Reaction rates are governed by the concentrations of reactants present during the reaction. Accordingly, to interpret any experiment properly, one needs to know the concentrations of enzyme present and the concentrations of products formed as a function of time. In most laboratories performing enzyme kinetics, it appears to be the convention to graph moles (nanomoles or picomoles) of product formed as a function of time and to provide details as to the number of moles of enzyme present in the reaction mixture. However, to interpret these results, the reader must determine the total volume of the reaction (if it is even given) and then calculate the concentration of product formed. It is much more straightforward to always operate in concentration units, referring to the concentrations of reactants after mixing to initiate the reaction and the concentrations of product formed during the reaction. Quantitative analysis of the data is greatly facilitated by maintaining this convention (Anderson and Johnson, 1990a; Anderson *et al.*, 1988a; Johnson, 1992).

V. Kinetics of DNA Polymerization

The pathway of DNA polymerization is shown in Fig. 2. The pathway was derived in studies on T7 DNA polymerase (Donlin *et al.*, 1991; Patel *et al.*, 1991; Wong *et al.*, 1991) and defines the steps in the pathway catalyzed by reverse transcriptase (Kati *et al.*, 1992) except for the absence of a proofreading exonuclease. Enzyme first binds DNA tightly, and then processive synthesis involves multiple rounds of polymerization without dissociation of the DNA.

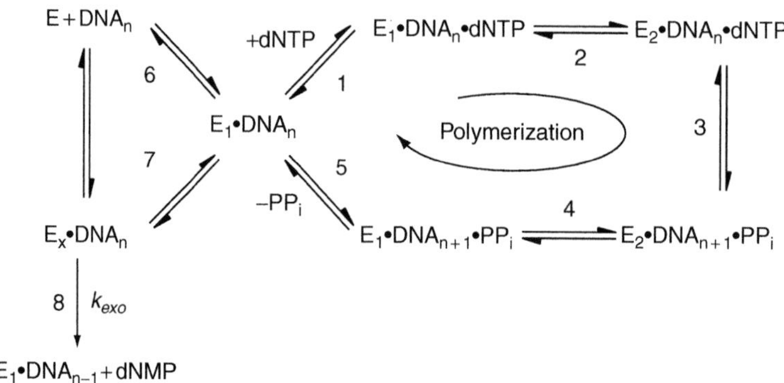

Fig. 2 Pathway of DNA polymerization. (Redrawn from Patel *et al.*, 1991.)

The binding of a correct deoxynucleoside triphosphate (dNTP) leads to a conformational change from an open to a closed state ($E_1 \cdot DNA_n \cdot dNTP \rightarrow E_2 \cdot DNA_n \cdot dNTP$) which is then followed by a fast chemical reaction. Relaxation of the closed state is then followed by release of pyrophosphate and translocation, all of which are much faster than the rate of product formation.

The important questions to address in studying polymerases center on understanding the extraordinarily high fidelity of DNA replication in terms of the elementary steps of the reaction sequence. The selectivity of the correct dNTP is a function of the steady-state kinetic parameter k_{cat}/K_m. However, measurement of k_{cat} and K_m does not define the mechanistic basis for fidelity; it only quantifies the problem. In fact, most often steady-state kinetic analysis of DNA polymerases is complicated by the nature of the processive synthesis whereby a single encounter of the polymerase with the DNA template results in multiple rounds of synthesis. Once the polymerase has filled in the template to which it is bound, the dissociation of the enzyme from the DNA is several orders of magnitude slower than polymerization. Accordingly, steady-state measurements of incorporation are a poorly defined average of the rate of synthesis and the rate of dissociation of the DNA according to the length of the DNA template. The magnitude of the artifact in attempting to measure k_{cat} and K_m by steady-state methods is approximately equal to the processivity of the enzyme.

Steady-state reaction measurements do provide an important check on the validity of measurement of partial reactions along the pathway by transient kinetic methods. For example, "processivity" is defined as the number of bases incorporated per encounter of the enzyme with the DNA. Therefore, processivity is simply the ratio of the rate of polymerization divided by the rate of dissociation of the enzyme-DNA complex. For T7 DNA polymerase, a maximum rate of polymerization of 300–500 s^{-1} and a dissociation rate of 0.2 s^{-1} predicts a processivity of 1500–2500, a value observed by bulk solution measurements.

To define the reactions leading to the high fidelity of replication, one needs to identify the rate-limiting step for incorporation and determine the contributions to fidelity for each step in the reaction sequence. This can be accomplished by analyzing the kinetics of single nucleotide incorporation using synthetic oligonucleotides.

A. Kinetics of Single Nucleotide Incorporation

Experiments can be performed using synthetic oligonucleotides to examine the kinetics of addition of a single nucleotide after the addition of only one dNTP to the reaction mixture. For example, the following 25/45-mer has a 25-base primer complexed to a 45-base template such that the addition of dATP results in the elongation by one residue only:

```
5′-GCCTCGCAGCCGTCCAACCAACTCA
   |||||||||||||||||||||||||
3′-CGGAGCGTCGGCAGGTTGGTTGAGTTCCAGCTAGGTTACGGCAGG-5′
```

The kinetics of elongation can be examined by mixing 50–100 nM enzyme with 200–300 nM DNA. The reaction shows a stoichiometric burst of product formation (26-mer) followed by slow steady-state turnover (see Fig. 3), which is limited by the release of the DNA product from the enzyme (Kati *et al.*, 1992). The reaction time course can then be analyzed by the burst equation, $Y = A \exp(-k_2 t) + k_{ss}t$. The concentration of E-DNA active sites is obtained from the amplitude, A, and the rate of product formation from the exponential rate constant, k_2. The steady-state rate, k_{ss}, divided by the amplitude, A, defines the rate constant for DNA release, k_6 (Fig. 2).

The analysis of the amplitude in such simple terms is dependent on the assumption that the rate of the reaction, k_2, is at least 20-fold faster than the rate of product release, k_3, for a simple three-step pathway (Johnson, 1992). In general, the amplitude of the burst is reduced owing to more rapid product release according to $A = [k_2/(k_2 + k_3)]$ (Johnson, 1993). In either case, the data can be fit to the burst equation and then interpreted according to the mechanism. It is not generally useful to include the equation for the amplitude in the fitting process because the amplitude may often be lower than that predicted because of dead enzyme or an internal equilibrium between substrate and product at the

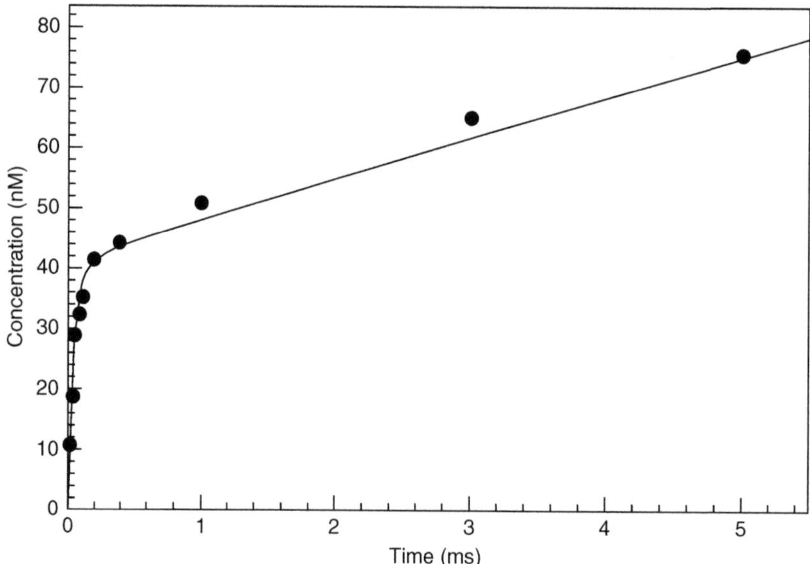

Fig. 3 Kinetics of a burst of DNA polymerization. The kinetics of single nucleotide incorporation were examined by reacting 40 nM reverse transcriptase (active-site concentration) with 200 nM DNA (25/45-mer synthetic oligonucleotide) for the times indicated. The reaction was initiated by the addition of Mg^{2+} (10 mM) and dATP (25 μM) and terminated by the addition of 0.3 M EDTA. Products were analyzed on a sequencing gel to quantify the formation of the product 26/45-mer. The data were fitted to a burst equation with rate constants of 20 and 0.18 s^{-1} for the rate of incorporation and release of DNA, respectively. (Reprinted with permission from Kati *et al.*, 1992.)

active site. Ultimately, data fitting based on computer simulation provides the most complete analysis to relate the reaction pathway directly to the observed kinetics (see Section VIII).

B. Active-Site Titration and Determination of K_d for DNA Binding of Polymerase Site

The rate of dissociation of the E DNA complex is slower than polymerization; therefore, when polymerization is measured on a fast time scale, only those enzyme molecules with DNA bound at the polymerase site will react to form product. Accordingly, the amplitude of the fast polymerization reaction provides a direct measurement of the concentration of E · DNA complexes at the start of the reaction. By measuring the variation in the burst amplitude with increasing concentration of DNA, one can establish the concentration of active enzyme sites and measure the dissociation constant for the binding of DNA. It is important to stress that this experiment provides a K_d for DNA binding in a productive mode at the active site of the enzyme because it is based on the ability of the DNA to be elongated rapidly.

Typically, the K_d values for DNA are in the range of 5–20 nM (Johnson, 1993; Kati *et al.*, 1992; Patel *et al.*, 1991). The experiment can then be performed at 50 nM enzyme and concentrations of DNA ranging from 10 to 250 nM, if the data are analyzed by fitting to the quadratic equation (see Section VIII). The hyperbolic equation is derived based on the assumption that substrate is always in excess of the enzyme concentration, and it is therefore not valid for this experiment.

C. dNTP Concentration Dependence

Analysis of the dNTP concentration dependence of the observed rate of the burst of incorporation provides an estimate of the K_d for ground state nucleotide binding for formation of the productive complex poised for catalysis. The reaction kinetics follow a two-step sequence:

$$\mathrm{E} \cdot \mathrm{DNA}_n + \mathrm{dNTP} \overset{K_d}{\rightleftharpoons} \mathrm{E} \cdot \mathrm{DNA}_n \cdot \mathrm{dNTP} \overset{K_{pol}}{\rightleftharpoons} \mathrm{E} \cdot \mathrm{DNA}_{n+1} \cdot \mathrm{PP_i}$$

Although the reaction pathway contains additional steps, the rate of incorporation is governed solely by the nucleotide binding followed by a rate-limiting step leading to elongation of the DNA (Kati *et al.*, 1992; Patel *et al.*, 1991). As illustrated by the data in Fig. 4A, the rate increases with increasing concentration of dNTP, reaching a maximum (Kati *et al.*, 1992). As shown in Fig. 4B, the fit of the concentration dependence of the rate to a hyperbola yields the maximum rate k_{pol} and the dissociation constant, K_d.

The fit to a hyperbola is dependent on two assumptions. First, the enzyme concentration must be much less than the K_d, so that the dNTP concentration added can be assumed to equal the free dNTP concentration. Second, the binding of dNTP must be a rapid equilibrium; that is, k_{-1} (the dNTP dissociation rate)

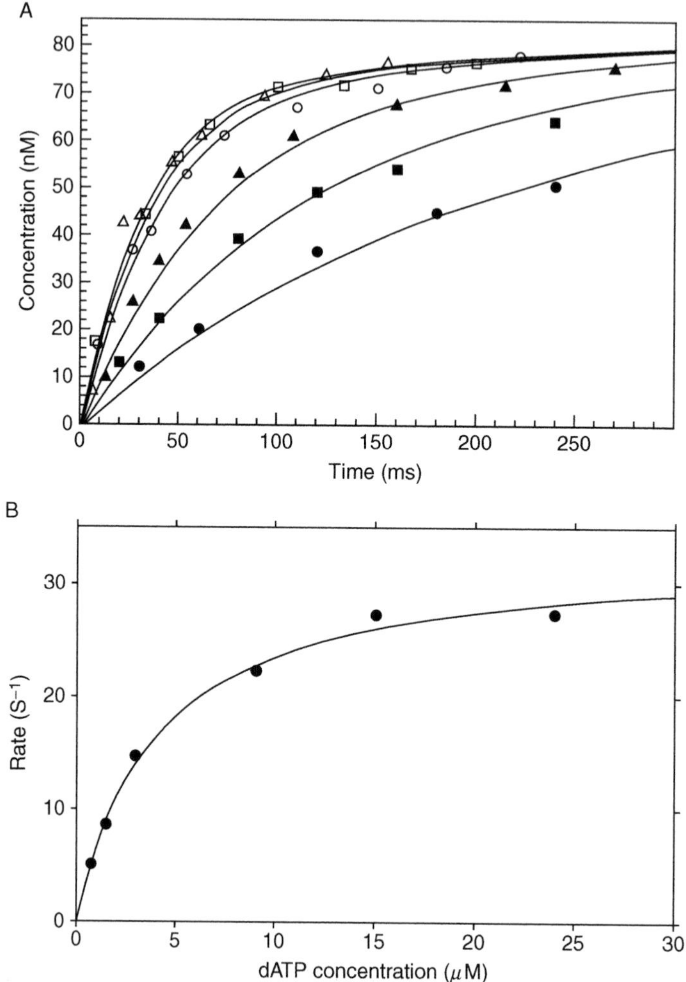

Fig. 4 Nucleotide concentration dependence of the rate of incorporation. (A) A solution of reverse transcriptase (80 nM) was incubated with 25/45-mer DNA (200 nM) and then mixed with Mg^{2+} (10 mM) and varying concentrations of dATP to initiate the reaction. The following dATP concentrations were used: 0.75 μM (●); 1.5 μM (■); 3.0 μM (▲); 9.0 μM (○); 15 μM (□); and 24 μM (Δ). The solid lines represent the best global fit to the data. (B) At each dATP concentration, the rate of polymerization was obtained by fit to a single exponential. The dATP concentration dependence was then fit to a hyperbola (solid line) to obtain a K_d of 4 ± 0.4 μM and a maximum rate of 33 ± 1.2 s^{-1}. (Reprinted with permission from Kati *et al.*, 1992.)

must be much greater than k_{pol}. It is not always easy to prove this experimentally. The rapid equilibrium assumption is supported by the observation that the polymerization reaction is a single exponential at all dNTP concentrations. A two-step irreversible mechanism predicts a lag phase clearly observable at low dNTP concentrations, most prominent at the half-maximal rate. In addition, pulse-chase type experiments can be used to examine the kinetics of formation of a tight nucleotide complex (Hsieh *et al.*, 1993; Patel *et al.*, 1991). Recent analysis has resolved the role of the nucleotide-induced conformational change based upon studies using a fluorescently labeled enzyme. These results establish that the conformational change commits a correct nucleotide to be incorporated, but binding of an incorrect nucleotide leads to an altered structure which promotes dissociation and attenuates incorporation of a mismatch (Johnson and Taylor, 1978).

D. Kinetics of Processive Synthesis

It is important to establish that the rate of the first enzyme turnover is not different than subsequent turnovers. Most often this can be accomplished by careful comparison of the steady-state parameters with the rates and mechanism determined by transient kinetic methods. In the case of DNA polymerases, one can perform an experiment to measure the kinetics of processive synthesis involving the incorporation of a small number of nucleotides. For example, using the synthetic oligonucleotide described above, the addition of dATP, dCTP, and dTTP (leaving out dGTP) results in the elongation of the DNA by 5 base pairs (see Fig. 5). The kinetics of processive elongation can then be examined by following the polymerization reaction on a sequencing gel and quantifying the formation and decay of each intermediate from $DNA_{25} \rightarrow DNA_{26} \rightarrow DNA_{27} \rightarrow DNA_{28} \rightarrow DNA_{29} \rightarrow DNA_{30}$. The reaction kinetics are complex and require analysis by numerical integration (Barshop *et al.*, 1983; Kati *et al.*, 1992; Patel *et al.*, 1991; Zimmerlie and Frieden, 1989). Nonetheless, using this analysis, the rates of polymerization for each step of elongation can be obtained (Kati *et al.*, 1992).

E. Pyrophosphorolysis

The kinetics of the reverse of polymerization, pyrophosphorolysis, can be analyzed by following the shortening of DNA on a sequencing gel if there is no exonuclease in the reaction mixture. If an exonuclease is present, then the kinetics of hydrolysis to produce dNMP can be distinguished from the pyrophosphorolysis to produce dNTP by labeling the 3-terminal base. It is necessary to run the reaction in the presence of excess unlabeled dNTP to trap any radiolabeled dNTP. Analysis of the kinetics of production of dNMP and dNTP can give the relative rates of pyrophosphorolysis and exonuclease hydrolysis. However, the time dependence of the reaction is a composite of both reactions such that the rate of pyrophosphorolysis, obtained as a fit to a single exponential, will be the sum of the rates of hydrolysis and pyrophosphorolysis even if care is taken to monitor the production of dNTP only (Patel *et al.*, 1991). Nonetheless, the rates of the two competing reaction can be resolved by analysis of the relative amplitudes of pyrophosphorolysis and hydrolysis.

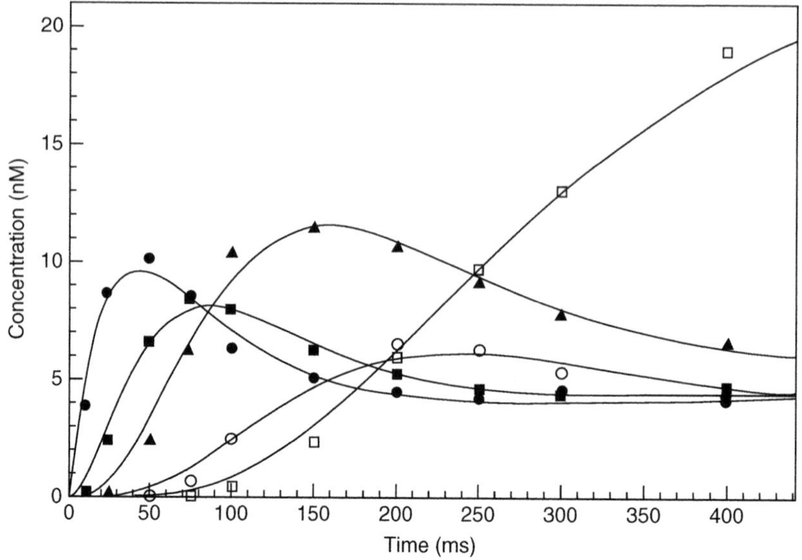

Fig. 5 Kinetics of processive polymerization. A solution of reverse transcriptase (40 nM) was preincubated with 25/45-mer DNA (100 nM) and then reacted with dATP, dCTP, and dTTP (20 μM each) with 10 mM Mg^{2+} for the times indicated before quenching with 0.3 M EDTA. The data show the formation of 26-mer (●), 27-mer (■), 28-mer (▲), 29-mer (○), and 30-mer (□). The solid lines represent the best fit to the data obtained by computer simulation using rates of 18, 31, 26, 9, and 13 s^{-1}, respectively, for the five consecutive nucleotide incorporation reactions and DNA dissociation rates of 1.8, 3.4, 3.5, 1.2, and 1.2 s^{-1}. (Reprinted with permission from Kati *et al.*, 1992.)

Another complication in measuring the kinetics of pyrophosphorolysis is due to the precipitation of magnesium pyrophosphate at concentrations required to saturate the rate of the reaction (above 2 mM). The problem is only partially overcome by initiating the reaction with Mg^{2+} in one syringe and pyrophosphate in the other. Precipitation begins almost immediately on mixing.

F. Exonuclease Kinetics

The kinetics of exonuclease hydrolysis can be easily measured by monitoring the shortening of DNA as observed on a sequencing gel with 5′-labeled DNA. Although multiple products will be observed owing to multiple rounds of hydrolysis, the quantification is based on the kinetics of disappearance of the starting material. In the case of T7 polymerase, different experimental designs have given different results that have been interpreted mechanistically. Single-stranded DNA is hydrolyzed quite rapidly (700–900 s^{-1}) (Donlin *et al.*, 1991). In contrast, duplex DNA is hydrolyzed at rates of 0.2 and 2.3 s^{-1} for correctly and incorrectly base-paired 3′ termini, respectively (Donlin *et al.*, 1991). These results have been taken to imply that the movement of the duplex DNA from the polymerase site to the exonuclease site is rate limiting and that the hydrolysis of single-stranded DNA represents the rate of hydrolysis of the DNA once it occupies the exonuclease site. In each case, the DNA is first incubated with the polymerase in the absence of Mg^{2+}, and the reaction is initiated by the addition of Mg^{2+}; the data can be fit to a single exponential (see Section VIII).

Biphasic kinetics have been observed when the reaction is initiated by mixing duplex DNA containing a mismatch with DNA (Donlin *et al.*, 1991). There is a fast initial hydrolysis of 50% of the DNA (with enzyme in excess), whereas the remainder of the DNA is hydrolyzed at a rate of 2.3 s^{-1}. These data have been interpreted in terms of a model whereby the DNA first binds to the exonuclease site and then partitions between hydrolysis and sliding into the polymerase site during the fast phase of the reaction. During the slow phase, the rate is limited by the sliding of the DNA from the polymerase site to the exonuclease site.

G. Complete Polymerase Pathway

The complete reaction pathway can be obtained by combining information from each experiment measuring the kinetics in the forward and reverse directions including estimates of the overall equilibrium constant for the reaction to constrain the fitting process. In the final fitting process, computer simulation can be used to model all of the data to a single model and a single set of rate constants (Barshop *et al.*, 1983; Johnson, 1992; Kati *et al.*, 1992; Patel *et al.*, 1991; Zimmerlie and Frieden, 1989).

VI. Adenosinetriphosphatase Mechanisms

In general, a mechanism for coupling ATP hydrolysis to do useful work can be analyzed in terms of the pathway shown in Fig. 6A. The enzyme exists in two states, E_1 and E_2, and the interconversion of the two forms is driven by the cycle of ATP hydrolysis coupled to do work. The pathway of coupling ATP hydrolysis to work is established by specifying the route through the maze shown in Fig. 6A by determining the kinetics of ATP binding, hydrolysis, and product release and the kinetics of the interconversion of the two enzyme forms. In the two examples given, actomyosin (Fig. 6B) and the microtubule-kinesin ATPase (Fig. 6C), the E_1 states are bound to the filament (A · M is actomyosin, Mt · K is microtubule-kinesin), and the E_2 states are dissociated from the filament (M is myosin, K is kinesin). Force is produced during conformational changes coupled to the changes in state while rebinding to the filament.

For actomyosin, ATP binding leads to rapid release of myosin from the actin (2000 s^{-1}), which is followed by hydrolysis of the ATP on the free myosin molecule (Johnson and Taylor, 1978; Lymn and Taylor, 1971). The myosin-products complex (M · ADP · P_i) rebinds to the actin, and the release of products (ADP and P_i) is coupled to force production. The dissociation of products from A · M · ADP · P_i

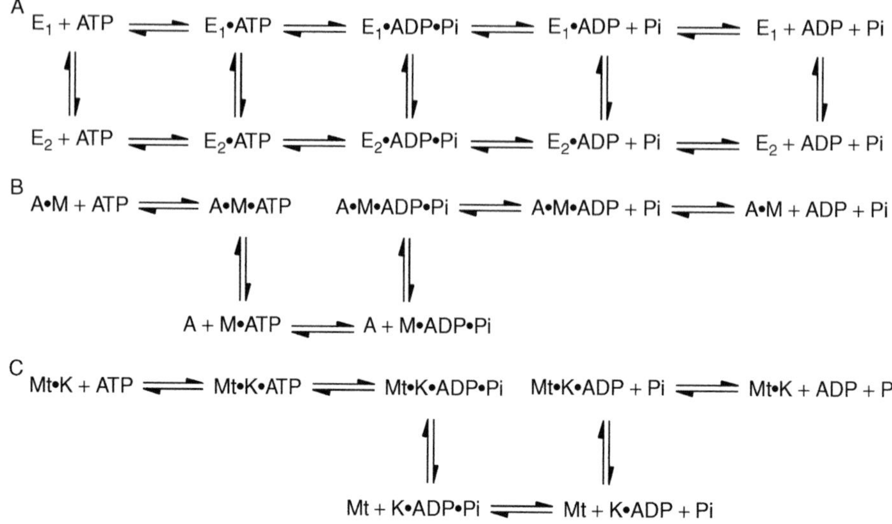

Fig. 6 Pathways of ATPase coupling. Pathways for ATPase coupling in general (A) and for actomyosin (B) and microtubule-kinesin (C) are shown. A, actin; M, myosin; K, kinesin; Mt, microtubules.

is the rate-limiting step (~20 s^{-1}) for the maximum actin-activated ATPase. A similar pathway has been observed for the dynein ATPase except phosphate release precedes rebinding to the microtubule (Johnson, 1983, 1985).

In contrast, in the kinesin pathway, ATP hydrolysis precedes the dissociation of the kinesin from the microtubule (Gilbert *et al.*, 1994). Phosphate release occurs from the free kinesin molecule, and the pathway is completed by the binding of the kinesin-ADP species (K · ADP) to the microtubule. Microtubules accelerate the release of ADP greater than 2000-fold. The release of kinesin from the microtubule following ATP hydrolysis is the rate-limiting step (20 s^{-1}). With dimeric kinesin, strain between the two heads, when simultaneously bound to a microtubule, leads to an alternating site ATPase cycle in which ATP binding to one head stimulates the release of ADP from the other. This pathway was resolved by stopped-flow measurements using a fluorescent nucleotide analog and the results were fit globally along with measurements by rapid quench methods to define the pathway (Auerbach and Johnson, 2005; Gilbert *et al.*, 1998).

In these two examples, the pathways are similar and the maximum rates of the fully activated ATPase are similar. In each case, however, a different step is rate limiting. For actomyosin, the extremely fast dissociation of the motor from the filament leads to a pathway where the motor spends very little of the duty cycle attached to the filament. In contrast, kinesin spends most of the time attached to the microtubules. These observations are in keeping with the biological roles of these motors wherein myosin operates in ordered arrays, whereas kinesin works to transport vesicles along tracks of individual microtubules through the cytoplasm.

The experiments described in the following subsections allow the pathway of coupling to be established by direct measurement of partial reactions.

A. Kinetics of ATP Hydrolysis

The kinetics of ATP hydrolysis can be measured in a rapid quench experiment by first mixing enzyme with ATP and then quenching with 1 N HCl. Because the acid liberates any products bound at the enzyme active site, the amount of ADP observed is given by the sum $[ADP]_{obs} = [E \cdot ADP \cdot P_i] + [E \cdot ADP] + [ADP]$. Accordingly, the time dependence of the formation of ADP follows burst kinetics. The burst phase provides a measurement of the rate of formation and amount of $E \cdot ADP \cdot P_i + E \cdot ADP$ at the active site.

To achieve a maximal burst rate and amplitude so as to measure the rate of hydrolysis at the active site, a high concentration of ATP must be mixed with enzyme to initiate the reaction such that ATP binding is not rate limiting. The experiment is difficult because the K_m for ATP is typically much larger than the concentrations of enzyme that can be obtained, and the signal is limited by the background of hydrolyzed ATP (1%) in the most purified preparations. An ATP concentration equal to five times the K_m provides only a lower limit on the concentration of ATP that may be required to achieve a binding rate faster than catalysis. The steady-state K_m only defines the concentrations required for the binding rate to equal the steady-state turnover rate. Because hydrolysis may be an order of magnitude faster than net turnover, a correspondingly higher concentration of ATP is required.

To satisfy the most stringent requirements for optimal signal-to-noise ratios (S/N), we have found that the kinetics of ATP hydrolysis can be monitored most accurately using [α-^{32}P]ATP and quantifying the appearance of [α-^{32}P]ADP following separation by TLC on PEI cellulose F plates (Gilbert and Johnson, 1994). The concentration of ADP formed is then calculated from the counts per minute in the ADP spot (cpm_{ADP}) relative to counts in the ATP spot (cpm_{ATP}):

$$[ADP] = \frac{[ATP]_0 cpm_{ADP}}{cpm_{ADP} + cpm_{ATP}}$$

The data can be fit to the burst equation (see Section VIII) to obtain estimates of the rate and amplitude of the hydrolysis reaction. However, if the kinetics of ATP binding are known and the rate is less than 10-fold faster than hydrolysis, then the hydrolysis kinetics will exhibit a lag phase which is a function of ATP binding. Accordingly, the rates of hydrolysis will be underestimated unless the lag phase predicted by the binding kinetics is included in the fitting process (Johnson, 1983). This is most easily accomplished by computer simulation (see Section VIII).

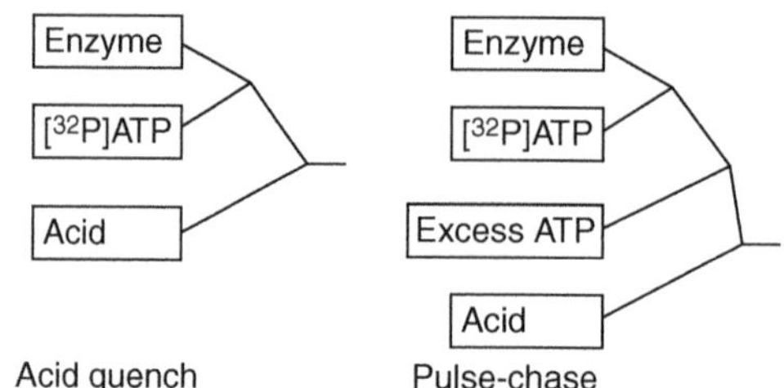

Fig. 7 Design a pulse-chase experiment.

B. Kinetics of ATP Binding: Pulse-Chase Experiments

The kinetics of ATP binding can be measured using a pulse-chase protocol outlined in Fig. 7. The reaction is initiated by mixing with [α-^{32}P]ATP and then, at various times, an excess of unlabeled ATP is added. The reaction is allowed to continue for 5–8 half-lives of the hydrolysis reaction, a time sufficient to convert all tightly bound ATP to ADP. The reaction is then terminated by the addition of acid. The amount of ADP formed is equal to the sum $[\mathrm{ADP}]_{\mathrm{obs}} = \theta[\mathrm{E \cdot ATP}] + [\mathrm{E \cdot ADP \cdot P_i}] + [\mathrm{E \cdot ADP}] + [\mathrm{ADP}]$, where θ is the fraction of bound ATP that is hydrolyzed before it is released from the enzyme. As described above, the data can be fit to a burst equation to a first approximation, and then fit more completely by computer simulation to the complete mechanism.

It is important that the chase period be as short as possible to prevent the accumulation of product owing to the binding and hydrolysis of the diluted radiolabeled ATP. A control experiment must be performed by mixing the enzyme with the diluted ATP for the time of the chase to ensure that the rate of dilution of the label is sufficient for the time required.

C. Kinetics of Phosphate Release

The kinetics of phosphate release from the enzyme can be measured by stopped-flow using a method newly developed by Webb using fluorescently modified phosphate-binding protein (Brune *et al.*, 1994). The details of this method, however, are beyond the scope of this review. Care must be taken to scrub the system of contaminating phosphate using a phosphate "mop." With appropriate attention to details given in the original manuscript, a burst of phosphate release of 50 nM can easily be observed at a concentration of 1 mM ATP.

D. Kinetics of ADP Release

The rate of ADP release can be measured by several methods. The easiest method relies on the use of fluorescently labeled ADP, mant-ADP (Sadhu and Taylor, 1994). On excitation at 350 nm, mant-ADP emits light at 440 nm. Mant-ADP fluorescence is quenched by solvent, and therefore there is a decrease in fluorescence on dissociation of the mant-ADP from the enzyme. When kinesin-ADP is mixed with microtubules, a single exponential decrease in fluorescence is observed, providing a direct measurement of the rate of ADP release from the microtubule-kinesin-ADP complex. On a much longer time scale, a second ADP is released from the second head of dimeric kinesin, but the rate is too slow to be physiologically significant. However, the binding of ATP greatly stimulates the second ADP release, and fitting of the data by simulation defined the reactions governing the alternating site mechanism (Auerbach and Johnson, 2005; Gilbert *et al.*, 1998).

E. Kinetics of Motor Dissociation and Rebinding to Filament

The kinetics of the dissociation and rebinding of the motor to the filament can be measured by stopped-flow light-scattering measurements. For example, the mixing of ATP with actomyosin or microtubule-kinesin leads to a reduction in light-scattering intensity that can be monitored in a stopped-flow apparatus. The method is relatively sensitive so that concentrations of 0.1 μM protein can be used and any concentration of ATP from 1 μM to 5000 mM can be employed. The data fit a single exponential. Plotting rate versus ATP concentration in the case of actomyosin yields a straight line, with the rate approaching 2000 s^{-1} (Johnson and Taylor, 1978). In contrast, the ATP concentration dependence observed for the microtubule-kinesin complex follows a hyperbola (Gilbert *et al.*, 1994).

F. Putting It All Together

The pathway of ATPase coupling is finally obtained by combining information from all of the partial reactions. For example, for kinesin, the rate of ATP hydrolysis is faster than the rate of dissociation of the kinesin from the microtubule, thereby establishing that ATP hydrolysis precedes kinesin-microtubule dissociation. Phosphate release occurs at a rate equal to the rate of kinesin-microtubule dissociation; therefore, one reaction must be much faster and follow the other. To complete the cycle without proposing additional intermediates, the most simple pathway involves the fast release of phosphate from kinesin after its dissociation (as K · ADP · P_i) from the microtubule. This conclusion is then supported by measurements of the rate of binding of kinesin-ADP (K · ADP) to the microtubules. Finally, measurement of the rate of ADP release from the microtubule-kinesin-ADP complex (Mt · K · ADP) shows that ADP release is fast, supporting the identification of the single rate-limiting step involving the dissociation of the kinesin from the microtubule. The overall pathway can then be modeled by computer simulation to confirm that it accounts for all of the observed reactions. Analysis of dimeric kinesin defined the alternating site mechanism by showing directly that ATP binding to one head stimulated release of ADP from the adjacent kinesin head (Auerbach and Johnson, 2005; Gilbert *et al.*, 1998).

VII. Detection of Enzyme Intermediates

The search for enzyme intermediates is often a frustrating and difficult task because when one begins the investigation one may not know the properties of the intermediate or the time and substrate concentrations required for its formation, and it may be formed in low amounts. Experiments based on single turnover kinetics overcome many of the difficulties and have led to the identification and isolation of several new enzyme intermediates (Anderson and Johnson, 1990a; Anderson *et al.*, 1988a,b, 1991a,b; Brown *et al.*, 1994; Marquardt *et al.*, 1993; Cho *et al.*, 1992).

We consider a single intermediate on the pathway from substrate to product:

$$E + S \rightleftharpoons E \cdot S \rightleftharpoons E \cdot I \rightleftharpoons E \cdot P \rightleftharpoons E + P$$

To prove the existence of the intermediate, I, it must be shown to be kinetically competent, and, if possible, it must be isolated to prove its structure. Kinetic competence is most precisely addressed by showing that the intermediate is formed and is broken down at rates sufficient to account for net conversion of substrate to product at the enzyme active site. A single turnover experiment with enzyme in excess over the substrate, S, provides the most sensitive means to look for intermediates and allows the kinetic competence of the intermediate to be established directly (Anderson and Johnson, 1990a,b; Anderson *et al.*, 1988a).

The kinetics of a single turnover of EPSP synthase is shown in Fig. 8A (Anderson *et al.*, 1988a). The reaction involves the combination of shikimate 3-phosphate (S3P) with phosphoenolpyruvate (PEP) to form a tetrahedral intermediate, which decays by elimination of phosphate to yield the product 5-enolpyruvoyl-shikimate 3-phosphate (EPSP). The reaction shown in Fig. 8A was initiated with 10 μM enzyme in excess over limiting [^{14}C]PEP (3.5 μM) in the presence of excess S3P (100 μM). The key to the success of the experiment lies in the use of enzyme excess over the limiting, radiolabeled substrate. This allowed the direct observation of the intermediate although it formed on only one-third of the enzyme sites. The kinetics of the reverse reaction are shown in Fig. 8B, again showing the formation of the intermediate from EPSP. The kinetics of the reactions could be fitted to the reaction sequence to obtain a unique set of rate constants describing the rates of formation and decay of the tetrahedral intermediate in each direction (Anderson *et al.*, 1988a). The kinetic analysis led directly to defining conditions under which the intermediate could be synthesized in quantities sufficient for structure determination (Anderson *et al.*, 1988b).

VIII. Data Analysis

Kinetic data can be fit to one of several simple equations involving either one or two exponential terms. The observed rates and amplitudes can then be interpreted in terms of the pathway of the reaction. As the pathway becomes more

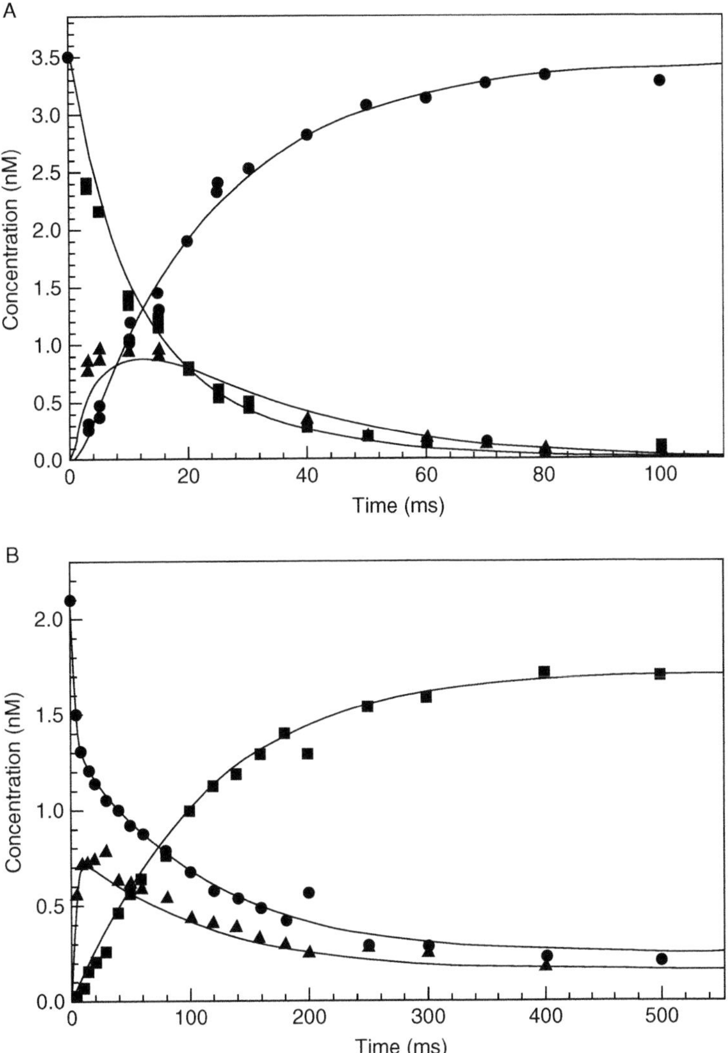

Fig. 8 Kinetics of a single turnover of EPSP synthase. (A) A single turnover in the forward reaction was observed by mixing a solution containing enzyme (10 μM) and S3P (100 μM) with [^{14}C]PEP (3.5 μM). The reaction was terminated by the addition of 1 N HCl, and then the solution was vortexed with chloroform and neutralized. Products were analyzed by ion-exchange HPLC. (B) Kinetics of the reverse reaction. The time course is shown for formation and disappearance of PEP (■), EPSP (•), and pyruvate (▲), a breakdown product of the intermediate. The curves were calculated by numerical integration according to the complete solution of the reaction pathway. (Reprinted with permission from Anderson *et al.* (1988). Copyright 1988 American Chemical Society.)

complex, then assumptions must be made to allow the use of the simplified equations. In general, there should be one exponential term for each step of the reaction (Johnson, 1992), except for steps in rapid equilibrium or, in general, when one step is much faster than neighboring steps. Although fitting data to analytical solutions of the rate equations is useful and is often a necessary first step, it is usually desirable to fit the data directly to the complete mechanism by numerical integration. Computer simulation allows all of the kinetic data to be fit directly to the enzyme reaction pathway by global analysis without the need for simplifying assumptions. This section first describes the fitting of data to standard solutions of simplified rate equations, then discusses the methods of global analysis.

A. Equilibrium Binding Measurements

It is often necessary and useful as part of the kinetic analysis to examine the equilibrium binding of substrates to the enzyme:

$$\mathrm{E} + \mathrm{S} \rightleftharpoons \mathrm{E} \cdot \mathrm{S}$$

If the binding is weak relative to the concentration of enzyme required to measure the binding, then the data can be fit to a hyperbola:

$$[\mathrm{E} \cdot \mathrm{S}] = \frac{[\mathrm{E}]_0[\mathrm{S}]_0}{[\mathrm{S}]_0 + K_\mathrm{d}}$$

where $[\mathrm{E}]_0$ and $[\mathrm{S}]_0$ are the total enzyme and substrate concentrations, respectively. Data should be fit by nonlinear regression, including $[\mathrm{E}]_0$, the starting enzyme concentration, as an unknown, if possible. If the data are in the form of an optical signal with an unknown extinction coefficient, a, then the signal must be fit including these additional unknowns, and the enzyme concentration must be assumed to be known:

$$Y = a[\mathrm{E} \cdot \mathrm{S}] + c = \frac{a[\mathrm{E}]_0[\mathrm{S}]_0}{([\mathrm{S}]_0 + K_\mathrm{d}) + c}$$

Several graphics programs containing nonlinear regression routines are available for personal computers (GraFit, Erithacus Software, Ltd., Staines, UK; KaleidaGraph, Synergy Software, Reading, PA, USA). After entering the appropriate equation and initial estimates of the parameters, the program will find the best fit by minimizing the sum of squares error between the calculated and observed curves. A failure to get the program to converge to a solution is most often due to errors in the initial estimates. Take note of the form of the equation and the values of y at $x = \infty$ to estimate the parameters from the graph.

If the binding is tight relative to the enzyme concentration, then the data must be fit to a quadratic equation:

$$[\mathrm{E}\cdot\mathrm{S}] = \frac{(K_\mathrm{d} + [\mathrm{E}]_0 + [\mathrm{S}]_0) - \mathrm{SQRT}((K_\mathrm{d} + [\mathrm{E}]_0 + [\mathrm{S}]_0)^2 - 4[\mathrm{E}]_0[\mathrm{S}]_0)}{2}$$

Again, care should be taken in considering the relationship between any optical signal and the concentration of E · S and in making initial estimates of the unknown parameters.

B. Single Exponential

Most kinetic experiments can be fit to a single exponential of the general form

$$Y = A\ \exp(-kt) + C$$

where A is the amplitude, k is the rate, and C is the end point. All three parameters should be taken as unknowns in fitting by nonlinear regression. In making initial estimates, note that a reaction with a signal increasing with time has a negative amplitude. At $t = 0$, $Y = A + C$, where as $t \to \infty$, $Y = C$.

C. Double Exponential

The equation for a double exponential follows the general form

$$Y = A_1\ \exp(-k_1 t) + A_2\ \exp(-k_2 t) + C$$

which is analogous to a single exponential except that there are two amplitude and rate terms. This equation is used whenever the reaction shows two phases, such as a fast and slow phase or a lag preceding a reaction.

D. Burst Kinetics

The kinetics of a presteady-state burst followed by linear steady-state turnover can be fit to the standard burst equation:

$$Y = A\ \exp(-k_1 t) + k_2 t + C$$

where A is the amplitude of the burst and k_1 is the rate of the burst (in units of s^{-1}), whereas k_2 is the steady-state rate (in units of concentration/s). Divide k_2 by the enzyme concentration or amplitude of the burst to convert it to a first-order rate constant defining product release rates.

E. Global Analysis

Ultimately the goal of kinetic analysis is to derive a single mechanism to account completely for the results without introducing steps or intermediates not required by the data. It is now possible to fit all of the kinetic experiments directly to the mechanism by nonlinear regression based on numerical integration of the rate equations (Barshop *et al.*, 1983; Zimmerlie and Frieden, 1989; Johnson *et al.*, 2009a,b). Computer simulation is also a useful learning tool that can aid in the design of experiments to optimize conditions to test opposing models.

In fitting data by numerical integration, no simplifying assumptions are needed, but it is necessary to be close to the best fit before beginning the nonlinear regression. In addition, it is necessary to allow only those constants to float which are sensitive parameters pertaining to each experiment. For example, if an experiment is done at high substrate concentration so the binding rate is fast and the observed reaction rate is a function of the chemistry step, then the binding rate must be fixed at a known value while floating the rate of the chemical reaction to obtain the best fit. Alternatively, for a series of experiments done as a function of substrate concentration, one may float the substrate binding rate while maintaining a fixed value for the rate of the chemical reaction. In the final analysis, the overall pathway can then be fit to all of the data and including in the fitting process estimates of rate constants even though they may pertain to different parts of the mechanism that are not tested by a given experiment. In this way, even subtle effects on the kinetics can be taken into consideration in the fitting process to achieve a global best fit.

Using these methods, the pathway of enzyme-catalyzed reactions can be established by direct measurement of individual steps of the reaction. By comprehensive analysis of the reaction kinetics, the partial reactions can be assembled to provide a simple, complete reaction pathway.

Recent advances in software now allow for the simultaneous fitting of multiple experiments to a single model and set of rate constants based upon numerical integration of rate equations (Johnson *et al.*, 2009a,b). Now data obtained using rapid quench-flow and by stopped-flow fluorescence methods can be fit simultaneously to provide the most rigorous analysis to distinguish alternative models. Moreover, fast dynamic simulation provides visual feedback to show the relationships between individual kinetic parameters and observable signals, which are critically important in exploring the range over which parameters can vary in fitting data as well as for finding starting estimates for nonlinear regression. Finally, confidence contour analysis provides a rigorous test of how well each of the kinetic parameters are constrained by the data (Johnson *et al.*, 2009b).

References

Anderson, K. A., Kati, W. M., Ye, C. Z., Liu, J., Walsh, C. T., Benesi, A. J., and Johnson, K. A. (1990). *J. Am. Chem. Soc.* **113,** 3198.

Anderson, K. S., and Johnson, K. A. (1990a). *Chem. Rev.* **90,** 1131.

Anderson, K. S., and Johnson, K. A. (1990b). *J. Biol. Chem.* **265,** 5567.

Anderson, K. S., Sikorski, J. A., and Johnson, K. A. (1988a). *Biochemistry* **27,** 7395.

Anderson, K. S., Sikorski, J. A., Benesi, A. J., and Johnson, K. A. (1988b). *J. Am. Chem. Soc.* **110,** 6577.

Anderson, K. S., Miles, E. W., and Johnson, K. A. (1991). *J. Biol. Chem.* **266,** 8020.

Auerbach, S. D., and Johnson, K. A. (2005). Alternating site ATPase pathway of rat conventional kinesin. *J. Biol. Chem.* **280,** 37048–37060.

Barshop, B. A., Wrenn, R. F., and Frieden, C. (1983). *Anal. Biochem.* **130,** 134.

Brown, E. D., Marquardt, J. L., Walsh, C. T., and Anderson, K. S. (1993). *J. Am. Chem. Soc.* **115,** 10398.

Brown, E. D., Marquardt, J. L., Lee, J. T., Walsh, C. T., and Anderson, K. S. (1994). *Biochemistry* **33,** 10638.
Brune, M., Hunter, J. L., Corrie, J. E. T., and Webb, M. R. (1994). *Biochemistry* **33,** 8262.
Cho, H., Krishnaraj, R., Kitas, E., Bannwarth, W., Walsh, C. T., and Anderson, K. S. (1992). *J. Am. Chem. Soc.* **114,** 7296.
Donlin, M. J., Patel, S. S., and Johnson, K. A. (1991). *Biochemistry* **30,** 538.
Froehlich, J. P., Sullivan, J. V., and Berger, R. L. (1976). *Anal. Biochem.* **73,** 331.
Gilbert, S. P., and Johnson, K. A. (1994). *Biochemistry* **33,** 1951.
Gilbert, S. P., Webb, M. R., Brune, M., and Johnson, K. A. (1994). *Nature* (*London*) (submitted for publication).
Gilbert, S. P., Moyer, M. L., and Johnson, K. A. (1998). Alternating site mechanism of the kinesin ATPase. *Biochemistry* **37,** 792–799.
Hanes, J. W., and Johnson, K. A. (2008). Real-time measurement of pyrophosphate release kinetics. *Anal. Biochem.* **372,** 125–127.
Hsieh, J. C., Zinnen, S., and Modrich, P. (1993). *J. Biol. Chem.* **268,** 24607.
Johnson, K. A. (1983). *J. Biol. Chem.* **258,** 13825.
Johnson, K. A. (1985). *Annu. Rev. Biophys. Biophys. Chem.* **14,** 161.
Johnson, K. A. (1992). *In* "The Enzymes" (P. D. Boyer, ed.), 4th edn., Vol. 20, p. 1. Academic Press, New York.
Johnson, K. A. (1986). *Method Enzymol.* **134,** p. 677.
Johnson, K. A. (1993). *Annu. Rev. Biochem.* **62,** 685.
Johnson, K. A., and Taylor, E. W. (1978a). *Biochemistry* **17,** 3432.
Johnson, K. A., and Taylor, E. W. (1978b). Intermediate states of subfragment 1 and actosubfragment 1 ATPase: Reevaluation of the mechanism. *Biochemistry* **17,** 3432–3442.
Johnson, K. A., Simpson, Z. B., and Blom, T. (2009a). FitSpace Explorer: An algorithm to evaluate multidimensional parameter space in fitting kinetic data. *Anal. Biochem.* **387,** 30–41.
Johnson, K. A., Simpson, Z. B., and Blom, T. (2009b). Global Kinetic Explorer: A new computer program for dynamic simulation and fitting of kinetic data. *Anal. Biochem.* **387,** 20–29.
Kati, W. M., Johnson, K. A., Jerva, L. F., and Anderson, K. S. (1992). *J. Biol. Chem.* **267,** 25988.
Lymn, R. W., and Taylor, E. W. (1970). *Biochemistry* **9,** 2975.
Lymn, R. W., and Taylor, E. W. (1971). *Biochemistry* **10,** 4617.
Patel, S. S., Wong, I., and Johnson, K. A. (1991). *Biochemistry* **30,** 511.
Sadhu, A., and Taylor, E. W. (1994). *J. Biol. Chem.* **267,** 11352.
Tsai, Y. C., and Johnson, K. A. (2006). A new paradigm for DNA polymerase specificity. *Biochemistry* **45,** 9675–9687.
Wiskind, H. K. (1964). *In* "Rapid Mixing and Sampling Techniques in Biochemistry" (B. Chance, R. H. Eisenhardt, Q. H. Gibson and K. K. Lonberg-Holm, eds.), p. 355. Academic Press, New York.
Wong, I., Patel, S. S., and Johnson, K. A. (1991). *Biochemistry* **30,** 526.
Zimmerlie, C. T., and Frieden, C. (1989). *Biochem. J.* **258,** 381.

CHAPTER 15

Presteady-State Kinetics of Enzymatic Reactions Studied by Electrospray Mass Spectrometry with Online Rapid-Mixing Techniques

Lars Konermann* and Donald J. Douglas†

*Department of Chemistry
University of Western Ontario
London, Ontario, Canada N6A 5B7

†Department of Chemistry
University of British Columbia, Vancouver
British Columbia, Canada V6T 1Z1

I. Introduction

Steady-state kinetic experiments on enzymatic reactions provide information such as the maximum turnover number (k_{cat}) and the Michaelis constant (K_m). Studies of this kind are also useful for distinguishing between different types of enzyme inhibition. However, K_m and k_{cat} are combinations of all the rate and equilibrium constants involved in an enzymatic reaction. Therefore, steady-state experiments usually provide little or no information on the actual reaction mechanism, that is, the number of transient intermediates, their chemical structure, and

CONTEMPORARY ENZYME KINETICS AND MECHANISM
Reprinted from *Methods in Enzymology*, Volume 354 (Elsevier Science (USA)).

DOI: 10.1016/B978-0-12-378608-1.00015-3

the rate constants or energy barriers of individual reaction steps (Fersht, 1999). A proper understanding of enzyme mechanisms requires kinetic experiments in the presteady-state regime. Immediately after initiating an enzymatic reaction there is a short time range (usually some milliseconds to seconds, depending on the rate constants involved) during which the system approaches steady-state conditions. The mechanism of the reaction dictates the order by which short-lived intermediates become populated successively during this period. Individual rate constants can be measured, from which the energy barriers along the reaction coordinate can be estimated (Anderson *et al.*, 1988; Fersht, 1999; Hiromi, 1979). The enzyme concentration required for presteady-state experiments is relatively high because the enzyme is a stoichiometric "reactant." Usually this is not a problem because recombinant techniques can be used to produce sufficiently large amounts of enzyme.

Presteady-state kinetic experiments often involve the use of stopped-flow optical spectroscopy (Johnson, 1992). In these experiments, two or more reactant solutions are expelled from a set of pneumatically or stepper motor-driven syringes. The solutions meet in a mixer where the reaction is initiated. The fresh reaction mixture is rapidly transferred to an observation cell while the previous contents of the cell are flushed out. The liquid flow is abruptly halted by a stopping syringe or a valve at the exit of the cell. The kinetics is monitored by recording an optical signal (e.g., fluorescence or absorbance) as a function of time as the reaction proceeds in the observation cell. Chemical quench-flow experiments (Johnson, 1995) are another method for studying fast enzyme kinetics. The reaction is initiated by rapid mixing of the reactants, followed by mixing with a quenching agent such as acid, base, or an organic solvent, after a specified period of time. The quenching step abruptly stops the reaction by denaturing the enzyme and liberates noncovalently bound substrates, intermediates, and products. Various methods can be used for the off-line analysis of the quenched reaction mixture. For example, the use of radiolabeled substrates allows the quantitation of individual reactive species at different reaction times by high-performance liquid chromatography (HPLC) (Anderson *et al.*, 1988). A relatively new and promising approach involves the off-line analysis of the quenched reaction mixture by matrix-assisted laser desorption/ionization (MALDI) mass spectrometry (MS). This elegant technique is the focus of another chapter in this volume and will not be discussed here (Gross and Frey, 2002).

Stopped- and quench-flow methods have provided a wealth of information on enzymatic reaction mechanisms. However, these techniques have a number of disadvantages. Stopped-flow optical absorption spectroscopy is limited to reactions that are associated with chromophoric changes. Most reactive species that are involved in the enzymatic conversion of "natural" substrates are not chromophoric and hence they cannot be directly studied by stopped-flow methods (although fluorescence of the enzyme can sometimes be monitored). For this reason most studies on enzyme kinetics involve the use of synthetic, chromophoric substrates. Unfortunately, the kinetics and the reaction mechanisms observed with these artificial compounds can be different from those of the natural enzyme substrates. Chromophoric substrate analogs can pose the further disadvantage of long and difficult chemical syntheses. Usually optical spectroscopy provides little information on the chemical transformations occurring in the reaction mixture and in most cases only one reactive species can be monitored. Chemical quench-flow experiments can also have problems. The quenching agent can induce degradation or chemical modification of labile species in the reaction mixture. The use of radiolabeled substrates in quench-flow studies not only is inconvenient but also can lead to unwanted artifacts arising from nonspecific entrapment of the radioactive label.

Here, we describe the use of electrospray ionization (ESI) MS in conjunction with online rapid mixing for studying the presteady-state kinetics of enzymatic reactions. It appears that this approach has the potential to overcome many of the limitations that are encountered with traditional stopped- and quench-flow methods. It seems likely that ESI-MS will soon become a standard technique in the laboratories of many enzyme kineticists.

II. Electrospray Ionization Mass Spectrometry

Electrospray is one of the most widely used ionization methods in biological MS (Fenn *et al.*, 1989). Analyte solution is pumped through a narrow capillary that is held at a high electric potential. Small, highly charged solvent droplets are emitted from the end of the ESI capillary at atmospheric pressure. These droplets rapidly shrink because of solvent evaporation, thus increasing the charge density on the droplet surface and eventually leading to the formation of offspring droplets via jet fission. Analyte gas-phase ions are generated, either as charged residues from nanometer-sized offspring droplets or by ion evaporation. This whole process of ion formation occurs on a time scale of roughly 1 ms (Kebarle and Ho, 1997). ESI is a very gentle process that can be used for ionizing a wide range of analytes, including small organic molecules, proteins, and even noncovalent ligand-protein or protein-protein complexes (Loo, 1997). The ions produced by ESI are transferred into the vacuum chamber of a mass spectrometer and analyzed according to their mass-to-charge ratio (m/z).

In enzymology, ESI-MS has been used for the detection of enzyme-substrate and enzyme-inhibitor complexes (Aplin *et al.*, 1990; Ashton *et al.*, 1991; Cheng *et al.*, 1995; Duncan *et al.*, 1999; Ganem *et al.*, 1991; Knight *et al.*, 1993; Menard *et al.*, 1991) and for monitoring the concentration of reactive species at different reaction times in quench-flow experiments (Boerner *et al.*, 1996; Brown *et al.*, 1996; Hsieh *et al.*, 1995; Ichiyama *et al.*, 2000; Lu *et al.*, 1999). All these studies were carried out on "stationary" samples, that is, the analyte solution was assumed to undergo no chemical changes during the analysis. In this chapter, we focus on recent developments that allow the direct online monitoring of enzyme-catalyzed reactions by ESI-MS. In these experiments, an enzyme/substrate mixture is injected directly into the ESI source of the mass spectrometer while the reaction proceeds in solution. With this approach it is possible to overcome many limitations of traditional kinetic methods (Simpson and Northrop, 2000). No chemical quenching steps are required because the analysis occurs online and is quasi-instantaneous.

MS does not rely on the presence of chromophores, and therefore, it is possible to monitor reactive species that cannot be detected optically. Radioactive labeling, artificial chromophoric substrate analogs, and coupled assays are not required. Instead, ESI-MS is a sensitive and selective method for studying the reactions of enzymes with their natural, biologically significant substrates. In principle, it is possible to monitor reactants, intermediates, and products simultaneously, as well as covalent and noncovalent enzyme-substrate complexes.

The enormous potential of ESI-MS for the online monitoring of reaction mixtures in kinetic experiments was first realized by Lee *et al.* (1989) who used this technique for studying the steady-state kinetics of lactase. They used a reaction vessel that was closely coupled to the ion source of an ESI mass spectrometer. Reactions were initiated by manual mixing of enzyme and substrate in the vessel. Similar approaches were used for steady-state experiments on a number of other enzymes (Bothner *et al.*, 2000; Fabris, 2000; Fligge *et al.*, 1999). These elegant studies demonstrated the versatility of ESI-MS as a direct method for monitoring enzyme kinetics. However, the time resolution in these experiments was relatively poor, on the order of several seconds to minutes, because manual mixing was used to initiate the reactions of interest. Although this is adequate for monitoring steady-state kinetics, it does not generally allow studies in the presteady-state regime. The following sections describe two different methods for the direct coupling of ESI-MS with rapid mixing devices, involving continuous- and stopped-flow systems, respectively. The time resolution that can be achieved by these techniques is orders of magnitude better than in manual mixing experiments. The results demonstrate that the coupling of rapid mixing devices with ESI-MS provides a new and powerful tool, not only for studying the presteady state of enzymatic reactions, but also for a wide range of other kinetic experiments in chemistry and biochemistry.

III. Presteady-State Kinetics of Xylanase Monitored by Electrospray MS with Online Continuous-Flow Mixing

Continuous-flow experiments with rapid online mixing are a simple and elegant method for studying biochemical reaction kinetics (Gutfreund, 1969). With optical detection, a time resolution in the microsecond range can be achieved (Shastry *et al.*, 1998). One limitation of many continuous-flow instruments is the substantial sample consumption; the original studies of Hartridge and Roughton (1923) required more than 3 l of sample for each experiment. Konermann *et al.* (1997a) developed a miniaturized continuous-flow mixing device and coupled it to the ESI source of a mass spectrometer. Through the use of narrow fused-silica capillaries, the sample consumption could be substantially reduced. This system allows extensive kinetic studies even if only a few milliliters of sample are available. The viability of this new experimental technique was amply demonstrated in a number of studies on protein folding kinetics (Konermann *et al.*, 1997a,b; Lee *et al.*, 1999; Sogbein *et al.*, 2000).

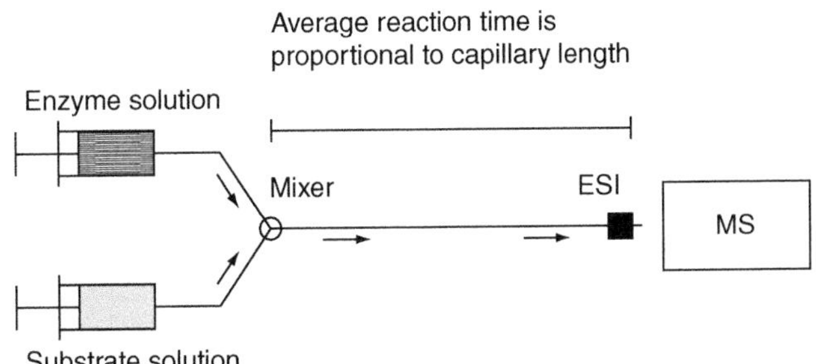

Fig. 1 Continuous-flow mixing setup for monitoring enzyme kinetics. The plungers of both syringes are advanced continuously by syringe pumps. The reaction is initiated by mixing the solutions from both syringes. The average reaction time is proportional to the length of the capillary downstream from the mixer. The end of the reaction capillary is connected to the ESI source of a mass spectrometer. Arrows indicate the direction of liquid flow. Reprinted with permission from Konermann *et al.* (1997). Copyright © 1997 American Chemical Society.

Here, we describe the application of this technique to the study of the presteady-state kinetics of enzymatic reactions. The setup used for these experiments is shown in Fig. 1 (Konermann *et al.*, 1997a). The plungers of two syringes (volume 1 ml each) are advanced simultaneously by syringe pumps. The syringes contain enzyme and substrate solution, respectively. The reaction is initiated by mixing the solutions in a custom-made mixing "tee." This tee has a volume of 3 nl which corresponds to a mixing time (dead time) of 5 ms at a total flow rate of 30 μl/min (Konermann *et al.*, 1997a). The outlet of the tee is connected to the ESI source of a quadrupole mass spectrometer via a reaction capillary (i.d. 75 μm). The average reaction time of the solution is determined by the flow rate and the length of the reaction capillary. For the flow rate given above, 1 cm of capillary corresponds to 81 ms. Shorter reaction times can be achieved by using higher flow rates. Typically, capillary lengths between 9 mm and 2 m are used. A miniaturized ion source with an overall length of only 7 mm had to be developed for this setup to accommodate the shortest reaction capillaries. Ions are generated by pneumatically assisted ESI at the outlet of the reaction capillary. Within one set of kinetic experiments (i.e., for a given substrate concentration), it is preferable to change the reaction time only by varying the length of the reaction capillary while leaving the flow rate constant. This is because different flow rates can change the appearance of ESI mass spectra, even if no chemical reaction occurs.

The apparatus of Fig. 1 was used for studying the presteady-state kinetics of an enzymatic reaction (Zechel *et al.*, 1998). The Y80F mutant of the extensively characterized *Bacillus circulans* xylanase (BCX, EC 3.2.1.8) was chosen as a model enzyme system. This enzyme cleaves the β-1,4-glycosidic bonds of xylan to release xylobiose. For this study, the chromophoric substrate 2,5-dinitrophenyl-β-xylobioside (DNP-X_2) was used, so the kinetics observed by ESI-MS could be independently confirmed by traditional stopped-flow optical spectroscopy. The Y80F mutant of BCX has a K_m of 60 μM for DNP-X_2. The mechanism for

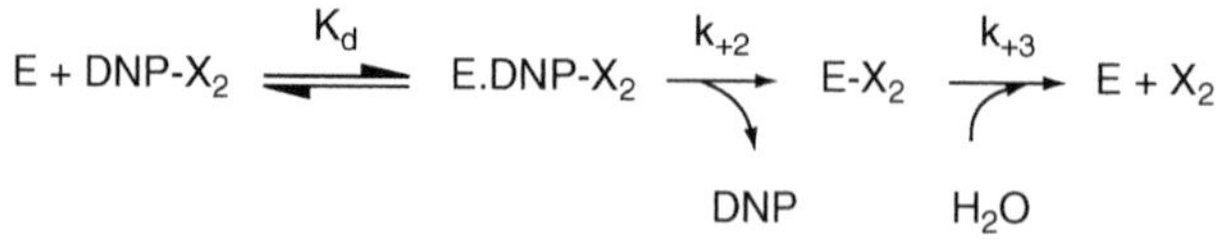

$$E + DNP\text{-}X_2 \underset{}{\overset{K_d}{\rightleftharpoons}} E.DNP\text{-}X_2 \xrightarrow[\text{DNP}]{k_{+2}} E\text{-}X_2 \xrightarrow[\text{H}_2\text{O}]{k_{+3}} E + X_2$$

Scheme 1 Mechanism for the hydrolysis of 2,5-dinitrophenyl-β-xylobioside (DNP-X_2) by *Bacillus circulans* xylanase (BCX). Reprinted with permission from Zechel *et al.* (1998). Copyright © 1998 American Chemical Society.

the hydrolysis of DNP-X_2 by BCX is described in Scheme 1 (Zechel *et al.*, 1998). The free substrate and the free enzyme are in equilibrium with a noncovalent enzyme-substrate complex E.DNP-X_2 (dissociation constant K_d). Subsequently, xylobiose (X_2) is covalently bound to the enzyme. The formation of this E-X_2 complex is accompanied by the release of 2,5-dinitrophenolate (DNP) which can be monitored optically at 440 nm. The last step of the reaction is the hydrolysis of the E-X_2 complex to release X_2 and to regenerate the free enzyme. As indicated in Scheme 1, k_{+2} and k_{+3} are the rate constants for glycosylation and deglycosylation of the enzyme, respectively. Wild-type BCX has not been amenable to presteady-state kinetic analysis by stopped-flow optical spectroscopy because the glycosylation step is rate limiting ($k_{+2} < k_{+3}$) with all known synthetic substrates. However, this is not the case for the Y80F mutant studied here. It shows a presteady-state phase which is characterized by a burst of DNP formation, indicating that deglycosylation is rate limiting ($k_{+2} > k_{+3}$) (Zechel *et al.*, 1998).

The apparatus of Fig. 1 was used for monitoring the time-resolved mass spectra of BCX Y80F at different times after mixing enzyme and substrate ("time-resolved ESI-MS"). All the concentrations given below refer to the final concentrations (i.e., after the mixing step). Ammonium acetate (5 mM, pH 6.3) was used as buffer system. The enzyme concentration was 4 μM, while the initial DNP-X_2 concentrations ranged from 0.035 to 2.2 mM. The limited solubility of DNP-X_2 precluded experiments at higher substrate concentration. The ESI mass spectrum of the mutant BCX Y80F in 5 mM ammonium acetate buffer and in the absence of substrate displayed a charge state distribution with a characteristic maximum of 8+ (data not shown). The m/z values of the peaks in the spectrum agreed with those expected from the calculated mass of the protein (20,384 Da). A very similar spectrum was recorded 0.5 s after mixing the enzyme with 0.11 mM DNP-X_2 (Fig. 2A) (Zechel *et al.*, 1998). However, after 4.2 s (Fig. 2B), each of the peaks in the spectrum displayed a pronounced satellite peak. On the basis of their m/z values, these satellite peaks could be assigned to the covalent xylobiosyl-enzyme intermediate E-X_2 (20,649 Da). The spectrum recorded after 48 s (Fig. 2C) displayed an even higher intensity for the E-X_2 peaks. Similar intensity ratios between E and E-X_2 peaks were observed for the different charge states for all reaction times and substrate concentrations. The intensity ratio of the dominant E 8+ and E-X_2 8+ peaks provides a direct measure of the relative concentrations of free enzyme (E) and covalent intermediate (E-X_2) in solution. Therefore, it is possible

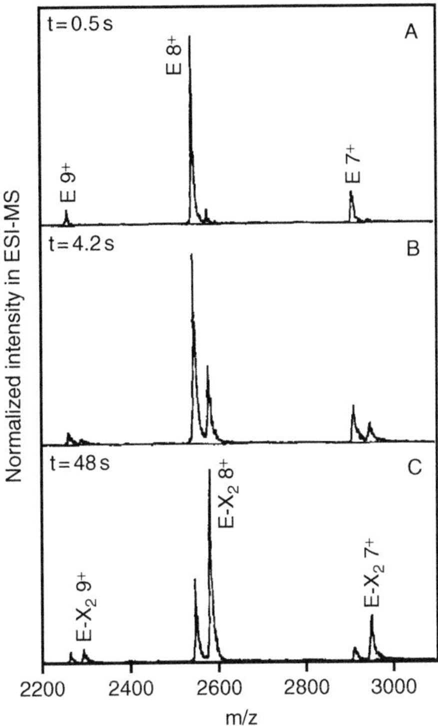

Fig. 2 ESI mass spectra recorded at 0.5 s (A), 4.2 s (B), and 48 s (C) after mixing a solution of the enzyme BCX Y80F (4 μM) with the substrate DNP-X_2 (110 μM). Peaks in the spectrum correspond to different charge states of the free enzyme (E) and the covalent xylobiosyl-enzyme intermediate (E-X_2). Reprinted with permission from Zechel *et al.* (1998). Copyright © 1998 American Chemical Society.

to monitor the presteady-state kinetics of this reaction by measuring these peak intensities as a function of time. The results of some representative kinetic measurements are shown in Fig. 3 (Zechel *et al.*, 1998). Consistent with the kinetic scenario in which $k_{+2} > k_{+3}$ (Scheme 1), the intensity of E-X_2 shows a fast increase before it levels off and reaches a constant value as the reaction approaches the steady-state regime. Higher substrate concentrations lead to an increased steady-state concentration of E-X_2, as expected. All the curves are well described by single-exponential fits with $I(t) = I_{SS}[1 - \exp(-k_{obs}t)]$, where $I(t)$ is the ion intensity as a function of time, I_{SS} is the steady-state intensity, and k_{obs} is the observed first-order rate constant (Fersht, 1999; Hiromi, 1979). A plot of k_{obs} versus substrate concentration is shown in Fig. 4 (Zechel *et al.*, 1998). Also shown in Fig. 4 are the values of k_{obs} that were measured by stopped-flow optical spectroscopy from the burst-phase kinetics of DNP formation (Zechel *et al.*, 1998). The datasets are in excellent agreement. According to Scheme 1, the first-order rate constant k_{obs} is given by Fersht (1999) and Hiromi (1979).

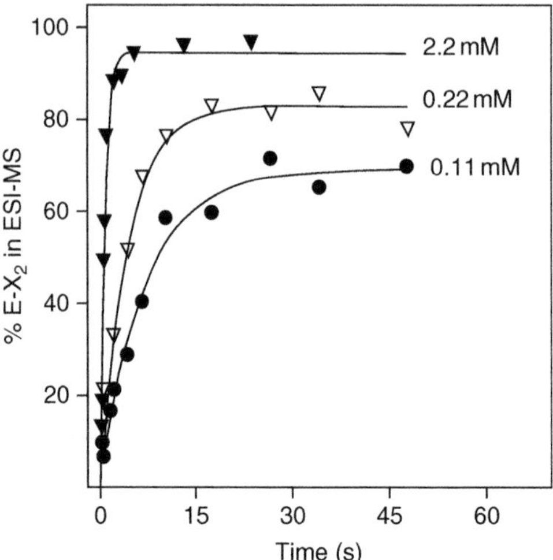

Fig. 3 Relative contribution of the 8+ ion generated from the covalent xylobiosyl-enzyme intermediate (E-X_2) in the ESI mass spectrum as a function of time. The substrate concentrations are as indicated. Circular and triangular symbols represent experimental data; solid lines are single-exponential fits to the data. Reprinted with permission from Zechel *et al.* (1998). Copyright © 1998 American Chemical Society.

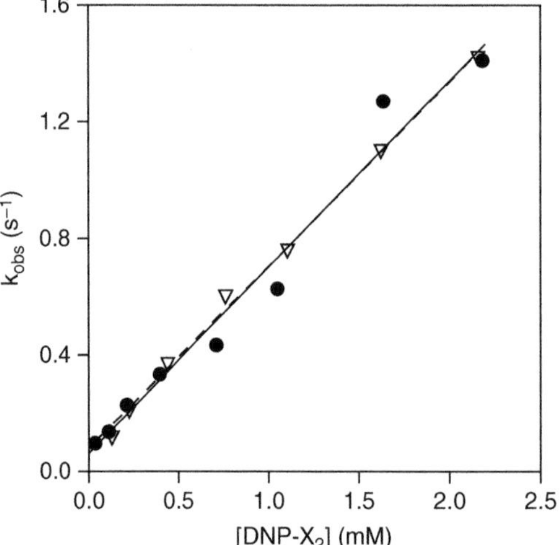

Fig. 4 First-order rate constant k_{obs} for the formation of the covalent xylobiosyl-enzyme intermediate E-X_2 measured by ESI-MS (circles) and for the formation of DNP measured by stopped-flow optical spectroscopy (triangles), as a function of substrate concentration. The solid and dashed lines are fits to the experimental ESI and optical data, respectively, according to Eq. (2). Reprinted with permission from Zechel *et al.* (1998). Copyright © 1998 American Chemical Society.

$$k_{obs} = k_{+3} + \frac{k_{+2}[S]}{K_d + [S]} \quad (1)$$

In this study, the low solubility of DNP-X_2 precluded saturation of the enzyme which indicates that K_d has to be in the high millimolar range ($K_d \gg [S]$). Under these conditions, Eq. (1) simplifies to a linear relationship.

$$k_{obs} = k_{+3} + \frac{k_{+2}}{K_d}[S] \quad (2)$$

The linear fits to the experimental data monitored by stopped-flow optical spectroscopy and by ESI-MS are virtually identical (Fig. 4). By extrapolating these fits to a substrate concentration of zero, k_{+3} is determined to be 0.08 s^{-1}. Unfortunately, the slope in Eq. (2) cannot be used to determine the individual values of K_d and k_{+2}. However, the ratio k_{+2}/K_d of 0.65 mM^{-1} s^{-1} from these presteady-state experiments agrees very well with the second-order rate constant k_{cat}/K_m of 0.70 mM^{-1} s^{-1} that was determined from steady-state experiments (Fersht, 1999; Zechel *et al.*, 1998). Unfortunately, the spectra in Fig. 2 do not show any peaks that could be assigned to noncovalent enzyme-substrate or enzyme-product complexes. Ganem *et al.* (1991) have shown previously that specific noncovalent complexes are only observed when the dissociation constant is in the micromolar range or lower. As noted above, K_d in this case is much higher.

These experiments on BCX Y80F marked the first time that ESI-MS in conjunction with online continuous-flow mixing was applied to study the presteady-state kinetics of an enzymatic reaction. This approach does not require radioactive labeling or chemical quenching. The chromophoric substrate DNP-X_2 was used only to validate this novel method by independently confirming the ESI-MS kinetics by traditional stopped-flow optical spectroscopy. The time resolution of the current setup is on the order of some tens of milliseconds. This resolution was sufficient for studying the relatively slow presteady-state kinetics of BCX Y80F. However, for many enzymes a higher time resolution will be required. This should be possible by using higher liquid flow rates, a mixing chamber with a lower dead volume, and capillaries with smaller inner diameters. Instead of using fused-silica capillaries, such a continuous-flow device could be incorporated into a microfluidic chip that is directly coupled to an ESI source. Through the use of hydrodynamic focusing, such a system could potentially reach a time resolution in the submillisecond range (Knight *et al.*, 1998). Although microfluidic chips for ESI-MS are now commonly used for various analytical applications (Oleschuk and Harrison, 2000), a fast kinetic mixer of this kind has not been developed yet.

Another experimental technique for studying the kinetics of enzymatic reactions has been developed by Northrop and Simpson (1998). It uses a continuous flow of carrier solution that flows past the interface of a membrane inlet mass spectrometer (MIMS). Pulses of reaction mixture are injected into the carrier stream and the reaction time is varied by using different lengths of delay line between the injection point and interface. This setup was used for monitoring the concentration of CO_2 during the conversion of carbon dioxide to bicarbonate by carbonic anhydrase. It appears that this technique could complement the ESI-based approach described above. ESI-MS allows the analysis of an extremely wide range of analytes. However, ESI cannot be used to ionize gases that are dissolved in a reaction mixture.

It is widely believed that continuous-flow systems can only produce meaningful kinetic data when they are operated in the turbulent flow regime (Fersht, 1999). Turbulent flow leads to continuous mixing of fast and slow liquid in a flow tube, therefore resulting in a relatively narrow distribution of solution "age" at every point along the tube, downstream from the mixer where the reaction is initiated. However, in narrow capillary devices with relatively low flow rates such as those used for the kinetic experiments by Zechel *et al.* (1998), the liquid flow is laminar (Konermann, 1999). Laminar flow is characterized by a parabolic velocity profile: the flow velocity in the center of the capillary is highest, whereas the liquid layer directly adjacent to the capillary wall is stationary. This velocity profile leads to a distribution of solution "age" at every point along the flow tube. It, therefore, seems somewhat surprising that reaction kinetics monitored in the laminar flow regime are in close agreement with those measured by traditional rapid mixing methods (see, e.g., Fig. 4). This apparent contradiction has been addressed in a study where computer modeling was used to simulate the liquid flow in continuous-flow systems (Konermann, 1999). It was shown that the distortion of reaction kinetics monitored under laminar flow conditions is surprisingly small. Radial diffusion in narrow flow tubes plays an important role since it transfers analyte between liquid layers of different velocity, thus narrowing the width of the age distribution of the analyte in the flow tube. These computer simulations have shown that the effects of laminar flow are almost undetectable for experiments such as the kinetic studies on BCX Y80F described above. The results of Konermann (1999) therefore clearly indicate the feasibility of kinetic continuous-flow experiments in the laminar flow regime.

IV. Stopped-Flow Electrospray MS

Stopped-flow optical spectroscopy is one of the most widely used methods for studying the presteady-state kinetics of enzymatic reactions. A variety of stopped-flow instruments are commercially available. They are easy to use and provide kinetic data in simple "push-button" experiments. It was pointed out earlier in this chapter that a significant limitation of conventional stopped-flow techniques is the requirement for chromophoric substrates. ESI-MS, on the other hand, does not require chromophores and allows direct monitoring of most reactive species in enzymatic assays with high sensitivity and selectivity. Therefore, there is considerable interest in the development of techniques that combine stopped-flow mixing with ESI-MS.

The first steps in this direction have been described by Paiva *et al.* (1997) who coupled a stopped-flow device to the ESI source of an ion trap mass spectrometer. This setup was used for the detection of a short-lived tetrahedral intermediate that was formed under single turnover conditions during the reaction catalyzed by 5-enolpyruvoyl-shikimate-3-phosphate synthase. Only one single time point (corresponding to 28 ms) could be monitored by this system. Therefore, a detailed kinetic study of the reaction was not possible. Ørsnes *et al.* (1998) developed a stopped-flow mass spectrometer that is capable of monitoring the concentration of reactive species at different reaction times. They coupled a stopped-flow mixer to a MIMS via a rotating ball inlet. The performance of this system was characterized by studying the concentrations of acetone and butanone in ketone-sulfite reactions. As demonstrated by Northrop and Simpson (1998) such an approach involving MIMS can be very useful for monitoring dissolved gases and small organic molecules in enzyme-catalyzed reactions. However, the use of ESI-MS would allow a much wider range of biochemical analytes to be monitored, most of which are not amenable to analysis by MIMS.

Kolakowski *et al.* (2000) have described a novel setup for stopped-flow ESI-MS that is capable of monitoring the concentration of multiple reactive species as a function of time (Kolakowski and Konermann, 2001). Thus far this new method has only been applied to bioorganic and protein folding reactions. Nevertheless, this novel setup will be briefly discussed in this chapter because it is anticipated that stopped-flow ESI-MS will become an important tool for studying the presteady state of enzymatic reactions. A schematic diagram of the experimental setup is shown in Fig. 5 (Kolakowski *et al.*, 2000). Two 20-ml syringes S1 and S2 and the mixer M are part of a commercial stopped-flow system (Bio-Logic, France). The plungers of S1 and S2 are operated by computer-controlled stepper motors. In its original configuration, two additional mixers and an optical observation cell are part of the instrument. These components are removed and replaced with a custom-made adapter which is installed directly downstream from M. S1 and S2 are used to deliver reactant solutions to M where the reaction is initiated. The bold lines downstream from M symbolize PEEK tubing (i.d. 0.76 mm). While the plungers of S1 and S2 are rapidly advanced, the valves V1 and V2 are open to permit rinsing of the reaction tube R with fresh reaction mixture. Subsequently, both valves are closed. The timing of this valve closing has to be adjusted carefully to optimize the time resolution of the experiment (Kolakowski and Konermann, 2001). While the reaction proceeds in R, the solvent from S3 pushes a steady "aliquot flow" of reaction mixture through a 1.5-cm-long fused-silica capillary (i.d. 75 μm) into a customized ESI source (Konermann *et al.*, 1997a). Analyte gas-phase ions are generated by pneumatically assisted ESI and analyzed by a triple quadrupole mass spectrometer (PE-SCIEX, Concord, Ont., Canada). Several reactants can be monitored quasi-simultaneously by operating the mass spectrometer in multiple ion mode, that is, by rapidly switching the mass analyzer between m/z values that correspond to different analyte ions.

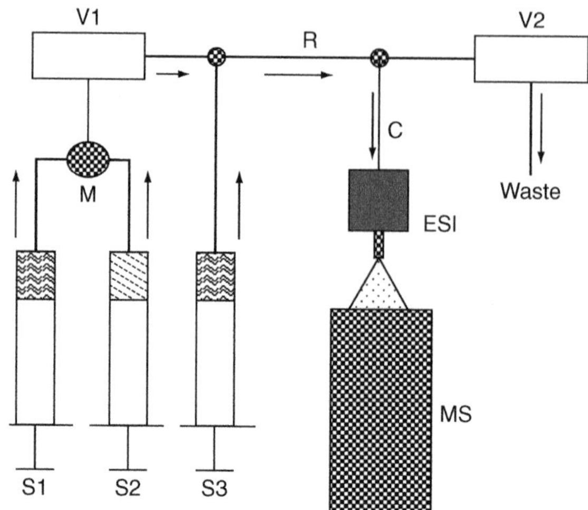

Fig. 5 Experimental setup for stopped-flow ESI-MS. S1 and S2, syringes for pulsed reactant injection; M, mixer; S3, syringe. S3 delivers solvent that pushes reaction mixture through the fused-silica capillary (C) to the ESI source after the reaction tube (R) has been filled with fresh reaction mixture; MS, quadrupole mass spectrometer; V1, V2, valves. Arrows indicate the direction of liquid flow. Reprinted with permission from Kolakowski *et al.* (2000). Copyright © 2000 John Wiley & Sons Ltd.

The performance of this apparatus can be demonstrated by using the acid-induced demetallation of chlorophyll *a* as a simple test reaction. This reaction is associated with a change of mass (loss of magnesium and gain of two protons) that can be easily monitored by ESI-MS; it is also associated with a color change, so that the ESI-MS results can be verified by traditional optical methods. Chlorophyll *a* (5 μM) was exposed to 14 mM HCl in methanol/water (90:10, v/v) and stopped-flow ESI-MS was used to monitor the intensity of molecular ions of chlorophyll (*m/z* of 894) as a function of time (Fig. 6). The experimental data are well described by an exponential fit with a pseudo-first-order rate constant of 1.9 s^{-1} which is in excellent agreement with the results of optical control experiments (Kolakowski and Konermann, 2001). The current time resolution that can be achieved with the stopped-flow ESI mass spectrometer shown in Fig. 5 is on the order of 100 ms. This time resolution will not be sufficient for studying the presteady-state kinetics of some enzymes. However, it is hoped that a substantial improvement of the time resolution will be possible by implementing a stopped-flow/ESI interface on a microfluidic chip, similar to the arrangement proposed for the continuous-flow approach described above.

Simpson and Northrop (2000) have suggested that stopped-flow ESI-MS could revolutionize enzyme kinetics because it will allow the direct and unambiguous determination of mechanistically important intermediates by their mass, at the same time providing detailed kinetic information. Different approaches can be used for monitoring the reaction mixture in stopped-flow ESI-MS. On slow-scanning mass spectrometers such as quadrupole instruments it will be most straightforward

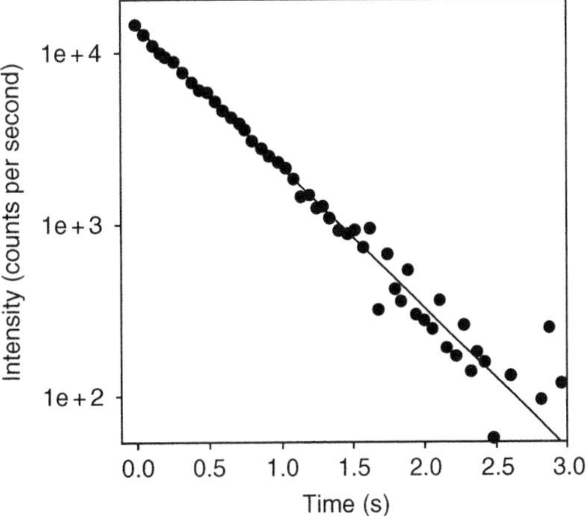

Fig. 6 Demetallation kinetics of chlorophyll *a* in a dilute solution of HCl monitored by stopped-flow ESI-MS. The circles represent the measured count rate of chlorophyll *a* ions (m/z 894) as a function of time. The solid line in this semilogarithmic plot is an exponential fit with a pseudo-first-order rate constant of 1.9 s^{-1}.

to monitor selected ions as a function of time in multiple ion mode. It will be possible to increase the sensitivity and the selectivity of the experiment by using tandem MS and multiple reaction monitoring (i.e., monitoring specific fragment ions of selected precursor species, instead of monitoring the precursor ions directly). By using mass spectrometers that allow a more rapid data acquisition, such as quadrupole time-of-flight instruments or quadrupole ion traps, it might be possible to monitor not only selected ion intensities, but also *entire mass spectra* as a function of reaction time. This could greatly facilitate the identification of transient intermediates and the elucidation of enzymatic reaction mechanisms. Such an approach is technically demanding because it would require the acquisition of high-quality mass spectra on a millisecond time scale. The feasibility of this experimental technique remains to be explored. However, spectral information can always be obtained, even with slow-scanning instruments, by using the continuous-flow methods described in this chapter. For practitioners who use quadrupole mass spectrometers it might be a good experimental strategy to (i) use continuous-flow methods to get an overview of the various species that are formed during the reaction and subsequently (ii) use stopped-flow ESI-MS to obtain detailed and accurate kinetic data by monitoring selected ion intensities as a function of reaction time.

V. Concluding Remarks

The new ESI-MS-based techniques described in this chapter clearly have the potential to become standard tools for studying enzymatic reactions in the pre-steady-state regime. They will also be useful for kinetic studies in other fields of chemistry and biochemistry. The current time resolution of these techniques (tens of milliseconds) is not as good as that of conventional quench- and stopped-flow experiments. However, it is anticipated that future developments such as the use of miniaturized flow-mixing devices will help to overcome this limitation. Certain restrictions will be encountered when choosing solvent systems that are compatible with online monitoring of the reaction mixture by ESI-MS. Salts, detergents, and pH buffers can interfere with the electrospray process, leading to low signal intensities, and the formation of adduct species. The electrospray process is compatible with pH buffers that are based on volatile components such as acetic acid, formic acid, ammonia, or piperidine. It might be feasible to develop microdialysis systems for kinetic experiments that allow the rapid desalting of the reaction mixture immediately before it is analyzed by ESI-MS (Xu *et al.*, 1998). At the current stage of development, experiments necessarily requiring high salt concentrations have to be carried out by other techniques such as chemical quench-flow methods in conjunction with off-line analysis by MALDI MS (Gross and Frey, 2002) or other analytical methods.

Acknowledgments

This work was supported by NSERC, an NSERC-SCIEX Industrial Chair (to D. J. D.), the Canada Foundation for Innovation, the Ontario Ministry of Energy, Science and Technology, the University of British Columbia, and the University of Western Ontario. We thank students, postdoctoral researchers, and colleagues who were involved in the work described and whose names appear in the reference list. The experimental data shown in Fig. 6 were recorded by Beata M. Kolakowski.

References

Anderson, K. S., Sikorski, J. A., and Johnson, K. A. (1988). *Biochemistry* **27,** 7395.

Aplin, R. T., Baldwin, J. E., Schofield, C. J., and Waley, S. G. (1990). *FEBS Lett.* **277,** 212.

Ashton, D. S., Beddell, C. R., Cooper, D. J., Green, B. N., Oliver, R. W. A., and Welham, K. J. (1991). *FEBS Lett.* **292,** 201.

Boerner, R. J., Kassel, D. B., Barker, S. C., Ellis, B., DeLacy, P., and Knight, W. B. (1996). *Biochemistry* **35,** 9519.

Bothner, B., Chavez, R., Wei, J., Strupp, C., Phung, Q., Schneemann, A., and Siuzdak, G. (2000). *J. Biol. Chem.* **275,** 13455.

Brown, R. P. A., Aplin, R. T., and Schofield, C. J. (1996). *Biochemistry* **35,** 12421.

Cheng, X., Chen, R., Bruce, J. E., Schwartz, B. L., Anderson, G. A., Hofstadler, S. A., Gale, D. C., Smith, R. D., Gao, J., Sigal, G. B., Mammen, M., and Whitesides, G. M. (1995). *J. Am. Chem. Soc.* **117,** 8859.

Duncan, C. W., Robertson, H. L., Hubbard, S. J., Gaskell, S. J., and Beynon, R. J. (1999). *J. Biol. Chem.* **274,** 1108.

Fabris, D. (2000). *J. Am. Chem. Soc.* **122,** 8779.

Fenn, J. B., Mann, M., Meng, C. K., Wong, S. F., and Whitehouse, C. M. (1989). *Science* **246,** 64.

Fersht, A. (1999). "Structure and Mechanism in Protein Science," W. H. Freeman & Co., New York.

Fligge, T. A., Kast, J., Bruns, K., and Przybylski, M. (1999). *J. Am. Soc. Mass Spectrom.* **10,** 112.

Ganem, B., Li, Y. T., and Henion, J. D. (1991). *J. Am. Chem. Soc.* **113,** 7818.

Gross, J. W., and Frey, P. A. (2002). *Methods Enzylmol.* **354**(2).

Gutfreund, H. (1969). *Methods Enzymol.* **16,** 229.

Hartridge, H., and Roughton, F. J. W. (1923). *Proc. R. Soc. (Lond.)* **A104,** 376.

Hiromi, K. (1979). "Kinetics of Fast Enzyme Reactions: Theory and Practice," John Wiley & Sons, New York.

Hsieh, F. Y. L., Tong, X., Wachs, T., Ganem, B., and Henion, J. (1995). *Anal. Biochem.* **229,** 20.

Ichiyama, S., Kurihara, T., Li, Y. F., Kogure, Y., Tsunasawa, S., and Esaki, N. (2000). *J. Biol. Chem.* **275,** 40804.

Johnson, K. A. (1992). *In* "The Enzymes" (D. S. Sigman, ed.), **20,** p. 1. Academic Press, New York.

Johnson, K. A. (1995). *Methods Enzymol.* **249,** 38.

Kebarle, P., and Ho, Y. (1997). *In* "Electrospray Ionization Mass Spectrometry" (R. B. Cole, ed.), p. 3. John Wiley & Sons, Inc., New York.

Knight, W. B., Swiderek, K. M., Sakuma, T., Calaycay, J., Shivley, J. E., Lee, T. D., Covey, T. R., Shushan, B., Green, B. G., Chabin, R., Shah, S., Mumford, R., *et al.* (1993). *Biochemistry* **32,** 2031.

Knight, J. B., Vishwanath, A., Brody, J. P., and Austin, R. H. (1998). *Phys. Rev. Lett.* **80,** 3863.

Kolakowski, B. M., and Konermann, L. (2001). *Anal. Biochem.* **292,** 107.

Kolakowski, B. M., Simmons, D. A., and Konermann, L. (2000). *Rapid Commun. Mass Spectrom.* **14,** 772.

Konermann, L. (1999). *J. Phys. Chem. A* **103,** 7210.

Konermann, L., Collings, B. A., and Douglas, D. J. (1997a). *Biochemistry* **36,** 5554.

Konermann, L., Rosell, F. I., Mauk, A. G., and Douglas, D. J. (1997b). *Biochemistry* **36,** 6448.

Lee, V. W. S., Chen, Y. L., and Konermann, L. (1999). *Anal. Chem.* **71,** 4154.

Lee, E. D., Mück, W., Henion, J. D., and Covey, T. R. (1989). *J. Am. Chem. Soc.* **111,** 4600.

Loo, J. A. (1997). *Mass Spectrom. Rev.* **16,** 1.

Lu, W. P., Sun, Y., Bauer, M. D., Paule, S., Koenigs, P. M., and Kraft, W. G. (1999). *Biochemistry* **38,** 6537.

Menard, R., Feng, R., Storer, A. C., Robinson, V. J., Smith, R. A., and Krantz, A. (1991). *FEBS Lett.* **295,** 27.

Northrop, D. B., and Simpson, F. B. (1998). *Arch. Biochem. Biophys.* **352,** 288.

Oleschuk, R. D., and Harrison, D. J. (2000). *Trends Anal. Chem.* **19,** 379.

Ørsnes, H., Graf, T., and Degn, H. (1998). *Anal. Chem.* **70,** 4751.

Paiva, A. A., Tilton, R. F., Crooks, G. P., Huang, L. Q., and Anderson, K. S. (1997). *Biochemistry* **36,** 15472.

Shastry, M. C. R., Luck, S. D., and Roder, H. (1998). *Biophys. J.* **74,** 2714.

Simpson, F. B., and Northrop, D. B. (2000). *In* "Mass Spectrometry in Biology and Medicine" (A. L. Burlingame, S. A. Carr and M. A. Baldwin, eds.), p. 329. Humana Press, Totowa, NJ.

Sogbein, O. O., Simmons, D. A., and Konermann, L. (2000). *J. Am. Soc. Mass Spectrom.* **11,** 312.

Xu, N., Lin, Y., Hofstadler, S. A., Matson, D., Call, C. J., and Smith, R. D. (1998). *Anal. Chem.* **70,** 3553.

Zechel, D. L., Konermann, L., Withers, S. G., and Douglas, D. J. (1998). *Biochemistry* **37,** 7664.

PART IV

Isotopic Probes of Enzyme Processes

CHAPTER 16

Isotope Exchange Methods for Elucidating Enzymic Catalysis

Daniel L. Purich and R. Donald Allison

Department of Biochemistry and Molecular Biology
University of Florida College of Medicine
Gainesville, Florida 32610-0245, USA

I. Introduction & Update

Isotope exchange methods constitute an entire domain of enzyme kinetics that until 45 years ago was largely concerned with net velocity measurements. The availability of isotopes and appropriate sensing devices (e.g., mass spectroscopy and scintillation counting) has encouraged the development of new approaches to understand enzyme catalysis and regulation. Among the early investigators concerned with exchange studies (Boyer *et al.*, 1959; Hass and Byrne, 1960; Rose, 1958), Boyer was probably the first to recognize their power and scope (Boyer, 1959; Boyer and Silverstein, 1963), and his recent review outlines many of the major findings (Boyer, 1978). At one time, equilibrium exchange studies were employed almost exclusively as an adjunct to initial-rate studies of the ordering of substrate

DOI: 10.1016/B978-0-12-378608-1.00016-5

binding and product release (Fromm *et al.*, 1964; Morrison and Cleland, 1966; Silverstein and Boyer, 1964a,b). Now, the approach has been considerably extended to involve rate measurement of loss, or exchange, of essentially all possible substrate atoms or functional groups of atoms, the determination of kinetic isotope effects, the definition of hitherto hidden stereochemical processes, and the examination of the interference of regulatory activators, inhibitors, and interconverting enzymes with the detailed chemical steps in the catalytic process. It is also interesting to note that oxidative phosphorylation and photophosphorylation, while outside the scope of this chapter, have also yielded important information through exchange studies.

To thoroughly examine each of the above aspects of exchange studies would require excessive space, and the reader is referred to several additional sources (Boyer, 1978; Cleland, 1970; Fromm, 1975). Generally, we will deal with representative applications of isotopic exchange at, or near, equilibrium. Here, one may find information on the binding and release of substrates, the rapidity of certain exchanges relative to each other and to the rate-limiting or rate-determining steps, the occurrence of abortive complexes, the likelihood of substrate synergism and possibly cooperativity, and the validity of the rapid-equilibrium assumption. We shall also examine exchange processes away from equilibrium, a condition of obvious importance when one studies irreversible or essentially irreversible processes. Again, valuable information frequently obtained in such cases includes the order of substrate binding and product release, the participation of covalent enzyme-substrate intermediates, and the nature of the irreversible step.

Although written nearly three decades ago, this chapter continues to define the underlying logic and best practice approaches for designing and implementing reliable isotope exchange-rate measurements.

Enzyme chemists continue to rely heavily on isotope exchange measurements to investigate enzyme catalysis and control. The related isotope flux technique has also gained in popularity, chiefly through the extensive work of John Albery and Jeremy Knowles on exchange reactions and isotope effects, as applied to the triose-phosphate isomerase and proline racemase reactions. An implicit advantage of the flux technique is that the experiments yield intrinsic rate parameters (i.e., ratios of rate constants) that tend to be less sensitive than traditional isotope exchange measurements to day-to-day variations in enzyme activity. A second advantage is that flux measurements can be conveniently carried at or away from thermodynamic equilibrium.

Finally, those interested in applying isotope exchange measurements in enzyme kinetic investigations are strongly encouraged to master the methods for deriving rate equations, as described in Chapter 1.

II. Systems at Equilibrium

Even at equilibrium enzymes relentlessly shuttle substrates and products forth and back, and the flux in each direction is essentially constant and balanced over the time period used in measurements. Inasmuch as the rate of enzymic catalysis depends upon the concentration of enzyme-substrate(s) and enzyme-product(s) complexes, the enzyme's behavior at equilibrium is a complex composite of the net reaction rates in the forward and reverse directions observed away from equilibrium. It is also true that the equilibrium flux in each direction depends upon enzyme concentration, and one may adjust the enzyme level to suit the limitations on the exchange experiment.

As noted earlier, introduction of a labeled substrate or product may be utilized to trace the course of the reaction quantitatively. Depending upon the position of the isotopic atom(s) in the labeled substrate or product, various exchanges may be examined. The types of exchanges subject to measurement can be illustrated by considering a bisubstrate reaction in the following form:

$$A - X + B - Y \rightleftharpoons A - Y + B - X \quad (1)$$

In all, one might expect that there are a number of exchange reactions to be examined. However, not all hypothetical exchanges actually involve the transfer of atoms (or functional groupings of atoms) between the various substrate-product partners; thus, only three exchanges occur. For example, in the hexokinase reaction it is possible to observe exchange reactions between glucose and glucose-6-phosphate, ADP and ATP, ATP and glucose-6-phosphate, but not glucose and ADP. Likewise, NAD-dependent dehydrogenases will never undergo exchange between the oxidized substrate and the oxidized coenzyme. Nonetheless, more complicated enzyme systems may have a number of exchange processes associated with the catalytic process, and the glutamine synthetase reaction provides an excellent example. Here, the following exchanges are possible (Stokes and Boyer, 1976; Wedler and Boyer, 1972).

$[^{14}C]$ATP $\leftrightarrows$ ADP
Glutamate $\leftrightarrows$ glutamine
ATP $\leftrightarrows$ P_i
NH_3 $\leftrightarrows$ glutamine
ATP $\leftrightarrows$ P_i
Glutamate $\leftrightarrows$ P_i
Glutamine $\leftrightarrows$ P_i
ATP $\leftrightarrows$ glutamate
ATP $\leftrightarrows$ glutamine
Glutamate $\leftrightarrows$ glutamine

The oxygen fluxes can provide additional probes of the dynamics of the reaction including stereochemical behavior of the enzyme. A detailed description of heavy-oxygen techniques is presented in Chapter 3, volume 64 of *Methods in Enzymology*.

To maintain equilibrium, attention must be given to the mass-action ratio, which is the product of each reaction product concentration divided by the product of each substrate.

$$K_{eq} = \frac{[P][Q]\ldots}{[A][B]\ldots} \tag{2}$$

Typically, the system is treated in terms of substrate-product pairs (e.g., $\alpha = [P]/[A]$ and $\beta = [Q]/[B]$). One substrate-product pair is held constant with respect to the absolute concentrations of each component; the other substrate-product pair may be adjusted to a variety of absolute levels. Nonetheless, the experimental conditions must allow the product of α and β to be K_{eq}. The nonvaried substrate-product pair may be prepared with buffer, salts, and modifiers (if desired), and enzyme is added to the complete reaction system to equilibrate the system. Since the experimenter has a good estimate of K_{eq}, the composition of the system will not change much from the preset conditions. A small aliquot of labeled reactant (substrate or product) is added, and the progress is followed by periodic sampling. Although the progress curve for the exchange is always first order, the magnitude of the rate constant depends upon the absolute concentrations of all reactants, level of enzyme, and the kinetic mechanism.

The most common application of such exchange experiments is the determination of the kinetic mechanism with respect to binding preference or order. Let us examine the Ordered Bi Bi mechanism with ternary complexes.

$$(E) \underset{k_2}{\overset{k_1[A]}{\rightleftharpoons}} (EA) \underset{k_4}{\overset{k_3[B]}{\rightleftharpoons}} (EAB) \underset{k_6}{\overset{k_5}{\rightleftharpoons}} (EPQ) \underset{k_8[P]}{\overset{k_7}{\rightleftharpoons}} (EQ) \underset{k_{10}[Q]}{\overset{k_9}{\rightleftharpoons}} (E) \tag{3}$$

The $A \leftrightharpoons Q$ exchange velocity as a function of the absolute levels of the [B]/[P] pair is the diagnostic test for compulsory ordered mechanisms wherein one substrate, A, must precede the other's binding and one product, P, must precede the other product's release. As the [B]/[P] pair is raised, the central complexes become the favored enzyme species, and the uncomplexed enzyme form becomes scarce. Exchange between A and Q requires that there be a form of the enzyme which may adsorb the labeled species. Since the uncomplexed enzyme form is required for A* to be converted to Q*, and vice versa, the $A \leftrightharpoons Q$ exchange will increase as we go from low to moderate levels of the [B]/[P] pair, but the $A \leftrightharpoons Q$ exchange will be depressed at high [B]/[P] absolute levels. The $A \leftrightharpoons Q$ exchange will not be depressed at high [A]/[Q] absolute levels since increasing A and Q concentrations favor the binding of A* or Q*. Interestingly, the $B \leftrightharpoons P$ exchange requires that (EA) or (EQ) be present for combination with B* or P*, respectively. Thus, raising the [A]/[Q] or [B]/[P] pair will result in increased $B \leftrightharpoons P$ exchange in a hyperbolic fashion consistent with rate saturation kinetics. In summary, the depression of the $A \leftrightharpoons Q$ exchange at high [B]/[P] serves to identify the ordered mechanism. However, in the random (i.e., noncompulsory) mechanism (Eq. (4)) we have alternative routes for exchange processes, and no exchanges will be inhibited at high levels of any substrate-product pair. This is true since A* may bind to (E) and (EB), and raising the [B]/[P] pair will not inhibit the combination of A* to the (EB) species. Because the mechanism is completely symmetrical, a similar statement is true for all exchanges.

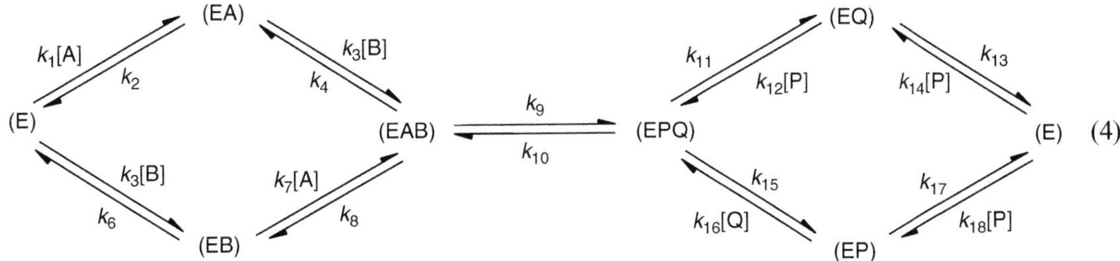

These qualitative conclusions will be justified later in terms of exchange-rate expressions, but it is significant to note that exchange kinetic studies may segregate substrate order into ordered and random processes. Likewise, it is possible to draw other conclusions based upon the relative rapidity of exchanges. For example, if ternary complex interconversion limits the rate of a random mechanism, then the maximal exchange rate of each exchange will be identical. One such case is provided by rabbit muscle creatine kinase at pH 8.0 (Morrison and Cleland, 1966). For yeast hexokinase, however, the nucleotide exchange is approximately twice the rate of the glucose $\leftrightarrows$ glucose-6-P exchange (Fromm *et al.*, 1964; Purich and Fromm, 1972a), and one must consider release of glucose or glucose-6-P as rate contributing. In the case of ovine brain or *Escherichia coli* glutamine synthetase (Allison *et al.*, 1977; Wedler, 1974a; Wedler and Boyer, 1972)

$$R_{\text{max,Glu}\leftrightarrows\text{Gln}} > R_{\text{max,NH}_3\leftrightarrows\text{Gln}} > R_{\text{max,P}_i\leftrightarrows\text{ATP}} \simeq R_{\text{max,ADP}\leftrightarrows\text{ATP}} \tag{5}$$

suggesting that nucleotide release contributes more to the overall rate than either interconversion or the association and dissociation of other substrates and products. On the other hand, with pea seed glutamine synthetase the $NH_3 \leftrightarrows$ glutamine exchange is slowest (Wedler, 1974b).

In summary, the power of exchange studies is quite considerable, as illustrated in Table I. Aside from the large body of information about the kinetic mechanisms of many enzymes, the references included in Table I contain a wealth of data on separating and measuring particular metabolites.

A. Rate Equations for Isotopic Exchange

There are a number of useful approaches to deriving equilibrium exchange-rate expressions. Boyer's method (Boyer, 1959; Boyer and Silverstein, 1963) is reasonably straightforward, and Fromm (1975) has a good step-by-step outline to obtain such rate laws. Morales *et al.* (1962) criticized this method and presented a more formal and generalized isotope distribution equation; however, Darvey, (1973) showed that Boyer's approach is sound. Numerous alternative routes for obtaining rate equations have been developed (Boyer, 1959; Boyer and Silverstein, 1963;

Table I
Summary of Enzyme Mechanisms Deduced by Isotope Exchange Data

Enzyme class	Specific reaction	Proposed kinetic mechanism	Diagnostic exchange(s)	References
Dehydrogenases	Lactate dehydrogenase (bovine heart and rabbit muscle)	Ordered (pH 7.9), partially ordered (pH 9.7)	$NAD^+ \leftrightharpoons$ NADH, pyruvate $\leftrightharpoons$ lactate, NADH $\leftrightharpoons$ lactate	Silverstein and Boyer (1964a), Yagil and Hoberman (1969)
	Alcohol dehydrogenase (yeast and equine liver)	Partially ordered	$NAD^+ \leftrightharpoons$ NADH, acetaldehyde $\leftrightharpoons$ ethanol	Silverstein and Boyer (1964b), Ainslie and Cleland (1972)
	Shikimate dehydrogenase (pea seedlings)	Ordered	$NADP^+ \leftrightharpoons$ NADPH, shikimate $\leftrightharpoons$ dehydroshikimate	Balinsky *et al.* (1971)
	Malate dehydrogenase (porcine heart)	Ordered (pH 8), partially ordered (pH 9)	$NAD^+ \leftrightharpoons$ NADH, oxaloacetate $\leftrightharpoons$ malate	Silverstein and Sulebele (1969)
	Glutamate dehydrogenase (bovine liver)	Partially ordered (pH 8)	$NAD(P)^+ \leftrightharpoons$ NAD(P)H, glutamate $\leftrightharpoons$ α-ketoglutarate	Silverstein and Sulebele (1973)
	Glutamate dehydrogenase, alanine deamination (bovine liver)	Random (pH 8)	$NAD^+ \leftrightharpoons$ NADH, alanine $\leftrightharpoons$ pyruvate	Silverstein and Sulebele (1974)
	$NADP^+$-isocitrate dehydrogenase (porcine heart)	Random	$NADP^+ \leftrightharpoons$ NADPH, $CO_2 \leftrightharpoons$ isocitrate	Uhr *et al.* (1974)
	20β-Hydroxysteroid dehydrogenase (*Streptomyces hydrogenans*)	Partially ordered	$NAD^+ \leftrightharpoons$ NADH	Betz and Taylor (1970)
	17β-Estradiol dehydrogenase (human placenta)	Random	$NAD(P)^+ \leftrightharpoons$ NAD(P)H, steroid ketone $\leftrightharpoons$ steroid alcohol	Betz (1971)
Kinases	Hexokinase (yeast)	Random	ATP $\leftrightharpoons$ ADP, glucose $\leftrightharpoons$ glucose-6-P	Fromm *et al.* (1964), Purich and Fromm (1972a)
	Nucleoside diphosphate kinase (yeast)	Ping-Pong	ATP $\leftrightharpoons$ ADP and UTP $\leftrightharpoons$ UDP	Garces and Cleland (1969)
	Adenylate kinase (myokinase) (yeast and rabbit muscle)	Random	ADP $\leftrightharpoons$ ATP and ADP $\leftrightharpoons$ AMP	Su and Russell (1968), Rhoads and Lowenstein (1968)
	Arginine kinase (sea crayfish)	Random (minor Ping-Pong pathway observed)	Arginine $\leftrightharpoons$ phosphoarginine and ATP $\leftrightharpoons$ ADP	Smith and Morrison (1969)
	Galactokinase (*Escherichia coli*)	Random	Galactose $\leftrightharpoons$ galactose-1-P and MgATP $\leftrightharpoons$ MgADP	Gulbinsky and Cleland (1968)
	Creatine kinase (rabbit muscle)	Random (rapid equilibrium) (pH 8)	Creatine $\leftrightharpoons$ creatine phosphate and ATP $\leftrightharpoons$ ADP	Morrison and Cleland (1966)

Table I (*continued*)

Enzyme class	Specific reaction	Proposed kinetic mechanism	Diagnostic exchange(s)	References
	Acetate kinase (*E. coli*)	Random ("activated Ping-Pong")	Acetate ⇋ acetyl phosphate and ATP ⇋ ADP	Skarstedt and Silverstein (1976)
	Aspartate kinase, lysine sensitive (*E. coli*)	Random on, ordered off	ADP ⇋ ATP and ATP ⇋ β-aspartyl phosphate	Shaw and Smith (1977)
	Fructokinase (bovine liver)	Partially ordered	Fructose ⇋ fructose-1-P and ATP ⇋ ADP	Raushel and Cleland (1977)
	Phosphofructokinase, nonallosteric (*Lactobacillus plantarum*)	Ordered	Fructose-6-P ⇋ fructose-1,6-diP and ATP ⇋ ADP	Simon and Hofer (1978)
Synthetases	Succinyl-CoA synthetase (*E. coli*)	Partially ordered (with other minor pathways)	ATP ⇋ ADP and succinate ⇋ succinyl-CoA	Bridger *et al.* (1968), Moffet and Bridger (1973)
	Asparagine synthetase (*E. coli*)	Ping-Pong (random addition of aspartate and ATP)	PP_i ⇋ ATP	Cedar and Schartz (1969)
	Short-chain fatty acyl-CoA synthetase (pine seeds)	Ping-Pong (ordered addition of ATP and acid)	PP_i ⇋ ATP and PP_i ⇋ dATP	Young and Anderson (1974)
	γ-Glutamylcysteine synthetase (toad and rat liver)	Ping-Pong (toad), sequential (rat)	Cysteine ⇋ γ-glutamylcysteine and glutamate ⇋ γ-glutamylcysteine	Davis *et al.* (1973)
	Arginyl tRNA synthetase (*E. coli*)	Random	PP_i ⇋ ATP, AMP ⇋ ATP, arginine ⇋ arg-tRNA, tRNA ⇋ arg-tRNA	Papas and Peterkofsky (1972)
	Isoleucine tRNA synthetase (*E. coli*)	Ping-Pong (random release)	PP_i ⇋ ATP	Cole and Schimmel (1970a,b)
	Glutamine synthetase (pea, *E. coli*, ovine brain)	Random	P_i ⇋ ATP, glutamate ⇋ glutamine, ATP ⇋ ADP	Boyer *et al.* (1959), Graves and Boyer (1962), Wedler and Boyer (1972), Wedler (1974a), Allison *et al.* (1977)
	Glutamine synthetase (ovine brain)	Partially ordered	ATP ⇋ ADP, ATP ⇋ P_i, glutamate ⇋ glutamine	Wedler (1974a)
Phosphorylases	Sucrose phosphorylase (*Pseudomonas saccharophila*)	Ping-Pong	Glucose-1-P ⇋ P_i	Doudoroff *et al.* (1947)
	Maltodextrin phosphorylase (*E. coli*)	Random	Glucose-1-P ⇋ P_i and glucose-1-P ⇋ maltoheptaose	Chao *et al.* (1969)

(*continues*)

Table I (*continued*)

Enzyme class	Specific reaction	Proposed kinetic mechanism	Diagnostic exchange(s)	References
	Glycogen phosphorylase *a* (rabbit muscle)	Random (rapid equilibrium)	Glucose-1-P ⇋ P_i and glucose-1-P ⇋ glycogen	Gold *et al.* (1970), Engers *et al.* (1970)
	Glycogen phosphorylase *b* (rabbit muscle)	Random (rapid equilibrium)	Glucose-1-P ⇋ P_i and glucose-1-P ⇋ glycogen	Engers *et al.* (1969)
Hydrolases	Glucose-6-phosphatase (rat liver)	Ordered off	Glucose ⇋ glucose-6-P	Hass and Byrne (1960)
	Phosphoserine phosphatase (rat and chicken liver)	Ordered off	Serine ⇋ phosphoserine	Neuhaus and Byrne (1959), Borkenhagen and Kennedy (1959)
	Pepsin	Ordered off (pH 4.7), Random (rapid equilibrium, pH 3.3)	*N*-Acetyl-l-phenylalanine-l-tyrosine ethyl ester ⇋ l-tyrosine ethyl ester	Ginodman and Lutsenko (1972)
Others	Aspartate transcarbamylase (*E. coli*)	Ordered	Aspartate ⇋ carbamyl aspartate, carbamyl phosphate ⇋ P_i	Wedler and Gasser (1974)
	Citrate cleavage enzyme (rat liver)	Mixed	Citrate ⇋ acetyl-CoA, citrate ⇋ oxaloacetate, P_i ⇋ ATP	Farrar and Plowman (1971)
	Phosphoglucomutase (rabbit muscle)	Random (Mg vs. sugar-phosphate)	Glucose-1-P ⇋ glucose-6-P	Ray *et al.* (1966)
	Galactose-1-P uridylyltransferase (*E. coli* and human erythrocyte)	Ping-Pong	UDP-glucose ⇋ glucose-1-P and galactose-1-P ⇋ UDP-galactose	Wong and Frey (1974), Marcus *et al.* (1977)
	Aldolase (yeast and rabbit muscle)	Ordered	Dihydroxyacetone-phosphate ⇋ water	Rose *et al.* (1965)
	CoA Transferase (porcine heart)	Ping-Pong	Succinate ⇋ succinyl-CoA and acetoacetate ⇋ acetaceyl-CoA	Hersh and Jencks (1967)

Britton, 1966; Cleland, 1967, 1975; Yagil and Hoberman, 1969) and there are two basic types of equations, depending upon whether the system is at, or significantly displaced from, equilibrium.

The derivation of exchange-rate laws is based upon the following assumptions:

1. The total enzyme concentration is much less than the concentration of any of its substrates or modifiers.

2. The concentration of the isotopically labeled compound is much less than that of the unlabeled compound. Thus, there is no significant change in the concentration of the unlabeled species.

3. All unlabeled species are present at their equilibrium concentrations prior to the addition of the labeled species (the enzyme system being reversible).

4. The pathway for the exchange of the labeled species is the same as that of the corresponding unlabeled species and this exchange is via the reaction path being studied. If there is a second enzyme present that catalyzes the same exchange, the value for R, the rate of exchange, will be a superimposition of two or more exchanges.

5. Kinetic isotope effects, if any, are negligible.

6. The labeled enzyme intermediates are at steady-state levels (the following results are not valid for the presteady state) (Darvey, 1973).

For a detailed description about the derivation of isotope rate expressions, the reader is referred to Chapter 1 in this volume. To conserve space in the presentation of rate laws, we have adopted the format used by Fromm (1975) and those who are interested in rewriting the equations in terms of individual rate constants should consult the appendix to his text. In Table II, we list the exchange-rate expressions for many typical enzyme kinetic mechanisms. These equations are relatively long and complex by comparison to initial-rate expressions, and it is generally cumbersome, if not completely impractical, to obtain kinetic constants for each step from isotope exchange data. Instead, one deals with the equations to obtain qualitative patterns of rate behavior, and the most frequently used method deals with the effect of raising particular substrate-product pairs. By far the most reliable way to do this uses L'Hospital's rule to determine the limits (Allison *et al.*, 1977). Treatment of the ordered mechanism provides a useful example. Let [Q]/[A] be α and [P]/[B] be β (or K_{eq}/α). From Table II, the rate of the A $\leftrightharpoons$ Q exchange becomes

$$R_{A\leftrightharpoons Q} = \frac{V_1[A][B]}{\left\{K_{ia}K_b + K_a[B] + \frac{K_{eq}K_{ia}K_bK_q[B]}{\alpha K_{iq}K_p}\left[1 + \frac{[A]}{K_{ia}}\left(1 + \frac{[B]}{K_{eb}}\right) + \frac{\alpha[A]}{K_{iq}}\right]\right\}} \tag{6}$$

Likewise,

$$R_{B\leftrightharpoons P} = \frac{V_1[B]}{K_b\left[1 + \frac{K_{ia}}{[A]} + \frac{[B]}{K_{eb}} + \frac{K_{ia}\alpha}{K_{iq}}\right]} \tag{7}$$

Now, we may establish the limiting value of $R_{A\leftrightharpoons Q}$ as the absolute concentration of the B-P pair becomes very large, that is, [B] $\to \infty$ in Eq. (6). In this case, R can be treated as a quotient of two functions: $f(B) = k[B]$ and $g(B) = k' + k''[B] + k'''[B]^2$; thus

$$\underset{\text{limiting}B-P}{R_{A\leftrightharpoons Q}} = \lim_{x\to\infty}\frac{f(B)}{g(B)} = \lim_{x\to\infty}\frac{f'(B)}{g'(B)} = \lim_{x\to\infty}\frac{k}{k'' + 2k'''[B]} = 0 \tag{8}$$

Table II
Isotope Exchange-Rate Expressions for Selected UNI and BI Substrate Systems[a]

1. Uni Uni

$$R_{A\rightleftharpoons P} = \frac{V_1[A]}{K_a[1 + [A]/K_{ia} + [P]/K_{ip}]}$$

2. Ordered Bi Uni

$$R_{A\rightleftharpoons P} = \frac{V_1[A][B]}{(K_{ia}K_b + K_a[B])[1 + [A]/K_{ia} + [P]/K_{ip}]}$$

$$R_{B\rightleftharpoons P} = \frac{V_1[B]}{K_b[1 + K_{ia}/[A] + K_{ia}[P]/K_{ip}[A]]}$$

3. Random Bi Uni (rapid equilibrium)

$$R_{all} = \frac{V_1}{1 + K_a/[A] + K_b/[B] + (K_{ia}K_b/[A][B])(1 + [P]/K_{ip})}$$

4. Random Bi Uni

$$R_{A\rightleftharpoons P} = \frac{[k_1k_3k_9[A][B] + k_7k_9[A](k_2 + k_3[B])[B]/K_{ib}]E_0}{[k_2(k_4 + k_8 + k_9) + k_3[B](k_8 + k_9)][1 + [A]/K_{ia} + [B]/K_{ib} + [A][B]/K_{ia}K_b]}$$

$$R_{B\rightleftharpoons P} = \frac{[k_5k_7k_9[A][B] + k_3k_9[B](k_6 + k_7[A])([A]/K_{ia})]E_0}{[k_6(k_4 + k_8 + k_9) + k_7[A](k_4 + k_9)][1 + [A]/K_{ia} + [B]/K_{ib} + [A][B]/K_{ia}K_b]}$$

5. Ordered Bi Bi*afasdasdf*

$$R_{A\rightleftharpoons Q} = \frac{V_1[A][B]}{[K_{ia}K_b + K_a[B] + K_{ia}K_bK_q[P]/K_{iq}K_p][1 + ([A]/K_{ia})(1 + [B]/K_{eb}) + [Q]/K_{iq}]}$$

$$R_{A\rightleftharpoons P} = \frac{V_1[A][B]}{[K_{ia}K_b + K_a[B]][1 + ([A]/K_{ia})(1 + [B]/K_{eb}) + [Q]/K_{iq}]}$$

$$R_{B\rightleftharpoons Q} = \frac{V_1[B]}{[K_b + K_bK_q[P]/K_{iq}K_p][1 + K_{ia}/[A] + [B]/K_{eb} + K_{ia}[Q]/K_{iq}[A]]}$$

$$R_{B\rightleftharpoons P} = \frac{V_1[B]}{K_b[1 + K_{ia}/[A] + [B]/K_{eb} + K_{ia}[Q]/K_{iq}[A]]}$$

6. Theorell-Chance Bi Bi

$$R_{A\rightleftharpoons Q} = \frac{V_1[A][B]}{[K_a[B] + K_{ia}K_b + K_{iq}[P]/K_{eq}][1 + [A]/K_{ia} + [Q]/K_{iq}]}$$

$$R_{A\rightleftharpoons P} = \frac{V_1[A][B]}{[K_{ia}K_b + K_a[B]][1 + [A]/K_{ia} + [Q]/K_{iq}]}$$

$$R_{B\rightleftharpoons Q} = \frac{V_1[B]}{[K_b + K_{iq}[P]/K_{ia}K_{eq}][1 + K_{ia}/[A] + K_{ia}[Q]/K_{iq}[A]]}$$

$$R_{B\rightleftharpoons P} = \frac{V_1[B]}{K_b[1 + K_{ia}/[A] + K_{ia}[Q]/K_{iq}[A]]}$$

7. Random Bi Bi (rapid equilibrium)

$$R_{all} = \frac{V_1}{1 + K_a/[A] + K_b/[B] + K_{ia}K_b/[A][B][1 + [P]/K_{ip} + [Q]/K_{iq} + [P][Q]/K_{ip}K_q]}$$

Table II (*continued*)

8. Random Bi Bi

$$R_{A\leftrightharpoons Q} = \frac{[k_9k_{11} + k_{13}(k_{10}[P] + k_{11})][k_1k_3[A][B] + k_7([A][B]/K_{ib})(k_2 + k_3[B])]E_0}{(Z)\{k_2[k_{10}[P](k_4 + k_8 + k_{13}) + k_{11}(k_4 + k_8 + k_9 + k_{13})] + k_3[B][k_{10}[P](k_8 + k_{13}) + k_{11}(k_8 + k_9 + k_{13})]\}}$$

$$R_{A\leftrightharpoons P} = \frac{[k_{13}k_{15} + k_9(k_{14}[Q] + k_{15})][k_1k_3[A][B] + k_7([A][B]/K_{ib})(k_2 + k_3[B])]E_0}{(Z)\{k_2[k_{14}[Q](k_4 + k_8 + k_9) + k_{15}(k_4 + k_8 + k_9 + k_{13})] + k_3[B][k_{14}[Q](k_8 + k_9) + k_{15}(k_8 + k_9 + k_{13})]\}}$$

$$R_{B\leftrightharpoons Q} = \frac{[k_9k_{11} + k_{13}(k_{10}[P] + k_{11})][k_5k_7[A][B] + k_3([A][B]/K_{ia})(k_6 + k_7[A])]E_0}{(Z)\{k_6[k_{10}[P](k_4 + k_8 + k_{13}) + k_{11}(k_4 + k_8 + k_9 + k_{13})] + k_7[A][k_{10}[P](k_4 + k_{13}) + k_{11}(k_4 + k_9 + k_{13})]\}}$$

$$R_{B\leftrightharpoons P} = \frac{[k_{13}k_{15} + k_9(k_{14}[Q] + k_{15})][k_5k_7[A][B] + k_3([A][B]/K_{ia})(k_6 + k_7[A])]E_0}{(Z)\{k_6[k_{14}[Q](k_4 + k_8 + k_9) + k_{15}(k_4 + k_8 + k_9 + k_{13})] + k_7[A][k_{14}[Q](k_4 + k_9) + k_{15}(k_4 + k_9 + k_{13})]\}}$$

where $Z = \{1 + [A]/K_{ia}(1 + [B]/K_b) + [B]/K_{ib} + [P]/K_{ip} + [Q]/K_{iq}\}$

9. Random on-Ordered off Bi Bi

$$R_{A\leftrightharpoons Q} = \frac{k_9k_{11}[k_1k_3[A][B] + k_7([A][B]/K_{ib})(k_2 + k_3[B])]E_0}{\{k_2[k_{10}[P](k_4 + k_8) + k_{11}(k_4 + k_8 + k_9)] + k_3[B][k_8k_{10}[P] + k_{11}(k_8 + k_9)]\}(Y)}$$

$$R_{A\leftrightharpoons P} = \frac{k_9[k_1k_3[A][B] + k_7([A][B]/K_{ib})(k_2 + k_3[B])]E_0}{\{k_3[B](k_8 + k_9) + k_2(k_4 + k_8 + k_9)\}(Y)}$$

$$R_{B\leftrightharpoons Q} = \frac{k_9k_{11}[k_5k_7[A][B] + k_3([A][B]/K_{ia})(k_6 + k_7[A])]E_0}{\{k_6[k_{10}[P](k_4 + k_8) + k_{11}(k_4 + k_8 + k_9)] + k_7[A][k_4k_{10}[P] + k_{11}(k_4 + k_9)]\}(Y)}$$

$$R_{B\leftrightharpoons P} = \frac{k_9[k_5k_7[A][B] + k_3([A][B]/K_{ia})(k_6 + k_7[A])]E_0}{\{k_7[A](k_4 + k_9) + k_6(k_4 + k_8 + k_9)\}(Y)}$$

where $Y = \{1 + ([A]/K_{ia})(1 + [B]/K_b) + [B]/K_{ib} + [Q]/K_{iq}\}$

10. Ping-Pong Bi Bi

$$R_{A\leftrightharpoons P} = \frac{V_1[A]}{K_a[1 + [A]/K_{ia} + K_{ip}[A]/K_{ia}[P] + [Q]/K_{iq}]}$$

$$R_{B\leftrightharpoons Q} = \frac{V_1[B]}{K_b[1 + [B]/K_{ib} + [P]/K_{ip} + K_{iq}[B]/K_{ib}[Q]]}$$

$$R_{A\leftrightharpoons Q} = \frac{V_1[A][B]}{K_a[B] + (K_{ia}K_b[P]/K_{ip}][1 + [A]/K_{ia} + [Q]/K_{iq} + K_{ip}[A]/K_{ia}[P]}$$

[a]The expressions are derived for systems at chemical equilibrium and in the absence of abortive complexes. All expressions, except for the Uni Uni and rapid-equilibrium cases, were derived assuming only one central catalytic complex.

On the other hand, $R_{A\leftrightharpoons Q}$ versus the A-Q pair will have its limit given by

$$\underset{\text{maximum}}{R_{A\leftrightharpoons Q}} = \frac{V_1[B]}{\frac{K_{eq}K_{ia}K_bK_q[B]}{\alpha K_{iq}K_P}\left[\frac{1}{K_{ia}}\left(1 + \frac{[B]}{K_{eb}}\right) + \frac{\alpha}{K_{iq}}\right]} \tag{9}$$

Thus, $R_{A\leftrightharpoons Q}$ falls to zero as the B-P pair is increased, but $R_{A\leftrightharpoons Q}$ achieves a finite plateau value as the A-Q pair is increased.

This approach is quite useful when a single substrate-product pair is increased; however, Wedler and Boyer (1972) introduced a protocol in which all substrates and products are increased simultaneously. Their qualitative conclusions about the exchange behavior under such conditions may be appreciated mathematically by the following treatment (Allison *et al.*, 1977). Consider again Eq. (6) for the $R_{A \leftrightharpoons Q}$ exchange in the Ordered Bi Bi mechanism. Let [B] = γ[A], then,

$$R_{A \leftrightharpoons Q} = \frac{\gamma V_1 [A]^2}{\left\{ K_{ia} K_b + K_a \gamma [A] + \frac{K_{eq} K_{ia} K_b K_q \gamma [A]}{\alpha K_{iq} K_p} \left[1 + \frac{[A]}{K_{ia}} \left(1 + \frac{\gamma [A]}{K_{eb}} \right) + \frac{\alpha [A]}{K_{iq}} \right] \right\}} \tag{10}$$

Now, we may again express $R_{A \leftrightharpoons Q}$ as the quotient of two functions: $f(A) = k[A]^2$ and $g(A) = k' + k''[A] + k'''[A]^2 + k''''[A]^3$. Applying L'Hospital's rule once again, we find that $R_{A \leftrightharpoons Q}$ reaches a limiting value of zero as [A], [B], [P], and [Q] are increased indefinitely. Thus, mechanisms with noncompetitive interactions will lead to depression in exchanges, but competitive interactions (even the formation of abortive complexes) will be balanced as [A], [B], [P], and [Q] are increased. The only major limitation with the Wedler-Boyer protocol is that it cannot be applied to reactions involving a different number of substrates and products. For example, a Uni Bi reaction has an equilibrium constant of [P][Q]/[A] and serial dilution will move the system away from equilibrium.

Examination of the equations presented in Table II indicates that there are a number of types of exchange-rate profiles. In terms of plots of R versus a substrate-product pair, we may observe hyperbolic (H), or sigmoidal (S) plots with no, partial, or complete depression of the exchange. These profiles are illustrated in Fig. 1, and it should be noted that variation of a single substrate-product pair should yield hyperbolic plots. The presence of sigmoidal plots may provide preliminary evidence for cooperativity; however, the Wedler-Boyer (1972) protocol in which all substrates and products are varied, should yield sigmoidal plots irrespective of the occurrence of cooperativity (of course, hyperbolic profiles in the Wedler-Boyer protocol may arise fortuitously, depending upon the magnitude of various kinetic parameters). In any case, the expected exchange-rate profiles for the systems defined in Table II are explicitly described in Table III. These do not take into account abortive complex effects, which are described later.

It may be valuable to note that the rapid-equilibrium ordered and rapid-equilibrium random mechanisms are virtually indistinguishable by isotope exchange data alone. Since the central complex interconversion rate is limiting in both cases, hyperbolic plots of R versus any substrate-product pair are to be expected. With the Wedler-Boyer method, R_{max} will be $V_1 V_2/(V_1 + V_2)$ where V_1 and V_2 are the forward and reverse V_{max} values obtained in initial-rate experiments. Under favorable conditions, a distinction might be possible by use of slope replots of $(1/R)$ versus the A-Q pair at various constant values of the B-P pair. This method would distinguish a rapid-equilibrium ordered mechanism by the intersection at the origin and the random at a minimal nonzero value of K_a/V_1. Of course, initial-rate studies may provide a clear indication of the mechanism if certain abortives do not form. Likewise, the steady-state mechanisms referred to as Ordered Bi Bi and Theorell-Chance Bi Bi may be distinguished by observation

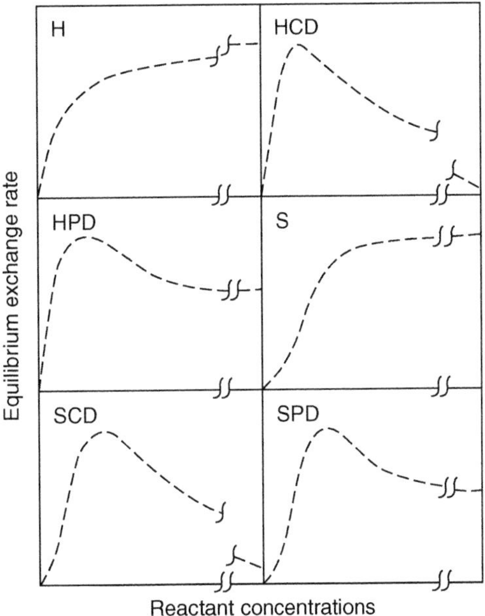

Fig. 1 Types of exchange-rate profiles for enzyme-catalyzed reactions: hyperbolic (H), hyperbolic with complete depression (HCD), hyperbolic with partial depression (HPD), sigmoidal (S), sigmoidal with complete depression (SCD), sigmoidal with partial depression (SPD), and linear (L).

of the linear dependence of the B $\leftrightharpoons$ P exchange versus the absolute concentration of the B-P pair. This is characteristic of the Theorell-Chance pathway because EA + B $\rightarrow$ EQ + P is a simple, bimolecular, process without ternary complex formation.

Finally, one may infer from the complexity of the equations in Table II that quantitative estimates of kinetic parameters, especially Michaelis and dissociation constants, are difficult to obtain from the exchange data. The Ping-Pong mechanisms are notable exceptions, and replots may provide valuable data.

B. Experimental Determination of Equilibrium Exchange Rates

The estimation of exchange rates may be achieved by adding an isotopically labeled substrate (or product) to a fully equilibrated reaction system containing enzyme and by subsequently determining the extent of interconversion to labeled product (or substrate) or periodic sampling. The basic equation for relating the rate of isotopic exchange (R) to the concentrations of two reactants, X and Y, undergoing exchange is given by

$$R = \frac{[X][Y][\ln(1 - F)]}{([X] + [Y])t} \tag{11}$$

Table III
Exchange Profiles for Enzyme Kinetic Mechanisms[a]

Mechanism	Exchange	Varied substrate-product pair(s)				
		A-P	B-P	A-Q	B-Q	A-B-P-Q[b]
1. Uni Uni	A ⇋ P	H				
2. Ordered Bi Uni	A ⇋ B	H	HCD			
	B ⇋ P	H	H			
3. Random Bi Uni (rapid equilibrium)	Any	H	H			
4. Random Bi Uni	A ⇋ P	H	H			
	B ⇋ P	H	H			
5. Ordered Bi Bi	A ⇋ P	H	HCD	H	HCD	SCD
	B ⇋ P	H	H	H	H	S
	A ⇋ Q	HCD	HCD	H	HCD	SCD
	B ⇋ Q	HCD	HCD	H	H	SCD
6. Theorell-Chance Bi Bi	A ⇋ P	H	H	H	HCD	S
	B ⇋ P	H	L	H	H	SL
	A ⇋ Q	HCD	H	H	HCD	S
	B ⇋ Q	HCD	H	H	H	S
7. Random Bi Bi (rapid equilibrium)	Any	H	H	H	H	S
8. Random Bi Bi	Any	H	H	H	H	S
9. Random on-Ordered off	A ⇋ P	H	H	H	H	S
	B ⇋ P	H	H	H	H	S
	A ⇋ Q	HCD	HCD	H	H	SCD
	B ⇋ Q	HCD	HCD	H	H	SCD

[a]See Table II for a parallel listing of rate equations. See Fig. 1 for types of exchange profiles.
[b]A-B-P-Q refers to the Wedler-Boyer protocol of varying all substrates and products in a constant ratio.

where "t" is the period elapsed between initiating and sampling the exchange process, and "F" is the fractional attainment of isotopic equilibrium (Frost and Pearson, 1961). Labeled species may be designated X* and Y*, and the label will be so distributed at isotopic equilibrium.

$$\frac{[X^*]}{[Y^*]} = \frac{[X]_e}{[Y]_e} \tag{12}$$

Using the subscripts "t" and "e" to indicate [X] and [Y] at their time-dependent and equilibrium values, respectively, "F" may be expressed as the dimensionless number as follows

$$F = \frac{[X^*]_t}{[X^*]_e} = \frac{[Y^*]_t}{[Y^*]_e} \tag{13}$$

In practice, $[X^*]_t$ and $[X^*]_e$ may be expressed in any consistent units that are proportional to their concentration or abundance (Norris, 1950). For radioactive

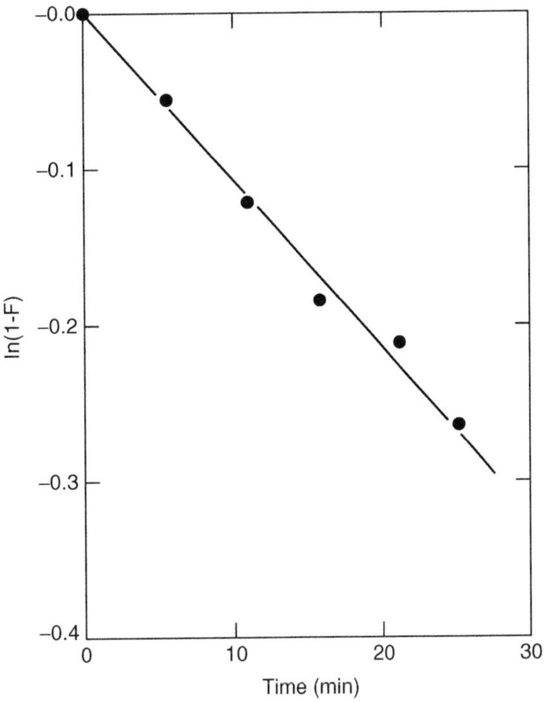

Fig. 2 Plot of $\ln(1-F)$, extent of isotopic equilibrium versus time of the [4A-^{3}H]NADH lactate exchange of bovine heart lactate dehydrogenase. The conditions of the assay were as follows: pH 8.03; $I = 0.2$ M Tris-KCl; $T = 28.5$ °C.

determinations, either total activity or specific radioactivity may be used; and for stable isotopes, mole fraction or gram atom excess may be employed.

The expression given above for R indicates that the exchange process is first order, and it will remain so provided tracer quantities are employed. It is also possible to roughly measure R by single-point determinations, but it is preferable to use several sampling times and to confirm the first-order rate law as shown in Fig. 2. In this case, the labeled NADH was added to a fully equilibrated lactate dehydrogenase reaction system (Yagil and Hoberman, 1969), and F was measured at various intervals. It will be noted that the concentrations of cosubstrate (pyruvate) and coproduct (NAD^+) do not enter into Eq. (11), but this does not mean that R is independent of the concentration of cosubstrate and coproduct. Indeed, R depends upon the fraction of enzyme capable of reacting with the labeled substrate (or product), and cursory examination of the enzyme exchange-rate expressions presented earlier in Table II indicates that all substrates and products affect R. This is true because the concentrations of substrate(s) and product(s) control the concentration of all Michaelis complexes as well as any "mixed" substrate-product-enzyme complexes resulting from abortive complex formation.

In some cases, Eq. (11) may be further simplified if the level of one exchange partner is considerably greater than the other. For example, consider the tritium exchange from labeled dihydroxyacetone phosphate to water in the aldolase reaction (Rose *et al.*, 1965) or the exchange from labeled malate to water in the fumarase reaction (Hansen *et al.*, 1969). In such cases, the rate of exchange will become

$$R = -\frac{[\ln(1-F)][\mathrm{X}][\mathrm{H_2O}]}{([\mathrm{X}]+[\mathrm{H_2O}])t} = -\frac{[\ln(1-F)][\mathrm{X}]}{t} \tag{14}$$

To establish valid conditions for estimating an equilibrium isotope exchange rate, one must have an accurate value for the apparent reaction equilibrium constant. This is necessary not only for estimating $[\mathrm{X}^*]_e$ and $[\mathrm{Y}^*]_e$, but also for manipulating the experimental conditions. One should note that the state of equilibrium is independent of the path by which it is achieved, and rarely does the practicing kineticist merely mix the substrate(s) and enzyme to attain equilibrium. Instead, it is more valuable to preset the values of substrate/product ratios as noted earlier, and this requires a value for K_{app}. If an incorrect value is chosen and an insufficient time is allowed for enzyme-catalyzed equilibration, the observed equilibrium rate may be incorrect because it will be the composite of the equilibrium rate plus or minus the net rate of reaction required to restore equilibrium. Such observed rate values nullify the validity of conclusions drawn from the experiment. One practical check on such problems may be accomplished by including the radiolabeled substrate or product at the time the enzyme is first added; the isotopic distribution at the time, which one would have otherwise added the isotope to initiate the exchange, should be consistent with the equilibrium value. A more strict condition is the demonstration of equal rates using X* or Y* to trace the exchange phenomena. Application of such controls may be especially helpful at the highest concentrations of reactants, where there may be a problem for the enzyme to adequately equilibrate the system. Another excellent procedure to monitor the equilibrium condition also permits the redetermination of K_{app} (Plowman, 1972; Segel, 1975). All products and substrates are present at known concentrations such that ([P][Q] ... /[A][B] ...) is near the anticipated value of the equilibrium constant. Following the addition of enzyme, the change in product or substrate concentration is measured after quenching the reaction and determining the analytical concentrations. If the amount of product has increased, then the initial value of K_{app} was less than the apparent equilibrium constant. If the amount of product decreased, the initial ratio was greater than K_{app}. Plotting ΔP (or ΔA) as a function of various initial mass-action ratios, the value of K_{app} becomes the point where $\Delta P = \Delta A = 0$. This is shown in Fig. 3 for a hypothetical case. In practice, the above protocol should be applied at several absolute concentrations of substrates and products because it is insufficient to assume that K_{app} will be completely independent of reactant concentrations. At very high levels of substrates, deviations from ideal behavior may be large. Likewise, metal ion and proton

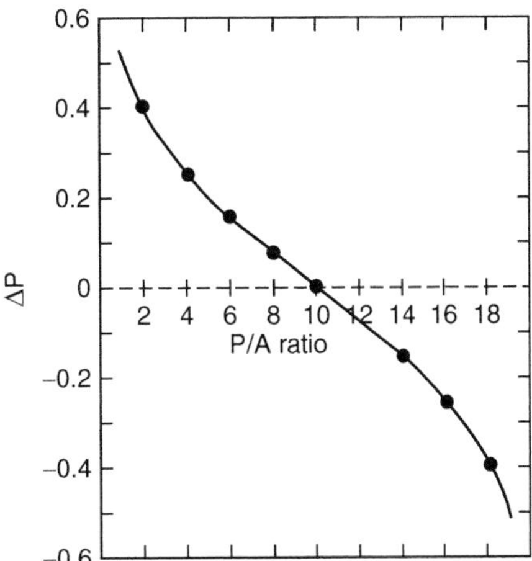

Fig. 3 Estimation of equilibrium constant by approximating the substrate(s)/product(s) ratio and measuring the direction of shift upon enzyme addition. Several initial ratios are prepared, and the direction of shift is determined. The intersection of ΔP or ΔA at zero establishes the equilibrium position.

concentration changes during equilibration may shift the mass-action ratio. Incidentally, some buffers may actually shift the equilibrium constant by reacting with substrates, products, or cofactors. Tris buffer, for example, reacts with some aldehydes, such as glyceraldehyde-3-phosphate and acetaldehyde, and pyrophosphate is a powerful metal ion-chelating agent. In this respect, K_{app} must be redetermined if the experimenter changes the pH, buffer components, metal ion concentrations, or reaction temperature. The experimenter may even elect to use the substrates and products as the pH buffer instead of adding additional ions (Allison *et al.*, 1977).

Before leaving equilibrium constants completely, it may be valuable again to note that the time required to attain equilibrium in the presence of enzyme may be deceptively long. Depending upon the enzyme levels, this may require several hours, but many investigators fail to allow adequate equilibration or to demonstrate the sufficiency of their protocol.

The range of substrate and product levels required adequately to distinguish ordered and random substrate addition must be experimentally determined. Ideally, the absolute levels of substrate and products should be varied to allow the observed value of R to fall between 0.5 and 20 times $R_{0.5}$, the concentrations of A, B, P, Q, etc., required to yield one-half the observed maximal exchange. Computer simulation studies indicate that the range of substrate(s) and product(s) concentration necessary for this condition may have little relation to the Michaelis constants obtained in initial-rate experiments. To observe inhibition of exchange at high "inner-pair" substrate-product levels, the upper range of concentrations should be extended to its practical limits. For rapid-equilibrium ordered cases, the depression of the A $\leftrightharpoons$ Q exchange as [P]/[B] is increased will never be observed. Thus, systems approaching this condition will fail to demonstrate inhibition even though the system does not involve random enzyme-substrate binding pathways.

As Fromm (1975) clearly indicates, the proper [P]/[A], [Q]/[A], [P]/[B], and [Q]/[B] ratios are also very important. These values determine the ultimate isotopic distribution of label at isotopic equilibrium, and it is frequently advantageous to use the isotopic label that corresponds to the least abundant reactant in a substrate-product pair. This permits the greatest redistribution of label and the most accurate estimate of F (One exception is the use of [^{18}O]H_2O in $H_2O \leftrightharpoons$ substrate exchanges. If one uses ^{18}O-labeled substrate, the atom excess in the water after exchange may not be appreciably above the natural abundance of the heavy-oxygen isotope). Ideally, the ratio of substrate to product should fall between 0.05 and 20. For values outside this range, one may encounter difficulty estimating accurately the rate of isotopic exchange. Of course, this limitation deals only with the substrate-product pair undergoing exchange (i.e., the substrate and product that becomes labeled at isotopic equilibrium). Thus, one may compensate for choosing a favorable ratio of this substrate-product pair by altering the ratio of the other substrate(s) or product(s). If the value of K_{app} is very large or very small, the experimenter may wish to alter the value of K_{app} by working at a different pH. For example, the glutamine synthetase equilibrium constant at pH 8 is approximately 10,000–12,000, but at pH 6.5 it is around 444. If one wanted a [glutamine]/[glutamate] ratio of 10 at pH 8, the product of the ratios of [ADP]/[ATP] and [P_i]/[NH_3] must equal 1000–1200. Under these conditions, even a slight redistribution of the system in the presence of the enzyme may alter the ratios of [ADP]/[ATP] and [P_i]/[NH_3] from their preset values. Thus, one would experience difficulties in obtaining meaningful plots of R versus the absolute levels of a particular substrate-product pair. In this respect, a change of pH can be useful to obtain more favorable experimental conditions, but it may also lead to a change in mechanism. Creatine kinase, for example, appears to be rapid-equilibrium random at pH 8 but equilibrium ordered at pH 7 (Morrison and Cleland, 1966; Schimerlik and Cleland, 1973). Another complication that results from lowering the pH is that stability constants for metal-substrate complexes may be appreciably altered. This may lead to a redistribution of substrate between active and inactive forms.

Another factor that is frequently overlooked in exchange studies is the occurrence of time-dependent changes in a chemical system supposedly at equilibrium. If an enzyme is not particularly stable, the fraction of active catalyst may be affected by the protective effects of ligands toward thermal denaturation. In this case, the exchange-rate profiles may yield completely ambiguous results. In the event of substrate instability, the equilibrium concentration of each component may slowly change, and the experimental conditions will be difficult to control. One example is provided by the need to examine the malate dehydrogenase reaction at 1 °C to avoid problems arising from decomposition of oxaloacetate (Silverstein and Sulebele, 1969). Such changes may be even more deceptive with polymerases and depolymerases that also catalyze minor transferase reactions. For example, the presence of debranching enzyme activity in the glycogen phosphorylase system may alter the distribution of available phosphorylase binding sites on the glycogen. Thus, the value of R and its dependence upon substrate/product ratios may be time dependent. Finally, there are even cases where a supposed allosteric modifier of a particular reaction turns out to be an alternative substrate for the reaction itself or some contaminating enzymic activity. If the first occurs, then the equilibrium distribution will change as the "modifier" concentration is altered. If the second case takes place, the observed exchange-rate behavior may be a complex composite of several effects. These situations may be surprisingly common for nucleotide-dependent enzymes where the presence of contaminating adenylate kinase, nucleoside diphosphate kinase, etc., may compromise the value of the experiment.

Upon choosing the ratios of products and substrates, the reaction samples may be prepared. Rather than laboriously adding each component to each reaction mix, it is advisable to prepare three stock solutions: one containing buffer, salts, and cofactors; the second containing one substrate-product pair, say [Q]/[A], at the highest desired concentrations for the experiment with allowance for dilution upon combination with the other stock solutions; and the third containing [P]/[B] in a similar fashion. Now, various dilutions may be made with the second and third stock solutions to obtain the desired absolute levels of each component, and this may be done without altering the ratios. Upon mixing the first stock solution with suitable dilutions of the others, the enzyme is added, and the reaction system is permitted to equilibrate. The isotope is then added to initiate the measurements of the exchange process. Another protocol, sometimes useful, involves the mixing of enzyme, substrates, and products at the desired level of the nonvaried substrate-product pair. After equilibration, appropriate aliquots of a stock solution containing the varied substrate-product pair are added at a ratio satisfying the equilibrium distribution. Then, isotope is added to begin the rate measurement. This protocol is fine provided the experimenter rigorously demonstrates that the system is at equilibrium prior to isotope addition. Examples of the two protocols are given for the hexokinase (Fromm *et al.*, 1964) and aspartokinase (Shaw and Smith, 1977) reactions, respectively.

It is advisable to give adequate thought to the method used for quenching the reaction. Additions of acid, base, phenol, ethanol, EDTA, ion exchange resin, sulfhydryl reagents, or rapid boiling or freezing have been utilized to stop the reaction with minimal uncertainty about the reaction period. Several methods should be used before one adopts a standard protocol, and the best method as regards convenience and accuracy should be adequately examined. By far, the best check is to stop the reaction after various periods and to demonstrate that the exchange obeys Eq. (11). The time required for quenching the process is an experimental parameter, and only trial and error will uncover the best protocol.

In the absence of subunit dissociation and association events, the exchange rate should be directly proportional to the enzyme concentration. This is an important point that should be verified. A typical example is provided by the work of Silverstein and Sulebele (1969) on the porcine heart malate dehydrogenase

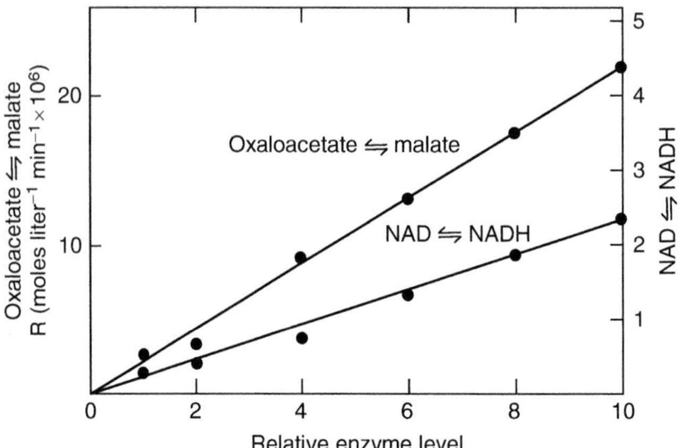

Fig. 4 Plot of the equilibrium reaction rate (R) versus enzyme concentration. The reaction mixture contained 0.33–3.3 μg/ml of pig heart mitochondrial malate dehydrogenase, 4.95 mM NAD^+, 48 μM NADH, 23.3 mM malate, 233 μM oxaloacetate, and 70 mM Tris-NO_3 at pH 8.0 and 1 °C.

reaction (Fig. 4). Here, the oxaloacetate ⇆ malate and NAD^+ ⇆ NADH exchange rates are linearly dependent upon enzyme concentration. It is also important to note that the levels of enzyme used in exchange reactions are frequently higher than in initial-rate studies; thus, contaminating enzyme activities may become a problem. This might lead to a nonlinear dependence of R upon enzyme level. A nonlinear relation may also be realized if the enzyme level is not vanishingly small by comparison to the levels of substrates and products. Quadratic terms compensating for the depletion of free substrate(s) or product(s) level by the bound forms will lead to this nonlinearity, and the experimenter will be confronted by the inadequacy of the rate laws in Table II to describe the observed behavior.

In the case of Ping-Pong mechanisms one may deal with three equilibrium conditions: A, B, P, and Q present; A and P only; and B and Q only. The last two cases are possible because the amount of enzyme linking two components of a half-reaction is very small compared to the substrate and product levels.

$$E + A \rightleftharpoons EX \rightleftharpoons E' + P \tag{15}$$

At any level of [A] and [P] the system will equilibrate by shifting the [E]/[E′] ratio. One experimental advantage of the Ping-Pong case is that more quantitative information is available. One experimental difficulty of measuring partial exchanges, however, is that the presence of a small contaminant amount of the second substrate or product can make a sequential mechanism appear to be catalyzing a partial reaction. For example, Switzer (1970) proposed the intermediary participation of a pyrophosphoryl enzyme in the PRPP synthetase reaction. However, later experiments discounted this idea (Switzer and Simcox, 1974). There are many such cases, and those with apparently slow exchanges are discussed later.

One source of the problem is substrate impurity, and the availability of high pressure liquid chromatography may mitigate this problem.

It may also be helpful to make some practical comments on the Wedler-Boyer (1972) method of raising all reaction components in an equilibrium ratio. This may be achieved by making a single stock solution at the highest analytical concentrations and then diluting this into buffer containing metal ions and cofactors whose free concentration is to remain constant. With metal-ligand complexes acting as substrates and products, it is necessary to estimate the amount of bound metal and to allow for this in the substrate-product stock solution. Likewise, the buffer should contain the estimated amount of uncomplexed metal ion with an allowance for dilution upon preparation of the complete reaction mix. The dilution of the substrate-product stock mix can then be easily accomplished with water, and upon combination with the buffer mix, the free uncomplexed metal ion level will be identical for each absolute level of substrates and products. One experimental limitation to the Wedler-Boyer technique is that large changes in ionic strength can affect stability constants and enzyme-rate behavior. The sensitivity of an enzyme to such changes may be evaluated by initial-rate experiments.

Finally, there are several practical considerations regarding isotope addition and exchange-rate assays. Since the addition of labeled reactant is used to initiate the exchange measurement, it is desirable that the label be of sufficient specific radioactivity to prevent a shift in the equilibrium distribution of reactants. In some cases, this may be difficult to achieve, especially if the labeled species cannot be prepared in high specific activity. In such situations, one may estimate the amount of perturbation on the basis of the specific activity. If this labeled reactant is of a low specific activity, one may mix it with nonlabeled product such that the [substrate*]/[product] ratio is close to the expected equilibrium distribution. When this mixture is added to the system at equilibrium, there should be little disturbance of the relative reactant concentrations. Likewise, all isotope additions should be made in a manner that obviates substantial dilution of the equilibrium samples. This is especially important in reaction systems having a different number of substrates and products (e.g., phosphatases, aldolases, and esterases). Such systems have equilibrium constants with units of molarity, and they will shift the distribution upon dilution. After the isotope is added, one may withdraw samples at various intervals for analysis. When it becomes impractical to withdraw identical aliquots, then one may use a ratio counting method. For example, if we were measuring an ADP* $\leftrightharpoons$ ATP exchange, the initial aliquots could be large, such that the number of ATP* counts was maximized. After separation on DEAE-ion exchange paper, the radioactivity of the ADP and ATP spots could be measured and normalized as (cpm ATP*)/[(cpm ATP*) + (cpm ADP*)], the percentage of radioactivity as ATP*. The same may be done after five to six half-lives of exchange to obtain the ratio at isotopic equilibrium. The quotient of these yields a value for *F*. This approach has the added advantage that no corrections are required for decreases in radioactivity, which are frequently a problem with ^{32}P.

C. Examples of Equilibrium Exchange-Rate Studies

Table I summarizes some of the systems that have been characterized by isotope exchange studies, but it is of value to consider several examples in more detail. The emphasis of this section will be on qualitative conclusions that may be drawn from exchange-rate profiles and the relative exchange velocities.

1. Lactate Dehydrogenase

Silverstein and Boyer (1964a) were the first to examine the rates of exchange between lactate and pyruvate as well as between NAD^+ and NADH in the lactate dehydrogenase system. Convenient ratios of [NADH]/[NAD^+] and [pyruvate]/[lactate] were chosen to satisfy the apparent equilibrium constant. They established that each exchange rate was proportional to time of reaction and proportional to enzyme concentration. They also demonstrated the equality of the pyruvate ⇋ lactate exchange rates with [1-^{14}C]pyruvate or [1-^{14}C]lactate to within 10%. To examine the substrate binding order, the effect of lactate and pyruvate concentrations on equilibrium reaction rates of the pyruvate ⇋ lactate and NAD^+ ⇋ NADH exchanges were examined. As shown in Fig. 5, the pyruvate ⇋ lactate exchange rises to an essentially stable plateau, but the coenzyme exchange is markedly depressed. This observation excludes random substrate addition (see Table III, Line 8), and indicates that the coenzymes form the binary Michaelis constants (i.e., NAD^+ is the first to bind and NADH is the last product to be released in each catalytic cycle).

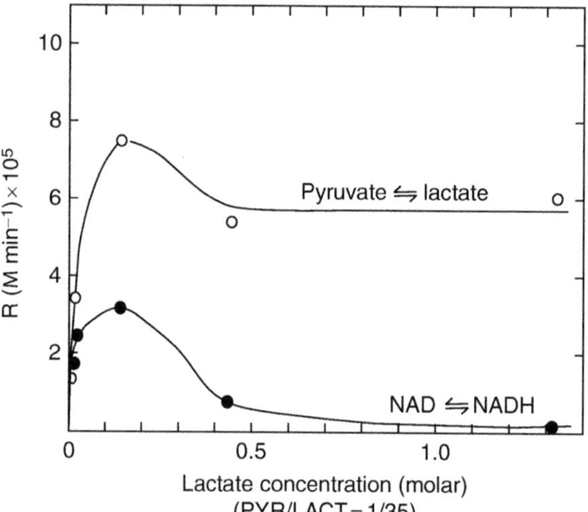

Fig. 5 The effect of lactate and pyruvate concentrations on equilibrium reaction rates with rabbit muscle lactate dehydrogenase at pH 7.9. Reaction mixtures contained 1.68–1.70 mM NAD^+, 30.2–45.5 μM NADH, and pyruvate and lactate as shown, in 145 mM Tris-NO_3 at pH 7.9 and 25 °C.

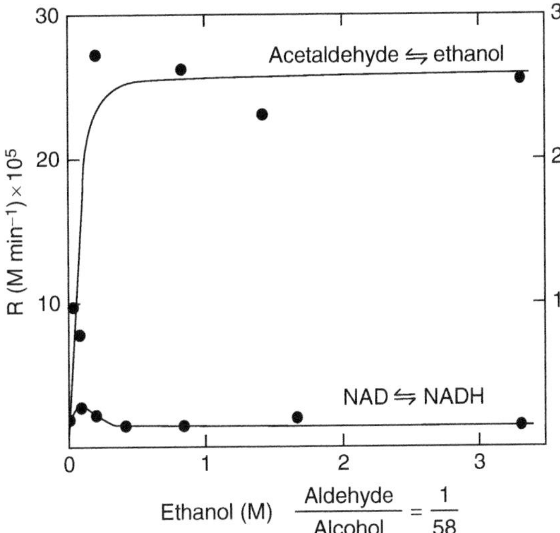

Fig. 6 Effect of ethanol and acetaldehyde concentrations on the equilibrium reaction rates with liver alcohol dehydrogenase. The scale for the $NAD^+ \leftrightharpoons NADH$ rate, given on the right ordinate, is 10 times that for the acetaldehyde ethanol exchange (at the left). Reaction mixtures contained, at 25 °C, 21 mM diethylbarbiturate buffer, pH 7.9, 18.2 mM $NaNO_3$, ethanol, and acetaldehyde as given in the figure, 1.55–1.49 mM NAD^+, and 51–115 μM NADH. In this and other experiments, it was not possible to hold NAD^+ and NADH ratios constant with a constant acetaldehyde to ethanol ratio because of shift in the equilibrium at the high concentrations of ethanol and acetaldehyde. The measured concentrations of NAD^+ for the various points given in the figure as ethanol was increased were 1.55, 1.56, 1.56, 1.56, 1.55, 1.54, 1.53, and 1.49 mM; the corresponding points for NADH were 56, 51, 52, 54, 60, 66, 80, and 115 μM. Thus only above 1 M ethanol was the equilibrium shift appreciable.

2. Alcohol Dehydrogenase

Liver alcohol dehydrogenase is among the most characterized bisubstrate enzyme systems. Sund and Theorell (1963) suggested that the liver enzyme may form all four binary complexes with substrates or products, and Silverstein and Boyer (1964b) demonstrated that the reaction does not proceed with a strictly compulsory substrate binding order. As shown in Fig. 6, the acetaldehyde $\leftrightharpoons$ ethanol and $NAD^+ \leftrightharpoons NADH$ exchanges have an exchange-rate profile very different from that in the analogous experiments with lactate dehydrogenase. The acetaldehyde $\leftrightharpoons$ ethanol exchange rises to a plateau value of about 2.6×10^{-4} M/min, but the $NAD^+ \leftrightharpoons NADH$ exchange rises to a plateau value about 100–160 times less rapid. This finding indicates that appreciable dissociation of enzyme • NADH • acetaldehyde to enzyme • ethanol + NAD^+ must occur, but the principal pathway of net reaction must exhibit coenzyme release as the slowest step. These investigators also recognized that the wide disparity of the two exchanges in the absence of a compulsory pathway eliminates the possibility that ternary complex interconversion is a slow step in the catalysis by the liver enzyme.

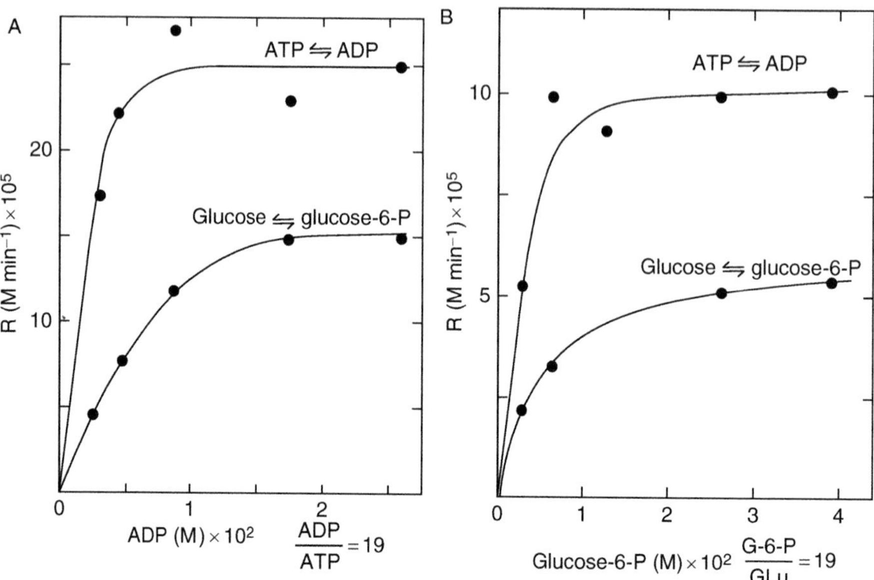

Fig. 7 (A) Effect of ATP and ADP concentrations on equilibrium reaction rates catalyzed by yeast hexokinase. The reaction mixtures contained, at 25 °C, 57.8 mM imidazole-NO_3, pH 6.5; 13 mM Mg $(NO_3)_2$; 2.5 mM glucose; 38.5 mM glucose-6-phosphate; 16.8 μg (29 Kunitz-McDonald units) of yeast hexokinase per milliliter; ATP and ADP as shown in the figure; and 0.34 mg of bovine serum albumin per milliliter. (B) Effect of glucose and glucose-6-phosphate concentrations on equilibrium reaction rates catalyzed by yeast hexokinase. The reaction mixtures contained, at 25 °C, 57.8 mM imidazole-NO_3, pH 6.5; 13 mM $Mg(NO_3)_2$; 0.99–2.2 mM ATP; 25.6 mM ADP; glucose and glucose-6-phosphate as in the figure; 7.83 μg (13.5 Kunitz-McDonald units) of yeast hexokinase per milliliter; and 0.624 mg of bovine serum albumin per milliliter.

3. Yeast Hexokinase

Isotope exchange studies played an important role in defining the kinetic mechanism of yeast hexokinase (Fromm *et al.*, 1964). The exchange work was carried out at pH 6.5 ($K_{app} = 490$) to obviate the experimental difficulties with a large K_{app} at higher pH. As shown in Figs. 7A, and B the ADP ⇋ ATP exchange exceeds the glucose ⇋ glucose-6-phosphate exchange rates by approximately 1.8-fold. Nonetheless, there is no evidence that the exchanges become depressed over the wide range of concentrations employed, and one must conclude that the mechanism is random but certainly not rapid-equilibrium random. This is to say, the interconversion of ternary complexes is relatively slow but not rate determining. This conclusion has been disputed by Noat and Ricard (1968), who favor an ordered mechanism with sugar binding as the obligatory first step, but Fromm (1969) has shown by computation that any depression in the sugar ⇋ sugar phosphate exchange would have been detected under the experimental regime used in the exchange studies. At present, the bulk of evidence accords with a random mechanism (Cleland, 1977; Purich *et al.*, 1974). This is illustrated in Fig. 8.

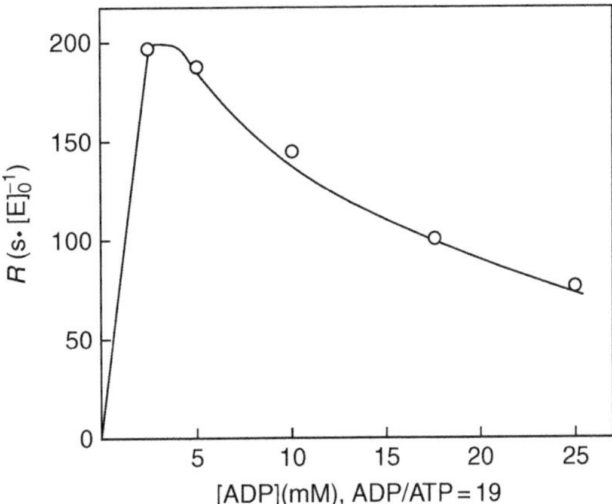

Fig. 8 The effect of ATP and ADP concentrations on equilibrium reaction rates (R) catalyzed by the enzyme yeast hexokinase. The points on the graph were calculated using the equation for an ordered mechanism, the rate constants provided by Noat and Ricard (1968), and the experimental conditions from Fig. 7A for the glucose ⇋ glucose-6-phosphate exchange.

It may be useful to point out several technical features of this report. The equilibrium constant at pH 6.5 is about 490. As noted earlier, this permits one far more latitude than the use of a very large equilibrium constant at higher pH. These investigators chose to set the [ADP]/[ATP] and [glucose-6-P]/[glucose] ratios at 19 and 15.4, respectively, giving an initial mass-action ratio of 293. After the addition of enzyme, the reaction mixtures were maintained at 25 °C for 1 h to achieve equilibration prior to isotope additions. If the system completely equilibrated, the [ADP]/[ATP] ratio at 10 and 25 mM ADP in Fig. 7A would have been 29 and 66, respectively. Thus, there is no doubt that the ratio of [ADP]/[ATP] varies considerably from the assumed constant value of 19. This flaw in experimental design is fairly common, and it illustrates the importance that must be attributed to satisfying the value of K_{app} in the initial design. Another experimental feature of this report is the great care taken to establish the radiopurity of the [^{14}C]glucose and [^{14}C]ATP. The calculation of exchange rates must always take radioimpurity into account, and it is best to rely upon enzymic assays because of their high specificity. Finally, these investigators obtained some information about the minimum values of dissociation constants from plots of R^{-1} versus S^{-1}. These approximate values were found to agree rather closely with K_{d} values obtained from initial-rate studies.

4. Creatine Kinase

Important mechanistic data about the rabbit muscle creatine kinase reaction was provided by Morrison and Cleland (1966). These investigators found that the initial rates of the ADP ⇋ ATP and creatine-P ⇋ creatine exchanges at

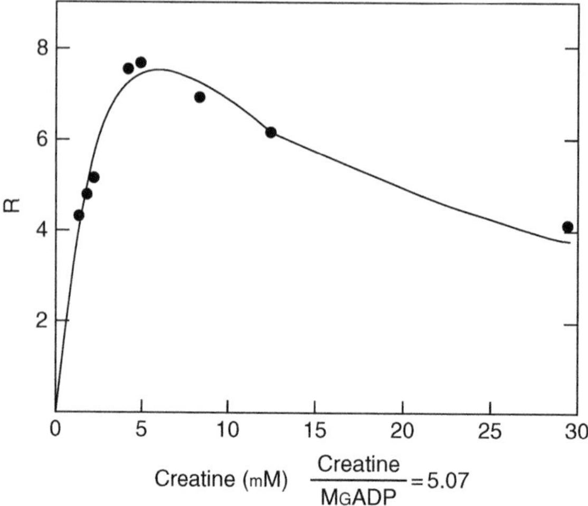

Fig. 9 Effect of increasing concentrations of the MgADP-creatine pair on the initial velocity of the ATP ⇋ ADP exchange. Basic reaction mixtures contained, in 0.5 ml; 0.1 M triethanolamine-HCl buffer (pH 8.0), 0.01 mM EDTA, 3.85 mM ATP, 0.323 mM ADP, 5.52 mM $MgCl_2$, 1.39 mM creatine, 0.756 mM phosphocreatine, and 1 μg of creatine kinase. The concentrations of $MgADP^-$ and creatine were increased as indicated. The exchange reaction was started by the addition of 40 μl (0.4 μCi); temperature, 30 °C. The exchange rate (*R*) is expressed as millimicromoles per minute per microgram of enzyme.

equilibrium are approximately equal, indicative of a rapid-equilibrium random mechanism. While the initial ADP ⇋ ATP exchange rate increased hyperbolically to a maximum value as the creatine-P-creatine pair was raised in concentration, higher concentrations of the MgATP-MgADP pair caused an inhibition of the exchange. This was shown to result from the inhibitory effect of NaCl that is obtained from magnesium chloride and sodium salts of the nucleotides. These investigators also provided excellent evidence for an enzyme•creatine•ADP abortive complex by observing the ADP ⇋ ATP exchange rate as a function of increases in the creatine-MgADP pair (see Fig. 9). The depression in the exchange is clearly evident, and this agrees with initial-rate studies by Morrison and James (1965). One technical advance presented in this exchange study was the use of initial rates of isotope exchange. These were obtained by the ratio (counts per minute of product per minute of reaction per microgram of enzyme)/(counts per minute of substrate per micromole), yielding exchange velocity (v^*) as micromoles per minute per microgram of enzyme. Here, a higher radiospecific activity is required to follow the course of the isotope exchange, but one does not need to use Eq. (11). Thus, the requirement for full attainment of equilibrium implicit in Eq. (11) is no longer necessary. It is also true that one need not rely upon accurate determination of the final distribution of label between substrate and product. Finally, Morrison and Cleland (1966) introduced the use of calculated theoretical curves to demonstrate the correspondence of theoretical and observed rate behavior.

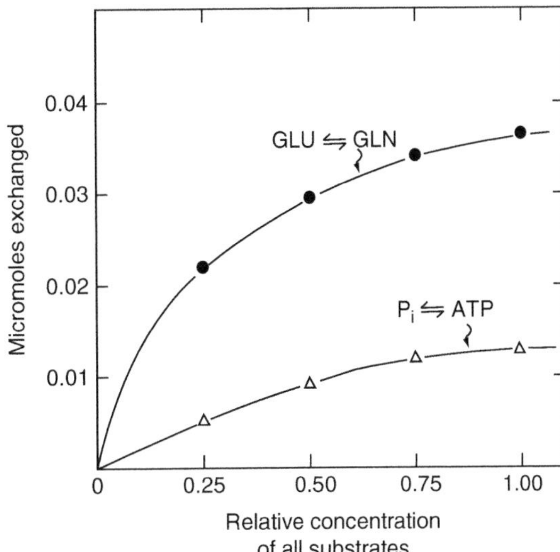

Fig. 10 The effects on equilibrium exchange rates of varying all substrate concentrations simultaneously and in constant ratio. The 1.0-ml reaction at pH 6.50, 37 °C with 1.00 relative concentration contained 2 mM NH_3, 2 mM glutamate, 1 mM ATP, 20 mM glutamine, 20 mM P_1, and 4 mM ADP. The reaction mixtures also contained 200 mM KCl, 50 mM β,β-dimethylglutarate buffer, 1 mM $MnCl_2$, $MgCl_2$ equal to nucleotide, plus 0.4 mg of *Escherichia coli* W glutamine synthetase, E_{10}.

5. Glutamine Synthetase

Isotope exchange studies of this enzyme are rather extensive, and only the Wedler-Boyer (1972) method is described here. This method is especially valuable for three-substrate systems in that the number of exchanges that may be dealt with by varying different substrate-product pairs is large. For distinguishing ordered and random addition pathways, the reader may already have noted that the former involve noncompetitive interactions but the latter cannot. By varying the levels of all substrates and products in a constant ratio, noncompetitive interactions in ordered systems will lead to a depression in the exchange. Competitive effects remain balanced under these conditions, and the relative concentrations of each enzyme species remain effectively constant. Thus, there is no tendency to change the availability of a particular enzyme form required for exchange. Figure 10 presents the results of varying all substrate levels simultaneously in a constant ratio corresponding to the equilibrium constant, as probed by the glutamate ⇋ glutamine and P_i ⇋ ATP exchanges. The maximum levels of substrates were raised considerably beyond the reported Michaelis constants for the substrates. These findings rule out compulsory binding orders or noncompetitive effects as being responsible for the previous inhibitions (Wedler and Boyer, 1972).

It is interesting to note that Wedler and Boyer (1972) did not obtain any evidence for partial exchange reactions between ADP and ATP in the presence of glutamate when ammonium ion was scrupulously omitted. This finding indicates that the following partial reaction is not kinetically significant in glutamine synthetase catalysis.

$$\text{Glutamate} + \text{ATP} \overset{\text{enzyme}}{\rightleftharpoons} \text{glutamyl} - \text{P} + \text{ADP} \quad (16)$$

Contrary to their conclusion that such evidence indicates that no enzyme-bound phosphorylated, enzyme-bound adenosine diphosphoryl, or γ-glutamyl-P participates in the reaction, there is sufficient chemical (Meister, 1974; Todhunter and Purich, 1975) and stereochemical (Midelfort and Rose, 1976) evidence for the latter. One alternative conclusion that may be drawn from the failure to observe a partial exchange reaction is that ADP release from the enzyme•ADP•γ-glutamyl-P complex is extremely slow, if it occurs at all.

6. Nucleoside Diphosphokinase

Of the kinase-type phosphotransferases, nucleoside diphosphokinase has drawn interest in terms of its stable phosphoenzyme intermediate. Initial velocity and product inhibition data are in full agreement with a Ping-Pong Bi Bi mechanism (Mourad and Parks, 1966a,b; Norman *et al.*, 1965). Garces and Cleland (1969) presented additional evidence for such a mechanism by examination of the MgADP $\leftrightarrows$ MgATP and MgUDP $\leftrightarrows$ MgUTP exchange processes. Initial velocities of each exchange were measured at various concentrations of the exchange partners in the absence of the other nucleoside diphosphate and triphosphate components. The MgADP $\leftrightarrows$ MgATP exchange is linear, but there is evidence for competitive substrate inhibition by MgADP. A similar effect was observed with MgUTP inhibition of the MgUDP $\leftrightarrows$ MgUTP exchange. To confirm the basic parallel nature of these exchange patterns by allowing for competitive substrate effects, Garces and Cleland (1969) found that variation in the absolute levels of the two reactants in a constant ratio (tri-/diphosphate = 14) gave completely linear reciprocal plots. Thus, the basic kinetic mechanism may be represented as follows:

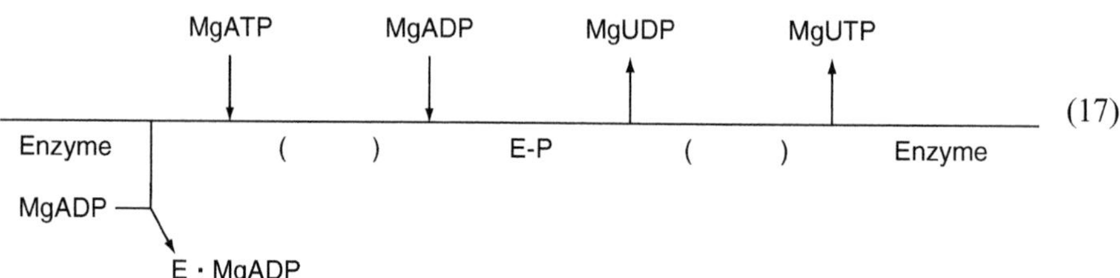

(17)

The viability of a phosphoryl enzyme intermediate was verified also with the bovine liver enzyme when the rates of phosphorylation and dephosphorylation were studied with a rapid mixing technique (Wålinder *et al.*, 1969). Finally, Cleland has presented cogent analyses of complex Ping-Pong mechanisms (Cleland, 1977; Santi *et al.*, 1974), and these reports are highly worthwhile reading, especially for those interested in three-substrate synthetase-type reactions.

III. Enzyme Systems Away From Equilibrium

Although most exchange studies have been desired for processes at equilibrium, the back exchange of labeled product while the reaction is proceeding in the forward direction can provide valuable information about enzymic catalysis. Interestingly, some of the first attempts to gain mechanistic insight about enzyme action were of this sort. Under favorable conditions, investigators may utilize such isotope exchange data to learn about the order of product release and the presence of covalent enzyme-substrate compounds. One of the first systems to be characterized in this way was glucose-6-phosphatase (Hass and Byrne, 1960), which has the following basic kinetic mechanism.

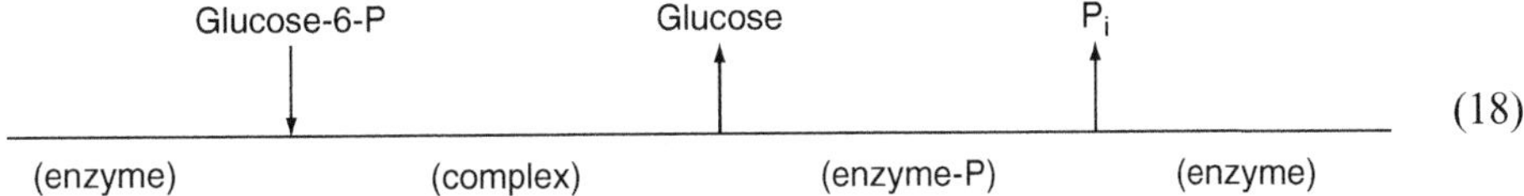

(18)

During the course of glucose-6-phosphate hydrolysis, radiolabeled glucose was added and the back exchange to form labeled glucose-6-P was examined (Likewise, the possibility of incorporation of labeled P_i into glucose-6-P was examined under identical conditions, but none was observed). As shown in Table IV, the rate of the glucose $\leftrightharpoons$ glucose-6-P exchange correlated rather well with the amount of glucose inhibition of the phosphatase. These findings led Hass and Byrne (1960) to conclude that glucose is the first product to leave and that the negative free energy of hydrolysis is preserved as a phosphoryl-enzyme compound. It is also interesting to note that Zatman *et al.* (1953) used such exchange phenomena away from equilibrium to synthesize [^{14}C]NAD$^+$ from [^{14}C]nicotinamide and NAD$^+$ with bovine spleen NADase. They also observed that labeled adenosine diphosphoribosylpyrophosphate (ADPR) failed to exchange in a similar fashion, and the following reaction sequence was favored.

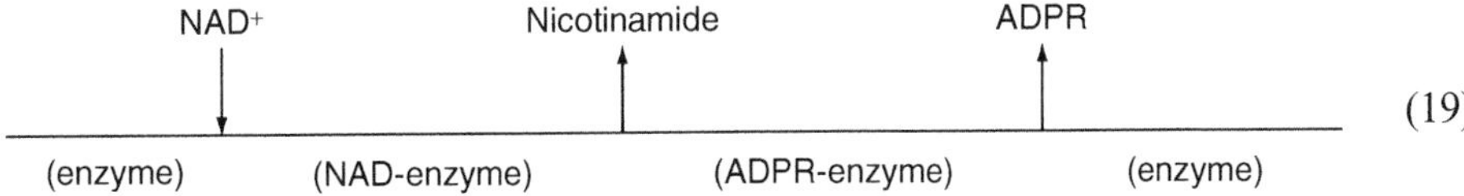

(19)

Such mechanisms involve the ordered release of products, and the scheme is formally of the Ping-Pong Bi Bi type with H_2O presumably entering the catalytic process after release of the first product. Labeled P can exchange back to A in the

Table IV
Examination of the Glucose ⇋ Glucose-6-Phosphate Exchange During Enzyme-Catalyzed Hydrolysis of Glucose-6-Phosphate[a]

Enzyme	Hydrolase activity (v)	Hydrolase activity with glucose (v_i)	Difference ($v-v_i$)	Exchange activity (v^*)
Normal rat liver enzyme	7.17	5.05	2.12	2.12
Diabetic rat liver enzyme	7.40	5.25	2.15	1.95

[a]Reaction samples at 37 °C (pH 6.0) contained 8.5 mM glucose-6-phosphate, and hydrolytic activity is presented as micromoles of P_i liberated per grams of liver per minute. Glucose, when present, was 72 mM. In the exchange measurements [^{14}C]glucose was added to a final specific activity of 4700 cpm/μmol, and the exchange rate is in the same units as above.

absence of Q only when there is a significant level of EQ in the steady state and P* is sufficiently high (i.e., equal or greater than the respective K_i for P). An exchange from Q* back to A may only occur when there is a significant level of P present. If Q* does exchange without P present, one must conclude that the order is noncompulsory (random) with adequate EP complex in the steady state to support significant exchange. In this respect, observation of a P* ⇋ A exchange is not strict proof of ordered release, and one may argue that a random mechanism with slow EQ breakdown to enzyme and Q is operative.

In the case of multisubstrate enzyme systems, one may gain evidence about the order of product release by adapting the experimental procedure to withdraw one product from participation in exchange. This approach was developed by Kosow and Rose (1970) to examine product release order in the hexokinase system. Glucose-6-P was rapidly depleted by excess NADP$^+$ and glucose-6-P dehydrogenase; labeled ADP was added to measure the ADP ⇋ ATP exchange. Without reference to the mode of substrate binding, we may write the following scheme:

( ) ⇌ E · Glu · MgATP ⇌ E · MgADP · Glu-P
E · MgADP · Glu-P ⇌ E · Glu-P (MgADP*) ; E · Glu-P ⇌ E (Glu-P, blocked)
E · MgADP · Glu-P ⇌ E · MgADP (Glu-P, blocked) ; E · MgADP ⇌ E (20)

If the enzyme • glucose-6-P complex is significant in the steady-state phosphorylation of glucose, then ADP* will combine with it and reform E • MgADP • glucose-6-P which will interconvert to form E • ATP* • glucose and lead to ATP* synthesis. Kosow and Rose (1970) found the rate of exchange to be 22% the rate of the forward reaction, where the forward reaction is inhibited 70% by ADP. The significance of this observed exchange may only be judged in qualitative terms (i.e., as evidence for enzyme•glucose-6-P complex). Unless the partition coefficient for E • MgADP • glucose-6-P interconversion to E • MgATP • glucose and E • MgADP is known, it becomes difficult to make any quantitative interpretations.

IV. Enzyme Interactions Affecting Exchange Behavior

A. Substrate Synergism

The cardinal feature of Ping-Pong mechanisms is the ability of the enzyme to catalyze partial exchange reactions as a result of the independence of the substrate's interactions with the enzyme. This is reflected in the fact that the second substrate is obliged to await the dissociation of the first product before it may bind to the enzyme. Multisubstrate enzymes frequently mediate such partial reactions which may be related to important steps in catalysis. One enzyme proposed to be of this sort is succinyl-CoA synthetase, but the partial reactions are relatively slow, and the participation of such reactions in catalysis becomes difficult to assess. Indeed, slow partial exchanges have been interpreted as proof of contamination of a particular enzyme with another enzyme or a small amount of the second substrate, an indication that the mechanism is not Ping-Pong, or that the presence of other substrates may markedly increase the rates of elementary reactions giving rise to the partial exchange. Bridger *et al.* (1968) proposed that the latter phenomenon be termed substrate synergism, and they examined this enzyme-substrate interaction by deriving appropriate raw laws for various exchanges. Their conclusion is that the rate of a partial exchange reaction must exceed the rate of the same exchange reaction in the presence of all the other substrates if the same catalytic steps and efficiencies are involved. If the opposite relation is observed, one must consider the possibility that synergism exists.

Lueck and Fromm (1973) also examined the significance of partial exchange-rate comparisons, and they focused on often misleading comparisons of exchange rates made with respect to initial velocity data. For the Ping-Pong Bi Bi mechanism, it can be shown that the following relationship pertains.

$$\frac{1}{R_{\mathrm{max,A}\leftrightharpoons\mathrm{P}}} + \frac{1}{R_{\mathrm{max,B}\leftrightharpoons\mathrm{Q}}} = \frac{1}{V_1} + \frac{1}{V_2} \qquad (21)$$

where $R_{\mathrm{max,A}\leftrightharpoons\mathrm{P}}$ and $R_{\mathrm{max,B}\leftrightharpoons\mathrm{Q}}$ refer to the maximal rates of the partial exchange reactions between the specified substrates, and V_1 and V_2 refer to the maximal velocity in the forward and reverse reactions, respectively. Thus, one may only evaluate the possibility of substrate synergism after these parameters are quantitated. For example, we may rewrite Eq. (21) to define a new parameter, Q_{syn}, the synergism quotient:

$$Q_{syn} = \frac{(R_{max,A\leftrightharpoons P})^{-1} + (R_{max,B\leftrightharpoons Q})^{-1}}{V_1^{-1} + V_2^{-1}} \tag{22}$$

Only when Q_{syn} is substantially greater than one, may the experimenter conclude that there is a possibility of substrate synergism or that the mechanism is not Ping-Pong. With yeast nucleoside diphosphokinase, Q_{syn} is about 1.3 based upon our replots of the data of Garces and Cleland (1969). Since the enzyme's specific activity may vary somewhat, a value of 1.3 is indicative of a Ping-Pong mechanism. On the other hand, the acetyl $\leftrightharpoons$ acetyl-P exchange in the *E. coli* acetate kinase reaction is quite feeble, and Q_{syn} is about 32 (Janson and Cleland, 1974). Thus, a Ping-Pong mechanism is not likely for acetate kinase but substrate synergism may be involved. Indeed, the so-called "activated Ping-Pong" mechanism of Skarstedt and Silverstein (1976) is a form of substrate synergism. Lueck and Fromm (1973) also presented another criterion for comparing partial exchange rates in the absence and in the presence of the substrate-product pair not involved in the isotope exchange. For the Ping-Pong Bi Bi mechanism, one may compare the rates of various exchanges using Eq. (23)

$$\frac{1}{R_{A\leftrightharpoons Q}} = \frac{1}{R_{A\leftrightharpoons P}} + \frac{1}{R_{B\leftrightharpoons Q}} \tag{23}$$

This expression indicates that the partial exchange rates must equal or exceed the overall exchange rate.

B. Abortive Complex Formation

As noted for creatine kinase, a variety of enzymes may form abortive complexes that are nonproductive forms of the enzyme. Such complexes form as a result of the adsorption of ligands under conditions where the enzyme may not carry out its usual chemistry. For example, the binding of ribulose and NAD^+ to ribitol dehydrogenase leads to the formation of an enzyme • NAD^+ • ribulose complex, which cannot allow for hydrogen transfer because both ligands are already in their oxidized states. The variety of possible binary and ternary abortive complexes (especially in three-substrate systems) presents a problem in analyzing exchange data in some cases. With regard to bisubstrate reactions and their exchange-rate behavior, Wong and Hanes (1964) were among the early investigators to describe several abortive effects that may limit the application of isotope exchange as an unambiguous tool for segregating ordered and random catalytic pathways. One case that attracted their attention involves the following ordered pathway with the release of A and Q from the central complexes to form EB and EP abortives.

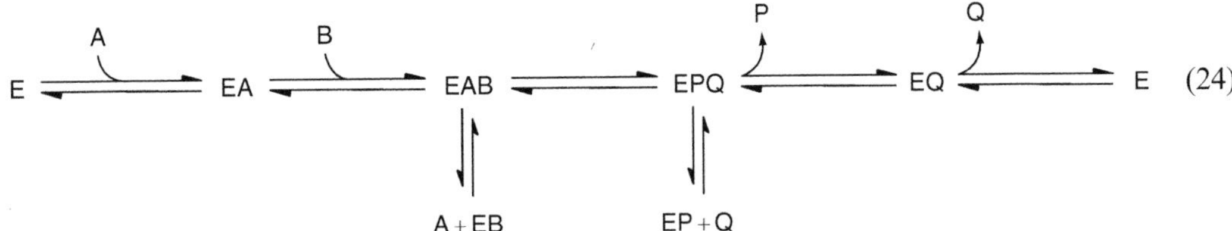

They reasoned that such abortive complex formation provides a second route of exchange between A and Q that may be active even when high levels of the B-P pair reduce the uncomplexed enzyme form (E). Thus, $R_{A \leftrightarrows Q}$ will vary in a hyperbolic fashion with respect to the B-P pair even though the kinetic mechanism is ordered.

Rudolph and Fromm (1971) considered yet another mechanism involving the formation of binary abortives of the sort described above. To illustrate their mechanism, it is helpful to discuss it in terms of a hypothetical dehydrogenase reaction

$$S_{red} + NAD^+ \rightleftharpoons P_{ox} + NADH \tag{25}$$

where S_{red} and P_{ox} represent the reduced and oxidized reactants. (This is done only for illustrative purposes, and the scheme may be applied to any type of enzyme reaction.) The detailed scheme, based on the Theorell-Chance mechanism, may be represented as follows:

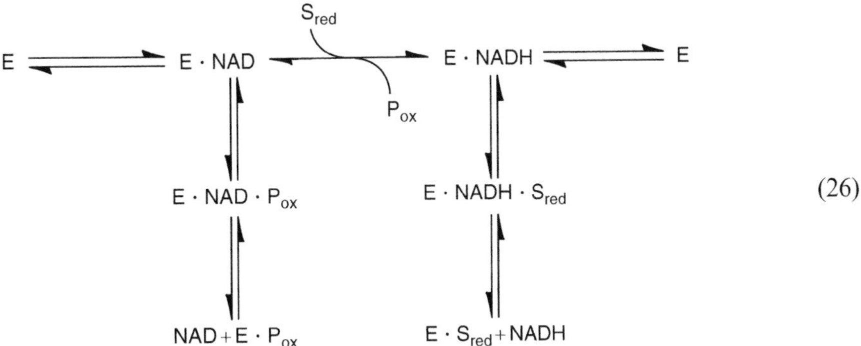

(26)

Here, the binding of NAD^+ (or NADH) facilitates the binding of P_{ox} (or S_{red}) to form the oxidized (or reduced) abortive ternary complexes, which may dissociate to give the oxidized (or reduced) abortive binary complexes E · P_{ox} and E · S_{red}. The complicated equilibrium exchange law for the $NAD^+ \leftrightarrows$ NADH exchange predicts that this exchange is depressed at high S_{red}-P_{ox}. Thus, this sort of abortive complex scheme does not behave differently from the ordered schemes considered in Table II.

Boyer (1978) stated that these mechanisms are somewhat unlikely but that such possibilities should be considered. Certainly, the catalytic efficiency of enzymes attests to their tendency to avoid unnecessary detours, but for systems at equilibrium (the condition under which the exchange is measured) these complexes will be present if these schemes apply. In this respect, the reader is well advised to remember that isotope exchange at equilibrium is only one of many kinetic tools and that there are limitations in the overall scope of each.

To detect the presence of abortive ternary complexes, the kineticist may raise the concentration of dissimilar substrate-product pairs (e.g., glucose-ADP and ATP-glucose-6-P in the hexokinase reaction). This will lead to the inhibition of all exchanges irrespective of the kinetic mechanism. Nonetheless, product inhibition is still unrivaled as the means for detecting abortive complex formation. From the generality of abortive complex formation, it becomes obvious that enzyme sites are fairly flexible. The physiologic conditions of the cell are such that abortive complexes probably have more mechanistic significance than regulatory value. Indeed, rather high concentrations of substrates and products are generally required for abortive complexation, but there has been some speculation on physiologic roles, especially for the dehydrogenases (Cahn *et al.*, 1962; Purich and Fromm, 1972b). It is also interesting to note that EB and EP abortive complexes might easily form in certain ordered schemes. One attractive example is postulated in the ligand exclusion model wherein A binds in a cleft and B binds over A to form the productive EAB complex (Cross and Fisher, 1970).

Frieden (1976) recently identified limitations on using initial velocity data to distinguish certain ordered and random mechanisms. He found that the rapid-equilibrium ordered mechanism with EB and EP abortive complexes cannot be distinguished from the random addition case except by equilibrium exchange measurements. Frieden considered the following mechanism, in which A is the leading substrate in the sequential formation of the productive ternary complex.

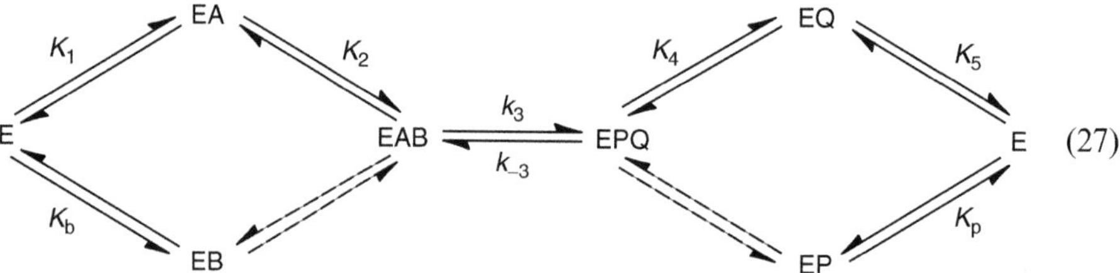

(27)

In this scheme, K_1, K_2, K_4, K_5, K_b, and K_p are dissociation constants, and the rate constant of the slow step is either k_3 or k_{-3}, depending upon the reaction direction considered. The dashed arrows represent the additional pathways that do not occur in this mechanism but are common to the random addition pathway. Frieden (1976) showed that the rate expressions in the absence (Eq. (28)) and in the presence (Eq. (29)) of EB abortive formation are

$$V_0 = \frac{V_{\mathrm{mf}}}{\left\{1 + \frac{K_2}{[\mathrm{B}]} + \frac{K_1K_2}{[\mathrm{A}][\mathrm{B}]}\right\}} \tag{28}$$

$$V_0 = \frac{V_{\mathrm{mf}}}{\left\{1 + \frac{K_2}{[\mathrm{B}]} + \frac{K_1K_2}{K_{\mathrm{b}}[\mathrm{A}]} + \frac{K_1K_2}{[\mathrm{A}][\mathrm{B}]}\right\}} \tag{29}$$

One cannot uniquely distinguish Eq. (29) from the random pathway by initial rates, Haldane relationships, Dalziel ϕ relationships, or the battery of inhibition techniques (including product inhibition, if EP forms) (Frieden, 1976).

When the rapid-equilibrium assumption is also applied in the derivation of isotope exchange-rate equations, Purich *et al.* (1977) find that this method also is incapable of providing a rigorous distinction. The rate law obtained for the $\mathrm{A} \leftrightarrows \mathrm{Q}$ exchange under the rapid-equilibrium assumption requires that

$$\lim_{\mathrm{B}\to\infty} R_{\mathrm{A}\leftrightarrows\mathrm{Q}} = \frac{k_3E_0}{\left\{\frac{1}{[\mathrm{A}]}\left[\frac{\beta}{K_{\mathrm{p}}} + \frac{1}{K_{\mathrm{b}}}\right] + \frac{1}{K_1K_2} + \frac{K_{\mathrm{eq}}}{K_4K_5}\right\}K_1K_2} \tag{30}$$

where $\beta = [\mathrm{P}]/[\mathrm{B}]$. This limit shows that as the level of B (and therefore P) is raised enormously high, the exchange rate reaches a maximum and will not decrease. One may anticipate this, since A* is in rapid equilibrium with the EA and EAB forms and the gross exchange rate depends only upon k_3[EAB].

Conclusions based on these exchange equations can in fact be somewhat misleading because these rapid-equilibrium cases are obtained by eliminating terms from the complete rate expressions. To circumvent this, Purich *et al.* (1977) numerically evaluated the complete equation for the ordered mechanism with EB and EP abortives. The basic idea stems from the fact that the rapid-equilibrium condition is valid only for certain values of the rate constants. For this reason, they began with the more general expressions for the rates at steady state (Eq. (31)) and equilibrium (Eq. (32)).

$$\frac{E_0}{v} = \frac{[\mathrm{E}] + [\mathrm{EA}] + [\mathrm{EAB}] + [\mathrm{EPQ}] + [\mathrm{EQ}] + [\mathrm{EB}] + [\mathrm{EP}]}{k_3[\mathrm{EAB}]} \tag{31}$$

$$\frac{R}{E_0} = \frac{k_1k_2k_3[\mathrm{A}][\mathrm{B}][\mathrm{E}]\left\{(k_{-1} + k_2[\mathrm{B}]) - \frac{k_2k_{-2}[\mathrm{B}]}{k_{-2}+k_3}\right\}^{-1}}{\{[\mathrm{E}] + [\mathrm{EA}] + [\mathrm{EAB}] + [\mathrm{EPQ}] + [\mathrm{EQ}] + [\mathrm{EB}] + [\mathrm{EP}]\}} \tag{32}$$

The determinants for the various enzyme species were obtained using the ENZ EQ program of Fromm (1975). These expressions can be examined by using values for rate constants to give rapid equilibration of all enzyme species except ternary complex interconversion. Bimolecular rate constants were set at $10^{7.5}\ \mathrm{M}^{-1}\ \mathrm{s}^{-1}$, which is a reasonable value for enzyme-substrate reactions (Hammes and Schimmel, 1970). With the exception of ternary complex interconversion, the unimolecular rate constants were set at $10^{3.5}$ s^{-1}, and thus all dissociation constants are 10^{-4} M. The ternary complex interconversion was described by various values. There are many combinations of rate constants that can be considered, but these are quite representative. It was noted that, for k_3 values of around 20 s^{-1} or less; initial-rate plots in the absence of EB and EP complex formation were characteristic of equilibrium ordered mechanisms (i.e., there was a characteristic convergence of all lines at the $1/v$ axis in plots of $1/v$ vs. 1/[B]). However, by including EB and EP abortives, all plots give convergence to the left of the $1/v$ axis, and this shows that the general model behaves as predicted by the rapid-equilibrium equation (Frieden, 1976). The important finding of this study is shown in Fig. 11, where the A $\leftrightharpoons$ Q equilibrium exchange rate and the initial velocity plots are compared at various values of k_3 and k_{-3}. When k_3 is fairly small (i.e., 10 s^{-1} or less), one must raise B and P to levels corresponding to 10–20 times K_b and K_p to observe any depression in the exchange. As k_3 gets greater than 30 or 50 s^{-1} the steady-state assumption becomes relevant, and the A $\leftrightharpoons$ Q exchange is depressed by raising B and P. Significantly, the initial velocity plots show a strong substrate inhibition, characteristic of EB abortive formation, at or above such k_3 values.

The applicability of the rapid-equilibrium assumption thus depends on the rate of the ternary complex step. By definition, the rate constant for this step is k_{cat} in either the rapid-equilibrium random or the Frieden mechanism. Since initial-rate or isotope exchange methods can easily distinguish these mechanisms for k_{cat} values above 30 or 50 s^{-1} (see Fig. 11), the problem pointed out by Frieden (1976) is not at all general. Indeed, the k_{cat} values of a number of enzyme systems are clearly considerably above this range (Purich *et al.*, 1977). Undoubtedly, there are a number of enzymes with lower catalytic constants, and these will not submit to a definitive distinction by these approaches. Nonetheless, knowledge of the k_{cat} value can be used in conjunction with the plots like those in Fig. 11. If the k_{cat} is above 30 or 50 s^{-1}, then the ordered mechanism with EB and EP abortives will be discernible by the characteristic substrate inhibition and depressed equilibrium exchange rates. Both phenomena will be measurable at experimentally accessible values of B and P. The key point is that the rapid-equilibrium assumption must be valid for Frieden's mechanism to be considered, and k_{cat} must be very small. Fromm (1976) suggested three experimental criteria for detecting the EB and EP abortives, and it will be of interest to determine their practical value.

C. Enzymes at High Concentrations

Because the intracellular concentration of enzyme may be higher than feasible in most steady-state experiments, the properties of enzymes at high concentrations becomes of theoretical and metabolic interest. Purich and Fromm (1972a) have noted that equilibrium exchange reactions provide a valuable means to investigate catalysis at high enzyme levels, thereby eliminating the need for coupled enzyme

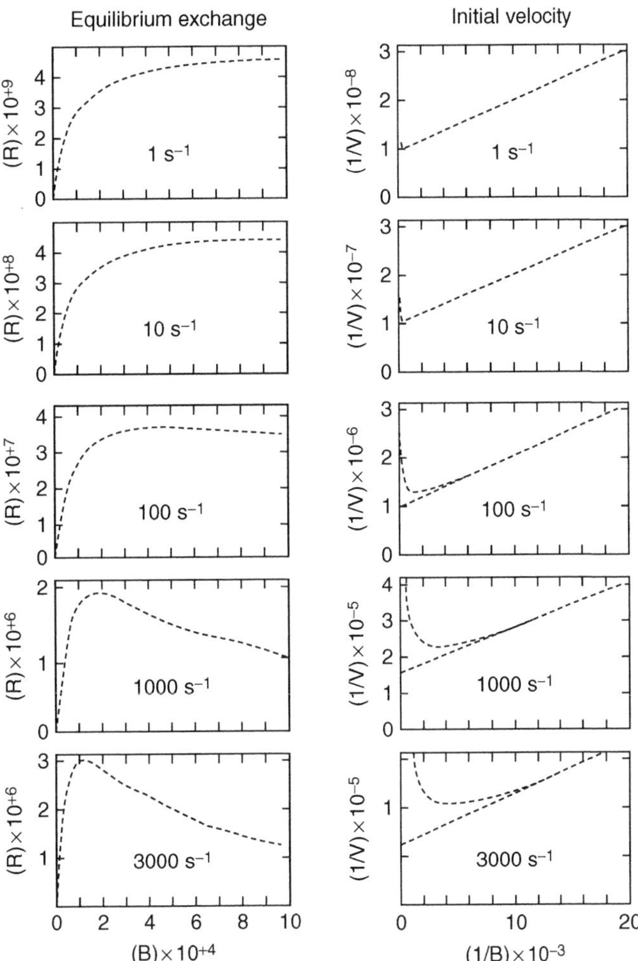

Fig. 11 Comparison of equilibrium exchange rates (R) and initial velocity (v) measurements for various values of the rate constants for ternary complex interconversion. Total enzyme was held at 20 nM; the bimolecular rate constants were $10^{7.5}$ M^{-1} s^{-1}, and the unimolecular constants were $10^{3.5}$ M^{-1} s^{-1}. For the equilibrium rate calculations, A and Q were maintained at 0.1 mM (K_m levels), and the absolute levels B and P were raised in a constant ratio as shown on the graph. For the initial-velocity comparisons, A was maintained at 0.05 mM (0.5 K_m), and P and Q were zero.

assays or for continuous direct assay (as in the case of the dehydrogenases). The initial-rate phase of most systems away from equilibrium is too short to be examined except by stopped-flow methods. Since equilibrium exchange is first order, the accuracy of such experiments is enhanced, and it is possible to study enzymes at moderately high levels even by manual mixing procedures. This is shown in Fig. 12 for the yeast hexokinase P-II isozyme. The linearity of the enzyme concentration dependence of R indicates that no kinetically important changes

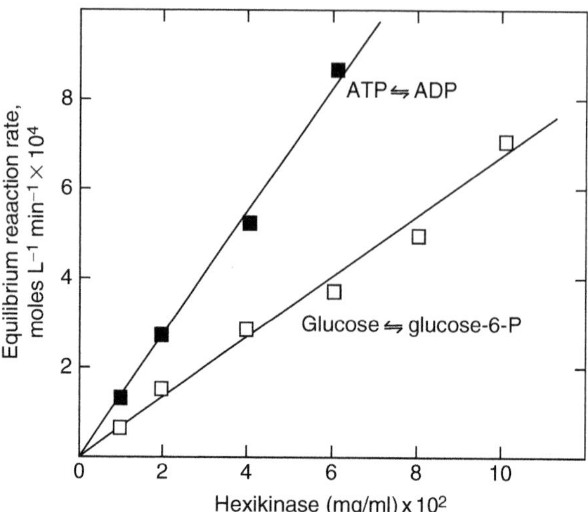

Fig. 12 Plot of the equilibrium exchange rate (R) of the glucose ⇋ glucose-6-P and ATP ⇋ ADP exchanges versus the concentration of yeast hexokinase. Reaction mixtures (final volume, 0.125 ml) contained 58 mM imidazole-NO_3 buffer (pH 6.5), 13 mM $Mg(NO_3)_2$, 50 mM glucose-6-P, 2.5 mM glucose, 38.0 mM ADP, 2.0 mM ATP, and a variable concentration of the P-II isozyme of yeast hexokinase. Reaction samples were preincubated for 75 min at 28 °C before the addition of approximately 200,000 cpm of either [^{14}C]glucose-6-P (specific radioactivity, 20 mCi/mmol).

occur in the catalytic power of the phosphotransferase. It was also possible to show that a steady-state random mechanism prevails under such conditions (Purich and Fromm, 1972a). Interestingly, the availability of rapid mixing-quenching devices may allow one to examine enzymes at concentrations in excess of 10 mg/ml with 1-s reaction periods (Purich and Fromm, 1972a). Provided the condition that total enzyme concentration is substantially below the least abundant substrate or product level, plots of R versus enzyme level may provide evidence about the catalytic potential of self-associating enzymes. A linear dependence is expected in the absence of significant kinetic differences between dissociated and oligomeric systems. If oligomerization increases activity, concave upward plots will be observed.

V. Concluding Remarks

The application of isotopic exchange methods to understand biological processes is truly broader than one may infer on the basis of this chapter. Such diverse topics as energy transduction, allosteric regulation, enzyme cooperativity, metabolic transport, and the behavior of supramolecular processes have all been enriched by the clever application of isotope exchange measurements. In this respect, the power and scope of the technique is still undergoing rapid expansion in theory and experiment. The only unfortunate aspect of this growth is the impossible task of adequately describing all the theoretical and technical advances in a single chapter. Obviously, our only choice was to limit the scope of this chapter, but we highly recommend that the reader examine Boyer's (1978) synopsis of the present status of isotope exchange methodology.

References

Ainslie, G. R., Jr., and Cleland, W. W. (1972). *J. Biol. Chem.* **247,** 946.
Allison, R. D., Todhunter, J. A., and Purich, D. L. (1977). *J. Biol. Chem.* **252,** 6046.
Balinsky, D., Dennis, A. W., and Cleland, W. W. (1971). *Biochemistry* **10,** 1947.
Betz, G. (1971). *J. Biol. Chem.* **246,** 2063.
Betz, G., and Taylor, P. (1970). *Arch. Biochem. Biophys.* **137,** 109.
Borkenhagen, L. F., and Kennedy, E. P. (1959). *J. Biol. Chem.* **234,** 849.
Boyer, P. D. (1959). *Arch. Biochem. Biophys.* **82,** 387.
Boyer, P. D. (1978). *Acc. Chem. Res.* **11,** 218.
Boyer, P. D., and Silverstein, E. (1963). *Acta Chem. Scand.* **17**(Suppl. 1), 195.
Boyer, P. D., Mills, R. C., and Fromm, H. J. (1959). *Arch. Biochem. Biophys.* **81,** 249.
Bridger, W. A., Millen, W. A., and Boyer, P. D. (1968). *Biochemistry* **7,** 3608.
Britton, H. G. (1966). *Arch. Biochem. Biophys.* **117,** 167.
Cahn, R. D., Kaplan, N. O., Levine, L., and Zwilling, E. (1962). *Science* **136,** 962.
Cedar, H., and Schartz, J. H. (1969). *J. Biol. Chem.* **244,** 4122.
Chao, J., Johnson, G. F., and Graves, D. J. (1969). *Biochemistry* **8,** 1459.
Cleland, W. W. (1967). *Annu. Rev. Biochem.* **36,** 77.
Cleland, W. W. (1970). *In* "The Enzymes" (P. Boyer, ed.), 3rd edn, Vol. 2, p. 1. Academic Press, New York.
Cleland, W. W. (1975). *Biochemistry* **14,** 3220.
Cleland, W. W. (1977). *Adv. Enzymol.* **45,** 273.
Cole, F. X., and Schimmerl, P. R. (1970a). *Biochemistry* **9,** 3143.
Cole, F. X., and Schimmerl, P. R. (1970b). *Biochemistry* **9,** 480.
Cross, D. G., and Fisher, H. F. (1970). *J. Biol. Chem.* **245,** 2612.
Darvey, I. G. (1973). *J. Theor. Biol.* **42,** 55.
Davis, J. S., Balinsky, J. B., Harington, J. S., and Shepherd, J. B. (1973). *Biochem. J.* **133,** 667.
Doudoroff, M., Baker, H. A., and Hassid, W. Z. (1947). *J. Biol. Chem.* **168,** 725.
Engers, H. D., Bridger, W. A., and Madsen, N. B. (1969). *J. Biol. Chem.* **244,** 5936.
Engers, H. D., Bridger, W. A., and Madsen, N. B. (1970). *Can. J. Biochem.* **48,** 755.
Farrar, Y. J. K., and Plowman, K. M. (1971). *J. Biol. Chem.* **246,** 3783.
Frieden, C. (1976). *Biochem. Biophys. Res. Commun.* **68,** 914.
Fromm, H. J. (1969). *Eur. J. Biochem.* **7,** 385.
Fromm, H. J. (1975). "Initial Rate Enzyme Kinetics." Springer-Verlag, Berlin and New York.
Fromm, H. J. (1976). *Biochem. Biophys. Res. Commun.* **72,** 55.
Fromm, H. J., Silverstein, E., and Boyer, P. D. (1964). *J. Biol. Chem.* **239,** 3645.
Frost, A. A., and Pearson, R. G. (1961). "Kinetics and Mechanism," 2nd edn. Wiley, New York.
Garces, E., and Cleland, W. W. (1969). *Biochemistry* **8,** 633.
Ginodman, L. M., and Lutsenko, N. G. (1972). *Biokhimya* **37,** 81.
Gold, A. M., Johnson, R. M., and Tseng, J. K. (1970). *J. Biol. Chem.* **245,** 2564.
Graves, D. J., and Boyer, P. D. (1962). *Biochemistry* **1,** 739.
Gulbinsky, J. S., and Cleland, W. W. (1968). *Biochemistry* **7,** 566.
Hammes, G. G., and Schimmel, P. R. (1970). *In* "The Enzymes" (P. Boyer, ed.), 3rd edn, Vol. 2, p. 67. Academic Press, New York.
Hansen, J. N., Dinovo, E. C., and Boyer, P. D. (1969). *J. Biol. Chem.* **244,** 6270.
Hass, L. F., and Byrne, W. L. (1960). *J. Am. Chem. Soc.* **82,** 947.

Hersh, L. B., and Jencks, W. P. (1967). *J. Biol. Chem.* **242,** 3468.
Janson, C. A., and Cleland, W. W. (1974). *J. Biol. Chem.* **249,** 2567.
Kosow, D. P., and Rose, I. A. (1970). *J. Biol. Chem.* **245,** 198.
Lueck, J. D., and Fromm, H. J. (1973). *FEBS Lett.* **32,** 184.
Marcus, H. B., Wu, J. W., Boches, F. S., Tedesco, T. A., Mellman, W. J., and Kallen, R. G. (1977). *J. Biol. Chem.* **252,** 5363.
Meister, A. (1974). *In* "The Enzymes" (P. Boyer, ed.), 3rd edn,Vol. 10, p. 699. Academic Press, New York.
Midelfort, C. F., and Rose, I. A. (1976). *J. Biol. Chem.* **251,** 5881.
Moffet, F. J., and Bridger, W. A. (1973). *Can. J. Biochem.* **51,** 44.
Morales, M. F., Horovitz, M., and Botts, J. (1962). *Arch. Biochem. Biophys.* **99,** 258.
Morrison, J. F., and Cleland, W. W. (1966). *J. Biol. Chem.* **241,** 673.
Morrison, J. F., and James, E. (1965). *Biochem. J.* **97,** 37.
Mourad, N., and Parks, R. E., Jr. (1966a). *J. Biol. Chem.* **241,** 271.
Mourad, N., and Parks, R. E., Jr. (1966b). *J. Biol. Chem.* **241,** 3838.
Neuhaus, F. C., and Byrne, W. L. (1959). *J. Biol. Chem.* **234,** 113.
Noat, G., and Ricard, J. (1968). *Eur. J. Biochem.* **5,** 71.
Norman, A. W., Wedding, R. T., and Black, M. K. (1965). *Biochem. Biophys. Res. Commun.* **20,** 703.
Norris, T. H. (1950). *J. Phys. Colloid Chem.* **54,** 777.
Papas, T. S., and Peterkofsky, A. (1972). *Biochemistry* **11,** 4602.
Plowman, K. M. (1972). "Enzyme Kinetics." McGraw-Hill, New York.
Purich, D. L., and Fromm, H. J. (1972a). *Biochem. Biophys. Res. Commun.* **47,** 916.
Purich, D. L., and Fromm, H. J. (1972b). *Curr. Top. Cell. Regul.* **6,** 131.
Purich, D. L., Fromm, H. J., and Rudolph, F. B. (1974). *Adv. Enzymol.* **39,** 249.
Purich, D. L., Allison, R. D., and Todhunter, J. A. (1977). *Biochem. Biophys. Res. Commun.* **77,** 753.
Raushel, F. M., and Cleland, W. W. (1977). *Biochemistry* **16,** 2176.
Ray, W. J., Jr., Roscelli, G. A., and Kirkpatrick, D. S. (1966). *J. Biol. Chem.* **241,** 2603.
Rhoads, D. G., and Lowenstein, J. M. (1968). *J. Biol. Chem.* **243,** 3963.
Rose, I. A. (1958). *Proc. Natl. Acad. Sci. USA* **44,** 10.
Rose, I. A., O'Connell, E. L., and Mehler, A. H. (1965). *J. Biol. Chem.* **240,** 1758.
Rudolph, F. B., and Fromm, H. J. (1971). *J. Biol. Chem.* **246,** 6611.
Santi, D. V., Webster, R. W., Jr., and Cleland, W. W. (1974). *Meth. Enzymol.* **29,** 49.
Schimerlik, M. I., and Cleland, W. W. (1973). *J. Biol. Chem.* **248,** 8418.
Segel, I. H. (1975). "Enzyme Kinetics." Wiley, New York.
Shaw, J. F., and Smith, W. G. (1977). *J. Biol. Chem.* **252,** 5304.
Silverstein, E., and Boyer, P. D. (1964a). *J. Biol. Chem.* **239,** 3901.
Silverstein, E., and Boyer, P. D. (1964b). *J. Biol. Chem.* **239,** 3908.
Silverstein, E., and Sulebele, G. (1969). *Biochemistry* **8,** 2543.
Silverstein, E., and Sulebele, G. (1973). *Biochemistry* **12,** 2164.
Silverstein, E., and Sulebele, G. (1974). *Biochemistry* **13,** 1815.
Simon, W. A., and Hofer, H. W. (1978). *Eur. J. Biochem.* **88,** 175.
Skarstedt, M. T., and Silverstein, E. (1976). *J. Biol. Chem.* **251,** 6775.
Smith, E., and Morrison, J. F. (1969). *J. Biol. Chem.* **244,** 4224.
Stokes, B. O., and Boyer, P. D. (1976). *J. Biol. Chem.* **251,** 5558.
Su, S., and Russell, P. J., Jr. (1968). *J. Biol. Chem.* **243,** 3826.
Sund, H., and Theorell, H. (1963). *In* "The Enzymes" (P. D. Boyer, H. Lardy and K. Myrbäck, eds.), 2nd edn, Vol. 7, p. 25. Academic Press, New York.
Switzer, R. L. (1970). *J. Biol. Chem.* **245,** 483.
Switzer, R. L., and Simcox, P. D. (1974). *J. Biol. Chem.* **249,** 5304.
Todhunter, J. A., and Purich, D. L. (1975). *J. Biol. Chem.* **250,** 3505.
Uhr, M. L., Thompson, V. W., and Cleland, W. W. (1974). *J. Biol. Chem.* **249,** 2920.
Wålinder, O., Zetterqvist, Ö., and Engström, L. (1969). *J. Biol. Chem.* **244,** 1060.

Wedler, F. C. (1974a). *J. Biol. Chem.* **249,** 5080.
Wedler, F. C. (1974b). *J. Biol. Chem.* **249,** 7715.
Wedler, F. C., and Boyer, P. D. (1972). *J. Biol. Chem.* **247,** 984.
Wedler, F. C., and Gasser, F. J. (1974). *Arch. Biochem. Biophys.* **163,** 57.
Wong, L. J., and Frey, P. A. (1974). *Biochemistry* **13,** 3889.
Wong, J. T. F., and Hanes, C. S. (1964). *Nature (Lond.)* **203,** 492.
Yagil, G., and Hoberman, H. D. (1969). *Biochemistry* **8,** 352.
Young, O. A., and Anderson, J. W. (1974). *Biochem. J.* **137,** 435.
Zatman, L. J., Kaplan, N. O., and Colowick, S. P. (1953). *J. Biol. Chem.* **200,** 197.

CHAPTER 17

Positional Isotope Exchange as Probe of Enzyme Action

Leisha S. Mullins and Frank M. Raushel

Department of Chemistry
Texas A&M University
College Station, Texas 77843

I. Introduction

The application and development of new mechanistic probes for enzyme-catalyzed reactions have significantly expanded our knowledge of the molecular details occurring at the active sites of many enzymes. Enzyme kinetic techniques have progressed from a simple determination of K_m and V_{max} values through a discrimination between sequential and ping-pong kinetic mechanisms to detailed evaluations of transition state structures by measurement of the very small differences in rate on isotopic substitution at reaction centers. The positional isotope exchange (PIX) technique, originally described by Midelfort and Rose (1976), is one of the newer techniques that has made a significant contribution to the elucidation of enzyme reaction mechanisms over the last decade.

CONTEMPORARY ENZYME KINETICS AND MECHANISM
Reprinted from *Methods in Enzymology*, Volume 249 (Academic Press, 1995).

DOI: 10.1016/B978-0-12-378608-1.00017-7

The PIX technique can be used as a mechanistic probe in any enzymatic reaction where the individual atoms of a functional group within a substrate, intermediate, or product become torsionally equivalent during the course of the reaction. This criterion is best illustrated with the example first presented by Midelfort and Rose (1976). Glutamine synthetase (glutamate-ammonia ligase) catalyzes the formation of glutamine via the overall reaction presented in Eq. (1):

$$\text{MgATP} + \text{glutamate} + \text{NH}_3 \rightleftharpoons \text{MgADP} + \text{glutamine} + \text{P}_i \quad (1)$$

The questions addressed by Midelfort and Rose concerned whether γ-glutamyl phosphate was an obligatory intermediate in the reaction mechanism and if this intermediate was synthesized at a kinetically significant rate on mixing of ATP, enzyme, and glutamate in the absence of ammonia. The putative reaction mechanism is illustrated in Scheme I.

NH_3^+ MgATP NH_3^+ O
-O O- -O O-P-O-
MgADP
O O O O O

NH_3^+ O NH_3^+
NH_3
-O O-P-O- -O NH_2
P_i
O O O- O O

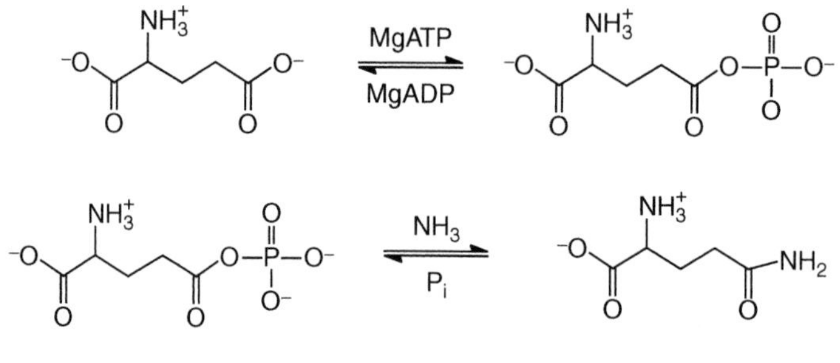

Scheme I

In the first step (Scheme I), glutamate is phosphorylated by MgATP at the terminal carboxylate group to form enzyme-bound MgADP and γ-glutamyl phosphate. In the second step, ammonia displaces the activating phosphate group to form the amide functional group of the product glutamine. Previous experiments with this enzyme had failed to detect an equilibrium isotope exchange reaction between ATP and ADP in the presence of enzyme and glutamate (Wedler and Boyer, 1972). Thus, if the γ-glutamyl phosphate intermediate was formed at the active site then the release of MgADP from the E · MgADP · intermediate complex must be very slow. The formation of the intermediate can be confirmed, however, when the oxygen atom between the β- and γ-phosphoryl groups in the substrate ATP is labeled with oxygen-18.

If the γ-glutamyl phosphate intermediate is formed and the β-phosphoryl group of the bound ADP is free to rotate then an isotopic label that was originally in the β, γ-bridge position will eventually be found 67% of the time in one of the two equivalent β-nonbridge positions of ATP as illustrated in Scheme II. This migration of the isotopic label can only occur if the γ-phosphoryl group of ATP is transferred to some acceptor and if the β-phosphoryl group of the enzyme-bound ADP is free to rotate torsionally. Midelfort and Rose found a PIX reaction

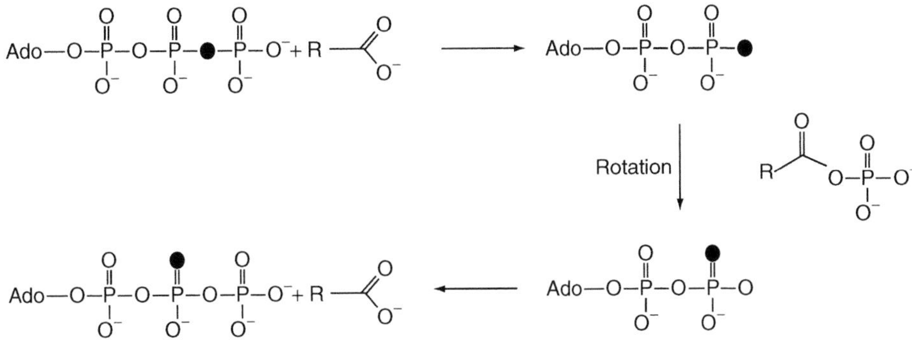

Scheme II

with glutamine synthetase, and thus γ-glutamyl phosphate is an obligatory intermediate in the enzymatic synthesis of glutamine.

II. Functional Groups for PIX

A variety of functional groups common to many substrates and products found in enzyme-catalyzed reactions are suitable for PIX analysis. Some of the generic examples are illustrated in Fig. 1. The two most common functional groups that have been utilized for PIX studies are esters of substituted carboxylic and phosphoric acids. Thus, all reactions involving nucleophilic attack at either the β- or γ-phosphoryl groups of nucleotide triphosphates are amenable to PIX analysis. Moreover, all reactions utilizing UDP-sugars for complex sugar biosynthesis can use these PIX techniques for mechanistic evaluation. Less common are examples that involve reactions at a substituted guanidino functional group. Specific examples include creatine kinase (Reddick and Kenyon, 1987) and argininosuccinate lyase (Raushel and Garrard, 1984). With these enzymes PIX of the two amino groups within arginine or creatine can be monitored by labeling with ^{14}N and ^{15}N and following the reaction with ^{15}N nuclear magnetic resonance (NMR) spectroscopy. Other cases include formation of radical or cationic centers at methylene carbons. Production of achiral acetaldehyde from chiral ethanolamine has been explained by intermediate formation of a C-2 radical during the course of the reaction (Retey *et al.*, 1974). Loss of stereochemistry in sp^2 centers can be exploited as a mechanistic probe when a methyl group is formed as an intermediate or product as for the case in the reaction catalyzed by pyruvate kinase.

III. Qualitative and Quantitative Approaches

The technique of PIX can be used for two interrelated probes of enzyme-catalyzed reactions. In the first, as exemplified by the example with glutamine synthetase, the experimenter is primarily interested in determining whether a

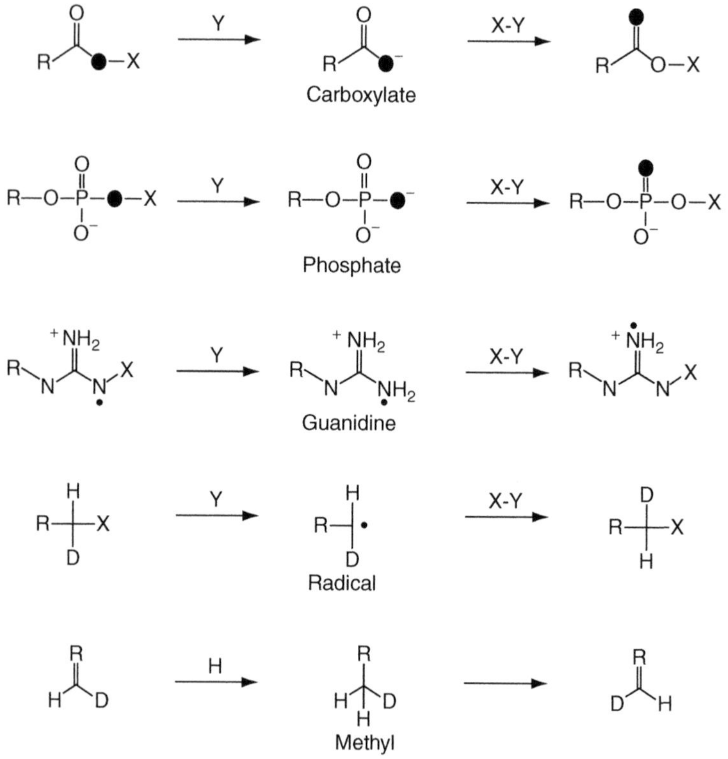

Fig. 1 Functional groups for positional isotope exchange analysis.

particular intermediate is formed during the course of the reaction. In the second approach, the primary interest is in a quantitative determination of the partition ratio of an enzyme-ligand complex. The partition ratio is defined here as the fraction of the enzyme-ligand complex that proceeds forward to form unbound products versus the fraction of the enzyme complex in question that returns to unbound substrate and free enzyme.

When one is interested in whether a specific intermediate or complex is formed the experiment is generally conducted with oe substrates absent from the reaction mixture. Take, for example, the simplified case of an enzyme reaction where two substrates are converted to two products via the covalent transfer of a portion of one substrate to the second substrate. This type of reaction is commonly found in many kinase reactions where ATP is used to phosphorylate an acceptor substrate. The generalized scheme is presented in Eq. (2):

$$A + B \rightleftharpoons P + Q \tag{2}$$

If the chemical mechanism involves the phosphorylation of the enzyme by ATP (substrate A) then a PIX reaction may proceed in the absence of the acceptor substrate (B) if the acceptor B is not required to be bound to the enzyme in order for the phosphorylation to occur. Because substrate B is not required to be bound to the protein in order for the torsional equilibration of the isotopic label to occur, then all that is required for the PIX reaction to be observed is the incubation of enzyme and the isotopically labeled substrate (A*). In such cases, the velocity of the PIX reaction can be derived from the model presented in Scheme III where A* represents the substrate as originally synthesized with the isotopic label and A^+ represents that fraction of A where the isotopic label is positionally equilibrated. It can be demonstrated that in such mechanisms the maximum value for the PIX reaction (V_{ex}) will be given by Eq. (3):

$$E \underset{k_2}{\overset{k_1}{\rightleftharpoons}} EA^* \xrightarrow{k_3} E\text{-}X\cdot P^* \underset{k_3}{\overset{k_4}{\rightleftharpoons}} EA^+ \xrightarrow{k_2} E + A^+$$

Scheme III

$$V_{ex} = \frac{k_2 k_3 k_4}{(k_2 + k_3)(k_3 + k_4)} \tag{3}$$

Rose has also demonstrated that in such mechanisms the minimum value for the PIX reaction is given by Eq. (4) (Rose, 1979):

$$V_{ex} \geq \frac{V_1 V_2}{V_1 + V_2} \tag{4}$$

where V_1 and V_2 are the maximal velocities of the steady-state reaction in the forward and reverse directions, respectively. This approach can be utilized with great utility to identify activated intermediates in many synthetase-type reactions where ATP is used to activate a second substrate molecule prior to condensation with a third substrate. This can be done either by phosphorylation or adenylylation of the second substrate. Equation (4) can be utilized to demonstrate that the presumed intermediate is formed rapidly enough to be kinetically competent.

When one is more interested in a quantitative analysis of the partition ratio of an enzyme-product complex, the PIX methodology can easily provide that information. Take, for example, a case where a substrate, A, is converted to a single product, P:

$$A \rightleftharpoons P \tag{5}$$

In this example, we will assume that the isotopic label originally in A becomes torsionally equivalent in the product, P.

The simplified kinetic scheme can be diagrammed as shown in Scheme IV. In this case, the bonds are broken and the isotopes positionally equilibrated in the EP complex. This complex can partition in one of two ways. Either the product P can

$$E \underset{k_2}{\overset{k_1 A}{\rightleftharpoons}} EA \underset{k_4}{\overset{k_3}{\rightleftharpoons}} EP \underset{k_6 P}{\overset{k_5}{\rightleftharpoons}} E$$

Scheme IV

dissociate irreversibly from the EP complex with a rate constant k_5 or it can reverse the reaction and form free enzyme and substrate A with a net rate constant of k'_4. The fraction of EP that partitions forward can be quantitated by the amount of P that is formed per unit time, whereas the fraction of EP that partitions backward toward free enzyme and unbound A can be quantitated by the rate of positional equilibration of the isotope labels within the pool of substrate A.

In the steady state the partition forward is given by

$$[\mathrm{EP}]k_5 = V_{\mathrm{chem}} \tag{6}$$

whereas the partition backward is given by

$$[\mathrm{EP}]k'_4 = [\mathrm{EP}]\frac{k_2 k_4}{k_2 + k_3} = V_{\mathrm{ex}} \tag{7}$$

and thus the ratio of $V_{\mathrm{ex}}/V_{\mathrm{chem}}$ is

$$\frac{V_{\mathrm{ex}}}{V_{\mathrm{chem}}} = \frac{k_2 k_4}{k_5(k_2 + k_3)} \tag{8}$$

Because the maximal velocity in the reverse direction, V_2/E_t, is given by

$$V_2 = \frac{k_2 k_4}{k_2 + k_3 + k_4} \tag{9}$$

then

$$\frac{k_2}{k_5} \geq \frac{V_{\mathrm{ex}}}{V_{\mathrm{chem}}} \geq \left(\frac{V_2}{E_t}\right) k_5 \tag{10}$$

Therefore, the lower limit for the experimentally determined partition ratio is given by the maximal velocity in the reverse direction divided by k_5 (the dissociation rate constant for the product P). The upper limit for the partition ratio is given by the relative magnitude for the off-rate constants for A and P from the enzyme. These values would be difficult to obtain directly by any other method.

Experimentally, the PIX rate, V_{ex}, is determined by measuring the fraction of positional isotope scrambling equilibrium (F) as a function of time (t) where $V_{\mathrm{ex}} = ([\mathrm{A}]/t)\ \ln(1 - F)^{-1}$. If the isotopically labeled substrate is being chemically depleted during the course of the PIX analysis then the corrected PIX rate is calculated from $V_{\mathrm{ex}} = [(X)/\ln(1-X)]\ (A_0/t)\ \ln(1-F)^{-1}$, where X is the fraction of substrate lost at time t and A_0 is the initial concentration of the labeled substrate. In the following section, we examine the quantitative effects induced by variation of the concentration of the unlabeled substrates and products on the PIX rates.

IV. Variation of Nonlabeled Substrates and Products

The PIX technique can be used to obtain information about the partitioning of enzyme complexes and the order of substrate addition and product release. This information is obtained by measuring the PIX rate in the enzyme-catalyzed reaction relative to the overall chemical transformation rate in the presence of variable amounts of added substrates or products. In many cases, it is possible with these methods to determine the microscopic rate constants for the release of substrates and products from the enzyme-ligand complexes. The following provides a presentation of how the PIX technique can be applied to the analysis of sequential and ping-pong kinetic reaction mechanisms.

A. Sequential Mechanisms

In sequential mechanisms, substrate addition and/or product release can be either ordered or random or a combination of various pathways. The PIX reaction can be utilized to distinguish between the possible kinetic mechanisms and to determine the net reaction flux through the various kinetic pathways. The effect of product addition can lead to an *enhancement* of the PIX reaction relative to overall rate (PIXE), whereas the variation of unlabeled substrates may reduce or *inhibit* the PIX reaction relative to overall chemical rate (PIXI).

B. PIX Enhancement

The simplest kinetic mechanism that can be written for an enzyme reaction with multiple products is a Uni Bi mechanism. In this mechanism, the release of the products can be either random (Uni Bi Random) or ordered (Uni Bi Ordered). A general model of a Uni Bi Random kinetic mechanism is shown in Scheme V, where A is designated as the substrate with the positionally labeled isotopic atoms and Q is the product in which the positional exchange occurs.

In Scheme V, the ratio of the PIX rate relative to the net chemical rate (V_{ex}/V_{chem}) is determined by the partitioning of the EPQ complex. Using the method of net rate constants, the partitioning of EPQ can be written as (Cleland, 1975):

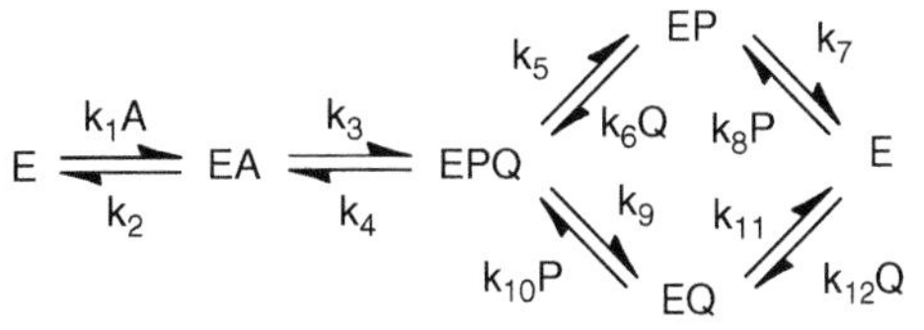

$$E \underset{k_2}{\overset{k_1A}{\rightleftharpoons}} EA \underset{k_4}{\overset{k_3}{\rightleftharpoons}} EPQ; \quad EPQ \underset{k_6Q}{\overset{k_5}{\rightleftharpoons}} EP \underset{k_8P}{\overset{k_7}{\rightleftharpoons}} E; \quad EPQ \underset{k_{10}P}{\overset{k_9}{\rightleftharpoons}} EQ \underset{k_{12}Q}{\overset{k_{11}}{\rightleftharpoons}} E$$

Scheme V

$$\frac{V_{ex}}{V_{chem}} = \frac{(k_2k_4)/(k_2+k_3)}{k_5+(k_9k_{11})/(k_{11}+k_{10}[\mathrm{P}])} \tag{11}$$

As can be seen in Eq. (11), the ratio V_{ex}/V_{chem} is dependent on the amount of P added to the reaction mixture. In a random mechanism, the ratio V_{ex}/V_{chem} increases as a function of [P] because the addition of P inhibits the flux through the lower pathway and then the ratio plateaus at a level that is determined by the net flux through the upper pathway. The net flux through each pathway can be determined by the ratio V_{ex}/V_{chem} as a function of [P]. At $[\mathrm{P}] = 0$, Eq. (11) becomes

$$\frac{V_{ex}}{V_{chem}} = \frac{k_2k_4/(k_2+k_3)}{k_5+k_9} \tag{12}$$

and at $[\mathrm{P}] = \infty$:

$$\frac{V_{ex}}{V_{chem}} = \frac{k_2k_4/(k_2+k_3)}{k_5} \tag{13}$$

Thus, the ratio of k_5 and k_9 can be determined by measurement of the PIX ratio at zero and saturating P.

Figure 2 also illustrates the change in the ratio V_{ex}/V_{chem} as a function of [P] as a sequential mechanism changes from random to ordered release of products. The ordered release of the products simplifies Eq. (11). If P is released first ($k_5 = 0$), then the equation becomes

$$\frac{V_{ex}}{V_{chem}} = \frac{k_2k_4/(k_2+k_3)}{k_9k_{11}/(k_{11}+k_{10}[\mathrm{P}])} \tag{14}$$

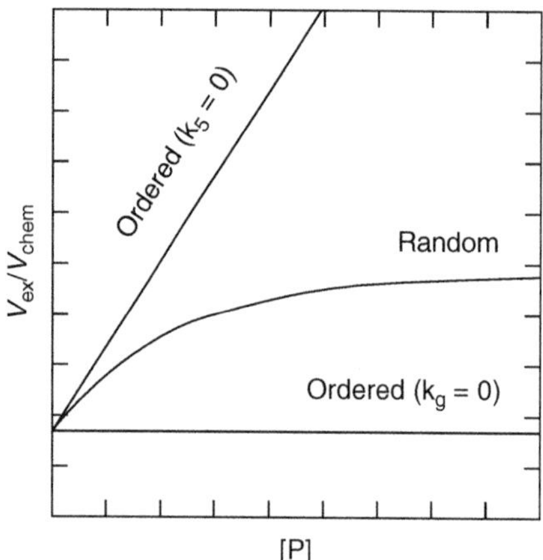

Fig. 2 Enhancement of the ratio of the positional isotope exchange rate and the net rate of chemical turnover as a function of added product inhibitor.

The ratio V_{ex}/V_{chem} is now linearly dependent on the concentration of added P. The intercept at [P] = 0 becomes

$$\frac{V_{ex}}{V_{chem}} = \frac{k_2k_4/(k_2+k_3)}{k_9} \tag{15}$$

It is possible to determine the lower limits for the off-rate constant from the ternary complex EPQ (k_9) and the binary complex EQ (k_{11}) relative to the turnover number in the reverse direction (V_2/E_t) as shown in Eqs. (16) and (17) since $V_2/E_t = (k_2k_4)/(k_2+k_3+k_4)$:

$$\frac{k_9}{V_2/E_t} = \frac{V_{chem}}{V_{ex}} + \frac{k_9}{k_2} \tag{16}$$

$$\frac{k_{11}}{V_2/E_t} = \left(\frac{1}{K_p}\right)\left(\frac{\text{intercept}+1}{\text{slope}}\right) \tag{17}$$

where K_p is the Michaelis constant for the product P.

If Q is the first product to be released ($k_9 = 0$) in an ordered mechanism, then there would be no dependence on V_{ex}/V_{chem} as [P] is varied, as apparent in Eq. (18):

$$\frac{V_{ex}}{V_{chem}} = \frac{k_2k_4}{k_2+k_4} \tag{18}$$

Again, the lower limit for the release of the product from the ternary complex can be determined relative to the turnover number in the reverse reaction:

$$\frac{k_5}{V_2/E_t} = \frac{V_{chem}}{V_{ex}} + \frac{k_5}{k_2} \tag{19}$$

PIX enhancement experiments can provide valuable information about enzyme kinetic mechanisms. Utilizing the PIXE technique, a simple inspection of a plot of V_{ex}/V_{chem} as a function of [P] readily identifies the kinetic mechanism (see Fig. 2). The PIXE experiments also allow the determination of the microscopic rate constants for product release from both the ternary and binary complexes. An often overlooked advantage of the PIXE experiment is that it can identify a "leaky" product in what otherwise would appear as an ordered product release.

C. PIX Inhibition

The concentration of unlabeled substrate also affects the ratio of the PIX rate relative to the net chemical rate (Raushel and Villafranca, 1988). An enzyme kinetic mechanism with random addition of substrates is shown in Scheme VI,

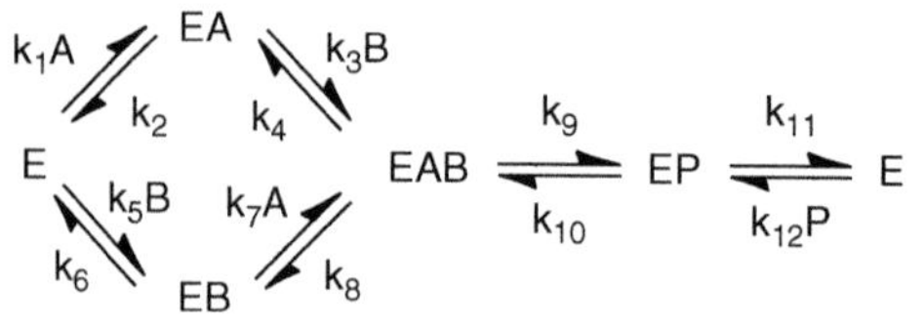

$$E \underset{k_2}{\overset{k_1A}{\rightleftharpoons}} EA \underset{k_4}{\overset{k_3B}{\rightleftharpoons}} EAB$$

$$E \underset{k_6}{\overset{k_5B}{\rightleftharpoons}} EB \underset{k_8}{\overset{k_7A}{\rightleftharpoons}} EAB$$

$$EAB \underset{k_{10}}{\overset{k_9}{\rightleftharpoons}} EP \underset{k_{12}P}{\overset{k_{11}}{\rightleftharpoons}} E$$

Scheme VI

where A is the substrate with the positionally labeled atoms and P is the product in which the positional exchange occurs.

If A is the first substrate to bind in an ordered kinetic mechanism ($k_8 = 0$), then increasing concentrations of B will inhibit the PIX rate relative to the net chemical rate (V_{ex}/V_{chem}) because saturation with B prevents A from dissociating from the enzyme. In contrast, if B must bind first ($k_4 = 0$), then its concentration will have no effect on V_{ex}/V_{chem}. This is because saturation with B would not affect the rate of dissociation of A from the EAB complex. If addition of the substrates to the enzyme is random, then saturation with B will reduce the value of V_{ex}/V_{chem} but not eliminate it entirely. A plot of V_{ex}/V_{chem} as a function of [B] will plateau at a value of V_{ex}/V_{chem} that equals the flux through the lower pathway and thus enables the determination of the ratio of k_4 and k_8. Figure 3 illustrates the plot of V_{ex}/V_{chem} as a function of added substrate. Substrate inhibition by the nonlabeled substrate can be quite diagnostic of the particular reaction mechanism.

In special cases, the investigation of substrate inhibition of PIX reactions can be used to determine the microscopic rate constants for the release of substrates and products from the enzyme complexes. A Bi Bi Ordered mechanism can be used to illustrate this application. The simplest mechanism that can be written for a Bi Bi

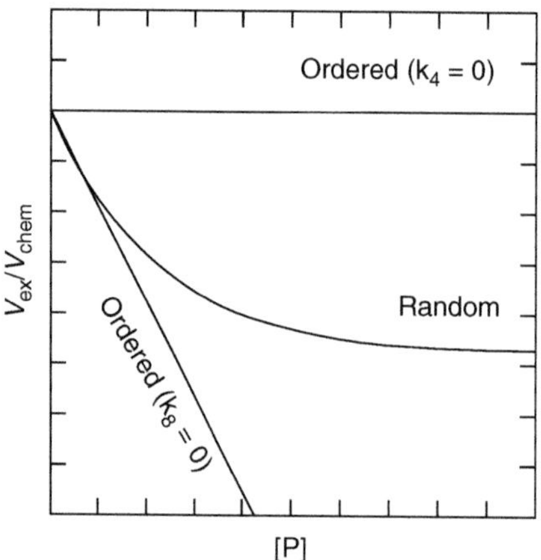

Fig. 3 Inhibition of the ratio of the positional isotope exchange rate and the net rate of chemical turnover as a function of added substrate.

Ordered reaction is shown in Scheme VII, where A is the substrate with the positional label and P is the product undergoing rotational exchange. For the forward reaction, the partitioning of the EPQ complex determines the PIX rate relative to the rate of net product formation. The partitioning of the EPQ complex can be written as

$$\mathrm{E} \underset{k_2}{\overset{k_1\mathrm{A}}{\rightleftharpoons}} \mathrm{EA} \underset{k_4}{\overset{k_3\mathrm{B}}{\rightleftharpoons}} \mathrm{EAB} \underset{k_6}{\overset{k_5}{\rightleftharpoons}} \mathrm{EPQ} \underset{k_8\mathrm{P}}{\overset{k_7}{\rightleftharpoons}} \mathrm{EQ} \underset{k_{10}\mathrm{Q}}{\overset{k_9}{\rightleftharpoons}} \mathrm{E}$$

Scheme VII

$$\frac{V_{\text{chem}}}{V_{\text{ex}}} = \frac{k_7(k_2k_4 + k_2k_5 + k_3k_5[\mathrm{B}])}{k_2k_4k_6} \tag{20}$$

The plot of $V_{\text{chem}}/V_{\text{ex}}$ as a function of the concentration of B yields a straight line. An analysis of the plot gives the following information:

$$\text{Intercept}_{[\mathrm{EPQ}]} = \frac{k_7(k_4 + k_5)}{k_4k_6} \tag{21}$$

$$\text{Slope}_{[\mathrm{EPQ}]} = \frac{k_3k_5k_7}{k_2k_4k_6} \tag{22}$$

If a PIX reaction can be followed for the reverse reaction, then analogous equations can be derived for the partitioning of the EAB complex:

$$\frac{V_{\text{chem}}}{V_{\text{ex}}} = \frac{k_4(k_6k_9 + k_7k_9 + k_6k_8[\mathrm{P}])}{k_5k_7k_9} \tag{23}$$

where the slope$_{[\mathrm{EAB}]}$ is equal to $(k_4k_6k_8)/(k_5k_7k_9)$ and the intercept$_{[\mathrm{EAB}]}$ is $k_4(k_6 + k_7)/(k_5k_7)$.

It can be shown that the microscopic rate constants for the release of the A and Q from the binary enzyme complexes can be determined from a combination of the respective slopes, intercepts, and the Michaelis constants as shown in Eqs. (24) and (25) (Raushel and Villafranca, 1988):

$$\frac{k_2}{V_1/E_t} = \left(\frac{1 + \text{intercept}_{[\mathrm{EPQ}]}}{\text{slope}_{[\mathrm{EPQ}]}}\right)\left(\frac{1}{K_b}\right) \tag{24}$$

$$\frac{k_9}{V_2/E_t} = \left(\frac{1 + \text{intercept}_{[\mathrm{EAB}]}}{\text{slope}_{[\mathrm{EAB}]}}\right)\left(\frac{1}{K_q}\right) \tag{25}$$

The lower limits for the release of B and P from the two ternary complexes can also be obtained:

$$\frac{k_7}{V_2/E_t} = \frac{k_7}{k_2} + \frac{k_7}{k_4} + \text{intercept}_{[EPQ]} \tag{26}$$

$$\frac{k_4}{V_1/E_t} = \frac{k_4}{k_9} + \frac{k_4}{k_7} + \text{intercept}_{[EAB]} \tag{27}$$

Therefore, the rate constants k_2 and k_9 relative to the net substrate turnover in the forward and reverse reactions are known from Eqs. (24) and (25), leaving two unknown rate constants, k_4 and k_7. Because there are two independent equations [Eqs. (26) and (27)] that contain both k_4 and k_7, both of the rate constants can be determined. Using this analysis of the PIX reaction, the rate constants for the release of all substrates and products from the enzyme complexes can be determined. If these results are combined with the steady-state kinetics parameters, V_1/E_t, V_2/E_t, K_a, K_b, K_p, and K_q, and the thermodynamic parameter, K_{eq}, then it is possible to estimate all 10 microscopic rate constants for the minimal Bi Bi Ordered kinetic mechanism shown in Scheme VII.

The analysis of PIX reactions as a function of added substrates and products in ordered kinetic mechanisms provides information that would otherwise be difficult to determine. It can also be used to corroborate classic steady-state experiments as well as to give information on the relative "stickiness" of a substrate or product.

D. Ping-Pong Mechanisms

In ping-pong reaction mechanisms, the first substrate binds and reacts covalently with the enzyme in the total absence of the second substrate to form a stable but modified enzyme-product complex. The second substrate binds and reacts with the modified enzyme to form the second product and regenerate the starting enzyme form. A simple Bi Bi Ping-Pong reaction is shown in Scheme VIII.

Because the first substrate can bind and react with the free enzyme in the absence of the second substrate, it is often assumed that the ping-pong reaction mechanism would be ideal for analysis by the PIX kinetic technique. The product bound to the modified enzyme could undergo torsional scrambling, and then reformation of the first substrate would yield a PIX reaction. However, this is a misconception because in reality no significant PIX reaction is expected to be observed because of the rapid dissociation of the first product from the modified enzyme (Hester and

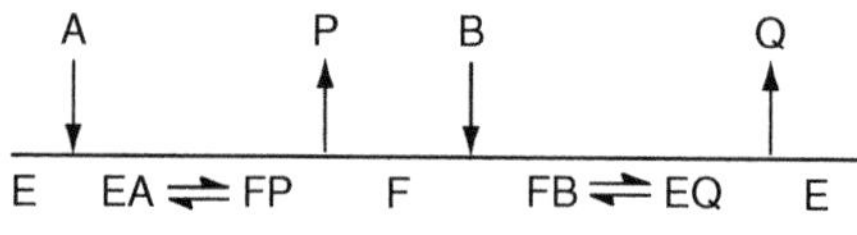

Scheme VIII

Raushel, 1987b). Essentially one enzyme equivalent of product is produced, leaving the enzyme in the modified form unable to process more substrate.

The minimal kinetic mechanism for a ping-pong reaction is shown in Scheme IX, where E is enzyme; A, substrate with the positional isotope label; P, product undergoing torsional scrambling; and F, stable modified enzyme. An analysis of the PIX reaction would give the partitioning of the FP complex shown in Eq. (28):

$$\mathrm{E} \underset{k_2}{\overset{k_1\mathrm{A}}{\rightleftharpoons}} \mathrm{EA} \underset{k_4}{\overset{k_3}{\rightleftharpoons}} \mathrm{FP} \underset{k_6\mathrm{P}}{\overset{k_5}{\rightleftharpoons}} \mathrm{F} \underset{k_8}{\overset{k_7}{\rightleftharpoons}} \mathrm{EQ} \underset{k_{10}\mathrm{Q}}{\overset{k_9}{\rightleftharpoons}} \mathrm{E}$$

Scheme IX

$$\frac{V_{\text{chem}}}{V_{\text{ex}}} = \frac{k_5(k_2 + k_3)}{k_2 k_4} \qquad (28)$$

However, for a PIX reaction to be observed, a method for reconverting the modified enzyme form F back to free enzyme E is required. This can be achieved with the addition of a large excess of unlabeled product P to the reaction mixture. The large excess of P returns the modified enzyme form F back to E, producing unlabeled A. This process provides a measure of the partitioning of the FP and F complexes. The excess level of P also dilutes the concentration of labeled P. This is important since the purpose of the PIX reaction is to determine the partitioning of the FP complex between F and E. To obtain an accurate determination for the partitioning of the FP complex, the formation of positionally exchanged substrate from the true PIX reaction must be distinguished from a pseudo-PIX reaction (caused by product dissociation and reassociation with the modified enzyme).

The ping-pong PIX experiment is initiated by addition of enzyme to labeled substrate A and unlabeled product P. The first-order equilibration of the positional label between substrate and product gives the partitioning between FP and F. Statistical considerations permit the determination of how much of the labeled product that is formed and released into solution will partition back to either the original labeled substrate or the positionally exchanged substrate. The rate of the pseudo-PIX mechanism can be calculated from the exchange rate of the label (positionally exchanged and nonexchanged) between product P and total substrate. The rate of the pseudo-PIX reaction can be used to correct the overall rate of formation of the positionally exchanged substrate to yield the true PIX rate for the interconversion of positionally labeled substrate. The correction factor will decrease as the ratio of unlabeled product relative to labeled substrate increases.

An analysis of the ping-pong PIX experiment is shown in Scheme X, where M, N, and O represent the positionally labeled, unlabeled, and positionally exchanged substrates, respectively. The equilibration of labeled and unlabeled substrates is represented by the interconversion of M $\rightarrow$ N $\rightarrow$ O. The formation of the

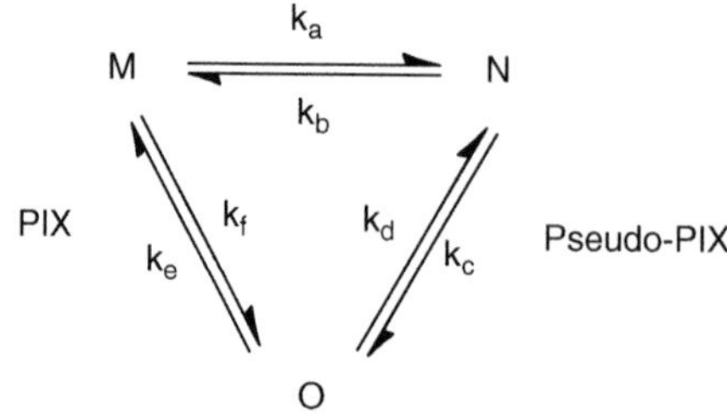

Scheme X

positionally exchanged substrate O by this pathway would be a pseudo-PIX reaction because of the disassociation and reassociation of the product. The interconversion of M $\rightarrow$ O without the formation of unbound labeled product provides the true PIX rate.

The rate constants in Scheme X can be defined as follows:

$$k_a = xy(1 + w) \tag{29}$$

$$k_b = x \tag{30}$$

$$k_c = xw \tag{31}$$

$$k_d = xy(1 + w) \tag{32}$$

$$k_e = xz \tag{33}$$

$$k_f = xzw \tag{34}$$

where x is proportional to the amount of enzyme used and therefore affects each step equally, y is the ratio of the initial concentration of unlabeled product P and labeled substrate A, w is the equilibrium ratio of positionally exchanged substrate and labeled substrate, and z is proportional to the true PIX reaction.

The factor x can be determined from an analysis of the plot of ([M] + [O])/([M] + [N] + [O]) versus time since the values of y and w are known. Once x is determined then z can be obtained by an analysis of the plot of ([O])/([M] + [O]) versus time. A numerical solution for the determination of the rate constants k_a through k_f is thus possible.

The time course for the equilibration of the labeled substrate and the positionally exchanged substrate, represented as a plot of [O]/([M] + [O]) versus time, is dependent on the ratio of the PIX and the pseudo-PIX pathways. The flux through the pseudo-PIX pathway can be diminished by increasing the ratio of unlabeled product to labeled substrate. This is represented in Fig. 4, where the curves are simulated with increasing values of z from 0.0 to 100. In the absence of a pathway

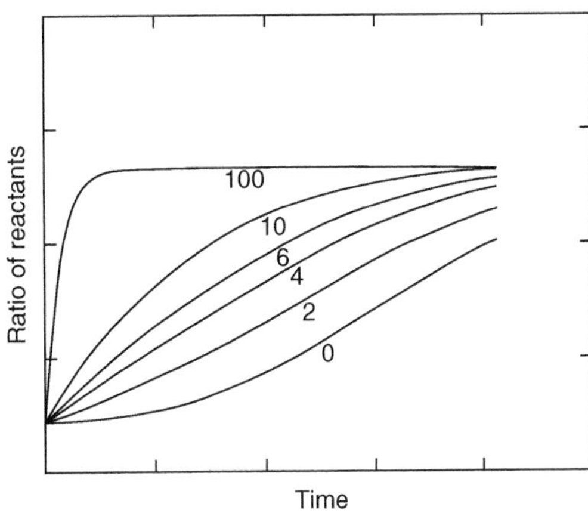

Fig. 4 Simulations for the interconversion of the labeled substrate (M) and the positionally exchanged substrate (O) as a function of time.

for the direct interconversion of M $\rightarrow$ O, there is a noticeable lag in the appearance of the positionally exchanged substrate.

To determine the partitioning of FP, the ratio V_{chem}/V_{ex} [Eq. (28)] must be expressed in terms of the rate constants k_a through k_f. The chemical rate for the conversion of substrate to product, V_{chem}, is $[X_0]k_a$, and the rate of PIX, V_{ex}, is $[X_0]$ $(k_e + k_f)$. The ratio V_{chem}/V_{ex} can now be expressed as

$$\frac{V_{chem}}{V_{ex}} = \frac{y}{z} \tag{35}$$

The lower limit for the conversion of FP to F, k_5, relative to the maximal velocity in the reverse direction, V_2/E_t, can be determined since V_2/E_t is less than or equal to $(k_2k_4)/(k_2 + k_5)$:

$$\frac{k_5}{V_2/E_t} = \frac{V_{chem}}{V_{ex}} \tag{36}$$

E. Specific Examples

Presented below are examples of five enzyme-catalyzed reactions that have been successfully studied by the PIX technique. Three of these enzymes, argininosuccinate lyase, UDPG pyrophosphorylase (UTP-glucose-1-phosphate uridylyltransferase), and galactose-1-phosphate uridylyltransferase (UTP-hexose-1-phosphate uridylyltransferase) give representative examples of the type of information available from a quantitative analysis of enzyme-ligand dissociation rates using PIX. The last two examples, D-alanine-D-alanine ligase and carbamoyl-phosphate synthase, are presented to illustrate how the PIX technique can be best utilized to identify reaction intermediates.

1. Argininosuccinate Lyase

The effect of added products on the observed PIX rate was first investigated by Raushel and Garrard (1984) on the reaction catalyzed by argininosuccinate lyase. Argininosuccinate lyase catalyzes the cleavage of argininosuccinate to arginine and fumarate. The effect of fumarate addition on the $^{15}N/^{14}N$ positional exchange reaction within argininosuccinate was measured by ^{15}N NMR spectroscopy. Scheme XI shows the PIX reaction that was followed with the enzyme. The two external nitrogens in the guanidino moiety of the product arginine are torsionally equivalent. This structural feature enables the scrambling of the ^{15}N and ^{14}N labels within argininosuccinate when the guanidino group is free to rotate about the C-N bond in the enzyme-arginine-fumarate complex.

Rotation

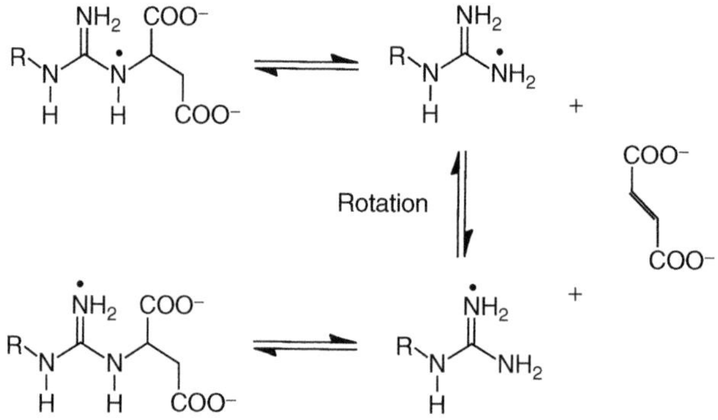

Scheme XI

Raushel and Garrard (1984) demonstrated that at zero fumarate there was no observable PIX reaction relative to the net formation of product ($V_{ex}/V_{chem} < 0.15$). However, at higher levels of added fumarate the ratio V_{ex}/V_{chem} increased until a plateau of 1.8 was reached. These results indicate quite clearly that the release of products from the enzyme-arginine-fumarate complex in the argininosuccinate lyase-catalyzed reaction is kinetically random. Because no exchange was observed at low fumarate, it can also be concluded that the release of at least one of the two products must be very fast. Moreover, the release of arginine from the ternary complex, relative to V_2/E_t, must be greater than or equal to 0.5 (based on the limiting value of 1.8 at saturating fumarate). Therefore, the release of fumarate from the ternary complex (see Scheme V) is very fast relative to the maximal velocity in the reverse direction [$k_9/(V_2/E_t) > 6$], and thus the release of fumarate from the enzyme-arginine-fumarate complex is at least 10 times faster than arginine release.

2. UDP-Glucose Pyrophosphorylase

Hester and Raushel (1987a) investigated the effects of substrate addition on the PIX reaction catalyzed by UDPG pyrophosphorylase. UDPG pyrophosphorylase catalyzes the conversion of UTP and glucose-1-phosphate to UDP-glucose and pyrophosphate. The reaction proceeds by the nucleophilic attack of glucose-1-phosphate on the α-phosphate of UTP. The kinetic mechanism has been previously shown to be ordered with UTP as the first substrate to bind and pyrophosphate the first product to be released (see Scheme VII) (Tsuboi *et al.*, 1969).

Hester and Raushel (1987a) showed that it was possible to follow a PIX reaction in both the forward and reverse reactions of the enzyme by utilizing the positionally labeled substrates shown in Scheme XII. The PIX reactions in the forward and reverse directions were able to be suppressed by increasing the concentration of the second substrate, thus confirming the strictly ordered nature of the reaction mechanism (see Fig. 5).

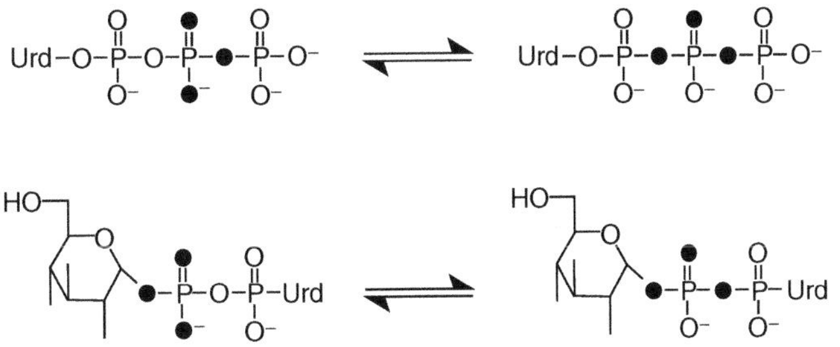

Scheme XII

By combining the quantitative information from the PIX reactions with the steady-state kinetic and thermodynamic parameters, Hester and Raushel (1987a) were able to obtain estimates for all of the microscopic rate constants in the UDPG pyrophosphorylase reaction. The values calculated for the rate constants (shown in Scheme XIII) reveal that the release of UDP-glucose is three times slower than the release of pyrophosphate and that the release of glucose-1-phosphate is five times slower than the release to UTP. The back-calculated kinetic constants are in excellent agreement with the experimental values. This example demonstrates quite clearly the potential for obtaining quantitative information about the binding and release of substrates and products from a thorough analysis of PIX reactions.

$$\mathrm{E} \underset{5.6\times10^{3}\ \mathrm{sec}^{-1}}{\overset{3.2\times10^{6}\ M^{-1}\ \mathrm{sec}^{-1}}{\rightleftharpoons}} \mathrm{E{\cdot}UTP} \underset{1.1\times10^{3}\ \mathrm{sec}^{-1}}{\overset{2.9\times10^{7}\ M^{-1}\ \mathrm{sec}^{-1}}{\rightleftharpoons}} \mathrm{E{\cdot}UTP{\cdot}Glu\text{-}1\text{-}P} \underset{3.0\times10^{4}\ \mathrm{sec}^{-1}}{\overset{2.3\times10^{4}\ \mathrm{sec}^{-1}}{\rightleftharpoons}} \mathrm{E{\cdot}PP_i{\cdot}UDPG} \underset{5.3\times10^{6}\ M^{-1}\ \mathrm{sec}^{-1}}{\overset{1.1\times10^{3}\ \mathrm{sec}^{-1}}{\rightleftharpoons}} \mathrm{E{\cdot}UDPG} \underset{2.4\times10^{6}\ M^{-1}\ \mathrm{sec}^{-1}}{\overset{3.4\times10^{2}\ \mathrm{sec}^{-1}}{\rightleftharpoons}} \mathrm{E}$$

Scheme XIII

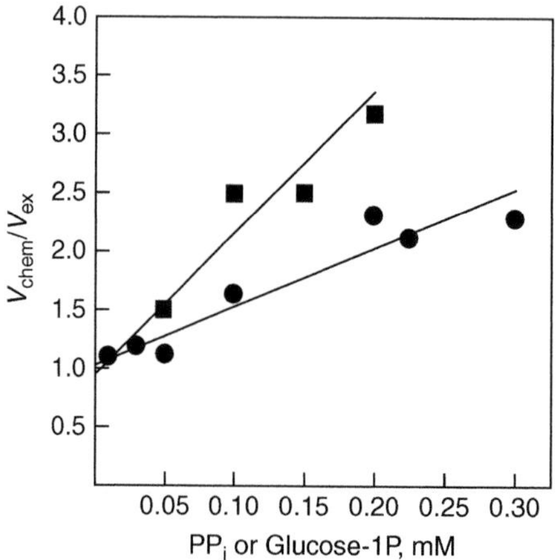

Fig. 5 Plot of the ratio of the net chemical turnover rate and the positional isotope exchange rate as a function of the concentration of added glucose-1-phosphate (●) or pyrophosphate (PP_i) (▪).

3. Galactose-1-Phosphate Uridylyltransferase

The PIX technique has rarely been applied to a ping-pong reaction mechanism. However, Hester and Raushel (1987b) have investigated the reaction catalyzed by galactose-1-phosphate uridylyltransferase. Galactose-1-phosphate uridylyltransferase catalyzes the transfer of the uridylyl group from UDP-glucose to galactose-1-phosphate. A covalent enzyme-uridylyl adduct is an intermediate in the reaction as determined by stereochemical and steady-state kinetic analysis (Sheu and Frey, 1978). The PIX experiment monitored only the first half-reaction of galactose-1-phosphate uridylyltransferase as shown in Scheme XIV.

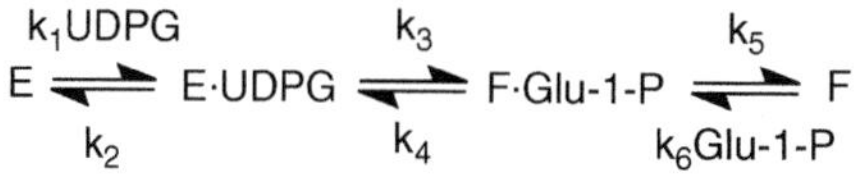

$$\mathrm{E} \underset{k_2}{\overset{k_1\mathrm{UDPG}}{\rightleftharpoons}} \mathrm{E{\cdot}UDPG} \underset{k_4}{\overset{k_3}{\rightleftharpoons}} \mathrm{F{\cdot}Glu\text{-}1\text{-}P} \underset{k_6\mathrm{Glu\text{-}1\text{-}P}}{\overset{k_5}{\rightleftharpoons}} \mathrm{F}$$

Scheme XIV

Unlabeled glucose-1-phosphate was included in the reaction mixture to recycle the enzyme so that the partitioning of the uridylyl enzyme-glucose-1-phosphate complex could be determined. The PIX reaction followed the torsional scrambling of $^{18}O_3$ within [β-$^{18}O_3$]UDP-glucose (Scheme XV) in the presence of variable amounts of unlabeled glucose-1-phosphate as a function of time. No PIX reaction was observed in the absence of glucose-1-phosphate as a function of time. No PIX reaction was observed in the absence of glucose-1-phosphate as expected for a

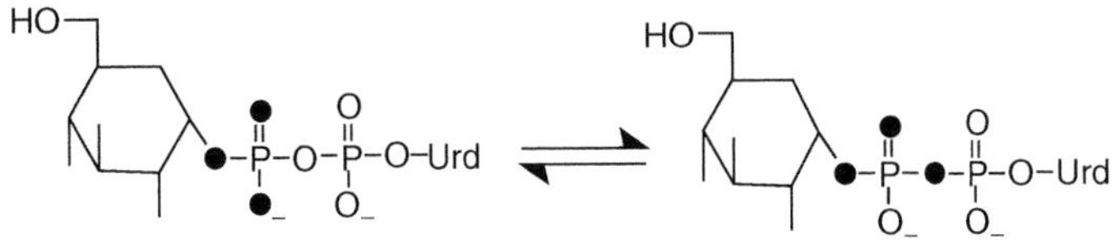

Scheme XV

ping-pong reaction. The partitioning of the uridylyl enzyme-glucose-1-phosphate (V_{chem}/V_{ex}) was determined to be 3.4. Therefore, the release of glucose-1-phosphate from the uridylyl enzyme complex is 3.4 times faster than the maximal velocity in the reverse direction.

4. D-Alanine-D-Alanine Ligase

Mullins *et al.* (1990) utilized the PIX technique to obtain kinetic evidence for the intermediate formation of D-alanyl phosphate in the reaction catalyzed by D-alanine-D-alanine ligase. The D-alanine-D-alanine ligase reaction is a difficult reaction to study mechanistically because it utilizes the same substrate twice. D-Alanine-D-alanine ligase catalyzes the formation of D-alanyl-D-alanine from ATP and two molecules of D-alanine. The reaction has been proposed to proceed through an acyl phosphate intermediate formed via the phosphorylation of the carboxyl group of the first D-alanine by the γ-phosphate of ATP (Scheme XVI) (Duncan and Walsh, 1988).

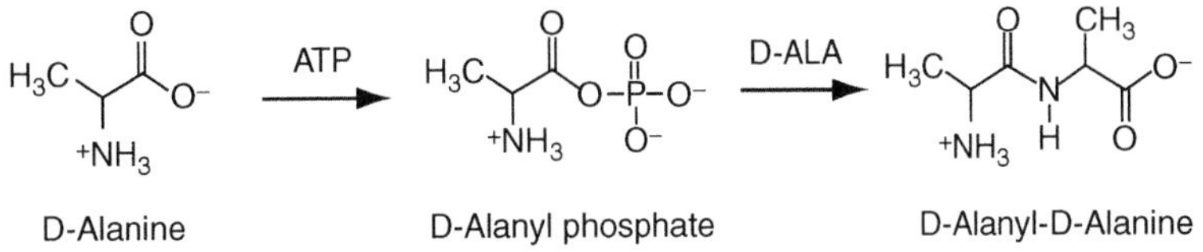

D-Alanine —ATP→ D-Alanyl phosphate —D-ALA→ D-Alanyl-D-Alanine

Scheme XVI

Mullins *et al.* followed the PIX reaction shown in Scheme XVII. In D-alanine-D-alanine ligase, the difficulty of investigating the kinetic mechanism is increased because the reaction cannot be artificially stopped after the addition of the first two substrates to probe for the existence of the intermediate. A PIX reaction is observed only in the presence of D-alanine, and this is consistent with the direct attack of the carboxyl group of one D-alanine on the γ-phosphate of ATP to give an acyl phosphate intermediate. Cleavage of the γ-phosphate of ATP could occur either after the first D-alanine binds or, alternatively, only after the binding of all three substrates. However, the observed PIX rate is diminished relative to the net substrate turnover as the concentration of D-alanine is increased. This is consistent with an ordered kinetic mechanism. In addition, the ratio of the PIX rate relative to the net chemical turnover of substrate (V_{ex}/V_{chem}) approaches a value of 1.4 as the concentration of D-alanine becomes very small. This ratio is 100 times larger than the ratio of the maximal reverse and forward chemical reaction velocities (V_2/V_1). This situation is only possible when the reaction mechanism in the forward direction proceeds in two distinct steps and the first step is much faster than the second step. Thus, it appears that formation of the acyl phosphate intermediate is faster than amide bond formation.

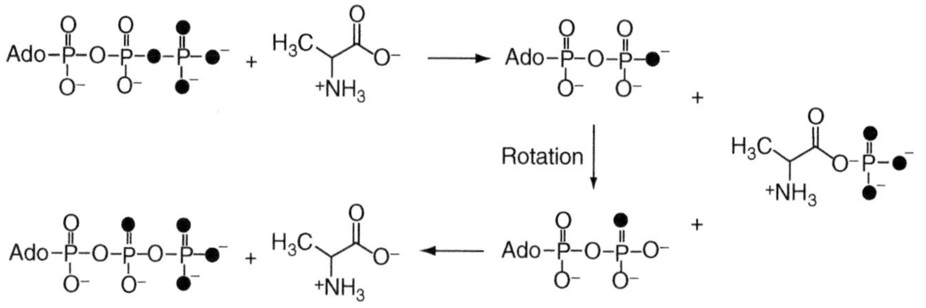

Scheme XVII

5. Carbamoyl-Phosphate Synthetase

The PIX technique has been used to identify two reactive intermediates in the reaction catalyzed by carbamoyl-phosphate synthetase (CPS). CPS catalyzes the following reaction:

$$\begin{aligned} 2\,\text{MgATP} + \text{HCO}_3^- + \text{glutamine} + \text{H}_2\text{O} &\rightarrow \text{carbamoyl phosphate} \\ &+2\,\text{MgADP} + \text{P}_i + \text{glutamine} \end{aligned} \tag{37}$$

The protein is heterodimeric consisting of a large (120 kDa) and a small (42 kDa) subunit (Trotta *et al.*, 1971). The proposed mechanism for the synthesis of carbamoyl phosphate (shown in Scheme XVIII) is composed of at least four individual steps (Anderson and Meister, 1965). Carboxyphosphate and carbamate have been postulated as the key intermediates in the overall reaction mechanism. The hydrolysis of glutamine to glutamate occurs on the small subunit while the remaining reactions occur on the large subunit. Therefore, the ammonia nitrogen must be transferred from the small subunit to the active site of the large subunit. Free ammonia can be utilized as the nitrogen source by the large subunit in the absence of an active glutaminase reaction from the small subunit. A partial reaction catalyzed by the large subunit is the bicarbonate-dependent hydrolysis of ATP.

This reaction is thought to result from the slow hydrolysis of the carboxyphosphate intermediate in the absence of a nitrogen source. In the reverse direction, the enzyme can utilize carbamoyl phosphate to phosphorylate MgADP. The other two products of the partial back reaction are CO_2 and NH_3, which probably arise from the decomposition of carbamate.

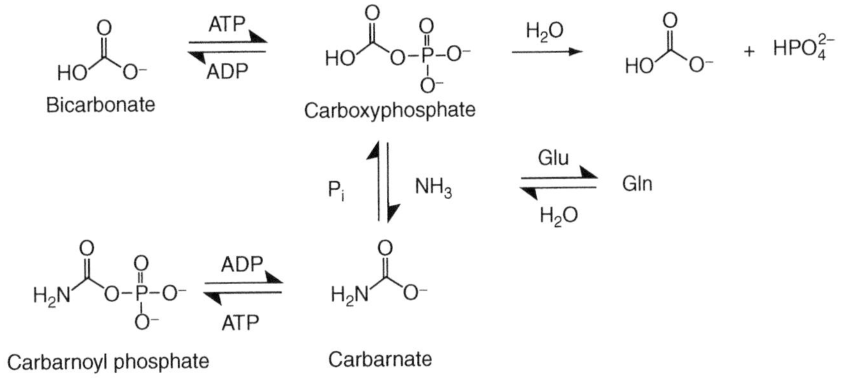

Bicarbonate
Carboxyphosphate
Carbarnoyl phosphate
Carbarnate

Scheme XVIII

Wimmer *et al.* (1979) utilized ^{18}O-labeled ATP to probe for a PIX reaction that would support the formation of carboxyphosphate as a reactive intermediate (see Scheme XIX). These workers found that in the presence of enzyme and bicarbonate CPS catalyzed a PIX within ATP which was 1.7 times as fast as the net hydrolysis of ATP. Therefore, the E · ADP · carboxyphosphate complex releases a product into solution about as fast as ATP is resynthesized and released into the bulk solution. These results have been confirmed by three other research groups (Meek *et al.*, 1987; Raushel and Villafranca, 1980; Reynolds *et al.*, 1983).

The second intermediate, carbamate, was probed by following the positional exchange of an ^{18}O label within carbamoyl phosphate (Raushel and Villafranca, 1980). Scheme XX shows that if carbamate is formed and stabilized at the active

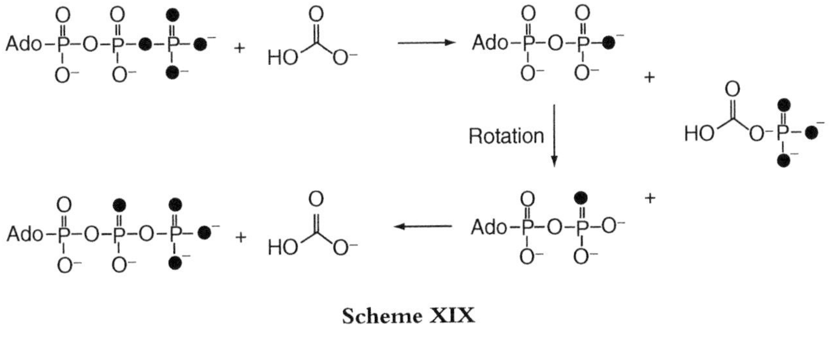

Rotation

Scheme XIX

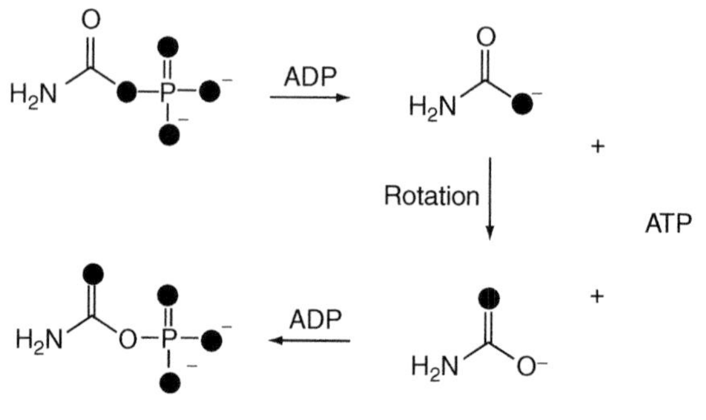

Scheme XX

site on mixing of ADP and carbamoyl phosphate, then a PIX reaction is possible via the torsional rotation of the carboxyl group of carbamate. The PIX rate was found to be four times faster than the net synthesis of ATP, and thus carbamate is a kinetically significant reactive intermediate in the carbamoyl-phosphate synthase reaction.

Work from the Meister laboratory has provided evidence that the glutaminase reaction catalyzed by the small subunit requires an essential cysteine residue (Pinkus and Meister, 1972). It was noted that when the glutaminase reaction was destroyed by incubation with a chloroketone analog of glutamine, the bicarbonate-dependent ATPase reaction was stimulated. Rubino *et al.* (1987) later identified the critical cysteine residue by site-directed mutagenesis to be Cys-269. The two mutants (C269G and C269S) made by Rubino *et al.* not only lost all glutaminase activity, but also more interestingly showed a significant enhancement of the bicarbonate-dependent ATPase reaction relative to wild-type enzyme. The bicarbonate-dependent hydrolysis of ATP normally occurs at approximately 5–10% of the rate of carbamoyl phosphate synthesis. The slow hydrolysis rate represents the protection of the unstable carboxyphosphate intermediate from water. Therefore, the stabilization of carboxyphosphate by the enzyme results from either exclusion of water from the active site or the very slow release of the reactive intermediate from the active site. Apparently, the two mutant enzymes (C269G and C269S) cannot adequately protect carboxyphosphate from water, thus permitting the partial hydrolysis reaction to compete with the overall synthesis of carbamoyl phosphate.

The energetics of the bicarbonate-dependent ATPase reaction catalyzed by the wild-type enzyme have been examined by Raushel and Villafranca (1979) using rapid quench and PIX experiments. Scheme XXI shows the kinetic mechanism for the bicarbonate-dependent ATPase reaction along with the kinetic barrier diagram for that mechanism, where ES represents the enzyme-HCO_3^--ATP complex and EI represents the enzyme-carboxyphosphate-ADP complex. At saturating levels of substrates the rate of the reaction shown in Scheme XXI is governed by Eq. (38), and the ratio of the PIX rate relative to the rate of net turnover of ATP (V_{ex}/V_{chem}) is given by the ratio of rate constants presented in Eq. (39):

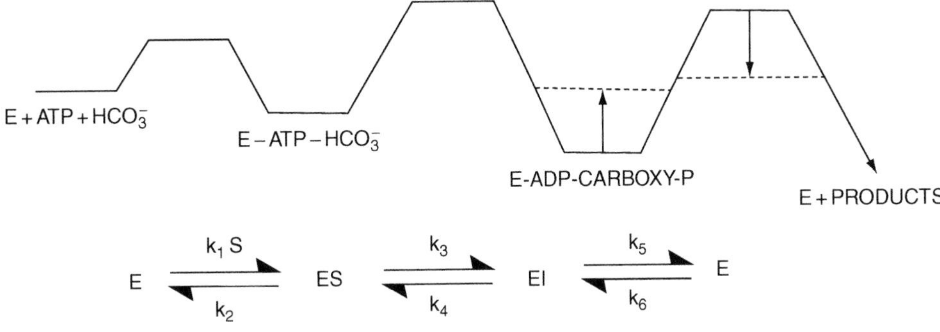

Scheme XXI

$$V\text{max} = \frac{k_3 k_5}{k_3 + k_4 + k_5} \tag{38}$$

$$\frac{V_{ex}}{V_{chem}} = \frac{k_2 k_4}{k_5(k_2 + k_3)} \tag{39}$$

The presteady-state time course exhibited "burst" kinetics with a rapid formation of acid-labile phosphate followed by a slower steady-state rate. Raushel and Villafranca measured the values of k_3, k_4, and k_5 as 4.2, 0.10, and 0.21 s^{-1}, respectively, from the rapid quench and previous PIX data. They also showed that k_2 is much greater than 3.1 s^{-1}. Therefore, the formation of carboxyphosphate is very fast, with the rate-limiting steps for net ATP hydrolysis involving either product release or hydrolysis of the intermediate in the active site.

Mullins *et al.* (1991) used the PIX technique to examine the changes in the reaction energetics of the ATPase reaction in the C269G and C269S mutants and the isolated large subunit. The PIX reaction they followed has been previously shown in Scheme XIX. They found the bicarbonate-dependent ATPase reaction for all three enzymes to be two- to threefold faster than the wild-type enzyme, but 4- to 10-fold faster if glutamine is present. As no increase is observed in the NH_3-dependent carbamoyl phosphate synthesis rate, k_3 is not changed in the mutants. The significant increase in the bicarbonate-dependent ATPase reaction of the mutants can be the consequence of two possible alterations in the kinetic barrier diagram. It could reflect the stabilization of the transition state for the reaction of carboxyphosphate with water. However, this stabilization would only affect k_5 and k_6 while leaving k_4 unchanged. Alternatively, the ground state for the enzyme-bound carboxyphosphate may be destabilized, resulting in an increase in k_4 and k_5 by the same factor. Mullins *et al.* showed that the two cases could be distinguished by measuring the partitioning of the E-ADP-carboxyphosphate complex. The ratio V_{ex}/V_{chem} simplifies to k_4/k_5 since $k_2 \gg k_3$ [Eq. (39)]. Therefore, if the ground state of the enzyme-carboxyphosphate complex is destabilized, then the ratio V_{ex}/V_{chem} will be identical for the mutants and the wild-type enzyme. If the transition state for the hydrolysis of carboxyphosphate is stabilized, however, the ratio V_{ex}/V_{chem} will be reduced in the mutants relative to the wild-type enzyme.

Mullins *et al.* (1991) observed that the ratio of the PIX rate relative to ATP turnover was identical for the wild type, C269G, C269S, and the isolated large subunit. Therefore, the alteration in the energetics is due to the destabilization of the ground state for the enzyme-bound carboxyphosphate-ADP complex. In the presence of NH_3, however, the value of V_{ex}/V_{chem} was reduced over 20-fold. Thus, the increased turnover in the presence of NH_3 results from the substantial stabilization of the transition state for the reaction of ammonia with carboxyphosphate relative to the reaction with water.

F. Other Examples

A number of other enzymes have been successfully analyzed using PIX methodology. Bass *et al.* (1984) have probed the reaction catalyzed by adenylosuccinate synthetase, which involves the synthesis of adenylosuccinate from GTP, IMP, and aspartate. Incubation of [β, γ-^{18}O]GTP, IMP, and enzyme resulted in scrambling of the label from the $\beta\gamma$-bridge position to the β-nonbridge positions. No scrambling was observed in the absence of IMP, and the addition of aspartate was not required for PIX to occur. This result has been interpreted to support a two-step mechanism for the synthesis of adenylosuccinate where the GTP phosphorylates the carbonyl oxygen of IMP to form a phosphorylated intermediate. The phosphorylated intermediate subsequently reacts with the α-amino group of aspartate to generate the final product, adenylosuccinate.

The Lowe laboratory has examined the reactions catalyzed by the aminoacyl-tRNA synthetases using the PIX technique (Lowe and Tansley, 1984; Lowe *et al.*, 1983a,b). The enzymes are responsible for the condensation of amino acids to the cognate tRNA. The reaction mechanism has been proposed to involve the formation of an aminoacyl adenylate intermediate from an amino acid and ATP. In support of this mechanism Lowe *et al.* have shown that the isoleucyl-, tyrosyl-, and methionyl-tRNA synthetases all catalyze a PIX reaction from the β-nonbridge position of ATP to the $\alpha\beta$-bridge position in the presence of the required amino acid. No exchange was observed in the absence of the amino acid, nor was any PIX reaction observed in the presence of the dead-end alcohol analogs of the amino acids.

The enzyme CTP synthase catalyzes the formation of CTP from glutamine, ATP, and UTP. The two most reasonable reaction mechanisms involve either the attack of ammonia at C-4 of UTP to form the carbinolamine or, alternatively, the phosphorylation of the carbonyl oxygen of UTP by the ATP. von der Saal *et al.* (1985a) have shown that the enzyme will catalyze the exchange of label from the $\beta\gamma$-bridge position to the β-nonbridge position in the presence of UTP. No ammonia or glutamine is required for the reaction to be observed. This result is consistent only with the phosphorylated UTP intermediate, and thus the carbinolamine intermediate can be discarded.

The enzymatic synthesis of GMP follows a reaction scheme that is analogous to the synthesis of CTP. The enzyme GMP synthetase utilizes ATP, XMP, and glutamine to construct the final bond in the formation of GMP. The proposed reaction mechanism involves the adenylation the carbonyl oxygen of XMP by ATP. von der Saal *et al.* (1985b) were able to demonstrate that on incubation of ^{18}O-labeled ATP and XMP a PIX reaction occurred which did not require the presence of glutamine or ammonia.

The enzyme pyruvate-phosphate dikinase catalyzes the formation of phosphoenolpyruvate (PEP) from ATP, phosphate, and pyruvate. The other two products of the reaction are AMP and pyrophosphate. The proposed reaction mechanism is thought to involve at least three separate reactions. ATP pyrophosphorylates the enzyme in the first reaction, and this intermediate subsequently phosphorylates phosphate to produce pyrophosphate and a phosphorylated enzyme intermediate. In the last step, the phosphorylated enzyme transfers the phosphoryl group to pyruvate to form the ultimate product, PEP. Wang *et al.* (1988) used a variety of ^{18}O-labeled ATP molecules to demonstrate that the observed PIX reactions were consistent with the proposed reaction mechanism.

The enzyme PEP carboxykinase catalyzes the formation of PEP and CO_2 from oxaloacetate and GTP. Chen *et al.* (1991) have utilized PIX methodology to examine the partitioning of enzyme-product complexes. No PIX was observed within the labeled GTP under initial velocity conditions when enzyme was mixed with oxaloacetate and [β, γ-^{18}O]GTP. These results have been interpreted to indicate that at least one of the products dissociates rapidly from the enzyme-GDP-PEP-CO_2 complex relative to the net rate of GTP formation from the complex.

V. Summary

The PIX technique has been found to be quite useful for the identification of reaction intermediates in enzyme-catalyzed reactions. For reactions where intermediates are not expected the method can be used with great utility for the quantitative determination of the partitioning of enzyme-product complexes. However, it must be remembered that it has been explicitly assumed that the functional group undergoing positional exchange is free to rotate. This assumption is not always valid since examples have been discovered where the functional group rotation is indeed hindered. For instance, in the reaction catalyzed by argininosuccinate synthetase a PIX reaction was not observed on incubation of ATP and citrulline even though a citrulline-adenylate complex has been identified from rapid quench experiments (Hilscher *et al.*, 1985).

Acknowledgments

The authors are grateful for financial support from the National Institutes of Health (DK30343, GM33894, and GM49706).

References

Anderson, P. M., and Meister, A. (1965). *Biochemistry* **4,** 2803.
Bass, M. B., Fromm, H. J., and Rudolf, F. B. (1984). *J. Biol. Chem.* **259,** 12330.
Chen, C. Y., Sato, Y., and Schramm, V. L. (1991). *Biochemistry* **30,** 4143.
Cleland, W. W. (1975). *Biochemistry* **14,** 3220.
Duncan, K., and Walsh, C. T. (1988). *Biochemistry* **27,** 3709.
Hester, L. S., and Raushel, F. M. (1987). *Biochemistry* **26,** 6465.
Hester, L. S., and Raushel, F. M. (1987). *J. Biol. Chem.* **262,** 12092.
Hilscher, L. W., Hanson, C. D., Russel, D. H., and Rauschel, F. M. (1985). *Biochemistry* **24,** 5888.
Lowe, G., Sproat, B. S., and Tansley, G. (1983a). *Eur. J. Biochem.* **130,** 341.
Lowe, G., Sproat, B. S., Tansley, G., and Cullis, P. M. (1983b). *Biochemistry* **22,** 1229.
Lowe, G., and Tansley, G. (1984). *Tetrahedron* **40,** 113.
Meek, T. D., Karsten, W. E., and DeBrosse, C. W. (1987). *Biochemistry* **26,** 2584.
Midelfort, C. G., and Rose, I. A. (1976). *J. Biol. Chem.* **251,** 5881.
Mullins, L. S., Lusty, C. J., and Raushel, F. M. (1991). *J. Biol. Chem.* **266,** 8236.
Mullins, L. S., Zawaszke, L. E., Walsh, C. T., and Raushel, F. M. (1990). *J. Biol. Chem.* **265,** 8993.
Pinkus, L. M., and Meister, A. (1972). *J. Biol. Chem.* **247,** 6119.
Raushel, F. M., and Garrard, L. J. (1984). *Biochemistry* **23,** 1791.
Raushel, F. M., and Villafranca, J. J. (1979). *Biochemistry* **18,** 3424.
Raushel, F. M., and Villafranca, J. J. (1980). *Biochemistry* **19,** 3174.
Raushel, F. M., and Villafranca, J. J. (1988). *Crit. Rev. Biochem.* **23,** 1.
Reddick, R. E., and Kenyon, G. L. (1987). *J. Am. Chem. Soc.* **109,** 4380.
Retey, J., Suckling, C. J., Arigoni, D., and Babior, B. (1974). *J. Biol. Chem.* **249,** 6359.
Reynolds, M. A., Oppenheimer, N. J., and Kenyon, G. L. (1983). *J. Am. Chem. Soc.* **105,** 6663.
Rose, I. A. (1979). *Adv. Enzymol. Relat. Areas Mol. Biol.* **50,** 361.
Rubino, S. D., Nyunoya, H., and Lusty, C. J. (1987). *J. Biol. Chem.* **262,** 4382.
Sheu, K. R., and Frey, P. A. (1978). *J. Biol. Chem.* **253,** 3378.
Trotta, P. P., Burt, M. E., Haschemeyer, R. H., and Meister, A. (1971). *Proc. Natl. Acad. Sci. USA* **68,** 2599.
Tsuboi, K. K., Fukunaga, K., and Petricciani, J. C. (1969). *J. Biol. Chem.* **244,** 1008.
von der Saal, W., Anderson, P. M., and Villafranca, J. J. (1985a). *J. Biol. Chem.* **260,** 14993.
von der Saal, W., Crysler, C. S., and Villafranca, J. J. (1985b). *Biochemistry* **24,** 5343.
Wang, H. C.h., Ciskanik, L., Dunaway-Mariano, D., von der Saal, W., and Villafranca, J. J. (1988). *Biochemistry* **27,** 625.
Wedler, F. C., and Boyer, P. D. (1972). *J. Biol. Chem.* **247,** 984.
Wimmer, M. J., Rose, I. A., Powers, S. G., and Meister, A. (1979). *J. Biol. Chem.* **254,** 1854.

CHAPTER 18

Enzymatic Transition-State Analysis and Transition-State Analogs

Vern L. Schramm
Department of Biochemistry
Albert Einstein College of Medicine
Bronx, New York 10461

CONTEMPORARY ENZYME KINETICS AND MECHANISM
Reprinted from *Methods in Enzymology*, Volume 308 (Academic Press, 1999).

DOI: 10.1016/B978-0-12-378608-1.00018-9

I. Introduction

Kinetic isotope effects (KIE) permit experimental access to the transition-state (TS) structure of enzymatic reactions and are the only method currently available to obtain detailed TS information. The theory and some methods of this approach have been discussed in two previous volumes of methods in enzymology, as well as in reviews and monographs (Cleland, 1982, 1995; Cleland *et al.*, 1977; Cook, 1991; Gandour and Schowen, 1978; Melander and Saunders, 1980). The application of isotope effects to establish TS structures for enzymatic reactions has evolved significantly since its last treatment in this series (Cleland, 1995). Some of the practical and theoretical advances during that period will be the focus of this section. In this and the following sections, three major advances since the mid-1990s are treated. This chapter deals with the methods for application of systematic KIE and methods to measure KIE in several complicated enzymatic systems. These include methods for (1) the enzymatic synthesis of substrates with specific and stereospecific labels in nucleotides, nucleosides, NAD^+, RNA, and DNA; (2) KIE analysis for enzymes with suppressed experimental KIE because of commitment factors; (3) the analysis of isotope effects in complex systems of protein covalent modification; (4) measuring KIE in large substrates typified by RNA and DNA; (5) the measurement of binding isotope effects and their influence on observed KIE; and (6) application of TS information for the design of TS inhibitors. In Chapter 19, recently developed methods are summarized for the computational modeling to match intrinsic KIE to bonding models of the TS (Berti, 2009). Computational approaches are increasingly accessible through the implementation of electron density functional theory, applied in the Gaussian computational chemistry programs, to estimate the bonding of nonequilibrium chemical states, which provide the models for the TS (Frisch *et al.*, 1995; Parr and Yang, 1989). These are matched to the KIE using current adaptations of bond-energy bond-order vibrational analysis programs (Huskey, 1993; Sims and Lewis, 1984; Sims *et al.*, 1977). Chapter 20 demonstrates the relationship of the TS structure to the TS binding energy, and the prediction of inhibitor binding energy by comparing its features to that of experimentally determined TS (Braunheim and Schwartz, 2009). These three chapters demonstrate that the combined use of KIE, computational chemistry, and TS inhibitor design have become powerful tools in understanding enzymatic TS. Knowledge of the TS is proving helpful in the design of TS inhibitors and has resulted in the most powerful inhibitors known for several different enzymes (Chen *et al.*, 1998; Li *et al.*, 1999; Miles *et al.*, 1998; Schramm, 1998; Schramm *et al.*, 1994).

II. Nature of Enzymatic TS

TS lifetimes are measured on the time scale of single bond vibrations, as the restoring force for the bond of interest is lost and is replaced by a translational motion, a negative restoring force (Sims and Lewis, 1984). Direct observations of the TS are problematic because of the short-time scale and have only been accomplished by laser spectroscopy or by supercooling of molecules in the gas phase (Liu *et al.*, 1993; Scherer *et al.*, 1987). Solution chemistry is largely intractable to direct TS observation because of the influence of the solvent on the spectroscopic methods to observe transient bonded states and also because of participation of the solvent in the organization or stabilization of the TS. In the usual construct of TS theory, the lifetime is within a single bond vibration or approximately 10^{-13} s. In TS investigations, we are concerned with the geometric and electronic features of this metastable structure. Many enzymes form unstable intermediates, but the TS is not considered to be an intermediate as intermediates have lifetimes of many bond vibrations. Intermediates in enzymatic reactions are often more closely related to the TS than substrates and therefore bind tightly, as do analogs of the tightly bound intermediate (Morrison and Walsh, 1988). Examples of this phenomenon are the sp^3-hybridized analogs of serine proteases that mimic the serine adduct at the carbonyl carbon (Radzicka and Wolfenden, 1995).

TS theory for enzymatic reactions proposes that the rate enhancement imposed by enzymes is solely due to the tight binding or "stabilization" of the activated complex (Pauling, 1948; Schowen, 1978). Theories that do not invoke tight binding of the TS complex have also been proposed (Cannon *et al.*, 1996). Knowledge of the TS for an enzymatic reaction can therefore provide information to design stable analogs as TS inhibitors. It is frequently impossible to estimate the rate of uncatalyzed biological reactions, but in the examples where enzymatic and nonenzymatic rates can be measured (e.g., in hydrolytic and decarboxylation reactions), it is well documented that the enzyme enhances the reaction rate by factors of 10^{10}–10^{18} (Radzicka and Wolfenden, 1995, 1996), TS constructs for enzyme-catalyzed reactions propose that the energetics of catalysis arise from binding the TS tighter than the substrate by a factor equivalent to the enzymatic rate enhancement. This means that the TS is bound 10^{10}–10^{18} times tighter than the substrate in the Michaelis complex. In summaries of TS and multisubstrate analogs, it is apparent that inhibitors with affinities of 10^{-9}–10^{-11} represent the majority of these inhibitors, with more tightly bound analogs being rare (Morrison and Walsh, 1988). Because Michaelis complexes have typical dissociation constants of 10^{-3}–10^{-7} M, enzymatic TS complexes are bound hypothetically with dissociation constants of 10^{-13}–10^{-25} M that correspond to −18 to −35 kcal/mol binding energy. These dissociation constants are astounding compared to those for known enzymatic inhibitors and indicate that many orders of magnitude of inhibitory potential are available beyond current inhibitor design methods. It may be possible to design improved inhibitors, provided that we have accurate information on the nature of the enzymatic TS and are capable of synthesizing closely related chemical mimics.

III. TS Binding Energy

Where does TS binding energy come from? The current proposal is that the enzyme conforms to the TS configuration in which amino acid side chains and the peptide backbone form hydrogen bonds to the TS complex, which are energetically more favorable than those in the Michaelis complex (Cleland, 1992; Gerlt and Gassman, 1993; Mildvan *et al.*, 1999; Shan and Herschlag, 1996). Hydrogen bond donor-acceptor pairs that contribute equally to electron sharing of an exchangeable hydrogen atom may form low-barrier hydrogen bonds that facilitate proton transfers and may be more energetic than average hydrogen bonds. The catalytic site becomes more constrained, water is excluded, the site becomes more hydrophobic, and the multiple hydrogen bonds shorten in the hydrophobic environment and lead to the TS (Shan and Herschlag, 1996, 1999). Enzymes with substrates of modest size (e.g., the size of glucose) can accommodate 15 or more hydrogen bonds from the enzyme (Aleshin *et al.*, 1998). Methods to detect low-barrier hydrogen bonds in enzymatic complexes are summarized in Chapter 10 of this volume (Mildvan *et al.*, 1999). Some enzymes make use of substrate neighboring group participation by conformational distortion of the bound substrate to provide catalytic groups that participate in TS formation (Degano *et al.*, 1998; Horenstein *et al.*, 1991; Muchmore *et al.*, 1998). At a relatively modest H-bond energy for optimally aligned bonds, the energy from 15 modest hydrogen bonds of only 1.4–3 kcal/mol can generate −23 to −45 kcal/mol to achieve the TS. A pertinent and recent example of this effect is a TS inhibitor bound to malarial hypoxanthine-guanine phosphoribosyltransferase (HGPRT). The complex demonstrates five hydrogen bonds with proton chemical shifts greater than 12 ppm downfield, none of which appear in a similar complex using substrate analogs (Li *et al.*, 1999). Based on the single criterion of the temperature dependence of the proton nuclear magnetic resonance (NMR) chemical shift for these bonds (Garcia-Viloca *et al.*, 1998), four of the five are low-barrier hydrogen bonds. Release of the TS binding energy following the reaction and on the time scale of catalysis is largely unexplored, but requires the loss of the hydrogen bonds that were aligned optimally at the TS. During the several milliseconds for the typical catalytic cycle, the enzyme collides with substrate, binds it into the weakly interacting Michaelis complex, increases the binding/catalytic energy to form the TS, relaxes the tight-binding energy of the TS, forms the weakly bound product complex (Michaelis complex for the reverse reaction), and dissociates the products.

Inhibitor design that takes advantage of the TS structure and binding energy, relies on accurate estimates of the electron distribution at the TS as deduced by the analysis of intrinsic KIE (Bagdassarian *et al.*, 1996b). This chapter emphasizes the procedure of measuring sufficient intrinsic kinetic and equilibrium isotope effects to construct a complete bonding and electronic structure of the TS for reactions not usually accessible by these methods.

IV. Chemistries Amenable to TS Analysis

KIE arise from changes in atomic vibrational states between the reactants and TS (Bigeleisen and Wolfsberg, 1958). In theory, all reactions can be probed by KIE, but in practice, electron transfers result in intrinsic KIE too small to be quantitated. In reactions where the atom of interest remains in the same bonding environment in the substrate and at the TS, there will be no KIE. Atoms that become vibrationally less constrained in the TS give normal KIE ($k_{normal}/k_{heavy} > 1$), with the heavier isotopic substrate reacting more slowly than that with natural abundance atoms. Conversely, atoms more constrained at the TS cause inverse KIE ($k_{normal}/k_{heavy} < 1$), with the heavy isotopically labeled substrate reacting more rapidly. These generalizations for the interpretation of KIE for some of the common reactions in biochemistry are summarized in Fig. 1. The atomic motions responsible for these effects have been described in the original accounts of Bigeleisen and Wolfsberg (1958) and have been reviewed by Melander and Saunders (1980), Suhnel and Schowen (1991), and Huskey (1991). The small size of KIE from heavy atom reactants ($^{14}C = 1.09$, $^{15}N = 1.04$, and $^{18}O = 1.07$), the necessity for the synthesis of labeled substrates, and the difficulty in establishing intrinsic KIE has prevented the application of KIE methods to most enzymatic

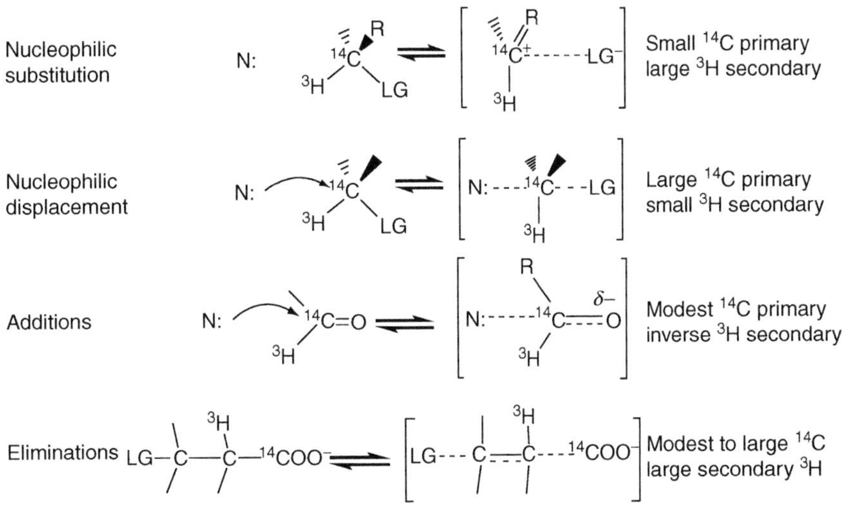

Fig. 1 Some common enzymatic reactions in biochemical pathways. The KIE expected for the primary ^{14}C and a secondary ^{3}H isotopic label are indicated. The pattern of these two simple KIE are sufficient to distinguish the mechanisms, and additional isotope effects are capable of providing quantitative information of bond orders at the transition state for each type of reaction.

reactions. Isotope effects from ^{13}C and ^{18}O in decarboxylases and phosphotransferases are small because the bond order to individual oxygen atoms changes only by a fraction of a bond order at the TS. Despite these problems, remote labeling and isotope ratio mass spectrometry techniques have permitted accurate measurements of these KIE (Weiss, 1991). Finally, the synthesis of isotopically labeled substrates must be accessible.

The most versatile method of measuring KIE is the competitive radiolabeled method. Two substrates are prepared, one with the labeled atom at a site expected to experience bonding changes at the TS and a second substrate with a different labeled atom at a site remote from the bond-breaking site, in a position expected to remain vibrationally unchanged at the TS (Dahlquist *et al.*, 1969). The double-label method depends on the ability to synthesize labeled substrates (Parkin *et al.*, 1984). With good techniques in dual-label radioactive counting, it is possible to measure isotope effects with standard errors of ± 0.002. Thus, isotope effects of 1.01 (1% difference in reaction rates between labeled and unlabeled substrates) can be measured with confidence. Isotope ratio mass spectrometry is more accurate by an order of magnitude, routinely to ± 0.0002. However, it is less versatile because the spectrometer measures only volatile samples (e.g., N_2, O_2, and CO_2) and requires more material for the analysis. Direct mass spectrometry methods provide accuracy similar to that of the dual-label radioactive counting method (Berti and Schramm, 1997; Berti *et al.*, 1997). Most of the atoms of interest in biochemical reactions are labeled readily with ^{3}H or ^{14}C and can be incorporated into substrates that also contain ^{2}H, ^{15}N, and ^{18}O. Double-label techniques provide the opportunity to measure virtually any of these KIE by the analysis of radioactive counting in samples of $^{3}H/^{14}C$. These isotopes can be counted with accuracies to 0.2% with small amounts of radiolabel, typically 1 μCi or less per experiment. Counting times decrease and accuracy improves with increased amounts of radioactivity (Parkin, 1991).

V. Experimental Procedure for Enzymatic TS Analysis and Inhibitor Design

Steps commonly used in this laboratory for the analysis of an enzymatic TS and the design of a TS inhibitor are to (Schramm, 1998) (1) select an enzyme expected to give significant experimental KIE; (2) synthesize substrates with isotopic labels at every position where bond changes are expected at the TS and at positions remote from the bonds where chemistry occurs; (3) measure experimental KIE; (4) establish the commitment factors for the enzyme, correct to intrinsic KIE; (5) match intrinsic KIE to the bond orders of a truncated TS state structure consistent with the KIE; (6) establish the molecular electrostatic potential (MEP) surface for substrate and TS; (7) design inhibitors with MEP surfaces similar to the TS; and (8) synthesize and test the inhibitors.

This chapter considers the practical steps in obtaining intrinsic KIE and interpretation of data to establish a TS structure that provides a practical guide to the design of TS inhibitors. Examples of inhibitor design projects using this experimental approach are described.

VI. Synthesis of Isotopically Labeled Substrates: Nucleoside and Nucleotide Metabolism

A. Purine Nucleotides and Nucleosides

A technical barrier to systematic KIE analysis is the requirement for the synthesis of as many as 10 different substrates, each labeled specifically with isotopes and usually where none are commercially available. Our approach to this problem is to use enzymatic pathways that synthesize the desired molecules *in vivo*. Enzymatic syntheses are specific and efficient and are increasingly accessible with the easy cloning and expression of most enzymes of bacterial and yeast metabolism. Once the conditions for synthesis are established, the same reaction mixtures, with different labeled precursors, can be used to produce the labeled substrates necessary for TS analysis. An example of the synthesis of a double-labeled [9-^{15}N, 5′-^{14}C]ATP illustrates this method (Fig. 2). Using the same reaction mixture, or with modifications of the mixture, the labeled ATP molecules summarized in Table I have been synthesized (Parkin and Schramm, 1987; Parkin *et al.*, 1984). Synthesis from a common precursor (glucose in the case of ATP) provides the benefit of commercially available labeled molecules as well as well-known enzymatic steps for the incorporation of additional isotopic labels. For example, the addition of excess phosphoglucomutase and $^{2}H_2O$ results in the incorporation of deuterium at the 2 position of glucose because of the solvent exchange catalyzed by the enzyme. ATP labeled in the indicated positions can be produced in yields of 50–90% from labeled precursors using procedures similar to that of Fig. 2. The synthetic procedures and the purification of the product nucleotides are made more efficient by combining the desired ^{3}H and ^{14}C precursors in the same reaction mixture to provide the product with the desired ratio of ^{3}H and ^{14}C product required for KIE measurements. For example, a combination of [2-^{14}C]glucose and [6-^{3}H]glucose in the same ATP synthesis mixture yields the [1′-^{14}C]ATP and [5′-^{3}H]ATP pair that is needed to measure the primary [^{14}C]KIE.

The labeled ATP molecules of Table I also provide a convenient starting material for the syntheses of other purine nucleotides and nucleosides. The conversions of ATP to dATP, ADP, dADP, AMP, IMP, dAMP, adenosine, D-adenosine, inosine, and D-inosine have all been accomplished in high yields using the steps shown in Fig. 3. Labeled AMP nucleotides used to establish the TS structure of AMP nucleosidase were synthesized by the combination of steps in Figs. 2 and 3 (Table II) (Parkin and Schramm, 1987). It is also possible to use the known stereochemistry of these reactions to incorporate additional isotopic labels. For example, ribonucleotide triphosphate reductase (RTR in Fig. 3) is known to

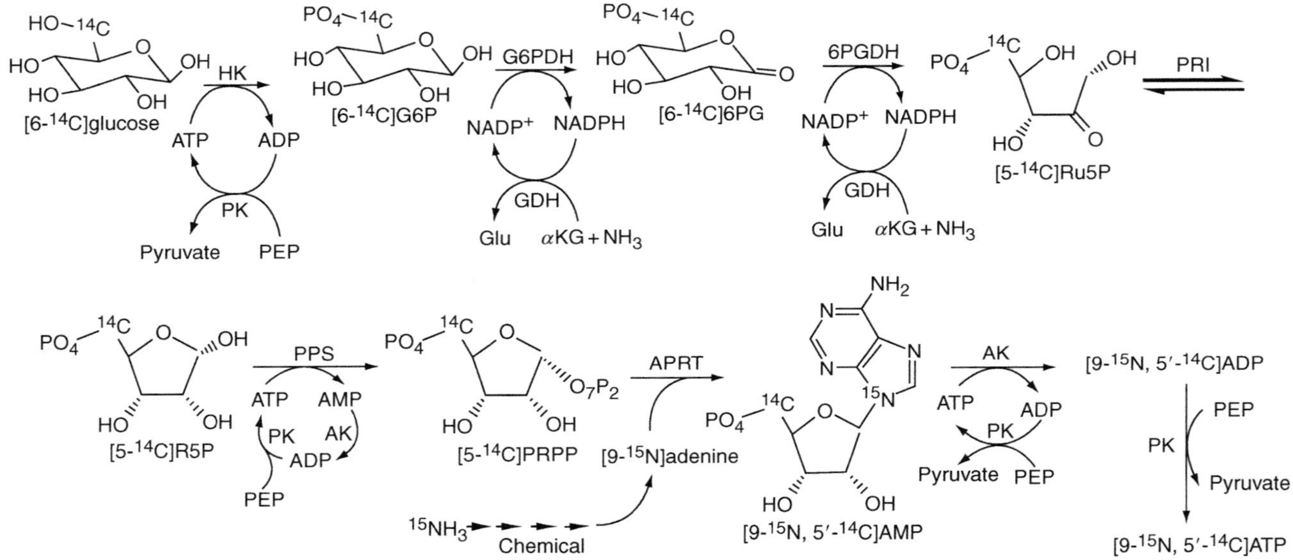

Fig. 2 Synthesis of labeled ATP using coupled enzymatic reactions with cycling of cofactors (Parkin *et al.*, 1984). Reaction mixtures (typically 1 ml) contain 50 mM potassium phosphate, 3.2 mM [9-^{15}N]adenine, 3 mM $MgCl_2$, 50 mM KCl, 5 mM dithiothreitol, 10 mM α-ketoglutarate (α-KG), 1 mM ATP, 0.1 mM $NADP^+$, 10 mM P-enolpyruvate, 1 mM [6-^{14}C]glucose, 0.1–2 U/ml hexokinase (HK), 2 U/ml pyruvate kinase (PK), 0.1–7 U/ml glucose 6-phosphate dehydrogenase (G6PDH), 0.5 U/ml glutamate dehydrogenase (GDH), 0.1–1 U/ml 6-phosphogluconate dehydrogenase (6PGDH), 6 U/ml phosphoriboisomerase (PRI), 0.5–2 U/ml adenylate kinase (myokinase, AK), 0.2–1.3 U/ml PRPP synthetase (PPS), 0.5 U/ml adenine phosphoribosyltransferase (APRT), and ammonium ion added as ammonium sulfate from the commercial preparations of enzymes stored in ammonium sulfate. With the exception of hexokinase, enzymes are mixed in the appropriate ratio and added to the otherwise complete reaction mixture. The reaction is initiated by the addition of hexokinase, incubated at 37 °C and is monitored for ATP formation by the luciferase reaction or by HPLC. Reactions are complete in 1–8 h, depending on the quantity of coupling enzymes. Tubes are placed in boiling water for 3 min to stop the reaction. Chromatography on semipreparative reverse-phase HPLC eluting with 50 mM triethylammonium acetate, pH 6.0, containing 5% methanol provides ATP purified sufficiently for RNA synthesis or for conversion to other nucleotides and nucleosides.

Table I
Synthesis of ATP for Kinetic Isotope Effects

Labeled substrate	Labeled ATP product	Changes from Fig. 1
[1-^{3}H]Ribose	[1′-^{3}H]ATP	Ribokinase replaces first four enzymes
[2-^{3}H]Ribose 5-PO_4	[2′-^{3}H]ATP	First four enzymes eliminated
[5-^{3}H]Glucose	[4′-^{3}H]ATP	
[6-^{3}H]Glucose	[5′-^{3}H]ATP	
[6-^{3}H]Glucose	[1′-^{2}H, 5′-^{3}H]ATP	Hexosephosphate isomerase in ^2H_2O
[6-^{3}H]Glucose	[2′-^{2}H, 5′-^{3}H]ATP	Phosphoribose isomerase in ^2H_2O
[2-^{14}C]Glucose	[1′-^{14}C]ATP	
[6-^{14}C]Glucose	[5′-^{14}C]ATP	
[6-^{14}C]Glucose	[9-^{15}N, 5′^{14}C]ATP	[9-^{15}N]Adenine replaces adenine
[2-^{14}C]Glucose	[1′-^{2}H, 1′-^{14}C]ATP	Hexosephosphate isomerase in ^2H_2O
[2-^{14}C]Glucose	[9-^{15}N, 1′-^{14}C]ATP	[9-^{15}N]Adenine replaces adenine

catalyze the reduction at C-2′ with the retention of configuration (Batterham *et al.*, 1967). Reactions conducted in the presence of ^2H_2O or ^3H_2O therefore incorporate isotopic hydrogen specifically in the pro-*R* position from which the hydroxyl is lost. It is possible to take advantage of this stereochemical specificity by producing [2′-^{2}H]ATP or [2′-^{3}H]ATP using the methods of Fig. 2, followed by conversion to dATP in the presence of normal water, to produce isotopic hydrogen specifically in the pro-*S* position. The availability of pro-*R* and -*S* isotopes by this synthetic method permits measurement of the stereochemical dependence of isotope effects for the group of enzymes that uses nucleosides, deoxynucleosides, nucleotides, and deoxynucleotides as substrates.

Any of the intermediates from Fig. 2 can be synthesized simply by truncating the reactions at the desired point. A special case is the production of 5-phosphoribosyl-1-pyrophosphate (PRPP), which is used as a substrate for at least a dozen enzymes of nucleotide and amino acid synthesis. The instability of this substrate requires that it be purified rapidly by DEAE ion-exchange chromatography and used promptly to prevent nonenzymatic breakdown. Another approach for production of the unstable PRPP is to include hypoxanthine and hypoxanthine-guanine phosphoribosyltransferase as a coupling enzyme following the step that produces PRPP. This result in IMP as the end product, which can be purified and stored. When PRPP is required, the reverse reaction—IMP + MgPPi → hypoxanthine + PRPP—can be accomplished, with hypoxanthine removal effected with xanthine oxidase to bring the reaction to completion. Coupling the reaction to remove hypoxanthine is necessary because of the unfavorable K_{eq} for PRPP production (Xu *et al.*, 1997).

The synthesis of molecules in addition to those just discussed can be inferred directly from the enzymatic reaction schemes. The addition of other isotopes in the starting materials, such as ^{2}H- or ^{13}C-labeled sugars or nitrogenous bases with additional labels, would lead to the expected products. The end product ATP can also be used as a precursor for other nucleosides, nucleotides, and th 2′-deoxynucleosides and -nucleotides. The example of Fig. 4 demonstrates the conversion of the

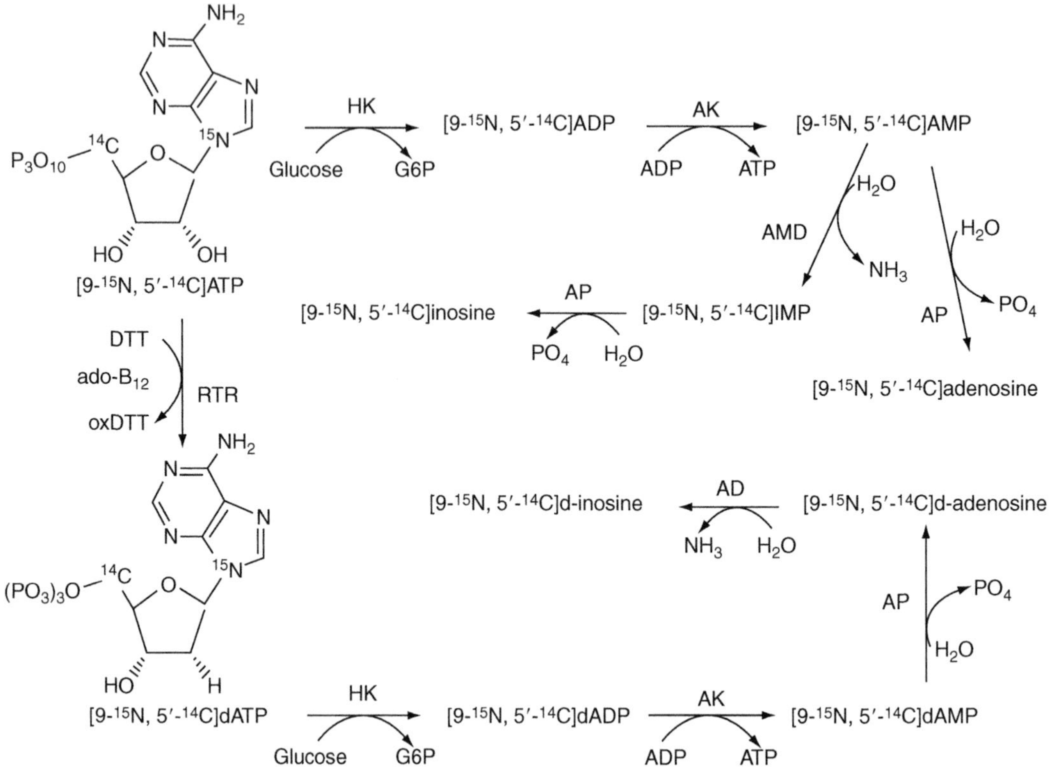

Fig. 3 Synthesis of other nucleotides and nucleosides from ATP (Horenstein *et al.*, 1991; Parkin *et al.*, 1984). Reaction mixtures for dATP production contained 50 mM potassium phosphate (pH 7.7), 100 mM sodium acetate (pH 7.2), 25 mM dithiothreitol (DTT), 0.1–1 mM labeled ATP, and 2 U/ml ribonucleotide triphosphate reductase (RTR). After incubation under argon, the anaerobic mixture was made to 20 μM in adenosylcobalamin (ado-B_{12}) in the dark. The reaction was incubated in the dark, under argon, for 3 h at 37 °C, placed at 95 °C for 2–3 min, cooled, and the labeled dATP purified by semipreparative HPLC as indicated in the legend to Fig. 2. Labeled ATP or dATP can be converted to monophosphates in the same reaction mixtures used to form them. Following denaturation of the ATP-forming enzymes by heating, the cooled reaction mixtures are made to 6 mM glucose, 25 U/ml hexokinase (HK), and 200 U/ml myokinase (AK). After a few minutes at 37 °C, the reactions are stopped by 3 min in boiling water. Conversions of AMP and dAMP to adenosine or D-adenosine can occur in the same reaction mixtures, after the previous enzymes are denatured. The reactions are made to 2 U/ml alkaline phosphatase (AP), incubated for 30 min at 37 °C, and boiled for 3 min. In the same reaction mixtures, the addition of 2 U/ml adenosine deaminase (AD) will convert adenosine or D-adenosine to the corresponding inosines in 30 min at 37 °C. Adenosine monophosphate deaminase (AMD) can be used in the same conditions to convert AMP to IMP, and IMP can be converted to inosine by alkaline phosphatase as indicated. Purification of the monophosphates or the nucleosides and deoxynucleosides can be accomplished by chromatography on Sephadex G-10 in 20 mM acetic acid, followed by HPLC C-18 reverse phase eluted with 0.1 M ammonium acetate, pH 5.0, containing 5% methanol.

Table II
Substrates for KIE Measurements: AMP Nucleosidase

Substrate	Kinetic isotope effect ^3H
[1′-^{3}H]AMP + [5′-^{14}C]AMP	α Secondary ^{3}H
[2′-^{3}H]AMP + [5′-^{14}C]AMP	β Secondary ^{3}H
[4′-^{3}H]AMP + [5′-^{14}C]AMP	Remote secondary ^{3}H
[5′-^{3}H]AMP + [5′-^{14}C]AMP	Remote secondary ^{3}H
[1′-^{2}H, 5′-^{3}H]AMP + [5′-^{14}C]AMP	α Secondary ^{2}H
[2′-^{2}H, 5′-^{3}H]AMP + [5′-^{14}C]AMP	β Secondary ^{2}H
[1′-^{14}C]AMP + [5′-^{14}C]AMP	Primary ^{14}C
[9-^{15}N, 5′-^{14}C]AMP + [5′-^{3}H]AMP	Primary ^{15}N
[1′-^{2}H, 1′-^{14}C]AMP + [5′-^{3}H]AMP	Dual primary ^{14}C and α secondary ^{2}H
[9-^{15}N, 1′-^{14}C]AMP + [5′-^{3}H]AMP	Dual primary ^{14}C and primary ^{15}N

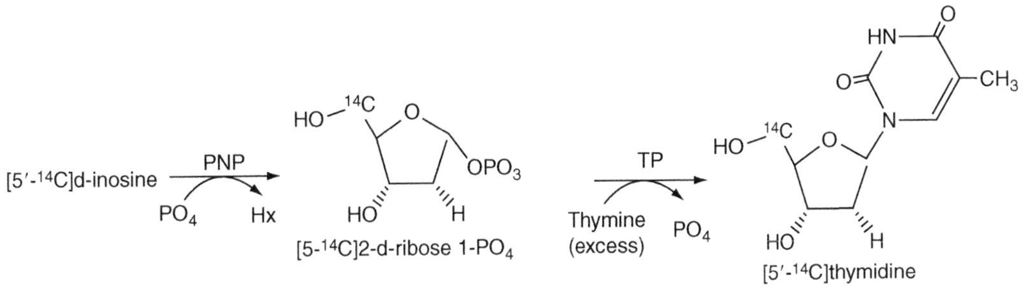

[5′-^{14}C]d-inosine → (PNP; PO_4 in, Hx out) → [5-^{14}C]2-d-ribose 1-PO_4 → (TP; Thymine (excess) in, PO_4 out) → [5′-^{14}C]thymidine

Fig. 4 Synthesis of labeled thymidine from D-inosine (M. Khalil and V. L. Schramm, unpublished experiments, 1998). Reaction mixtures contained 2 mM potassium phosphate, pH 7.7, 30 μM D-inosine, 10 mM thymine, 0.4 U/ml purine nucleoside phosphorylase (PNP), 1 U/ml thymidine phosphorylase (TP), and 0.1 U/ml xanthine oxidase (converts hypoxanthine (Hx) to uric acid). The product is purified by HPLC C-18 reverse phase, eluted with 50 mM ammonium acetate, pH 5.0.

[5′-^{14}C]2′-deoxyribosyl group of [5′-^{14}C]deoxyinosine (Fig. 3) to thymidine, a 2′-deoxypyrimidine nucleoside. A direct extension of this method would provide other pyrimidine nucleosides, nucleotides, and the 2′-deoxy compounds.

B. Nicotinamide Adenine Dinucleotide and Related Nucleotides

NAD^+ is a central metabolite used in hydride transfer and was one of the first substrates used to measure enzymatic KIE. The large KIE for deuteride ($^2H^-$) transfer reactions is measured easily by direct initial rate techniques, a consequence of the relatively large mass increase from ^1H to ^2H (Blanchard and Wong, 1991). NAD^+ plays additional roles in metabolism that are of increasing interest. These include the ADP-ribosylation reactions, which are catalyzed by the ADP-ribosylating bacterial toxins; the formation and use of poly-ADP-ribose in DNA repair reactions and in maintaining chromatin structure; and the formation of cyclic ADP-ribose as a signal causing calcium release from the endoplasmic reticulum (Aktories, 1992; Lee *et al.*, 1989; Miwa *et al.*, 1983). The reaction mechanisms of these enzymes involve scission of the N1-ribosidic bond to nicotinamide followed by transfer of the resulting ADP-ribose to the acceptor molecule. Investigation of the TS for these reactions requires labeling the NMN^+-ribose group of NAD^+. An example of the synthesis of [1′-^{3}H]NAD^+, with the labels being incorporated in the NMN^+-ribose, is provided in Fig. 5 (Rising and Schramm, 1994). Synthesis of NAD^+ is accomplished in two steps. In the first, all reagents needed for nicotinic acid adenine dinucleotide ($NAAD^+$) synthesis are combined to convert labeled precursors to this end product. The reaction is stopped by brief heating, the $NAAD^+$ is purified from the reagents of the first step, and the NAD^+ synthetase reaction is accomplished in a second incubation (Fig. 5). In this procedure, a component of $NAAD^+$ synthesis inhibits NAD^+ synthetase and must be removed prior to NAD^+ formation. The same precursors described in Table I can be used to synthesize a similar group of specifically labeled NAD^+ molecules (Table III). By combining the synthesis of specifically labeled ATP and that of NAD^+, it is possible to synthesize NAD^+ with specific atomic labels in any one or more of the nonexchangeable atoms in the entire molecule. The specificity of the synthetic enzymes permits obvious extensions of the methods, for example, the conversion of labeled NAD^+ molecules to nicotinamide guanine dinucleotide (NGD^+) and the production of labeled NMN^+ molecules by hydrolyzing the phosphodiester of labeled NAD^+ (Fig. 6). Both NGD^+ and NMN^+ are substrates for CD38, an NAD^+, NMN^+ nicotinamide *N*-ribohydrolase, and ADP-ribosyl cyclase ($NAD^+ \rightarrow$ cyclic ADP-ribose + nicotinamide) (Sauve *et al.*, 1998).

C. RNA and DNA Oligonucleotides

Isotopically labeled nucleotides and deoxynucleotides (Figs. 2 and 3) serve as substrates for incorporation into oligonucleotide polymers. Figure 7 demonstrates the synthesis of RNA and DNA polymers using [9-^{15}N, 5′-^{14}C]ATP and [9-^{15}N, 5′-^{14}C]dATP for incorporation into oligonucleotide sites coding for adenine nucleotides. RNA synthesis uses T7 RNA polymerase and a double-stranded DNA primer. Under these conditions the product RNA begins with an uncoded GTP, followed by faithful transcription of the remaining DNA primer to the final base. The product RNA terminates with a free 3′-hydroxyl group and contains a triphosphate at the 5′ terminus. The primer is not consumed in the reaction, and T7 polymerase is capable of rebinding and synthesis of another RNA strand, displacing the previous RNA strand. Thus, small amounts of the DNA primer can produce large amounts of the labeled RNA product. This process has been used to produce unlabeled stem-loop RNA structures for the adenine depurination reaction catalyzed by the ricin A-chain (Chen *et al.*, 1998), as well as labeled RNA for KIE measurements with this reaction (X.-Y. Chen and V. L. Schramm, unpublished results, 1998).

Synthesis of labeled DNA oligonucleotides uses a hairpin primer-template, the desired labeled and unlabeled deoxynucleotide triphosphates, and Klenow fragment DNA polymerase (Fig. 7). The addition of the first nucleotide to the primer

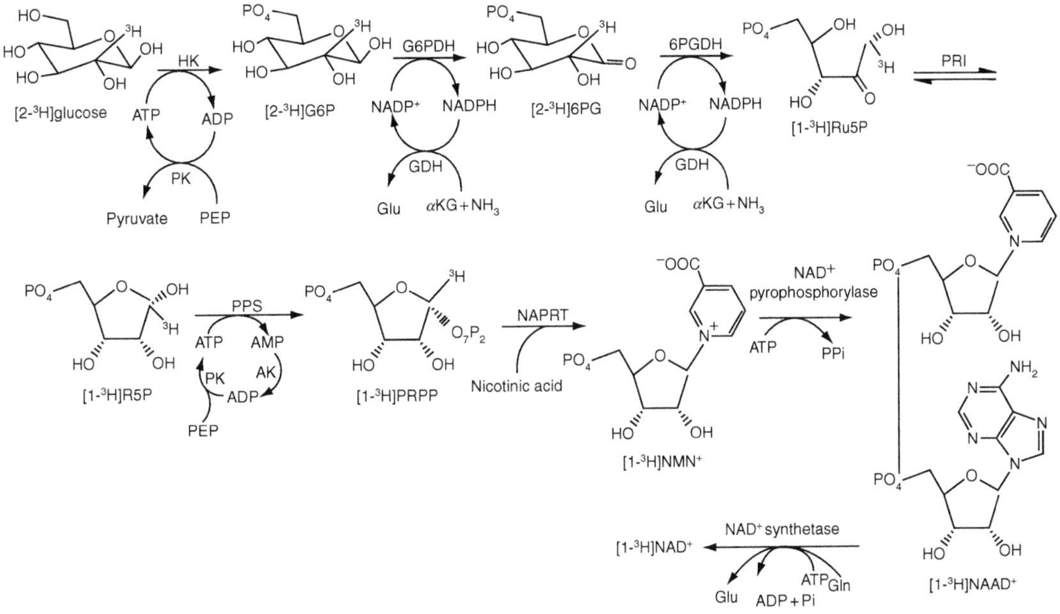

Fig. 5 Synthesis of isotope-labeled NAD^+ (Rising and Schramm, 1994). Reaction mixtures of 1 ml contained 50 mM potassium phosphate, pH 7.5, 3 mM $MgCl_2$, 50 mM KCl, 5 mM dithiothreitol, 2 mM ATP, 10 mM P-enolpyruvate, 10 mM α-ketoglutarate, 0.1 mM $NADP^+$, 2 mM nicotinic acid, pH 6.2, and 1 mM glucose containing 20–100 μCi [^{3}H] or [^{14}C]glucose. Enzymes (as supplied) were diluted into 50 mM potassium phosphate and added to give final concentrations of 0.1 U/ml hexokinase, 2 U/ml pyruvate kinase, 0.1 U/ml glucose 6-phosphate dehydrogenase, 0.5 U/ml glutamate dehydrogenase, 0.1 U/ml 6-phosphogluconate dehydrogenase, 6 U/ml phosphoriboisomerase, 0.5 U/ml myokinase, 0.25 U/ml NAD^+ pyrophosphorylase, 0.2 U/ml PRPP synthetase, and 0.02 U/ml nicotinate phosphoribosyltransferase (NAPRT). All enzymes, except hexokinase, were mixed and added together. The other mixture components were added in the order listed. The reaction was initiated by the addition of hexokinase. Synthesis was monitored for 4 h at 37 °C for $NAAD^+$ production by periodic HPLC analysis on a C-18 reverse-phase column (7.8 × 300 mm) eluted with 0.1 M ammonium acetate, pH 5.0. The reaction was terminated by placing the sealed reaction mixture (in a plastic vial) in a heating block at 120 °C for 1.5 min. The sample was centrifuged and the supernate subjected to HPLC using the same method to provide $NAAD^+$. Purified $NAAD^+$ was freeze dried and reconstituted in a reaction mixture of 0.5 ml containing 50 mM potassium phosphate, pH 7.5, 50 mM KCl, 3 mM $MgCl_2$, 5 mM dithiothreitol, 4 mM ATP, 20 mM glutamine, and 0.2 U/ml yeast NAD^+ synthetase. Incubation for 2 h at 37 °C converted $NAAD^+$ to NAD^+. At the end of the incubation, the entire reaction mixture was injected directly onto the HPLC column (7.8 × 300 mm) and eluted as indicated earlier. In this HPLC system, NAD^+ is the last component eluted from the column. Fractions containing NAD^+ are freeze dried and stored in 50% ethanol at −70 °C. Under these conditions the labeled NAD^+ is stable (>99%) for at least 12 months. Synthesis of [2′-^{3}H]NAD^+ was initiated with [2-^{3}H]R5P. In this synthesis, only the components for PPS and the following steps are used. [2-^{3}H]R5P is prepared by the phosphoribose isomerase (PRI) exchange of ^3H_2O into an equilibrium mixture of R5P and Ru5P.

Table III
Synthesis of NAD$^+$ for Kinetic Isotope Effects

Labeled substrate	Labeled NAD$^+$ product	Changes from Fig. 4
[2-^{3}H]Glucose	[1′-^{3}H]NAD$^+$	
[2-^{3}H]Ribose 5-PO_4	[2′-^{3}H]NAD$^+$	First four enzymes eliminated
[5-^{3}H]Glucose	[4′-^{3}H]NAD$^+$	
[6-^{3}H]Glucose	[5′-^{3}H]NAD$^+$	
[2-^{14}C]Glucose	[1′-^{14}C]NAD+	
[6-^{14}C]Glucose	[5′-^{14}C]NAD$^+$	
[5-^{18}O]Glucose	[4′-^{18}O]NAD$^+$	
[6-^{14}C]Glucose	[1-^{15}N, 5′-^{14}C]NAD$^+$[1-^{15}N]nicotinamide	
[2-^{14}C]Glucose	[1-^{15}N, 1′-^{14}C]NAD$^+$[1-^{15}N]nicotinamide	
[8-^{14}C]ATP	[8-^{14}C]NAD$^+$ NMN$^+$ + [8-^{14}C]ATP → [8-^{14}C]NAD$^+$	
[5-^{18}O]Glucose	[4′-^{18}O, 8-^{14}C]NAD$^+$[8-^{14}C]ATP	

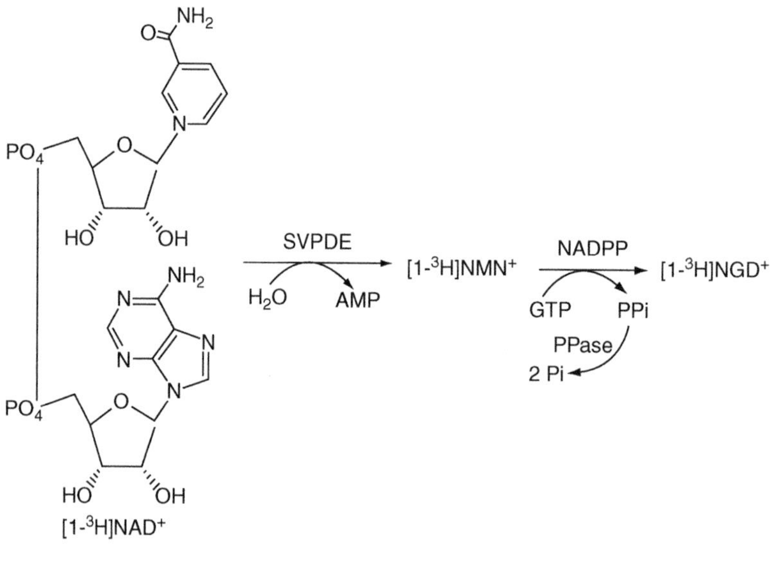

Fig. 6 Synthesis of labeled NGD$^+$ from NAD$^+$ (Sauve *et al.*, 1998). The reaction mixture for the synthesis of NAD$^+$ (legend to Fig. 5) was treated with snake venom phosphodiesterase (SVPDE) (~0.5 U/ml for 30 min 37 °C) to produce NMN$^+$, heated to 110 °C in a heating block for 40 s, centrifuged, and purified on C-18 reverse-phase HPLC with 0.1% trifluoroacetic acid containing 1% methanol. After freeze drying, the NMN$^+$ was made to 2 mM with 50 mM potassium phosphate, pH 7.5, containing 5 mM $MgCl_2$, 4 mM GTP, 1 U NAD$^+$ pyrophosphorylase (NADPP), and 0.1 U pyrophosphatase (PPase). After 6 h at 37 °C, another addition of enzymes was followed by a further 6-h incubation at 37 °C. The reaction was terminated by 20 s at 110 °C, centrifuged, and the supernate purified on preparative C-18 HPLC eluted with 50 mM ammonium acetate, pH 5.0.

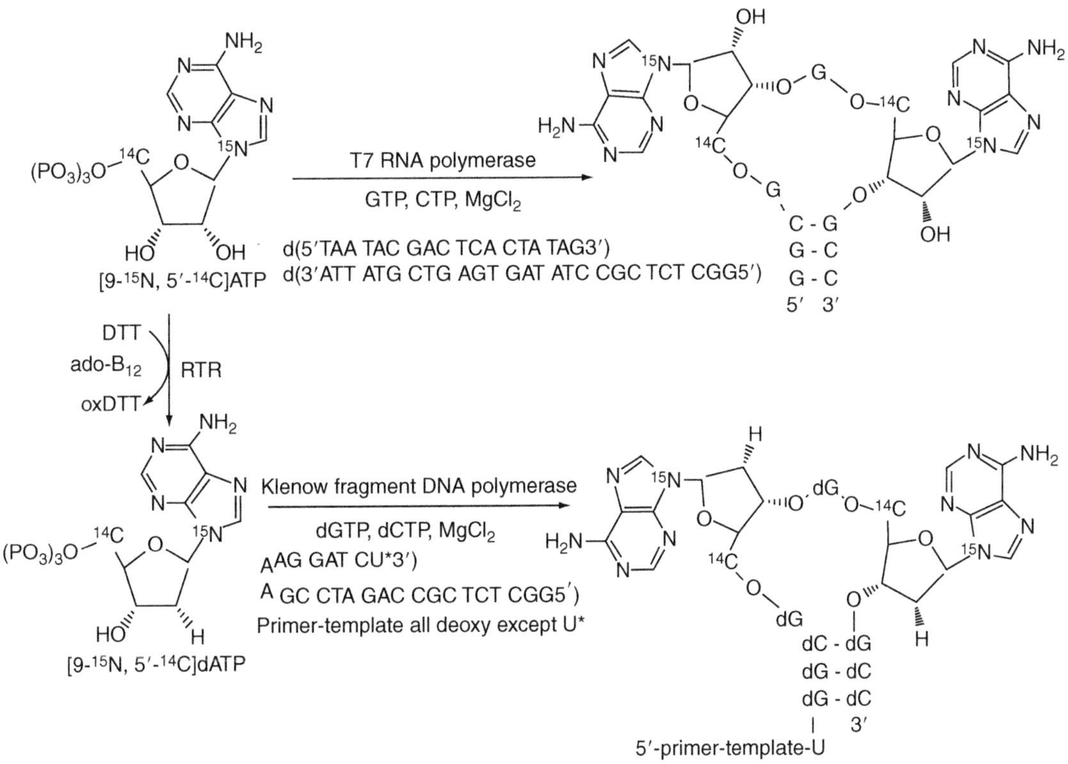

Fig. 7 Synthesis of labeled stem-loop RNA and DNA from labeled nucleotides and deoxynucleotides. Reaction mixtures of 0.4 ml for RNA synthesis contained 0.22 mM ATP, including the desired ^3H and ^{14}C, 4.5 mM GTP, 4.5 mM CTP, 22 mM $MgCl_2$, 25 mM NaCl, 2 mM spermidine, 5 mM dithiothreitol, 40 mM Tris-HCl, pH 8.0, 0.2 μM DNA primer-template, 40 U Rnasin (Promega), and 6000 U T7 RNA polymerase. Reactions were incubated at 37 °C overnight, made to 50 mM EDTA and 0.4 M sodium acetate, and the RNA precipitated with 2.5 volumes of ethanol followed by cooling to −70 °C. RNA 10-mers were isolated by 24% denaturing polyacrylamide gel electrophoresis, located by UV shadowing, eluted into 1 M ammonium acetate, and desalted on a C-18 Sep-Pac column from Waters. Reaction mixtures of 1.0 ml for DNA synthesis contained 10 mM Tris-HCl, pH 7.5, 5 mM $MgCl_2$, 50 μM hairpin primer-template, 3 mM dGTP, 3 mM dCTP, 50 μM dATP containing the desired ^3H and ^{14}C, and 100 U 3′–5′ exonuclease-deficient mutant of Klenow fragment DNA polymerase I (US Biochemical). Note that the 3′-terminal U in the primer is the ribosyl nucleotide. After incubation at 37 °C overnight, the mixture was made to 0.3 N NaOH, followed by incubation at 55 °C for 2 h to cleave at the uridine site. The mixture was neutralized with 0.36 M acetic acid, and the DNA was precipitated with ethanol and isolated as described for RNA.

results in a covalent phosphodiester bond that must be specifically removed if it is intended to remove the primer-template from the finished product. This is accomplished by the synthesis of a primer-template that incorporates 2′-deoxynucleotides at every position except at the 5′-terminal, where a uridine nucleotide is incorporated. This introduces a site for specific chemical cleavage by base (0.3 M NaOH) without hydrolysis of the DNA primer or the DNA product. Base hydrolysis of the ribonucleotide phosphodiester bond in the product DNA/RNA hybrid yields the replicated DNA strand and regenerates the 3′-phosphorylated primer-template that can be reused following dephosphorylation at the 3′ terminus with alkaline phosphatase. The replicative synthesis by this process is stoichiometric, but with the method described here, it can be used in a repetitive fashion following recycling of the primer-template. The methods of Fig. 7 have been used to synthesize stem-loop RNA and DNA containing most of the labeled ATP and dATP described in Table I. For the specific stem-loop RNA and DNA of Fig. 7, the labeled nucleotides are incorporated into both adenylate positions of the oligonucleotide.

D. Purification of Isotopically Labeled Substrates

Readers will realize that these complex mixtures of enzymes, cofactors, and buffers represent an empirical approach to isotopically labeled synthesis. The complexity of the reaction mixtures and the incomplete conversion of isotopically labeled intermediates present a purification challenge to yield the desired product free of reaction mixture components that would interfere with subsequent KIE determinations. The sequential application of two or three rapid purification steps has provided efficient purifications for all of the small molecule substrates. Following denaturation of the enzymes by brief treatment near 100°, or their removal by ultrafiltration, the first step for nucleosides and nucleotides is molecular exclusion chromatography on Sephadex G-10 or G-25 in 20 mM acetic acid to remove denatured protein and salts. This is followed by anion-exchange chromatography on DEAE Sephadex A-25 developed with gradients of acetic acid, ammonium bicarbonate, or ammonium acetate. In many cases, these two steps are adequate. If not, the product is freeze dried to remove salts followed by HPLC chromatography. NAD^+ is more labile and is purified directly in a single step by chromatography of the reaction mixtures on HPLC using a reverse-phase C-18, 7.8 × 300-mm column, developed with 0.1 M ammonium acetate, pH 5.0. The stainless-steel components of HPLC systems catalyze the hydrolysis of NAD^+ to ADP-ribose, resulting in an approximately 5% hydrolysis during the purification (Rising and Schramm, 1994). NAD^+ hydrolysis is avoided by the use of coated HPLC systems designed to prevent solvent metal contact.

Purification of DNA and RNA oligonucleotides is accomplished by precipitation of the product oligonucleotide from reaction mixtures with 2.5 volumes of ethanol, cooling on dry ice, centrifugation, ethanol wash, and chromatography on 24% denaturing polyacrylamide gel electrophoresis. The appropriate product is located with UV shadowing using known oligonucleotides as chromatography standards. Segments containing the desired oligonucleotides are excised, eluted into 1 M ammonium acetate, and desalted on a C-18 Sep-Pack column (Waters) (Chen *et al.*, 1998).

VII. Experimental Measurement of KIE

A. Methods for *N*-Ribosyl Hydrolases and Transferases

Partial conversion of the isotopically labeled substrates to products generates the mixture of isotopically labeled products that is used to establish the KIE. Quantitative analysis of this ratio requires accurate analysis of (1) the isotopic ratio in the substrate, (2) the fraction of substrates converted to product, (3) separation of the unreacted substrates from the products without altering the ratio of the products, and (4) analysis of the isotopic ratio in the products. An example of the experimental approach for a KIE study for the nucleoside hydrolase reaction is provided in Fig. 8. The isotopic ratio in the substrate (labeled inosines) is measured following complete conversion of inosine to ribose with excess enzyme. It is also possible to determine the ratio of $^3H/^{14}C$ in substrate inosines directly; however, analysis of the $^3H/^{14}C$ ratio in the products eliminates the possibilities that quenching is different when counting inosine and ribose or that the substrate contains a small fraction of impurity as ^3H or ^{14}C that is not converted to product and thus gives an inaccurate $^3H/^{14}C$ ratio in inosine. Chromatographic or other physical means of complete separation of unreacted substrate from product is essential. In the case of nucleoside hydrolase (Fig. 8), the product ribose is resolved efficiently from inosine by chromatography on small columns of acid-washed, powdered charcoal-cellulose in a mixture to give good flow rates, and is packed in disposable Pasteur pipettes. Ratios of dry charcoal to cellulose between 1:1 and 1:4 are satisfactory to provide the desired flow rate of approximately 0.5 ml/min. Elution with buffered, unlabeled ribose provides complete yields of product ribose, and the complete retention of inosine. In our experience, the resolution of substrates from products on charcoal columns is superior to other chromatographic mediums and should be used when substrates and products separate by this method. In some cases, it is possible to quantitatively convert the products of a reaction to compounds that can be analyzed conveniently by charcoal columns.

B. KIE in Enzymes of RNA and DNA Processing

The conversion of RNA and DNA oligonucleotides to products with a single adenine removed presents a challenge in the quantitative separation and recovery of the oligonucleotide substrate from the product oligonucleotide. This problem is solved by the conversion of both substrate and product RNA and DNA to the corresponding nucleosides with methods that generate free ribose or 2-deoxyribose only at the site of depurination. The substrate sites where depurination has not occurred generate adenosine from this position. The ultimate products of the reactions yield a mixture of nucleosides from all sites not depurinated and ribose or 2-deoxyribose from all sites that have been depurinated (Fig. 9). These products are resolved conveniently on charcoal columns and analyzed for total counts and

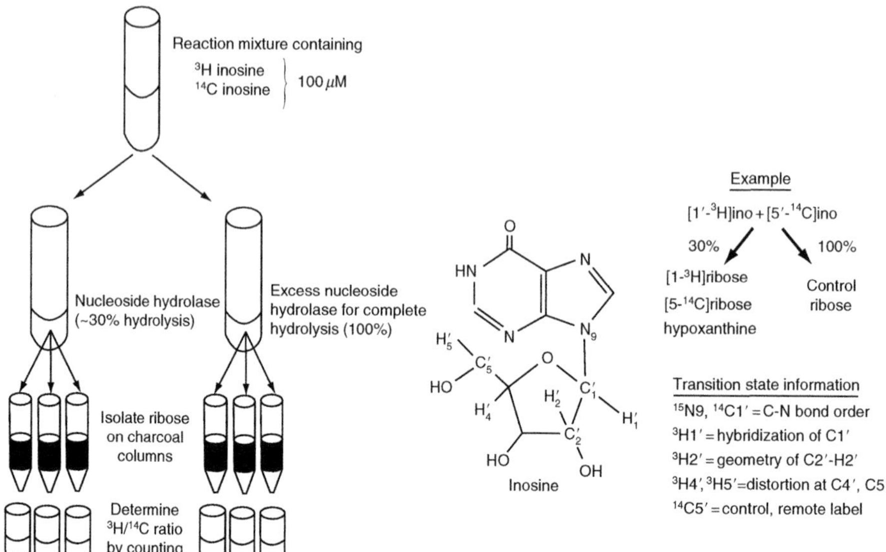

Fig. 8 The charcoal column method to measure KIE and transition-state information from specific isotopic labels (Parkin, 1991; Parkin and Schramm, 1987; Parkin *et al.*, 1984). The reaction mixture (upper tube) is typically 1.5 ml of buffered reaction mixture containing 0.1–1 mM substrate and the desired ^{3}H and ^{14}C substrates to provide equal counts in both isotopes. The mixture is divided into portions to provide equal counts in the product ribose after the reactions have occurred (1.0 ml for the partial reaction (left tube) and 0.4 ml for the complete reaction (right tube)). Following the enzymatic conversions to product, the reactions are quenched with acid (or EDTA for reactions dependent on divalent cations), and equal aliquots are applied to three or four charcoal-cellulose columns (three are shown) equilibrated previously with the reaction product to be collected. For nucleoside hydrolases, the columns are equilibrated with 50 or 100 mM ribose. For AMP nucleosidase (product is ribose 5-phosphate), the columns were equilibrated with 10 mM ribose 5′-phosphate. A control to establish that all substrate is retained on the column uses an additional column and the reaction mixture prior to reaction with the enzyme. The columns are eluted with the equilibration solutions, and 1.0-ml fractions are collected into scintillation vials, made to 20 ml with scintillation fluid, and counted to quantitate ^{3}H and ^{14}C at ~30 and 100% conversion to products. The observed ^{3}KIE = (^{3}H/^{14}C for 100%)/(^{3}H/^{14}C for 30%). Preparation of multiple samples from a single reaction mixture provides statistical analysis of the procedures for chromatography, sample preparation, and counting. Samples are counted for 10-min cycles until sufficient counts are accumulated to provide small counting errors, usually 10–15 cycles per sample set.

the ^{3}H/^{14}C ratio. Collection and counting of the product ^{3}H/^{14}C ratio in the partial reaction compared to that of the samples from the complete conversion gives the observed KIE.

C. KIE in Enzymes of Protein Covalent Modification

The action of bacterial exotoxins on GTP-binding proteins (G-proteins) has been developed as the first example in which the TS structure for protein covalent modification has been attempted using KIE methods. The ADP-ribosylation of

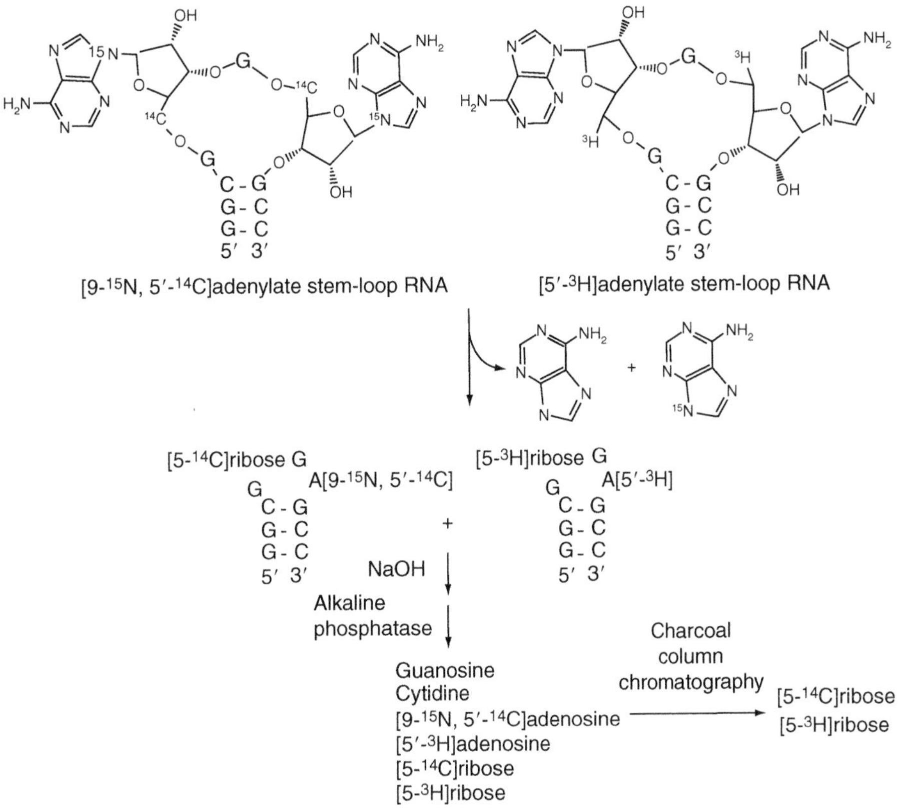

Fig. 9 Method to measure the ^{15}N KIE for and RNA depurination reaction (X.-Y. Chen and V. L. Schramm, unpublished observations, 1998). Stem-loop RNA with the labels in the site of depurination (left stem-loop RNA) or in a remote position (right stem-loop RNA) are incubated in companion reaction mixtures with ricin A-chain to provide ~30 or 100% hydrolysis of the susceptible bond, using the method described in Fig. 8. Adenine or [9-^{15}N]adenine is released in the ratio reflecting the ^{15}N KIE, leaving [5-^{14}C]ribose and [5-^{3}H]ribose at the depurination sites. Treatment with 0.3 M NaOH at 37 °C overnight followed by neutralization to pH 8 and alkaline phosphatase yields nucleosides for every base that has not been depurinated and ribose for every depurination site. The samples are then processed as in Fig. 8 to determine the KIE.

G-proteins by NAD^+ occurs with bacterial exotoxins, including cholera, diphtheria, and pertussis exotoxins. Covalent modification of other mammalian proteins by ADP-ribose is also common, but is not as well understood. The reactions catalyzed by bacterial exotoxins are

$$NAD^+ + G\text{-protein} \rightarrow \text{nicotinamide} + \text{ADP-ribosylated-G-protein}$$

where the ADP-ribosylation site is variable. Reactions catalyzed by cholera, diphtheria, and pertussis toxins ADP-ribosylate a specific arginine, diphthamide (a modified histidine), or cysteine residues. The reaction of pertussis toxin involves

ADP-ribosylation of the $G_{i\alpha 1}$ subunit of the trimeric GTP-binding complex. The $G_{i\alpha 1}$ subunit has been overexpressed in *Escherichia coli* and was purified for use at substrate levels to measure the KIE. Substrate trapping studies with labeled NAD^+ and a variable $G_{i\alpha 1}$ subunit demonstrated a negligible forward commitment for the reaction. Reactions to measure KIE were conducted with excess isotopically labeled NAD^+ (Table IV) and a limiting $G_{i\alpha 1}$ subunit. This procedure permits isotopic fractionation of the labeled NAD^+ substrate until the $G_{i\alpha 1}$ subunit is largely ADP-ribosylated. The resolution of the ADP-ribosyl-$G_{i\alpha 1}$ subunit from the labeled NAD^+ that had not reacted is accomplished by precipitation of the protein with 15% trichloroacetic acid. The pellet is washed with TCA to remove all unincorporated NAD^+. The pellets were dissolved in sodium dodecyl sulfate (SDS) and counted by scintillation counting to determine the KIE. The KIE measured for the ADP-ribosylation of the $G_{i\alpha 1}$ subunit by pertussis toxin are intrinsic and were used to solve the first TS for the covalent modification of a protein (Scheuring *et al.*, 1998).

D. Correction of Observed KIE for Isotopic Depletion

The $^3H/^{14}C$ ratios obtained by the procedures just described are those for the total amount of labeled substrate that has been converted to products. The experimental KIE is dependent on the fractional conversion to product, as the $^3H/^{14}C$ ratio in the substrate changes as a function of the extent of the reaction. Comparison of the total product ribose counts in the 30–40% conversion to the 100% conversion gives an accurate value of the fractional conversion to product. $KIE_{experimental}$ values are converted to KIE_{actual} values that would occur at 0% of substrate depletion using the expression (Bigeleisen and Wolfsberg, 1958)

$$KIE_{actual} = \frac{\ln[1 - (KIE_{experimental}) \times (\text{fraction converted})]}{\ln[1 - \text{fraction converted}]}.$$

Table IV
Isotopically Labeled NAD^+ Mixtures for Kinetic Isotope Effects

Labeled NAD^+ substrate	Kinetic isotope effect
$[1'\text{-}^3H]NAD^+ + [5'\text{-}^{14}C]NAD^+$	α Secondary ^3H
$[2'\text{-}^3H]NAD^+ + [5'\text{-}^{14}C]NAD^+$	β Secondary ^3H
$[4'\text{-}^3H]NAD^+ + [5'\text{-}^{14}C]NAD^+$	Remote ^3H
$[5'\text{-}^3H]NAD^+ + [5'\text{-}^{14}C]NAD^+$	Remote ^3H
$[1'\text{-}^{14}C]NAD^+ + [4'\text{-}^3H]NAD^+$	Primary ^{14}C
$[4'\text{-}^{18}O]NAD^+$	Direct α secondary ^{18}O
$[1\text{-}^{15}N, 5'\text{-}^{14}C]NAD^+ + [4'\text{-}^3H]NAD^+$	Primary ^{15}N
$[1\text{-}^{15}N, 1'\text{-}^{14}C]NAD^+ + [4'\text{-}^3H]NAD^+$	Combined primary $^{14}C + ^{15}N$
$[4'\text{-}^{18}O, 8\text{-}^{14}C]NAD^+ + [4'\text{-}^3H]NAD^+$	α Secondary ^{18}O

Details of the $^{3}H/^{14}C$ scintillation counting procedures to minimize errors and to accuracies of 0.2–0.3% have been described (Parkin, 1991; Parkin and Schramm, 1987). The value of KIE_{actual} can vary from unity (no KIE) to the intrinsic KIE, defined as the KIE for the TS with no contribution from steps other than the chemical structure of the TS. The most common problem faced in the determination of intrinsic KIE for enzyme-catalyzed reactions is forward and reverse commitments (Northrop, 1975, 1977, 1981).

VIII. Measurement of Commitment Factors and Binding Isotope Effects

With the potential for KIE to provide direct information of the enzymatic TS, why has the application been limited to a relatively few enzymatic reactions? Enzymes achieve catalytic power by multiple steps in the reaction coordinate, often of comparable energetic barriers, so that the chemical step which gives the intrinsic KIE may be buried between enzymatic steps that include forward and reverse commitment, rate-limiting product release, or rate-limiting enzymatic conformations (Fig. 10). These problems were recognized and quantitated by Northrop (1975, 1977, 1981) at the inception of the application of KIE analysis for enzymatic reactions. Because most of the first KIE measurements were accomplished with $^{1}H/^{2}H$, methods for correcting commitment factors were proposed primarily for dehydrogenases. Hydride transfer represents the most optimal case in KIE measurements, as isotope effects from ^{2}H or ^{3}H transfer as hydride ions give large isotope effects related by the Swain-Schadd relationship (^{3}H-KIE = ^{2}H-KIE$^{1.44}$) in classical mechanics of hydride transfer. Deviations from this relationship or from the temperature dependence for KIE in hydride transfer reactions provide indications of commitment factors or quantum mechanical tunneling (Bahnson and Klinman, 1995; Northrop, 1981). When tunneling can be eliminated, the deviation from the Swain-Schadd relationship can be used to establish the intrinsic KIE for hydride transfers. However, the decision that establishes or excludes tunneling is difficult. In the case of quantum mechanical tunneling, experimental KIE for hydride ion transfer can be large or small and may deviate significantly from the Swain-Schadd relationship (Antoniou and Schwartz, 1997). Large KIE are not required in hydride tunneling, and it has been proposed that in cases of coupled motion at the TS, the KIE for hydride transfer may appear to follow classical limits of bond vibrational isotope effects, but that α secondary effects, coupled to the tunneling atom, may show deviations from the Swain-Schadd relationship (Bahnson and Klinman, 1995; Huskey and Schowen, 1983). Tunneling complications for hydride transfer reactions have caused the KIE that are the easiest to measure to become the most difficult to interpret. Quantum mechanical tunneling has been discussed in several

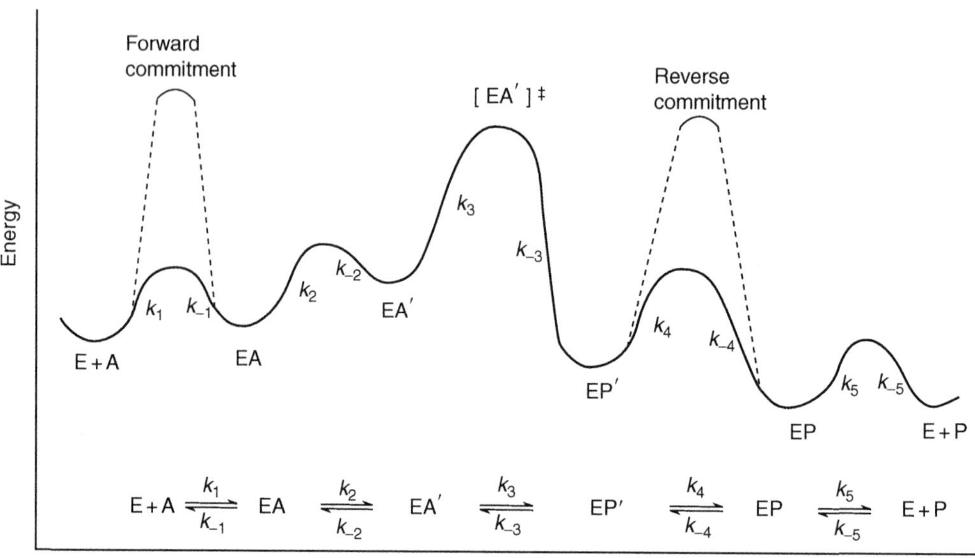

Fig. 10 Reaction coordinate diagram demonstrating forward and reverse commitments. The reaction coordinate diagram formed by the solid line demonstrates a reaction with chemistry as the highest energetic barrier for the catalytic cycle. This pattern yields intrinsic KIE. Forward commitment is indicated by the dashed line, where substrate molecules (A), which cross the binding barrier k_1, are more likely to react (k_3) than to return to E + A. Reverse commitment is indicated by the energetic barriers for EP′, which is more likely to be converted back to EA′ than to EP. Reprinted with permission, from the *Annual Review of Biochemistry*, Volume 67, © 1998, by Annual Reviews: http://www.annualreviews.org (Schramm, 1998).

reviews and reports and will not be discussed further here. Reactions without hydride transfer are not subject to tunneling effects and are interpreted more readily in the framework of classic bond vibrational theory.

Apart from the complications in the special case of quantum mechanical tunneling, the commitment factors can be measured by the following experimental methods. These can then be used to establish the intrinsic KIE using the relationship of Northrop (1981):

$$\frac{{}^{x}V}{K} = \frac{{}^{x}k + C_{\mathrm{f}} + C_{\mathrm{r}} \cdot {}^{x}K_{\mathrm{eq}}}{1 + C_{\mathrm{f}} + C_{\mathrm{r}}} \quad (1)$$

where x is the isotopic label (^{3}H, ^{14}C, etc.), $^{x}V/K$ is the experimental competitive KIE, which is KIE on the V/K kinetic constants when competing labels are used, ^{x}k is the intrinsic KIE on the chemical step k, C_{f}, and C_{r} are forward and reverse commitments, and ^{x}Keq is the isotope effect on the equilibrium constant for the chemical equilibrium. Analysis of ^{x}k is required for TS analysis based on KIE. Therefore, C_{f} and C_{r} must be known to solve for ^{x}k.

A. Forward Commitment and KIE

Forward commitment arises from exceptions to the case that a thermodynamic equilibrium is established between the pool of free substrate molecules and those that populate the TS (Northrop, 1981). The procedure to detect forward commitment depends on measuring the fraction of substrate bound in the Michaelis complex, which is converted to product and that which is released unchanged from the catalytic site for each catalytic event. Forward commitment is the probability function for the fate of each substrate molecule in the Michaelis complex. A reaction coordinate diagram illustrates the forward commitment problem (Fig. 10). The substrate trapping procedure established by Rose (1995) provides a convenient solution to this problem. The enzyme is saturated with labeled substrate, followed by mixing with a large excess of the unlabeled substrate followed by several catalytic turnovers and quenching of catalytic activity. Analysis of the fraction of bound, labeled substrate converted to products gives the commitment factor.

The experimental protocol and the results of a substrate trapping experiment for the determination of forward commitment are shown schematically in Fig. 11. The fraction of bound substrate [bound substrate = free substrate concentration/(K_d + free substrate concentration)] is converted to product following five or more catalytic cycles (to permit equilibration of bound substrate with the new pool of diluted substrate and conversion of bound substrate to product) and provides the commitment factor. Enzymes with two or more substrates, bound randomly, are analyzed easily by this approach, as the binary enzyme-substrate complex forms but does not react until the second substrate is added together with the large excess of unlabeled first substrate. Purine nucleoside phosphorylase (PNP) demonstrated large forward commitment with bound inosine when phosphate was added, and the KIE for the phosphorolysis reaction were too small to be measured accurately (Fig. 11) (Kline and Schramm, 1993, 1995).

Single substrate enzymes begin reaction as soon as the enzyme and labeled substrate are mixed, and the label/dilution steps must therefore occur on the time scale of the catalytic turnover rate. Measurement of forward commitment under these conditions requires rapid mix, chemical quench approaches. These methods have been used to determine forward commitments for NAD^+ in CD-38 and for ATP in *S*-adenosylmethionine synthetase (Markham *et al.*, 1987; Sauve *et al.*, 1998).

Forward commitment is the most common factor to obscure intrinsic KIE in reactions with an irreversible chemical step, for example, in hydrolytic reactions with large negative free energy values. In enzymes where the conversion of enzyme-bound products to bound substrates does not occur, Eq. (1) reduces to ${}^{x}V/K = ({}^{x}k + C_f)/(1 + C_f)$. The observed isotope effects vary between intrinsic (${}^{x}k$) and unity as forward commitment increases from negligible to dominant. Measurement of the chemical isotope effect and the forward commitment is sufficient to establish the nature of the intrinsic KIE unless the experimental isotope effect is so

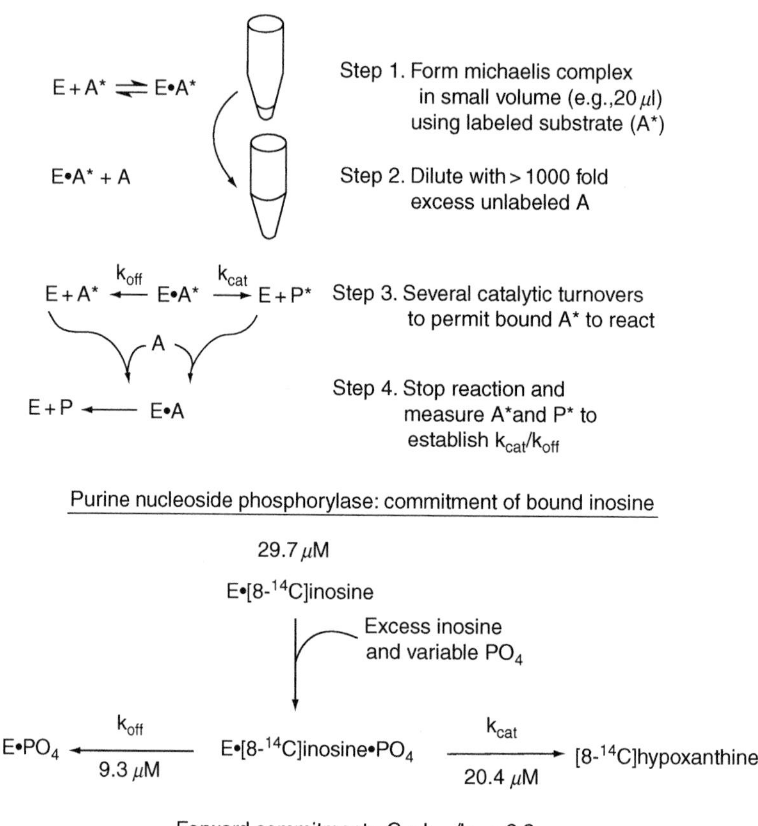

Fig. 11 Experimental measurement of substrate commitment factor (Kline and Schramm, 1993; Rose, 1995). In step 1, 30 μM PNP was incubated for 10 s with 400 μM [8-^{14}C]inosine in 50 mM triethanolamine, pH 7.5, in a volume of 20 μl. In step 2, 1 ml was added containing 5 mM inosine and variable concentrations of phosphate (100, 125, 200, 2000, and 4000 μM). Following 15 s, the reactions were quenched with 100 μl 1 N HCl and analyzed by HPLC (C-18 reverse phase eluted with 5% methanol in H_2O) to establish the fraction of bound substrate converted to product (ribose 1-phosphate). The commitment factor is determined at saturating phosphate concentration.

near unity that corrections are numerically unsound. In KIE measurements, the assumption is made that binding isotope effects are negligible. Binding isotope effects are discussed in later.

B. Reverse Commitment

Reactions that are energetically reversible often exhibit reverse commitment (C_r in Eq. (1)) in which the enzyme-bound products react to reform enzyme-bound substrate before product release. Reverse commitment can also occur in reactions with irreversible overall reactions provided that the equilibrium between enzyme-bound substrates and products permits reversal. The equilibrium constant for enzyme-bound substrates and products is often nearer unity than the chemical reaction, thus reverse commitment must be considered for all reactions. Reverse commitment alone obscures the intrinsic KIE by the expression

$$\frac{{}^{x}V}{K} = \frac{{}^{x}k + C_r \cdot {}^{x}K_{eq}}{1 + C_r}$$

where ${}^{x}k$ is the intrinsic isotope effect. The value of the observed KIE, ${}^{x}V/K$, can vary between the intrinsic KIE and ${}^{x}K$eq. Reverse commitments for chemically reversible reactions can be measured with the same techniques as forward commitment, using the substrates for the reverse reaction. If present, the ${}^{x}K$eq must also be established to interpret the intrinsic KIE. Methods to measure ${}^{x}K$eq are simple and have been outlined previously in this series (Cleland, 1982). With experiments to measure both forward and reverse commitments, it has become experimentally feasible to measure intrinsic isotope effects for many enzymes.

C. Enzymes with Large Commitment Factors

Substrates in the Michaelis complex of enzymes with large forward and reverse commitment factors are converted efficiently to enzyme-bound products. Bound products react to form bound substrates with a high probability relative to the release as free products. Large commitment factors are independent of the chemical thermodynamics of the reaction because the equilibrium constants for bound substrates may vary considerably from that in solution. This phenomenon has been well established by the NMR measurement of the equilibrium concentrations of substrates and products bound at catalytic sites (Cohn and Reed, 1982). Thus, reverse commitment can occur in enzymes with freely reversible and in experimentally irreversible reactions. If the experimental KIE are near unity, even accurate measurement of the commitment factors does not permit an accurate analysis of the intrinsic isotope effect, as the correction factor is the product of the observed isotope effect and the correction factors (Eq. (1)). An experimental solution for large commitments is to measure presteady-state KIE on partial reactions.

D. Overcoming Large Commitment Factors

A highly committed reaction coordinate diagram is shown in Fig. 12 and will be used to illustrate presteady-state reaction methods to overcome large commitments, even when the observed KIE are unity in the steady state. The reaction is initiated from a complex of enzyme with appropriately labeled substrates (E*A). In a two substrate reaction, this complex is nonreactive, and the reaction is initiated by rapid mixing with sufficient B to form the E*AB complex. The reaction is permitted to proceed until approximately 30% of the E*AB complex is converted to the E*PQ complex. This mixture is quenched chemically or physically before a

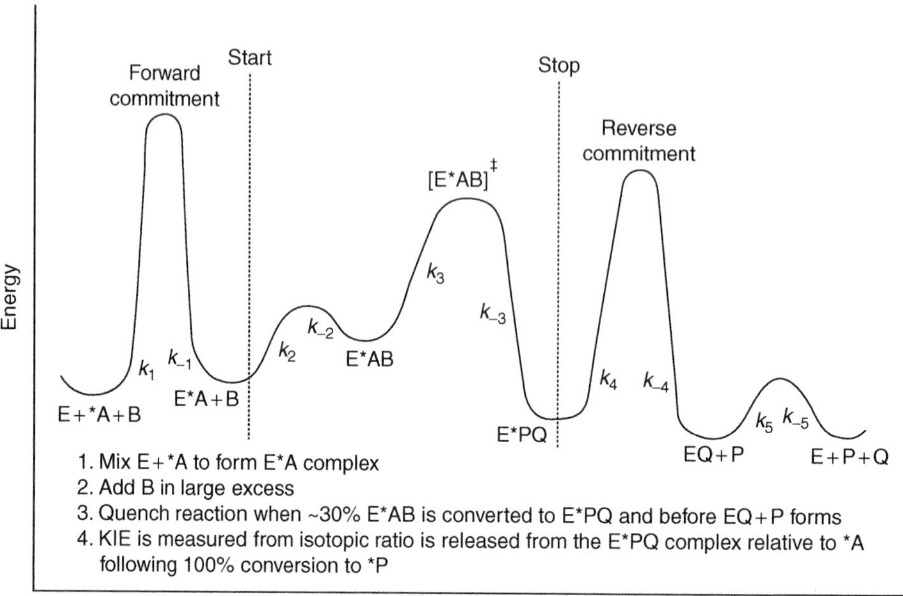

Fig. 12 Presteady-state analysis to circumvent commitment factors. The reaction coordinate diagram illustrates the unfavorable case of strong forward and reverse commitments in a bisubstrate enzymatic reaction. KIE for the chemical step can be measured by forming the E*A complex (where A indicates the mix of isotopically labeled substrates). Assuming no binding isotope effect on k_1 (usually diffusion controlled), the E*AB complex is formed by mixing with B. After a small fraction (~20%) is converted to E*PQ, the reaction is quenched. The product is analyzed in the same way as for normal KIE. Analysis of KIE for a series of reactions with conversions to E*PQ from 10% to 40% of E*AB can be used to extrapolate to the KIE at 0% conversion, the intrinsic KIE for the chemical step.

significant fraction of E*PQ reverses to E*AB or dissociates to products. Under these conditions the KIE arises from isotopic fractionation of the E*AB complex, which contains a mixture of *A with the appropriate isotopic labels. Both forward and reverse commitments are avoided in this method, which uses the enzyme as a stoichiometric reactant rather than a steady-state catalyst. This method measures specific steps in the reaction coordinate and contrasts to steady-state measurements where many steps in the cycle contribute to the reaction rate and therefore to the *V*/*K* KIE. The presteady-state approach to enzymatic TS analysis is certain to be widely applicable in complex enzymatic reactions, but only a few examples have been reported. The approach requires relatively large amounts of enzyme, knowledge of the reaction coordinate barriers, and sufficient knowledge of the reaction rates to stop the reaction when the desired fraction of labeled substrate is converted to enzyme-bound product. This approach has been used to determine the TS structure for the hydrolysis of inosine by PNP, where reverse commitment and rate-limiting product release obscured intrinsic KIE (Kline and Schramm, 1995).

Single substrate enzymes can therefore be analyzed by partial catalytic turnover techniques to avoid the problem of reverse commitment. Transient KIE have been measured most often with dehydrogenases, where deuterium isotope effects and the easily measured spectral changes make the approach most accessible. The relatively limited number of studies available at that time were summarized by McFarland (1991).

E. Binding Isotope Effects

It has been assumed that binding equilibrium isotope effects are small relative to the magnitude of KIE and have been disregarded in most studies of enzymatic KIE (Northrop, 1998). However, it is apparent from inspection of the reaction coordinate diagram that steady-state KIE studies are influenced by binding isotope effects arising from the interactions of substrate and enzyme in the Michaelis complex (e.g., Fig. 12; competitive KIE report on the isotopic responses for every rate constant included in the *V*/*K* expression for the specific mechanism). Because kinetic constants in this term include every step between free substrate and the first irreversible step, *V*/*K* KIE always include substrate binding effects. Equilibrium binding isotope effects between free and bound substrate ($E + {}^*A \leftrightarrow E^*A$) occur if the enzyme-induced distortion of the substrate is sufficient to cause bond vibration differences at the atomic positions labeled with isotopes for the KIE experiments. Binding isotope effects are well documented in a small number of enzymatic reactions. For example, the binding isotope effect for [4′-^{3}H]NAD$^+$ has been measured to be approximately 1.10 (10%) (LaReau *et al.*, 1989). Although an isotope effect of 10% is measured easily, the intrinsic KIE for hydride transfer reactions relative to tritium ion transfer in dehydrogenase reactions are much larger, with an upper limit near 13 (1300%), and are commonly found to be 3–6 (300–600%). For these reactions, the assumption that binding isotope effects are negligible is valid (Northrop, 1998). Reactions with C–C, C–N, C–S, and C–O bond scission at the TS typically cause small isotope effects of 1–13% for the primary ^{14}C atom and α and β secondary ^{3}H effects from 1% to 25%, depending on the nature of the TS. In these cases, a ^{3}H-binding isotope effect of 10% would be a major correction required in the calculation of the intrinsic KIE.

Experimental binding isotope effects can be measured by binding equilibrium experiments. Ultrafiltration or equilibrium dialysis followed by the analysis of bound and free isotopically labeled substrates provides a simple and direct measurement of isotope effects (Fig. 13). In ordered kinetic reaction mechanisms, trapping of the first substrate by extrapolation to a saturating concentration of the second has also been used to analyze binding isotope effects (Gawlita *et al.*, 1995). This method does not provide binding equilibrium effects because return over the substrate binding barrier (k_{-1} in Fig. 12) is prevented by saturation with the second substrate. The method therefore provides the KIE for the bimolecular formation (k_1 in Fig. 12) of the Michaelis complex (the TS for substrate binding). Binding isotope effects are not subject to commitment factors or any other

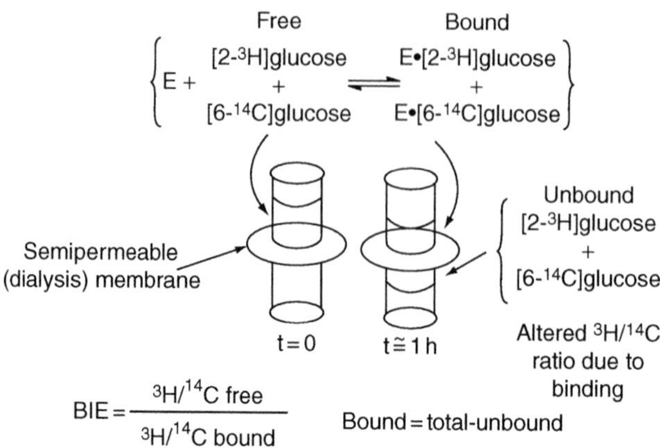

Fig. 13 The ultrafiltration method used to measure substrate binding equilibrium isotope effects (B. Lewis and V. L. Schramm, unpublished observations, 1998). A solution (100 μl is typical) of buffered enzyme containing the appropriate ratio of labeled substrates (glucose in the example) is placed above an ultrafiltration membrane that is freely permeable to ligand but retains the enzyme. The conditions are selected to permit approximately 50% of the total ligand to bind. The upper chamber is placed under N_2 pressure, approximately 50% of the volume is forced through the membrane, and the bottom chamber is sampled for unbound ^{3}H and ^{14}C substrate. This is compared to the isotopic label in the original mixture. In the example shown here, the ^{14}C is the control as no equilibrium binding isotope effect is expected with the carbon. Control experiments move the ^{14}C to other positions, keeping the ^{3}H at the 2 position to demonstrate that the binding isotope effect is independent of the placement for the remote (^{14}C) label.

obscuring effects. Recent experiments have tested the hypothesis that substrate binding to hexokinase causes binding isotope effects in the Michaelis complex of enzyme-glucose. Binding of specific [^{3}H]glucose substrates indicates that the remote ^{3}H-binding isotope effects can be as large as 9% (Lewis and Schramm, 1998). These early results suggest that binding isotope effects may need to be considered in obtaining accurate intrinsic KIE in many enzymatic reactions. The results of isotope-edited difference Raman spectroscopy (Chapter 8, this volume) lead to the same conclusion (Callender and Deng, 1994). The Raman shifts observed in several Michaelis complexes indicate altered bond vibrational frequencies that correspond to significant binding isotope effects. It is expected that data from Raman or infrared-red spectroscopy and binding isotope effects will be complementary, as the physical basis of all three effects is the change in bond vibrational modes for the atoms of interest when free and when bound to the enzyme. In those cases where binding isotope effects are demonstrated, additional terms can be incorporated into the intrinsic KIE expression (Eq. (1)) to correct for the binding isotope effects.

Equation (2) indicates that substrate binding isotope effects will influence the calculation of the intrinsic binding isotope effects both by the equilibrium effect and by isotopic discrimination at the TS for binding (the effect on k_1 discussed earlier) (Lewis and Schramm, 1998):

$$\frac{{}^xV}{K} = \frac{{}^xk{}^xK_a + {}^xk_1C_f + C_r{}^xK_{eq}{}^xK_a}{1 + C_f + C_r}.$$

In addition to the terms described in Eq. (1), ${}^xK_a$ is the substrate binding equilibrium isotope effect and ${}^xK_1$ is the effect on the initial binding of the substrate. Each of the terms in Eq. (2) can be evaluated experimentally. The binding expressions for a variety of mechanisms and the experimental impact of binding isotope effects require further evaluation.

F. Summary of Commitment Factors and Their Influence on KIE and TS Analysis

The commitment and binding problems discussed earlier can be overcome experimentally in most enzymes. Enzymatic reactions with multiple chemical steps where the energies are approximately the same require presteady-state KIE approaches. Similar to other technological advances, there has been a period of experimental development to establish the techniques that permit the measurement of intrinsic KIE for enzymes of ever-increasing mechanistic complexity. Experimental and computational methods are now available to overcome most of the commitment factor barriers imposed by slow enzymatic steps and binding isotope effects for a broad range of important and interesting enzymatic reactions.

IX. Matching of Intrinsic KIE to TS Structures

The relative atomic vibrational environments in the reactant and TS molecules are the major factors that determine the magnitude of the intrinsic KIE. Conversely, the intrinsic KIE for an enzymatic reaction permits reconstruction of the atomic vibrational environment of the TS if sufficient isotope effects have been measured. This section provides brief examples of three experimentally determined intrinsic KIE and the steps that lead to a construction of a TS structure for the enzyme-stabilized TS. The steps outlined here were developed with computationally modest resources and are semiquantitative estimates of the TS structures. This approach systematically compares many possible TS structures in reaction coordinate space and assumes a linear relationship among bond orders, angles, and changes in atomic rehybridization (Markham *et al.*, 1987; Mentch *et al.*, 1987). Despite the semiquantitative approach, the TS methods have captured the major features that are altered between substrate and TS and give rise to the intrinsic KIE. Chapter 19 provides the most recent evolution of the systematic computational approach. A systematic quantum chemical computational approach matches intrinsic KIE to atomic structures of the TS. This new approach incorporates nonlinear features of bond order and geometric changes at the TS. TS imbalance is a relatively recently recognized property of enzymatic TS, and future applications of TS determination will incorporate this feature of chemical bond changes. Coupling these methods to a systematic search of reaction coordinate space improves selection of the TS that is most consistent with intrinsic KIE (Berti, 2009).

A. AMP Deaminase

Determination of the TS structure for AMP deaminase had the purpose of comparing the nature of the experimentally determined TS with the structure of coformycin 5′-phosphate. Coformycin is a natural product TS inhibitor that binds 10^7 more tightly than substrate to adenosine deaminase, and the 5′-phosphate shares an equal affinity for AMP deaminase. Deamination of AMP occurs by the enzymatic activation of a water nucleophile that attacks C-6 of the purine ring (Fig. 14) (Merkler *et al.*, 1993). The metalloenzyme contains a tightly bound Zn^{2+} that chelates the water and lowers the pK_a to create a catalytic site hydroxide ion. The catalytic mechanism is common to adenosine deaminase and AMP deaminase, as all amino acids in contact with the Zn^{2+} and purine ring are conserved in both amino acid sequences (Merkler and Schramm, 1993). The sequence homology outside the catalytic site is limited, and AMP deaminase is more than double the size of adenosine deaminase (352 and 810 amino acids, respectively) because of the presence of allosteric activation and inhibition regulatory domains (Meyer *et al.*, 1989). The hydroxide oxygen reacts and displaces the amino-leaving group to convert the adenine ring to hypoxanthine. Nucleophilic substitutions at an aromatic ring are well known from chemical precedent and involve the formation of an unstable intermediate (the Meisenheimer complex) with the carbon under nucleophilic attack being converted from sp^2 to sp^3 hybridization, followed by protonation of the leaving group and its elimination, regenerating the sp^2 center (Schramm and Bagdassarian, 1999). The natural product inhibitor, coformycin, resembles this intermediate, thus it was proposed that the TS complex must be closely related to the Meisenheimer complex (Kline and Schramm, 1994).

Intrinsic KIE were established from the observations that (1) the KIE were the same for AMP over a wide range of *V*/*K* values with allosteric activator, allosteric inhibitor, and alternative slow substrates and (2) the [^{14}C]KIE was near the theoretical limit for any possible degree of bond formation to the attacking oxygen nucleophile (Merkler *et al.*, 1993). The TS structure was determined from the KIE summarized in Fig. 14. In the first model of the TS, the extent of oxygen nucleophile to C-6 bond order was varied systematically and fixed at values of 0.3–0.8 with a full bond retained to the NH_2, which was fixed as the NH_2 and not permitted to become NH_3. The hybridization of C-6 was converted incrementally from sp^2 to sp^3 as the C-6–oxygen bond was formed. With fixed values for each C-6–O bond distance, the remainder of the molecule was minimized using MOPAC 6.0. The optimized structure from MOPAC 6.0 was used to establish the KIE that each TS structure would be expected to generate using BEBOVIB-IV. This method introduces only one variable, the distance to the nucleophile, with the remainder of the structure assigned by computational optimization. Two additional

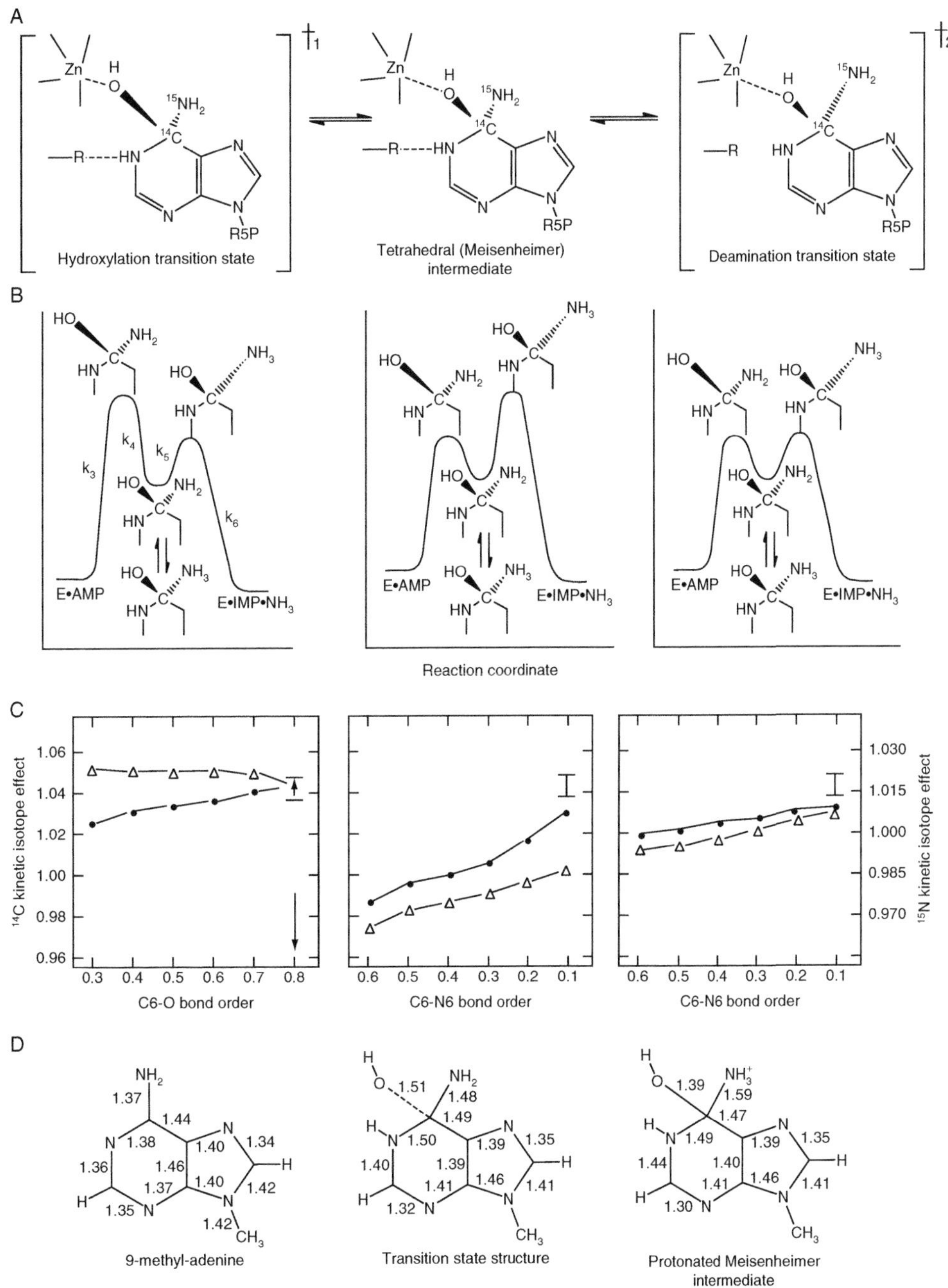

Fig. 14 Reaction coordinate, KIE, and transition-state structure for yeast AMP deaminase. (A) The three steps in the aromatic nucleophilic substitution to deaminate NH_3 from AMP. (B) Possible reaction coordinate diagrams with rate-limiting OH attack (left), rate-limiting NH_3 departure (middle),

TS models were generated by fixing C-6 as an sp^3 center with fully bonded oxygen and varying the bond order to the leaving group NH_3. Finally, a mechanism was tested in which two equal TS, one for hydroxide oxygen attack and one for NH_3 departure, contributed equally in two distinct TS separated by the Meisenheimer intermediate. The isotope effects calculated for each of these TS models were compared to the experimental isotope effects for ^{14}C-6 and for ^{15}N-6 as the leaving group. Only the TS with the attacking hydroxide ion and fully bonded NH_2 provided agreement with the experimental KIE and agreement was consistent with a bond order of 0.8 to the attacking oxygen. This TS is late in the nucleophilic attack portion of an $A_N + D_N$ mechanism and occurs just before formation of the Meisenheimer intermediate. Bond lengths for the MOPAC-6 optimized structures of the reactant, TS, and Meisenheimer complex for AMP deaminase are shown in Fig. 14.

B. Nucleoside Hydrolase

Nucleoside hydrolases are enzymes found in protozoan parasites, which contribute to the essential pathway of purine salvage in these purine auxotrophs (Parkin *et al.*, 1991). The enzymes produce purine or pyrimidine bases and ribose and are considered targets for novel inhibitor design as mammals have no nucleoside hydrolases. The nonspecific nucleoside hydrolase from *Crithidia fasciculata* was used to measure the KIE for labeled inosine substrates synthesized as outlined in Figs. 2 and 3. In addition to extensive kinetic studies, the KIE for each isotopically labeled position indicated that the enzyme expressed intrinsic KIE (Fig. 15) (Horenstein *et al.*, 1991). A match between experimentally measured intrinsic KIE and a TS structure was obtained by the systematic variation of the C-1′–N-9 bond order with varying degrees of nucleophilic participation by the attacking water nucleophile. The reactant-state structure for inosine was taken from the crystal structure with the bond lengths for C–H bonds obtained by AMPAC calculations using the AM1 parameter set (Dewar *et al.*, 1985). The initial structure for the TS ribosyl was the X-ray crystal structure coordinates for ribonolactone. An imidazole was used as the leaving group, and the bond changes, which characterize the leaving group, were estimated by AMPAC calculations as the C-1′–N-9 bond was broken. The reaction coordinate was generated by coupling the stretching motion

and equivalent rates for OH attack and NH_3 departure (right). (C) The KIE expected for each case is plotted as a function of attacking OH bond order (left, C-6–C bond order) or departing NH_3 bond order (middle and right, C-6–N-6 bond order). Superimposed on the predicted KIE as a function of bond order are the actual KIE values for ^{14}C and ^{15}N. Ordinate and abscissa scales have been adjusted so that both ^{14}C and ^{15}N KIE are represented by the same value, shown with their standard error bars. The only reaction coordinate diagram consistent with the intrinsic KIE is rate-limiting OH attack followed by rapid protonation and departure of the NH_3. Agreement between experimental and calculated KIE was best at a C-6–O bond order of 0.8, indicated by the arrow. (D) 9-Methyl truncations of the substrate, transition state, and protonated Meisenheimer complexes. The bond lengths are shown for the optimized structures.

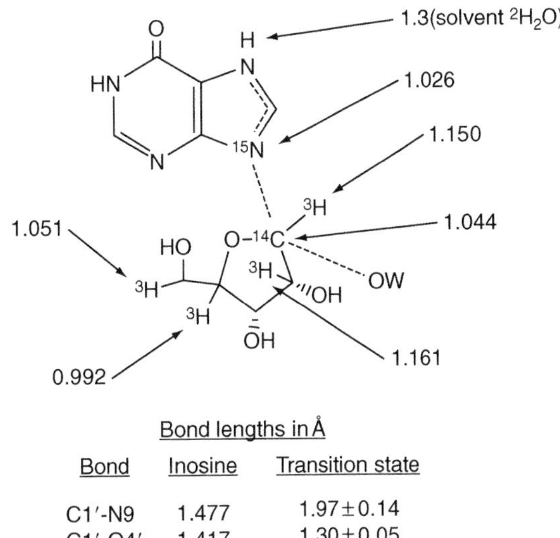

Bond lengths in Å

Bond	Inosine	Transition state
C1′-N9	1.477	1.97±0.14
C1′-O4′	1.417	1.30±0.05
C1′-C2′	1.530	1.502±0.001
C2′-H2′	1.128	1.160±0.001
N9-C8	1.372	1.332±0.006
C8-N7	1.308	1.346±0.006
C1′-WO	---	3.0±0.4

Fig. 15 Intrinsic KIE and transition-state bond lengths for the nucleoside hydrolase reaction. Arrows indicate the positions at which the intrinsic KIE were measured. Isotopic labels are indicated in the molecule; however, in practice, each KIE was measured with the isotopic pairs as indicated in Table V. The table indicates bond lengths in substrate and transition state.

of the breaking C-1′–N-9 bond with the C-1′–oxygen nucleophile bond. Inversion of configuration at C-1′ was incorporated into the TS model with an interaction force constant of ±0.05 between the stretching modes for the oxygen nucleophile–C-1′ and C1′–N-9 with the angle bending modes for all bond angles with C1′ at the center (six bond angles of the type OW–C-1′–H-1′ and N-9–C-1′–O-4′). Reaction coordinates for the TS were compatible with the experimental KIE only in TS structures that had low bond order in the C-1′–N-9 bond and low but significant bond order to OW, the attacking nucleophile. The match between hypothetical TS structures and experimental KIE employed the BEBOVIB program restricted to 25 atoms for both reactant and TS models. Truncations used in the calculations included every atom within two bonds of the C-1′–N-9 bond, which defines the reaction coordinate. It is generally assumed that KIE (perturbations in bond vibrational environments) will not be significant at a distance of more than two atoms away from the site of bond scission at the TS (Sims and Lewis, 1984).

Calculated KIE for atoms of the labeled substrate as a function of the nature of the TS are shown in Fig. 16. The experimental KIE are compared to calculate KIE for dissociated (S_N1) TS, nucleophilic (S_N2) TS, and a mixed TS with weak

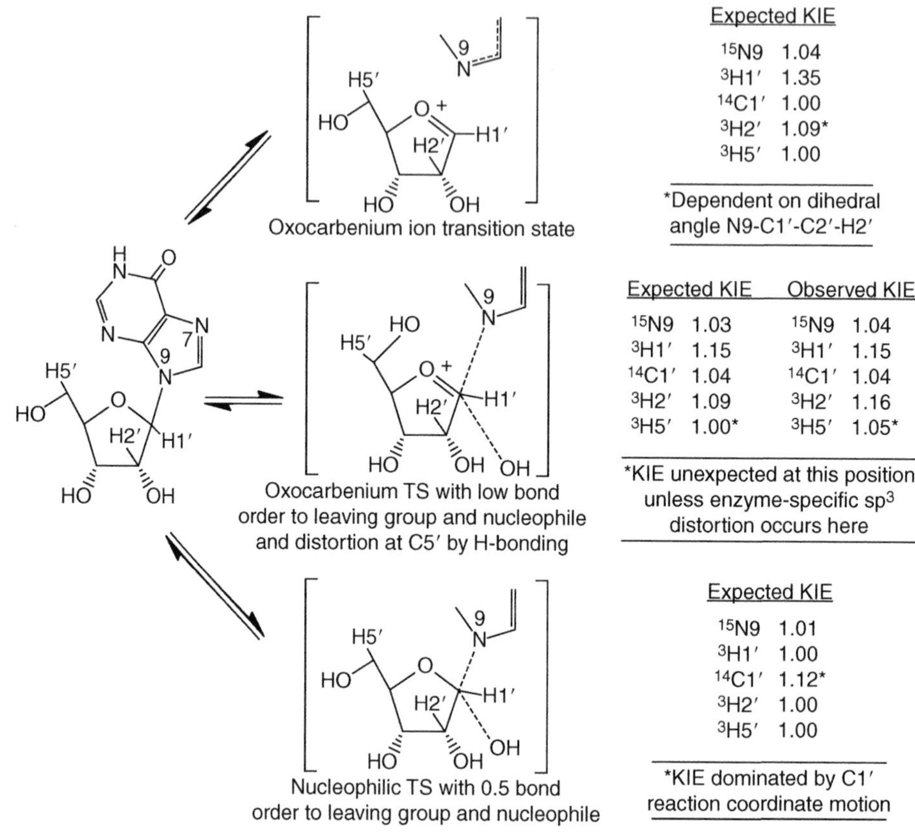

Fig. 16 Calculated KIE for three potential reaction mechanisms compared to an intrinsic KIE dataset. The upper and lower TS structures represent the extremes of dissociative and associative mechanisms calculated for the gas phase. The expected KIE were calculated from BEBOVIB calculations (Horenstein *et al.*, 1991). The intrinsic KIE ruled out the limiting mechanisms and are in good agreement with the structure of the intermediate TS. Exact matches of the ^{3}H2′ and ^{3}H5′ are accommodated by the hyperconjugation and H-bonding effects specific to the enzymatic catalytic site.

participation of both the attacking nucleophile and the significant bond order to the leaving group. Matches of the calculated KIE and the experimental values indicate agreement of the mixed TS. However, the 5.1% KIE at the 5′-^{3}H position was unpredicted because it is four bonds away from the reaction center, and any reasonable chemical model of the TS would not perturb this bond vibrational environment. The substrate specificity of nucleoside hydrolase indicated that the 5′-hydroxyl is essential for catalytic activity, but not for substrate binding (Parkin *et al.*, 1991). Together, the steady-state kinetic analysis and KIE results indicated that formation and stabilization of the TS require enzymatic interaction with the 5′-hydroxymethyl group in such a way that the sp^{3} hybridization is distorted to give rise to a 5.1% normal KIE at this position. Normal isotope effects indicate increased vibrational freedom for the isotopic atom in the TS relative to the reference state. The most likely TS structure that could account for the remote KIE was obtained by rotation of the 5′-hydroxyl group to form a dihedral angle between O-5′–C-5′–C-4′–O-4′ that locates the 5′-O near the ring oxygen, O-4′. In this conformation, the lone pair of O-5′ is directed toward the ring oxygen, the site of ribooxocarbenium cation formation at the TS. A normal hydrogen bond to the hydrogen of the 5′-hydroxyl group in this geometry can easily account for the 5.1% KIE measured experimentally. This unexpected and unprecedented KIE predicted the unusual nucleoside TS configuration that was later established to have a dihedral angle of 336° when the X-ray crystal structure was solved with a TS inhibitor (Degano *et al.*, 1998). This geometry is unknown in nucleoside and nucleotide chemistry, where the average dihedral angle at this position is 233° (Gelbin *et al.*, 1996). The TS structure, which is in agreement with all of the experimental KIE, is indicated in Figs. 15 and 16. The C-1′–N-9 ribosidic bond is nearly broken at 1.97 ± 0.14 Å, and the attacking water nucleophile oxygen is at 3.0 ± 0.4 Å, in van der Waals contact at the TS. Loss of the bonding electrons without compensation from the attacking nucleophile leaves the ribosyl group with an electron deficiency and nearly a full positive charge. This introduces double bond character into the ribosyl ring, primarily between C-1′ and O-4′, the site of the ribooxocarbenium cation. Departure of the bonding electrons into the leaving group gives it a negative charge, and neutralization of the charge is accomplished by the protonation of N-7 at the TS. Because the *N*-ribosidic bond is not completely broken at the TS, it is conceivable that the protonation of N-7 is in progress at the TS. To test this hypothesis, the solvent deuterium KIE was measured to establish the role N-7 protonation plays in leaving group activation. The solvent deuterium KIE is small, 1.3 for k_{cat} and 0.99 for k_{cat}/K_m, whereas a TS that demonstrates proton transfer at the TS gives a KIE >2.0 (Quinn and Sutton, 1991). It was concluded that proton transfer to N-7 is complete at the TS, as the ^{15}N-9 KIE required imidazole bonding consistent with N-7 protonation.

Atomic structures for the truncated molecules used in the BEBOVIB analysis were converted into the full atomic structures. Atoms for the substrate inosine were taken from the X-ray crystal structure with the hydrogens positioned using Gaussian 92 with the STO-3G basis set (Frisch *et al.*, 1992). The TS state structure was computed with the 3′-exo configuration of the ribose, as required by the β secondary ^{3}H KIE from 2′-^{3}H. All bonds defined in the final BEBOVIB-TS model (the structure that recreates the experimentally observed intrinsic KIE) were fixed, and the positions of the remaining bonds were minimized energetically with Gaussian 92 and the STO-3G basis set. The enzyme-bound product α-D-ribose was modeled as the 3-exo conformation based on its relationship to the structure of the TS and least-motion principles to achieve the product. Likewise, the enzyme-bound hypoxanthine product was modeled using the same computational parameters, with the site of ring protonation at N-7 corresponding to that of the TS rather than the preferred protonation at N-9 as occurs in solution. These parameters were selected to preserve the features of the TS imprinted onto the products in the next few bond vibrations that occur following TS separation and before release from the catalytic site. This approach permitted analysis of substrate, TS, and products to provide an electronic model of the reaction coordinate pathway (Horenstein and Schramm, 1993a).

C. Purine Nucleoside Phosphorylase

The phosphorolysis of inosine catalyzed by PNP plays a unique role in the development of the human immune system, as infants with a genetic deficiency of the enzyme develop a specific and ultimately fatal T-cell deficiency within 1–2 years of birth (Hershfield and Mitchell, 1995). This physiological role has made PNP the target of inhibitor discovery efforts for agents to modulate T-cell disorders and malignancies of T-cell origin (Niwas *et al.*, 1994; Walsh *et al.*, 1994). The PNP reaction was selected for TS analysis in an effort to apply TS practice to the design of TS inhibitors.

KIE for the reaction using the same labeled inosines as for nucleoside hydrolase were near unity and could not be corrected to reliable values because of large forward and reverse commitment factors, the most unfavorable of the commitment cases described earlier (Table V). The problem of reverse commitment was solved by replacing phosphate with its analog arsenate, which reacts at a rate similar to phosphate in the PNP reaction, but forms α-D-ribose-1-arsenate as the chemically unstable product (Kline and Schramm, 1993). This product hydrolyzes to form ribose and arsenate with a rate constant greater than the reverse reaction and thus prevents the reverse reaction from occurring. The forward commitment factor was determined experimentally using the method of inosine substrate trapping described in Fig. 11. The commitment factor permits the calculation of intrinsic KIE from the experimental KIE measured in the arsenolysis reaction. The experimentally measured and the intrinsic KIE for the arsenolysis reaction (Table V) were used to generate a truncated structure of the TS with the bond lengths around the N-9–C-1′ bonds defined (Fig. 17). A comparison of the bond lengths for the

Table V
KIE for Arsenolysis by PNP

Substrate	Type KIE	Experimental KIE	Intrinsic KIE
[1′-^{3}H]- + [5′-^{14}C]inosine	α Secondary	1.118 ± 0.003	1.141 ± 0.004
[2′-^{3}H]- + [5′-^{14}C]inosine	β Secondary	1.128 ± 0.003	1.152 ± 0.003
[1′-^{14}C]- + [5′-^{3}H]inosine	Primary	1.022 ± 0.005	1.026 ± 0.006
[1′-^{14}C]- + [4′-^{3}H]inosine	Primary	1.020 ± 0.006	1.024 ± 0.007
[9-^{15}N, 5′-^{14}C]- + [5′-^{3}H]inosine	Primary	1.009 ± 0.004	1.010 ± 0.005
[4′-^{3}H]- + [5′-^{14}C]inosine	γ Secondary	1.007 ± 0.003	1.008 ± 0.004
[5′-^{3}H]- + [5′-^{14}C]inosine	δ Secondary	1.028 ± 0.004	1.033 ± 0.005

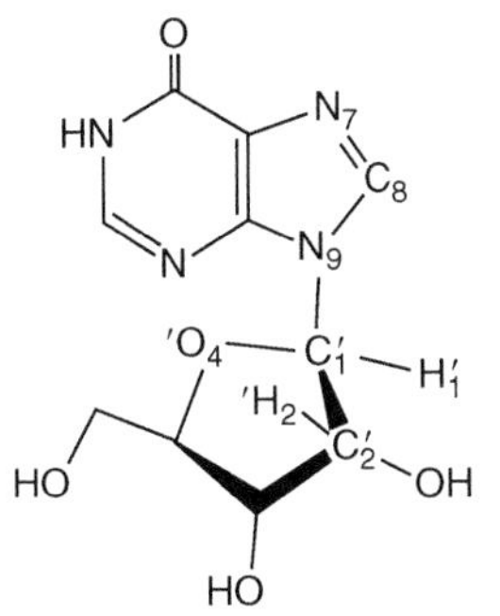

Bond	Å bond length, inosine	Bonds arsenolysis transition state	Bonds hydrolysis transition state
C1′-N9	1.47	1.77 ± 0.04	1.90 ± 0.12
C1′-O4′	1.42	1.29 ± 0.02	1.32 ± 0.06
C1′-H1′	1.13	1.12 ± 0.01	1.12 ± 0.001
C1′-C2′	1.53	1.50 ± 0.02	1.50 ± 0.003
C2′-H2′	1.12	1.16 ± 0.01	1.16 ± 0.002
C1′-OAs	---	3.01 ± 0.02	2.96 ± 0.4
C8-N9	1.37	1.32 ± 0.02	1.33 ± 0.005
C4-N9	1.37	1.33 ± 0.02	1.39 ± 0.004
C8-N7	1.31	1.31 ± 0.02	1.34 ± 0.003

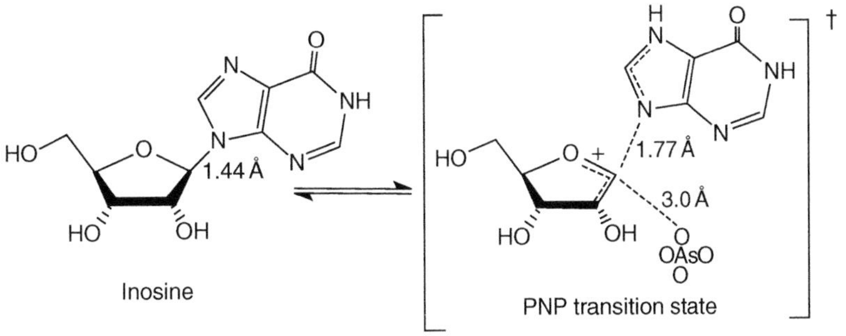

Fig. 17 Transition-state structures for arsenolysis and hydrolysis transition states of purine nucleoside phosphorylase. *Top*: bond lengths at transition states for the reactions. Errors in bond length relate to the error limits of the KIE used to establish the bond length. The transition state for hydrolysis is more advanced in the reaction coordinate than that for arsenolysis, as demonstrated by the increased C-1′–N-9 bond length for hydrolysis. The position of the arsenate or water nucleophile (indicate as C-1′–OAs) is the same for both transition states. *Bottom*: features that distinguish the substrate from the transition state.

reactant inosine and the TS indicates that the C-1′–N-9 bond retains approximately 0.3 bond order at the TS whereas the attacking arsenate anion is 3.01 Å away, in weak van der Waals contact. Although the arsenate is not bonded, the presence of arsenate at this distance is required to match the TS structure to the intrinsic KIE. The significance of these low bond orders in TS stabilization and on the values of the intrinsic KIE is discussed in more detail in Chapter 19 (Berti, 2009). The residual 0.3 bond order to C-1′ indicates that the net charge on the ribooxocarbenium ion is near 0.7 and is apparent in the shortened C-1′–O-4′ bond from 1.42 to 1.29 Å. The value of the ^{15}N-9 KIE is consistent with the protonation of N-7 at the TS as would be expected for the charge separation leading to TS formation. These features are informative for TS inhibitor design, for example, the weak interaction with the attacking anion nucleophile establishes that this group is not required as a covalent group in the design of a TS inhibitor (see later).

D. Summary of KIE and TS Relationships

The information content from KIE is greatest for large isotope effects, where the error is a small fraction of the experimental isotope effect. In this case, there is less ambiguity as to whether a small experimental KIE is a consequence of small bond differences at the TS or an unfavorable commitment factor. The magnitude of intrinsic KIE for three reactions for substitution at carbon are shown in Fig. 16. KIE at the central atom of the reaction are dominated by reaction coordinate motion at the TS: the motion of the central atom away from the leaving group and toward the acceptor. An example of this is in nucleophilic (S_N2) substitutions at carbon, where at the TS the carbon atom is in a coordinated motion away from the leaving group and toward the attacking nucleophile. The displacement of the tripolyphosphate of ATP by a sulfur nucleophile in the *S*-adenosylmethionine synthetase mechanism results in a [5′-^{14}C]ATP KIE of 12.8% (Fig. 18) (Markham *et al.*, 1987). This KIE is near the theoretical limit for the relative masses of ^{12}C/^{14}C and provides unambiguous proof that the TS involves symmetric nucleophilic participation of the attacking group and departure of the leaving group and that the commitment factors for the reaction are small. In this most favorable of cases, the single KIE provides nearly complete information about the bonding nature of the TS.

The reaction examples of Fig. 16 demonstrate that reactions involving hybridization changes at the TS provide strong signals in the measurement of KIE. The lack of isotope effects in reactions where chemistry is known to dominate the reaction coordinate can also provide useful information. At least four distinct types of TS information are available from intrinsic KIE.

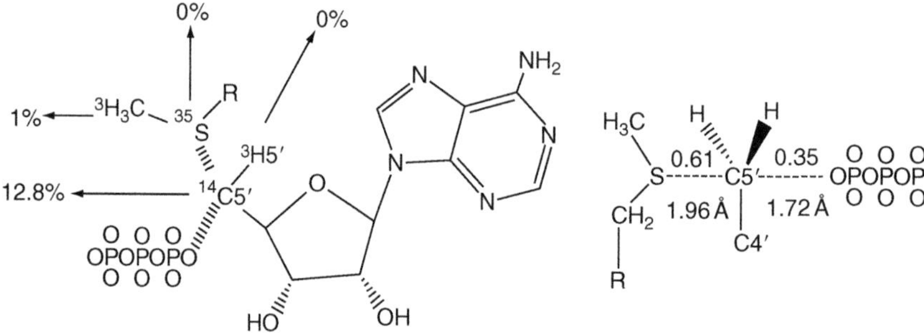

Fig. 18 Kinetic isotope effects and transition-state structure in the nucleophilic displacement catalyzed by *S*-adenosylmethionine synthetase. Individual intrinsic KIE are demonstrated in the structure on the left, which indicates the transition-state attack of the S of methionine on C-5′ of ATP to displace the tripolyphosphate. Isotope effects were measured individually using labeled ATP pairs, as indicated in Table I, or with labeled methionine pairs. Note the dominant isotope effect from ^{14}C-5′ at the reaction center. The bond orders (above) and bond lengths (below) at the TS are shown in the structure on the right. The mass difference between S and O causes the S–C-5′ and C-5′–O bonds to be vibrationally isoenergetic at the bond orders indicated in the figure, a characteristic of S_N2 reaction chemistry.

1. The primary ^{14}C isotope effect gives reaction coordinate motion at the TS, is large for nucleophilic displacements, is smaller in dissociative mechanism where some bond order remains to the leaving group at the TS, and is negligible in dissociative mechanisms where the TS is a fully developed carbenium ion. In this case, the reaction coordinate motion for carbon has ceased at the TS.

2. α Secondary KIE are dominated by the out-of-plane bending mode for the C–H bond. A change of hybridization from sp^3 to sp^2 results in increased vibrational freedom for the hydrogen atom and a resulting large normal isotope effect. The magnitude for this out-of-plane vibrational mode ^3H KIE can be calculated (in the gas phase) to be about 40% (1.40) as a limiting value; however, the experimental values are smaller, with α secondary KIE of 1.20 being common for dissociative mechanisms, but rarely larger than 1.3. The largest of these are observed where rehybridization to sp^2 is nearly complete at the TS, but where there is no significant participation of the attacking nucleophile. The out-of-plane C–H bending mode is sensitive to crowding by nearby heteroatoms so that even the weak participation of nucleophiles (e.g., at 3.0 Å) causes a significant change in the KIE (Mentch *et al.*, 1987). In solution, there is chemical evidence that solvent atoms organize around a charged TS to dampen the out-of-plane motion, thereby decreasing the isotope effect relative to the gas phase (Berti and Schramm, 1997; Burton *et al.*, 1977). Heavy atom α secondary isotope effects are useful in establishing bond changes at the TS. In the case of glucoside or riboside bond scission with dissociative and oxocarbenium-like TS, atoms that are neighbors to the anomeric carbon accumulate double bond character as the TS is reached. In sugars, the ring oxygen in the reactant molecule has a net bond order of 2, and in a fully developed sugar oxocarbenium ion, the oxygen in increasingly bonded to the anomeric carbon to give a net bond order approaching 3. This stiffening of the bond vibrational environment for the oxygen results in an inverse isotope effect, with the ^{18}O-labeled substrate reacting more rapidly than the ^{16}O substrate. Hydrolysis of the *N*-ribosidic bond to the nicotinamide of NAD^+ by acid-catalyzed solvolysis or by diphtheria toxin gives an inverse ^{18}O-4′ KIE of 1.2%, whereas that of methyl β-glucoside is inverse 0.9%. Calculated ^{18}O isotope effects for this position can be inverse 2% (Bennet and Sinnott, 1986; Berti and Schramm, 1997; Berti *et al.*, 1997). Information from the ring ^{18}O KIE complements that from the primary ^{14}C and ^{15}N KIE in establishing the extent of rehybridization of the anomeric carbon at the TS. It should be apparent from this analysis that the α secondary heavy atom KIE will be negligible in nucleophilic displacements, as is demonstrated in the reaction in Fig. 18.

3. The β secondary KIE provides both geometric and bonding information for the TS. KIE in geometrically constrained reactants and analysis by computational chemistry have established that the dihedral angle between hydrogen isotopes β to the vacant p-orbital of the leaving group determines the magnitude of the β isotope effect (Mentch *et al.*, 1987; Schramm and Bagdassarian, 1999; Sunko *et al.*, 1977). Hyperconjugative interactions favor dihedral angles near 0° and 180° as these angles provide maximal stabilization of the electron-deficient TS of the type

shown in Fig. 16 (middle panel). Of particular interest are cases in which the β KIE can be measured for both pro-*R* and pro-*S* stereochemistries. In this case, the relative magnitude of the KIE can be used to solve the equation

$$\ln {}^{x}\mathrm{H}(\mathrm{KIE}_{\mathrm{exp}}) = \cos^2\theta \ \ln[{}^{x}\mathrm{H}(\mathrm{KIE}_{\mathrm{max}})] + {}^{x}\mathrm{H}(\mathrm{KIE}_{\mathrm{ind}})$$

which describes the angular dependence of the β KIE in terms of the dihedral angle θ, the experimental KIE, the maximum KIE for a dihedral of 0° or 180°, and the inductive isotope effect, which is 0.96 for ^{2}H (Sunko *et al.*, 1977). The geometric dependence of this β isotope effect is one of the most powerful tools for probing TS geometry, but has been used in a relatively few cases in enzymology.

4. KIE decrease rapidly from the site of covalent bond modification at the TS. The general rule for solution or gas-phase chemistry is that the isotope effects are insignificant beyond the β position in molecules where the atoms are connected by single bonds. More remote isotope effects occur in conjugated ring systems where changes in conjugation alter bond order to groups. For example, the ionization of *p*-nitrophenol alcohol is influenced by the presence of ^{15}N in the nitro group, as electronic changes between alcohol and NO_2 are transmitted through the conjugated ring (Cleland, 1995). Remote KIE are rarely observed in solution chemistry in singly bonded reactants. In enzyme-catalyzed reactions, remote ^{3}H KIE are becoming common. In KIE studies with both nucleoside hydrolase and PNP, the [5′-^{3}H]inosine label was designed as a remote position, because it is four bonds away from the reactive center. Nevertheless, it was discovered that this remote position causes a 5.1% (1.051) KIE in nucleoside hydrolase and a 3.3% KIE in PNP, whereas the same isotopically labeled molecule hydrolyzed by acid in solution gave small or insignificant KIE (Horenstein *et al.*, 1991; Kline and Schramm, 1993; Parkin *et al.*, 1984). These observations forced a reexamination of remote KIE in enzyme-catalyzed reactions. It is apparent from these studies that hydrogen bonds to groups at sp^3 centers cause sufficient geometric distortion to change the out-of-plane modes of adjacent hydrogens. A normal energy hydrogen bond is sufficient to cause angular distortion of the tetrahedral geometry at an anchored sp^3 center and give rise to a KIE of 5.1% (1.051) (Horenstein and Schramm, 1993a). Based solely on the KIE at this remote center, an unprecedented TS geometry was predicted for the nucleoside hydrolase reaction. Remote 5′-^{3}H KIE have now been documented in the hydrolysis of inosine by nucleoside hydrolases (two different enzymes), the arsenolysis and hydrolysis reactions catalyzed by PNP, and the hydrolysis of NAD$^+$ by cholera, diphtheria, and pertussis toxins (Schramm, 1998). In every case, the nonenzymatic solvolysis of the same labeled molecules has demonstrated smaller KIE at the remote position. These enzymes enforce distortion at remote positions from the chemistry of the TS.

X. MEP Surfaces of Substrates and TS

MEP surface of an enzymatic TS obtained from KIE was first reported in 1993 for nucleoside hydrolase (Horenstein and Schramm, 1993a). As outlined in that report, the utility of the MEP permits comparison of the electron potential at all points in space for the TS and the substrate. The van der Waals surfaces of the molecules are of the most interest, as this distance from nuclei contains the electron interactions involved in H-bonding and ionic interactions, the major forces in protein-substrate interaction and TS stabilization. The catalytic potential of enzymes has been attributed to binding the TS complex more tightly than the substrate by the factor of the catalytic rate enhancement. Therefore, the difference between MEP surfaces of the substrate and TS contain the answer to the tight binding of the TS. Analysis of the MEP surfaces of enzymatic TS has only been possible since enzymatic TS structures have become available from KIE investigations. The MEP has been widely used to describe the properties of stable molecules (Politzer and Truhlar, 1981; Sjoberg and Politzer, 1990; Venanzi *et al.*, 1992). It represents the energy imposed on a point of unit charge positioned at the site of interrogation, a defined distance from the nuclei (Srebrenik *et al.*, 1973). The procedure to obtain the MEP uses crystallographic data for the substrate and KIE data for the TS. Computational refinement is used to locate H atoms prior to calculation of the MEP for both substrate and TS (Horenstein and Schramm, 1993a). The wave function for the molecule is determined in a point calculation with the bond angles and lengths fixed at the values determined from the KIE analysis. These calculations can be done at modest basis sets, for example, STO-3G, as calculations using a range of basis sets have demonstrated that the MEP at the van der Waals surface is not strongly dependent on the basis set. For the few TS structures currently available, this level of MEP is adequate for visual representation and for the prediction of hydrogen bonding sites that differ between substrate and the TS. The MEP is determined from the wave function for the molecule using the CUBE function of Gaussian. For purposes of calculation and visualization, the van der Waals surfaces are considered to be at an electron density of $0.002e/a_0^3$ (Horenstein and Schramm, 1993a). Results of the CUBE function analysis were visualized with the AVS-Chemistry viewer (Advanced Visual Systems Inc. and Molecular Simulations Inc.). Since the original report of 1993, a variety of alternative molecular visualization programs have become available.

The MEP for substrate and inhibitor molecules can be compared to those of the TS following MEP analysis of the molecules. Substrates and inhibitor molecules are adjusted to be in conformations similar to that of the TS, and a single-point calculation is performed to determine the wave function of each molecule using Gaussian. The MEP for all molecules is determined from the CUBE function as described earlier for the TS. These results provide quantitative data for visual and numerical comparison of the electrostatic features for substrate, inhibitors, and the TS. This relationship is now being used as a means to predict the ability of unknown molecules to bind as analogs of the TS (Bagdassarian *et al.*, 1996a,b). Chapter 20 discusses the use of substrate, TS, and inhibitor MEP to train neural networks for the analysis of inhibitor binding energy.

An example of the relationship between the MEP of the TS and an established TS inhibitor is provided by the AMP deaminase study (Bagdassarian *et al.*, 1996b). The TS for AMP deamination was characterized by the KIE and BEBOVIB analysis in Fig. 14. Substrate and TS were compared to the TS inhibitor coformycin 5′-phosphate, which is known to bind with an equilibrium dissociation constant of 10 pM relative to the substrate K_d of 0.5 mM under the same assay conditions. Thus, the inhibitor binds 5×10^7 more tightly than the substrate (Kline and Schramm, 1994). For purposes of MEP analysis, the ribose 5-phosphate group was truncated to a methyl group, as the sugar phosphate is common to substrate, TS, and inhibitor, but is chemically inert in the reaction. AMP deamination is a relatively simple and convenient reaction for TS analysis. The rigid purine ring system fixes neighboring atoms in the TS, and the attack of the hydroxide ion from the pro-*R* face of the purine ring defines the reaction coordinate quite closely. The results demonstrate a striking similarity between the MEP surfaces of the TS and the TS inhibitor (Fig. 19, upper panels; see color insert). Both the TS and the TS inhibitor are different from the MEP of the substrate. This difference defines the essential property of the TS that accounts for the 5×10^7 tighter binding of the TS analog than the substrate and the TS binding energy that leads to the facile deamination of AMP at a rate of 1200 s^{-1} (Michels *et al.*, 1986). This result provided evidence that electrostatic features of the TS, when incorporated into a stable analog, result in powerful inhibition of the target enzyme.

The similarity among TS, substrates, and TS analogs can be evaluated by comparing the properties of the MEP surfaces (Bagdassarian *et al.*, 1996b). The molecules are first oriented to align the geometries of the van der Waals surfaces as closely as possible. The MEP surfaces are then compared for electrostatic similarity using an electrostatic grid at the van der Waals surfaces with typically 17 points per atom at the van der Waals surface. At each point on the TS, the corresponding point on the test molecule (substrate or inhibitor analog) is evaluated for distance and similarity of MEP. Identical molecules will score equivalent properties of geometry and MEP at every point. Close relatives of the TS will be similar for the sum of these measures, and distant relatives of the TS will be dissimilar at each point. The sum of these comparisons provides a similarity index that is scaled from a value of 1.0 for exact similarity to a value of 0 for molecules with little similarity. The inhibition of AMP deaminase by TS inhibitors correlates to this measure of similarity to the TS, as do inhibitors for adenosine deaminase and AMP nucleosidase (Bagdassarian *et al.*, 1996b). The correlation of binding energy with similarity to the TS electrostatics emphasizes that the complete molecular description is contained in the MEP at the van der Waals surface and the geometry of the molecules. These are the features used to design analogs as TS inhibitors. The TS state structure and its similarity to a TS inhibitor for PNP are discussed in the following section.

XI. Designing TS Inhibitors

The MEP surfaces of the TS most important for tight-binding interactions involve the formation of new H-bond donor and acceptor sites as substrate is converted to the TS. A comparison of AMP at the Michaelis complex of AMP

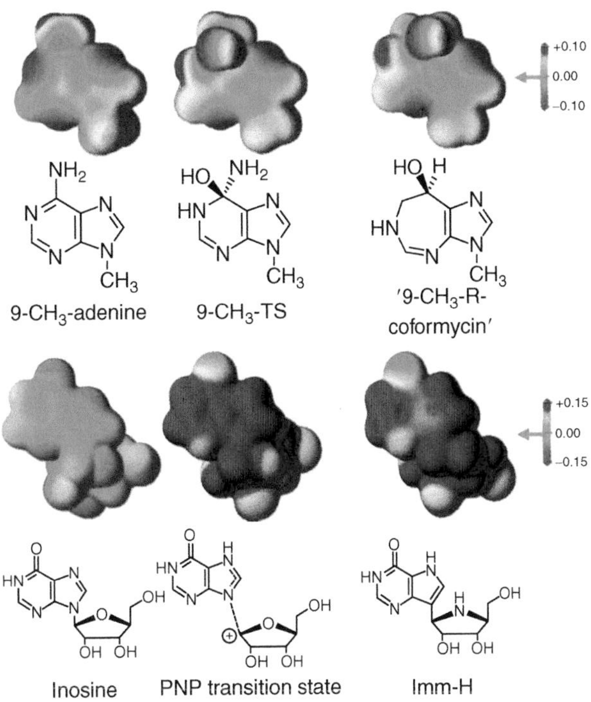

Fig. 19 Molecular electrostatic potential surfaces for the substrates, TS, and TS inhibitors for AMP deaminase and PNP. *Top*: the electrostatic potential for N-9–CH_3 truncations of AMP, AMP TS, and TS inhibitor coformycin 5′-phosphate. Convential structures are shown below in the same geometry. The calibration colors are shown on the right and indicate the relative charge a unit point charge experiences at the molecular electrostatic potential surfaces corresponding to the van der Waals distance from the nuclei. The NH_2 group of adenine is planar with the conjugated ring. Conjugation is broken at the transition state with the attacking OH and the formation of sp^3 carbon at C-6. The NH_2 has not yet accepted a proton at the transition state, but now protrudes behind the ring and is rehybridized to expose the lone pair electrons. Loss of the conjugated ring results in the protonation of N-1, which changes the electrostatic potential. The TS inhibitor for AMP deaminase binds approximately 10^7 more tightly than the substrate. The electrostatic similarity between the TS and the TS state inhibitor is apparent, as is their common difference from the substrate. The computational analysis is described in the text. *Bottom*: a similar analysis for the reaction catalyzed by PNP. The ribooxocarbenium ion formed at the transition state is the site of attack by the phosphate anion. At the transition state, the phosphate oxygen is 3 Å away and is therefore not bonded. The structures are shown below without the phosphate anion. Note the similarity between the transition state and the transition-state inhibitor. This inhibitor binds approximately 10^6 more tightly than substrate. (See Plate no. 1 in the Color Plate Section.)

deaminase and at the TS permits the identification of several new H-bond donor or acceptor sites available at the TS that were not present in the substrate or are relocated to a new geometry. The catalytic site amino acids reposition from the Michaelis complex to form optimal contacts to the TS. Even six modest H bonds of 3 kcal/mol provide 18 kcal/mol energy for TS stabilization. Relocation of existing H-bond donor/acceptors from the substrate to the TS can be exploited in the design of TS inhibitors. The relocation of H-bond sites in the conversion of AMP to the Meisenheimer-like TS complex of AMP is illustrated in Fig. 20. The example of AMP deaminase does not qualify as inhibitor design, as the natural product coformycin was designed by natural selection. However, it provides an example of the tight binding that occurs when the electrostatics of the inhibitor approach those of the TS. We can apply the same logic to TS solved by the methods of KIE analysis.

A. Purine Nucleoside Phosphorylase

TS analysis of the arsenolysis reaction catalyzed by PNP established a TS with ribooxocarbenium character and protonation of the leaving group (Fig. 17). The attacking phosphate anion lags far behind bond breaking to the purine ring. Two major features of the purine nucleoside at the TS are the positive charge of the ribooxycarbenium ion and the protonation of the leaving group. KIE analysis implicated N-7 as the site of protonation, although the X-ray crystal complex indicates the presence of an asparagine near N-7 (Erion *et al.*, 1997a,b). The low bond order to the attacking arsenate (and phosphate) nucleophile indicated that the TS analog need not incorporate the phosphate group, as the phosphate anion site can readily fill with free phosphate and accomplish the bond order of the TS through van der Waals and ionic interactions. The substrate specificity of PNP includes hypoxanthine and guanine purine bases, and these features were included in the synthesis of TS analogs. The positive charge of the ribooxocarbenium was

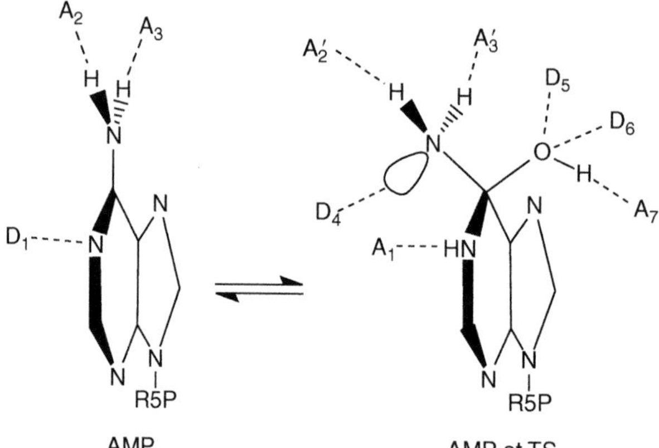

Fig. 20 An example of the new H bonds that can be formed at reacting atoms of the TS for AMP deaminase. Bonds altered in position and new sites for H-bond formation are key sites for transition-state interactions. The H-bond acceptors A_2 and A_3 are altered in geometry and H-bond donor D_1 is altered to an acceptor (A_1) at the transition state. New H-bond donor/acceptor sites are indicated by D_4–D_6 and by acceptor A_7. The sum of the ΔG for these new interactions is proposed to be the major contribution to TS formation. The crystal structure of adenosine deaminase has not yet been solved with an amino group in the position of the leaving group amino, therefore these contacts are hypothetical.

incorporated with a 4′-imino group replacing the 4′-oxygen. This group has a pK_a of 6.5 in the related phenyliminoribitol (Horenstein and Schramm, 1993b). Chemical stability was introduced with a carbon bond to the purine base using 9-deazahypoxanthine and 9-deazaguanine. Carbon substitution at the 9 position also changes the pK_a of the N at N-7. Inosine and guanosine have pK_a values of approximately 2 at this position. As the TS is formed, the pK_a increases at N-7 and is proposed to be protonated or in a low-barrier hydrogen bond as the TS is achieved (Kline and Schramm, 1995). The N-7 position of immucillin-H (Imm-H) and immucillin-G has a pK_a >8.5, resembling the elevated pK_a of the TS (Fig. 20, lower panel) (Miles *et al.*, 1998).

Both Imm-H and immucillin-G are slow-onset, tight-binding inhibitors of PNP from calf spleen or from human erythrocytes. Equilibrium dissociation constants (K_i^*) are in the range of 23–72 pM with this combination of inhibitors and enzymes. Complete inhibition of PNP occurs with a single catalytic site of the PNP homotrimer occupied by inhibitor, and binding at one site prevents equivalent binding of inhibitor molecules at the remaining two sites. Inhibitor binding stoichiometry indicates that the enzyme works in an obligate sequential catalytic fashion. Dissociation of the enzyme-inhibitor complex is slow, for example, $t_{1/2} = 4.8$ h for the complex of bovine PNP and Imm-H. The geometries and MEP surfaces of substrate, TS, and TS inhibitor were compared. The relationship between the binding energy ($\Delta G/RT$) and the MEP surface similarity $\langle S_e \rangle$ were correlated in the same manner that was used for the interactions of TS inhibitors with AMP deaminase, AMP nucleosidase, and adenosine deaminase (Fig. 21) (Bagdassarian *et al.*, 1996a,b). In this comparison, the geometric and electrostatic similarity of inosine to the TS is 0.483, whereas that of Imm-H is 0.723. The TS inhibitor binds 7×10^5 more tightly than substrate (Miles *et al.*, 1998). The inhibitors have potential in the treatment of T-cell proliferative disorders and provide a practical example of the application of TS structure to TS inhibitor design.

B. Nucleoside Hydrolase

One of the protozoan nucleoside hydrolases (IU-NH) is nonspecific for the aglycone and accepts both purine and pyrimidine nucleosides as substrate. KIE were used to establish a ribooxocarbenium ion character at the TS with protonation of the purine leaving group as a secondary interaction for TS stabilization (see Fig. 16). The major factor in TS stabilization is ribooxocarbenium ion interaction and was demonstrated by the inhibition with iminoribitol, which binds 38-fold more tightly than substrate, even without any elements of the leaving group (Fig. 22). The addition of hydrogen bond acceptor or hydrophobic groups at the 1 position of iminoribitol provided inhibitors binding 120 and 12,000 times more tightly than substrate (Parkin *et al.*, 1997). The nonspecific nature of the leaving-group interactions for TS formation are demonstrated by the requirement for a ribooxycarbenium mimic in tight-binding analogs. In contrast, a variety of substituent groups can be used to replace the protonated purine of the TS while

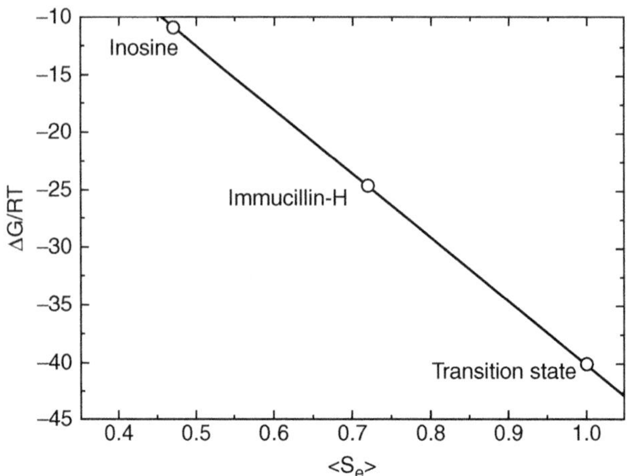

Fig. 21 The electronic similarity $\langle S_e \rangle$ of substrate and TS inhibitor to the TS for PNP (Miles *et al.*, 1998). The dimensionless binding free energy ($\Delta G/RT$) is determined experimentally for substrate inosine and TS inhibitor and is estimated for the transition state using the Wolfenden thermodynamic box principle for transition-state binding affinity (Radzicka and Wolfenden, 1995). The calculations for $\langle S_e \rangle$ are based on a combination of geometric and molecular electrostatic potential surface comparisons as indicated in the text. Reprinted with permission from the American Chemical Society.

retaining nM binding ability. A large number of inhibitors for IU-NH have been used to train neural networks to recognize features important to the tight-binding interactions of the TS (see Chapter 20).

The specificity of TS analogs is illustrated by the differing specificities with two isozymes of nucleoside hydrolases. The IU-NH has low-leaving group specificity, high ribose specificity, and is inhibited by most iminoribitol analogs. Inhibitor binding is improved by hydrophobic groups or by hydrogen bond acceptors as purine or pyrimidine analogs (Fig. 23). In contrast, the purine-specific IAG-NH catalyzes the same reaction, but half of the TS binding energy is involved in purine contacts to activate the purine-leaving group (Miles *et al.*, 1998).

TS inhibitors designed specifically for the nonspecific-leaving group of the IU-NH bind poorly to the IAG-NH. By incorporating features of leaving groups that are accepted by both of the nucleoside hydrolase isozymes, TS inhibitors have been designed and synthesized that bind tightly to both nucleoside hydrolase isozymes (I and II in Fig. 23). The hydrophobic and nonspecific-leaving group properties of IU-NH permit tight binding of III and IV (Fig. 23), which are inert for IAG-NH. Conversely, precise leaving-group interactions with lower specificity for the ribooxycarbenium ion of the TS permit V and VI in Fig. 23 to bind tightly to IAG-NH while being inert for IU-NH (Miles, Tyler, Evans, Furneaux, and Schramm, unpublished results, 1998). TS information permits the design of isozyme-specific inhibitors. These inhibitors have potential for species-specific interaction with target enzymes, an ultimate goal of logically designed TS inhibitors.

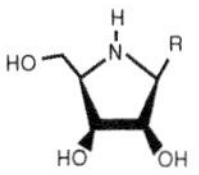

R	IU-nucleoside hydrolase[a] K_i	IU-nucleoside hydrolase[a] K_m/K_i	IAG-nucleoside hydrolase[b] K_i	IAG-nucleoside hydrolase[b] K_m/K_i	$(K_m/K_i$ IU-NH) / $(K_m/K_i$ IAG-NH)
(I) H	4.5±0.4 μM	25	44±4 μM	0.41	60
(II) 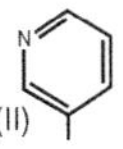	7.9±0.6 μM	14	>240[c] μM	<0.075	>187
(III) 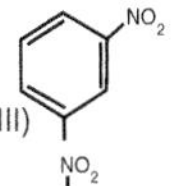	7.5±0.5 μM	15	>360[c] μM	<0.05	>296
(IV) 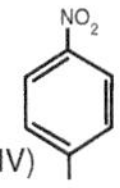	1.1±0.1 μM	101	>360[c] μM	<0.05	>2000
(V) 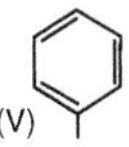	300±27 nM	370	180±15 μM	0.1	3700
(VI) 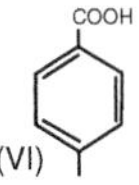	96±7 nM	1160	>480[c] μM	<0.04	>30,800
(VII) 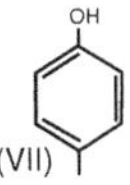	75±4 nM	1480	35±2 μM	0.51	2900
(VIII) 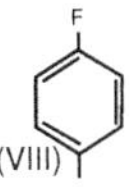	57±5 nM	1950	205±14 μM	0.088	22,180
(IX) 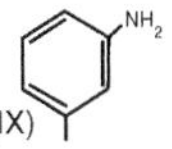	51±4 nM	2180	38±4 μM	0.47	4590
(X) 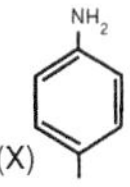	30±2 nM	3700	12±1 μM	1.5	2470
(XI) 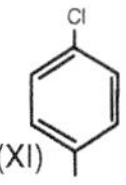	30±1.4 nM	3700	190±8 μM	0.095	39,000
(XII) 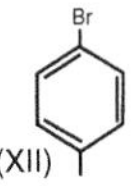	28±4 nM	3960	113±6 μM	0.16	24,890

[a]The K_m for inosine was 111 ± 17 μM under these assay conditions. K_m/K_i values are for inosine as substrate. [b]The K_m for inosine was 18±1 μM under these assay conditions. K_m/K_i values are for inosine as substrate. [c]No inhibiton was observed at 80, 120, 120, and 160 μM II, III, IV, and VI, respectively, when assayed at 75 μM inosine. The indicated inhibitor constants are lower limits of the constants based on the sensitivity of the assay.

Fig. 22 Isozyme specificity of TS inhibitors for IU-nucleoside hydrolase and IAG-nucleoside hydrolase. The dissociation constants for inhibitor (K_i) and the factor by which the inhibitor binds relative to substrate (K_m/K_i) are compared for the nonspecific IU-nucleoside hydrolase and the purine-specific IAG-nucleoside hydrolase. From Parkin *et al.* (1997). Reprinted with permission from the American Chemical Society.

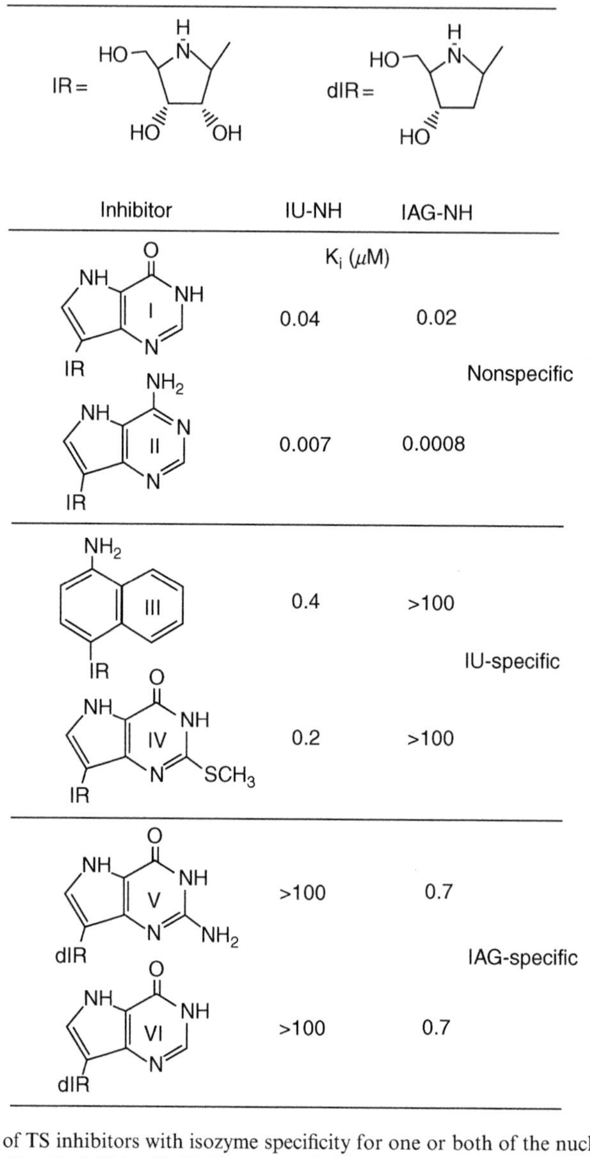

IR = ; dIR =

Inhibitor	IU-NH	IAG-NH	
	K_i (μM)		
I (IR)	0.04	0.02	Nonspecific
II (IR)	0.007	0.0008	
III (IR)	0.4	>100	IU-specific
IV (IR)	0.2	>100	
V (dIR)	>100	0.7	IAG-specific
VI (dIR)	>100	0.7	

Fig. 23 Design of TS inhibitors with isozyme specificity for one or both of the nucleoside hydrolase isozymes (R. W. Miles, P. C. Tyler, R. H. Furneaux, G. Evans, and V. L. Schramm, unpublished data, 1998). Inhibitors I and II contain purine-leaving-group contacts and the ribooxocarbenium ions mimic and bind tightly to both enzymes. Inhibitors III and IV are unable to be leaving group activated, a major energetic interaction for IAG-NH, and are IU specific. Inhibitors V and VI lack the ribosyl-specific interactions required for TS formation in IU-NH by containing all purine activation groups and thus are IAG-NH specific.

XII. Conclusions

Experimental methods are now available for obtaining TS information for most enzymatic reactions. Intrinsic KIE are established and used to calculate TS bonding geometry. The bonding structure provides information for calculation of the MEP surface. Differences between the MEP surfaces of the substrate and the TS indicate the hydrogen bonding and charge patterns to incorporate into synthetic TS analogs. This approach has now been used for the design of several powerful TS inhibitors. The combination of quantum computational chemistry, enzymatic mechanisms, and KIE provides a new level of information for enzymatic catalysis. The reliable success of the approach provides confidence that enzymatic TS structures from KIE analysis provide a sufficiently accurate structure of the TS for the practical purpose of TS inhibitor design.

References

Aktories, K. (ed.) (1992). "Current Topics in Microbiology and Immunology: ADP-Ribosylating Toxins." Springer-Verlag, Berlin.

Aleshin, A. E., Zeng, C., Bartunik, H. D., Fromm, H. J., and Honzatko, R. B. (1998). *J. Mol. Biol.* **282,** 345.

Antoniou, D., and Schwartz, S. D. (1997). *Proc. Natl. Acad. Sci. USA* **94,** 12360.

Bagdassarian, C. K., Brauheim, B. B., Schramm, V. L., and Schwartz, S. D. (1996a). *Int. J. Quant. Chem.* **60,** 73.

Bagdassarian, C. K., Schramm, V. L., and Schwartz, S. D. (1996b). *J. Am. Chem. Soc.* **118,** 8825.

Bahnson, B. J., and Klinman, J. P. (1995). *Methods Enzymol.* **249,** 374.

Batterham, T. J., Ghambeer, R. K., Blakley, R. L., and Brownson, C. (1967). *Biochemistry* **6,** 1203.

Bennet, A. J., and Sinnott, M. L. (1986). *J. Am. Chem. Soc.* **108,** 7287.

Berti, P. (2009). *Methods Enzymol.* Chapter 19.

Berti, P. J., and Schramm, V. L. (1997). *J. Am. Chem. Soc.* **119,** 12069.

Berti, P. J., Blanke, S. R., and Schramm, V. L. (1997). *J. Am. Chem. Soc.* **119,** 12079.

Bigeleisen, J., and Wolfsberg, M. (1958). *Adv. Chem. Phys.* **1,** 15.

Blanchard, J. S., and Wong, K. K. (1991). *In* "Enzyme Mechanism from Isotope Effects" (P. F. Cook, ed.), p. 341. CRC Press, Boca Raton, FL.

Braunheim, B. B., and Schwartz, S. D. (2009). *Methods Enzymol.* Chapter 20.

Burton, G. W., Sims, L. B., Wilson, J. C., and Fry, A. (1977). *J. Am. Chem. Soc.* **99,** 3371.

Callender, R., and Deng, H. (1994). *Annu. Rev. Biophys. Biomol. Struct.* **23,** 215.

Cannon, W. R., Singleton, S. F., and Benkovic, S. J. (1996). *Nat. Struct. Biol.* **3,** 821.

Chen, X. Y., Link, T. M., and Schramm, V. L. (1998). *Biochemistry* **37,** 11605.

Cleland, W. W. (1982). *Methods Enzymol.* **87,** 625.

Cleland, W. W. (1992). *Biochemistry* **31,** 317.

Cleland, W. W. (1995). *Methods Enzymol.* **249,** 341.

Cleland, W. W., O'Leary, M. H., and Northrop, D. B. (1977). "Isotope Effects on Enzyme-Catalyzed Reactions," University Park Press, Baltimore, MD.

Cohn, M., and Reed, G. H. (1982). *Annu. Rev. Biochem.* **51,** 365.

Cook, P. F. (ed.) (1991). "Enzyme Mechanisms from Isotope Effects," CRC Press, Boca Raton, FL.

Dahlquist, F. W., Rand-Meier, T., and Raftery, M. A. (1969). *Biochemistry* **8,** 4214.

Degano, M., Almo, S. C., Sacchettini, J. C., and Schramm, V. L. (1998). *Biochemistry* **37,** 6277.

Dewar, M. J. S., Zoebisch, E. F., Healy, J. J. P., and Stewart, J. (1985). *J. Am. Chem. Soc.* **107,** 3902.

Erion, M. D., Stoeckler, J. D., Guida, W. C., Walter, R. L., and Ealick, S. E. (1997a). *Biochemistry* **36,** 11735.

Erion, M. D., Takabayashi, K., Smith, H. B., Kessi, J., Wagner, S., Honger, S., Shames, S. L., and Ealick, S. E. (1997b). *Biochemistry* **36,** 11725.

Frisch, M. J., Trucks, G. W., Hend-Gordon, M., Gill, P. M. W., Wong, M. W., Foresman, J. B., Johnson, B. G., Schlegel, H. B., Robb, M. A., Replogle, E. S., Gomperts, R., Andres, J. L., *et al.* (1992). Gaussian 92 User's Guide Gaussian Inc., Pittsburgh.

Frisch, M. J., Trucks, G. W., Schlegel, H. B., Gill, P. M. W., Johnson, B. G., Robb, M. A., Cheeseman, J. R., Keith, T., Petersson, G. A., Montgomery, J. A., Raghavachari, K., Al-Laham, M. A., *et al.* (1995). Gaussian 94, Revision C.2 Gaussian, Inc., Pittsburgh.

Gandour, R. D. and Schowen, R. L. (eds.) (1978). "Transition States of Biochemical Processes," Plenum Press, New York.

Garcia-Viloca, M., Gelabert, R., Gonzalez-Lafont, A., Moreno, M., and Lluch, J. M. (1998). *J. Am. Chem. Soc.* **120,** 10203.

Gawlita, E., Caldwell, W. S., O'Leary, M. H., Paneth, P., and Anderson, V. E. (1995). *Biochemistry* **34,** 2577.

Gelbin, A., Schneider, B., Clowney, L., Hsieh, S. H., Olson, W. K., and Berman, H. M. (1996). *J. Am. Chem. Soc.* **118,** 519.

Gerlt, J. A., and Gassman, P. G. (1993). *J. Am. Chem. Soc.* **115,** 11552.

Hershfield, M. S., and Mitchell, B. S. (1995). *In* "The Metabolic Basis of Inherited Disease" (C. R. Scriber, A. L. Beaudet, W. S. Sly and D. Valle, eds.), 7th edn. p. 1725. McGraw-Hill, New York.

Horenstein, B. A., Parkin, D. W., Estupiñán, B., and Schramm, V. L. (1991). *Biochemistry* **30,** 10788.

Horenstein, B. A., and Schramm, V. L. (1993a). *Biochemistry* **32,** 7089.

Horenstein, B. A., and Schramm, V. L. (1993b). *Biochemistry* **32,** 9917.

Huskey, W. P. (1991). *In* "Enzyme Mechanism from Isotope Effects" (P. F. Cook, ed.), p. 37. CRC Press, Boca Raton, FL.

Huskey, W. P. (1993). *J. Am. Chem. Soc.* **118,** 14.

Huskey, W. P., and Schowen, R. L. (1983). *J. Am. Chem. Soc.* **105,** 5704.

Kline, P. C., and Schramm, V. L. (1993). *Biochemistry* **32,** 13212.

Kline, P. C., and Schramm, V. L. (1994). *Biochemistry* **34,** 1153.

Kline, P. C., and Schramm, V. L. (1995). *Biochemistry* **34,** 1153.

LaReau, R., Wah, W., and Anderson, V. (1989). *Biochemistry* **28,** 3619.

Lee, H. C., Walseth, T. F., Bratt, G. T., Hayes, R. N., and Clapper, D. L. (1989). *J. Biol. Chem.* **264,** 1608.

Lewis, B., and Schramm, V. (1998). unpublished results.

Li, C., Tyler, P. C., Furneaux, R. H., Kicska, G., Xu, Y., Grubmeyer, C., Girvin, M. E., and Schramm, V. L. (1998). *Nat. Struct. Biol.* **6.**

Liu, K., Polanyi, J. C., and Yang, S. (1993). *J. Chem. Phys.* **98,** 5431.

Markham, G. D., Parkin, D. W., Mentch, F., and Schramm, V. L. (1987). *J. Biol. Chem.* **262,** 5609.

McFarland, J. T. (1991). *In* "Enzyme Mechanism from Isotope Effects" (P. F. Cook, ed.), p. 151. CRC Press, Boca Raton, FL.

Melander, L., and Saunders, W. J., Jr. (1980). "Reaction Rates of Isotopic Molecules," Wiley, New York.

Mentch, F., Parkin, D. W., and Schramm, V. L. (1987). *Biochemistry* **26,** 921.

Merkler, D. J., Kline, P. C., Weiss, P., and Schramm, V. L. (1993). *Biochemistry* **32,** 12993.

Merkler, D. J., and Schramm, V. L. (1993). *Biochemistry* **32,** 5792.

Meyer, S. L., Kvalnes-Krick, K. L., and Schramm, V. L. (1989). *Biochemistry* **28,** 8734.

Michels, P. A. M., Polisczak, A., Osinga, K. A., Misset, O., Van Beeumen, J., Wierenga, R. K., Borst, P., and Opperdoes, F. R. (1986). *EMBO J.* **5,** 1049.

Mildvan, A. S., Harris, T. K., and Abeygunawardana, C. (1999). *Methods Enzymol.* **308,** Chapter 10.

Miles, R. W., Tyler, P. C., Furneaux, R. H., Bagdassarian, C. K., and Schramm, V. L. (1998). *Biochemistry* **37,** 8615.

Miwa, M., Hayaishi, O., Shall, S., Smulson, M. and Sugimura, T. (eds.) (1983). "ADP-Ribosylation. DNA Repair and Cancer," Japan Scientific Societies Press, Tokyo.

Morrison, J. F., and Walsh, C. T. (1988). *Adv. Enzymol. Relat. Areas Mol. Biol.* **61,** 201.

Muchmore, C. R., Krahn, J. M., Kim, J. H., Zalkin, H., and Smith, J. L. (1998). *Prot. Sci.* **7,** 39.

Niwas, S., Chand, P., Pathak, V. P., and Montgomery, J. A. (1994). *J. Med. Chem.* **37,** 2477.

Northrop, D. B. (1975). *Biochemistry* **14,** 2644.

Northrop, D. B. (1977). *In* "Isotope Effects on Enzyme-Catalyzed Reactions" (W. W. Cleland, M. H. O'Leary and D. B. Northrop, eds.), p. 122. University Park Press, Baltimore, MD.

Northrop, D. B. (1981). *Annu. Rev. Biochem.* **50,** 103.

Northrop, D. (1998). *J. Chem. Ed.* **75,** 1153.

Parkin, D. W. (1991). *In* "Enzyme Mechanism from Isotope Effects" (P. F. Cook, ed.), p. 269. CRC Press, Boca Raton, FL.

Parkin, D. W., Leung, H. B., and Schramm, V. L. (1984). *J. Biol. Chem.* **259,** 9411.

Parkin, D. W., Horenstein, B. A., Abdulah, D. R., Estupiñán, B., and Schramm, V. L. (1991). *J. Biol. Chem.* **266,** 20658.

Parkin, D. W., Limberg, G., Tyler, P. C., Furneaux, R. H., Chen, X. Y., and Schramm, V. L. (1997). *Biochemistry* **36,** 3528.

Parkin, D. W., and Schramm, V. L. (1987). *Biochemistry* **26,** 913.

Parr, R. G., and Yang, W. (1989). "Density Functional Theory of Atoms and Molecules," Oxford University Press, New York.

Pauling, L. (1948). *Am. Sci.* **36,** 50.

Politzer, P. and Truhlar, G. (eds.) (1981). "Chemical Applications of Atomic and Molecular Electrostatic Potentials," Plenum Press, New York.

Quinn, D. M., and Sutton, L. D. (1991). *In* "Enzyme Mechanism from Isotope Effects" (P. F. Cook, ed.), p. 72. CRC Press, Boca Raton, FL.

Radzicka, A., and Wolfenden, R. (1995). *Methods Enzymol.* **249,** 284.

Radzicka, A., and Wolfenden, R. (1996). *Science* **267,** 90.

Rising, K. A., and Schramm, V. L. (1994). *J. Am. Chem. Soc.* **116,** 6531.

Rose, I. A. (1995). *Methods Enzymol.* **249,** 315.

Sauve, A. S., Munshi, C., Lee, H. C., and Schramm, V. L. (1998). *Biochemistry* **37,** 13239.

Scherer, N. F., Khundkar, L. R., Bernstein, R. B., and Zewail, A. H. (1987). *J. Chem. Phys.* **87,** 1451.

Scheuring, J., Berti, P. J., and Schramm, V. L. (1998). *Biochemistry* **37,** 2748.

Schowen, K. B. J. (1978). *In* "Transition States of Biochemical Processes" (R. D. Gandour and R. L. Schowen, eds.), p. 225. Plenum Press, New York.

Schramm, V. L. (1998). *Annu. Rev. Biochem.* **67,** 693.

Schramm, V. L., and Bagdassarian, C. (1999). *In* "Comprehensive Natural Product Chemistry" (C. D. Poulter, ed.), Vol. 5, pp. 71–100. Pergamon Press, Elsevier Science, New York.

Schramm, V. L., Horenstein, B. A., and Kline, P. C. (1994). *J. Biol. Chem.* **269,** 18259.

Shan, S. O., and Herschlag, D. (1996). *Proc. Natl. Acad. Sci. USA* **93,** 14474.

Shan, S. O., and Herschlag, D. (1999). *Method Enzymol.* **308,** Chapter 11.

Sims, L. B., Burton, G. W., and Lewis, D. E. (1977). BEBOVIB-IV, QCPE No. 337 Quantum Chemistry Program Exchange, Department of Chemistry, University of Indiana, Bloomington.

Sims, L. B., and Lewis, D. E. (1984). *In* "Isotopes in Organic Chemistry" (E. Buncel and C. C. Lee, eds.), p. 161. Elsevier, New York.

Sjoberg, P., and Politzer, P. (1990). *J. Phys. Chem.* **94,** 3959.

Srebrenik, S., Weinstein, H., and Pauncz, R. (1973). *Chem. Phys. Lett.* **20,** 419.

Suhnel, J., and Schowen, R. L. (1991). *In* "Enzyme Mechanism from Isotope Effects" (P. F. Cook, ed.), p. 3. CRC Press, Boca Raton, FL.

Sunko, D. E., Szele, I., and Hehre, W. J. (1977). *J. Am. Chem. Soc.* **99,** 5000.

Venanzi, C. A., Plant, C., and Vananzi, T. J. (1992). *J. Med. Chem.* **35,** 1643.

Walsh, G. M., Reddy, N. S., Bantia, S., Babu, Y. S., and Montgomery, J. A. (1994). *Hematol. Rev.* **8,** 87.

Weiss, P. M. (1991). *In* "Enzyme Mechanism from Isotope Effects" (P. F Cook, ed.), p. 291. CRC Press, Boca Raton, FL.

Xu, Y., Eads, J. C., Sacchettini, J. C., and Grubmeyer, C. (1997). *Biochemistry* **36,** 3700.

CHAPTER 19

Determining Transition States from Kinetic Isotope Effects

Paul J. Berti

Department of Chemistry
Department of Biochemistry & Biomedical Science McMaster University
Hamilton, Ontario, Canada

Abbreviations

BEBOVA	bond-energy/bond-order vibrational analysis
EIE	equilibrium isotope effect
EXC	contribution from vibrationally excited states to an IE
GVFF	general valence force field
IE	isotope effect
KIE	kinetic isotope effect
LSC	liquid scintillation counting

CONTEMPORARY ENZYME KINETICS AND MECHANISM
Reprinted from *Methods in Enzymology*, Volume 308 (Academic Press, 1999).

DOI: 10.1016/B978-0-12-378608-1.00019-0

MMI	contribution from the change in mass and moments of inertia to an IE
NAD^+	nicotinamide adenine dinucleotide
PDB	Protein Databank file format
RHF	restricted Hartree-Fock
SVFF	simple valence force field
TS	transition state
TSI	transition-state imbalance
ZPE	contribution from zero point energy to an IE

I. Introduction

Kinetic isotope effects (KIEs) have long been used for qualitative analyses of enzyme mechanisms. Qualitative analysis of KIEs has been a cornerstone of enzymology in showing, for example, the orderedness of reactions, the identification of rate-limiting steps, and identifying which bonds are being formed or broken at the transition state (TS). It is only since the mid-1980s that it has become feasible to routinely use precise measurements of KIEs to quantitatively determine the TSs of enzymatic reactions. The quantitative analysis of KIEs has become possible due to a number of technical advances that have made it practical to routinely synthesize complex molecules with isotopic labels at any one of many specific positions, to measure KIEs accurately and precisely, and to analyze the experimental KIEs via the structure interpolation approach to bond-energy/bond-order vibrational analysis (BEBOVA) to yield precise, quantitative determinations of experimental TSs. This contribution will describe the process of proceeding from observed KIEs to the experimental TS using structure interpolation and BEBOVA.

The biggest change in TS determination since Sims and Lewis's comprehensive description of BEBOVA (Sims and Lewis, 1984) has been the huge increase in computational power available to the average biochemist. At one time, the molecular models used in BEBOVA analyses were truncated as far as possible to make them computationally tractable. Today, desktop computers can routinely handle molecules of 100 atoms. Another advance has been the introduction of the structure interpolation technique of generating many test TS structures automatically. This approach allows many more structures to be tested with greater accuracy than was possible using manually adjusted TS models. In the past, many simplifying assumptions were made in the modeling process. Given the greater accuracy of TS determination that is now routinely possible, these assumptions can be reexamined. The theoretical and practical complexities in determining a TS are still formidable, but the invaluable increase in our knowledge of enzyme mechanisms from each experimental TS, and the impossibility of achieving these results by any other technique, fully justify the effort.

The method for deriving an experimental TS from experimental KIEs will be described, as far as possible, in terms of general principles, although reference will often be made to the specific case where the approach is most developed:

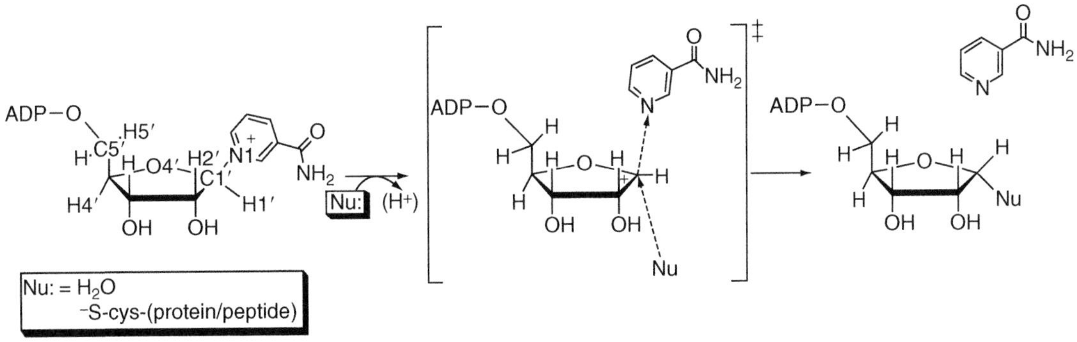

Fig. 1 Nucleophilic substitutions on NAD^+. (Left) Sites of isotopic labels (^3H, ^{14}C, ^{15}N, ^{18}O) are indicated in the reactant. (Middle) The reactions pass through an ANDN TS before forming the products (right).

nucleophilic substitutions on NAD^+ (Fig. 1), including hydrolysis (Berti and Schramm, 1997; Berti *et al.*, 1997; Rising and Schramm, 1997; Scheuring and Schramm, 1997a) and ADP-ribosylation reactions of peptide substrates (Scheuring and Schramm, 1997b; Scheuring *et al.*, 1998). Our interest in nucleophilic substitutions on NAD^+ stems from its role as a substrate in bacterial toxin-catalyzed ADP-ribosylation reactions. All the hydrolytic TSs follow bimolecular, A_ND_N (Guthrie and Jencks, 1989) (S_N2) mechanisms with highly dissociative TSs, similar to the TSs for other *N*-glycoside hydrolysis reactions examined in this laboratory (Horenstein *et al.*, 1991; Kline and Schramm, 1993, 1995; Mentch *et al.*, 1987; Merkler *et al.*, 1993; Parkin and Schramm, 1987; Parkin *et al.*, 1991b). In these TSs, the *N*-glycosidic bond is almost completely broken, with low but significant bonding to the incoming water nucleophile. TSs for the ADP-ribosylation reactions catalyzed by pertussis toxin are more concerted than the hydrolytic reactions.

II. Theory

A. Transition State and Its Structure

In any reaction proceeding between two stable species, the reacting molecules must pass through higher energy states between reactants and products. One of these states is of particular interest, the TS. The TS is the highest energy point on the lowest energy path between reactants and products. The significance of the TS for enzymologists was noted by Pauling (1946) who pointed out that enzymes catalyze reactions by binding to and stabilizing TSs in preference to either substrates or products. Furthermore, the extent to which an enzyme increases the rate over the uncatalyzed reaction is equal to the extent to which it binds the TS more tightly than the substrates or products. There are now several enzymes that have been shown to catalyze reactions with rate enhancements of ca. 10^{17}-fold (Miller *et al.*, 1998; Radzicka and Wolfenden, 1995; Wolfenden *et al.*, 1998a,b), including orotate decarboxylase. This rate enhancement, taken with the equilibrium dissociation constants for the substrate, implies that the equilibrium dissociation constant for the enzyme · TS complex would be 10^{-24} M (Radzicka and Wolfenden, 1995). If it were possible to capture a small fraction of that binding energy in a TS mimic—a stable molecule that reproduces geometric and charge features of the TS—a potent inhibitor would result.

The potential energy surface of a reaction is often likened to a mountain pass between two valleys. Both the hiker and the molecule seek the lowest energy route between the valleys (reactants and products), passing over the lowest barrier (i.e., the TS), rather than over a mountain peak. Topologically, the TS is at a saddle point on the potential energy surface, with the energy minimized in every dimension except the reaction coordinate, where it is a maximum. The TS is a stationary state, that is, the structure is at equilibrium, with no net forces acting on any atom. In a regular bond, any distortion results in a force tending to restore the structure back to a minimum energy structure. In contrast, any distortion in the reaction coordinate direction causes forces *away* from the TS that increase with increasing displacement from the stationary point. The definition used here of the reaction coordinate is that it is a normal vibrational mode that has an imaginary frequency (Buddenbaum and Shiner, 1976). Displacement of atoms along this normal mode results in a decrease in energy and movement of the molecule either forward toward products, or backward toward reactants.

Because of the emphasis on understanding enzyme-substrate interactions that promote catalysis, and on designing TS mimics as inhibitors, analysis of TSs tends to focus on their structure. It is intrinsic to the process of determining TS structures, however, to also characterize the reaction coordinate. For example, in the highly dissociative A_ND_N TSs of glycoside nucleophilic substitutions, the residual bond order to the leaving group and the incipient bond order to the incoming nucleophile (both <0.05) are so low that it seems they should have little or no observable effect on the KIEs. The question then arises whether these groups are present in the TS at all. It is the contribution of the leaving group and the nucleophile to the reaction coordinate frequency (and its associated contribution to the primary KIEs) that makes it possible to unambiguously judge whether they are present at the TS, that is, that the reaction has a bimolecular, A_ND_N mechanism, not a stepwise, $D_N + A_N$ (S_N1) mechanism.

B. Balls-and-Springs Vibrational Models and Geometry

The physical origin and significance of KIEs have been very well described in several reviews (Buddenbaum and Shiner, 1977; Huskey, 1991; Suhnel and Schowen, 1991) and will not be covered in detail here. Isotope effects (IEs) are primarily a vibrational phenomenon, in the same sense as infrared (IR) and Raman spectroscopies. KIEs reflect a *change* in the vibrational environment of a given atom between the reactant and TS of a reaction. If there is a change in molecular structure in the vicinity of an isotopic label on going from the reactant to the TS, there will be changes in its vibrational environment and therefore in the frequencies of normal vibrational modes. These frequency changes lead to an IE. There is also a (usually small, but not always) contribution from the isotopic label to a change in mass and moments of inertia (MMI) of the molecule.

IEs are an inherently quantum phenomenon; they arise as a consequence of the application of the Heisenberg uncertainty principle to vibrational frequencies. An atom in a molecule cannot be at rest, even at 0 K, because that would violate the principle that the position and momentum of a particle cannot be perfectly known. Thus, molecules possess a vibrational energy even at 0 K, the zero point energy (ZPE). The ZPE is a function of the vibrational frequency, which is in turn a function of the mass of the vibrating atoms and the force constant (Fig. 2). There is also a small (near room temperature) contribution from normal modes that are vibrationally excited (not in the vibrational ground state).

Vibrational energy (E_v) is a simple function of frequency:

$$E_v = \left(v + \frac{1}{2}\right)h\omega \tag{1}$$

where v is the vibrational quantum number, h is Planck's constant, and ω is the angular frequency ($=2\pi v$). Thus, the lowest vibrational energy an oscillator may have is the ZPE, when $v = 0$ and $E_v = h\omega/2$. Changes in ZPEs between molecules are generally the largest contributor to IEs. The other factors that contribute to IEs are contributions from excited-state energies (EXCs), the change in MMI,

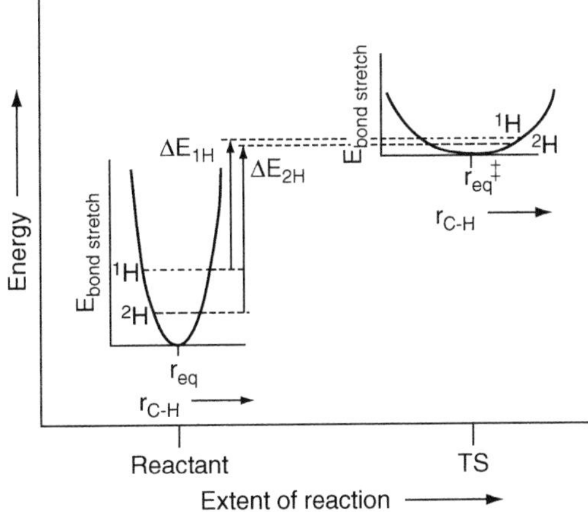

Fig. 2 Each inset shows the energy of changing the length of a C–H bond from its equilibrium length in the reactant (r_{eq}) and the TS ($r^{\ddagger}_{eq}$). Zero point energies for ^{1}H and ^{2}H (dashed horizontal lines) depend on the mass and the bond-stretching force constant. For a normal IE, the zero point energy decre7ases *more* for the light isotope (^{1}H) than the heavy isotope (^{2}H) when the bond-stretching force constant decreases at the TS. This leads to a difference in activation energies ($\Delta\Delta E\ddagger = \Delta E_{2H} - \Delta E_{1H}$), which is the source of the ZPE contribution to the IE. This illustration is for a secondary IE; for a primary KIE, the reaction coordinate normal mode at the TS has no zero point energy because it is an imaginary frequency and therefore has no restoring force. For a primary KIE, $\Delta\Delta E\ddagger = E_{2H,\text{ reactant}} - E_{1H,\text{ reactant}}$.

transmission effects, and tunneling (Cha *et al.*, 1989; Kohen and Klinman, 1998). The contribution from transmission coefficient effects is in the realm of variational TS theory, where barrier recrossing can lead to transmission coefficients <1 (Truhlar and Garrett, 1980; Truhlar *et al.*, 1983; Tucker *et al.*, 1985). In general, barrier recrossing becomes significant for reactions in the gas phase, at high temperature, where small atoms are transferred between much larger groups. There is no evidence in the literature indicating transmission coefficients <1 for enzymatic reactions. Tunneling through an energetic barrier, rather than passing over it, has been observed only in reactions involving electron, hydron, or hydride transfer, reactions that will not be considered here.

The expression for calculating equilibrium isotope effects (EIEs) from vibrational frequencies was formulated by Bigeleisen and Goeppert Mayer (1947) with additions by Bigeleisen and Wolfsberg (1958) to calculate KIEs. The contributions to the IEs are multiplicative:

$$\text{KIE} = \text{MMI}\cdot\text{ZPE}\cdot\text{EXC} \tag{2}$$

where each factor is the contribution to the KIE from the change in MMI, ZPE, and EXC.

In a stable nonlinear molecule, there are 3*N*-6 normal modes. In the TS, however, one of these modes becomes the reaction coordinate with an imaginary frequency. The imaginary frequency is sometimes expressed as a negative number, although it is correct to write a frequency of $100i$ cm^{-1}, rather than -100 cm^{-1}. The reaction coordinate normal mode has no restoring force; it is aperiodic. It therefore makes no contribution to ZPE or EXC. Thus, for a KIE, ZPE and EXC are calculated using 3*N*-7 normal modes for the TS molecule.

The expression for each term is

$$\text{MMI} = \frac{\left[\left(\frac{{}^{\text{light}}M}{{}^{\text{heavy}}M}\right)^{3/2} \frac{{}^{\text{light}}I_x\,{}^{\text{light}}I_y\,{}^{\text{light}}I_z}{{}^{\text{heavy}}I_x\,{}^{\text{heavy}}I_y\,{}^{\text{heavy}}I_z}\right]_{\text{TS}}}{\left[\left(\frac{{}^{\text{light}}M}{{}^{\text{heavy}}M}\right)^{3/2} \frac{{}^{\text{light}}I_x\,{}^{\text{light}}I_y\,{}^{\text{light}}I_z}{{}^{\text{heavy}}I_x\,{}^{\text{heavy}}I_y\,{}^{\text{heavy}}I_z}\right]_{\text{reactant}}} \tag{3a}$$

where ${}^{\text{light}}M$ is the molecular mass with the light isotope and ${}^{\text{light}}I_x$ is the moment of rotational inertia in the *x*-axis.

$$\text{ZPE} = \frac{\left[\prod_i^{3N-7\ddagger} e^{-({}^{\text{light}}u_i - {}^{\text{heavy}}u_i)/2}\right]_{\text{TS}}}{\left[\prod_i^{3N-6} e^{-({}^{\text{light}}u_i - {}^{\text{heavy}}u_i)/2}\right]_{\text{reactant}}} \tag{3b}$$

where ${}^{\text{light}}u_i$ is the *i*th normal mode, u_i is $h\nu_i/k_BT$ (*h* is Planck's constant, ν_i is frequency, k_B is the Boltzman constant, and *T* is absolute temperature) and 3*N*-7‡ is the number of normal modes in a nonlinear TS, where *N* is the number of atoms. For linear molecules, use 3*N*-6‡ normal modes. 3*N*-6 is the number of normal modes in a nonlinear stable molecule. For linear molecules, use 3*N*-5.

$$\mathrm{EXC} = \frac{\left[\prod_i^{3N-7\ddagger} \frac{1-\mathrm{e}^{\left(-^{\mathrm{heavy}}u_i\right)}}{1-\mathrm{e}^{\left(-^{\mathrm{light}}u_i\right)}}\right]_{\mathrm{TS}}}{\left[\prod_i^{3N-6} \frac{1-\mathrm{e}^{\left(-^{\mathrm{heavy}}u_i\right)}}{1-\mathrm{e}^{\left(-^{\mathrm{light}}u_i\right)}}\right]_{\mathrm{reactant}}} \quad (3c)$$

MMI can be calculated directly, as in Eq. (3a). For EIEs, it can also be calculated from the vibrational product (VP) by application of the Teller-Redlich rule:

$$\mathrm{MMI} = \mathrm{VP} = \frac{\left(\prod_i^{3N-6} \frac{^{\mathrm{light}}\nu_i}{^{\mathrm{heavy}}\nu_i}\right)_{\mathrm{final}}}{\left(\prod_i^{3N-6} \frac{^{\mathrm{light}}\nu_i}{^{\mathrm{heavy}}\nu_i}\right)_{\mathrm{initial}}}. \quad (4)$$

For KIEs, the reaction coordinate mode contributes to the MMI, which can also be expressed as

$$\mathrm{MMI} = \frac{^{\mathrm{light}}\nu*}{^{\mathrm{heavy}}\nu*} \times \mathrm{VP} \quad (5)$$

where $^{\mathrm{light}}\nu*$ is the reaction coordinate frequency for the light isotope and VP is the VP calculated using 3N-6 normal modes for the reactant state and 3N-7‡ for the TS.

Even though IEs are a quantum phenomenon, ZPE and EXC factors are simple functions of vibrational frequencies. These frequencies can be calculated by using purely mechanical models of the molecules. To a very good approximation, the vibrational frequencies of a molecule can be described in terms of a balls-and-springs harmonic oscillator model. Atoms are treated as balls of defined mass, and bonds are treated as ideal massless springs with defined strengths. At its equilibrium length, the spring exerts no force. If a displacement of the atoms changes the length of the spring, it will exert a restoring force proportional to the displacement, following Hook's law:

$$F = -k\cdot\Delta x \quad (6)$$

where F is the restoring force, k is the spring force constant (in units of force per unit displacement), and Δx is the displacement from equilibrium spring length.

The simplest vibrational model is a diatomic molecule, with two balls connected by a spring. With this model, the harmonic bond-stretching frequency can be calculated using

$$\nu = \frac{1}{2\pi}\sqrt{\frac{k}{\mu}} \quad (7)$$

where ν is the vibrational frequency, k is the spring force constant, μ is the reduced mass of the oscillator, $m_1m_2/(m_1 + m_2)$, where m_1, m_2 = mass of the spheres.

A diatomic molecule has only one vibrational motion, bond stretching. In polyatomic molecules, molecular vibrations do not occur as isolated vibrations of individual bonds. Rather, individual bond vibrations become coupled, depending on symmetry and frequency, into a discrete number of complex motions, the normal modes of vibration (Harris and Bertolucci, 1978; Wilson *et al.*, 1955). Normal modes correspond to the observable vibrational frequencies detected by IR or Raman spectroscopies (although not all normal modes are IR or Raman active, depending on symmetry). There are 3*N*-6 normal modes in nonlinear molecules and 3*N*-5 in linear molecules. In normal coordinate analysis, the frequencies and motions of individual atoms for each normal mode are determined. Vibrational frequency is synonymous with vibrational energy; changes in vibrational frequency caused by an isotopic label translate into changes in the relative energies of the reactant and TS, that is, a change in activation energy, and therefore a change in reaction rate. Thus, the change in mass associated with an isotopic label leads to the change in relative reaction rates: the KIE.

Thus, it is possible to calculate KIEs based on balls-and-springs vibrational models of molecules by applying Eq. (3) to those frequencies. The next step is to make the connection between molecular structures and vibrational frequencies.

The relationship between structure and frequency arises from the relationship between bond length and the force required to distort a bond. The relation between bond length and stretching force constants is described by the Pauling bond order and Badger's rule. Pauling bond orders relate bond length to bond order:

$$n_{ij} = e^{(r_1 - r_{ij})/0.3} \tag{8}$$

where n_{ij} is the Pauling bond order between atoms i and j, r_1 is the length of a single bond of that type, and r_{ij} is the length of the bond between atoms i and j.

Badger's rule describes the relationship between bond order and bond strength, that is, the force constant for that bond:

$$F_{ij} = F_1 \cdot n_{ij} \tag{9}$$

where F_{ij} is the force constant for the bond between atoms i and j and F_1 is the force constant for a single bond between those atom types.

Force constants are expressed in units of force per unit displacement, the same units as k in Hook's law, generally mdyne/Å. Evidence supporting the use of the Pauling and Badger rules in BEBOVA analysis have been reviewed (Burton *et al.*, 1977; Johnston, 1966; Sims and Lewis, 1984). This establishes the connection between geometry and vibrational frequencies through the variation in bond-stretching force constants with bond length. Thus, the chain of causality is changes in bond lengths between reactant and TS lead to changes in bond-stretching force constants, which leads to changes in vibrational frequencies, which leads to IEs.

Structural changes also affect bond-bending force constants, although the relationship between bond angles and bending force constants is not as clear cut. Sims and Lewis (1984) developed a series of functions to describe the variation in bending force constants as a function of bond angle (see later).

C. Competitive KIEs

Large IEs (ca. >1.5) can be determined with acceptable precision from the steady-state kinetic constants for isotopically pure species. This is the direct, or noncompetitive, method. For smaller IEs, such as secondary hydrogen or heavy atom IEs, competitive, or isotope discrimination, methods are used to obtain measurements precise enough to be useful for quantitative TS analysis. In competitive IE measurements, a mixture of isotopically labeled and unlabeled substrates is allowed to react with the enzyme. The isotopologs compete with each other as substrates, which mean that their relative rates of reaction are determined by the relative values of k_{cat}/K_M, the "specificity constant" (Fersht, 1985). Competitive KIEs measure IEs on k_{cat}/K_M. This is true regardless of the concentration of each isotopic label, that is, whether the isotopic labels are trace, radioactive isotopes, or nonradioactive labels present as a large molar fraction of the total substrate concentration. This is contrary to occasional statements in the literature that this is true only for trace (i.e., radioactive) labels (Parkin, 1991). There are many techniques for measuring competitive IEs, including isotope ratio mass spectrometry (O'Leary, 1980; Weiss, 1991), whole molecule mass spectrometry (Bahnson and Anderson, 1991; Berti *et al.*, 1997; Gawlita *et al.*, 1995), liquid scintillation counting of radioisotopes (Parkin, 1991), and nuclear magnetic resonance (NMR) (Singleton and Thomas, 1995).

D. Irreversible Step(s) and Commitment to Catalysis

In the simplest case, the kinetic constant k_{cat}/K_M (or V/K) reflects partitioning between the free reactant in solution and TS of the *first irreversible step* of the enzymatic reaction. In this case, KIEs on k_{cat}/K_M reflect the IEs between these same species, the reactant in solution, and the TS for the first irreversible step. The meaning of k_{cat}/K_M, and therefore the meaning of KIEs, may change if there are intermediates in the chemical mechanism; that is, if there are chemical species with finite lifetimes that can partition either forward or backward in the reaction pathway. If these intermediates partition mostly back to the substrate and are therefore in equilibrium with it, then the meaning of the KIEs is unchanged. If there is significant partitioning of these intermediates both forward and backward, then the TS leading to each intermediate will contribute to the KIEs. The case of one intermediate is discussed further later.

There is no reason why the first irreversible step must be a chemical step. If every time a substrate molecule binds to the enzyme it reacts to give products, that is, if there is full commitment to catalysis (C_f), then the first irreversible step is association of the substrate and enzyme to form the Michaelis complex. In this case, the observed KIEs would be for diffusion of the enzyme and substrate together, which would carry no information on the chemical mechanism. Ideally one seeks to find reaction conditions where the commitment to catalysis is zero, that is, where $k_2 \gg k_3$ in Eq. (10)

$$E + S \underset{k_2}{\overset{k_1}{\rightleftharpoons}} E{\cdot}S \xrightarrow{k_3} E{\cdot}P \xrightarrow{k_5} E + P \tag{10}$$

where E is the enzyme, S is the substrate, E · S is the Michaelis complex, and k_n is the intrinsic rate constant on step n. This can be accomplished by using alternate substrates, adjusting reaction conditions, or using a mutant enzyme with increased k_2/k_3. When commitment is zero, the observable KIEs are those of the chemical step(s).

If conditions of zero commitment cannot be found, it is possible to measure the commitment and derive the intrinsic KIEs of the chemical steps from the observed KIEs using Eq. (11) (Northrop, 1981; Rose, 1980):

$$\mathrm{KIE_{obs}} = \frac{\mathrm{KIE_{intrinsic}} + C_f + \mathrm{EIE}{\cdot}C_r}{1 + C_f + C_r} \tag{11}$$

where $\mathrm{KIE_{obs}}$ is the observed KIE, $\mathrm{KIE_{intrinsic}}$ is the intrinsic KIE on the chemical step, C_f is the commitment to catalysis in the forward direction, C_r is the commitment to catalysis in the reverse direction, and EIE is the EIE between reactants and products. The reverse commitment to catalysis, C_r, may be neglected if the reaction is run under conditions that are effectively irreversible.

If a nonchemical step, such as a protein conformational change or product release, is the irreversible step, then once again the observable KIEs will not include a contribution from the chemical step. If this situation is encountered, then a change in reaction conditions may increase the rate of the nonchemical step or slow the chemical step, thereby allowing the KIEs on the chemical step to be expressed (Kline and Schramm, 1995; Scheuring *et al.*, 1998).

So far, we have only considered the case where one step is cleanly the irreversible step. In the case where one intermediate is formed, formation and/or breakdown of the intermediate will contribute to the observable KIEs, depending on the rate constants. Let us consider a kinetic mechanism with one intermediate:

$$E + S \underset{k_2}{\overset{k_1}{\rightleftharpoons}} E{\cdot}S \underset{k_4}{\overset{k_3}{\rightleftharpoons}} [E{\cdot}S'] \xrightarrow{k_5} E{\cdot}P \xrightarrow{k_7} E + P \tag{12}$$

where E · S′ is the enzyme-bound intermediate with a finite lifetime.

The k_{cat}/K_M on this reaction will be

$$\frac{k_{cat}}{K_M} = \frac{k_1k_3k_5k_7}{k_2k_4k_6\left[1 + \frac{k_7}{k_6}\left(1 + \frac{k_5}{k_4}\left(1 + \frac{k_3}{k_2}\right)\right)\right]}. \tag{13}$$

The observable competitive KIEs will be

$$\mathrm{KIE} = \frac{\frac{\alpha_1\alpha_3}{\alpha_2}\left(\frac{\alpha_5}{\alpha_4} + \frac{k_5}{k_4}\right)}{1 + \frac{k_5}{k_4}} \tag{14}$$

where α_n is the intrinsic IE on step n.

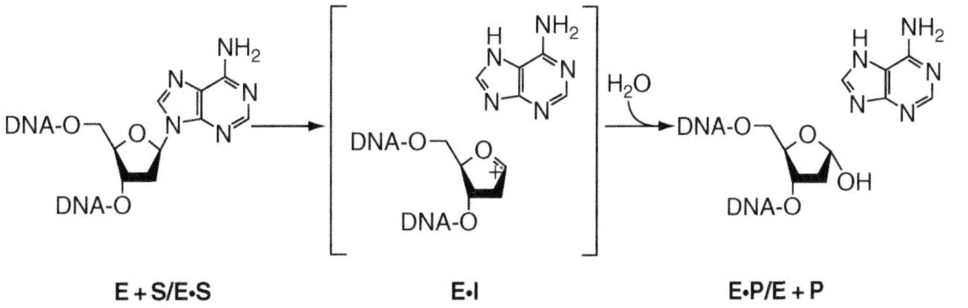

Fig. 3 Ricin-catalyzed DNA hydrolysis. This reaction follows the kinetic scheme in Eq. (12). Ricin depurinates 28S RNA *in vivo*.

The observable IE depends on k_5/k_4, the partitioning of the enzyme-bound intermediate, $E \cdot S'$. If $k_5/k_4 \gg 1$, then k_3 is the first irreversible step, and $KIE = (\alpha_1\alpha_3)/\alpha_2$. The KIE will be that between the reactant in solution and the TS of the first chemical step. If k_5/k_4 » 0, then k_5 will be the first irreversible step, and $KIE = (\alpha_1\alpha_3\alpha_5)/(\alpha_2\alpha_4)$. This is the KIE between the reactant and the TS for the second chemical step. Alternately, and equivalently, it is the product of the EIE for $E \cdot S'$ formation $(\alpha_1\alpha_3)/(\alpha_2\alpha_4)$ and the KIE of $E \cdot S'$ breakdown, α_5. At intermediate values of k_5/k_4, the observable KIE varies monotonically between those for the extreme values of k_5/k_4. Whether it is possible to determine k_5/k_4 by BEBOVA will depend on the particular system.

In the case of ricin-catalyzed depurination of DNA (Chen *et al.*, in preparation) (Fig. 3), experimental KIEs showed that the TS did not involve a bimolecular nucleophilic A_ND_N displacement conforming to Eq. (10). Rather, they showed that an oxocarbenium ion · enzyme complex with a finite lifetime (more than several bond vibrations, ca. 10^{-12} s) was being formed, that is, the reaction was proceeding through a $D_N + A_N$ mechanism. The experimental primary $1'$-^{14}C KIE was 1.015 ± 0.001, compared with the calculated KIEs for $k_5/k_4 \gg 1$ (KIE = 1.018) and k_5/k_4 » 0 (1.010). At the same time, the primary 1-^{15}N KIE was 1.023 ± 0.004, compared with the calculated KIEs of 1.028 for $k_5/k_4 \gg 1$, and 1.024 for k_5/k_4 » 0. In this case, the calculated KIEs were in good agreement with the experimental KIEs, but were not sufficiently different from each other to make it possible to estimate the value of k_5/k_4 from the experimental KIEs. The ricin example is a reminder that although competitive KIEs always measure the IEs up to the first irreversible step, one does not know beforehand (and sometimes afterward) which step that is.

III. Route from KIEs to TS Structures

A. Overall Strategy

It is not possible to calculate a TS directly from experimental KIEs. Rather, it is necessary to create a test TS, then calculate what the KIEs would be for that TS. The calculated KIEs are then compared with the experimental ones. KIEs for the test TS structure are calculated using the BEBOVA approach. BEBOVA attempts to recreate the vibrational frequencies of the reactant and test TS based on simple rules relating geometry and force constants (see later). Test TS structures for BEBOVA have been created using the structure interpolation method, an improvement on the old, *ad hoc* approach.

In the old, *ad hoc* method used in this laboratory, one searches for the TS by adjusting the structure of the TS manually. Structural parameters that are expected to contribute to the KIEs are adjusted and the KIEs calculated. The structure is then readjusted iteratively to arrive at a structure that gives calculated KIEs in agreement with experimental KIEs. The limitations of this approach include the fact that it is possible to create structures that are subtly incorrect or to fail to make important adjustments to the structure because they are expected (incorrectly) not to affect the KIEs. A second limitation is that only one TS is determined, without any indication whether there is another structure, or group of structures, that would match the experimental KIEs. Third, the process is very time-consuming. Finally, because one does not know what the KIEs should be for a given reaction, it is not possible to recognize enzyme-caused or other unusual effects. The transition-state imbalance (TSI) effects observed for pertussis toxin-catalyzed ADP-ribosylation of protein $G_{i\alpha 1}$ (Scheuring *et al.*, 1998) were recognized as such only because the experimental KIEs were different from those expected based on the NAD^+ hydrolysis reactions (Berti and Schramm, 1997; Berti *et al.*, 1997).

For these reasons, the structure interpolation approach was developed (Berti and Schramm, 1997). Rather than trying to reach one possible TS, an algorithm is used to generate many test TS structures throughout reaction space (Fig. 4), and KIEs are calculated for all the test TSs. Plots are made showing the areas in reaction space where calculated KIEs match experimental KIEs for each isotopic label (Fig. 5). If the modeling process has been successful, there will be a point in reaction space where calculated KIEs match experimental KIEs for all of the isotopic labels. This is the experimentally determined TS. If more than one TS is consistent with experimental KIEs that will be obvious also.

The structure interpolation approach is similar to systematic searches of reaction space that have been described previously (Lewis *et al.*, 1980; Rodgers *et al.*, 1982), with several changes. The two reference structures in structure interpolation are either X-ray crystal structures or high-level quantum mechanically optimized structures, rather than structures based on ideal bond lengths and angles. All the internal coordinates of the test TS structures are varied between the values of the first and second reference structures, rather than relying on simplified models of structure variation throughout reaction space, and the structure interpolation algorithm was based on the results of quantum mechanical optimizations of intermediate structures. This allowed accurate modeling of the structures throughout reaction space.

The BEBOVA approach achieves its greatest power if a *unified model* of a given reaction type can be created. Nucleophilic substitutions on NAD^+ are a case in point. TSs of nonenzymatic (Berti and Schramm, 1997) and three enzyme-

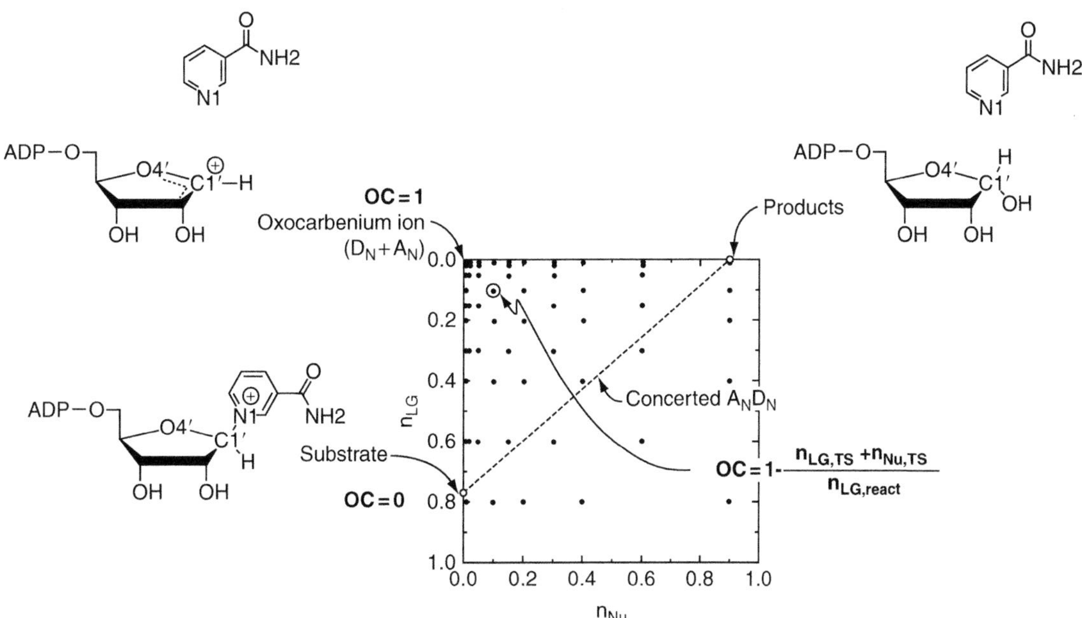

Fig. 4 Reaction space for hydrolysis of NAD^+ to ADP-ribose and nicotinamide. The main processes of the reaction are breaking the bond to the leaving group ($n_{LG} \equiv n_{C\text{-}1'\text{-}N\text{-}1}$) and forming the bond to the nucleophile ($n_{Nu} \equiv n_{C\text{-}1'\text{-}Nu}$). A classical $D_N + A_N$ mechanism involves departure of the leaving group ($n_{LG} = 0$) to form an oxocarbenium ion intermediate plus the first product, nicotinamide. In the second step, the nucleophile approaches to form the C-1′–Nu bond. In a fully concerted A_ND_N mechanism, the formation of n_{Nu} exactly matches the loss of n_{LG}, as shown by the diagonal dashed line. The dots in reaction space represent the 85 test TS structures generated by structure interpolation. OC is the oxocarbenium ion character described in the text.

catalyzed hydrolysis reactions (Berti *et al.*, 1997 and unpublished results), plus the two ADP-ribosylation TSs (Scheuring and Schramm, 1997b; Scheuring *et al.*, 1998),were determined using a unified model of NAD^+ nucleophilic substitution. A unified model differs from any other TS determination only in that the same calculation system has been used successfully to determine several different TSs of that type, lending support to the correctness of each step in the process. Thus, the same reference structures, same interpolation algorithm, same force constants, and same vibrational model were used for all these TS determinations.

A unified model of a reaction yields several advantages. (a) *Speed*: Once the calculations are complete, the TS for any reaction of that type can be determined simply by matching the already calculated KIEs to the new experimental ones. (b) *Reliability*: If many TSs can be determined from a single unified model, this is a strong indication of the validity of that model. (c) *Sensitivity*: BEBOVA detects differences between TSs with much greater precision than the absolute values of bond orders for a single TS. This characteristic is especially useful in understanding

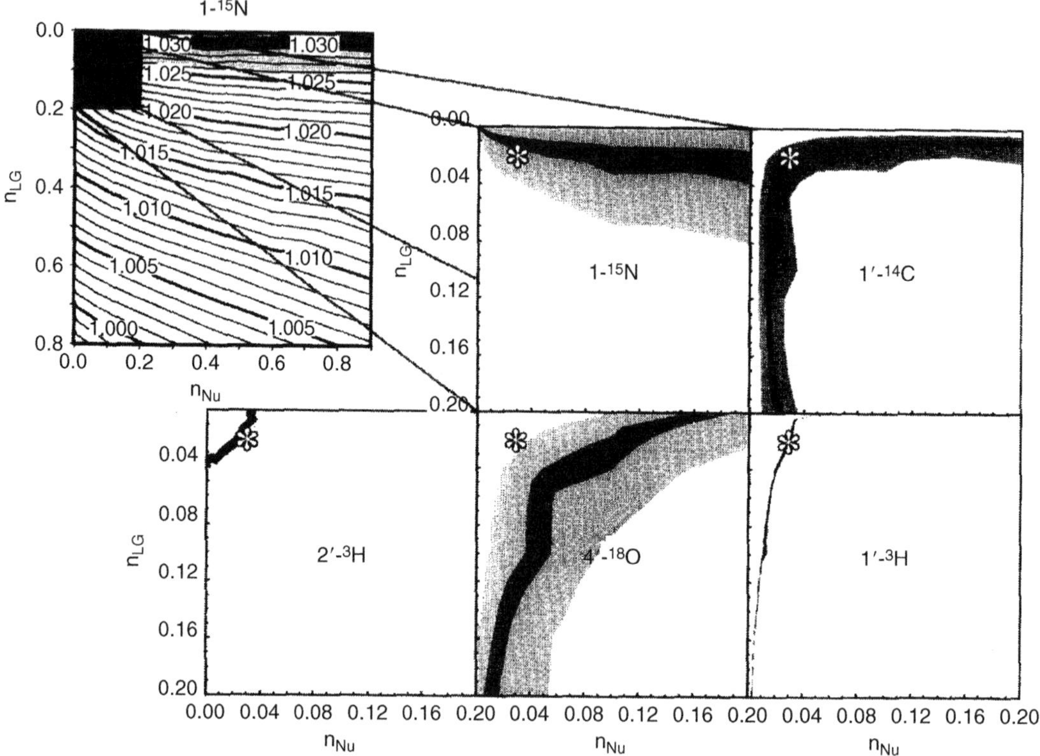

Fig. 5 Match of calculated to experimental KIEs in reaction space for diphtheria toxin-catalyzed hydrolysis of NAD^{+4}. (Top left) Contour plot of calculated 1-^{15}N KIEs throughout reaction space. (Top middle) Expansion of the dissociative section of reaction space for 1-^{15}N KIEs. The shading shows the areas in reaction space where the calculated KIEs match the experimental KIEs exactly (dark gray) or within the 95% confidence interval (light gray). The asterisk is the experimentally determined TS. Isotopic labels as shown in Fig. 1.

enzyme mechanisms because the difference between TSs for the enzymatic and nonenzymatic reactions reflects the interaction between the inherent reaction pathway of the substrate and the enzyme's use of binding energy to lower the energetic barrier to the TS.

The description of the process will be in two parts, the first covering BEBOVA and the second covering structure interpolation.

B. Bond-Energy/Bond-Order Vibrational Analysis

The program used for all the following BEBOVA work was BEBOVIBIV, written by Sims *et al.* (1977). It is available through the Quantum Chemistry Program Exchange and is the primary program for BEBOVA studies. There is one other program for doing bond vibrational analysis, VIBIE (Casamassina and

Huskey, 1993; Huskey, 1996). VIBIE does not automatically vary force constants as a function of structure, as BEBOVIB does, so they must be included in the input directly.

BEBOVA versus quantum mechanical KIEs. As an alternative to BEBOVA, it is sometimes possible to use an entirely quantum mechanical approach to TS determination. Increases in computational power have made it possible to perform quantum mechanical calculations at usefully high levels of theory on molecules of biologically relevant size, although only in a vacuum or a continuum dielectric reaction field. Using electronic structure programs, the reactant and TS structures for a reaction may be determined. Calculating the vibrational frequencies of these structures allows KIEs to be calculated and compared with experimental ones. If the calculated KIEs match the experimental ones, that will, in effect, confirm the computational result and make BEBOVA calculations unnecessary. It continues to be the general case, however, that the computed and experimental KIEs (and therefore the TSs) do not match, despite sometimes heroic efforts. At this point, if one is limited to using only a quantum mechanical approach, one is left knowing what the TS is not, which is not generally a useful result. Quantum mechanical calculations are a crucial component in solving TSs via the structure interpolation approach, but are unlikely on their own to yield the correct TS. Quantum mechanical calculations are still constrained by the limits of computer power and of theory. These limitations are likely to remain significant factors into the foreseeable future.

Using BEBOVIB. The principles of BEBOVA discussed by Sims and Lewis (1984) in their excellent description of the program BEBOVIB have not changed, but certain details of the process have. BEBOVA work by attempting to recreate the vibrational frequencies of the molecules involved. Once the frequencies have been calculated, the vibrational (ZPE and EXC) contributions to the KIE can be calculated, as well as the contribution from the change in MMI. The functioning of BEBOVIB is illustrated schematically in Fig. 6.

The input required for BEBOVA analysis are (a) structures of the initial and final states, (b) list of internal coordinates, (c) force constants for each internal coordinate, (d) (optionally) interaction or "off-diagonal" force coefficients or constants, (e) list of atomic masses, (f) temperature, and (g) the reaction coordinate (implicit to (c) and (f)).

a. *Structures.* Structures of the initial and final states (for KIE modeling, this is the reactant and the TS) are required. Because IEs report on the *change* in vibrational environment, structures of the reactant and the TS are equally important, and have equal effects on calculated KIEs. The creation of models of the reactant and test TS structures will be discussed in Section III.C.

It has been a general practice in BEBOVA analyses to use cutoff models; this includes the NAD^+ substitution reactions discussed here. In a cutoff model, atoms more than a couple of bonds away from an isotopic label are deleted from the model. The justification for using cutoff models is the assumption that IEs are local

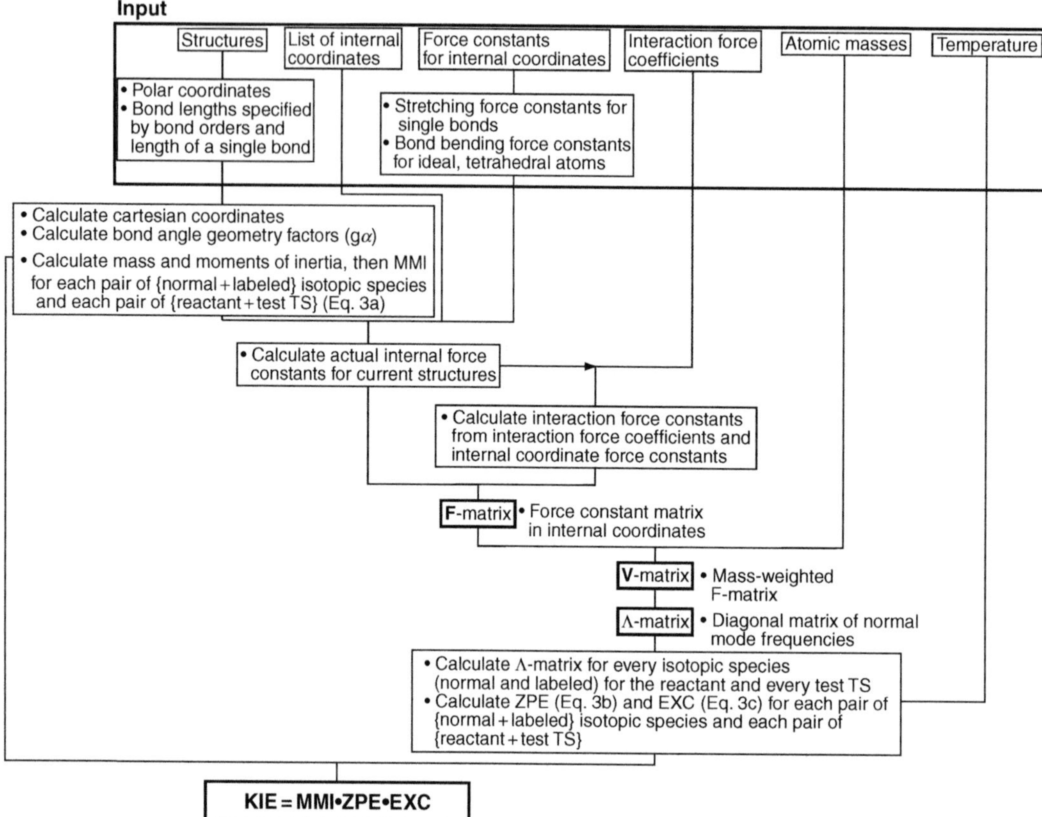

Fig. 6 Schematic description of BEBOVIB operation. The route from input, at the top, to the calculated KIE, at the bottom, is illustrated. Procedures are in light outline and objects are in dark outline. This scheme assumes that the internal coordinate force constants for each structure are being calculated from the force constants for single bonds and ideal, tetrahedral bond angles in the input. It also assumes that interaction force coefficients are supplied. It is possible to provide internal coordinate force constants and/or interaction force constants directly without BEBOVIB recalculating them.

effects, that the vibrational environment of any atom is not affected significantly by atoms more than a few bonds away. In the past, cutoff models were used because computers were not powerful enough and did not have enough memory to handle large molecules. In 1999, even desktop computers can easily handle molecules with 100 atoms, so this limitation is no longer important and the use of cutoff models is less necessary.

Stern and Wolfsberg (1966) (see also Sims and Lewis, 1984) described the conditions under which cutoff models could be made. Models consisting of all atoms within three bonds of any site of isotopic substitution will give acceptable cutoff models by the criteria of Stern and Wolfsberg. If certain conditions are met, cutoff models consisting of all atoms within two bonds of any isotopic label, or even one bond, are possible.

The use of cutoff models can add an artifact to calculated KIEs, however, in the form of inaccurate MMI factors. For heavy atom KIEs, the contribution of the MMI factor can be large, up to one-half of the total IE (Huskey, 1991). If the size of the model molecules is significantly smaller than the real molecules, there is the risk of introducing an artifact in the MMI factor. The use of cutoff models should be evaluated for each reaction studied.

The format for structure input in BEBOVIB is a polar coordinate system. Bond lengths are not given directly, rather the bond order between two connected atoms is defined, along with the length of a single bond of that type. Structure input in VIBIE is in Cartesian coordinates or in a *Z*-matrix (Frisch *et al.*, 1995) of internal coordinates.

b. *List of internal coordinates.* Once the structures have been defined, it is necessary to define the connectivities between atoms and which types of vibrational motions may occur. There are five types of internal coordinates: bond stretch, bond angle bend, out-of-plane bend, linear bend, and torsion (Fig. 7). Out-of-plane bends are used in any structure containing three atoms bonded to a central atom where all four atoms are coplanar. Only one out-of-plane internal coordinate needs to be defined for a set of four coplanar atoms because displacing any of the terminal atoms out of the plane is equivalent to any other (because a plane can be defined with the central atom and any two of the three terminal atoms). Linear bends are defined for each set of three colinear atoms. For any three colinear atoms, two linear bends are defined by two virtual atoms connected to the

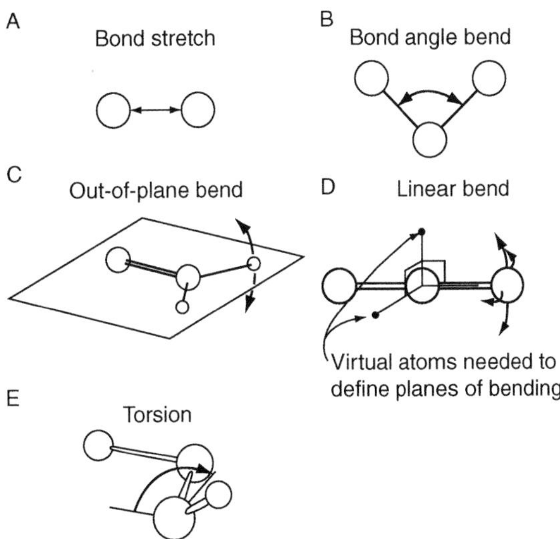

Fig. 7 Internal coordinates used in BEBOVA.

central atom with bonds that are perpendicular to the linear bonds and perpendicular to each other. Acetylene contains two overlapping sets of three colinear atoms and requires two sets of two linear bends to be defined. The torsional force constants are for librations of the torsional angles, not free rotation.

Force constants may be supplied to the program as either force constants to be used directly or force constants for single bonds (or "ideal" bonds for bends), which BEBOVIB then uses to calculate the actual force constants for a given structure. The treatment to be used for the force constants is included in the list of internal coordinates, although the values of the force constants are listed separately. The stretching force constant is a simple function of the bond order, $F_{ij} = F_1 \cdot n_{ij}$ (Eq. (9)).

Several treatments for the force constants for bond angle bends are possible. Sims and Lewis (1984) and Burton *et al.* (1977) recommend:

$$F_{ijk} = (n_{ij}n_{jk})^x \cdot g_\alpha \cdot F_1 \tag{15}$$

where F_{ijk} is the bending force constant for the angle, α, formed by bonds *i–j* and *i–k*; x is 0.5 or 1.0 (see later); and g_α is the hybridization factor, $g_\alpha = 1.39 + 1.17 \cdot \cos \alpha$.

The value of x should usually be 0.5, except in the case of large changes in n_{ij} or n_{jk}, where 1.0 is preferable. The hybridization factor, g_α, attempts to reproduce observed trends in bond-bending force constants in certain hydrocarbons and other types of compounds (see Burton *et al.*, 1977; Sims and Lewis, 1984). For an ideal sp^3 carbon atom, with $\alpha = 109.5°$, $g_\alpha = 1.0$; it decreases as α increases, to 0.81 at 120° and 0.22 at 180°.

Only 3*N*-6 nonredundant internal coordinates are required to specify the normal modes of a molecule. A significant disadvantage of using nonredundant coordinates is that identical bond angle bending coordinates are treated unequally. For an sp^3 atom, for example, methane, there are six possible bond angle bends, but one of those is redundant (Fig. 8). If five bending modes for methane are used, the normal modes must be described as linear combinations of the five (unequal) bending force constants, which would be represented in BEBOVIB as off-diagonal elements of the **F** matrix (see later). In BEBOVIB calculations, it is simpler to use six bending modes for methane, including one redundant one, all with equal force constants. BEBOVIB can calculate vibrational frequencies from an arbitrarily large number of internal force constants (i.e., >3*N*-6). It should be remembered when using force constants from the literature that they may have been derived from experimental vibrational frequencies using nonredundant sets of internal coordinates.

c. *Force constants for each internal coordinate*. Internally, BEBOVIB builds an **F** matrix that contains, as its diagonal elements, the force constants for each internal coordinate (see later), that is, **F** is a square matrix of dimension *n*, where *n* is the number of internal coordinates, with F_{ii} the force constant for internal coordinate *i*. The off-diagonal elements will be discussed later. An **F** matrix containing only diagonal elements is referred to as a simple valence force field (SVFF, as distinct from a general valence force field, GVFF, which includes off-diagonal elements).

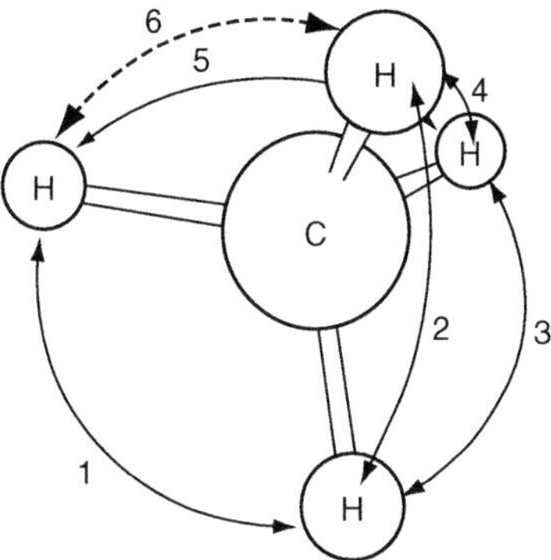

Fig. 8 Redundant bond-bending coordinates in methane. With five bond-bending internal coordinates defined, the sixth (dashed line) is redundant for specifying atom positions and vibrational frequencies, but is still useful. The order of numbering is arbitrary.

Generally, SVFFs are used in BEBOVA analyses, with the off-diagonal forces used to generate the reaction coordinate if necessary.

Values of force constants designed specifically for use in vibrational analyses can be found in several sources in the literature (Burton *et al.*, 1977; Sims and Lewis, 1984; Sims *et al.*, 1977; Wilson *et al.*, 1955). For NAD^+ TS determinations, the force constants for bond stretches and bond angle bends were taken from the AMBER force field (Cornell *et al.*, 1995; Weiner *et al.*, 1984). This force field was derived by reference to the normal modes of model compounds and contains force constants for a large variety of atom types, making it a good source of force constants for use with BEBOVIB.

d. *Interaction force coefficients or constants (optional)*. Force constants for the internal coordinates described in the previous section are the diagonal components of the **F** matrix. They represent the restoring force exerted upon distortion of a bond. Interaction force constants are the off-diagonal components of the **F** matrix. They represent the interactions between internal coordinates. If, for example, stretching a given bond (internal coordinate i) reduces the force constant for a bond angle bend (internal coordinate j), then there is an interaction between those internal coordinates. This interaction constant is given by the value of matrix element F_{ij} (or F_{ji}; the matrix is symmetrical). A *positive* interaction force constant couples an *increase* in the value of one internal coordinate with a *decrease* in the other. A negative interaction force constant has the opposite effect. Coupling a bond stretch with a bond angle bend with a positive interaction force constant means that stretching the bond is coupled with closing the bond angle, and vice versa. Coupling two bond stretches with a positive interaction force constant means that the stretching of one bond is coupled with compression of the other. Reaction coordinate normal modes for bimolecular substitutions are generated using interaction force constants (see later).

Interaction force constants may be defined directly as a force constant, although, more commonly, they are defined as an interaction force coefficient (a_{ij}). This allows the interaction force constant to vary along with the internal coordinate force constants as they are adjusted as a function of the structure. The interaction force coefficient (a_{ij}) gives an interaction force constant based on the equation:

$$F_{ij} = a_{ij} \cdot \sqrt{F_{ii} F_{jj}}. \tag{16}$$

In general, it has been found that acceptable accuracy in IE calculations can be achieved using SVFFs for stable molecules (Burton *et al.*, 1977; Sims and Lewis, 1984). Off-diagonal elements can be used to generate the reaction coordinate mode, as described later.

e. *List of atomic masses.* Each isotopically labeled species is defined by listing the atomic mass of each atom in the molecule for initial and final structures. Many isotopologs may be defined in a single file so IEs at every labeled position can be calculated in one run.

f. *Temperature.* Vibrational frequencies are independent of temperature, but IEs are temperature dependent because the proportion of vibrationally excited molecules increases with temperature, which changes the contribution from the EXC term (Bigeleisen and Goeppert Mayer, 1947; Bigeleisen and Wolfsberg, 1958; Stern *et al.*, 1968, 1970). IEs at up to four temperatures can be calculated in a given run with BEBOVIB.

g. *Reaction coordinate.* As discussed earlier, the reaction coordinate is the one normal mode in the TS that has an imaginary frequency. The reaction coordinate is generated through appropriate values of the internal coordinate force constants and/or interaction force constants discussed in c and d, respectively, but is treated separately here.

The reaction coordinate can be broken down conceptually into the main processes, and coupled processes. The main processes are the bond-making and bond-breaking events in the reaction. For a nucleophilic substitution, the main processes depend on the mechanism. For an A_ND_N mechanism, they are the concerted departure of the leaving group and approach of the nucleophile. In a $D_N + A_N$ mechanism, it is either the unimolecular departure of the leaving group or the nucleophilic attack on the electrophilic intermediate, depending on which is the first irreversible step. One coupled process for nucleophilic substitutions is the inversion of stereochemistry of the electrophilic carbon in an umbrella-like motion called the Walden inversion. In the NAD^+ example, this was achieved in part by introduction of a positive interaction force coefficient between the N-1–C-1′ bond stretch and the N-1–C-1′–X bond bends, which meant that stretching the N-1–C-1′ bond made decreasing the N-1–C-1′–X angles more favorable, facilitating Walden inversion. The inversion of stereochemistry was coupled with the main processes by using interaction force coefficients between bond-bending and bond-stretching forces (see later).

The first decision in modeling a reaction coordinate is to identify the main processes and decide which forces should be coupled to those main processes. For an A_ND_N mechanism, the main process is the concerted movement of the leaving group and nucleophile. The Walden inversion is coupled to this. Should there be other motions included in the reaction coordinate? Let us consider the nucleophilic attack of water on NAD^+ as an example.

As illustrated in Fig. 9A, a number of structural changes occur to NAD^+ on forming an oxocarbenium ion-like A_ND_N TS. A number of bonds, including C-1′–O-4′ and C-1′–C-2′, become stronger in compensation for the positive charge on the ribosyl ring. In the nicotinamide ring, the N-1–C-2 and N-1–C-6 bonds become stronger to compensate for the loss of the N-1–C-1′ bond. Because these structural changes occur as the reaction proceeds from the reactants to the TS, should they not be included in the reaction coordinate motion? Let us consider three bond lengths in turn (Fig. 9B, top). Electronic structure calculations show that $n_{\text{C-1}'\text{–O-4}'}$ increases from 0.96 to 1.59 as the ribosyl ring becomes more oxocarbenium ion like (Berti *et al.*, 1997). Because the TS is bimolecular, the bond order from the electrophilic carbon to the nucleophile water, $n_{\text{C-1}'\text{–Nu}}$ (or n_{Nu}), is increasing as the bond order to the leaving group, $n_{\text{C-1}'\text{–N-1}}$ (or n_{LG}), decreases. At some point, which we expect to be at or near the TS, the increase in $n_{\text{C-1}'\text{–Nu}}$ balances the

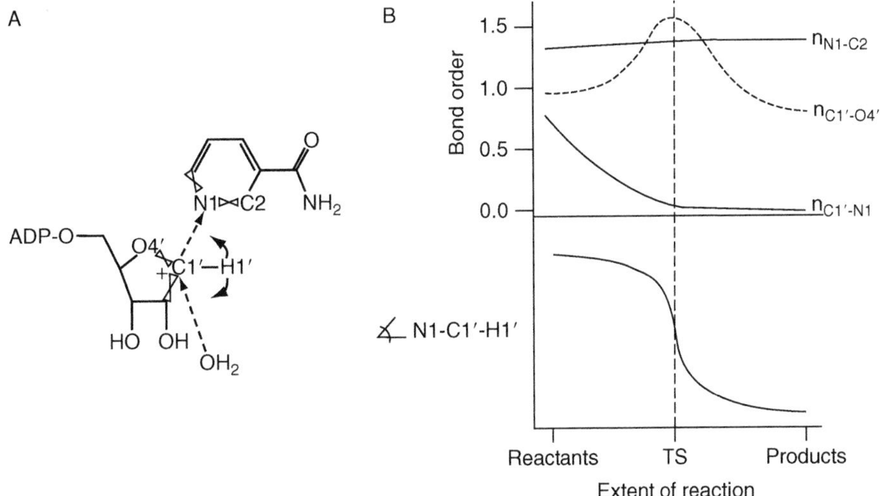

Fig. 9 Selection of structural changes to be coupled to the main processes in forming the reaction coordinate. (a) The main processes of this A_ND_N reaction are shown with dark arrowheads. Associated structural changes are shown as open arrowheads. (b) Changes in selected bond orders (top) or the ∠N-1–C-1′–H-1′ bond angle through the course of the reaction (bottom). Curve shapes are approximate, based on electronic structure calculations (Berti and Schramm, in preparation).

decrease in $n_{C\text{-}1'\text{–}N\text{-}1}$. At this point the ribosyl ring achieves its maximal oxocarbenium ion character (OC), that is, the C-1′–O-4′ bond is at its shortest length and will start to lengthen as $n_{C\text{-}1'\text{–}Nu}$ continues to increase. Thus, at or near the TS, the length of the C-1′–O-4′ bond stops changing and is constant as the reaction coordinate proceeds through the TS toward products. Therefore, the C-1′–O-4′ bond stretch should not be coupled to the reaction coordinate normal mode. The bond order $n_{N\text{-}1\text{–}C\text{-}2}$ (and $n_{N\text{-}1\text{–}C\text{-}6}$) in the nicotinamide ring continues to increase as the TS is passed and a nicotinamide molecule is formed. In this case, however, the relative change in bond order[1] is small, going from $n_{N\text{-}1\text{–}C\text{-}2} = 1.32$ in the reactant to 1.39 at the TS and 1.40 in nicotinamide, so its contribution to the reaction coordinate will be small and is therefore neglected. The Walden inversion, as followed by the N-1–C-1′–H-1′ angle, proceeds monotonically from reactants, through the TS, to the products (Fig. 9B, bottom). It is included in the reaction coordinate motion. In summary, the reaction coordinate motion for an A_ND_N mechanism will include the main processes, the bonds being made and broken, plus the coupled process, Walden inversion of the electrophilic carbon.

Main process reaction coordinate motions are generated using appropriate values of the internal coordinate force constants and/or the interaction force constants. The simplest reaction coordinate motion would be for a unimolecular dissociation. In this case the imaginary frequency of the reaction coordinate is generated by giving the internal coordinate for the breaking bond a zero or small negative force constant. Nucleophilic attack on the intermediate of a $D_N + A_N$ reaction is equivalent to a dissociation (in the reverse direction) and is treated in the same way.

For an A_ND_N mechanism, the main process involves a concerted motion of the leaving group and the nucleophile. The reaction coordinate mode with an imaginary frequency is generated by coupling the bond stretches from the electrophilic carbon to the leaving group and the nucleophile. In NAD^+ hydrolysis, the C-1′–N-1 and C-1′–Nu bond stretches were coupled with a positive interaction force coefficient, so a stretch of one bond made compression of the other more favorable. By making the interaction force coefficient greater than 1 (1.1 in this case), the asymmetric Nu $\rightarrow$ C-1′ $\rightarrow$ N-1 stretch acquired an imaginary frequency. An interaction force coefficient, rather than an interaction force constant, is used in these situations so that the proper reaction coordinate motion is generated automatically for different test TS structures with different values of $n_{C\text{-}1'\text{–}Nu}$ and $n_{C\text{-}1'\text{–}N\text{-}1}$. One advantage of generating a reaction coordinate through interaction forces, as opposed to simply assigning negative force constants to the Nu–C-1′ and C-1′–N-1 bond stretches, is that the symmetric Nu $\rightarrow$ C-1′ $\leftarrow$ N-1 bond stretch remains a harmonic vibration, as it should. In addition, by using an interaction

[1] As discussed in Berti and Schramm, 1997, IEs are a nonlinear function of bond order, being more closely correlated with the relative change in bond order. Thus, the IEs associated with a 1.06-fold increase in the N-1–C-2 bond order will be very much smaller than the 40-fold decrease in the C-1′–N-1 bond order.

force coefficient, the interaction force constant for a given test TS is calculated depending on the internal coordinate force constants for leaving group and nucleophile bonds. Because of this, the reaction coordinate imaginary frequency will have the desired characteristic of depending on $n_{C\text{-}1'\text{-}N\text{-}1}$ and $n_{C\text{-}1'\text{-}Nu}$, being higher for concerted and lower for less concerted (more dissociative) TSs (Buddenbaum and Shiner, 1976). A general method for generating a preselected imaginary frequency in a preselected normal vibrational motion has been described (Buddenbaum and Shiner, 1977; Buddenbaum and Yankwich, 1967).

The Walden inversion motion is generated by coupling the C-1′–Nu and C-1′–N-1 bond stretches to the bond angle bends centered on C-1′, that is, X–C-1′–Y, where X = N-1 and Nu, and Y = H-1′, C-2′, and O-4′ (Fig. 1). An interaction force coefficient of 0.05 for coupling the C-1′–N-1 bond stretch to bends with X = N-1 and −0.05 to bends with X = Nu were used (Berti and Schramm, 1997; Berti *et al.*, 1997), based on the work of Horenstein *et al.* (1991) and Markham *et al.* (1987) (Fig. 10). Similarly, interaction force coefficients were used to couple the C-1′–Nu bond to bends where X = Nu (0.05) and X = N-1 (−0.05). A general method for generating interaction force constants has been described (Buddenbaum and Shiner, 1977; Buddenbaum and Yankwich, 1967).

In an $A_{xh}D_HD_N$ (E_2) elimination mechanism, where two groups depart in concert, a negative interaction force coefficient would be used so that the concerted departure of both groups would have an imaginary frequency, whereas an asymmetric stretch would be a harmonic motion.

BEBOVIB calculates KIEs as KIE = MMI · ZPE · EXC, where MMI is calculated directly (Eq. (3a)). The contribution to the KIE from the reaction coordinate is obtained by calculating MMI and VP (Eq. (4)) directly, then using ${}^{light}\nu^{*}/{}^{heavy}\nu^{*}$ = MMI/VP (see Eq. (5)). Individual reaction coordinate frequencies

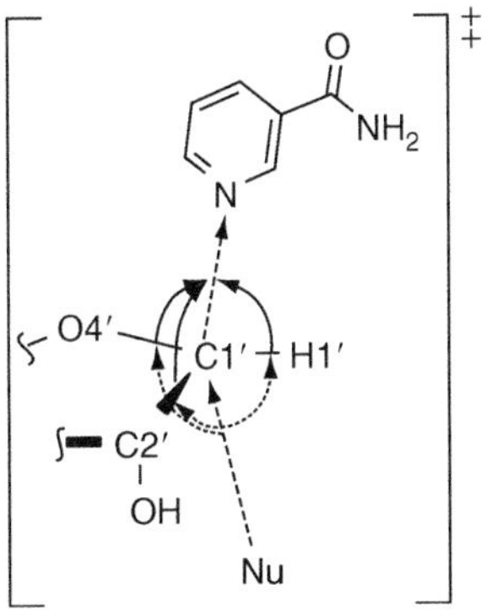

Fig. 10 Coupling of bond bending forces to the C-1′–N-1 bond stretch to create the Walden inversion motion. (Solid curves) Bond bends N-1′–C-1′–Y (Y = H-1′, O-4′, C-2′) are coupled with positive interaction force coefficients. (Dashed curves) Bond bends Nu–C-1′–Y are coupled with negative interaction force coefficients to the C-1′–N-1 bond stretch. These same bending internal coordinates are coupled through interaction force coefficients to the C-1′–Nu stretch, but with the opposite sign.

are also calculated for each isotopolog from the eigenvalues and reported. In principle, the ratio of frequencies from eigenvalues should be the same as the ratio calculated from MMI/VP. In practice, rounding and other arithmetic errors mean that the ratios differ slightly.

C. Structure Interpolation

The structure interpolation method involves generating many test TS structures for which KIEs are calculated and compared with the experimental KIEs. It works by interpolating test TS structures throughout reaction space based on two reference structures. One of the reference structures will be the reactant or product. The other will be a high energy (possibly hypothetical) intermediate that is close in structure to the expected TS structure, such as an oxocarbenium ion for glycoside hydrolyses or a tetrahedral intermediate for amide hydrolysis.

CTBI (Cartesian coordinates to TS structures to BEBOVIB format using internal coordinates). Structure interpolation is performed using the program CTBI: Cartesian coordinates to TS structures to BEBOVIB format, using internal coordinates (written in QBasic, available from the author). The input includes the two reference structures and a list of values for the independent variables (n_{LG} and n_{Nu} for a nucleophilic substitution). The algorithm for determining interpolated structures is written into the program and would have to be changed for other reactions.

CTBI performs the following procedures: (a) Read in reference structures in Cartesian coordinates, including atom connectivities (Protein Databank (PDB) format) (Bernstein *et al.*, 1977). (b) Parse structures to (i) find which atoms belong to the leaving group, the nucleophile, and the rest of the molecule and (ii) find cyclic fragments. (c) Convert the molecular definitions from Cartesian to internal coordinates. (d) Using an interpolation algorithm, create test TS structures throughout reaction space. (e) Convert the resulting test TS structures back to Cartesian coordinates and write the structures in the polar coordinate format for BEBOVIB.

The reference structures (step a), the parsing process (step b), and the interpolation process (step d) are described in some detail. Steps c and e are straightforward matrix and arithmetic calculations and will not be discussed further.

Reaction space and interpolation philosophy. Reaction space (Fig. 4) is defined by the two independent variables: the internal coordinates that describe the "main processes" of the reaction, that is, the bonds that are being made or broken. In a nucleophilic substitution, the main processes are breaking the bond to the leaving group and forming the bond to the nucleophile. Thus, the independent variables in reaction space are n_{LG} and n_{Nu}. In an elimination reaction, the main processes would be breaking the two leaving group bonds. All other structural changes that occur in the reaction are calculated as a function of the independent variables. For example, in NAD^+ hydrolysis, the reaction space was defined as a function of n_{LG} and n_{Nu}, and the reorganization of the ribosyl ring structure in oxocarbenium ion formation was a dependent variable, governed by the values of n_{LG} and n_{Nu}.

The structure interpolation method attempts to create a reasonable test TS structure for any values of n_{LG} and n_{Nu}, whether or not those values are themselves reasonable. For example, it is clearly unreasonable to have a pentavalent TS, with $n_{LG} = n_{Nu} = 1$, but CTBI would extrapolate from known structures in an attempt to create a "reasonable" pentavalent structure. In essence, structure interpolation answers the question: If the TS occurred at this arbitrary point in reaction space (i.e., these values of n_{LG} and n_{Nu}), what would the structure be? The question of which values of n_{LG} and n_{Nu} constitute the correct TS is then answered by calculating KIEs for each isotopic label for each test TS structure. There is no attempt to limit the independent variables to reasonable values because what is reasonable is not clearly known until after the TS has been determined. There is no information on the reaction coordinates in the structure interpolation-derived structures; this is a function of the internal coordinate and/or interaction forces in BEBOVIB.

Advantages of the structure interpolation approach over the old, *ad hoc* approach have already been discussed. It also has the advantages of speed and accuracy over one possible alternative that of using electronic structure optimizations at fixed values of n_{LG} and n_{Nu}. The speed advantage arises because structure interpolation is simply an arithmetic manipulation of structures. The accuracy advantage arises from the fact that interpolated structures are derived from high accuracy reference structures to which empirical adjustments can be made to account for experimentally known phenomena, such as hyperconjugation, that are not modeled adequately in electronic structure calculations (see later).

Structure interpolation in internal coordinates. Structure interpolations are performed in internal coordinates. Molecular structures are defined in the same *Z*-matrix internal coordinates (Fig. 7) as used by the electronic structure program GAUSSIAN (Frisch *et al.*, 1995). These internal coordinates are of the same type as used for BEBOVA, but without the implicit motion. Thus, a bond stretching in BEBOVA becomes a bond length when the structure is being defined. Similarly, bond angles and torsions are used. Out-of-plane bends and linear bends are not used in defining molecular structure.

The reason for using internal coordinates for interpolations is illustrated in Fig. 11. Interpolation between the atom at position 4 and 4′ is accomplished by increasing the torsional angle d_{3214}. No other internal coordinates are changed. If the structures were interpolated in Cartesian coordinates, the atom 4 would move in a straight line (shown by the dashed line and shaded circle), an unrealistic motion that would result in a decrease in r_{14} at intermediate values. Reference structure input is in Cartesian coordinates, and the BEBOVIB formatted output is in polar (i.e., a variation on Cartesian) coordinates, so CTBI converts the structures to and then back from internal coordinates.

The interpolation algorithm applied to NAD^+ hydrolysis. The structure interpolation technique is best explained in the context of a concrete example: NAD^+ hydrolysis reactions. The details of interpolation will vary with each

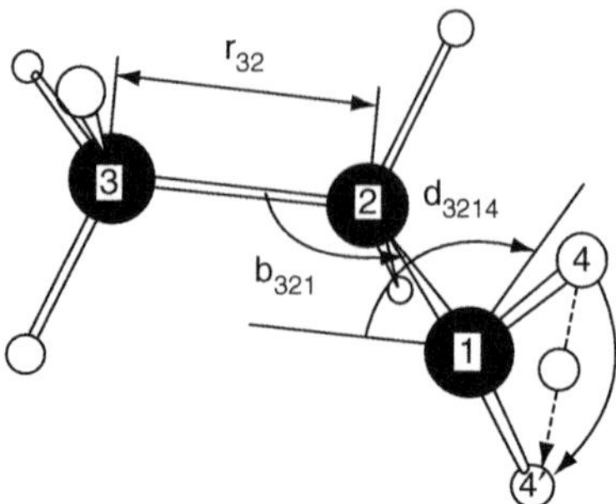

Fig. 11 Internal coordinates used in defining molecular structures and internally by CTBI during structure interpolation. The consequences of performing structure interpolations in Cartesian coordinates are illustrated for atom 4 (see text). With internal coordinates, this movement is accomplished by rotation about the C-2–O-1 bond by varying the dihedral angle d_{3214}.

reaction type, so this can only be considered a template, or archetypal, interpolation.

Based on other glycoside hydrolytic reactions studied in this and other laboratories (Kline and Schramm, 1995; Mazzella *et al.*, 1996; Mentch *et al.*, 1987; Parkin *et al.*, 1991a,b; Schramm, 1991; Sinnott, 1990), NAD^+ hydrolysis was expected to proceed through an asynchronous, oxocarbenium ion-like A_ND_N TS. The emphasis, therefore, was on modeling the oxocarbenium ion-like section of reaction space accurately (the top left in Fig. 4). The first reference structure was the reactant, NAD^+. The second reference structure was the hypothetical oxocarbenium ion, plus free nicotinamide. These would be the intermediates in a $D_N + A_N$ mechanism (Fig. 4). The ribosyl ring, leaving group, and nucleophile were each treated differently.

For the ribosyl ring, the simplifying assumption was made that the structure of the ribosyl ring is a function of one parameter, the oxocarbenium ion character (OC), that is, the similarity between the ribose ring in a given structure and the pure oxocarbenium ion. By definition, OC = 0 in the reactant NAD^+ molecule, and OC = 1 in the oxocarbenium ion. OC is defined as a simple function of n_{LG} and n_{Nu} in the test TS structure ($n_{\mathrm{LG,TS}}$ and $n_{\mathrm{Nu,TS}}$) and in the reactant ($n_{\mathrm{LG,reactant}}$):

$$\mathrm{OC} = 1 - \frac{n_{\mathrm{Nu,TS}} + n_{\mathrm{LG,TS}}}{n_{\mathrm{LG,reactant}}} \tag{17}$$

In the simplest model, each internal coordinate in the ribose ring would be the linear combination of the first and second reference structures. When TSI (Bernasconi, 1992a,b) effects in the ribosyl ring were recognized and elucidated further by electronic structure calculations (Berti and Schramm, in preparation), the ribosyl ring structure was made an exponential function of OC, that is, $\mathrm{OC}^{1.71}$. For example, for a bond length in the ribosyl ring ($r_{\mathrm{ribosyl,TS}}$):

$$r_{\text{ribosyl,TS}} = r_{\text{ribosyl,reactant}} + \text{OC}^{1.71} \cdot (r_{\text{ribosyl,oxocarbenium}} - r_{\text{ribosyl,reactant}}). \quad (18)$$

Internal coordinates in the nicotinamide leaving group (r_{nic} for bond lengths) were a function of n_{LG} only:

$$r_{\text{nic,TS}} = r_{\text{nic,reactant}} + (1 - n_{\text{LG,TS}}/n_{\text{LG,reactant}})(r_{\text{nic,nic}} - r_{\text{nic,reactant}}) \quad (19)$$

where $r_{\text{nic,nic}}$ is the corresponding bond length in the crystal structure of nicotinamide (Wright and King, 1954).

To a first approximation, the internal structure of the nicotinamide ring should not depend on n_{Nu}, and the internal structure of the nucleophile should not depend on n_{LG}. In NAD$^+$ hydrolysis reactions, the nucleophile water was a single atom of mass = 18 amu, with no internal structure (i.e., O–H bonds) to adjust.

The interpolation algorithm: The general case. The NAD$^+$ hydrolysis case was relatively straightforward because it was known beforehand that the TS would occur somewhere in the top left corner of reaction space (Fig. 4). In other cases, it may be necessary to use more than one pair of reference structures for different parts of reaction space. For example, in a phosphorolysis reaction, it is possible, *a priori*, that an associative mechanism is followed, with nucleophile approach leading leaving group departure. In this case, a pentacoordinate phosphorus structure may be needed as a reference structure. In elimination reactions, where $D_N + A_{xh}D_H$ (E_1), $A_{xh}D_HD_N$ (E_2), and $A_{xh}D_H + D_N$ (E_{1cB}) mechanisms are all possible, several pairs of reference structures may be needed.

Reference structures. Two reference structures are required for a structure interpolation. The first reference structure is the reactant molecule. The preferred source for the reactant molecule structure would be a crystallographic structure. As the reactant is a stable molecule, it is possible that there will be structures of that molecule, or a closely related one, available through one of the crystallographic databases (Allen *et al.*, 1991; Bernstein *et al.*, 1977). Because the majority of crystal structures do not include the coordinates of hydrogen atoms, these have to be added and their positions optimized using electronic structure calculations.

If several sources of structural information are available, all those structures should be considered in creating the first reference structure. For instance, there were several crystallographic structures of NAD$^+$ available from both small molecule and protein crystal structures. All of the crystal structures fell into two categories, having either 3′-*endo* or 2′-*endo* ribose ring puckers (Berti and Schramm, 1997). Unexpectedly, and unlike other (e.g., adenylyl) nucleotides (Gelbin *et al.*, 1996; Moodie and Thornton, 1993), the conformation of the ribose ring attached to the nicotinamide has a significant effect on bond orders. Therefore, there was a significant difference in KIEs calculated using the different conformations, and it was not possible to find a TS that matched the experimental KIEs using the 2′-*endo* conformer, or a combination of 3′-*endo* and 2′-*endo* conformers, as the reactant model. Only the 3′-*endo* reactant gave calculated KIEs that matched the measured KIEs. This was unexpected because, in general, ribose rings in nucleotides have weak conformational preferences, they interchange conformations readily, and there is little effect of ribose ring conformation on the overall structure (Altona and Sundaralingam, 1973; Moodie and Thornton, 1993; Thornton and Bayley, 1977). Coupling constants from solution NMR offered a resolution to this apparent dilemma by showing that the distribution of ring conformers in solution is skewed toward 3′-*exo* and 2′-*exo* (Oppenheimer, 1982, 1987), conformers that have not been observed in the solid state. These conformers are expected to have, like the 3′-*endo* conformer, longer C-1′–N-1 and C-4′–O-4′ bonds than the 2′-*endo* conformer, due to the anomeric effect (Berti and Schramm, 1997; Briggs *et al.*, 1984; Perrin *et al.*, 1994). Thus, it was necessary to consider structural information from a variety of sources to arrive at an appropriate first reference structure for NAD^+ hydrolysis. Because KIEs reflect a *change* in vibrational environment, the reactant molecule is as important in determining KIEs as the TS.

The second reference structure will be a computationally derived TS or a high energy (or hypothetical) intermediate structure for which electronic structure calculations are needed. Crystal structures of similar molecules may be useful. For example, earlier models (Horenstein *et al.*, 1991) of the oxocarbenium ion of the ribose ring were based in part on the structure of ribonolactone, a molecule that, like the oxocarbenium ion, contains a planar, ribose-like ring structure (Kinoshita *et al.*, 1981).

For the NAD^+ hydrolysis studies, the second reference structure consisted of two parts, the oxocarbenium ion and the nicotinamide molecule. The nicotinamide model was from a crystal structure (Wright and King, 1954), with hydrogens added computationally. The oxocarbenium ion was an *ab initio* structure, optimized at the RHF/6–31G** level of theory. The large magnitude of 2′-^{3}H KIEs of NAD^+ hydrolysis are due to hyperconjugation (Ashwell *et al.*, 1992; Berti and Schramm, 1997; Horenstein *et al.*, 1991; Sunko *et al.*, 1977). The RHF-optimized structure produced KIEs much lower than the experimental KIEs because Hartree-Fock optimizations are inherently unable to reproduce electron correlation effects, including hyperconjugation. Empirical adjustments to the oxocarbenium ion were made to reproduce the experimental 2′-^{3}H KIEs, as discussed later.

Reference structure input in CTBI. Reference structure input is in Cartesian coordinates, specifically in PDB format (Bernstein *et al.*, 1977) . The PDB format is used because all the bonds in a molecule are listed in the "CONECT" section of the file. A full connectivity list of all bonded atoms is needed for parsing and pruning the molecule. No explicit bonding information is included in the *Z*-matrix internal coordinate format as atoms may be defined with respect to atoms to which they are not bonded. Also, there is no ring closure information (see later). Finally, *Z*-matrix structures are not unique in that the same structure may be defined with different lists of internal coordinates. CTBI requires that the atom numbering and connectivities be identical in the first and second reference structures.

Both reference structures must contain all the atoms that will be present in the test TS structures. For instance, the first reference structure includes the nucleophile. The bond length is unimportant as this will be adjusted according to the list of n_{Nu}'s provided, but its orientation relative to the ribosyl ring must be correct.

In internal coordinate definitions of molecular structures, and the polar coordinate format of BEBOVIB, each successive atom in the structure is defined by reference to a previously defined atom (the first atom is the origin; its position is not defined). Strictly speaking, these methods of defining structures give only atomic positions, with no information on bonding, so atoms can, in principle, be defined in any order. However, it is common in these methods, and it is required by CTBI, that each successive atom in a structure be specified by reference to a lower numbered atom to which it is bonded (Fig. 12).

The order of atom numbering is important in CTBI for two reasons. First, in the BEBOVIB structure definition, each atom must be bonded to an already defined atom. Thus, atom 2 must be bonded to 1. Atom 3 may be bonded to 1 or 2, and so on. Second, this numbering scheme makes possible the pruning procedure for finding cyclic structures described later.

Empirical adjustments to reference structures. The example of hyperconjugation cited earlier is a specific case of a general problem: the limited accuracy of electronic structure calculations in representing real molecules. One of the major strengths of the structure interpolation approach is its ability to include alternate sources of information in creating the reference structures.

Evidence from several sources was incorporated in modeling hyperconjugation: theoretical considerations of the angular dependence (Sunko *et al.*, 1977), molecular orbital theory to explain the source and effects of hyperconjugation (Hehre, 1975), experimental KIEs from previous studies (Ashwell *et al.*, 1992; Horenstein *et al.*, 1991; Rising and Schramm, 1997), and density functional theory calculations that take some electron correlation effects into account (unpublished results). Hyperconjugation arises from the interaction of the occupied π-symmetry orbital of the β-carbon, C-2′, with the developing vacant p-orbital of the anomeric carbon as the leaving group departs (Hehre, 1975) (Fig. 13). This leads to a lengthening of the C-2′–H-2′ bond, which in turn causes the large 2′-^{3}H KIE. Because the cause

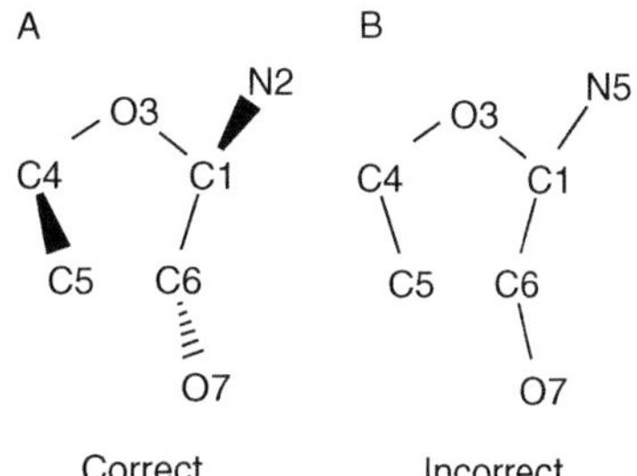

Fig. 12 Atom serial numbers in a fragment of an *N*-riboside structure for CTBI input. (a) Correct numbering, with each successive atom specified in reference to existing atoms. The C-5–C-6 bond is implicit because atom C-5 is specified by atoms C-4, O-3, and C-1, and atom C-6 is specified by atoms C-1, O-3, and C-4 (or C-1, N-2, and O-3). (b) Incorrect numbering, with C-2 specified in reference to C-1, to which it is not bonded.

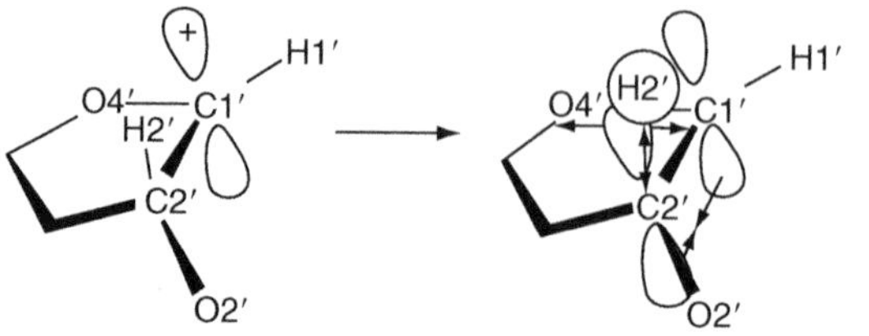

Fig. 13 Hyperconjugation of the C-2′–H-2′ bond (see text).

and many of the characteristics of KIEs arising from hyperconjugation are understood, it was possible to model this effect by adjusting the structure of the oxocarbenium ion in the second reference structure. This was done by decreasing $n_{\text{C-2}'\text{–H-2}'}$ in the oxocarbenium ion, with an equal increase in $n_{\text{C-1}'\text{–C-2}'}$ and decrease in $n_{\text{C-1}'\text{–O-4}'}$ to conserve the total bond order to C-1′. The adjustment was increased until the calculated 2′-^{3}H KIE at the TS matched the experimental KIE. The fact that the match of calculated to experimental 2′-^{3}H KIEs is excellent in all the NAD$^+$ hydrolysis and ADP-ribosylation reactions supports this approach.

Structure parsing. Because the interpolation algorithm treats the leaving group, ribosyl ring, and nucleophile differently, CTBI parses the structures to find which atoms belong to each fragment of the molecule. The program assumes that atom 1 is the electrophilic carbon (C-1′) and that atom 2 is the leaving group nitrogen (N-1). Atoms other than 1 that are bound to atom 2 are part of the leaving group; all others are part of the ribosyl moiety. At present, the nucleophile is the last atom defined so it must be only one atom.

Structure pruning to find rings. The significance of cyclic structures is that even though all the bond connectivities must be included to create a reasonable vibrational model of the molecule, one of the bonds in a cyclic structure is not defined when the structure is converted from Cartesian to internal coordinates. When a molecular structure is defined in terms of internal coordinates, one of the bonds in the ring is implicit; it is not needed to define the position of any atom (see Fig. 12). Which bond is the "ring closure" bond is somewhat arbitrary, being governed by the atoms' serial numbers, but it is important because (a) the ring closure bond must be added to the list of internal coordinates to generate a complete vibrational model, (b) BEBOVIB requires that ring closure bonds be included in the structure definition, and (c) because of the possibility of distortion of nonplanar rings (see later).

The ring closure bond. A *ring closure* bond is defined as a bond that is not required for specifying atomic positions (i.e., it is redundant for defining the molecular structure), but that is specified to close a cyclic structure. This bond is required to generate a complete list of internal coordinates (i.e., stretching, different types of bending, and torsions) for the vibrational model of the molecule.

CTBI uses a pruning algorithm to recognize the presence of a ring (or rings) and to decide which is the ring closure bond. Because a list of all bonds is provided along with the Cartesian coordinates in the input file, it is simply a matter of finding which bond is implicit when the molecular structure is defined in internal coordinates. CTBI recognizes rings by "pruning" from the terminal atoms toward the middle. Terminal atoms (with no other atoms bound to them) are deleted iteratively until there are no more terminal atoms. At this point, the molecule contains only ring atoms, or internal atoms connecting rings. Because of the way that atoms in a molecule are numbered, any atom connected *only* to lower numbered atoms is (a) a ring atom and (b) one atom in the ring closure bond. The next highest numbered atom is the other atom in the ring closure bond. In a fused ring system (e.g., purines), two or more ring closure bonds must be defined. This is done automatically by CTBI.

Distortion of nonplanar rings. Some distortion is inevitable at the site of ring closure in nonplanar rings. The problem is illustrated more easily than explained. Consider a ribosyl ring undergoing a conformational change where its ring pucker flips from 3-*endo* (Fig. 14) to 3-*exo*. Given the serial numbers of the atoms as shown in Fig. 14, the bond connecting C-5–C-6 is the ring closure bond. If a structure is generated that is halfway between the two stable conformers, then atom C5 will be moved to position C-5′ by changing the dihedral angle ∠C-1–O-3–C-4–C-5 to 0°. The C-4–C-5 bond length remains unchanged, but the implicit bond C-5–C-6 is shortened from 1.53 to 1.35 Å, giving an increase in the Pauling bond order from 0.98 to 1.80. Because bond-stretching force constants are directly proportional to bond order, in this case the bond stretch force constant for the C-5–C-6 bond will be almost doubled whereas there will be no change in the C-4–C-5 bond stretch. This will affect the vibrational frequencies involving that bond, and therefore the calculated IEs. This is an extreme example, but any structure interpolation of nonplanar rings will exhibit this type of distortion to some extent.

To avoid this problem altogether would require performing a full geometric optimization of each test TS structure, which would be too expensive computationally. Alternatively, the problem can be alleviated to a large extent by using redundant atoms. Two C-5 atoms are defined in each reference model, both with the same Cartesian coordinates. One, C-5a, is bonded to C-4, but not C-6, and its

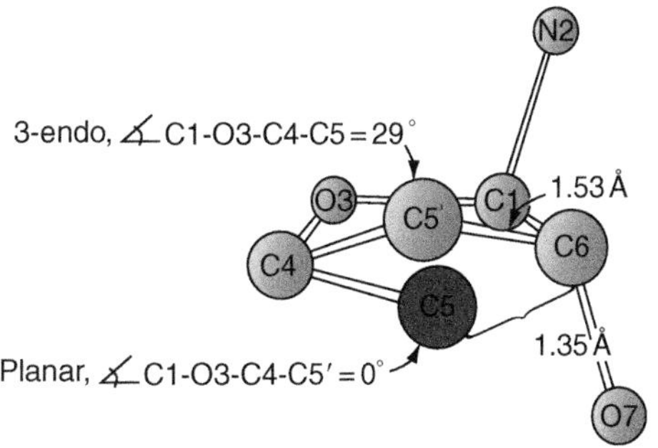

Fig. 14 Distortion of a nonplanar ring structure in structure interpolation (see text).

internal coordinates are specified relative to C-4. The other, C-5b, is bonded to C-6 (not to C-4 or C-5a), with its internal coordinates specified in relation to C-6. Force constants and vibrational frequencies involving C-4 are calculated using C-5a, similarly with C-6 and C-5b. Because neither C-4 nor C-6 is subject to the types of distortion seen in the example in Fig. 14, secondary ^{3}H KIEs can be calculated with better accuracy for hydrogens bound to C-4 and C-6.

D. Matching Test TS Structures to Experimental KIEs

We can now use the structure interpolation method to generate many test TS structures and BEBOVA analysis to calculate the KIE for each isotopic label for each test TS structure. The last step is to interpret the calculated KIEs, that is, to find which test TS structure gives calculated KIEs that most closely match the measured, experimental KIEs.

Contour diagrams. Finding the experimental TS is accomplished most conveniently by plotting contour diagrams of the calculated KIEs (Fig. 5). The top left panel of Fig. 5 shows a plot of the calculated 1-^{15}N KIE as a function of position in reaction space. The 1-^{15}N KIE depends most strongly on n_{LG} and less on n_{Nu}. The top middle panel of Fig. 5 is an expansion of the top left corner of reaction space, in the region of highly dissociative TSs. The gray shading in Fig. 5 represents the areas in reaction space where the calculated KIEs match the experimental values exactly (dark gray) or within the 95% confidence interval (light gray). Similar plots are shown for the other isotopically labeled positions. If both the KIE measurements and the modeling process were accurate, there should be (at least) one point in reaction space where the calculated and experimental KIEs match for all the labeled positions. This was the case for diphtheria toxin-catalyzed hydrolysis of NAD$^+$ (Fig. 5). The position of the experimental TS in reaction space is marked by an asterisk in Fig. 5. The calculated KIEs for the TS were within experimental error for all the isotopic positions, except for the 4′-^{18}O label, where it fell just outside of experimental error. Having found a TS that matches the experimental KIEs, it remains to (a) assess the accuracy of the structure, (b) add atoms back to the cutoff model, if necessary, and (c) analyze the structure with respect to geometry, charge distribution, and so on with an eye to explaining enzyme mechanisms.

E. TS Accuracy and Sources of Error

Broadly speaking, there are two sources of uncertainty in assessing the accuracy of the experimental TS: errors in measuring KIEs and errors in the modeling process.

Many of the issues concerning the accuracy and precision of KIE measurements are those of analytical chemistry in general, whereas some are specific to IEs, such as whether greater accuracy can be achieved by measuring isotope ratios in the products or residual substrates (Duggleby and Northrop, 1989), or the quenching of counts. Many of these issues have been discussed by Parkin (1991).

Assuming that the experimental KIEs are correct, the second source of uncertainty is in the modeling process. There is no theory to allow a systematic evaluation of the accuracy of TSs determined from experimental KIEs; one must consider each case individually to find evidence to corroborate the KIE-based TS. The accuracy of TSs derived from KIEs is difficult to assess because KIEs provide data on TSs that is not available from other sources. Approaches such as linear-free energy relationships, inhibition constants for TS analogs or crystallographic structures, do provide information on the TS, but only on a qualitative level that will support or contradict the general nature of the TS determined by KIEs, but not address the more precise, quantitative conclusions that can be achieved with KIEs.

Based on TSs solved in this laboratory, several strong enzyme inhibitors have been designed, lending support to KIE-derived TSs. These include formycin monophosphate with AMP nucleosidase ($K_d = 43$ nM) (DeWolf *et al.*, 1979), a *p*-nitrophenylamidrazone compound with nucleoside hydrolase IU ($K_d = 2$ nM) (Boutellier *et al.*, 1994), and immucillin-H with purine nucleoside phosphorylase ($K_d = 23$ pM) (Miles *et al.*, 1998). The structure of nucleoside hydrolase IU cocrystallized with a TS mimic inhibitor (Degano *et al.*, 1998) strongly corroborated the TS determined from KIEs (Horenstein and Schramm, 1993a,b; Horenstein *et al.*, 1991). These studies substantiated the conclusion that all these enzymes catalyze reactions passing through oxocarbenium ion-*like* TSs, but were unable to provide evidence to differentiate between A_ND_N and $D_N + A_N$ TSs; this is the unique domain of KIEs. These examples all represent *post facto* support for KIE-derived TSs, and none address the finer, quantitative distinctions of which KIEs are capable.

In some studies, estimates of the range of possible TSs that were consistent with the experimental KIEs were made by fitting TSs to the upper and lower limits of the error range of the experimental KIEs (Horenstein *et al.*, 1991; Markham *et al.*, 1987; Mentch *et al.*, 1987; Merkler *et al.*, 1993; Parkin *et al.*, 1991b). This approach provides an indication of how sensitive the experimental TS is to variation in the KIEs. This approach is useful because it gives some indication of the range of TSs that are possible, but it should be remembered that it assumes that the only source of error is in the KIE measurements and that the calculated KIEs for a given TS are exactly correct. The error range is also used in the structure interpolation approach in that the contour plots contain information on the exact measured KIEs (dark gray, Fig. 5) and the confidence interval (light gray). The significance of the confidence interval for a given single experimental KIE is not as clear, however, when the TS is defined as the intersection of the contour bands for many KIEs.

One promising approach that has not yet been used would take advantage of the high accuracy of calculated frequencies possible with many *ab initio* methods, particularly density functional theory methods (Pople *et al.*, 1993; Scott and Radom, 1996; Wong, 1996), to act as a reference for the BEBOVIB vibrational model. *Ab initio* optimized reactant and TS structures would not necessarily be expected to be correct, but the vibrational frequencies calculated for those structures are likely to be highly accurate. If the BEBOVIB vibrational model could duplicate the *ab initio*-based KIEs for the *ab initio* optimized reactant and TS structures that would support the accuracy of the BEBOVIB KIE calculations. If the vibrational model used with BEBOVIB is known to be accurate, that would be evidence for the accuracy of the (different) TS determined by matching BEBOVIB-based KIEs to the experimental KIEs.

Other evidence that may be considered includes the consistency of the TS with existing chemical reactivity results and the consistency of results between related TSs in a unified model. With the TS for solvolytic hydrolysis of NAD^+(Berti and Schramm, 1997), even though the experimental KIEs indicated an extremely dissociative TS with very little nucleophile bond order, it was possible to be confident of the mechanism because a large body of reactivity data on NAD^+ and related ribosides supported an A_ND_N mechanism rather than $D_N + A_N$ (Berti and Schramm, 1997). The combination of KIE and reactivity data on their own gave good confidence for the TS of the solvolytic hydrolysis reaction. The unified model used to determine the TSs of five other reactions[2] with no changes to the reference structures, interpolation algorithm, or vibrational model gave a strong indication that all are essentially correct.

Another source of confidence in the results is if the experimental KIEs force one to reach previously unexpected conclusions about the TS that can be shown to be true, or at least reasonable. Three examples are given. The first has already been discussed, the effect of 3′-*endo* versus 2′-*endo* ring conformation in the reactant on calculated KIEs. A second example is in the cholera (Rising and Schramm, 1997) and pertussis (Scheuring and Schramm, 1997a) toxin-catalyzed hydrolyses of NAD^+, for which TSs were determined before the introduction of structure interpolation, using the *ad hoc* method, without using quantum chemically derived reference structures for guidance. In those cases, it was possible to match calculated to experimental KIEs only when the total bond order to the anomeric carbon, C-1′, *increased* between the reactant and the highly dissociative TSs. The same result was observed with the TSs solved by the structure interpolation method, namely those for diphtheria toxin-catalyzed (Berti *et al.*, 1997) and solvolytic hydrolysis (Berti and Schramm, 1997). A variety of electronic structure calculations on the oxocarbenium ion support the conclusion that the π-bonding interactions formed from C-1′ to O-4′ and C-2′ at the TS are greater in sum than the loss of the *N*-glycosidic bond order (Berti and Schramm, 1997 and unpublished results). The third example is the observation of TSI in the ADP-ribosylation of

[2] The TSs for diphtheria toxin catalyzed hydrolysis (Berti *et al.*, 1997) and ADP-ribosylation of protein $G_{i\alpha1}$ (Scheuring *et al.*, 1998) have been published. Reanalysis of the KIEs hydrolyses catalyzed by cholera (Rising and Schramm, 1997) and pertussis (Scheuring and Schramm, 1997) toxins gave TSs very similar to that for diphtheria toxin reaction, and reanalysis of the KIEs for pertussis toxin-catalyzed ADP-ribosylation of a 20-mer peptide (Scheuring and Schramm, 1997) gave a TS identical to that for ADP-ribosylation of $G_{i\alpha1}$.

protein $G_{i\alpha 1}$ (Scheuring *et al.*, 1998). TSI is relatively well known in some contexts, such as proton transfer (Bernasconi, 1985, 1992a,b; Bernasconi and Wenzel, 1994), but was not expected in this case. It was impossible to match the experimental KIEs using the structure interpolation method without accounting for TSI effects in the interpolation algorithm, even though the structural differences were very small. The differences in bond lengths in the TS if TSI was neglected in the interpolation algorithm were $\Delta r_{\text{C-1}'\text{–O-4}'} = -0.028$ Å, $\Delta r_{\text{C-1}'\text{–C-2}'} = -0.012$ Å, $\Delta r_{\text{C-1}'\text{–H-1}'} = 0.006$ Å. The smallness of these changes illustrates an important property of the unified model: that changes in TS structure can be detected more sensitively than the absolute structure. In other words, the TS structures are not accurate to within 0.01 Å. They are based on reference structures derived from crystal structures and calculations, both of which have >0.01 Å errors in atom position. Whatever the accuracy may be of any one of the TS structures, deviation by one structure from the unified model is detected with great precision. The TSI effects were unexpected, but electronic structure calculations and a consideration of the changes in molecular orbital interactions on TS formation confirm their existence. The fact that such small structural changes could be detected highlights the accuracy and precision possible in TS determinations.

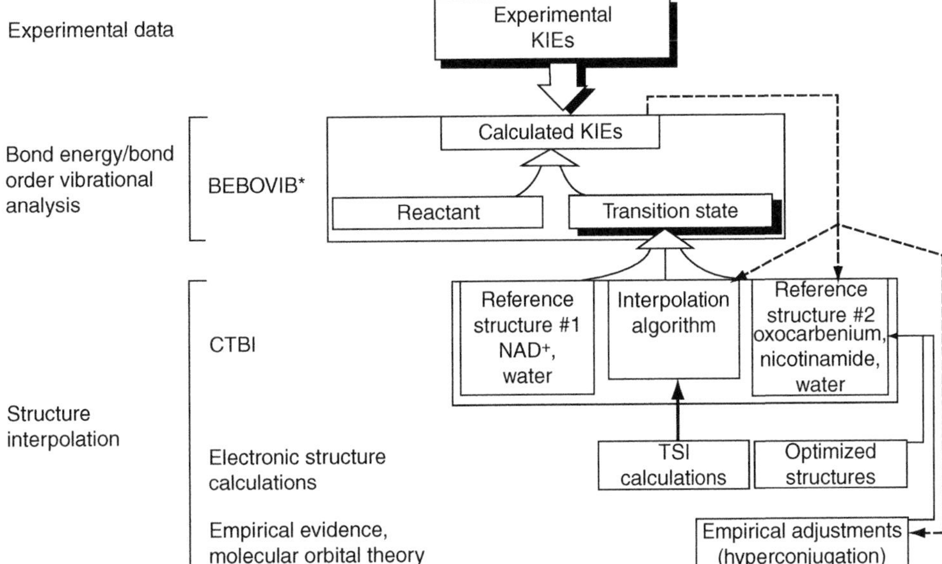

Fig. 15 Schematic depiction of the process of determining a TS. The goal of the process is to find a TS giving calculated KIEs that match the experimental KIEs. The arrows represent, roughly, the flow of information during the process. Dashed lines represent "feedback" from calculated KIEs to earlier stages in the modeling process. The left column shows the three separate parts of the process. The middle column shows the computer program or other source of information. *: The function of BEBOVIB is illustrated in Fig. 6.

IV. Summary

BEBOVA-based TS determination has been very successful in elucidating enzyme mechanisms at a level of detail that would be otherwise inaccessible. The resulting TS structures have been used successfully as the basis for designing TS mimics as enzyme inhibitors with dissociation constants to 10^{-11} M. The structure interpolation approach has systematized the process of finding a TS, increasing both the speed and accuracy of TS determination.

The combination of information from several TSs into a unified model increases the accuracy of the process significantly and results in an extremely sensitive probe of changes in TS with varying reaction conditions (i.e., enzymatic vs. nonenzymatic reactions, different enzymes, or different nucleophiles). The TS determination process is summarized in Fig. 15.

Acknowledgments

The author thanks Vern Schramm for support and advice throughout this work and for the freedom to explore new directions in TS analysis, as well as other members of the Schramm laboratory whose work is discussed here: Drs Xiangyang Chen, Robert Miles, Kathy Rising, and Johannes Scheuring. The author also thanks Drs Phil Huskey of Rutgers University and Alvan Hengge of Utah State University for their critical reading and thoughtful comments on the manuscript. This work was supported by NIH Research Grant AI34342 to Vern L. Schramm and by a postdoctoral fellowship from the Natural Sciences and Engineering Research Council (Canada) to P. J. B.

References

Allen, F. H., Davies, J. E., Galloy, J. J., Johnson, O., Kennard, O., Macrae, C. F., Mitchell, E., Mitchell, G. F., Smith, J. M., and Watson, D. G. (1991). *J. Chem. Inf. Comput. Sci.* **31,** 187.

Altona, C., and Sundaralingam, M. (1973). *J. Am. Chem. Soc.* **95,** 2333.

Ashwell, M., Guo, X., and Sinnott, M. L. (1992). *J. Am. Chem. Soc.* **114,** 10158.

Bahnson, B. J., and Anderson, V. E. (1991). *Biochemistry* **30,** 5894.

Bernasconi, C. F. (1985). *Tetrahedron* **41,** 3219.

Bernasconi, C. F. (1992a). *Acc. Chem. Res.* **25,** 9.

Bernasconi, C. F. (1992b). *Adv. Phys. Org. Chem.* **27,** 119.

Bernasconi, C. F., and Wenzel, P. J. (1994). *J. Am. Chem. Soc.* **116,** 5405.

Bernstein, F. C., Koetzle, T. F., Williams, G. J. B., Meyer, E. F., Jr., Brice, M. D., Rodgers, J. R., Kennard, O., Shimanouchi, T., and Tasumi, M. (1977). *J. Mol. Biol.* **112,** 535.

Berti, P. J., and Schramm, V. L. (1997). *J. Am. Chem. Soc.* **119,** 12069.

Berti, P. J., and Schramm, V. L. (1997).

Berti, P. J., Blanke, S. R., and Schramm, V. L. (1997). *J. Am. Chem. Soc.* **119,** 12079.

Bigeleisen, J., and Goeppert Mayer, M. (1947). *J. Chem. Phys.* **15,** 261.

Bigeleisen, J., and Wolfsberg, M. (1958). *Adv. Chem. Phys.* **1,** 15.

Boutellier, M., Horenstein, B. A., Semenyaka, A., Schramm, V. L., and Ganem, B. (1994). *Biochemistry* **33,** 3994.

Briggs, A. J., Glenn, R., Jones, P. G., Kirby, A. J., and Ramaswamy, P. (1984). *J. Am. Chem. Soc.* **106,** 6200.

Buddenbaum, W. E., and Shiner, V. J., Jr. (1976). *Can. J. Chem.* **54,** 1146.

Buddenbaum, W. E., and Shiner, V. J., Jr. (1977). *In* "Isotope Effects on Enzyme-Catalyzed Reactions" (W. W. Cleland, M. H. O'Leary and D. B. Northrop, eds.), 1University Park Press, Baltimore, MD.

Buddenbaum, W. E., and Yankwich, P. E. (1967). *J. Phys. Chem.* **71,** 3136.
Burton, G. W., Sims, L. B., Wilson, J. C., and Fry, A. (1977). *J. Am. Chem. Soc.* **99,** 3371.
Casamassina, T. E., and Huskey, W. P. (1993). *J. Am. Chem. Soc.* **115,** 14.
Cha, Y., Murray, C. J., and Klinman, J. P. (1989). *Science* **243,** 1325.
Chen, X. Y., Berti, P. J., and Schramm, V. L. Manuscript in preparation.
Cornell, W. D., Cieplak, P., Bayly, C. I., Gould, I. R., Merz, K. M., Jr., Ferguson, D. M., Spellmeyer, D. C., Fox, T., Caldwell, J. W., and Kollman, P. A. (1995). *J. Am. Chem. Soc.* **117,** 5179.
Degano, M., Almo, S. C., Sacchettini, J. C., and Schramm, V. L. (1998). *Biochemistry* **37,** 6277.
DeWolf, W. E., Jr., Fullin, F. A., and Schramm, V. L. (1979). *J. Biol. Chem.* **254,** 10868.
Duggleby, R. G., and Northrop, D. B. (1989). *Bioorg. Chem.* **17,** 177.
Fersht, A. R. (1985). "Enzyme Structure and Mechanism." Freeman, New York.
Frisch, M. J., Trucks, G. W., Schlegel, H. B., Gill, P. M. W., Johnson, B. G., Robb, M. A., Cheeseman, J. R., Keith, T., Petersson, G. A., Montgomery, J. A., Raghavachari, K., Al-Laham, M. A., *et al.* (1995). "Gaussian 94, Revision C.2, D.4." Gaussian, Pittsburgh, PA.
Gawlita, E., Paneth, P., and Anderson, V. E. (1995). *Biochemistry* **34,** 6050.
Gelbin, A., Schneider, B., Clowney, L., Hsieh, S. H., Olson, W. K., and Berman, H. M. (1996). *J. Am. Chem. Soc.* **118,** 519.
Guthrie, R. D., and Jencks, W. P. (1989). *Acc. Chem. Res.* **22,** 343.
Harris, D. C., and Bertolucci, M. D. (1978). "Symmetry and Spectroscopy: An Introduction to Vibrational and Electronic Spectroscopy." Dover Publications, New York.
Hehre, W. J. (1975). *Acc. Chem. Res.* **8,** 369.
Horenstein, B. A., and Schramm, V. L. (1993a). *Biochemistry* **32,** 7089.
Horenstein, B. A., and Schramm, V. L. (1993b). *Biochemistry* **32,** 9917.
Horenstein, B. A., Parkin, D. W., Estupinan, B., and Schramm, V. L. (1991). *Biochemistry* **30,** 10788.
Huskey, W. P. (1991). *In* "Enzyme Mechanism from Isotope Effects" (P. F. Cook, ed.), p. 37. CRC Press, Boca Raton, FL.
Huskey, W. P. (1996). *J. Am. Chem. Soc.* **118,** 1663.
Johnston, H. S. (1966). "Gas Phase Reaction Rate Theory." Ronald Press, New York.
Kinoshita, Y., Ruble, J. R., and Jeffrey, G. A. (1981). *Carbohydr. Res.* **92,** 1.
Kline, P. C., and Schramm, V. L. (1993). *Biochemistry* **32,** 13212.
Kline, P. C., and Schramm, V. L. (1995). *Biochemistry* **34,** 1153.
Kohen, A., and Klinman, J. P. (1998). *Acc. Chem. Res.* **31,** 397.
Lewis, D. E., Sims, L. B., Yamataka, H., and McKenna, J. (1980). *J. Am. Chem. Soc.* **102,** 7411.
Markham, G. D., Parkin, D. W., Mentch, F., and Schramm, V. L. (1987). *J. Biol. Chem.* **262,** 5609.
Mazzella, L. J., Parkin, D. W., Tyler, P. C., Furneaux, R. H., and Schramm, V. L. (1996). *J. Am. Chem. Soc.* **118,** 2111.
Mentch, F., Parkin, D. W., and Schramm, V. L. (1987). *Biochemistry* **26,** 921.
Merkler, D. J., Kline, P. C., Weiss, P., and Schramm, V. L. (1993). *Biochemistry* **32,** 12993.
Miles, R. W., Tyler, P. C., Furneaux, R. H., Bagdassarian, C. K., and Schramm, V. L. (1998). *Biochemistry* **37,** 8615.
Miller, B. G., Traut, T. W., and Wolfenden, R. (1998). *J. Am. Chem. Soc.* **120,** 2666.
Moodie, S. L., and Thornton, J. M. (1993). *Nucleic Acids Res.* **21,** 1369.
Northrop, D. B. (1981). *Annu. Rev. Biochem.* **50,** 103.
O'Leary, M. H. (1980). *Methods Enzymol.* **64,** 83.
Oppenheimer, N. J. (1982). *In* "The Pyridine Nucleotide Coenzymes" (J. Everse, B. Anderson and K.-S. You, eds.), p. 51. Academic Press, New York.
Oppenheimer, N. J. (1987). *In* "Pyridine Nucleotide Coenzymes" (D. Dolphin, R. Poulson and O. Avramovic, eds.), p. 185. Wiley-Interscience, New York.
Parkin, D. W. (1991). *In* "Enzyme Mechanism from Isotope Effects" (P. F. Cook, ed.), p. 269. CRC Press, Boca Raton, FL.
Parkin, D. W., and Schramm, V. L. (1987). *Biochemistry* **26,** 913.

Parkin, D. W., Horenstein, B. A., Abdulah, D. R., Estupinan, B., and Schramm, V. L. (1991a). *J. Biol. Chem.* **266,** 20658.
Parkin, D. W., Mentch, F., Banks, G. A., Horenstein, B. A., and Schramm, V. L. (1991b). *Biochemistry* **30,** 4586.
Pauling, L. (1946). *Chem. Eng. News* **24,** 1375.
Perrin, C. L., Armstrong, K. B., and Fabian, M. A. (1994). *J. Am. Chem. Soc.* **116,** 715.
Pople, J. A., Scott, A. P., Wong, M. W., and Radom, L. (1993). *Isr. J. Chem.* **33,** 345.
Radzicka, A., and Wolfenden, R. (1995). *Science* **267,** 90.
Rising, K. A., and Schramm, V. L. (1997). *J. Am. Chem. Soc.* **119,** 27.
Rodgers, J., Femec, D. A., and Schowen, R. L. (1982). *J. Am. Chem. Soc.* **104,** 3263.
Rose, I. W. (1980). *Methods Enzymol.* **64,** 47.
Scheuring, J., and Schramm, V. L. (1997a). *Biochemistry* **36,** 4526.
Scheuring, J., and Schramm, V. L. (1997b). *Biochemistry* **36,** 8215.
Scheuring, J., Berti, P. J., and Schramm, V. L. (1998). *Biochemistry* **37,** 2748.
Schramm, V. L. (1991). *In* "Enzyme Mechanism from Isotope Effects" (P. F. Cook, ed.), p. 367. CRC Press, Boca Raton, FL.
Scott, A. P., and Radom, L. (1996). *J. Phys. Chem.* **100,** 16502.
Sims, L. B., and Lewis, D. E. (1984). *In* "Isotope Effects: Recent Developments in Theory and Experiment" (E. Buncel and C. C. Lee, eds.), **6,** p. 161. Elsevier, New York.
Sims, L. B., Burton, G. W., and Lewis, D. E. (1977). "BEBOVIB-IV, QCPE No. 337." Quantum Chemistry Program Exchange, Department of Chemistry, University of Indiana, Bloomington, IN.
Singleton, D. A., and Thomas, A. A. (1995). *J. Am. Chem. Soc.* **117,** 9357.
Sinnott, M. L. (1990). *Chem. Rev.* **90,** 1171.
Stern, M. J., and Wolfsberg, M. (1966). *J. Chem. Phys.* **45,** 4105.
Stern, M. J., Spindel, W., and Monse, E. U. (1968). *J. Chem. Phys.* **48,** 2908.
Stern, M. J., Spindel, W., and Monse, E. U. (1970). *J. Chem. Phys.* **52,** 2022.
Suhnel, J., and Schowen, R. L. (1991). *In* "Enzyme Mechanism from Isotope Effects" (P. F. Cook, ed.), p. 3. CRC Press, Boca Raton, FL.
Sunko, D. E., Szele, I., and Hehre, W. J. (1977). *J. Am. Chem. Soc.* **99,** 5000.
Thornton, J. M., and Bayley, P. M. (1977). *Biopolymers* **16,** 1971.
Truhlar, D. G., and Garrett, B. C. (1980). *Acc. Chem. Res.* **13,** 440.
Truhlar, D. G., Hase, W. L., and Hynes, J. T. (1983). *J. Phys. Chem.* **87,** 2664.
Tucker, S. C., Truhlar, D. G., Garrett, B. C., and Isaacson, A. D. (1985). *J. Chem. Phys.* **82,** 4102.
Weiner, S. J., Kollman, P. A., Case, D. A., Singh, U. C., Ghio, C., Alagona, G., Profeta, S., Jr., and Weiner, P. (1984). *J. Am. Chem. Soc.* **106,** 765.
Weiss, P. M. (1991). *In* "Enzyme Mechanism from Isotope Effects" (P. F. Cook, ed.), p. 291. CRC Press, Boca Raton, FL.
Wilson, E. B., Jr., Decius, J. C., and Cross, P. C. (1955). "Molecular Vibrations: The Theory of Infrared and Raman Vibrational Spectroscopy." McGraw-Hill, New York.
Wolfenden, R., Lu, X., and Young, G. (1998a). *J. Am. Chem. Soc.* **120,** 6814.
Wolfenden, R., Ridgway, C., and Young, G. (1998b). *J. Am. Chem. Soc.* **120,** 833.
Wong, M. W. (1996). *Chem. Phys. Lett.* **256,** 391.
Wright, W. B., and King, G. S. D. (1954). *Acta Crystallogr.* **7,** 283.

CHAPTER 20

Computational Methods for Transition State and Inhibitor Recognition

Benjamin B. Braunheim* and Steven D. Schwartz†

*Department of Physiology and Biophysics
Albert Einstein College of Medicine
Bronx, New York 10461

†Departments of Physiology and Biophysics, and Biochemistry
Albert Einstein College of Medicine
Bronx, New York 10461

I. Introduction

The discovery of enzymatic inhibitors for therapeutic use is often accomplished through random searches. Hundreds of thousands of molecules are tested with refined enzyme for signs of inhibition. This process is expensive and time-consuming, costing a great deal of money and taking many years. The development of computational and theoretical methods for prediction of the binding constant of a putative inhibitor, prior to synthesis and testing, would facilitate the discovery of novel inhibitors.

CONTEMPORARY ENZYME KINETICS AND MECHANISM
Reprinted from *Methods in Enzymology*, Volume 308 (Academic Press, 1999).

DOI: 10.1016/B978-0-12-378608-1.00020-7

The first attempts to develop computational methods to predict the potency of an inhibitor prior to synthesis are described broadly as quantitative structure-activity relationship (QSAR) studies. QSARs are polynomial equations with n terms, where n corresponds to the number of ways regions of a molecule can be altered (Ariens, 1989). The value of these n coefficients are varied depending on how changes at those regions affect binding (Ariens, 1989). It is believed that once enough enzyme inhibition experiments have been done and the polynomial has been adjusted properly, it could be used to predict the inhibition constant of molecules that have not been tested experimentally (Ariens, 1989). The problems with these techniques lie in their need for the user to define a functional relationship between molecular structure and action. In the QSAR approach, or any approach where a person is charged with adjusting a mathematical model, the investigator must use variations in the structure of a molecule as the motivation for changing the value of coefficients in the model. It is not possible to predict *a priori all* the effects a change to a substrate molecule will have on enzymatic action. An example of this is the fact that variations in double bond conjugation can have only slight effects on the size and electrostatic potential of a group, but it can have a large effect on the electrostatic potential of the molecule as a whole, particularly if the enzyme polarizes the substrate on formation of the transition state. Polarizable groups of molecules can have a large effect on the electrostatic potential of a molecule, even if the only field acting on the molecule is that of the molecule itself. The following statements were made about the limitations of traditional QSAR: "With drugs one usually isn't even sure of the atom from which to base steric effects. Second, in obscurity are electronic effects. Again, with drugs part of the problem is to decide which is the key atom from which substituent effects should be measured.... To summarize, a big problem of traditional QSAR is describing molecules to the computer. This is the most time-consuming and ambiguous part of a QSAR analysis." (Martin *et al.*, 1996).

This chapter describes an approach that is different from that of a QSAR. We do not use classically derived quantities such as volume, hydrophobicity, or number of specific groups to describe molecules. We use *ab initio* quantum mechanics to describe molecules. In addition, we describe molecules as coincidentally oriented surfaces that vary in geometry and electrostatic potential. QSARs describe molecules as a collection of features that are independent of orientation. As a result of this simplification, libraries can be searched more easily with the QSAR approach, but subtle information about the bonding capability of a molecule is lost. The energy of noncovalent interactions such as ionic interactions and hydrogen bonds drops off with $1/r$. Van der Waals interactions drop off with $1/r^{12}$. As a result of these distance dependencies, the relative geometric position information of groups is vital to the task of simulating molecular recognition.

The structure of this chapter is as follows: it begins with a brief description of transition-state structures and enzyme stabilization. The concept of a transition-state analog will be central to our discovery process. We then describe how quantum mechanics and the electrostatic potential are used to discover putative inhibitors. We next describe the use of mathematical similarity measures for quantum systems in the prediction of binding strength. Application to three enzyme systems will be used to demonstrate the strength and weaknesses of the method. We then discuss the application of neural computation networks to the recognition problem. The method will be demonstrated through a study of an enzyme system not amenable to similarity measure analysis. This chapter concludes with future directions.

A. Enzyme-Stabilized Transition-State Structures

Along the reaction coordinate from reactant to product, the reactant reaches its most unstable configuration at the transition state. Transition-state stabilization theory says that enzymes increase reaction rates by stabilizing the transition-state configuration. The theory suggests that enzymes bind the transition-state structure with high energy contacts. The atomic and electronic structure of the transition state gives information about the enzyme-active site when the enzyme is in the transition-state stabilizing structure. Molecules in the transition-state structure are used in our approach because they represent the structure that the enzyme evolved to bind most tightly. With this information one can design stable molecules that mimic the geometry and electrostatic potential of the enzyme-stabilized transition state. If one were to find a stable molecule that could bind to the enzyme in the same way as the transition state, that molecule would bind strongly to the enzyme through slow onset inhibition and destroy enzymatic action.

The most basic description of a molecule is that of the quantum mechanical wave function. The quantum properties of an inhibitor are in fact what an enzyme-active site will recognize; we therefore use quantum mechanics as the descriptor of molecules. Quantum mechanics is important for simulating molecular recognition because the molecular interactions that define recognition are sensitive to subtle variations caused by intra- and intermolecular polarizations. Polarizations across conjugated bonds can have a large effect on binding energy. Polarizations of large atoms such as Br or I can have profound effects on binding. We will argue that to be effective as a recognition algorithm, any attempt to simplify the description of a molecule before presentation to a comparison algorithm must not remove information concerning the electrostatic potential on the surface of the entire molecule and the relative position of points on the surface of the molecule.

We create quantum descriptions of molecules in the following way: First, the molecular structures are energy minimized using semiempirical methods. Molecules with many degrees of freedom are configured such that they all have their flexible regions in the same relative position. Then the wave function for the molecule is calculated with the program Gaussian 94 (Frisch *et al.*, 1995). From the wave function, the electrostatic potential is calculated at all points around and within the molecule. The electron density, the square of the wave function, is also calculated at all points around and within the molecule. With these two pieces of information the electrostatic potential at the van der Waals surface can be generated. Such information sheds light on the kinds of interactions a given molecule can have with the active site (Horenstein and Schramm, 1993). Regions with electrostatic potentials close to zero are likely to be capable of van der Waals interactions, regions with a partial positive or negative charge can serve as hydrogen bond donor or acceptor sites, and regions with even greater positive or negative potentials may be involved in coulombic interactions. The electrostatic potential also conveys information concerning the likelihood that a particular region can undergo electrophilic or nucleophilic attack (Evans *et al.*, 1975). Since molecules described by quantum mechanics have a finite electron density in all space, a reasonable cutoff is required to define a molecular geometry. One choice is the van der Waals surface, within which 95% of the electron density is found. One can approximate the van der Waals surface closely by finding all points around a molecule where the electron density is close to $0.002 \pm \delta$ electrons/bohr (Frisch *et al.*, 1995). δ is the acceptance tolerance. δ is adjusted so that about 15 points per atom are accepted, creating a fairly uniform molecular surface, as shown previously (Bagdassarian *et al.*, 1996b). The information about a given molecular surface is thus described by a matrix with dimensions of $4 \times n$, where n is the number of points for the molecule and the row vector of length 4 contains the x, y, z coordinates of a given point and the electrostatic potential at that point.

II. Similarity Measures

Having defined the necessary inputs to our recognition algorithms, the quantum similarity measure will now be described. Our work with similarity measures is based on the principle that stable molecules that are similar in structure to the transition state make good transition-state inhibitors. The structure of a molecule is defined for this application as the electrostatic potential at the van der Waals surface of a molecule (Bagdassarian *et al.*, 1996a,b). The molecules are oriented for maximum geometric coincidence and the degree of similarity (both in electrostatic potential and in geometry) of the surfaces is used to generate an output. Two molecules that are quite similar will produce an output close to one, whereas two different molecules will generate an output close to zero (Bagdassarian *et al.*, 1996a, b). This method is most useful when searching for transition-state mimics. If the transition-state structure is known from heavy atom kinetic isotope effect experiments (Mentch *et al.*, 1987; Merkler *et al.*, 1993; Parkin *et al.*, 1991), the similarity measure can be used to compare ground-state molecules to the transition state. Ground-state molecules that are similar to the enzyme-induced transition state are potent inhibitors (Bagdassarian *et al.*, 1996b). A strong inhibitor can be effective even if the substrate is more concentrated in the solution by a factor of 10^6.

Molecular similarity measures derive from the idea that different molecules can be compared and the degree to which they both share the same qualities can be measured (Mezey, 1993). We have defined a similarity measure as

$$S = \frac{\int \varepsilon_A(r)\varepsilon_B(r)\mathrm{d}r}{\sqrt{\int \varepsilon_A(r)\mathrm{d}r}\sqrt{\int \varepsilon_B(r)\varepsilon_B(r)\mathrm{d}r}} \tag{1}$$

Let ε_A and ε_B be some quality of interest that the two molecules posses, such as electron density or electrostatic potential; r is the position vector. Regions within the van der Waals surface of a molecule are not important in noncovalent interactions at biological temperatures, thus this integral is computationally expensive and compares regions of the molecules that are irrelevant from a biological point of view.

We thus define a discretized similarity measure that compares points at the van der Waals surfaces of two molecules:

$$S = \frac{\sum_{i=1}^{nA}\sum_{j=1}^{nB} \varepsilon_i^A \varepsilon_j^B \exp(-\alpha r_{ij}^2)}{\sqrt{\sum_{i=1}^{nA}\sum_{j=1}^{nA} \varepsilon_i^A \varepsilon_j^A \exp(-\alpha r_{ij}^2)}\sqrt{\sum_{i=1}^{nB}\sum_{j=1}^{nB} \varepsilon_i^B \varepsilon_j^B \exp(-\alpha r_{ij}^2)}} \tag{2}$$

Here ε_i^A and ε_j^B are the electrostatic potential at point i on molecule A and point j on molecule B; r_{ij} is the distance between two points. Unlike the similarity measure that uses a point-by-point integration over all space, this similarity measure compares points on surfaces that are not coincident in space. Thus, each comparison of n points must be weighted. This is accomplished by an exponential decay. The term in the exponent is negative so all surface point comparisons that occur over large distances have less effect than those of small distances. The term is squared to reduce computation time. α acts to modulate the effects of the distance term r_{ij}; when α is small, points separated by large distances have little effect on the similarity output whereas the opposite is true for large values of α. Figure 1 shows how the similarity measure output of a molecular surface point comparison varies with α (Bagdassarian *et al.*, 1996a). This graph shows the similarity output for five inhibitors of two enzymes. (*R*)- and (*S*)-coformycin are inhibitors of AMP deaminase; they along with AMP are compared to the transition state of this reaction. Formycin and AMP are compared to the transition state of AMP nucleosidase. The important point of this graph is that the lines for any two molecules' S never cross. For any physically reasonable value of α the similarity output for two molecules will remain in the same relative order. Physically reasonable values of α range from 0.1 to 0.5 bohr^{-2}. With $\alpha = 0.1, 0.3$, and 0.5 bohr^{-2}, the exponential decays to e^{-1} for distance of 1.67, 0.97, and 0.75 Å, respectively (Bagdassarian *et al.*, 1996b). Use of similarity measures requires that the molecules be oriented for maximum geometric coincidence. This can be accomplished in a variety of different ways. In our original work (Bagdassarian *et al.*, 1996a,b), we defined a geometric similarity measure in which ε was replaced by unity. We demonstrated that maximization of this measure maximized geometric coincidence. To orient two molecules with a geometric similarity measure one assigns a unit mass to all points nA and nB and translates both molecule centers of mass to the origin. Then one assigns unit charge to all points and allows one of the molecules to undergo center of mass rotations until the similarity is maximized. One hundred thousand random reorientations were sufficient to locate the optimal orientation. This approach is computationally impractical for large molecules and does not work well for molecules that are dissimilar in structure. In such cases one

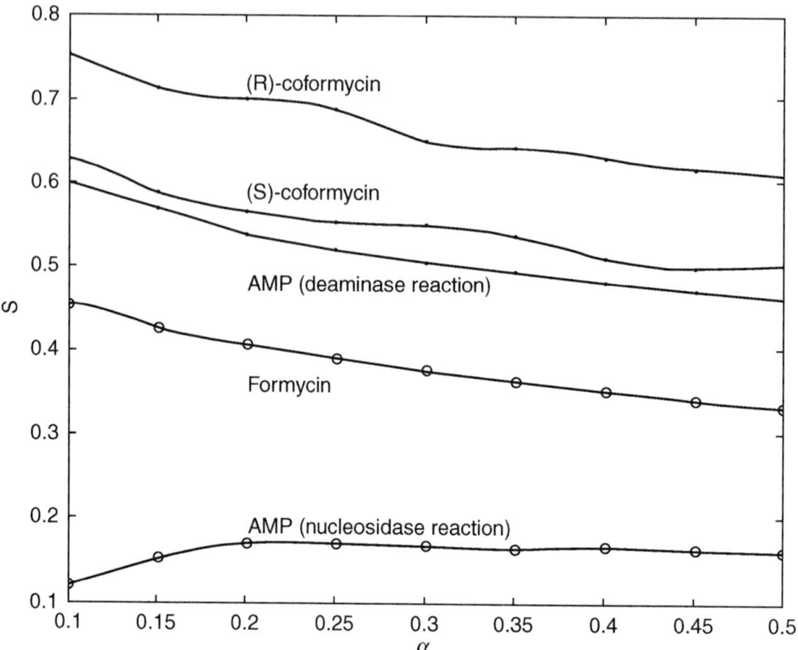

Fig. 1 Graph of similarity (S) as a function of α.

can orient molecules by choosing three common atoms in the backbone of the molecule and label them a, b, and c. a is translated to the origin and this translation is performed on b and c and all the surface points. The basis set is rotated such that b is on the positive z axis. Then the basis set is rotated so that c is in the positive x, z plane.

We now report on the application of the molecular surface point similarity measure to three different enzyme systems: adenosine deaminase, AMP nucleosidase, and cytidine deaminase.

A. Adenosine Deaminase

This enzyme facilitates the chemical hydrolysis of an amine group from adenosine converting it to inosine. The three inhibitors used here are hydrated purine ribonucleoside, 1,6-dihydropurine ribonucleoside, and (*R*)-coformycin. Stick figures of methyl derivatives of adenosine, the transition-state structure, and inhibitors are shown in Fig. 2. Enzymatic hydroxylation at C-6 and protonation at N-1 of adenosine represent the highest energetic barrier to catalysis and yield the transition-state structure. Because AMP deaminase, for which the transition-state structure is known from kinetic isotope effects (Kline and Schramm, 1994), has similar chemistry and catalytic site structure to adenosine deaminase that

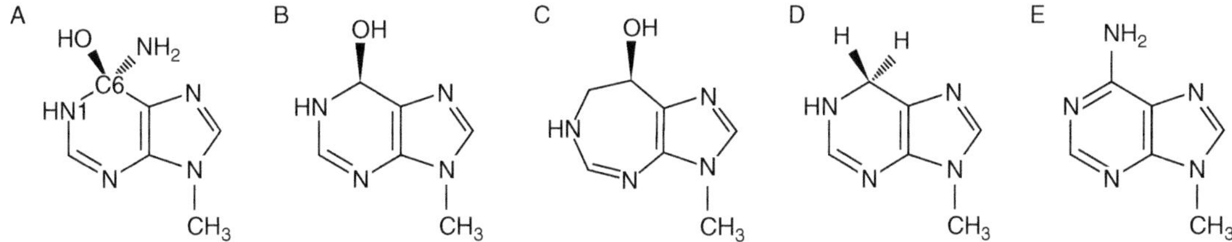

Fig. 2 Adenosine deaminase transition state (A), hydrated purine ribonucleoside (B), (*R*)-coformycin (C), 1,6-dihydropurine ribonucleoside (D), and adenosine (E).

transition state is constructed through analogy with the AMP deaminase transition state. All molecules in Fig. 2 have methyl groups in place of the ribose 5′-PO_4 (at the N-9 positions of adenosine and the transition state, at the N-3 position of (*R*)-coformycin, and at the analogous loci in the remaining two inhibitors). Those residues are constant and are not involved in the chemistry of the reaction and are assumed to affect the electrostatic potentials and geometries of the five molecules equally. Equilibrium constants for the binding of the molecules to adenosine deaminase are found in Radzicka and Wolfenden (1995) who review the original work. $K_{TX} = 1.5 \times 10^{-17}$ M for the transition state; $K_i = 3.0 \times 10^{-13}$ M for the hydrate of purine ribonucleoside; (*R*)-coformycin binds with an inhibition constant of $K_i = 1 \times 10^{-11}$ M; and 1,6-dihydropurine ribonucleoside is the weakest binding inhibitor with $K_i = 5.4 \times 10^{-6}$ M. Finally, $K_m = 3.0 \times 10^{-5}$ M for the substrate adenosine. For a visual representation of the molecular electrostatic potential surfaces for the structures shown in Fig. 2, see Bagdassarian *et al.* (1996b) and Kline and Schramm (1994).

B. AMP Nucleosidase

Figure 3 shows AMP, the transition state for the nucleosidase reaction, and three inhibitors, formycin 5′-PO_4, 4-aminopyrazolo-(3,4-*d*)pyrimidine-1-ribonucleotide, and tubercidin 5′-PO_4. The chemistry of this reaction proceeds with a protonation at the N-7 position of AMP and the inclusion of an attacking hydroxyl nucleophile at the C-1′ carbon to give the transition state. The C-1′ to N-9 bond is partially broken with a bond order of 0.2 at the transition state (Ehrlich and Schramm, 1994). Orientation of the ribose with respect to the purine across the breaking glycosidic bond is important for this reaction, so both the ribose and the phosphate group are included in the calculations. The phosphate groups are neutralized by ionic interactions in the bound form of these molecules, and quantum mechanical calculations to give the electrostatic potentials are performed with protonated phosphates (Bagdassarian *et al.*, 1996b). The equilibrium constants are (DeWolf *et al.*, 1979) $K_{TX} = 2 \times 10^{-17}$ M for the transition state of AMP; $K_i = 4.3 \times 10^{-8}$ M for formycin 5′-PO_4; $K_i = 1.0 \times 10^{-5}$ M for aminopyrazolo pyrimidine ribonucleotide; $K_i = 5.1 \times 10^{-5}$ for tubercidin 5′-PO_4; and

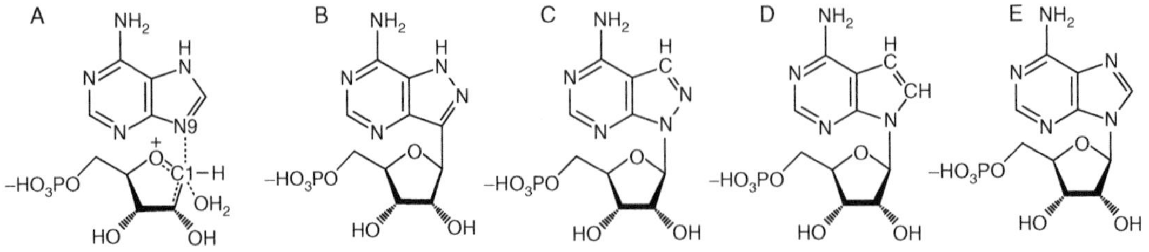

Fig. 3 AMP nucleosidase transition state (A), formycin 5′-phosphate (B), aminopyrazolo pyrimidine ribonucleoside (C), tubercidin 5′-phosphate (D), and AMP (E).

$K_m = 1.2 \times 10^{-4}$ M for the substrate AMP. More detailed biochemistry and the electrostatic potential surfaces can be found by Bagdassarian *et al.* (1996b) and Ehrlich and Schramm (1994).

C. Cytidine Deaminase

This enzyme catalyzes the hydrolysis of the amine group on cytidine to give uridine and ammonia as products. For this system, more inhibitors have been studied. Twelve molecules are used here to analyze binding to this enzyme, and the usefulness of the method is explored more fully. The detailed structure of the transition state for this enzyme has yet to be determined by isotope effect methods; nonetheless, the crystal structure of the enzyme complexed with the transition-state analog 5-fluoropyrimidin-2-one ribonucleoside is known and serves as a starting point for the construction of a model transition-state structure (Betts *et al.*, 1994). Figure 4A shows a schematic of this transition-state model for cytidine. The reaction mechanism is assumed to be similar to that for adenosine deaminase. The C-4 to O (of the attacking –OH) bond distance is constrained at 1.67 Å, corresponding to that found in the crystal structure of the enzyme-inhibitor complex. The angles to the attacking –OH and to the leaving group $-NH_2$ are modeled by analogy to the AMP (or adenosine) deaminase reaction where a detailed transition-state structure is available. The rest of the molecule is energy minimized as described later. Methyl derivatives are used again for the transition states, substrates, and inhibitors of this system. The 12 molecules in Fig. 4 are in order of decreasing binding strength to cytidine deaminase. The transition state for cytidine has a binding constant of Frick *et al.* (1989) $K_{TX} = 4 \times 10^{-16}$ M. Cytidine, the substrate, has a binding constant of Frick *et al.* (1989) $K_m = 5 \times 10^{-5}$ M. 5,6-Dihydrocytidine is another substrate with Xiang *et al.* (1995) $K_m = 1.1 \times 10^{-4}$ M, and the transition state resulting from it with $K_{TX} = 7.9 \times 10^{-10}$ M (calculated from information given in Frick *et al.* (1989). The products corresponding to these substrates are uridine and 5,6-dihydrouridine with $K_i = 2.5$ and 3.4×10^{-3} M, respectively (Frick *et al.*, 1989). The remaining inhibitors are pyrimidine-2-one ribonucleoside, which binds as the hydrate (Betts *et al.*, 1994) with a K_i estimated

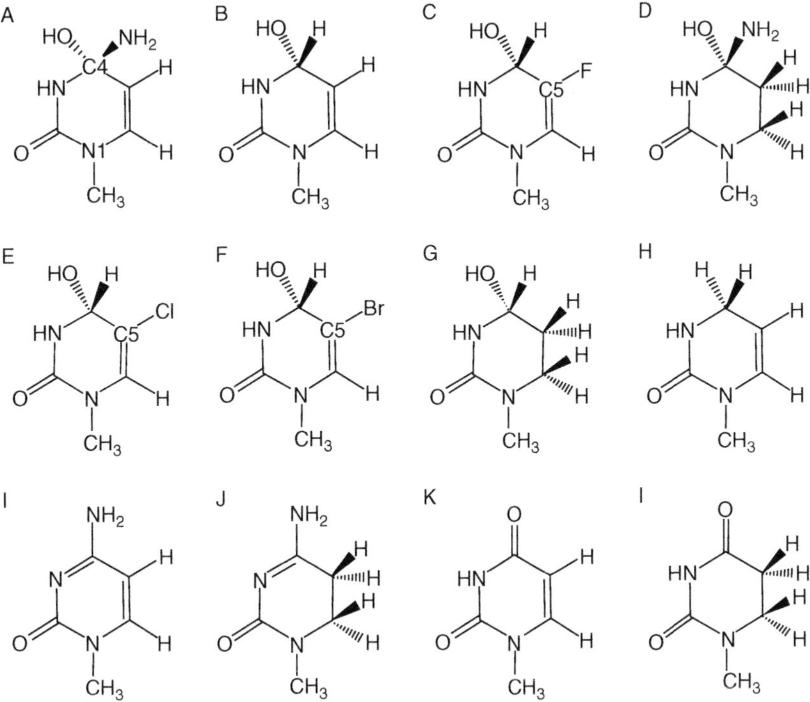

Fig. 4 Cytidine deaminase series: transition state for cytidine (A), hydrated pyrimidine-2-one ribonucleoside (B), hydrated 5-fluoropyrimidine-2-one ribonucleoside (C), transition state for 5,6-dihydrocytidine (D), hydrated 5-chloropyrimidine-2-one ribonucleoside (E), hydrated 5-bromopyrimidine-2-one ribonucleoside (F), 3,4,5,6-tetrahydrouridine (G), 3,4-dihydrozebularine (H), cytidine (I), 5,6-dihydrocytidine (J), uridine (K), and 5,6-dihydrouridine (L).

to be 2×10^{-13} M (Frick *et al.*, 1989). The bound form (hydrate) of 5-fluoropyrimidine-2-one ribonucleoside has an equilibrium constant of 3.9×10^{-11} M (Carlow *et al.*, 1996). We applied the same water addition constant offered in Carlow *et al.* (1996) to calculate the inhibition constants of 5-chloropyrimidine-2-one ribonucleoside, $K_i = 7.1 \times 10^{-9}$ M (McCormack *et al.*, 1980), and 5-bromopyrimidine-2-one ribonucleoside (McCormack *et al.*, 1980), $K_i = 1.8 \times 10^{-8}$ M. Finally, 3,4,5,6-tetrahydrouridine has $K_i = 1.8 \times 10^{-17}$ M (the average of the two quantities given by Frick *et al.*, 1989) and 3,4-dihydrozebularine has (Xiang *et al.*, 1995) $K_i = 3.0 \times 10^{-5}$ M. For a comprehensive review of cytidine deaminase and the related adenosine deaminase, see Radzicka and Wolfenden (1995).

Constrained energy minimizations for all inhibitors for the three enzyme systems were required. For both AMP nucleosidase and deaminase, on which adenosine deaminase is modeled, structures for the transition states were derived from heavy atom kinetic isotope effects (Mentch *et al.*, 1987; Merkler *et al.*, 1993; Parkin *et al.*, 1991). All inhibitor and substrate structures for these systems are energy minimized to accommodate the bonds defined by kinetic isotope methods and are influenced by the transition-state geometry. For example, the –OH group of (*R*)-coformycin is positioned and constrained to mimic the experimentally derived orientation of the transition-state –OH group. The *ab initio* calculations are done with the 3-21G basis set at the Hartee-Fock level of theory. The transition-state structure for cytidine bound to cytidine deaminase is deduced from the closely related transition state of adenosine deaminase. The remaining 11 molecules in this series are energy minimized except for substituents, which are analogous to that defined by the transition-state geometry.

Results of the application of Eq. (2) are shown in Tables I and II. The names of the compounds are listed on the far left; the experimentally determined binding free energies are listed in the second column from the left. These values are divided by the gas constant and absolute temperature, making them dimensionless. The middle column shows the output of the similarity measure. The second column from the right shows the binding energy as predicted by the similarity measure. For both AMP nucleosidase and adenosine deaminase, the binding free energies of the transition states and of the substrates are defined by experimental values. Predicted values of $\Delta G/RT$ from S are made by linear extrapolation between these values to give values for the inhibitors (Bagdassarian *et al.*, 1996b). In cases where there are few inhibitors with known affinities for the enzyme, the similarity measure was extremely effective.

Table I
Prediction of Ligand-Binding Energy by a Molecular Surface Point Similarity Measure

Enzyme/molecule	$\Delta G/RT$ (experimental)	S	$\Delta G/RT(S)$
AMP nucleosidase			
Transition state	−39	1.000	−39
Formycin	−17	0.434	−18
Aminopyrazolopyrimidine ribonucleotide	−12	0.310	−14
Tubercidin	−9.9	0.298	−13
AMP	−9.0	0.173	−9.0
	Average deviation		1.2
Adenosine deaminase			
Transition state	−39	1.000	−39
Hydrated purine ribonucleoside	−29	0.765	−27
(R)-Cofomycin	−25	0.604	−19
1,6-Dihydropurine ribonucleoside	−12	0.677	−23
Adenosine	−10	0.428	−10
	Average deviation		3.8

Table II
Prediction of Ligand-Binding Energy by a Well-Trained Neural Network for Cytidine Deaminase

Enzyme/molecule	$\Delta G/RT$ (experimental)	S	$\Delta G/RT(S)$
Cytidine deaminase			
Transition state for cytidine	−36	1.00	−36
Hydrated pyrimidine-2-one ribonucleoside	−27	0.87	−30
Hydrated 5-fluoropyrimidine-2-one ribonucleoside	−24	0.78	−26
Transition state for 5,6-dihydrocytidine	−21	0.88	−30
Hydrated 5-chloropyrimidine-2-one ribonucleoside	−19	0.70	−22
Hydrated 5-bromopyrimidine-2-one ribonucleoside	−18	0.68	−21
3,4,5,6-Tetrahydrouridine	−16	0.76	−25
3,4-Dihydrozebularine	−10	0.68	−22
Cytidine	−9.9	0.39	−9.9
5,6-Dihydrocytidine	−9.1	0.28	−5.3
Uridine	−6.0	0.58	−18
5,6-Dihydrouridine	−5.7	0.45	−12
	Average deviation		5.3

III. Similarity Measure Results and Discussion

For AMP nucleosidase, the errors in $\Delta G/RT$ as predicted by the similarity measure are 0.0 for the transition state (by construction), 1.0 for formycin, 2.0 for aminopyrazolo pyrimidine ribonucleotide, 3.1 for tubercidin, and 0.0 for AMP (again, by construction). The average deviation for the similarity measure when used in the study of these AMP nucleosidase inhibitors is 1.2. The average deviation for the adenosine deaminase inhibitors for the similarity measure is 3.8.

The similarity measure predictions for the enzyme/inhibitor system of cytidine deaminase are reported in column 3 of Table II. It is unable to give accurate predictions. Many of the molecules, especially the 5,6-hydrogenated ones, when compared with the transition state (the one for cytidine), give large errors in the predicted binding energy. The products uridine and 5,6-dihydrouridine are scored with 12 and 6.3 $\Delta G/RT$ deviation, respectively, by the similarity measure.

The study of cytidine deaminase has shown that it is crucial that the comparison algorithm be able to learn the binding rules of the system because binding energy is not always a linear function of similarity to the transitions state. The product uridine, a poor inhibitor, was predicted to be a good inhibitor. This is an example of the fact that sometimes products look like the transition state, but enzymes have evolved to bind products weakly to minimize back reactions. The similarity measure is not capable of determining complex sets of rules that govern binding energy. The study of cytidine deaminase led our group to the development of methods that could determine such rules. These learning approaches are broadly called computational neural networks and are described in Section IV.

IV. Neural Networks

A computational neural network is a computer algorithm that, during its training process, can learn features of input patterns and associate these with an output. Neural networks learn to approximate the function defined by the input/output pairs. The function is never specified by the user. After the learning phase, a well-trained network should be able to predict an output for a pattern not in the training set. In the context of the present work, the neural net is trained with a set of molecules—the transition state, substrate, and inhibitors for a given enzyme—until it can associate with every quantum mechanical description of the molecules in this set, a free energy of binding (which is the output). Then the network is used to *predict* the free energy of binding for unknown molecules.

The use of neural networks for the prediction of binding free energies becomes crucial in the case where the geometric and electrostatic similarity are poor predictors of binding energy. Enzymes interact with specific regions of the molecules they bind. Similarity measures that compare inhibitors to the transition state usually weigh all regions of the inhibitor as being equally important to the binding process. A difference between the inhibitor and transition state distant to any region involved in binding will affect the similarity output adversely, leading to a lowered measure, but this difference will not affect binding. The neural network is conditioned to *ignore* such regions. It is also desirable to use a neural network when the enzyme might be modifying the inhibitors chemically in the first stages of binding (e.g., protonation). If this is the case, the molecular features of the inhibitor that bind to the enzyme might be altered from the entity used in the calculation of similarity. In the study of cytidine deaminase the similarity measure was ineffective because the products looked like the transition state. A neural network could learn to identify product-like features in the same way that the enzyme has and to predict poor binding.

Neural networks have been used previously for the task of simulating biological molecular recognition. Gasteiger *et al.* (1994) have used Kohonen self-organizing networks to preserve the maximum topological information of a molecule when mapping its three-dimensional surface onto a plane. Wagener *et al.* (1995) have used autocorrelation vectors to describe different molecules. In that work (Wagener *et al.*, 1995), the molecular electrostatic potential at the molecular surface was collapsed onto 12 autocorrelation coefficients. Neural networks were used by Weinstein *et al.* (1992) to predict the mode of action of different chemotherapeutic agents. The effectiveness of the drugs on different malignant tissues served as descriptors, and the output target for the network was the mode of action of the drug (e.g., alkylating agent, topoisomerase I inhibitor). Tetko *et al.* (1994) used an approach similar to the autocorrelation vectors. So and Richards (1992) used networks to learn and to predict biological activity from QSAR descriptions of molecular structure. Neural networks were used by Thompson *et al.* (1995) to predict the amino acid sequence that the HIV-1 protease would bind most tightly, and this information was used to design HIV protease inhibitors.

The present work is a departure from all previous work because the quantum mechanical electrostatic potential at the van der Waals surface of a molecule is used as the physicochemical descriptor. The *entire* surface for each molecule, represented by a discrete collection of points, serves as the input to the neural network. To preserve the geometric and electrostatic integrity of the training molecules, a collapse onto a lower dimensional surface is avoided. After alignment of the inhibitor molecule for maximal geometrical overlap with the transition-state structure, the electrostatic potentials on the inhibitor surface are mapped onto a reference surface. Input patterns for a neural network are presented in the form of a vector with entries $(I_1, I_2, \ldots, I_n)$. Because the molecules are represented by a $4 \times n$ matrix, a method is needed to discard the x, y, z coordinates but to maintain spatial information. This is accomplished by mapping the surface points of every molecule onto a reference surface. Mapping ensures that similar regions on different molecules enter the same part of the neural network. To minimize the amount of geometric information loss, we augment the network with geometric information. Geometrical information is generated by a reference surface that was larger, in all regions, than the surfaces used in the study (a sphere of diameter larger than the largest molecule). During the nearest neighbor mapping, the shortest distance between each point on the reference and each point on the inhibitor is calculated. The input to the neural network includes the electrostatic potential at many points and the distance of that point from the nearest point on the reference surface. Using a reference surface larger than the other surfaces permits a uniform outward mapping. In the limit, with an infinite number of points, all mappings are normal to the surface of the inhibitor, and the mapping distances will be as small as possible. To approach this limit, a 10-fold excess of points was selected to describe the molecules. The surfaces of the molecules are described by 150 points per atom. The reference sphere that the points are mapped onto is described by a smaller number of points, 15 times the average number of atoms in the molecules of the study. As a result of mapping to the reference sphere, all molecules are described by the smaller number of points.

Computational neural networks are composed of many simple units operating in parallel. These units and the aspects of their interaction are inspired by biological nervous systems. The network function is determined largely by the interactions between units. Networks learn by adjusting the values of the connections between elements (Fausett, 1994). The neural network employed in this study is a feed forward with back propagation of error network that learns with momentum. The basic construction of a back propagation neural network has three layers: input, hidden, and output. The input layer is where input data are transferred. The link between the layers of the network is one of multiplication by a weight matrix, where every entry in the input vector is multiplied by a weight and sent to every hidden-layer neuron so that the hidden-layer weight matrix has the dimensions $n \times m$, where n is the length of the input vector and m is the number of hidden-layer neurons. A bias is added to the hidden- and output-layer neurons; it scales all the arguments before they are input into the transfer function (Fausett, 1994).

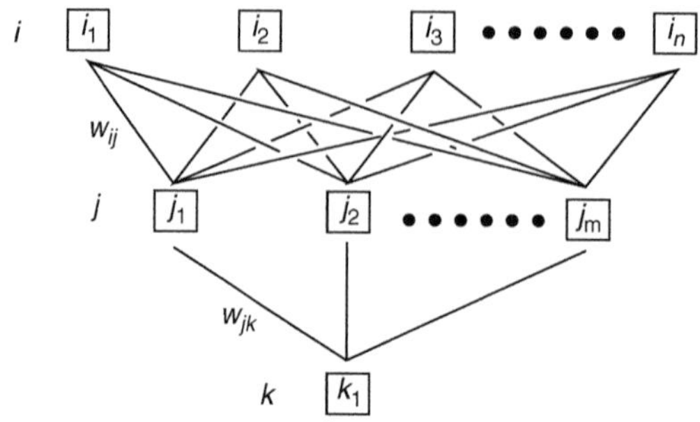

Fig. 5 Schematic of a three-layer neural network with an n dimensional input layer and a single output layer neuron.

Referring to the schematic in Fig. 5, the input layer is represented by the squares at the top of the diagram. The weights are represented by the lines connecting the layers: w_{ij} is the weight between the ith neuron of the input layer and jth neuron of the hidden layer, and w_{jk} is the weight between the jth neuron of the hidden layer and kth neuron of the output layer. In this diagram the output layer has only one neuron because the target pattern is a single number $\Delta G/RT$. The hidden-layer input from pattern number 1 for neuron j, $h_i^I(1)$, is calculated:

$$h_j^I(1) = b_j + \sum_{i=1}^{n} x_i^o(1) \times w_{\mathrm{ij}} \tag{3}$$

where $x_i^o(1)$ is the output from the ith input neuron, w_{ij} is the element of the weight matrix connecting input from neuron i with hidden-layer neuron j, and b_j is the bias on the hidden-layer neuron j. This vector h_j^I is sent through a transfer function, f. This function is nonlinear and usually sigmoidal, taking any value and returning a number between −1 and 1 (Fausett, 1994). A typical example is

$$f(h_j^{\mathrm{I}}) = \frac{1}{1 + \mathrm{e}^{-h_j^{\mathrm{I}}}} - 1 \equiv h_j^{\mathrm{o}} \tag{4}$$

The hidden-layer output, h_j^{o}, is then sent to the output layer. The output layer input o_k^{I} is calculated for the kth output neuron:

$$o_k^{\mathrm{I}} = b_k + \sum_{j=1}^{m} h_j^{\mathrm{o}} w_{\mathrm{jk}} \tag{5}$$

where w_{jk} is the weight matrix element connecting hidden-layer neuron j with output layer neuron k. The output layer output, o_k^{o}, is calculated with the same transfer function given earlier. The calculation of an output concludes the feed forward phase of training. Back propagation of error is used in conjunction with learning rules to increase the accuracy of predictions. The difference between $o_k^{\mathrm{o}}(1)$ and the target value for input pattern number 1, $t_k(1)$, determines the size and sign of the corrections to the weights and biases during back propagation. The relative change in weights and biases is proportional to a quantity δ_k:

$$\delta_k = (t_k - o_k^{\mathrm{o}}) \times f'(o_k^{\mathrm{I}}) \tag{6}$$

where f' is the first derivative of Eq. (4). Corrections to the weights and biases are calculated:

$$\Delta w_{\mathrm{jk}} = \alpha \delta_k h_j^{\mathrm{o}} \tag{7}$$

$$\Delta b_k = \alpha \delta_k \tag{8}$$

The size corrections are moderated by α, the learning rate; this number ranges from 0 to 1 exclusive of the end points. The same learning rule is applied to the hidden-layer weight matrix and biases. In most adaptive systems, learning is facilitated with the introduction of noise. In neural networks this procedure is called learning with momentum. The correction to the weights of the output layer at iteration number τ is a function of the correction of the previous iteration, $\tau - 1$, and μ, the momentum constant:

$$\Delta w_{\mathrm{jk}}(\tau) = \alpha \delta_k h_j^{\mathrm{o}} + \mu \Delta w_{\mathrm{jk}}(\tau - 1) \tag{9}$$

$$\Delta b_k(\tau) = \alpha \delta_k + \mu \Delta b_k(\tau - 1) \tag{10}$$

The same procedure is applied to hidden-layer weights and biases. The correction terms are added to the weights and biases, concluding the back propagation phase of the iteration. The network can train for hundreds to millions of iterations, depending on the complexity of the function defined by the input/output pairs. This type of back propagation is a generalization of the Widrow-Hoff learning rule applied to multiple-layer networks and nonlinear differentiable transfer functions (Rumelhart *et al.*, 1986).

Input vectors and the corresponding output vectors are used to train until the network can approximate a function (Rumelhart *et al.*, 1986). The strength of a back propagation neural network is its ability to form internal representations through the use of a hidden layer of neurons. For example, the "exclusive or" problem demonstrates the ability of neural networks, with hidden layers, to form internal representations and to solve complex problems. Suppose we have four input patterns [(0, 1) (0, 0) (1, 0) (1, 1)] with output targets [1, 0, 1, 0], respectively. A perceptron or other single layer system would be unable to simulate the function described by these four input/output pairs. The only way to solve this problem is to learn that the two types of inputs work together to affect the output. In this case the least similar inputs cause the same output, and the more similar inputs have different outputs (Rumelhart *et al.*, 1986). The ability required to solve this problem is not unlike that required to find the best inhibitor when it does not share all the same quantum features of the transition state. It is this inherent ability of neural networks to solve complex puzzles that makes them well conditioned for the task of simulating biological molecular recognition. As a result of the mapping during input preparation, each input neuron is given information (electrostatic potential or geometry) about the nearest point on the surface of the inhibitor. It is as though each input neuron is at a fixed point on the sphere around the inhibitors, judging each inhibitor in the same way the enzyme-active site would.

The left-most columns of Table III show the results of the neural network approach in the study of cytidine deaminase. The (11) column shows the results of a neural network that was trained with 11 of the 12 molecules used in this study. The neural network predicted the binding energy of only one of the molecules and its value is reported in this column. The (7) column reports the results of neural network predictions when the network was trained with seven molecules. Molecules left out of the training set were chosen at random such that they represented a large range in binding energy and that they were spaced evenly throughout this range. A situation similar to the "exclusive or" problem was encountered in the study of cytidine deaminase. The products uridine and 5,6-dihydrouridine are similar structurally to the transition state and found to be so by the similarity measure. This makes the similarity measure predictions error prone because binding energy is not a linear function of similarity to the transition state. As with the "exclusive or" problem, the network was able to learn that two similar inputs can cause very different outputs.

An even better example of the use of the neural network recognition method is found in a study of the inosine-uridine-preferring nucleoside hydrolase (IU-NH). Protozoan parasites lack *de novo* purine biosynthetic pathways and rely on the ability to salvage nucleosides from the blood of their host for RNA and DNA

Table III
Prediction of Ligand-Binding Energy by a Well-Trained Neural Network for Cytidine Deaminase

				$\Delta G/RT$ (neural net)	
Enzyme/molecule	$\Delta G/RT$ (experimental)	S	$\Delta G/RT$ (S)	(11)	(7)
Cytidine deaminase					
Transition state for cytidine	−36	1.00	−36	−30	
Hydrated pyrimidine-2-one ribonucleoside	−27	0.87	−30	−27	−26
Hydrated 5-fluoropyrimidine-2-one ribonucleoside	−24	0.78	−26	−19	
Transition state for 5,6-dihydrocytidine	−21	0.88	−30	−23	
Hydrated 5-chloropyrimidine-2-one ribonucleoside	−19	0.70	−22	−19	−18
Hydrated 5-bromopyrimidine-2-one ribonucleoside	−18	0.68	−21	−17	
3,4,5,6-Tetrahydrouridine	−16	0.76	−25	−15	
3,4-Dihydrozebularine	−10	0.68	−22	−12	−13
Cytidine	−9.9	0.39	−9.9	−8.3	
5,6-Dihydrocytidine	−9.1	0.28	−5.3	−7.5	
Uridine	−6.0	0.58	−18	−6.1	−9.9
5,6-Dihydrouridine	−5.7	0.45	−12	−6.2	
	Average deviation		5.3	1.7	2.2

synthesis (Hammond and Gutteridge, 1984). The IU-NH from *Crithidia fasciculata* is unique and has not been found in mammals (Degano *et al.*, 1998). This enzyme catalyzes the *N*-ribosyl hydrolysis of all naturally occurring RNA purines and pyrimidines (Degano *et al.*, 1998). The active site of the enzyme has two binding regions: one binding ribose and other the base. The inosine transition state requires $\Delta\Delta G = 17.7$ kcal/mol activation energy: 13.1 kcal/mol are used in activation of the ribosyl group and only 4.6 kcal/mol are used for activation of the hypoxanthine-leaving group (Parkin *et al.*, 1997). Analogs that resemble the inosine transition state have proved to be powerful competitive inhibitors of this enzyme both geometrically and electronically and could be used as antitrypanosomal drugs (Degano *et al.*, 1998).

The transition state for these reactions feature an oxocarbenium ion achieved by the polarization of the C-4′ oxygen C-1′ carbon bond of ribose. The C-4′ oxygen is in proximity to a negatively charged carboxyl group from Glu-166 during transition-state stabilization (Degano *et al.*, 1998). This creates a partial double bond between the C-4′ oxygen and C-1′ carbon, causing the oxygen to have a partial positive charge and the carbon to have a partial negative charge. Nucleoside analogs with iminoribitol groups have a secondary amine in place of the C-4′ oxygen of ribose and have proven to be effective inhibitors of IU-NH.

IU-NH acts on all naturally occurring nucleosides (with C-2′ hydroxyl groups). The lack of specificity for the leaving groups results from the small number of amino acids in this region to form specific interactions: Tyr-229, His-82, and His-241 (Degano *et al.*, 1998). The only crystal structure data available concerning the configuration of bound inhibitors were generated from a study of the enzyme bound to *p*-aminophenyliminoribitol (pAPIR) (Fig. 6, t4). Tyr-229 relocates during binding and moves above the phenyl ring of pAPIR. The side chain hydroxyl group of Tyr-229 is directed toward the cavity that would contain the six-member ring of a purine, were it bound (Degano *et al.*, 1998). His-82 is 3.6 Å from the phenyl ring of pAPIR and in the proper position for positive charge-π interactions to occur (Degano *et al.*, 1998). His-241 has been shown to be involved in leaving-group activation in the hydrolysis of inosine, presumably as the proton donor in the creation of hypoxanthine (Degano *et al.*, 1998).

A similarity measure does not work well in the prediction of binding energy of inhibitors of IU-NH because the enzyme has many substrates. The chemistry involved in catalyzing the scission of the *N*-ribosidic bonds of these substrates is quite different. Therefore, the transition states for the various substrates are quite different. Because there is only one known transition-state structure, that of inosine, it would be unwise to presume that the binding energy of all inhibitors would be a linear function of similarity only in the inosine transition state. The neural network has the capacity to learn the binding rules of this enzyme if enough information is present in the training set.

To study this enzyme, inhibitors with different base and ribose analogs were used. The crystal structure of the enzyme bound to pAPIR (Degano *et al.*, 1998) shows the relative position of the phenyl group with the iminoribitol group. This

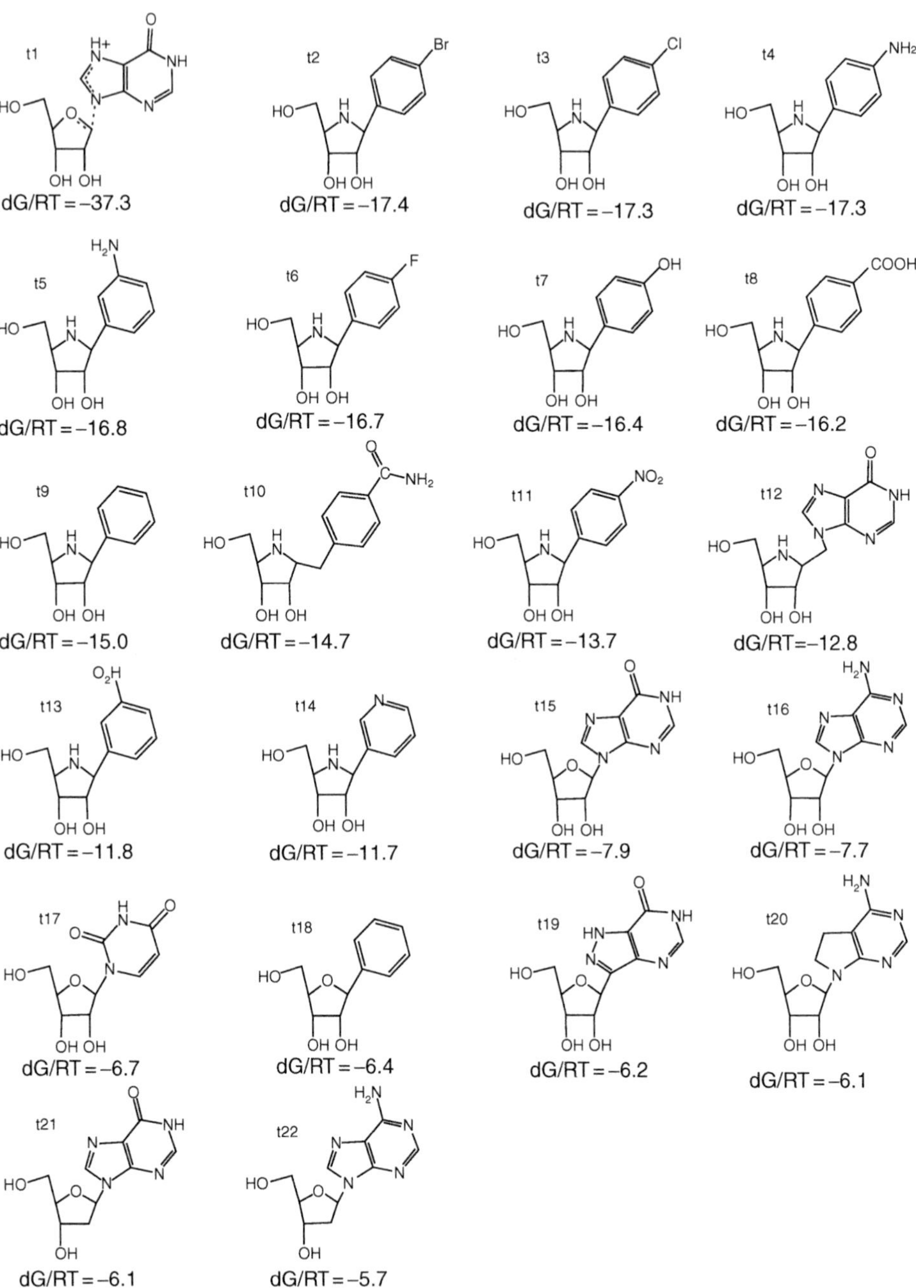

Fig. 6 Molecules in the training set, binding energies in dimensionless units of $\Delta G/RT$. Most of these values are in Ref. 37, the others have yet to be published.

compound is identical on rotation of 180°. This being the case, the structure offers little information as to how molecules that are ambiguous on rotation should be oriented. Molecule t1, the transition state, probably binds to the enzyme in a particular configuration with the hypoxanthine group extended either over the C-2′ carbon of the ribose group or in the opposite direction, but there is no way to tell which of these is correct from existing data. Subtle orientation discrepancies are unlikely to introduce sufficient misinformation (if the orientation is conserved for all of the molecules) for this to be a problem for the neural network. Both orientations are used to demonstrate that the neural network is not dependent on the selection of a "correct" orientation. The neural network was trained with the patterns of each molecule in both orientations. The network was trained twice with the patterns of molecules that are identical on rotation of 180°; this prevents the network from being biased to the other patterns.

In addition to using both configurations about the ribosidic bond, there are other degrees of freedom that have to be consistent for all molecules. We assume that the enzyme will bind all molecules in a similar low energy conformation. The neural network need not know the configuration as long as the conformation of all molecules we present to the neural network is consistent. The known crystal structure of pAPIR bound to IU-nucleoside hydrolase is used as the model conformation.

The inosine transition-state structure is stabilized by a negatively charged carboxyl group within the active site 3.6 Å from the C-4′ oxygen (Horenstein *et al.*, 1991). To simulate this aspect of the active site, we included a negatively charged fluoride ion (at the same relative position of the nearest oxygen of the carboxyl group) in the calculations of the electrostatic potential at the van der Waals surface.

In an attempt to simulate a real-world application of the use of our method in drug discovery, the structures of 41 molecules were used but the binding energy of only 22 of the molecules was known at the time of study. The binding energies of the other 19 molecules were yet to be determined. The 22 molecules in the training set included eight substrates and substrate analogs, the transition state of inosine, 11 inhibitors from a previous publication, and 2 inhibitors with known but unpublished binding energies. The neural network was trained with 22 molecules as inputs. The output (training target) is the binding free energy of the molecule with the enzyme.

Training a neural network requires variation of four adjustable parameters: number of hidden-layer neurons, learning rate, momentum constant, and number of training iterations. The only way to tell that a network is well trained is to minimize the training set prediction error. This can be calculated by taking the difference between the target_i value for a molecule (experimentally determined binding energy) and the number the neural network predicted for that pattern_i and summing the absolute value of this number for all the molecules in the training set. As training progresses, the training set prediction error will decrease. Minimizing training set error is not without negative consequence; overtraining occurs when the network trains for too many iterations, has too many hidden-layer neurons, has too large of a learning rate, or has too small of a momentum constant. The only way to tell that a neural network has not been overtrained is to have it make a prediction for a pattern not in the training set, that is, see if the network can generalize from the information contained in the input/output pairs of the training set and apply that information to a molecule it has not trained with. In accommodation of this fact we employ 1 of the 22 training set molecules as an adjuster molecule. This molecule is left out of the training set during training and is used to check if the neural network was overtrained. The procedure is to train the neural network until the prediction set error has decreased into a plateau. We then end training and test the neural network with the adjuster molecule. If the neural network predicts the binding energy of the adjuster molecule within 5%, then the construction of that neural network is saved; if the prediction is more than 5% off, a new construction is chosen. This procedure is repeated until a construction is found that allows the neural network to predict the binding energy of the adjuster molecule within 5%. This is done for all 22 molecules in the training set, and these 22 best neural network constructions are used to predict binding energies for the 19 molecules in the predicting set. The final reported result is the average of the predictions for each of the network constructions.

A. IU-NH Neural Network Results

Figure 6 shows the structure of the 22 molecules in the training set with their experimental binding energies with IU-NH. Figure 7 shows the 19 molecules in the predicting set, with their experimentally determined binding energies (*E*) and their predicted binding energies (*P*). Figure 8 shows a graphic representation of data: prediction versus experimentally determined binding energy. It is clear that the method is effective over the entire range of binding energies. The method is able to distinguish between inhibitors that bind with a little, medium, and large amount of energy. The inhibitors in the predicting set were tested only experimentally in ranges below -9.81 $\Delta G/RT$ binding energy, so any predictions above this number are considered to have zero deviation if the experimental binding energy is greater than -9.81 $\Delta G/RT$. The experimental dimensionless binding energy range is 31.6 $\Delta G/RT$, the average deviation is 0.90 $\Delta G/RT$, and the largest deviation is 2.69 $\Delta G/RT$.

The accuracy of the results leads us to believe that the method was able to create a mathematics model that describes the function defined by the input-output pairs. There are specific successes that are the product of synthetic reasoning on the part of the neural network that warrant discussion. The prediction for prediction set molecule number one (p1 in Fig. 7) was an example of such synthetic reasoning. Training set molecule number eleven (t11 in Fig. 6) and t13 have nitro groups at the *para* and *meta* positions, respectively. Nitro groups, such as fluoro groups (like that of p1), are capable of electron withdrawing and little electron donating. The neural network could have learned about the negative effects electron-withdrawing substitutions to pyrimidine analogs have on binding from nitro-substituted inhibitors in the training

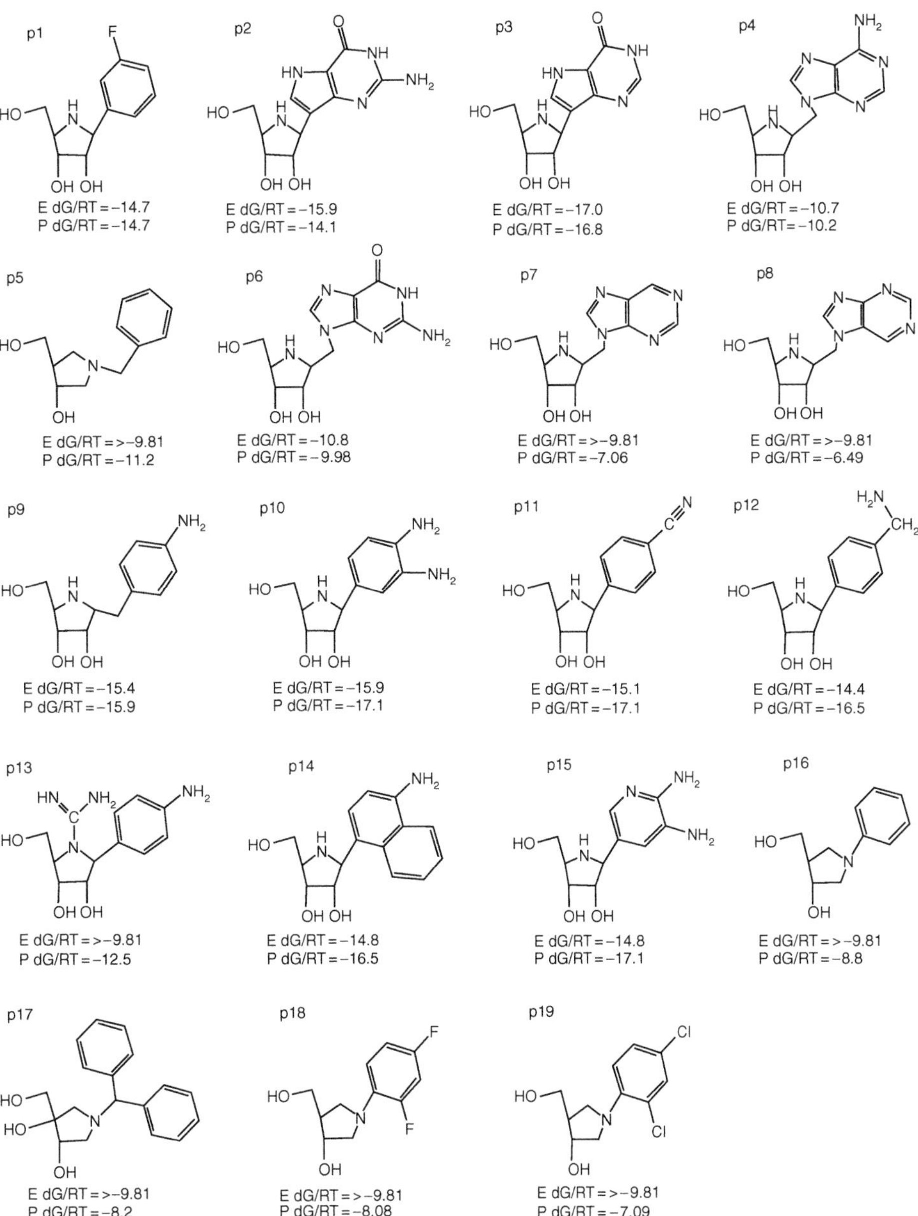

Fig. 7 Molecules in the predicting set. Experimentally determined binding affinities (*E*) and prediction (*P*).

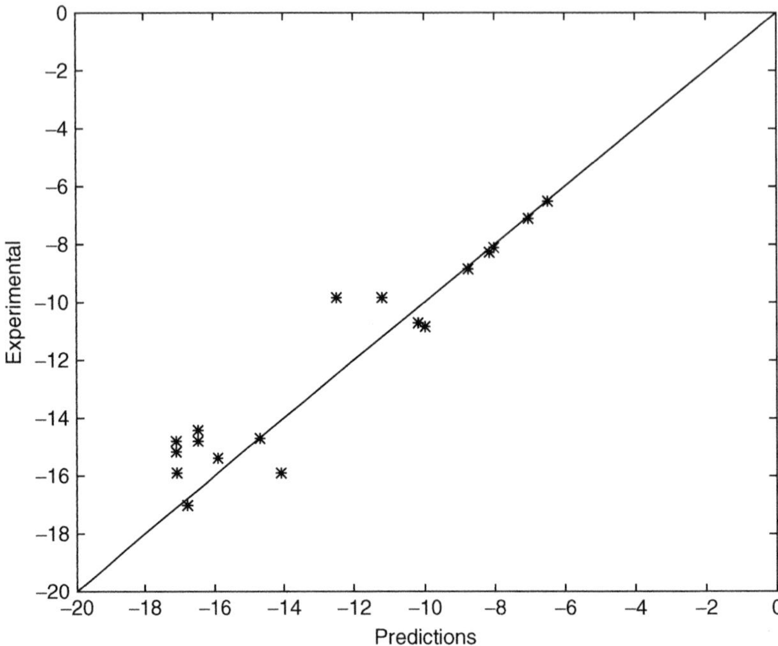

Fig. 8 Neural network predictions versus experimental data for IU-nucleoside hydrolase.

set. The algorithm also learned of the negative effects of *meta* versus *para* substitutions from studying the nitro inhibitors. Because *p*-flourophenyliminoribitol (t6) was in the training set, the network was able to combine these two pieces of information and subtract about 2 $\Delta G/RT$ from the binding energy of t6 when generating a prediction of *m*-fluorophenyliminoribitol (p1). The prediction of p2 is slightly more complicated, as there are no guanosine analogs in the training set, but the neural network could have learned about the negative effects of electron-donating groups to purine analogs from the implied structure-activity relationship present in the training set: t1 > t15 > t16. Prediction set molecule number 2 has an electron-donating amine group at C-4 of the purine six-member ring. This group is the only difference between p2 and p3; notice the difference between the predictions. The prediction of p3 was also generated by synthetic reasoning. p3 has three features that are conducive to tight binding: an electron-withdrawing group at C-6, a partially positive charged hydrogen at N-7, and an iminoribitol group. The neural network combined these pieces of information and predicted a large binding energy. The prediction of binding energy for p4 also required a combination of facts. Prediction set molecule 4 has two qualities that make it a poor inhibitor of IU-NH: (1) p4 has a substituted C-1′–N-9 glycosidic bond, the neural network could have learned about the negative effects of this feature from t10 and t12 and (2) p4 has an electron-donating group at C-6. As already mentioned, these groups decrease the binding of purine analogs. Prediction set molecule 4 does have an iminoribitol group that is a tight binding group, but overall the molecule is a poor inhibitor and is identified as such by the neural network.

These results show that this quantum neural network technology can make completely blind predictions of enzyme/inhibitor-binding constants in a complex biochemical system. To highlight the nature of this accomplishment, we examined the different nature of transition-state structures for the two different kinds of substrates: purines and pyrimidines. The transition state of inosine is the only one for which there is a determined structure, which is shown in Fig. 6 (t1). The transition-state structure for inosine is created by polarization of the ribosyl group across the C-4′ oxygen C-1′ bond and protonation of N-7 of the purine group. This protonation would be impossible when dealing with the pyrimidine uridine, t17, as there is no place for this group to receive a proton (the electrons of N-3 are involved in the ring conjugation). Therefore, it is clear that these two types of substrates have quite different transition-state structures and that the rules of binding pyrimidine analogs are quite different from those of binding purines. For pyrimidine analogs, the binding energy tends to decrease with increasingly electron-withdrawing substitutions. The opposite trend is seen with purine analogs. From the accuracy of the results, it is clear that the neural network was able to distinguish between purine and pyrimidine analogs and to apply the correct rules when evaluating the two families in the predicting set.

V. Molecular Importance and Hidden-Layer Magnitudes

We have devised a way to probe trained neural networks and get information about what is most important about the molecules and what gives them their efficacy. This method can aid in the design of new inhibitors. This technique probes part of a trained neural network directly and reveals which parts of the molecules the neural network found most important.

This technique relies on the fact that the weights in the hidden layer associated with regions of the input important in binding have large absolute values. The network is presented with an input pattern and an output; to minimize the error the network must recognize regions that change and affect the binding energy relative to those regions that change and do not affect binding energy. This recognition occurs when weights of the neural network are adjusted so that important regions are multiplied by large weights and unimportant regions are multiplied by small weights. Documentation of this behavior is made by inspection of the absolute values of every number in the hidden-layer weight matrix of a trained network. The matrix is collapsed into a vector V_i by summing on j where $j = 1, \ldots, m$ and m is the number of hidden-layer neurons:

$$V_i = \sum_{j=1}^{n} |w_{ij}| \tag{11}$$

and where i refers to the input surface points. Large values for V_i represent regions found to be important to the neural network, whereas, small values represent regions found to be unimportant. Figure 9 (see color insert) shows a colored graphic representation of the hidden-layer weight matrix of a neural network trained with the molecules of the cytidine deaminase series. All the patterns were generated by mapping the electrostatic potential of different molecules onto a 5-bromo transition-state surface (instead of a sphere as in the IU-NH study). The common 5-bromo transition-state geometry is used to identify those regions on the molecules found as most important by the neural net. This is represented by coloring points on a van der Waals surface with large V_i values yellow and regions with small values blue. Regions on molecular surfaces with intermediate weights have a mixture of the two colors.

Figure 9A (see color insert) shows the first half of the hidden-layer weight matrix of the neural network associated with the electrostatic potential used in the study of cytidine deaminase. Figure 9B (see color insert) shows the bottom half of that

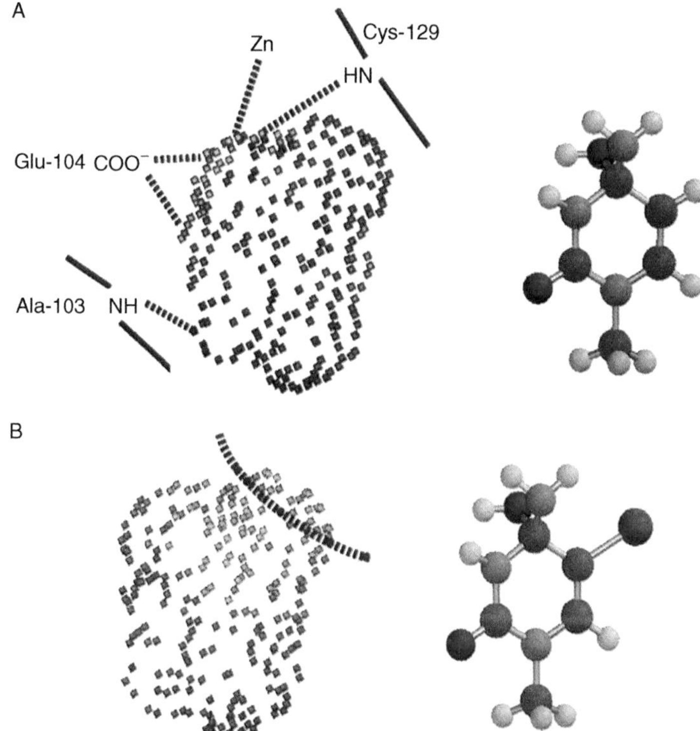

Fig. 9 Representation of the hidden-layer weight matrix of a trained neural network. It provides evidence that the neural network found the regions of the inhibitors, which are important in binding, to be important in the calculation. The range of these values is from 0.118 to 15.70. (See Plate no. 2 in the Color Plate Section.)

weight matrix associated with the geometry of the van der Waals surface. On inspection of Fig. 9A, the attacking hydroxyl nucleophile and protonated N-3 were both characterized by large weights and therefore were deemed important to the neural network. It should be noted that C-6 has large weights, which is likely caused by the fact that the bond between C-6 and C-5 is saturated and unsaturated in various inhibitors and is responsible for large effects on binding energy. Figure 9B shows that the geometry of the C-5 region is the only region that affects binding greatly. This is consistent with crystallographic data, as there is a non-binding region of the enzyme that interferes sterically with inhibitors that have large substitutions at that region. Inhibitors with increasingly bulky substitutions at C-5 have decreasing binding energies.

VI. Conclusions

The very accurate predictions of inhibitor potency of IU-nucleoside hydrolase could not have been done by any method that did not use quantum mechanically derived features. In a QSAR study the IU-NH inhibitors would have been described to the computer in terms of their volumes, hydrophobicity, H-bonding donor and acceptor sites, and other experimentally or classically derived features. These descriptions would not have provided the information necessary to model the binding preferences of IU-NH. This is demonstrated by inspection of the weaker inhibitors in the training set. 3-Pyridineiminoribitol (t14) is the weakest binding substituted iminoribitol: it binds with less energy than iminoribitol alone (iminoribitol, $\Delta G/RT = -12.3$; 3-pyridineiminoribitol, $\Delta G/RT = -11.7$). It is not likely that the pyridine group is interacting with the leaving-group binding region with a positive $\Delta\Delta G$. It is more likely that the quantum mechanical nature of pyridine is affecting the electrostatic potential of the iminoribitol group and decreasing its ability to bind the enzyme. Because subtle quantum mechanical differences between the molecules dictate the binding preferences of IU-NH, an approach that uses quantum mechanically derived features is required.

The molecular surface point similarity measure is most effective when there are few inhibitors with known inhibition constants for the enzyme. The small training sizes of adenosine deaminase and AMP nucleosidase made it difficult to train the neural network, and neural network predictions for these systems were less accurate than those generated by the similarity measure. The IU-nucleoside hydrolase enzyme system and its inhibitors were studied by our group with the molecular surface point similarity measure. The similarity measure worked when purine analogs were compared to the inosine transition state. The binding strengths of pyrimidine analogs were not predicted accurately by the similarity measure when the analogs were compared to the inosine transition state. The reason is that the binding rules for this enzyme are different for purines and pyrimidines.

The most important result is that even though we do not know *a priori* what important molecular quantum features are needed to bind IU-NH, the information necessary to know this was presented adequately to the neural network during training, and this information was used by the neural network to make accurate predictions. Our method uses quantum mechanical descriptions of molecules, and with this information the neural network functions as the ultimate structure/activity relationship machine. It is able to determine what quantum structural features are of importance and form a highly accurate association of these with binding energy.

This is not a surprise from a mathematic sense, as the most common application for neural networks in engineering is in pattern classification and the creation of priority trees. The neural network first identifies that there are two or more groups of molecules that have one mutually exclusive characteristic that has an effect on the output. For our purposes, this characteristic could be the fact that purines have a double ring structure and pyrimidines have a single ring. Once two distinct groups of molecules have been identified, rules for evaluating the two can be learned, and the rules of one group can be the same or different from the rules of the other group. Examples of this from the IU-nucleoside hydrolase study are methyl substitutions that lengthen the glycosidic bond, decrease binding energy for purines and pyrimidines, but extremely electronegative groups (such as protons and double-bonded oxygens) increase the binding energy for purines ($\Delta G/RT$ for t1 $>$ t19 $>$ t20) but electron-withdrawing groups decrease the binding energy of pyrimidines ($\Delta G/RT$ for t2 $<$ t6 $<$ t11). As shown earlier with the "exclusive or" problem, the ability of a neural network to solve a problem when it appears to contain contradictory information is not a new finding. What is new is that this ability is being employed to solve a real-world problem in biochemical engineering.

The ability of the neural network to model the system defined by the input/output pairs is potentially useful for the search of chemical libraries. It might also be useful to aid in the design of new drugs. In the cytidine deaminase system, the weight matrix of the network was adjusted such that regions important in binding were owned by large weights. This was found to be true for the two kinds of information in the input. This technique could be helpful to researchers, giving them the ability to know what regions are to be avoided sterically and what regions are important in terms of electrostatic potential.

References

Ariens, E. J. (1989). "QSAR: Quantitative Structure-Activity Relationships in Drug Design." p. 3. A. R. Liss, New York.

Bagdassarian, C. K., Braunheim, B. B., Schramm, V. L., and Schwartz, S. D. (1996a). *Int. J. Quant. Chem. Quant. Biol. Symp.* **23,** 73.

Bagdassarian, C. K., Schramm, V. L., and Schwartz, S. D. (1996b). *J. Am. Chem. Soc.* **118,** 8825.

Betts, L., Xiang, S., Short, S. A., Wolfenden, R., and Carter, C. W., Jr. (1994). *J. Mol. Biol.* **235,** 635.

Carlow, D. C., Short, S. A., and Wolfenden, R. (1996). *Biochemistry* **35,** 948.

Degano, M., Almo, S. C., Sacchettini, J. C., and Schramm, V. L. (1998). *Biochemistry* **37,** 6277.

DeWolf, W. E., Fullin, F. A., and Schramm, V. L. (1979). *J. Biol. Chem.* **254,** 10868.

Ehrlich, J. L., and Schramm, V. L. (1994). *Biochemistry* **33,** 8890.

Evans, B. E., Mitchell, G. N., and Wolfenden, R. (1975). *Biochemistry* **14,** 621.
Fausett, L. (1994). "Fundamentals of Neural Networks." Prentice-Hall, New Jersey.
Frick, L., Yang, C., Marquez, V. E., and Wolfenden, R. (1989). *Biochemistry* **28,** 9423.
Frisch, M. J., Trucks, G. W., Schlegel, H. B., Gill, P. M. W., Johnson, B. G., Robb, M. A., Cheeseman, J. R., Keith, T., Petersson, G. A., Montgomery, J. A., Raghavachari, K., Al-Laham, M. A., *et al.* (1995). "Gaussian 94, Revision C.2" Gaussian, Pittsburgh, PA.
Gasteiger, J., Li, X., Rudolph, C., Sadowski, J., and Zupan, J. (1994). *J. Am. Chem. Soc.* **116,** 4608.
Hammond, D. J., and Gutteridge, W. E. (1984). *Mol. Biochem. Parasitol.* **13,** 243.
Horenstein, B. A., Parkin, D. W., Estupinan, B., and Schramm, V. L. (1991). *Biochemistry* **30,** 10788.
Horenstein, B. A., and Schramm, V. L. (1993). *Biochemistry* **32,** 7089.
Kline, P. C., and Schramm, V. L. (1994). *J. Biol. Chem.* **269,** 22385.
Martin, Y., Kim, K., and Lin, T. C. (1996). "Advances in Quantitative Structure-Property Relationships," p. 3. JAI Press, Greenwich, CT.
McCormack, J. J., Marquez, V. E., Liu, P. S., Vistica, D. T., and Driscoll, J. S. (1980). *Biochem. Pharmacol.* **29,** 830.
Mentch, F., Parkin, D. W., and Schramm, V. L. (1987). *Biochemistry* **26,** 921.
Merkler, D. J., Kline, P. C., Weiss, P., and Schramm, V. L. (1993). *Biochemistry* **32,** 12993.
Mezey, P. G. (1993). "Shape in Chemistry: An Introduction to Molecular Shape Topology." VCH, New York.
Parkin, D. W., Mentch, F., Banks, G. A., Horenstein, B. A., and Schramm, V. L. (1991). *Biochemistry* **30,** 4586.
Parkin, D. W., Limberg, G., Tyler, P. C., Furneau, R. H., Chen, X. Y., and Schramm, V. L. (1997). *Biochemistry* **36**(12), 3528.
Radzicka, A., and Wolfenden, R. (1995). *Methods Enzymol.* **249,** 284.
Rumelhart, D. E., Hinton, G. E., and Williams, R. J. (1986). "Parallel Distributed Processing." Vol. 1. MIT Press, Cambridge, MA.
So, S. S., and Richards, W. G. (1992). *J. Med. Chem.* **35,** 3201.
Tetko, I. V., Tanchuk, V. Y., Chentsova, N. P., Antonenko, S. V., Poda, G. I., Kukhar, V. P., and Luik, A. I. (1994). *J. Med. Chem.* **37,** 2520.
Thompson, T. B., Chou, K. C., and Zheng, C. (1995). *J. Theor. Biol.* **177,** 369.
Wagener, M., Sadowski, J., and Gasteiger, J. (1995). *J. Am. Chem. Soc.* **117,** 7769.
Weinstein, J. N., Kohn, K. W., Grever, M. R., Viswanadhan, V. N., Rubinstein, L. V., Monks, A. P., Scudiero, D. A., Welch, L., Koutsoukos, A. D., Chiausa, A. J., and Paull, K. D. (1992). *Science* **258,** 447.
Xiang, S., Short, S. A., Wolfenden, R., and Carter, C. W., Jr. (1995). *Biochemistry* **34,** 4516.

APPENDIX

Selected Exercises and Problems

Daniel L. Purich

Professor of Biochemistry and Molecular Biology
University of Florida, College of Medicine
Gainesville, Florida, USA

I. Exercises and Problems

1. The following basic terms are frequently used in enzyme kinetics and mechanism, and you should be sufficiently prepared to define and provide examples of each: (a) elementary reaction; (b) reaction mechanism; (c) collision frequency; (d) encounter; (e) reaction order; (f) zero-order kinetics; (g) series first-order process; (h) bimolecular rate constant; (i) rate-limiting step; (j) general acid catalysis; (k) buffer catalysis; (l) preequilibrium kinetics; (m) Arrhenius plot; (n) in-line displacement; (o) dissociative mechanism; (p) pH-function; (q) initial velocity; (r) van't Hoff plot; (s) diffusion limit; (t) turnover number; (u) activation energy; (v) transition state; (w) kinetic isotope effect; (x) Swain relationship; (y) partition function; and (z) zero-point energy.

2. The generalized form of a chemical reaction is

$$aA + bB \rightarrow cC + dD$$

It shows that the velocity can be expressed as

$$-\frac{1}{a}\frac{d[A]}{dt} = -\frac{1}{b}\frac{d[B]}{dt} = \frac{1}{c}\frac{d[C]}{dt} = \frac{1}{d}\frac{d[D]}{dt}$$

DOI: 10.1016/B978-0-12-378608-1.00022-0

3. Derive an integrated rate equation for two parallel first-order reactions:

$$A \xrightarrow{k_1} P, A \xrightarrow{k_2} Q$$

If you assume that [P] and [Q] are initially zero, then your rate equation from the first part of this problem should reveal an interesting relationship between [P] and [Q] as the reaction progresses.

4. Consider the following overall system at equilibrium:

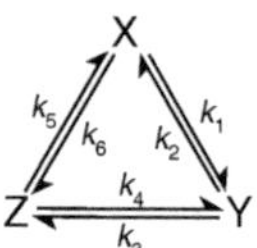

What does the principle of detailed balance tell one about the levels of each intermediate and their relationship to the other chemical species?

5. The stoichiometry of a bimolecular reaction is such that the levels of both reactants A and B are depleted simultaneously, and this introduces the variable x to account for the level of product formed.

$$\frac{dx}{dt} = k(a - x)(b - x)$$

(a). Integrate this equation between the limits of 0 and x from time 0 to t. (b) Is the solution of the integration valid for $a = b$?

6. Suppose that one is studying the autocatalytic activation of a protease zymogen by its active enzyme form:

$$A + B \xrightarrow{k} 2A$$

where A is the active enzyme and B is the inactive form. Derive an integrated rate law for the time-course of this autoactivation process using $[A]_0$ and $[B]_0$ to represent the initial active and inactive enzyme concentrations, respectively.

7. Consider a mechanism whereby two different reactants produce a common product:

$$A \xrightarrow{k_1} C, B \xrightarrow{k_2} C$$

Write a rate law for each reaction and derive the integrated rate law which has the following form:

$$\log([C]_\infty - [C]) = \log\left([A]_0 e^{-k_1 t} + [B]_0 e^{-k_2 t}\right)$$

What does a plot of $\log([C]_\infty - [C])$ versus t look like if (a) $k_1 = k_2$; (b) $k_1 = 5k_2$; and (c) $k_1 = 100k_2$?

8. A protein (15 μM) has three reactive thiol groups which react with Ellman's reagent to give a colored product which may be quantitatively measured. From the following time-course tabulation, calculate the number of thiol groups reacting in each kinetically discernible class and the corresponding rate constant.

Reaction time (min)	Unreacted thiol (μM)	Reaction time (min)	Unreacted thiol (μM)
0	45.0	2.00	9.65
0.25	32.3	4.00	5.53
0.50	24.3	6.00	3.35
0.75	19.1	8.00	2.03
1.00	15.7	12.0	0.75
1.50	11.8	16.0	0.27

9. The limiting magnitude of a typical bimolecular rate constant for macromolecule interactions with small ligands is about 10^7–10^8 M^{-1} s^{-1}. One notable exception is the binding of the *Lac* repressor molecule to the defined region of bacterial DNA known as the *Lac* operator, and the rate constant has been estimated at about 10^{10} M^{-1} s^{-1} from studies with the repressor and purified DNA. How might one interpret such a finding?

10. As noted in the previous problem, the magnitude of bimolecular rate constants for macromolecule–ligand interactions proceeding at the diffusion limit range from 10^7 to 10^8 M^{-1} s^{-1}. Frequently, however, observed bimolecular rate constants appear to fall into the range 10^4–10^5 M^{-1} s^{-1}, and occasionally they are even smaller. How can one interpret these rate constants in a physically realistic manner?

11. The Michaelis–Menten rate equation describes the following simplistic pathway:

$$\mathrm{E} + \mathrm{S} \overset{K_m}{\rightleftharpoons} \mathrm{ES} \overset{k}{\rightarrow} \mathrm{E} + \mathrm{P}$$

(a). State each of the assumptions on which the Michaelis–Menten equation is based. (b) Derive the rate law for this process.

12. Briggs and Haldane first derived the steady-state solution for the one-substrate, one-intermediate, one-product mechanism in the previous problem. Using k_1 and k_2 to represent the on and off rate constants for ES formation and k_3 for the conversion of ES to E and product, derive a steady-state rate expression for this process. By what assumptions about the magnitude of rate constants will the Briggs–Haldane solution reduce to the Michaelis–Menten equation?

13. Repeat the previous problem for the reverse reaction (i.e., P $\rightarrow$ S without any S present). By what assumptions about the magnitude of rate constants will the Briggs–Haldane solution reduce to the Michaelis–Menten equation?

14. Is it possible for a given enzyme system to obey the Michaelis–Menten rate laws in both directions? Explain.

15. The enzyme "Xase" converts substrate X to product Y, and the following rate data were obtained at an enzyme concentration of 10^{-8} M:

[X] mM	5.00	1.67	1.00	0.71	0.56	0.46		
V (mol/min)	$\times$	10^7	2.59	1.95	1.67	1.41	1.24	1.10

Graphically determine K_m and V_{max}, and then estimate k_{cat}.

16. R. A. Alberty and W. H. Pierce (*J. Am. Chem. Soc.* **79,** 1526 (1957)) analyzed the fumarase reaction mechanism in terms of the rate constants obtainable from studies of the forward and backward reactions. For the reaction at pH 6, the following constants apply: k_1, 3×10^8 M^{-1} s^{-1}; k_2, 2.6×10^2 s^{-1}; k_3, 1.45×10^3 s^{-1}; and k_4, 3.8×10^8 M^{-1} s^{-1} in the following scheme:

$$\mathrm{E} + \mathrm{F} \underset{k_2}{\overset{k_2}{\rightleftharpoons}} \mathrm{EX} \underset{k4}{\overset{k_3}{\rightleftharpoons}} \mathrm{E} + \mathrm{M}$$

(a). Calculate the K_m values and dissociation constants for both fumarate and malate. (b) If the subunit molecular weight is 50,000, calculate the specific activity under conditions of saturating fumarate for the forward reaction rate. (c) Calculate the reaction equilibrium constant. (d) If the half-life for the transient phase preceding the steady state is given by the relation $t_{1/2} = 0.693/(k_1\,[\mathrm{Fum}_{\mathrm{initial}}] + k_2 + k_3)$, estimate the duration of the presteady-state phase. (e) Would much product accumulate prior to the steady state and limit the ease of measuring the steady-state rate?

17. The plots given in the accompanying figures illustrate the time-course of a one-substrate, two-intermediate, one-product enzyme reaction as computed by an interactive microprocessor-based reaction kinetics simulator (D. Shindell, C. Magagnosc, and D. L. Purich, *J. Chem. Educ.* **55,** 708 (1978)). The following rate constants and reaction conditions apply: $k_{+1} = 10^6$ M^{-1} s^{-1}; $k_{-1} = 10$ s^{-1}; $k_{+2} = 9$ s^{-1}; $k_{-2} = 1$ s^{-1}; $k_{+3} = 10$ s^{-1}; $k_{-3} = 10^5$ M^{-1} s^{-1}; $[\mathrm{E}_{\mathrm{tot}}] = 10^{-6}$ M; $[\mathrm{S}]_{\mathrm{init}} = 0.15$ mM; $[\mathrm{P}]_{\mathrm{init}} = 0$. Identify the presteady-state, steady-state, poststeady-state, and equilibrium phases of the reaction. Roughly estimate the amount of substrate consumption during each phase.

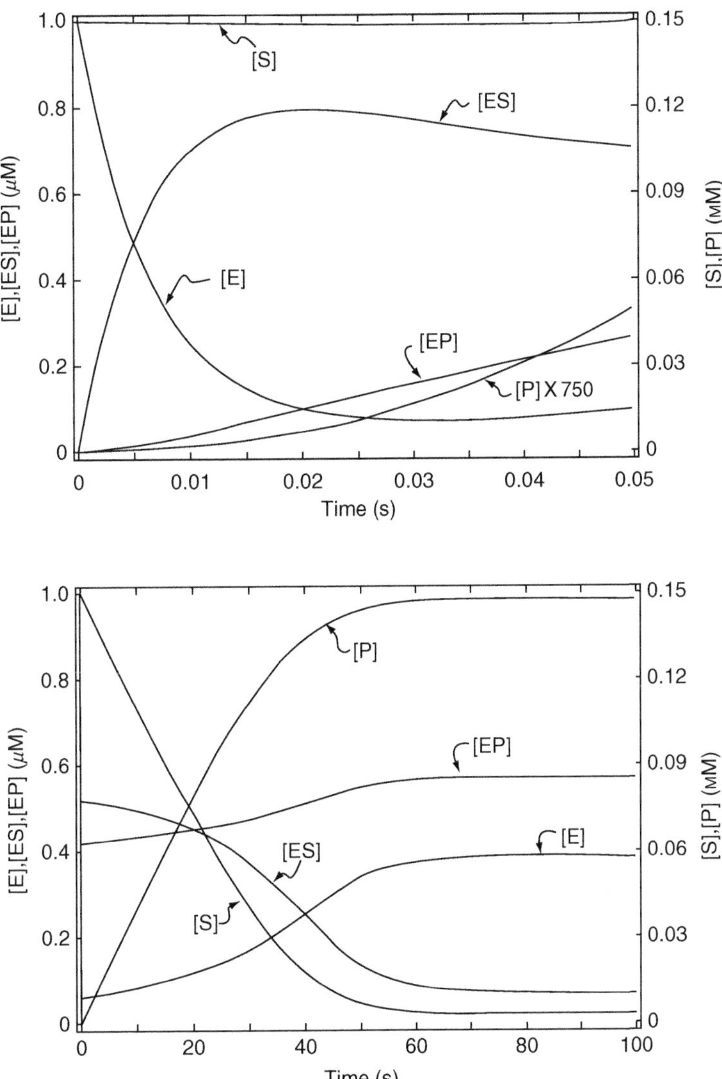

18. Derive the initial rate law for the following mechanism using the systematic approach described in Chapter 1 and do not assume [P] = 0:

$$E + S \underset{k_2}{\overset{k_1}{\rightleftharpoons}} EX \underset{k4}{\overset{k_3}{\rightleftharpoons}} EY \underset{k6}{\overset{k_5}{\rightleftharpoons}} E + P$$

For comparing the ease of derivation, try to use the method of simultaneous equations, the King–Altman method, or the determinant method.

19. One possible scheme for the action of a bisubstrate enzyme in the presence of substrates A and B plus an alternative substrate for B (designated below as B′) is as follows:

$$E + A \rightleftharpoons EA, \qquad EAB \rightleftharpoons EQ + P$$

$$EA + B \rightleftharpoons EAB, \qquad EA'B \rightleftharpoons EQ + {}'P$$

$$EA + {}'B \leftarrow EA'B, \qquad EQ \leftarrow E + Q$$

(a). Arrange this mechanism as an enclosed geometrical figure with each enzyme form at a vertex interconnected to those other forms as described by the mechanism. (b) Write out all the King–Altman patterns. (c) Confirm part (b) by calculating the total number of patterns with $(n - 1)$ lines joining the enzyme forms.

20. Occasionally, the combined equilibrium-steady-state approach of S. Cha (*J. Biol. Chem.* **243,** 820 (1968)) provides a valuable, simplifying route for deriving special rate equations. As an exercise, use Cha's method with the mechanisms in Schemes 6a or 7a of Chapter 1 to obtain the corresponding special rate law.

21. The parameter V_{max}/K_m consists of rate constants up to and including the first irreversible step of an enzymatic reaction whereas V_{max} consists of the rate constants of catalysis and all subsequent steps through pr oduct release. Derive expressions for V_{max}/K_m and V_{max} for the following reaction scheme:

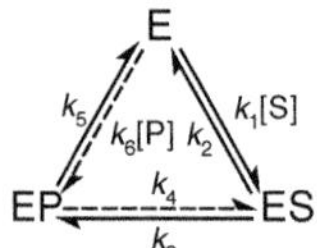

Let $k_4 = k_6[P] = 0$.

22. Generally, the linearity of the reaction rate has been taken as one additional indication that the initial rate assumption is valid, and the shape of v versus $[E]_{total}$ plots can be helpful in discerning complications in rate assays. Describe how the following might lead to upward curvature in such plots: (a) the assay medium itself contains a small amount of potent inhibitor; (b) the enzyme preparation contains an activator; (c) the enzyme undergoes self-association to a more active form; and (d) the enzyme undergoes time-dependent activation.

23. How does the level of substrate relative to the level of enzyme catalyst affect the validity of the steady-state assumption?

24. Derive a rate expression for the action of a noncompetitive inhibitor in the following scheme:

$$\begin{array}{ccccc} E + S & \overset{K_m}{\rightleftharpoons} & ES & \overset{k}{\longrightarrow} & E + P \\ \updownarrow & & \updownarrow & & \\ EI & & ESI & & \end{array}$$

where $K_i = [E][I]/[EI]$ and $K_{ii} = [ES][I]/[ESI]$. Next, determine the point of intersection of inhibition data plotted as a Lineweaver–Burk type plot (i.e., v^{-1} versus $[S]^{-1}$ in the presence and absence of I) for the case where (a) $K_i = K_{ii}$; (b) $K_i > K_{ii}$; and (c) $K_i < K_{ii}$. (Hint: Start with your equation in double-reciprocal form, and let [I] = 0 and [I] = x to obtain two equations which may then be simultaneously solved.)

25. Show that your equation from the previous exercise will reduce to the competitive and uncompetitive inhibition cases depending on K_i and K_{ii} values.

26. Derive a rate expression for the following mechanism involving participation of an essential activator:

$$E + A \rightleftharpoons EA$$

$$EA + S \rightleftharpoons EAS \rightarrow EA + P$$

where $K_a = [E][A]/[EA]$.

27. Write the *net* rate constants for the following three-intermediate enzyme pathway:

$$E_1 \underset{k_2}{\overset{k_1[A]}{\rightleftharpoons}} E_2 \underset{k4}{\overset{k_3}{\rightleftharpoons}} E_3 \underset{k6}{\overset{k_5}{\rightleftharpoons}} E_4 \overset{k_7}{\rightarrow} E_1$$

28. Repeat Problem (21) using the net rate constant method to get V_{max}/K_m and V_{max}.

29. Consider the following monomer–dimer interaction in terms of near-equilibrium relaxation approaches:

$$2E \underset{k_{-1}}{\overset{k_{+1}}{\rightleftharpoons}} E_2$$

(a). Derive an appropriate expression for the relaxation time τ. (b) Suggest a plotting method to obtain values of k_{+1} and k_{-1}.

30. G. G. Hammes and P. R. Schimmel ("The Enzymes" (P. D. Boyer, ed.), 3rd. ed., Vol. 2, pp. 67–114. Academic Press, New York (1970)) have provided an excellent overview of rapid reaction and transient state kinetic studies as applied to enzyme behavior. Following their step-by-step outline of relaxation kinetic derivations, compare the one-step and two-step cases for enzyme–substrate interactions.

31. Suppose that you wish to determine the anomeric specificity of an enzyme acting on a sugar substrate. The spontaneous rate of anomerization can, of course, be too fast for the usual initial rate methods to be used. Design a series of experiments which could allow one to define the anomeric specificity of an enzyme.

32. The dependence of reaction rate on temperature or pH can offer valuable mechanistic information, but irreversible enzyme instability can lead to complicated observed rate behavior. What type of control experiments are necessary to validate such approaches?

33. K. D. Gibson (*Biochim. Biophys. Acta* **10,** 221 (1953)) was among the first to recognize the need to determine both K_m and V_m parameters in experiments designed to measure the energy of activation of enzyme systems. Why is the

determination of K_m essential even when one is only interested in the temperature dependence of V_{max}?

34. Derive a rate law for the action of aldolase on fructose 1,6-bisphosphate for the mechanism described below, and include one product at a time to obtain two expressions, each describing the action of a single product.

$$E + A \underset{k_2}{\overset{k_1}{\rightleftharpoons}} EA$$

$$EA \underset{k_4}{\overset{k_3}{\rightleftharpoons}} EQ + P$$

$$EQ \underset{k_6}{\overset{k_5}{\rightleftharpoons}} E + Q$$

How would you determine whether glyceraldehyde-3-P or dihydroxyacetone-P is the first product to be released?

35. C. Frieden (*Biochim. Biophys. Res. Commun.* **68,** 914 (1976)) presented an interesting example leading to mechanistic ambiguity. Consider the following *rapid equilibrium* ordered pathway:

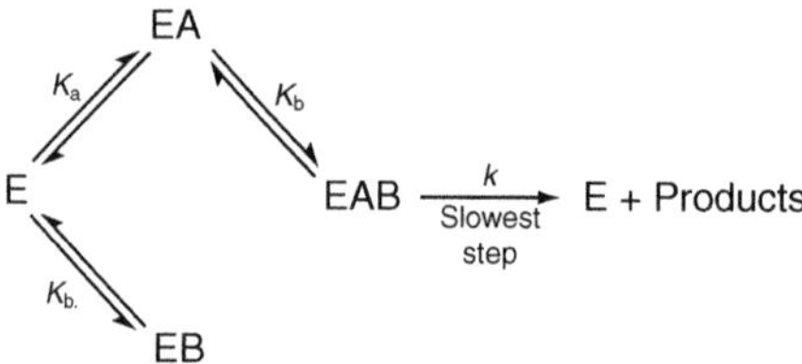

Here, EB is an abortive binary complex which is inactive and must dissociate prior to the binding of substrate A. (a) Derive the initial rate equation for the above mechanism *without* including EB abortive complex formation. (b) Repeat the derivation with EB formation. (c) Rearrange both to double-reciprocal form and compare plots of v^{-1} versus $[A]^{-1}$ (at different constant levels of B) and v^{-1} versus $[B]^{-1}$ (at different levels of A) for both cases. (d) Compare with other sequential mechanisms.

36. H. J. Fromm (*Biochim. Biophys. Acta* **52,** 199 (1961)) was one of the first to recognize that bisubstrate enzymes can form abortive ternary complexes from the binding of substrate and product such as enzyme · NAD^+·pyruvate or enzyme · NADH · lactate in the lactate dehydrogenase reaction. In such cases, high substrate levels in the presence of a product can lead to potent substrate inhibition. (a) Derive a steady-state rate law for the following dehydrogenase mechanism with ribitol present:

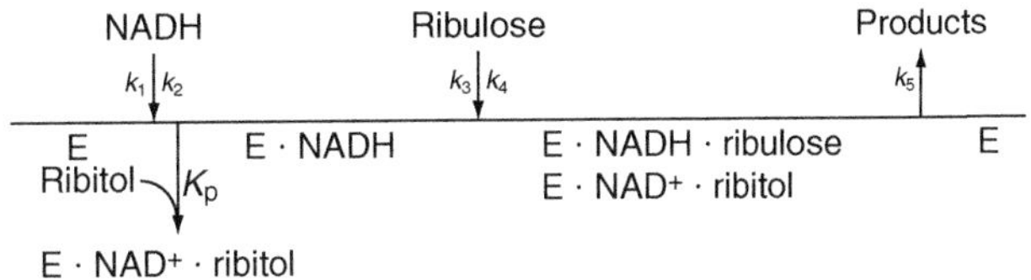

Assume that the NAD^+ level remains negligible. (b) Make plots of v^{-1} versus $[\text{ribulose}]^{-1}$ in the presence and absence of NAD^+.

37. Substrate inhibition may also occur with so-called ping pong mechanisms, which are kinetically observable double-displacement pathways. Although substrate B is normally obliged to bind only after A binds and P is released, elevated levels of B can lead to abortive complex formation. Another possibility is that at high levels of substrate A one may find that A combines a second time with the enzyme, but the second molecule of A binds to the same enzyme form as B. How might these cases be distinguished on the basis of plots of v^{-1} versus $[A]^{-1}$ at several constant levels of substrate B?

38. An enzymologist is characterizing the inhibition of an ADP-requiring enzyme with a very low K_m for ADP. He finds that AMPP(CH_2)P completely inhibits the enzyme in a competitive fashion, but with ATP the inhibition reaches an intermediate plateau activity level. Even at saturating ATP levels, the inhibition remains incomplete. Propose an explanation for such behavior and defend it with an equation.

39. For some enzymatic systems there exists the distinct possibility that the substrate is contaminated with a potent competitive inhibitor. Let $\alpha = [I]/[S]$, and insert this relationship into the competitive inhibitor expression for a one-substrate system to observe the effect of such contamination on the rate behavior.

40. Derive an initial rate expression for the following mechanism and show that it can be used to explain the bell-shaped curves frequently found in pH-rate profiles of enzymes.

$$\begin{array}{ccccc} EH_2 & & EH_2S & & \\ \Big\Updownarrow K_a & & \Big\Updownarrow K_a & & \\ EH + S & \overset{K_b}{\rightleftharpoons} & EHS & \xrightarrow{k} & EH + P \\ \Big\Updownarrow K_b & & \Big\Updownarrow K_b & & \\ ES & & ES & & \end{array}$$

For simplicity, assume that all acid–base reactions are very rapid and do the same for substrate–enzyme interactions.

41. Why is the maintenance of solution ionic strength so important in the experimental characterization of pH-dependent changes of enzyme kinetic parameters?

42. Equation (31) in Chapter 5 is the logarithmic form of Eq. (25), and it allows one to readily determine pK values. Rearrange Eqs. (26) and (27) into the same eneral form as Eq. (31); then verify that the plots in Fig. 8 can be related to each of your new equations.

43. How will the following affect the determination of pK values from pH kinetic studies? (a) The substrate(s) is "sticky" and does not rapidly desorb. (b) The substrate(s) is ionic but the dissociation does not change over the pH range under study. (c) The substrate(s) is ionic and the degree of dissociation changes over the pH range under study. (d) The enzyme requires an essential metal ion for activity.

44. Frequently, one would like to confirm the results of pH-kinetic studies by using an independent method. List three other approaches and discuss their advantages and limitations.

45. Draw plots of v/V_{max} versus [substrate] for a one-substrate enzyme which acts only on the metal ion–substrate complex. Let the K_m be 0.5 mM and assume that the uncomplexed substrate is a competitive inhibitor ($K_i = 0.5$ mM). The stability constant for the formation of the metal ion–substrate complex should be assumed to be 10,000 M^{-1}. The total level of substrate should be varied in the range from 0.02 to 1.0 mM, and consider the following cases: (a) metal ion is varied in a one-to-one fashion with substrate; (b) metal ion is added such that the free, uncomplexed metal ion concentration is 5 mM. (This problem is most readily handled with a programmable hand calculator to solve the quadratic equations.)

46. L. Tabatabai and D. J. Graves (*J. Bio. Chem.* **253,** 2196 (1978)) reported that phosphorylase kinase exhibits sequential bisubstrate kinetics when studied with Mg–ATP^{2-} and a tetradecapeptide corresponding to the phosphoryl acceptor site on glycogen phosphorylase *b*. They also noted that AMPP(CH_2)P was a linear competitive inhibitor relative to the nucleotide substrate, but it was a noncompetitive inhibitor relative to the peptide substrate. Another peptide (a threonine-containing heptapeptide) acted as a nonlinear (S-parabolic) competitive inhibitor relative to the peptide substrate, and it served as a nonlinear (I-parabolic, S-parabolic) noncompetitive inhibitor with respect to ATP. (a) What do these results indicate about the order of substrate addition. (b) Why would anyone wish to substitute a small peptide substrate in place of the physiologic protein substrate?

47. What special property of chromium(III) and cobalt(III) complexes of nucleotides renders them so valuable as mechanistic probes of nucleotide-dependent enzymes?

48. Active-site-directed irreversible enzyme inhibitors frequently follow the kinetic pathway presented below:

$$E + I \overset{K_1}{\rightleftharpoons} E{\cdot}I \overset{k}{\rightarrow} E - I$$

where E · I and E–I represent the noncovalent and covalent species arising from the interaction of enzyme and inhibitor. (a) Derive a rate equation for the initial rate of loss of enzyme activity. (b) How might the presence of the corresponding substrate influence the rate of inactivation. (c) How might the presence of another substrate in a multisubstrate enzyme mechanism influence the process?

49. A. H. Mehler, J.-J. Park Kim, and A. A. Olsen (*Arch. Biochem. Biophys.* **212,** 475 (1981)) observed that the periodate-cleavage product of ATP (hereafter dialdehydo-ATP) inactivates the partial reactions of aminoacyl-tRNA synthetases in which the amino acid is transferred from the aminoacyl adenylate to the tRNA. Yet, these investigators noted that the dialdehydo-ATP interaction with the enzyme had no effect on the other partial-exchange reactions in which ATP itself is a substrate or product. What does this series of experiments suggest regarding the commonly held assumptions about the specificity of enzyme interactions with enzyme inactivators?

50. Inactivation of horse liver alcohol dehydrogenase by diethyl pyrocarbonate (ethoxyformic anhydride) does not proceed with pseudo-first-order kinetics as indicated by the nonlinearity of a plot of the logarithm of remaining enzymatic activity versus period of reaction. Indeed, the rate of the reaction appears to decrease with time. A possible explanation is that the diethyl pyrocarbonate hydrolyzes during the above reaction, and, in fact, this reagent has a half-life of 20 min at pH 8 and 25 °C. On the assumptions that the enzyme has a single group which is involved in the inactivation process and that there is an excess of reagent over enzyme (sufficient to allow a pseudofirst-order assumption), derive an integrated rate equation which could describe the observed inactivation kinetics.

51. When chemical modification of a single amino acid residue of an enzyme leads to complete or almost complete loss of enzymatic activity, it is often concluded that the modified residue is "essential" for activity. In many cases, however, there is some residual activity, and it is important to know if this activity is due to some unmodified enzyme or if it is the result of conversion of fully modified enzyme to a less active form. Describe what procedures you might use to distinguish between these two possibilities.

52. The inactivation of chymotrypsin by the active-site-directed reagent phenylmethanesulfonylfluoride, which reacts with Ser-195, proceeds most rapidly when a group on the enzyme with a pK of 7 is unprotonated. Similarly, the inactivation of the enzyme by tosyl-L-phenylalanylchloromethane, which alkylates His-57, proceeds most rapidly when a group on chymotrypsin with a pK of 6.8 is in the unprotonated form. Explain these results on the basis of known properties of chymotrypsin's catalytic center from the studies of the analogous enzyme α-lytic protease by W. W. Bachovchin and J. D. Roberts (*J. Am. Chem. Soc.* **100,** 8041 (1978)).

53. NAD^+ analogs containing bromoacetyl functions on either the nicotinamide ring or the adenine ring have been used to study the inactivation of horse liver and yeast alcohol dehydrogenase. Modification occurs at cysteine residues which are ligated to the active-center zinc ion and proximal to the nicotinamide

binding pocket as determined by X-ray diffraction studies of the complexes of the horse liver enzyme with various coenzyme analogs by H. Eklund, J.-P. Samama, L. Wallén, C.-V. Bränden, Å. Åkeson, and T. A. Jones (*J. Mol. Biol.* **146,** 561 (1981)). How might you explain the basis for a reagent containing a bromoacetyl group in the adenine ring reacting with cysteine residues near the nicotinamide binding site? (The data are presented in Table I (item 38) of Chapter 12.)

54. Suicide substrates are mechanism-based inactivators which frequently obey the following reaction scheme:

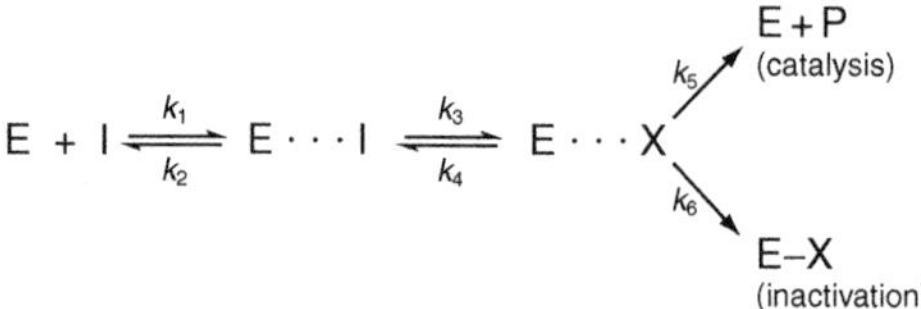

The E···X species can undergo partitioning between catalytic turnover to product or covalent inactivation, and the ratio k_5/k_6 defines the partition ratio. How is this value determined?

55. State the basic assumptions of the sequential model for cooperative ligand binding by an oligomeric protein. How do these compare with the concerted (or symmetry conserving) model?

56. Scatchard plots and Hill plots have become standard graphical methods for probing macromolecule interactions with ligands. (a) Derive the Scatchard and Hill equations. (b) Describe how positive cooperativity is measured by each. (c) Describe how negative cooperativity is determined by each.

57. The saturation function for the symmetrical, concerted nonexclusive ligand binding model of Monod, Wyman, and Changeux is given by the following equation:

$$\bar{Y} = \frac{Lc\alpha(1 + c\alpha)^{n-1} + \alpha(1 + \alpha)^{n-1}}{L(1 + \alpha)^n + (1 + \alpha)^n}$$

(a). Define the parameters $\bar{Y}$, L, c, α, and n. (b) For the case of a dimer ($n = 2$), make a plot of $\bar{Y}$ versus α with $c = 0$, $L = 100$. Repeat part (b) with $L = 0$ to see how much the cooperativity depends on the value of L. (c) Show that the saturation function presented above can reduce to a hyperbolic saturation function when a monomer is considered. (d) Why is the parameter α a constant?

58. The following equation defines $\bar{R}$, the fraction of protein molecules in the R-state in the MWC model:

$$\bar{R} = \frac{(1 + \alpha)^n}{(1 + \alpha)^n + L(1 + \alpha)^n}$$

Compare $\bar{R}$ and $\bar{Y}$ for the parameters given in Problem 57. Why might it be difficult to use comparisons of $\bar{Y}$ and $\bar{R}$ to distinguish between the MWC and KNF models?

59. Many investigators have analyzed enzyme cooperativity by using kinetic data alone and by assuming that the extent of ligand saturation may be evaluated by the ratio of the initial velocity and the maximal velocity. Is this a reliable practice? Explain.

60. C. Y. Huang and D. J. Graves (*Biochemistry* **9,** 660 (1970)) presented the theory for determining the equilibrium constant for dimer–tetramer interactions and the specific activities of each oligomeric form of glycogen phosphorylase *a* from rate data and light scattering intensity data. Outline a series of experiments which might be of value in distinguishing among the cases of active dimer only, active tetramer only, and both species having catalytic activity.

61. V. L. Schramm and J. F. Morrison (*Biochemistry* **9,** 671 (1970)) carried out an illuminating comparative study of freshly prepared rat liver nucleoside diphosphatase and the same enzyme after prolonged storage at $-10\,°C$. They found that such cold storage resulted in linear plots of v^{-1} versus [Mg-IDP] while freshly prepared enzyme exhibited nonlinear (concave-up) plots. Yet, both preparations have about the same molecular weight, and both are also activated by Mg-ATP^{2-}, an allosteric enzyme effector. Propose a likely explanation for such behavior.

62. B. S. Hartley and B. A. Kilby (*Biochem. J.* **56,** 288 (1954)) observed the time-course of chymotrypsin-catalyzed hydrolysis of *p*-nitrophenylethyl carbonate by measuring the accumulation of *p*-nitrophenol. In this classical study, they observed a presteady-state burst of the nitrophenol product, and the amplitude of the burst corresponded to about 63% of the enzyme concentration at several enzyme concentrations. What does the existence of such a burst say about the reaction?

63. P. R. Krishnaswamy, V. Pamiljans, and A. Meister (*J. Biol. Chem.* **237,** 2932 (1962)) reported on a novel isotope experiment which has led to much wider application in recent years. Glutamine synthetase was preincubated with [^{14}C]glutamate and ATP in the strict absence of ammonia or hydroxylamine. This solution was then mixed with a second solution containing hydroxylamine and a large excess of unlabeled glutamate. After the reaction was quenched and the γ-glutamylhydroxamate was separated, the specific radioactivity of this product was determined to be much higher than the specific radioactivity estimated by taking both labeled and unlabeled glutamate into account. What does this say about the "stickiness" of the glutamate and the course of the glutamine synthetase reaction?

64. Although aminoacyl-AMP compounds are isolated from aminoacyl-tRNA synthetase, it was not clear for some time whether the aminoacyl adenylates served as covalent intermediates on the reaction pathway. Suggest several experiments using rapid-mixing techniques which could settle this issue.

65. Schimmel's laboratory has provided valuable insights into the mode of recognition and interaction of aminoacyl-tRNA synthetases with their nucleic acid substrates. When purine bases of the tRNA molecule are exposed to tritiated water solvent, the rates of exchange at the H-8 of the purine ring can be measured in the absence and presence of the synthetase. Sequential T1 and T2 digestion of the tRNA can be used to localize the sites of exchange and to evaluate their rate constants. This work also led to the finding that the enzyme catalyzes an H-5 exchange in the pyrimidine ring of uridine-8 of the tRNA. Moreover, R. M. Starzyk, S. W. Koontz, and P. R. Schimmel (*Nature* (*London*) **298,** 136 (1982)) found that 5-bromouridine interacts with *E. coli* alanyl-tRNa synthetase in a manner leading to a single site of covalent modification after a first-order loss ($k = 0.017\ \text{min}^{-1}$) of enzyme activity. Exposure of the modified enzyme to 10 mM dithiothreitol led to reversal of the activity loss. These investigators also observed that incubation with the bromouridine and enzyme resulted in inactivation of the ATP-PP_i exchange reaction. (a) Propose chemical mechanisms for the H-8 and H-5 exchange reactions to explain the observed behavior. (b) Propose a scheme to explain the bromouridine results.

66. J. R. Knowles (*Ann. Rev. Biochem.* **49,** 896 (1980)) has presented a valuable discussion of the internal thermodynamics of substrates and reaction intermediates on the surface of an enzyme as opposed to their solution thermodynamics. How does one gain information about the internal thermodynamics of central complexes in enzymatic catalysis?

67. Investigations on transition-state analogs over the past decade are in excellent agreement with L. Pauling's postulates (*Am. Scientist* **36,** 51 (1958)) that an enzyme's surface is somewhat complementary to the transition-state configuration of a substrate. Indeed, he specifically stated that "the picture even presents us with ideas as to the nature of substances which would be effective inhibitors—they should resemble the activated complex." Write a thermodynamic cycle which accounts for the extraordinary affinity of an enzyme for a transition-state analog by considering the corresponding nonenzymatic reaction scheme as well.

68. Transition-state analogs and related multisubstrate-geometrical inhibitors may also be used to examine the order of substrate binding to enzymes. For example, K. Collins and G. R. Stark (*J. Biol. Chem.* **246,** 6599 (1971)) observed that *N*-(phosphonacetyl)-L-aspartate acts as a competitive inhibitor relative to carbamyl-P and as a noncompetitive inhibitor relative to aspartate in the reaction catalyzed by *E. coli* aspartate transcarbamoylase. In contrast, D. L. Purich and H. J Fromm (*Biochim. Biophys. Acta* **276,** 563 (1972)) found that P^1,P^4-di(adenosine-5′)-tetraphosphate served as a linear competitive inhibitor relative to Mg-ATP^{2-} or AMP in the adenylate kinase reaction. What do these results indicate about the kinetic mechanisms of enzyme–substrate interactions?

69. List six basic assumptions which must be satisfied to derive equilibrium-exchange reaction rate laws for enzymic systems. (As an extension of this exercise, chose one of the papers corresponding to references 7–9 or 10 of Chapter 16,

and verify that each assumption in your list from above is satisfied for the system under study.)

70. Derive the general expression for isotopic exchange for the reaction

$$\mathrm{A-X+B-X^{*} \rightleftharpoons A-X^{*}+B-X}$$

where the asterisk indicates the isotopically labeled forms of A–X and B–X. Assume that the unlabeled and labeled forms do not differ in their reactivity, and assume that the amount of the isotopically labeled form is always small compared to the unlabeled species. Use the symbol R to represent the *gross* rate of exchange (i.e., the rate of exchange of all atoms X whether they represent like or different isotopes).

71. For practice, try to identify as many of the exchanges which may be anticipated for the following enzyme reactions: (a) the nucleoside diphosphate kinase reaction (ping pong); (b) the creatine kinase reaction (rapid equilibrium random); and (c) the lactate dehydrogenase reaction (ordered sequential). (*Hint*: Some exchanges depend on the stereochemical course of the reaction.)

72. Derive the equilibrium exchange rate expression for the A $\rightleftharpoons$ P exchange in the absence of substrate B and product Q for the ping pong mechanism. (*Hint*: You will only need to consider the following reaction scheme:

$$\mathrm{E+A} \underset{k_2}{\overset{k_1}{\rightleftharpoons}} \mathrm{EX} \underset{k_4}{\overset{k_3}{\rightleftharpoons}} \mathrm{E'+P}$$

73. For the process described in the last problem, is it possible to vary one reactant (say A) at various fixed levels of the other (say P)? Explain.

74. Derive the equilibrium exchange rate equation for the B $\rightleftharpoons$ P exchange in the ordered ternary complex mechanism:

$$\mathrm{E} \underset{k_{-1}}{\overset{k_1\mathrm{A}}{\rightleftharpoons}} \mathrm{EA} \underset{k_{-2}}{\overset{k_2\,(\mathrm{B}^*)}{\rightleftharpoons}} \mathrm{EAB}^* \underset{k_{-3}}{\overset{k_3}{\rightleftharpoons}} \mathrm{EP^*Q} \underset{k_{-4}}{\overset{k_4\,(\mathrm{P}^*)}{\rightleftharpoons}} \mathrm{EQ} \underset{k_{-5}}{\overset{k_5}{\rightleftharpoons}} \mathrm{E}$$

75. Derive the exchange-rate equation for the A $\rightleftharpoons$ P exchange for the following random bi bi mechanism:

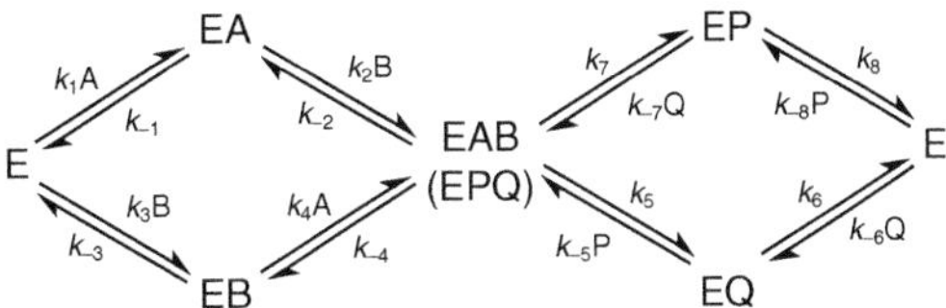

76. How are the rate parameters v^* and R related to each other?

77. Frequently, isotope-exchange equations are used to study exchange experiments wherein one substrate and one product are held in a constant ratio relative to each other and the absolute levels are raised or lowered. For practice, rearrange

the rate law derived in Problem 74 to account for the effect of raising substrate A and product Q in a constant ratio (say $\alpha = [Q]/[A]$).

78. Rederive the exchange rate law in Problem 74 by the steady-state method.

79. Convert the answer in Problem 78 into the answer for Problem 74 by making equilibrium assumptions.

80. The equilibrium constant of the hexokinase reaction at pH 6.5 with excess magnesium ion present may be written as follows:

$$K = \frac{[\text{Glc} - 6 - \text{P}][\text{ADP}]}{[\text{Glc}][\text{ATP}]} = 490$$

Suppose the following substrate levels were used in an isotope-exchange experiment: 1 mM Glc; 1 mM ATP; 24.5 mM Glc-6-P; and 20 mM ADP. (a) If complete equilibration of exchange occurs after the addition of 25,500 cpm of labeled Glc-6-P, how many counts will be distributed into the Glc and Glc-6-P pools? (b) If complete equilibration occurs after the addition of 25,000 cpm of labeled ATP, how many counts will be distributed into the ADP and ATP pools? (c) Ten minutes after the addition of 25,500 cpm of labeled Glc to the reaction system in the presence of hexokinase, the reaction was quenched and further analysis demonstrated that 4900 cpm of Glc-6-P was formed. Calculate R for this rate process.

81. For the ping pong bi bi mechanism, show that

$$R^{-1}_{\text{max,A-P}} + R^{-1}_{\text{max,B-Q}} = V^{-1}_{\text{max,forward}} + V^{-1}_{\text{max,reverse}}$$

82. For both yeast hexokinase and bacterial galactokinase, there is very good evidence that the substrates bind in a random order. From equilibrium isotopic-exchange studies, however, the sugar $\rightleftharpoons$ sugar-P exchange is about 1.5- to 2.5-fold slower than the ADP $\rightleftharpoons$ ATP exchange. What does this indicate about the reaction mechanism?

83. How does the equilibrium exchange rate of an enzyme-catalyzed reaction depend on the concentration of enzyme?

84. In 1953, L. J. Zatman, N. O. Kaplan, and S. P. Colowick (*J. Biol. Chem.* **200,** 197 (1953)) devised a very useful scheme for the synthesis of [^{14}C]NAD$^+$ based on the properties of the bovine spleen nucleotidase reaction. When [^{14}C]nicotinamide and NAD$^+$ are incubated with this enzyme (which catalyzes the hydrolysis of NAD$^+$) a very significant amount of radiolabeled NAD$^+$ is formed. What does the observation of such an exchange reaction indicate about the NADase mechanism?

85. C. F. Midelfort and I. A. Rose (*J. Biol. Chem.* **251,** 5881 (1976)) developed an ingenious method to probe the glutamine synthetase mechanism by observing the "scrambling" or positional isotope exchange of ^{18}O atoms from the $\beta\gamma$-bridge position to the β-non-bridge position with ATP and glutamate in the strict absence of ammonia: (O = ^{16}O and Ø = ^{18}O)

$$\text{Adenosine}\text{—O—}\underset{\text{O}}{\overset{\text{O}}{\text{P}}}\text{—O—}\underset{\text{O}}{\overset{\text{O}}{\text{P}}}\text{—Ø—}\underset{\text{O}}{\overset{\text{O}}{\text{P}}}\text{—O} \xrightleftharpoons[\text{Enzyme (NH}_3\text{-free)}]{\text{L-glutamate}} \text{Adenosine—O—}\underset{\text{O}}{\overset{\text{O}}{\text{P}}}\text{—O—}\underset{\text{O}}{\overset{\text{Ø}}{\text{P}}}\text{—O—}\underset{\text{O}}{\overset{\text{O}}{\text{P}}}\text{—O}$$

(a). How does such an observation add credence to the Meister mechanism which involves an acyl-P intermediate? (b) What kinetic information can be obtained from this method?

86. Many enzymes catalyze the exchange of hydrogen atoms from substrates to solvent during the course of enzymatic reactions involving isomerization reactions. What can be said of such processes when $R_{\text{S·H} \rightleftharpoons \text{HOH}}$ is (a) greater or (b) less than the rate of the overall reaction?

87. Rewrite the solution to Problem 70 for substrate $\rightleftharpoons$ solvent exchanges of hydrogen atoms. Is it a safe assumption that S–^{1}H, S–^{2}H, and S–^{3}H exchange protium, deuterium, and tritium with the solvent at the same rate? Explain.

88. W. J. Albery and J. R. Knowles (*Biochemistry* **15,** 5588, 5627, and 5631 (1976)) have presented a detailed analysis of the triose-P isomerase reaction based upon isotope-exchange studies. In addition, they have discussed some of the constraints on enzyme efficiency, aspects regarding potential improvements in catalytic efficiency, and the conceptual basis of the "perfect" enzyme. As an exercise, the reader should carefully examine Knowles' exhaustive analysis of the triose-P isomerase catalytic pathway in these papers and the accompanying reports. It would also be instructive to examine the means by which the free-energy profile of this reaction was elucidated.

89. Draw a schematic reaction coordinate diagram which will explain the physical basis of the primary kinetic isotope effect.

90. The primary isotope effect on reaction rate of isotopic isomers A–B and A–B′ can be estimated by the following expression.

$$\frac{k_{\text{A-B}}}{k_{\text{A-B}'}} = \exp\left\{\frac{hv}{2kT}\left(1 + \left[\frac{m_{\text{B}}(m_{\text{A}} + m_{\text{B}'})}{m_{\text{B}'}(m_{\text{A}} + m_{\text{B}})}\right]^{1/2}\right)\right\}$$

Here, h, v, k, and m_i are Planck's constant, stretching frequency, Boltzmann's constant, and the mass of each atom, respectively. Calculate $k_{\text{C-H}}/k_{\text{C-D}}$ using a stretching frequency for a C—H vibration near 2900 cm^{-1} and that for C–D stretching at 2100 cm^{-1}. Let the temperature be 298°K.

91. Estimate the $k_{\text{C-H}}/k_{\text{C-T}}$ from the Swain relationship and your result from Problem 90.

92. How strongly does the magnitude of a primary deuterium kinetic isotope effect depend on temperature? (*Hint*: Use the equation in Problem 90.)

93. Occasionally, the observed primary isotope effects ($k_{\text{H}}/k_{\text{D}}$) for chemical reactions exceed the limiting values predicted on the basis of the classical Hooke's law theory presented in Problem 90. Indeed, at low temperatures, the ratios of k_{H} to k_{D} can become surprisingly large. What is the basis for such large kinetic isotope effects?

94. D. B. Northrop (*Biochemistry* **14,** 2645 (1975)) developed an insightful approach for evaluating the kinetic isotope effects in enzymatic reactions. Using the scheme presented earlier in Problem 21, Northrop obtained (V_H/V_D) and $(V/K)_H/(V/K)_D$ from the earlier problem (except that here k_3 is the rate constant for the isotopically sensitive step, and it is k_{3H} and k_{3D} depending on which isotopic isomer is the substrate):

$$\frac{V_H}{V_D} = \frac{k_{3H}/k_{3D} + (k_3/k_5)_H}{(k_3/k_5)_H + 1}$$

$$\frac{(V/K)_H}{(V/K)_D} = \frac{k_{3H}/k_{3D} + (k_3/k_2)_H}{(k_3/k_2)_H + 1}$$

Derive these equations.

95. For the preceding problem, suppose that k_{3H}/k_{3D} is about 8. Evaluate V_H/V_D for the case where k_3 is essentially rate-determining and for the case where k_5 is rate-determining. Also, evaluate $(V/K)_H/(V/K)_D$ for the case where k_3 is very small relative to k_2 and vice versa.

96. A commonly observed phenomenon in enzyme dynamics is the strong dependence of reaction rate on the rate of product desorption. As noted in the previous example (k_5 very small relative to k_3), the observed isotope effect on V can be rather minor. B. V. Plapp, R. L. Brooks, and J. D. Shore (*J. Biol. Chem.* **248,** 3470 (1973)) found that chemical modification of the lysyl residue side-chains of liver alcohol dehydrogenase resulted in an increased rate of coenzyme dissociation. Considering that most dehydrogenases have an ordered kinetic mechanism, how might this be reflected in the primary kinetic isotope effects on this reaction.

97. M. I. Schimerlik, J. E. Rife, and W. W. Cleland (*Biochemistry* **14,** 5347 (1975)) first reported on an intriguing phenomenon observed in experiments on the malic enzyme mechanism. When the enzyme was added to a solution containing $NADP^+$, malate-2-D, CO_2, pyruvate, and NADPH (all calculated to maintain the reaction at equilibrium), the NADPH concentration decreased at first and then gradually reappeared until the equilibrium position was attained. The shape of the curve resembled an $A \rightarrow B \rightarrow C$ series first-order process. (a) Rationalize the origin of this displacement. (b) What would have occurred if malate-2-H and NADPH had been utilized? (c) Would it matter in (b) whether the site of deuteration was the A or B position in NADPD?

98. Kinetic studies of solvent-isotope effects in a series of mixtures of HOH and DOD have provided important information about factors contributing to solvent-isotope effects on enzymic processes. Describe the basis and use of the "proton inventory" method for analyzing enzyme mechanisms.

99. As an exercise, describe the origin and magnitude of secondary kinetic isotope effects in S_N1 and S_N2 reaction mechanisms. Give several examples for enzyme-catalyzed reactions.

100. D. V. Santi's group has probed the thymidylate synthetase mechanism by measuring the secondary kinetic isotope effect on the rate of desorption of 5-fluoro-2′-deoxyUMP from the enzyme-FdUMP binary complex. Using a mixture of [2-^{14}C]FdUMP and [6-^{3}H]FdUMP to make the complex, they followed the course of the complex's dissociation and obtained a value of 1.24 for k_T/k_H, which is an inverse isotope effect. Explain the mechanistic basis of this inverse secondary kinetic isotope effect.

II. Answers and/or Comments

The reader should note that a few of the problems take the form of exercises, and answers to these are genereally unnecessary and are not included.

1. These terms are well defined in standard textbooks on physical chemistry, organic mechanism, or chemical kinetics.

3. (a) $[P] = [P]_0 + k_1[A]_0(1 - e^{-kt})/k$, and $[Q] = [Q]_0 + k_2[A]_0(1 - e^{-kt})/k$, where $k = (k_1 + k_2)$; (b) If $[P]_0 = [Q]_0 = 0$, the products will accumulate in a constant ratio relative to each other, independent of the time and $[A]_0$. The relative rate constants can be readily evaluated by measuring the relative P and Q concentrations as the reaction progresses.

4. The principle of detailed balance requires that if the overall system is at equilibrium, so must the individual steps be at equilibrium. (The inverse need not be true.) From detailed balance, $[Y]_e/[X]_e = k_1/k_2$; $[Z]_e/[Y]_e = k_3/k_4$; and $[X]_e/[Z]_e = k_5/k_6$, where the subscript e designates the equilibrium concentrations.

5. (a) The integrated rate law is

$$\frac{1}{(a-b)} \text{In} \frac{b(a-x)}{a(b-x)} = kt$$

(b) When $a = b$, the above expression is undefined because $(a - b)$ appears in the denominator.

6. $([A]0 + [B]_0)^{-1} \ln([A]_0[B]/[B]_0[A]) = kt$.

7. (a) Linear. (b) Linear at the most initial point; curvilinear at the intermediate phase; and linear when in the time domain of the slower process. When the two processes have almost equal rate constants, the fitting of the rate data becomes difficult.

8. The first two thiols react quickly ($k = 2\ \text{min}^{-1}$) and the third reacts more slowly ($k = 0.25\ \text{min}^{-1}$). In practice, one could compare the rates of such protein modification reactions with model compounds under identical solution conditions to determine whether the reaction rates on the protein are greater or less than the model reactions, and this provides information on the environment around the protein-bound groups.

9. The best current explanation is that the repressor molecule first forms a loose, nonspecific interaction with virtually any region of the DNA and subsequently migrates along the DNA helix to the operator region. This reduction in dimensionality of the diffusion process could speed up the complexation reaction and reconcile what would be an otherwise baffling interaction requiring long-distance recognition of approaching chemical species.

10. The reaction rate constant may be only an apparent bimolecular rate constant resulting from any of several multistep, rather than single-step, processes. The ligand may bind initially in a very loose configuration which must isomerize on a slower time scale to a more tightly bound species. The macromolecule may exist in several conformational states or even states of ionization, and only a relatively less abundant form of the macromolecule reacts with the ligand at the diffusion limit. In some cases, the substrate may also exist in several states and exhibit similar rate behavior. Table I in the chapter by G. G. Hammes and P. R. Schimmel ("The Enzymes" (P. D. Boyer, ed.), 3rd ed., Vol. II, p. 109. Academic Press, New York (1970)) is a very good source for examining the factors affecting bimolecular rate constants.

11. (a) There are several implicit assumptions: that the enzyme and substrate are in rapid equilibrium; that the total enzyme concentration is negligible in relation to the total substrate concentration; that the reverse rate is zero; that the amount of product is negligible *or* that the accumulation of product does not affect the reaction rate; and that the breakdown of ES is slow relative to its formation, thereby justifying the first assumption. (b) $v = V_m[S]/(K_m + [S])$.

12. Setting the differential equation for EX (i.e., $d[ES]/dt = k_1[E][A] - (k_2 + k_3)[EX]$) equal to zero, one solves for [E] in terms of [ES]. Thus, K_m becomes equal to $(k_2 + k_3)/k_1$, whereas V_m remains $k[E]_{total}$. Note that K_m has units of molarity, but it is not simply a measure of enzyme affinity because it is a kinetic constant and not an equilibrium constant. Also, note that the steady-state and equilibrium solutions to the mechanism become identical when k_3 is insignificant relative to k_2.

13. Note that the pathway in the previous problem is symmetrical to the pathway for the reverse reaction. In such symmetrical cases, one may use the following shorthand notation:

S	P
k_1	k_4
k_2	k_3

If one takes the forward reaction scheme which contains constants in the left column and then replaces them with the neighboring constants or chemical species in the right column, the reverse rate equation will be obtained. This method will work for any symmetrical reaction scheme. For example, the ordered ternary complex mechanism and the shorthand notation may be given as follows:

$$E + A \underset{k_2}{\overset{k_1}{\rightleftharpoons}} EA$$

$$EA + B \underset{k_4}{\overset{k_3}{\rightleftharpoons}} EAB$$

$$EAB \underset{k_6}{\overset{k_5}{\rightleftharpoons}} EPQ$$

$$EPQ \underset{k_8}{\overset{k_7}{\rightleftharpoons}} EQ + P$$

$$EQ \underset{k_{10}}{\overset{k_9}{\rightleftharpoons}} E + Q$$

A	Q
B	P
k_1	k_{10}
k_2	k_9
k_3	k_8
k_4	k_7
k_5	k_6

Finally, it should be clear that the inequality $k_2 \gg k_3$ for the previous problem will now become $k_3 \gg k_2$ for the reverse reaction case, and this too could be obtained using the above shorthand method.

14. Both k_2 and k_3 are unimolecular rate constants, and the levels of substrate and product have no influence on their effective magnitude. Thus, no enzyme can simultaneously satisfy the conflicting inequalities between k_1 and k_2 for the forward and reverse directions.

15. $V_m = 3 \times 10^{-7}$ mol/min; $K_m = 0.8$ mM; and $k = 0.5\ s^{-1}$.

16. (a) $K_{mF} = 5.7 \times 10^{-6}$ M; $K_{dF} = 8.7 \times 10^{-7}$ M; $K_{mM} = 4.5 \times 10^{-6}$ M; $K_{dM} = 3.8 \times 10^{-7}$ M.

(b). The specific activity (in units of μmol/min/mg) is 1740 in the forward direction and 312 in the reverse direction. (c) $K_{eq} = k_1k_3/k_2k_4 = 4.4$. (d) 4×10^{-4} s. (e) Not really, because the state preceding the steady state would only persist for a time equal to about 5–10 times the value given in (d).

18. The solution is presented in Chapter 1 on p. 4. Note that the form of the final equation is the same as that with only one intermediate; this result should remind one that the steady-state rate laws can be derived with the least number of internal isomerization steps without affecting the form of the resulting rate equation.

19. See Scheme 2 in Chapter 1 and the expression on p. 9.

21. The solution from Problem 18 can be used if terms containing k_4 or k_6[P] are eliminated. $V/K = k_1k_3[E]_{tot}/(k_2 + k_3)$ and $V = k_3k_5/(k_3 + k_5)$.

22. If you cannot rationalize how each of these leads to upward curvature, then it may be helpful to read pp. 48–55 of "Enzymes" by M. Dixon and E. C. Webb, Academic Press, New York (1979). This updated treatise on enzyme action also presents several explanations of downward curvature.

23. J. T.-F. Wong (*J. Am. Chem. Soc.* **87,** 1788 (1965)) analyzed the one-substrate, one-intermediate case when $k_4 = 0$. He found that the steady-state assumption becomes increasingly more valid as the ratio of substrate to enzyme is increased.

24. $v^{-1} = V_m^{-1}(1 + [I]/K_{ii}) + (K_m/V_m)(1 + [I]/K_i)[S]^{-1})$. The point of intersection in a double-reciprocal plot will be (a) on the $[S]^{-1}$ axis; (b) below the $[S]^{-1}$ axis (third quadrant); and (c) above $[S]^{-1}$ axis (second quadrant).

25. If $K_{ii} \rightarrow \infty$, the equation becomes that for competitive inhibitor action. If $K_i \rightarrow \infty$, the equation becomes that for a noncompetitive inhibitor.

26. $v = V_m/\{1 + (K_m/[S])[1 + (K_a/[A])]\}$.

27. The net rate constant for any step has been defined by W. W. Cleland as the rate constant for the forward reaction multiplied by a partition factor which describes how much of the particular enzyme form in a step goes on to give products, as opposed to undergoing reversal. For E_4 to E_1 (the step farthest to the right in the mechanism), the net rate constant (designated by a prime) is

$$k_7' = k_7 \text{ (note that the } k_7 \text{ is irreversible in the mechanism)}$$

For E_3 to E_4 (the next to the farthest right-hand term), the net rate constant is

$$k_5' = k_5\left(\frac{k_7'}{k_6 + k_7'}\right) = \frac{k_5 k_7}{k_6 + k_7}$$

For E_2 to E_3, the net rate constant is

$$k_3' = k_3\left(\frac{k_5'}{k_4 + k_5'}\right) = \frac{k_3 k_5 k_7}{k_4(k_6 + k_7) + k_5 k_7}$$

For E_1 to E_2, the net rate constant is

$$k_1' = k_1[A]\left(\frac{k_3'}{k_2 + k_3'}\right) = \frac{k_1 k_3 k_5 k_7 [A]}{k_2[k_4(k_6 + k_7) + k_5 k_7] + k_3 k_5 k_7}$$

(Note that all the bimolecular rate constants are converted to effective unimolecular rate constants by multiplying them by the associated reactant. Thus, in the case listed above $k_1[A]$ is used.)

28. One can follow W. W. Cleland's approach as given on p. 276 of his article in *Adv. Enzymol.* **45,** 273 (1977). As in Problem 27, one first obtains the net rate constants: $k_5' = k_5, k_3' = k_3$; and $k_1' = k_1 k_3[A]/(k_2/k_3)$ because $k_4 = k_6[P] = 0$. The velocity of the reaction is $[E]_{tot}$ divided by the sum of the reciprocals of all the rate constants for the mechanism which is now a series of first-order processes with net rate constants for each:

$$v = \frac{[E]_{tot}}{1/k_1' + 1/k_3' + 1/k_5'} = \frac{[E]_{tot}}{1/k_1' + 1/k_3 + 1/k_5}$$

For V_m, let $[A] \rightarrow \infty$, then $1/k_1' \rightarrow 0$, and

$$V_m = \frac{[E]_{tot}}{1/k_3 + 1/k_5} = \frac{k_3 k_5 [E]_{tot}}{k_3 + k_5}$$

For V_m/K_m, let $[A] \rightarrow 0$, then $1/k_1$ becomes far larger than $1/k_3$ or $1/k_5$, and

$$v = \frac{V_m}{K_m}[A] = \frac{[E]_{tot}}{(1/k_1')} = k_1'[E]_{tot}$$

Thus, equating the second and fourth terms of this series of equations give

$$\frac{V_m}{K_m} = \frac{k_1'[E]_{tot}}{[A]} = \frac{k_1 k_3 [E]_{tot}}{k_2 + k_3}$$

29. (a) $\tau^{-1} = (4k_1[A] + k_{-1})$. (b) $k_1 = (\text{slope})/4$ and $k_{-1} = (\text{intercept})$.

30. For the one-step process, $\tau^{-1} = \{k_1([E]_e + [S]_e) + k_{-1}\}$. For the two-step process, $\tau_1^{-1} = \tau_{\text{one-step}}^{-1}$; and $\tau_2^{-1} = \{k_2/[1 + (k_{-1}/k_1)([E]_e + [S]_e)] + k_{-2}\}$

31. See Chapter 14 by S. J. Benkovic in Vol. 63 of *Methods in Enzymology*.

32. See the discussions on pp. 134 and 169.

33. The true temperature dependence of V_{max} will only be obtained when the enzyme is fully saturated by substrate at each temperature surveyed. Likewise, special care must be taken to ensure that those enzymes which require cofactors, coenzymes, and metal ions are properly assayed at each temperature so that these additional ligands are also never limiting the observed rate.

34. See Eq. (12) and Table I in Chapter 8.

35. (a) $v^{-1} = V_m^{-1}\{1 + K_b/[B] + K_a K_b/[A][B]\}$. (b) $v^{-1} = V_m^{-1}\{1 + K_a K_b/[A] + K_b/[B] + K_a K_b/[A][B]\}$. (c) The first equation predicts that a plot of v^{-1} versus $[B]^{-1}$ will converge on the v^{-1} axis. (d) The result in (b) suggests that initial rates alone cannot help one to distinguish the rapid-equilibrium ordered mechanism with EB abortive from other sequential mechanisms. For further consideration of this case see pp. 442–445.

36. See Eqs. (26) and (27) in Chapter 8, and eliminate all steps beyond the k_5 step by deleting terms containing k_6, k_7, k_8, and K_{IB}. For the actual rate experiments on ribitol dehydrogenase, see the early paper by H. J. Fromm and D. R. Nelson (*J. Biol. Chem.* **237,** 215 (1962)).

37. See pp. 254–256.

38. Adenosine triphosphate from commercial sources contains trace levels of ADP; if the enzyme under study has a low K_m for ADP, then the inhibition by ATP will never be total. If the ratio of ATP/ADP is α for the "ATP" used in the experiment, then one can substitute α[ADP] for [ATP] in the equation describing competitive inhibition:

$$v^{-1} = \frac{1}{V_{max}} + \frac{K_m}{V_{max}}\left(1 + \frac{[ATP]}{K_i}\right)[ADP]^{-1}1$$

$$= \frac{1}{V_{max}} + \frac{K_m}{V_{max}}\left(1 + \frac{\alpha[ATP]}{K_i}\right)[ADP]^{-1}1$$

$$= \frac{1 + K_m\alpha/K_i}{V_{max}} + \frac{K_m}{V_{max}}[ADP]^{-1}1$$

39. Recall the approach used in the preceding problem.

40. See Eq. (23) in Chapter 5.

41. The ionic strength provides a quantitative means for maintaining the bulk solvent-ionic environment. Reactions between ions in solution and those reactions on the enzyme which involve charge neutralization or separation will be especially sensitive to the ionic nature of the solvent. K. J. Ellis and J. F. Morrison (*Methods in Enzymology*, Vol. 87, p. 405). Discuss many of the important factors which must be considered in developing buffers of constant ionic strength for studying pH-dependent processes.

43. In addition to the information in Chapter 5, the reader is encouraged to examine the section entitled "Determination of Catalytic Groups by pH-Variation Studies" in the paper cited in Answer 29.

44. Same as Answer 28.

45. You should find that varying the substrate and metal ion concentration in a 1:1 ratio can be quite misleading and result in apparent cooperativity. See Chapters 11 and 12 in Vol. 63 of *Methods in Enzymology* and consult the report by D. L. Purich and H. J. Fromm (*Biochem. J.* **130,** 63 (1972)).

46. (a) See p. 246. (b) Many proteins which serve as substrates of other enzymes will quite naturally have ligand binding sites of their own; thus, one must devise alternative substrate approaches such as described in Chapter 13 of Vol. 64 of *Methods in Enzymology* to fully analyze such enzymic systems.

47. Certain nucleotide complexes of tervalent chromium and cobalt are inert to ligand exchange, which is not the case for divalent metal ions. Thus, one can prepare stable complexes which are valuable as potent inhibitors and as stereochemical probes. The interested reader may wish to consult Chapters 11 and 12 of Vol. 87 of *Methods in Enzymology*.

48. (a) The initial rate equation is analogous to the Michaelis–Menten equation because the rate of inactivation should display saturation kinetics with respect to the irreversible inhibitor: $-d[E]/dt = k[E][I]/(K_I + [I])$. (b) The substrate in the one-substrate case will serve as a competitive inhibitor of inactivation. (c) The effect of a second substrate in a bisubstrate reaction will depend on the kinetic mechanism. For example, an affinity label analogous in structure to the second substrate in an ordered ternary complex mechanism should require the presence of the first substrate, but it may not if abortive binary complexes such as discussed in Problem 35 are formed.

49. The facile synthesis of dialdehyde derivatives of ribonucleosides and ribonucleotides with periodate has led to their widespread application as affinity labeling agents. All too frequently, experimentalists seeking active-site-directed irreversible enzyme inhibitors assume that inactivation by a reagent which resembles a substrate reflects action at the substrate binding site. With aminoacyl-tRNA synthetases one might imagine that the dial-ATP could act at the active center and also at locations which represent recognition sites for the tRNA, and the latter might block productive binding of tRNA. It is noteworthy that Mehler's group

found that the inactivation was nonspecific with regard to the substituents on the dialdehyde and with regard to the enzymes susceptible to inactivation. For example, dial-GTP and dial-uridine react as well as dial-ATP with the synthetase, and all three reagents also react with rabbit muscle aldolase. Another somewhat related point was made by B. Boettcher and A. Meister (*J. Biol. Chem.* **256,** 5977 (1981)), who found that dial-UMP binding to carbamyl-P synthetase leads to enzyme activation whereas UMP is a well-recognized allosteric inhibitor. Thus, one cannot presume that the affinity label analog will behave as the natural ligand.

50. The scheme for these reactions can be written as follows:

$$\mathrm{E} + \mathrm{R} \xrightarrow{k_1} \mathrm{E} - \mathrm{X}$$
$$\mathrm{R} \xrightarrow{k_2} \mathrm{Q}$$

where E, R, E–X, and Q are active enzyme, active reagent, inactivated enzyme, and inactivated reagent, respectively. The differential equations may be written as

$$\frac{-\mathrm{d}E}{\mathrm{d}t} = k_1[\mathrm{R}][\mathrm{E}] \text{ and } \frac{-\mathrm{d}R}{\mathrm{d}t} = k_2[\mathrm{R}]$$

Integrating the second equation first, we get $[\mathrm{R}] = [\mathrm{R}]_0 \mathrm{e}^{-k_2 t}$; thus, we may substitute this expression for [R] in the first differential equation to obtain a rate law allowing for the inactivation of R. Then, after separating variables and integrating once more, we obtain the desired rate expression:

$$\ln\left(\frac{[\mathrm{E}]_t}{[\mathrm{E}]_0}\right) = \left(\frac{k_1[\mathrm{R}]_0}{k_2}\right)(\mathrm{e}^{-k_2 t} - \mathrm{e}^{-0})$$

or

$$\frac{[\mathrm{E}]_t}{[\mathrm{E}]_0} = e - \left(\frac{k_1[\mathrm{R}]_0}{k_2}\right)(1 - \mathrm{e}^{-k_2 t})$$

51. As described in References 9–11 on p. 322, there are several experiments to be carried out. First, repeated modification of the enzyme should be attempted to see if the enzyme activity can be reduced to much less than 1% of the original activity. Second, the modified enzyme can be purified from the reaction mixture, and one should try to demonstrate that it is chemically distinct from the unmodified enzyme. Third, the steady-state kinetic characteristics of the enzyme should be determined for forward and reverse reaction directions and at several pH values to ascertain that the modified enzyme has different kinetic properties. If these characteristics are unique, this constitutes good evidence that the modified amino acid residues are not absolutely essential for catalytic activity. (In the case of liver alcohol dehydrogenase, cysteine-46 and lysine-228 are in the active site, but they are not directly involved in catalysis; rather, it appears that modification of these residues disturbs substrate binding and the efficiency of catalysis without completely eliminating activity.) Fourth, it is sometimes helpful to use a less bulky alkylating reagent because the modified enzyme with such a bulky reagent

attached may partly block access of substrate to the catalytic center. Last, it is useful to try different substrates with the modified enzyme to determine whether there is a change in substrate specificity between unmodified and modified enzyme.

52. Bachovchin and Roberts studied the α-lytic protease from histidine-auxotrophs of myxobacterium because the growth conditions of the organism may be manipulated to allow enrichment of the single histidine residue of this enzyme with ^{15}N in the imadazole ring of the histidine. Like chymotrypsin, in which serine-195 and the histidine-57 are part of the catalytic triad at the active center, the α-lytic protease behaves as many other "serine" proteases. The serine hydroxyl and the histidine imidazole groups constitute a *system* which ionizes with a pK value of about 7, and this controls the rate of reaction of the active-site-directed reagents. Note that the pK for Ser-195 is probably much higher than 7, whereas the pK of the His-57 imidazole is approximately 7, as shown by ^{15}N nuclear magnetic resonance studies.

53. Although we often think that substrates and inhibitors only bind in one way to the active site, this case clearly shows that NAD can bind with the modified adenine ring in the nicotinamide site. This type of nonproductive binding may also occur in the interaction of normal substrates with the enzyme active site. In any case, such alternative binding modes of active-site-directed inactivators can render some active-site mapping less certain in the absence of information as provided by X-ray studies. (*Hint*: Recall the result from Problem 3.) For a thorough discussion of suicide inhibition, see the articles by R. H. Abeles and A. Maycock (*Acc. Chem. Res.* **9,** 313 (1976)) and by C. T. Walsh (*Horizons Biochem. Biophys.* **3,** 36 (1977)).

54. See p. 296.

55. See pp. 272–275. The interested reader will also wish to consult the article by F. W. Dahlquist in *Methods in Enzymology*, Vol. 48, p. 271.

56. See p. 294.

57. For the MWC model, $\bar{R} > \bar{Y}$ except where the two parameters converge at saturating ligand; for the KNF model, $\bar{R} = \bar{Y}$. In practice, however, this distinction may be difficult to achieve, especially with regard to the ability of a particular observable to really measure $\bar{R}$ accurately. J. Wyman (*Curr. Top. Cell Regul.* **6,** 209–226 (1972)) has presented a useful discussion of cooperativity in terms of a general model of which the MWC and KNF models are special cases.

58. For studies of cooperativity, initial rate kinetic approaches for estimating $\bar{Y}$ are rarely as diagnostic and reliable as equilibrium binding data. Complications which frequently compromise the kinetic approach include the following: (a) the uncertainty over whether substrate binding is sufficiently fast as to justify the rapid equilibrium assumption; (b) the inability to evaluate V_{max} accurately in cases of multisubstrate mechanisms: and (c) the difficulty in establishing the degree of substrate saturation in multisubstrate–enzyme systems in which the order of ligand addition increases the kinetic complexity of the process.

59. See the Discussion and Appendix sections of the report by Huang and Graves. The interested reader may also wish to consult the analysis of B. I. Kurganov, Z. S. Kagan, A. I. Dorozhko, and V. A. Yakovlev (*J. Theorei. Biol.* **47**, 1 (1974)).

60. Many enzymes are subject to cold inactivation or cold modification, and oligomeric enzymes appear to especially susceptible. Presumably, the enzyme can become "locked" in a conformation which does not readily isomerize to the active form.

61. The attainment of a steady-state rate requires that the enzyme undergo catalytic turnover. Prior to achieving the steady-state rate, the efficiency of catalysis with respect to bond-breaking and bond-making steps may far exceed the rate of product desorption. With chymotrypsin, *p*-nitrophenol is formed at a rate which is faster than acyl-enzyme hydrolysis, and this gives rise to the burst. Actually, the magnitude of enzyme bursts can be used to evaluate a variety of phenomena such as the relative saturation of the enzyme by substrate in the state preceding the steady state, the number of catalytically active subunits in an oligomeric enzyme, and the role of covalent intermediates in catalysis.

62. This experiment was interpreted as providing further proof for the formation of a tightly bound covalent acyl-P intermediate, and ammonia or hydroxylamine is omitted from the preincubation medium to permit the intermediate to accumulate. I. A. Rose (*Methods in Enzymology*, Vol. 64 of Chapter 2) describes how this pulse-chase type of protocol can be extended into a general method for examining the "stickiness" of enzyme–substrate complexes in multisubstrate reactions. Of course, this method does require high concentrations of enzyme, and one must confirm that the enzyme concentration does not alter the catalytic properties.

63. A. R. Fersht and M. M. Kaether (*Biochemistry* **15,** 818 and 3342 (1976)) describe an elegant series of kinetic experiments demonstrating the catalytic competence of such intermediates.

64. (a) P. R. Schimmel and H. J. P. Schoemaker ("RNA and Protein Synthesis" (K. Moldave, ed.), p. 227. Academic Press, New York (1981)) describe the mechanism of H-8 exchange as an ylid-type reaction involving protonation at N-7, followed by removal of the proton at C-8 to generate an ylid, reprotonation with solvent protons, and deprotonation at N-7 to complete the cycle. The concentration of ylid formed will be quite low, and this leads to small rate constants for exchange. The low rates of exchange are crucial for the analytical protocols to evaluate sites of exchange and to measure the rate constants. With H-5 exchange at uridine-8 in the tRNA, transient formation of an enzyme-S-adduct at the C-6 of the pyrimidine, akin to the sequence of 5-fluorodUMP interactions with thymidylate synthetase (see Problem 100). (b) Again, as in the FdUMP interaction with thymidylate synthetase, the addition of an enzyme-SH might add across the double bond between the C-5 and C-6 positions. It is also noteworthy that all tRNAs have a uridine at the eight-position, suggesting a common theme in enzyme–tRNA interactions.

65. See reports by N. Rao, F. J. Kayne, and M. Cohn (*J. Biol. Chem.* **254,** 2689 (1979)), W. J. Ray and J. W. Long (*Biochemistry* **15,** 4018 (1976)), and K. D. Wilkinson and I. A. Rose (*J. Biol. Chem.* **254,** 12567 (1979)).

66. R. Wolfenden (*Methods in Enzymology*, Vol. 46, p. 15) presents the following scheme comparing the nonenzymatic activation process (upper reaction) and the corresponding enzyme-facilitated step (lower reaction):

$$\begin{array}{ccc} S & \overset{K_{S\ddagger}}{\rightleftharpoons} & S\ddagger \\ E \updownarrow K_S & & E \updownarrow K_{TX} \\ ES & \underset{}{\overset{K_{ES\ddagger}}{\rightleftharpoons}} & ES\ddagger \end{array}$$

where the double-dagger symbol represents the transition state and K_{TX} is the equilibrium constant for the interaction of the substrate in the transition-state configuration with the enzyme [$K_{TX} = K_S \ddagger K_S/K_{ES} \ddagger \simeq 10^{-7}K_S$]. Thus, while one cannot observe this interaction directly, the scheme serves to explain why some inhibitors might be bound with very high affinity.

67. The results with aspartate transcarbamoylase are consistent with ordered binding in which carbamyl-P binding precedes aspartate. The results with adenylate kinase are in agreement with a rapid equilibrium random kinetic mechanism.

68. See pp. 407 and 412.

69. See p. 311 of "Kinetics and Mechanism" by J. W. Moore and R. G. Pearson, J. Wiley & Sons, New York (1981).

70. (a) ATP $\rightleftharpoons$ ADP; ATP $\rightleftharpoons$ GTP; GTP $\rightleftharpoons$ GDP; and positional isomerizations arising from the interaction of ATP or GTP with the enzyme to form a covalent E–P intermediate; inversion at the phosphorus atom of the transferred phosphoryl group upon E–P formation. (b) No exchange between ATP and ADP, ATP and creatine-P, or creatine-P and creatine *unless* all substrates and products are present; inversion at the terminal phosphoryl group of ATP. (c) Same as (b) except with LDH substrates.

71. See Table II for a collection of isotope-exchange rate equations for selected mechanisms on p. 414.

72. The equilibrium constant for the A $\rightleftharpoons$ P exchange is

$$K_{A-B} = \frac{[P][E']}{[A][E]}$$

Note that the amount of enzyme catalyst is insignificant in relation to the level of P or A. Thus, when A is added alone, the above mass action ratio will adjust to a new equilibrium level for each concentration of A, but the amount of P formed will always be less than or equal to the amount of enzyme.

73. See p. 24.

74. See p. 26.

75. With equations expressed in terms of v^*, one is actually stating the rate of labeled isotope exchange, whereas R is the gross rate of exchange which is the exchange of all atoms X (see Problem 70) whether they are of like or different isotopic composition. If a labeled substrate (say A*) is described in terms of some measure of specific activity (e.g., cpm radioactivity per unit of A concentration), then A* can be replaced by A and likewise v^* by R.

76. See pp. 412–413.

77. See p. 26.

78. See p. 29.

79. (a) 24,500 cpm Glc-6-P and 1000 cpm Glc. (b) 22,727 cpm ADP and 2273 cpm ATP. (c) 0.203 mmol/min.

80. For the ping pong mechanism the following kinetic parameters apply: $R_{max,A\text{–}P} = k_2k_3[E]_{tot}/(k_2 + k_3)$; $V_{m,forward} = k_3k_7[E]_{tot}/(k_3 + k_7)$; $R_{max,B\text{–}Q} = k_6k_7[E]_{tot}/(k_6 + k_7)$; $V_{m,reverse} = k_2k_6[E]_{tot}/(k_2 + k_6)$. (Recall that R_{max} parameters are readily obtained from expressions for the exchange rate expressed in reciprocal form, letting the concentration of A and P approach infinity and thereby eliminating those terms containing $[A]^{-1}$ and $[P]^{-1}$, which go to zero.)

81. Neither kinetic reaction mechanism is truly rapid equilibrium random, and the dissociation of sugar and/or sugar-P from these enzymes during catalysis must partially limit the rate of the reaction.

82. See p. 445.

83. See p. 436.

84. (a) The scrambling reaction suggests the internal equilibrium between [Enz · ATP · glutamate] and [Enz · ADP · P–X], where P–X is very likely the glutamyl-P or phosphorylated tetrahedral intermediate of glutamate which forms prior to ammonia attack and glutamine formation. (b) The rates of scrambling were found to exceed the ADP $\rightleftharpoons$ ATP exchange which had been measured by Boyer's group for the entire reaction system at equilibrium. For details of other approaches to such positional exchange processes see Chapter 20 of Vol. 87 of *Methods in Enzymology*.

85. See the excellent chapter by I. A. Rose in "The Enzymes" (P. D. Boyer, ed.), 3rd ed., Vol. 2, Academic Press, New York (1970) and Chapter 20 by the same author in Vol. 87 of *Methods in Enzymology*.

86. (a) $R = -\{[S\text{–}H]\ln(1 - F)\}/t$. (b) No, one can anticipate, in general, that there will be associated primary kinetic isotope effects which will be reflected in the observed rates of exchange.

87. See the excellent chapter by J. H. Richards in "The Enzymes" (P. D. Boyer, ed.) 3rd. ed., Vol. 2, Academic Press, New York (1970) for a clear account of primary isotope effects. The advanced student may wish to consult the following monographs: "Isotope Effects on Enzyme-Catalyzed Reactions" (W. W. Cleland, M. H. O'Leary, and D. B. Northrop, eds.), University Park Press, Baltimore (1977); "Reaction Rates of Isotopic Molecules" by L. Melander and W. H.

Saunders, Jr., Wiley-Interscience, New York (1980); and "Transition States of Biochemical Reactions" (R. D. Gandour and R. L. Schowen, eds.), Plenum Press, New York (1978).

88. Your k_H/k_D value will be around 7–8.

89. The so-called Swain relationship results from changes in the reduced mass when deuterium is replaced by tritium: $(k_H/k_T) = (k_H/k_D)^{1.44}$. (The interested reader can confirm this by rewriting the expression in Problem 90 twice, first for k_H/k_D and second for k_H/k_T, and then by writing an equation $(k_H/k_T) = (k_H/k_D)^n$ and solving for n.) For a value of k_H/k_D corresponding to 8 in the previous problem, k_H/k_T will be $8^{1.44}$ or about 20.

90. From the expression in Problem 90, one can estimate k_{C-H}/k_{C-D} to be about 8.2, 6.9, and 4.6 at 0, 25, and 100°.

91. The effect is the result of the Heisenberg uncertainty principle, and the wavelength of very small particles can be estimated by the de Broglie wavelength. It is known that small particles such as hydrogen atoms, protons, hydride ions, and electrons have a reasonably good probability of existing in regions excluded on the grounds of classical mechanics. Thus, depending on the width of a potential energy barrier to chemical reaction, such particles can tunnel through, rather than surmount, the reaction barrier. At low temperature, the width of such barriers narrows, and this contributes to increased efficiency of tunneling at lower temperatures. For proton-transfer reactions, the proton's wavelength can readily be in the range of the energy barrier widths even at room temperature.

92. For V_H/V_D, first write V_H and V_D expressions by using the answer from Problem 21 or 28 by substituting k_{3H} and k_{3D} for the isotopically sensitive k_3; then, letting all other constants remain unaffected by isotopic substitution, divide V_H by V_D and rearrange. The same procedure will yield the isotope effect on V/K.

93. For $k_3 = 0.01\, k_5$, $V_H/V_D \cong 8$. For $k_3 = 100\, k_5$, $V_H/V_D \cong 1.08$. Note also that V/K contains terms up to and including the first irreversible step; so $(V/K)_H/(V/K)_D$ is insensitive to the relative magnitude of k_5. The apparent isotope effect on V/K varies inversely with k_3/k_2 to give values between 1 (i.e., no observed isotope effect) and k_{3H}/k_{3D} (i.e., the true isotope effect).

94. Coenzyme release in dehydrogenase reactions is frequently the last step in an ordered catalytic cycle, and the coenzyme release is often sufficiently sluggish to limit partially the reaction rate. Thus, anything which would increase the celerity of coenzyme desorption should also tend to increase the observed kinetic isotope effect. Indeed, R. T. Dworshack and B. V. Plapp (*Biochemistry* **16,** 2716 (1977)) found values of V_H/V_D of 3.1, 3.6, and 2.4 for the action of hydroxybutyrimidylated liver ADH action on pentadeuterioethanol, [1,1-D_2]benzyl alcohol, and a series of *p*-substituted benzaldehydes. There were no observable kinetic isotope effects with any of these substrates when tested with the unmodified dehydrogenase.

95. (a) A primary kinetic isotope effect reduces the rate of conversion of malate-2-D to pyruvate, CO_2, and NADPH; however, the rate in the reverse direction is unimpeded. This leads to a temporary displacement in the reaction equilibrium, and the kinetics resemble a series first-order process. (b) The opposite direction of displacement would be observed. (c) Transfer only occurs to and from the A side (or *re*-face) of the coenzyme with the pigeon liver enzyme.

96. K. B. Schowen and R. L. Schowen (*Methods in Enzymology*, Vol. 87, p. 551) present a lucid account of the proton inventory method along with other protocols and theory for understanding solvent-isotope effects.

97. See the references cited in the answer to Problem 89.

98. A cardinal feature of the mechanism is that the enzyme-bound FdUMP is a covalent, sp^3-hybridized adduct at the six-position of the pyrimidine ring, and the inverse secondary isotope effect reflects the change in hybridization (sp^3 to sp^2) upon ligand desorption. For more details, see Chapter 6 of Vol. 64 of *Methods in Enzymology*.

INDEX

D

E

K

L

Q

R

S